Digital Signal Processing

Digital Signal Processing
Fundamentals and Applications

Second edition

Li Tan
Purdue University North Central

Jean Jiang
Purdue University North Central

ELSEVIER

AMSTERDAM • BOSTON • HEIDELBERG • LONDON
NEW YORK • OXFORD • PARIS • SAN DIEGO
SAN FRANCISCO • SINGAPORE • SYDNEY • TOKYO
Academic Press is an Imprint of Elsevier

Academic Press is an imprint of Elsevier
225 Wyman Street, Waltham, MA 02451, USA
The Boulevard, Langford Lane, Kidlington, Oxford, OX5 1GB, UK

First edition 2007
Second edition 2013

Notices

Knowledge and best practice in this field are constantly changing. As new research and experience broaden our understanding, changes in research methods, professional practices, or medical treatment may become necessary.

Practitioners and researchers must always rely on their own experience and knowledge in evaluating and using any information, methods, compounds, or experiments described herein. In using such information or methods they should be mindful of their own safety and the safety of others, including parties for whom they have a professional responsibility.

To the fullest extent of the law, neither the Publisher nor the authors, contributors, or editors, assume any liability for any injury and/or damage to persons or property as a matter of products liability, negligence or otherwise, or from any use or operation of any methods, products, instructions, or ideas contained in the material herein.

Library of Congress Cataloging-in-Publication Data
A catalog record for this book is available from the Library of Congress

British Library Cataloguing in Publication Data
A catalogue record for this book is available from the British Library

ISBN: 978-0-12-415893-1

For information on all Academic Press publications
visit our website at elsevier.com

Printed and bound in the United States of America
13 10 9 8 7 6 5 4 3 2 1

Working together to grow
libraries in developing countries

www.elsevier.com | www.bookaid.org | www.sabre.org

ELSEVIER   BOOK AID International   Sabre Foundation

Contents

Preface xiii

CHAPTER 1 Introduction to Digital Signal Processing 1
   1.1. Basic Concepts of Digital Signal Processing 1
   1.2. Basic Digital Signal Processing Examples in Block Diagrams 3
       1.2.1. Digital Filtering 3
       1.2.2. Signal Frequency (Spectrum) Analysis 3
   1.3. Overview of Typical Digital Signal Processing in Real-World
          Applications 5
       1.3.1. Digital Crossover Audio System 5
       1.3.2. Interference Cancellation in Electrocardiography 5
       1.3.3. Speech Coding and Compression 7
       1.3.4. Compact-Disc Recording System 7
       1.3.5. Vibration Signature Analysis for Defective Gear Teeth 9
       1.3.6. Digital Photo Image Enhancement 9
   1.4. Digital Signal Processing Applications 12
   1.5. Summary 13

CHAPTER 2 Signal Sampling and Quantization 15
   2.1. Sampling of Continuous Signal 15
   2.2. Signal Reconstruction 21
       2.2.1. Practical Considerations for Signal Sampling: Anti-Aliasing Filtering 25
       2.2.2. Practical Considerations for Signal Reconstruction: Anti-Image
              Filter and Equalizer 30
   2.3. Analog-to-Digital Conversion, Digital-to-Analog Conversion,
          and Quantization 35
   2.4. Summary 47
   2.5. MATLAB Programs 48
   2.6. Problems 49

CHAPTER 3 Digital Signals and Systems 57
   3.1. Digital Signals 57
       3.1.1. Common Digital Sequences 58
       3.1.2. Generation of Digital Signals 61
   3.2. Linear Time-Invariant, Causal Systems 63
       3.2.1. Linearity 63
       3.2.2. Time Invariance 65
       3.2.3. Causality 66
   3.3. Difference Equations and Impulse Responses 67
       3.3.1. Format of the Difference Equation 67
       3.3.2. System Representation Using Its Impulse Response 68

3.4. Bounded-In and Bounded-Out Stability.. 71
3.5. Digital Convolution 72
3.6. Summary 79
3.7. Problem 80

CHAPTER 4   Discrete Fourier Transform and Signal Spectrum 87
4.1. Discrete Fourier Transform 87
    4.1.1. Fourier Series Coefficients of Periodic Digital Signals............... 88
    4.1.2. Discrete Fourier Transform Formulas... 91
4.2. Amplitude Spectrum and Power Spectrum.. 97
4.3. Spectral Estimation Using Window Functions 107
4.4. Application to Signal Spectral Estimation... 116
4.5. Fast Fourier Transform 123
    4.5.1. Decimation-in-Frequency Method 123
    4.5.2. Decimation-in-Time Method... 128
4.6. Summary 132
4.7. Problem 132

CHAPTER 5   The z-Transform 137
5.1. Definition 137
5.2. Properties of the z-Transform.. 140
5.3. Inverse z-Transform 144
    5.3.1. Partial Fraction Expansion Using MATLAB 150
5.4. Solution of Difference Equations Using the z-Transform...................... 152
5.5. Summary 156
5.6. Problems 156

CHAPTER 6   Digital Signal Processing Systems, Basic Filtering Types,
                 and Digital Filter Realizations 161
6.1. The Difference Equation and Digital Filtering...................................... 161
6.2. Difference Equation and Transfer Function... 166
    6.2.1. Impulse Response, Step Response, and System Response................ 169
6.3. The z-Plane Pole-Zero Plot and Stability 172
6.4. Digital Filter Frequency Response 178
6.5. Basic Types of Filtering 186
6.6. Realization of Digital Filters... 192
    6.6.1. Direct-Form I Realization 193
    6.6.2. Direct-Form II Realization... 193
    6.6.3. Cascade (Series) Realization... 195
    6.6.4. Parallel Realization... 196
6.7. Application: Signal Enhancement and Filtering.................................... 199
    6.7.1. Pre-Emphasis of Speech... 200
    6.7.2. Bandpass Filtering of Speech... 203
    6.7.3. Enhancement of ECG Signal Using Notch Filtering.................... 205

6.8. Summary 206

6.9. Problem 208

CHAPTER 7   Finite Impulse Response Filter Design.................................. 217

   7.1. Finite Impulse Response Filter Format.. 217

   7.2. Fourier Transform Design 219

   7.3. Window Method 230

   7.4. Applications: Noise Reduction and Two-Band Digital Crossover 253

      7.4.1. Noise Reduction 253

      7.4.2. Speech Noise Reduction.. 256

      7.4.3. Noise Reduction in Vibration Signals................................. 257

      7.4.4. Two-Band Digital Crossover 258

   7.5. Frequency Sampling Design Method 262

   7.6. Optimal Design Method 269

   7.7. Realization Structures of Finite Impulse Response Filters............... 280

      7.7.1. Transversal Form 280

      7.7.2. Linear Phase Form.. 281

   7.8. Coefficient Accuracy Effects on Finite Impulse Response Filters.... 282

   7.9. Summary of FIR Design Procedures and Selection of FIR Filter Design
        Methods in Practice 285

  7.10. Summary 288

  7.11. MATLAB Programs 288

  7.12. Problems 290

CHAPTER 8   Infinite Impulse Response Filter Design.............................. 301

   8.1. Infinite Impulse Response Filter Format.. 302

   8.2. Bilinear Transformation Design Method 303

      8.2.1. Analog Filters Using Lowpass Prototype Transformation 304

      8.2.2. Bilinear Transformation and Frequency Warping 308

      8.2.3. Bilinear Transformation Design Procedure 314

   8.3. Digital Butterworth and Chebyshev Filter Designs......................... 318

      8.3.1. Lowpass Prototype Function and Its Order 318

      8.3.2. Lowpass and Highpass Filter Design Examples.................. 322

      8.3.3. Bandpass and Bandstop Filter Design Examples 331

   8.4. Higher-Order Infinite Impulse Response Filter Design Using the
        Cascade Method.. 338

   8.5. Application: Digital Audio Equalizer 341

   8.6. Impulse-Invariant Design Method... 345

   8.7. Pole-Zero Placement Method for Simple Infinite Impulse Response
        Filters 351

      8.7.1. Second-Order Bandpass Filter Design 352

      8.7.2. Second-Order Bandstop (Notch) Filter Design................... 354

      8.7.3. First-Order Lowpass Filter Design....................................... 355

      8.7.4. First-Order Highpass Filter Design...................................... 357

8.8. Realization Structures of Infinite Impulse Response Filters 358
    8.8.1. Realization of Infinite Impulse Response Filters in Direct-Form I and Direct-Form II.. 358
    8.8.2. Realization of Higher-Order Infinite Impulse Response Filters via the Cascade Form 361
8.9. Application: 60-Hz Hum Eliminator and Heart Rate Detection Using Electrocardiography 362
8.10. Coefficient Accuracy Effects on Infinite Impulse Response Filters 369
8.11. Application: Generation and Detection of DTMF Tones Using the Goertzel Algorithm.. 373
    8.11.1. Single-Tone Generator 374
    8.11.2. Dual-Tone Multifrequency Tone Generator.................................... 375
    8.11.3. Goertzel Algorithm 377
    8.11.4. Dual-Tone Multifrequency Tone Detection Using the Modified Goertzel Algorithm 383
8.12. Summary of Infinite Impulse Response (IIR) Design Procedures and Selection of the IIR Filter Design Methods in Practice 388
8.13. Summary... 391
8.14. Problem.. 392

CHAPTER 9   Hardware and Software for Digital Signal Processors 405
9.1. Digital Signal Processor Architecture.. 406
9.2. Digital Signal Processor Hardware Units 408
    9.2.1. Multiplier and Accumulator 408
    9.2.2. Shifters 409
    9.2.3. Address Generators.. 409
9.3. Digital Signal Processors and Manufacturers 411
9.4. Fixed-Point and Floating-Point Formats 411
    9.4.1. Fixed-Point Format.. 412
    9.4.2. Floating-Point Format.. 419
    9.4.3. IEEE Floating-Point Formats 423
    9.4.5. Fixed-Point Digital Signal Processors 426
    9.4.6. Floating-Point Processors 427
9.5. Finite Impulse Response and Infinite Impulse Response Filter Implementations in Fixed-Point Systems... 429
9.6. Digital Signal Processing Programming Examples 434
    9.6.1. Overview of TMS320C67x DSK 434
    9.6.2. Concept of Real-Time Processing.. 438
    9.6.3. Linear Buffering 440
    9.6.4. Sample C Programs 445
9.7. Summary... 448
9.8. Problems 449

CHAPTER 10 Adaptive Filters and Applications **453**

    10.1. Introduction to Least Mean Square Adaptive Finite Impulse Response
        Filters 453

    10.2. Basic Wiener Filter Theory and Least Mean Square Algorithm.................. 457

    10.3. Applications: Noise Cancellation, System Modeling, and Line
        Enhancement... 462

        10.3.1. Noise Cancellation.. 462

        10.3.2. System Modeling 468

        10.3.3. Line Enhancement Using Linear Prediction 473

    10.4. Other Application Examples 476

        10.4.1. Canceling Periodic Interferences Using Linear
              Prediction 476

        10.4.2. Electrocardiography Interference Cancellation.............................. 476

        10.4.3. Echo Cancellation in Long-Distance Telephone Circuits................. 479

    10.5. Laboratory Examples Using the TMS320C6713 DSK............................. 480

    10.6. Summary 485

    10.7. Problems.. 486

CHAPTER 11 Waveform Quantization and Compression **497**

    11.1. Linear Midtread Quantization 497

    11.2. μ-law Companding 501

        11.2.1. Analog μ-Law Companding 501

        11.2.2. Digital μ-Law Companding... 504

    11.3. Examples of Differential Pulse Code Modulation (DPCM), Delta
        Modulation, and Adaptive DPCM G.721... 509

        11.3.1. Examples of Differential Pulse Code Modulation and Delta
              Modulation 509

        11.3.2. Adaptive Differential Pulse Code Modulation G.721....................... 512

    11.4. Discrete Cosine Transform, Modified Discrete Cosine Transform,
        and Transform Coding in MPEG Audio 519

        11.4.1. Discrete Cosine Transform 519

        11.4.2. Modified Discrete Cosine Transform 522

        11.4.3. Transform Coding in MPEG Audio 525

    11.5. Laboratory Examples of Signal Quantization Using the TMS320C6713
        DSK.. 528

    11.6. Summary 533

    11.7. MATLAB Programs.. 533

    11.8. Problems.. 548

**CHAPTER 12 Multirate Digital Signal Processing, Oversampling
                   of Analog-to-Digital Conversion, and Undersampling
                   of Bandpass Signals** **555**

    12.1. Multirate Digital Signal Processing Basics...................................... 555

        12.1.1. Sampling Rate Reduction by an Integer Factor............................ 556

12.1.2. Sampling Rate Increase by an Integer Factor.....................................562
12.1.3. Changing the Sampling Rate by a Noninteger Factor *L/M*..............567
12.1.4. Application: CD Audio Player...571
12.1.5. Multistage Decimation...574
12.2. Polyphase Filter Structure and Implementation...578
12.3. Oversampling of Analog-to-Digital Conversion...585
12.3.1. Oversampling and Analog-to-Digital Conversion Resolution...........586
12.3.2. Sigma-Delta Modulation Analog-to-Digital Conversion..................592
12.4. Application Example: CD Player..601
12.5. Undersampling of Bandpass Signals...603
12.6. Sampling Rate Conversion Using the TMS320C6713 DSK..........................608
12.7. Summary...613
12.8. Problems...613

CHAPTER 13   Subband- and Wavelet-Based Coding..621
13.1. Subband Coding Basics..621
13.2. Subband Decomposition and Two-Channel Perfect Reconstruction
        Quadrature Mirror Filter Bank..626
13.3. Subband Coding of Signals..635
13.4. Wavelet Basics and Families of Wavelets..638
13.5. Multiresolution Equations..650
13.6. Discrete Wavelet Transform..655
13.7. Wavelet Transform Coding of Signals...664
13.8. MATLAB Programs..668
13.9. Summary...672
13.10. Problems...673

CHAPTER 14   Image Processing Basics...683
14.1. Image Processing Notation and Data Formats...684
14.1.1. 8-Bit Gray Level Images...684
14.1.2. 24-bit Color Images...686
14.1.3. 8-Bit Color Images..687
14.1.4. Intensity Images...688
14.1.5. Red, Green, and Blue Components and Grayscale Conversion........688
14.1.6. MATLAB Functions for Format Conversion.....................................690
14.2. Image Histogram and Equalization...692
14.2.1. Grayscale Histogram and Equalization..692
14.2.2. 24-Bit Color Image Equalization...695
14.2.3. 8-Bit Indexed Color Image Equalization...700
14.2.4. MATLAB Functions for Equalization...702
14.3. Image Level Adjustment and Contrast..704
14.3.1. Linear Level Adjustment..704
14.3.2. Adjusting the Level for Display...707
14.3.3. MATLAB Functions for Image Level Adjustment..............................707

14.4.   Image Filtering Enhancement.. 707

14.4.1. Lowpass Noise Filtering... 709

14.4.2. Median Filtering 712

14.4.3. Edge Detection... 715

14.4.4. MATLAB Functions for Image Filtering 718

14.5.   Image Pseudo-Color Generation and Detection.............................. 722

14.6.   Image Spectra 725

14.7.   Image Compression by Discrete Cosine Transform 728

14.7.1. Two-Dimensional Discrete Cosine Transform................ 729

14.7.2. Two-Dimensional JPEG Grayscale Image Compression Example ... 731

14.7.3. JPEG Color Image Compression.......................... 735

14.7.4. Image Compression Using Wavelet Transform Coding 738

14.8.   Creating a Video Sequence by Mixing Two Images 745

14.9.   Video Signal Basics 746

14.9.1. Analog Video... 747

14.9.2. Digital Video... 753

14.10. Motion Estimation in Video 755

14.11. Summary 757

14.12. Problems 758

Appendix A: Introduction to the MATLAB Environment 767

Appendix B: Review of Analog Signal Processing Basics 775

Appendix C: Normalized Butterworth and Chebyshev Functions 805

Appendix D: Sinusoidal Steady-State Response of Digital Filters 813

Appendix E: Finite Impulse Response Filter Design Equations by the Frequency Sampling Design Method.. 817

Appendix F: Wavelet Analysis and Synthesis Equations 821

Appendix G: Some Useful Mathematical Formulas...................................... 825

Answers to Selected Problems 831

References.. 857

Index 861

Preface

Technology such as microprocessors, microcontrollers, and digital signal processors have become so advanced that they have had a dramatic impact on the disciplines of electronics engineering, computer engineering, and biomedical engineering. Engineers and technologists need to become familiar with digital signals and systems and basic digital signal processing (DSP) techniques. The objective of this book is to introduce students to the fundamental principles of these subjects and to provide a working knowledge such that they can apply DSP in their engineering careers.

The book is suitable for a two-semester course sequence at the senior level in undergraduate electronics, computer, and biomedical engineering technology programs. Chapters 1 to 8 provide the topics for a one-semester course, and a second course can complete the rest of the chapters. This textbook can also be used in an introductory DSP course in an undergraduate electrical engineering program at traditional colleges. Additionally, the book should be useful as a reference for undergraduate engineering students, science students, and practicing engineers.

The material has been tested for two consecutive courses in a signal processing sequence at Purdue University North Central in Indiana. With the background established from this book, students will be well prepared to move forward to take other upper-level courses that deal with digital signals and systems for communications and control.

The textbook consists of 14 chapters, organized as follows:

- Chapter 1 introduces concepts of DSP and presents a general DSP block diagram. Application examples are included.
- Chapter 2 covers the sampling theorem described in the time domain and frequency domain and also covers signal reconstruction. Some practical considerations for designing analog anti-aliasing lowpass filters and anti-image lowpass filters are included. The chapter ends with a section dealing with analog-to-digital conversion (ADC) and digital-to-analog conversion (DAC), as well as signal quantization and encoding.
- Chapter 3 introduces digital signals, linear time-invariant system concepts, difference equations, and digital convolutions.
- Chapter 4 introduces the discrete Fourier transform (DFT) and digital signal spectral calculations using the DFT. Methods for applying the DFT to estimate the spectra of various signals, including speech, seismic signals, electrocardiography data, and vibration signals, are demonstrated. The chapter ends with a section dedicated to illustrating fast Fourier transform (FFT) algorithms.
- Chapter 5 is devoted to the z-transform and difference equations.
- Chapter 6 covers digital filtering using difference equations, transfer functions, system stability, digital filter frequency responses, and implementation methods such as direct-form I and direct-form II.
- Chapter 7 deals with various methods of finite impulse response (FIR) filter design, including the Fourier transform method for calculating FIR filter coefficients, window method, frequency sampling design, and optimal design. Chapter 7 also includes applications that use FIR filters for noise reduction and digital crossover system design.

- Chapter 8 covers various methods of infinite impulse response (IIR) filter design, including the bilinear transformation (BLT) design, impulse-invariant design, and pole-zero placement design. Applications using IIR filters include audio equalizer design, biomedical signal enhancement, dual-tone multifrequency (DTMF) tone generation, and detection with the Goertzel algorithm.
- Chapter 9 introduces DSP architectures, software and hardware, and fixed-point and floating-point implementations of digital filters.
- Chapter 10 covers adaptive filters with applications such as noise cancellation, system modeling, line enhancement, cancellation of periodic interferences, echo cancellation, and 60-Hz interference cancellation in biomedical signals.
- Chapter 11 is devoted to speech quantization and compression, including pulse code modulation (PCM) coding, mu-law compression, adaptive differential pulse code modulation (ADPCM) coding, windowed modified discrete cosine transform (W-MDCT) coding, and MPEG audio format, specifically MP3 (MPEG-1, layer 3).
- Chapter 12 covers topics pertaining to multirate DSP and applications, as well as principles of oversampling ADC, such as sigma-delta modulation. Undersampling for bandpass signals is also examined.
- Chapter 13 introduces a subband coding system and its implementation. Perfect reconstruction conditions for a two-band system are derived. Subband coding with an application of data compression is demonstrated. Furthermore, the chapter covers the discrete wavelet transform (DWT) with applications to signal coding and denoising.
- Finally, Chapter 14 covers image enhancement using histogram equalization and filtering methods, including edge detection. The chapter also explores pseudo-color image generation and detection, two-dimensional spectra, JPEG compression using DCT, image coding using the DWT, and the mixing of two images to create a video sequence. Finally, motion compensation of the video sequence is explored, which is a key element of video compression used in MPEG.

MATLAB programs are listed whenever they are possible. Therefore, a MATLAB tutorial should be given to students who are new to the MATLAB environment.

- Appendix A serves as a MATLAB tutorial.
- Appendix B reviews key fundamentals of analog signal processing. Topics include Fourier series, Fourier transform, Laplace transform, and analog system basics.
- Appendixes C, D, and E review Butterworth and Chebyshev filters, sinusoidal steady-state responses in digital filters, and derivation of the FIR filter design equation via the frequency sampling method, respectively.
- Appendix F details the derivations of wavelet analysis and synthesis equations.
- Appendix G offers general useful mathematical formulas.

In this new edition, MATLAB projects dealing with practical applications are included in Chapters 2, 4, 6, 7, 8, 10, 12, and 13.

Instructor support, including solutions, can be found at http://textbooks.elsevier.com. MATLAB programs and exercises for students, plus Real-time C programs can be found at booksite.elsevier.com/ 9780124158931.

Thanks to all the faculty and staff at Purdue University North Central in Westville, Indiana, for their encouragement. In particular, the authors wish to thank Professors Thomas F. Brady, Larryl Matthews,

Christopher J. Smith, Alain Togbe, Edward Vavrek, Nuri Zeytinoglu, and Shengyong Zhang for their support and suggestions. We are also indebted to all former students in our DSP classes at Purdue University North Central for their feedback over the years, which helped refine this edition.

Special thanks go to Joseph P. Hayton (Publisher), Chelsea Johnston (Editorial Project Manager), and Renata Corbani (Project Manager) at Elsevier for their encouragement and guidance in developing the second edition.

The book has benefited from many constructive comments and suggestions from the following reviewers and anonymous reviewers. The authors take this opportunity to thank them for their significant contributions. We would like to thank the following reviewers for the second edition:

Professor Oktay Alkin, Southern Illinois University Edwardsville
Professor Rabah Aoufi, DeVry University-Irving, TX
Dr. Janko Calic, University of Surrey, UK
Professor Erik Cheever, Swarthmore College
Professor Samir Chettri, University of Maryland Baltimore County
Professor Nurgun Erdol, Florida Atlantic University
Professor Richard L Henderson, DeVry University, Kansas City, MO
Professor JeongHee Kim, San Jose State University
Professor Sudarshan R. Nelatury, Penn State University, Erie, PA
Professor Javad Shakib, DeVry University in Pomona, California
Dr.ir. Herbert Wormeester, University of Twente, The Netherlands
Professor Yongpeng Zhang, Prairie View A&M University

In addition we would like to repeat our thanks to the reviewers for the first edition: Professor Mateo Aboy, Oregon Institute of Technology; Professor Jean Andrian, Florida International University; Professor Rabah Aoufi, DeVry University; Professor Larry Bland, John Brown University; Professor Phillip L. De Leon, New Mexico State University; Professor Mohammed Feknous, New Jersey Institute of Technology; Professor Richard L. Henderson, DeVry University; Professor Ling Hou, St. Cloud State University; Professor Robert C. (Rob) Maher, Montana State University; Professor Abdulmagid Omar, DeVry University; Professor Ravi P. Ramachandran, Rowan University; Professor William (Bill) Routt, Wake Technical Community College; Professor Samuel D. Stearns, University of New Mexico; Professor Les Thede, Ohio Northern University; Professor Igor Tsukerman, University of Akron; Professor Vijay Vaidyanathan, University of North Texas; and Professor David Waldo, Oklahoma Christian University.

Li Tan
Jean Jiang

Introduction to Digital Signal Processing

CHAPTER OUTLINE

1.1 Basic Concepts of Digital Signal Processing 1
1.2 Basic Digital Signal Processing Examples in Block Diagrams 3
    1.2.1 Digital Filtering 3
    1.2.2 Signal Frequency (Spectrum) Analysis 3
1.3 Overview of Typical Digital Signal Processing in Real-World Applications 5
    1.3.1 Digital Crossover Audio System 5
    1.3.2 Interference Cancellation in Electrocardiography 5
    1.3.3 Speech Coding and Compression 7
    1.3.4 Compact-Disc Recording System 7
    1.3.5 Vibration Signature Analysis for Defective Gear Teeth 9
    1.3.6 Digital Photo Image Enhancement 9
1.4 Digital Signal Processing Applications 12
1.5 Summary 13

OBJECTIVES:

This chapter introduces concepts of digital signal processing (DSP) and reviews an overall picture of its applications. Illustrative application examples include digital noise filtering, signal frequency analysis, speech and audio compression, biomedical signal processing such as interference cancellation in electrocardiography, compact-disc recording, and image enhancement.

1.1 BASIC CONCEPTS OF DIGITAL SIGNAL PROCESSING

Digital signal processing (DSP) technology and its advancements have dramatically impacted our modern society everywhere. Without DSP, we would not have digital/Internet audio and video; digital recording; CD, DVD, and MP3 players; iPhone and iPad; digital cameras; digital and cellular telephones; digital satellite and TV; or wired and wireless networks. Medical instruments would be less efficient or unable to provide useful information for precise diagnoses if there were no digital electrocardiography (ECG) analyzers, digital X-rays, and medical image systems. We would also live in many less efficient ways, since we would not be equipped with voice recognition systems, speech synthesis systems, and image and video editing systems. Without DSP, scientists, engineers, and technologists would have no powerful tools to analyze and visualize the data necessary for their designs, and so on.

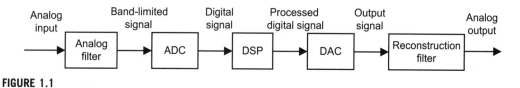

FIGURE 1.1

A digital signal processing scheme.

The basic concept of DSP is illustrated by the simplified block diagram in Figure 1.1, which consists of an analog filter, an analog-to-digital conversion (ADC) unit, a digital signal (DS) processor, a digital-to-analog conversion (DAC) unit, and a reconstruction (anti-image) filter.

As shown in the diagram, the analog input signal, which is continuous in time and amplitude, is generally encountered in the world around us. Examples of such analog signals include current, voltage, temperature, pressure, and light intensity. Usually a transducer (sensor) is used to convert the nonelectrical signal to the analog electrical signal (voltage). This analog signal is fed to an analog filter, which is applied to limit the frequency range of analog signals prior to the sampling process. The purpose of filtering is to significantly attenuate *aliasing distortion*, which will be explained in the next chapter. The band-limited signal at the output of the analog filter is then sampled and converted via the ADC unit into the digital signal, which is discrete both in time and in amplitude. The DS processor then accepts the digital signal and processes the digital data according to DSP rules such as lowpass, highpass, and bandpass digital filtering, or other algorithms for different applications. Notice that the DS processor unit is a special type of digital computer and can be a general-purpose digital computer, a microprocessor, or an advanced microcontroller; furthermore, DSP rules can be implemented using software in general.

With the DS processor and corresponding software, a processed digital output signal is generated. This signal behaves in a manner according to the specific algorithm used. The next block in Figure 1.1, the DAC unit, converts the processed digital signal to an analog output signal. As shown, the signal is continuous in time and discrete in amplitude (usually a sample-and-hold signal, to be discussed in Chapter 2). The final block in Figure 1.1 is designated as a function to smooth the DAC output voltage levels back to the analog signal via a reconstruction (anti-image) filter for real-world applications.

In general, the analog signal process does not require software, an algorithm, ADC, and DAC. The processing relies wholly on the electrical and electronic devices such as resistors, capacitors, transistors, operational amplifiers, and integrated circuits (ICs).

DSP systems, on the other hand, use software, digital processing, and algorithms; thus they have a great deal of flexibility, less noise interference, and no signal distortion in various applications. However, as shown in Figure 1.1, DSP systems still require minimum analog processing such as the anti-aliasing and reconstruction filters, which are musts for converting real-world information into digital form and digital signals back into real-world information.

Note that there are many real-world DSP applications that do not require DAC, such as data acquisition and digital information display, speech recognition, data encoding, and so on. Similarly, DSP applications that need no ADC include CD players, text-to-speech synthesis, and digital tone generators, among others. We will review some of them in the following sections.

1.2 BASIC DIGITAL SIGNAL PROCESSING EXAMPLES IN BLOCK DIAGRAMS

We first look at digital noise filtering and signal frequency analysis, using block diagrams.

1.2.1 Digital Filtering

Let us consider the situation shown in Figure 1.2, depicting a digitized noisy signal obtained from digitizing analog voltages (sensor output) containing a useful low-frequency signal and noise that occupies all of the frequency range. After ADC, the digitized noisy signal $x(n)$, where n is the sample number, can be enhanced using digital filtering.

Since our useful signal contains the low-frequency component, the high-frequency components above that of our useful signal are considered noise, which can be removed by using a digital lowpass filter. We set up the DSP block in Figure 1.2 to operate as a simple digital lowpass filter. After processing the digitized noisy signal $x(n)$, the digital lowpass filter produces a clean digital signal $y(n)$. We can apply the cleaned signal $y(n)$ to another DSP algorithm for a different application or convert it to the analog signal via DAC and the reconstruction filter.

The digitized noisy signal and clean digital signal, respectively, are plotted in Figure 1.3, where the top plot shows the digitized noisy signal, while the bottom plot demonstrates the clean digital signal obtained by applying the digital lowpass filter. Typical applications of noise filtering include acquisition of clean digital audio and biomedical signals and enhancement of speech recording, among others (Embree, 1995; Rabinar and Schafer, 1978; Webster, 1998).

FIGURE 1.2

The simple digital filtering block.

1.2.2 Signal Frequency (Spectrum) Analysis

As shown in Figure 1.4, certain DSP applications often require that time domain information and the frequency content of the signal be analyzed. Figure 1.5 shows a digitized audio signal and its calculated signal spectrum (frequency content), that is, the signal amplitude versus its corresponding frequency for the time being, obtained from a DSP algorithm, called the *fast Fourier transform* (FFT), which will be studied in Chapter 4. The plot in Figure 1.5(a) is a time domain display of the recorded audio signal with a frequency of 1,000 Hz sampled at 16,000 samples per second, while the frequency content display of plot (b) displays the calculated signal spectrum versus frequency, in which the peak amplitude is clearly located at 1,000 Hz. Plot (c) shows a time domain display of an audio signal consisting of one signal of 1,000 Hz and another of 3,000 Hz sampled at 16,000 samples per second. The frequency content display shown in plot (d) gives two locations (1,000 Hz and 3,000 Hz) where the peak amplitudes reside, hence the frequency content display presents clear frequency information of the recorded audio signal.

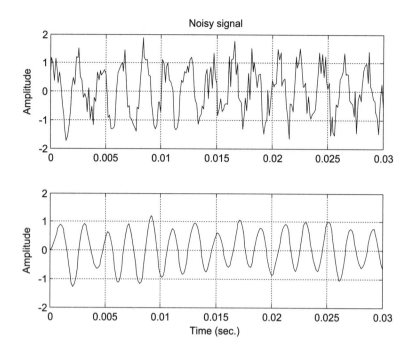

FIGURE 1.3

(Top) Digitized noisy signal. (Bottom) Clean digital signal using the digital lowpass filter.

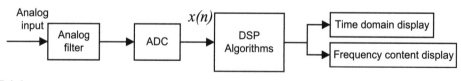

FIGURE 1.4

Signal spectral analysis.

As another practical example, we often perform spectral estimation of a digitally recorded speech or audio (music) waveform using the FFT algorithm in order to investigate spectral frequency details of speech information. Figure 1.6 shows a speech signal produced by a human in the time domain and frequency content displays. The top plot shows the digital speech waveform versus its digitized sample number, while the bottom plot shows the frequency content information of speech for a range from 0 to 4,000 Hz. We can observe that there are about ten spectral peaks, called *speech formants,* in the range between 0 and 1,500 Hz. Those identified speech formants can be used for applications such as speech modeling, speech coding, speech feature extraction for speech synthesis and recognition, and so on (Deller et al., 1993).

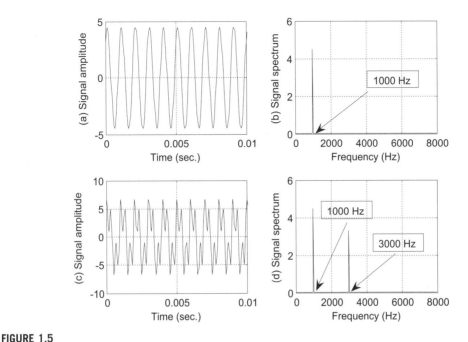

FIGURE 1.5

Audio signals and their spectrums.

1.3 OVERVIEW OF TYPICAL DIGITAL SIGNAL PROCESSING IN REAL-WORLD APPLICATIONS

1.3.1 Digital Crossover Audio System

An audio system is required to operate in an entire audible range of frequencies, which may be beyond the capability of any single speaker driver. Several drivers, such as the speaker cones and horns, each covering a different frequency range, are used to cover the full audio frequency range.

Figure 1.7 shows a typical two-band digital crossover system consisting of two speaker drivers: a woofer and a tweeter. The woofer responds to low frequencies, while the tweeter responds to high frequencies. The incoming digital audio signal is split into two bands by using a digital lowpass filter and a digital highpass filter in parallel. Then the separated audio signals are amplified. Finally, they are sent to their corresponding speaker drivers. Although the traditional crossover systems are designed using the analog circuits, the digital crossover system offers a cost-effective solution with programmability, flexibility, and high quality. This topic is taken up in Chapter 7.

1.3.2 Interference Cancellation in Electrocardiography

In ECG recording, there often is unwanted 60-Hz interference in the recorded data (Webster, 1998). The analysis shows that the interference comes from the power line and includes magnetic induction,

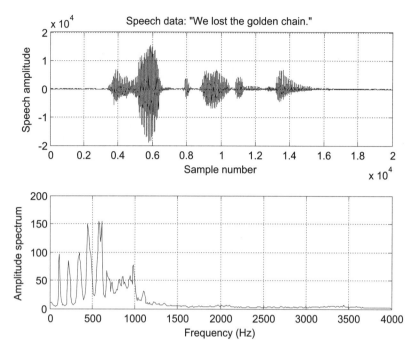

FIGURE 1.6

Speech samples and speech spectrum.

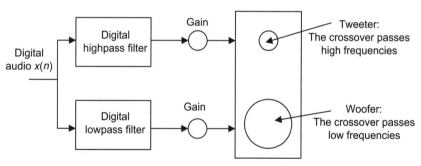

FIGURE 1.7

Two-band digital crossover.

displacement currents in leads or in the body of the patient, effects from equipment interconnections, and other imperfections. Although using proper grounding or twisted pairs minimizes such 60-Hz effects, another effective choice can be use of a digital notch filter, which eliminates the 60-Hz interference while keeping all the other useful information. Figure 1.8 illustrates a 60-Hz interference eliminator using a digital notch filter. With such enhanced ECG recording, doctors in clinics could give accurate diagnoses for patients.

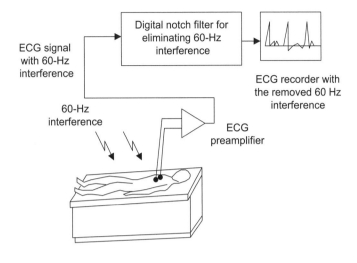

FIGURE 1.8

Elimination of 60-Hz interference in electrocardiography (ECG).

This technique can also be applied to remove 60-Hz interference in audio systems. This topic is explored in depth in Chapter 8.

1.3.3 Speech Coding and Compression

One of the speech coding methods, called *waveform coding*, is depicted in Figure 1.9A, describing the encoding process, while Figure 1.9B shows the decoding processing. As shown in Figure 1.9A, the analog signal is first sent through an analog lowpass filter to remove high frequency noise components and is then passed through the ADC unit, where the digital values at sampling instants are captured by the DS processor. Next, the captured data are compressed using data compression rules to reduce the storage requirements. Finally, the compressed digital information is sent to storage media. The compressed digital information can also be transmitted efficiently, since compression reduces the original data rate. Digital voice recorders, digital audio recorders, and MP3 players are products that use compression techniques (Deller et al., 1993; Li and Drew, 2004; Pan 1985).

To retrieve the information, the reverse process is applied. As shown in Figure 1.9B, the DS processor decompresses the data from the storage media and sends the recovered digital data to DAC. The analog output is acquired by filtering the DAC output via the reconstruction filter.

1.3.4 Compact-Disc Recording System

A compact-disc (CD) recording system is described in Figure 1.10A. The analog audio signal is sensed from each microphone and then fed to the anti-aliasing lowpass filter. Each filtered audio signal is sampled at the industry standard rate of 44.1 kilo-samples per second, quantized, and coded to 16 bits for each digital sample in each channel. The two channels are further multiplexed and encoded, and extra bits are added to provide information such as playing time and track number for the listener. The encoded

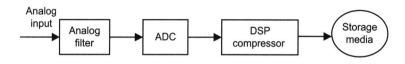

FIGURE 1.9A

Simplified data compressor.

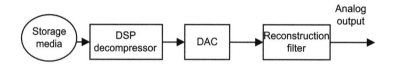

FIGURE 1.9B

Simplified data expander (decompressor).

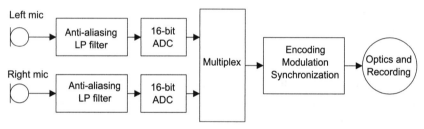

FIGURE 1.10A

Simplified encoder of the CD recording system.

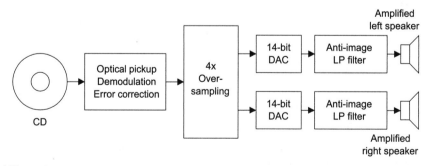

FIGURE 1.10B

Simplified decoder of the CD recording system.

data bits are modulated for storage, and more synchronized bits are added for subsequent recovery of sampling frequency. The modulated signal is then applied to control a laser beam that illuminates the photosensitive layer of a rotating glass disc. When the laser turns on and off, the digital information is etched on the photosensitive layer as a pattern of pits and lands in a spiral track. This master disc forms the basis for mass production of the commercial CD from the thermoplastic material.

During playback, as illustrated in Figure 1.10B, a laser optically scans the tracks on a CD to produce a digital signal. The digital signal is then demodulated. The demodulated signal is further oversampled by a factor of 4 to acquire a sampling rate of 176.4 kHz for each channel and is then passed to the 14-bit DAC unit. For the time being, we can consider the oversampling process as interpolation, that is, adding three samples between every two original samples in this case, as we shall see in Chapter 12. After DAC, the analog signal is sent to the anti-image analog filter, which is a lowpass filter to smooth the voltage steps from the DAC unit. The output from each anti-image filter is fed to its amplifier and loudspeaker. The purpose of the oversampling is to relieve the higher-filter-order requirement for the anti-image lowpass filter, making the circuit design much easier and economical (Ambardar, 1999).

Software audio players installed on computer systems that play music from CDs, such as Windows Media Player and RealPlayer, are examples of DSP applications. These audio players often have many advanced features, such as graphical equalizers, which allow users to change audio through techniques such as boosting low-frequency content or emphasizing high-frequency content (Ambardar, 1999; Embree, 1995; Ifeachor and Jervis, 2002).

1.3.5 Vibration Signature Analysis for Defective Gear Teeth

Gearboxes are widely used in industry and vehicles. During their extended service lifetimes, the gear teeth will inevitably be worn, chipped, or go missing. Hence, with DSP techniques, effective diagnostic methods can be developed to detect and monitor the defective gear teeth in order to enhance the reliability of the entire machine before any unexpected catastrophic events occur. Figure 1.11(a) shows the gearbox; two straight bevel gears with a transmission ratio of 1.5:1 inside the gearbox are shown in Figure 1.11(b). The number of teeth on the pinion is 18. The gearbox input shaft is connected a sheave and driven by a "V" belt drive. The vibration data can be collected by a triaxial accelerometer installed on the top of the gearbox, as shown in Figure 1.11(c). The data acquisition system uses a sampling rate of 12.8 kHz. Figure 1.11(d) shows that a pinion has a missing tooth. During the test, the motor speed is set to 1,000 RPM (revolutions per minute) so the meshing frequency is determined as $f_m = 1000(\text{RPM}) \times 18/60 = 300\,\text{Hz}$ and input shaft frequency is $f_i = 1000(\text{RPM})/60 = 16.17\,\text{Hz}$. The baseline signal and spectrum (excellent condition) from the x-direction of the accelerometer are displayed in Figure 1.12, where we can see that the spectrum contains the meshing frequency component of 300 Hz and a sideband frequency component of $283.33\ (300 - 16.67)$ Hz. Figure 1.13 shows the vibration signature for the damaged pinion in Figure 1.11(d). For the damaged pinion, the sidebands ($f_m \pm f_i, f_m \pm 2f_i \ldots$) become dominant. Hence, the vibration failure signature is identified. More details can be found in Randall (2011).

1.3.6 Digital Photo Image Enhancement

Digital image enhancement is another example of signal processing in two dimensions. Figure 1.14(a) shows a picture of an outdoor scene taken by a digital camera on a cloudy day. Due to the weather

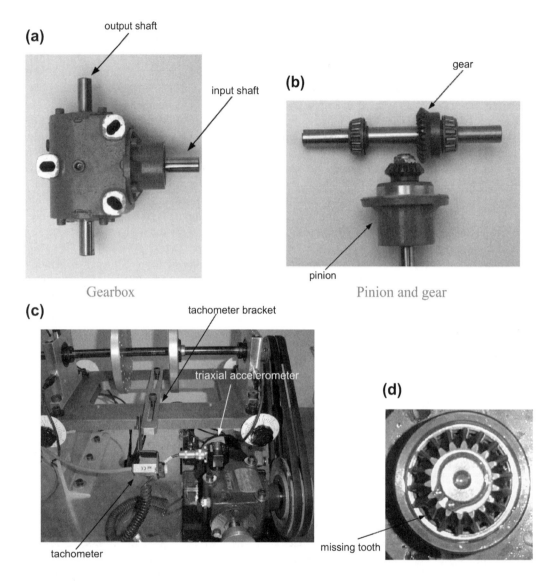

Spectra Quest's Gearbox Dynamics Simulator (GDS)   Damaged pinion

FIGURE 1.11

Vibration signature analysis of the gearbox.

(Courtesy of SpectaQuest, Inc.)

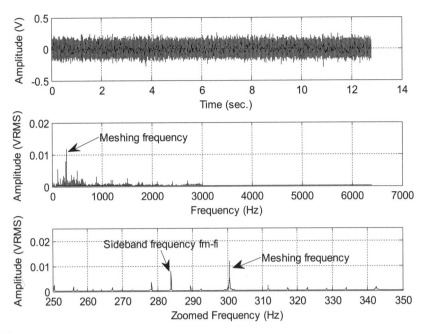

FIGURE 1.12

Vibration signal and spectrum from the gearbox in good condition.

(Data provided by SpectaQuest, Inc.)

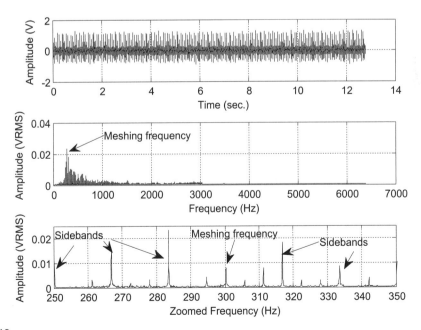

FIGURE 1.13

Vibration signal and spectrum from the damaged gearbox.

(Data provided by SpectaQuest, Inc.)

(a) Original image   **(b)** Enhanced image

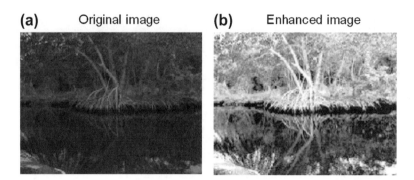

FIGURE 1.14

Image enhancement.

conditions, the image was improperly exposed in natural light and came out dark. The image processing technique called *histogram equalization* (Gozalez and Wintz, 1987) can stretch the light intensity of an image using the digital information (pixels) to increase image contrast so that detailed information in the image can easily be seen, as we can see in Figure 1.14(b). We will study this technique in Chapter 14.

1.4 DIGITAL SIGNAL PROCESSING APPLICATIONS

Applications of DSP are increasing in many areas where analog electronics are being replaced by DSP chips, and new applications are depending on DSP techniques. With the cost of DS processors decreasing and their performance increasing, DSP will continue to affect engineering design in our modern daily life. Some application examples using DSP are listed in Table 1.1.

Table 1.1 Applications of Digital Signal Processing	
Digital audio and speech	Digital audio coding such as CD players and MP3 players, digital crossover, digital audio equalizers, digital stereo and surround sound, noise reduction systems, speech coding, data compression and encryption, speech synthesis and speech recognition
Digital telephone	Speech recognition, high-speed modems, echo cancellation, speech synthesizers, DTMF (dual-tone multifrequency) generation and detection, answering machines
Automobile industry	Active noise control systems, active suspension systems, digital audio and radio, digital controls, vibration signal analysis
Electronic communications	Cellular phones, digital telecommunications, wireless LAN (local area networking), satellite communications
Medical imaging equipment	ECG analyzers, cardiac monitoring, medical imaging and image recognition, digital X-rays and image processing
Multimedia	Internet phones, audio and video, hard disk drive electronics, iPhone, iPad, digital pictures, digital cameras, text-to-voice and voice-to-text technologies

However, the list in the table by no means covers all DSP applications. Engineers and scientists are exploring many new potential applications. DSP techniques will continue to have a profound impact and improve our lives.

1.5 SUMMARY

1. An analog signal is continuous in both time and amplitude. Analog signals in the real world include current, voltage, temperature, pressure, light intensity, and so on. The digital signal contains the digital values converted from the analog signal at the specified time instants.
2. Analog-to-digital signal conversion requires an ADC unit (hardware) and a lowpass filter attached ahead of the ADC unit to block the high-frequency components that ADC cannot handle.
3. The digital signal can be manipulated using arithmetic. The manipulations may include digital filtering, calculation of signal frequency content, and so on.
4. The digital signal can be converted back to an analog signal by sending the digital values to DAC to produce the corresponding voltage levels and applying a smooth filter (reconstruction filter) to the DAC voltage steps.
5. Digital signal processing finds many applications in the areas of digital speech and audio, digital and cellular telephones, automobile controls, vibration signal analysis, communications, biomedical imaging, image/video processing, and multimedia.

Signal Sampling and Quantization

2

CHAPTER OUTLINE

2.1 Sampling of Continuous Signal 15
2.2 Signal Reconstruction 21
    2.2.1 Practical Considerations for Signal Sampling: Anti-Aliasing Filtering 25
    2.2.2 Practical Considerations for Signal Reconstruction: Anti-Image Filter and Equalizer 30
2.3 Analog-to-Digital Conversion, Digital-to-Analog Conversion, and Quantization 35
2.4 Summary 47
2.5 MATLAB Programs 48

OBJECTIVES:

This chapter investigates the sampling process, sampling theory, and the signal reconstruction process. It also includes practical considerations for anti-aliasing and anti-image filters and signal quantization.

2.1 SAMPLING OF CONTINUOUS SIGNAL

As discussed in Chapter 1, Figure 2.1 describes a simplified block diagram of a digital signal processing (DSP) system. The analog filter processes the analog input to obtain the band-limited signal, which is sent to the analog-to-digital conversion (ADC) unit. The ADC unit samples the analog signal, quantizes the sampled signal, and encodes the quantized signal level to the digital signal.

Here we first develop concepts of sampling processing in the time domain. Figure 2.2 shows an analog (continuous-time) signal (solid line) defined at every point over the time axis (horizontal line) and amplitude axis (vertical line). Hence, the analog signal contains an infinite number of points.

It is impossible to digitize an infinite number of points. The infinite points cannot be processed by the digital signal (DS) processor or computer, since they require an infinite amount of memory and infinite amount of processing power for computations. Sampling can solve such a problem by taking samples at a fixed time interval as shown in Figure 2.2 and Figure 2.3, where the time T represents the sampling interval or sampling period in seconds.

As shown in Figure 2.3, each sample maintains its voltage level during the sampling interval T to give the ADC enough time to convert it. This process is called *sample and hold*. Since there exits one amplitude level for each sampling interval, we can sketch each sample amplitude level at its corresponding sampling time instant shown in Figure 2.2, where 14 samples at their sampling time instants are plotted, each using a vertical bar with a solid circle at its top.

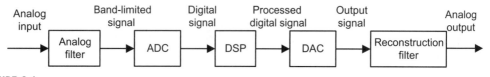

FIGURE 2.1

A digital signal processing scheme.

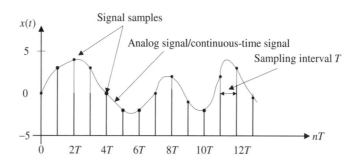

FIGURE 2.2

Display of the analog (continuous) signal and the digital samples versus the sampling time instants.

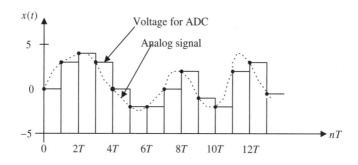

FIGURE 2.3

Sample-and-hold analog voltage for ADC.

For a given sampling interval T, which is defined as the time span between two sample points, the sampling rate is therefore given by

$$ f_s = \frac{1}{T} \text{ samples per second (Hz)} $$

For example, if a sampling period is $T = 125$ microseconds, the sampling rate is $f_s = 1/125\mu s = 8,000$ samples per second (Hz).

After obtaining the sampled signal whose amplitude values are taken at the sampling instants, the processor is able to process the sample points. Next, we have to ensure that samples are collected at a rate high enough that the original analog signal can be reconstructed or recovered later. In other words, we are looking for a minimum sampling rate to acquire a complete reconstruction of the analog signal from its sampled version. If an analog signal is not appropriately sampled, *aliasing* will occur, which causes unwanted signals in the desired frequency band.

The sampling theorem guarantees that an analog signal can be in theory perfectly recovered as long as the sampling rate is at least twice as large as the highest-frequency component of the analog signal to be sampled. The condition is described as

$$f_s \geq 2f_{\max}$$

where $f_{\max}$ is the maximum-frequency component of the analog signal to be sampled. For example, to sample a speech signal containing frequencies up to 4 kHz, the minimum sampling rate is chosen to be at least 8 kHz, or 8,000 samples per second; to sample an audio signal possessing frequencies up to 20 kHz, at least 40,000 samples per second, or 40 kHz, of the audio signal are required.

Figure 2.4 illustrates sampling of two sinusoids, where the sampling interval between sample points is $T = 0.01$ second, and the sampling rate is thus $f_s = 100$ Hz. The first plot in the figure displays a sine wave with a frequency of 40 Hz and its sampled amplitudes. The sampling theorem condition is satisfied since $2f_{\max} = 80 < f_s$. The sampled amplitudes are labeled using the circles shown in the first

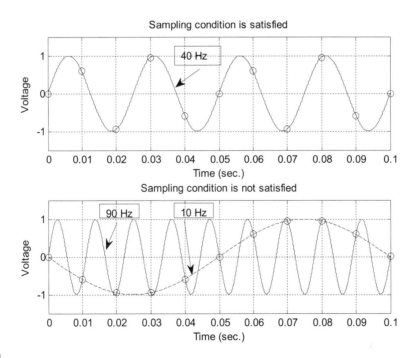

FIGURE 2.4

Plots of the appropriately sampled signals and nonappropriately sampled (aliased) signals.

plot. We notice that the 40-Hz signal is adequately sampled, since the sampled values clearly come from the analog version of the 40-Hz sine wave. However, as shown in the second plot, the sine wave with a frequency of 90 Hz is sampled at 100 Hz. Since the sampling rate of 100 Hz is relatively low compared with the 90-Hz sine wave, the signal is undersampled due to $2f_{max} = 180 > f_s$. Hence, the condition of the sampling theorem is not satisfied. Based on the sample amplitudes labeled with the circles in the second plot, we cannot tell whether the sampled signal comes from sampling a 90-Hz sine wave (plotted using the solid line) or from sampling a 10-Hz sine wave (plotted using the dot-dash line). They are not distinguishable. Thus they are *aliases* of each other. We call the 10-Hz sine wave the aliasing noise in this case, since the sampled amplitudes actually come from sampling the 90-Hz sine wave.

Now let us develop the sampling theorem in frequency domain, that is, the minimum sampling rate requirement for sampling an analog signal. As we shall see, in practice this can help us design the anti-aliasing filter (a lowpass filter that will reject high frequencies that cause aliasing) that will be applied before sampling, and the anti-image filter (a reconstruction lowpass filter that will smooth the recovered sample-and-hold voltage levels to an analog signal) that will be applied after the digital-to-analog conversion (DAC).

Figure 2.5 depicts the sampled signal $x_s(t)$ obtained by sampling the continuous signal $x(t)$ at a sampling rate of f_s samples per second.

Mathematically, this process can be written as the product of the continuous signal and the sampling pulses (pulse train):

$$x_s(t) = x(t)p(t) \tag{2.1}$$

where $p(t)$ is the pulse train with a period $T = 1/f_s$. From spectral analysis, the original spectrum (frequency components) $X(f)$ and the sampled signal spectrum $X_s(f)$ in terms of Hz are related as

$$X_s(f) = \frac{1}{T} \sum_{n=-\infty}^{\infty} X(f - nf_s) \tag{2.2}$$

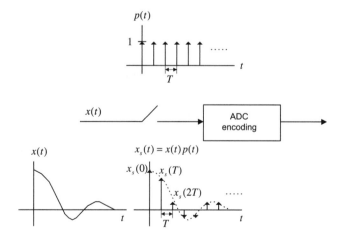

FIGURE 2.5

The simplified sampling process.

where $X(f)$ is assumed to be the original baseband spectrum while $X_s(f)$ is its sampled signal spectrum, consisting of the original baseband spectrum $X(f)$ and its replicas $X(f \pm nf_s)$. Since Equation (2.2) is a well-known formula, the derivation is omitted here and can be found in well-known texts (Ahmed and Nataranjan, 1983; Ambardar, 1999; Alkin, 1993; Oppenheim and Schafer, 1975; Proakis and Manolakis, 1996).

Expanding Equation (2.2) leads to the sampled signal spectrum in Equation (2.3):

$$X_s(f) = \cdots + \frac{1}{T}X(f + f_s) + \frac{1}{T}X(f) + \frac{1}{T}X(f - f_s) + \cdots \tag{2.3}$$

Equation (2.3) indicates that the sampled signal spectrum is the sum of the scaled original spectrum and copies of its shifted versions, called *replicas*. Three possible sketches based on Equation (2.3) can be obtained. Given the original signal spectrum $X(f)$ plotted in Figure 2.6(a), the sampled signal spectrum according to Equation (2.3) is plotted in Figure 2.6(b), where the replicas $\frac{1}{T}X(f)$, $\frac{1}{T}X(f - f_s)$, $\frac{1}{T}X(f + f_s)$, ..., have separations between them. Figure 2.6(c) shows that the baseband spectrum and its replicas, $\frac{1}{T}X(f)$, $\frac{1}{T}X(f - f_s)$, $\frac{1}{T}X(f + f_s)$, ..., are just connected, and finally, in Figure 2.6(d), the original

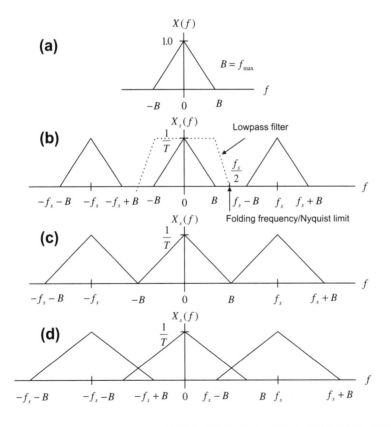

FIGURE 2.6

Plots of the sampled signal spectrum.

spectrum $\frac{1}{T}X(f)$ and its replicas $\frac{1}{T}X(f - f_s)$, $\frac{1}{T}X(f + f_s)$, ..., are overlapped; that is, there are many overlapping portions in the sampled signal spectrum.

From Figure 2.6, it is clear that the sampled signal spectrum consists of the scaled baseband spectrum centered at the origin, and its replicas centered at the frequencies of $\pm nf_s$ (multiples of the sampling rate) for each of $n = 1, 2, 3, \dots$.

If applying a lowpass reconstruction filter to obtain exact reconstruction of the original signal spectrum, the following condition must be satisfied:

$$f_s - f_{max} \geq f_{max} \tag{2.4}$$

Solving Equation (2.4) gives

$$f_s \geq 2f_{max} \tag{2.5}$$

In terms of frequency in radians per second, Equation (2.5) is equivalent to

$$\omega_s \geq 2\omega_{max} \tag{2.6}$$

This fundamental conclusion is well known as the **Shannon sampling theorem**, which is formally described below:

> For a uniformly sampled DSP system, an analog signal can be perfectly recovered as long as the sampling rate is at least twice as large as the highest-frequency component of the analog signal to be sampled.

We summarize two key points here.

1. The sampling theorem establishes a minimum sampling rate for a given band-limited analog signal with highest-frequency component f_{max}. If the sampling rate satisfies Equation (2.5), then the analog signal can be recovered via its sampled values using the lowpass filter, as described in Figure 2.6(b).
2. Half of the sampling frequency $f_s/2$ is usually called the *Nyquist frequency* (Nyquist limit) or *folding frequency*. The sampling theorem indicates that a DSP system with a sampling rate of f_s can ideally sample an analog signal with a maximum frequency that is up to half of the sampling rate without introducing spectral overlap (aliasing). Hence, the analog signal can be perfectly recovered from its sampled version.

Let us study the following example.

EXAMPLE 2.1

Suppose that an analog signal is given as

$$x(t) = 5\cos(2\pi \cdot 1,000t), \text{ for } t \geq 0$$

and is sampled at the rate 8,000 Hz.

a. Sketch the spectrum for the original signal.
b. Sketch the spectrum for the sampled signal from 0 to 20 kHz.

Solution:

a. Since the analog signal is sinusoid with a peak value of 5 and frequency of 1,000 Hz, we can write the sine wave using Euler's identity:

$$5\cos(2\pi \times 1,000t) = 5 \cdot \left(\frac{e^{j2\pi \times 1,000t} + e^{-j2\pi \times 1,000t}}{2}\right) = 2.5e^{j2\pi \times 1,000t} + 2.5e^{-j2\pi \times 1,000t}$$

which is a Fourier series expansion for a continuous periodic signal in terms of the exponential form (see Appendix B). We can identify the Fourier series coefficients as

$$c_1 = 2.5 \text{ and } c_{-1} = 2.5$$

Using the magnitudes of the coefficients, we then plot the two-side spectrum as shown in Figure 2.7A.

b. After the analog signal is sampled at the rate of 8,000 Hz, the sampled signal spectrum and its replicas centered at the frequencies $\pm nf_s$, each with a scaled amplitude of $2.5/T$, are as shown in Figure 2.7B:

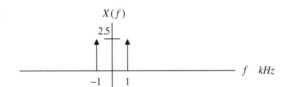

FIGURE 2.7A

Spectrum of the analog signal in Example 2.1.

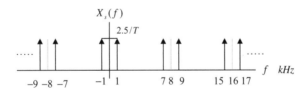

FIGURE 2.7B

Spectrum of the sampled signal in Example 2.1.

　　　Notice that the spectrum of the sampled signal shown in Figure 2.7B contains the images of the original spectrum shown in Figure 2.7A; that the images repeat at multiples of the sampling frequency f_s (for our example, 8 kHz, 16kHz, 24kHz, ...); and that all images must be removed, since they convey no additional information.

2.2 SIGNAL RECONSTRUCTION

In this section, we investigate the recovery of analog signal from its sampled signal version. Two simplified steps are involved, as described in Figure 2.8. First, the digitally processed data $y(n)$ are converted to the ideal impulse train $y_s(t)$, in which each impulse has amplitude proportional to digital output $y(n)$, and two consecutive impulses are separated by a sampling period of T; second, the analog

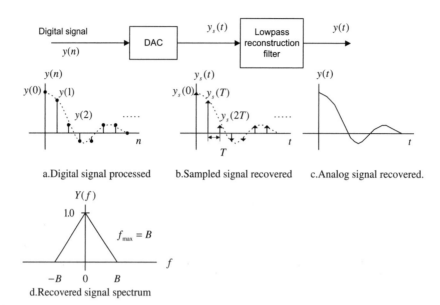

a.Digital signal processed   b.Sampled signal recovered   c.Analog signal recovered.

d.Recovered signal spectrum

FIGURE 2.8

Signal notations at the reconstruction stage.

reconstruction filter is applied to the ideally recovered sampled signal $y_s(t)$ to obtain the recovered analog signal.

To study the signal reconstruction, we let $y(n) = x(n)$ for the case of no DSP, so that the reconstructed sampled signal and the input sampled signal are ensured to be the same; that is, $y_s(t) = x_s(t)$. Hence, the spectrum of the sampled signal $y_s(t)$ contains the same spectral content of the original spectrum $X(f)$, that is, $Y(f) = X(f)$, with a bandwidth of $f_{max} = B$ Hz (described in Figure 2.8d) and the images of the original spectrum (scaled and shifted versions). The following three cases are discussed for recovery of the original signal spectrum $X(f)$.

Case 1: $f_s = 2f_{max}$

As shown in Figure 2.9, where the Nyquist frequency is equal to the maximum frequency of the analog signal $x(t)$, an ideal lowpass reconstruction filter is required to recover the analog signal spectrum. This is an impractical case.

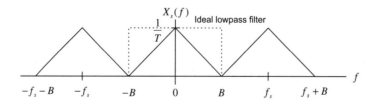

FIGURE 2.9

Spectrum of the sampled signal when $f_s = 2f_{max}$.

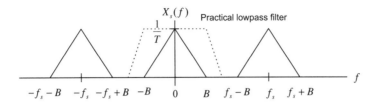

FIGURE 2.10

Spectrum of the sampled signal when $f_s > 2f_{max}$.

Case 2: $f_s > 2f_{max}$

In this case, as shown in Figure 2.10, there is a separation between the highest-frequency edge of the baseband spectrum and the lower edge of the first replica. Therefore, a practical lowpass reconstruction (anti-image) filter can be designed to reject all the images and achieve the original signal spectrum.

Case 3: $f_s < 2f_{max}$

Case 3 violates the condition of the Shannon sampling theorem. As we can see, Figure 2.11 depicts the spectral overlapping between the original baseband spectrum and the spectrum of the first replica and so on. Even when we apply an ideal lowpass filter to remove these images, in the baseband there are still some foldover frequency components from the adjacent replica. This is aliasing, where the recovered baseband spectrum suffers spectral distortion, that is, it contains an aliasing noise spectrum; in the time domain, the recovered analog signal may consist of the aliasing noise frequency or frequencies. Hence, the recovered analog signal is incurably distorted.

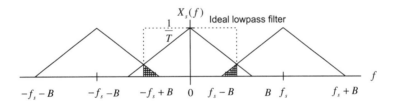

FIGURE 2.11

Spectrum of the sampled signal when $f_s < 2f_{max}$.

Note that if an analog signal with a frequency f is undersampled, the aliasing frequency component f_{alias} in the baseband is simply given by the following expression:

$$f_{alias} = f_s - f$$

The following examples give a spectrum analysis of the signal recovery.

EXAMPLE 2.2

Assume that an analog signal is given by

$$x(t) = 5\cos(2\pi \cdot 2,000t) + 3\cos(2\pi \cdot 3,000t), \text{ for } t \geq 0$$

and is sampled at the rate of 8,000 Hz.

a. Sketch the spectrum of the sampled signal up to 20 kHz.
b. Sketch the recovered analog signal spectrum if an ideal lowpass filter with a cutoff frequency of 4 kHz is used to filter the sampled signal ($y(n) = x(n)$ in this case) to recover the original signal.

Solution:
a. Using Euler's identity, we get

$$x(t) = \frac{3}{2}e^{-j2\pi \cdot 3,000t} + \frac{5}{2}e^{-j2\pi \cdot 2,000t} + \frac{5}{2}e^{j2\pi \cdot 2,000t} + \frac{3}{2}e^{j2\pi \cdot 3,000t}$$

The two-sided amplitude spectrum for the sinusoid is displayed in Figure 2.12:

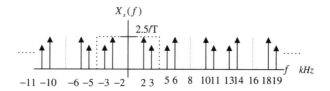

FIGURE 2.12

Spectrum of the sampled signal in Example 2.2.

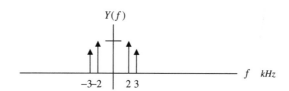

FIGURE 2.13

Spectrum of the recovered signal in Example 2.2.

b. Based on the spectrum in (a), the sampling theorem condition is satisfied; hence, we can recover the original spectrum using a reconstruction lowpass filter. The recovered spectrum is shown in Figure 2.13.

EXAMPLE 2.3

Assume an analog signal is given by

$$x(t) = 5\cos(2\pi \times 2,000t) + 1\cos(2\pi \times 5,000t), \text{ for } t \ge 0$$

and is sampled at a rate of 8,000 Hz.

a. Sketch the spectrum of the sampled signal up to 20 kHz.
b. Sketch the recovered analog signal spectrum if an ideal lowpass filter with a cutoff frequency of 4 kHz is used to recover the original signal ($y(n) = x(n)$ in this case).

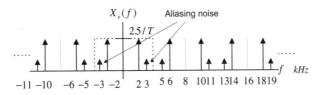

FIGURE 2.14

Spectrum of the sampled signal in Example 2.3.

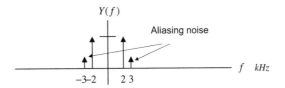

FIGURE 2.15

Spectrum of the recovered signal in Example 2.3.

Solution:
a. The spectrum for the sampled signal is sketched in Figure 2.14.
b. Since the maximum frequency of the analog signal is larger than that of the Nyquist frequency—that is, twice the maximum frequency of the analog signal is larger than the sampling rate—the sampling theorem condition is violated. The recovered spectrum is shown in Figure 2.15, where we see that aliasing noise occurs at 3 kHz.

2.2.1 Practical Considerations for Signal Sampling: Anti-Aliasing Filtering

In practice, the analog signal to be digitized may contain other frequency components in addition to the folding frequency, such as high-frequency noise. To satisfy the sampling theorem condition, we apply an anti-aliasing filter to limit the input analog signal, so that all the frequency components are less than the folding frequency (half of the sampling rate). Considering the worst case, where the analog signal to be sampled has a flat frequency spectrum, the band limited spectrum $X(f)$ and sampled spectrum $X_s(f)$ are depicted in Figure 2.16, where the shape of each replica in the sampled signal spectrum is the same as that of the anti-aliasing filter magnitude frequency response.

Due to nonzero attenuation of the magnitude frequency response of the anti-aliasing lowpass filter, the aliasing noise from the adjacent replica still appears in the baseband. However, the amount of aliasing noise is greatly reduced. We can also control the aliasing noise by either using a higher-order lowpass filter or increasing the sampling rate. For illustrative purpose, we use a Butterworth filter. The method can also be extended to other filter types such as the Chebyshev filter. The Butterworth magnitude frequency response with an order of n is given by

$$|H(f)| = \frac{1}{\sqrt{1 + \left(\dfrac{f}{f_c}\right)^{2n}}} \tag{2.7}$$

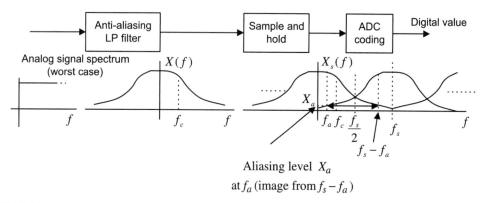

FIGURE 2.16

Spectrum of the sampled analog signal with a practical anti-aliasing filter.

For a second-order Butterworth lowpass filter with unit gain, the transfer function (which will be discussed in Chapter 8) and its magnitude frequency response are given by

$$H(s) = \frac{(2\pi f_c)^2}{s^2 + 1.4141 \times (2\pi f_c)s + (2\pi f_c)^2} \tag{2.8}$$

$$|H(f)| = \frac{1}{\sqrt{1 + \left(\dfrac{f}{f_c}\right)^4}} \tag{2.9}$$

A unit gain second-order lowpass filter using a Sallen-Key topology is shown in Figure 2.17. Matching the coefficients of the circuit transfer function to that of the second-order Butterworth lowpass transfer function in Equation (2.10) gives the design formulas shown in Figure 2.17, where for a given cutoff

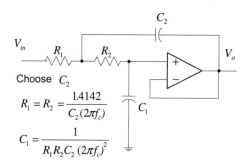

FIGURE 2.17

Second-order unit gain Sallen-Key lowpass filter.

frequency of f_c in Hz, and a capacitor value of C_2, we can determine the other elements using the formulas listed in the figure.

$$\frac{\frac{1}{R_1R_2C_1C_2}}{s^2 + \left(\frac{1}{R_1C_2} + \frac{1}{R_2C_2}\right)s + \frac{1}{R_1R_2C_1C_2}} = \frac{(2\pi f_c)^2}{s^2 + 1.4141 \times (2\pi f_c)s + (2\pi f_c)^2} \qquad (2.10)$$

As an example, for a cutoff frequency of 3,400 Hz, and by selecting $C_2 = 0.01$ microfarad (μF), we get

$$R_1 = R_2 = 6,620\ \Omega, \text{ and } C_1 = 0.005\ \mu F$$

Figure 2.18 shows the magnitude frequency response, where the absolute gain of the filter is plotted. As we can see, the absolute attenuation begins at the level of 0.7 at 3,400 Hz and reduces to 0.3 at 6,000 Hz. Ideally, we want the gain attenuation to be zero after 4,000 Hz if our sampling rate is 8,000 Hz. Practically speaking, aliasing will occur anyway with some degree. We will study achieving the higher-order analog filter via Butterworth and Chebyshev prototype function tables in Chapter 8. More details of the circuit realization for the analog filter can be found in Chen (1986).

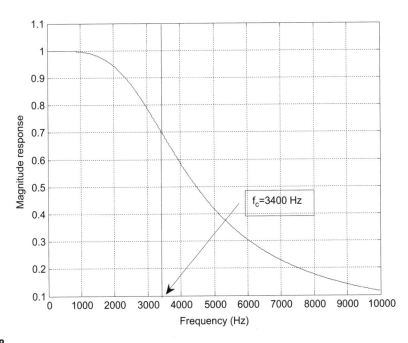

FIGURE 2.18

Magnitude frequency response of the second-order Butterworth lowpass filter.

According to Figure 2.16, we can derive the aliasing level percentage using the symmetry of the Butterworth magnitude function and its first replica. It follows that

$$\text{aliasing level \%} = \frac{X_a}{X(f)|_{f=f_a}} = \frac{|H(f)|_{f=f_s-f_a}}{|H(f)|_{f=f_a}} = \frac{\sqrt{1 + \left(\frac{f_a}{f_c}\right)^{2n}}}{\sqrt{1 + \left(\frac{f_s-f_a}{f_c}\right)^{2n}}} \quad \text{for } 0 \leq f \leq f_c \quad (2.11)$$

With Equation (2.11), we can estimate the aliasing noise percentage, or choose a higher-order anti-aliasing filter to satisfy the requirement for the aliasing level percentage.

EXAMPLE 2.4

Given the DSP system shown in Figures 2.16 to 2.18, where a sampling rate of 8,000 Hz is used and the anti-aliasing filter is a second-order Butterworth lowpass filter with a cutoff frequency of 3.4 kHz, determine

a. the percentage of aliasing level at the cutoff frequency;
b. the percentage of aliasing level at a frequency of 1,000 Hz.

Solution:

$$f_s = 8,000, \ f_c = 3,400, \ \text{and} \ n = 2$$

a. Since $f_a = f_c = 3,400$ Hz, we compute

$$\text{aliasing level \%} = \frac{\sqrt{1 + \left(\frac{3.4}{3.4}\right)^{2\times2}}}{\sqrt{1 + \left(\frac{8-3.4}{3.4}\right)^{2\times2}}} = \frac{1.4142}{2.0858} = 67.8\%$$

b. With $f_a = 1,000$ Hz, we have

$$\text{aliasing level \%} = \frac{\sqrt{1 + \left(\frac{1}{3.4}\right)^{2\times2}}}{\sqrt{1 + \left(\frac{8-1}{3.4}\right)^{2\times2}}} = \frac{1.03007}{4.3551} = 23.05\%$$

Let us examine another example with an increased sampling rate.

EXAMPLE 2.5

Given the DSP system shown in Figures 2.16 to 2.18, where a sampling rate of 16,000 Hz is used and the anti-aliasing filter is a second-order Butterworth lowpass filter with a cutoff frequency of 3.4 kHz, determine the percentage of aliasing level at the cutoff frequency.

Solution:

$$f_s = 16,000, \ f_c = 3,400, \ \text{and} \ n = 2$$

Since $f_a = f_c = 3,400$ Hz, we have

$$\text{aliasing level } \% = \frac{\sqrt{1 + \left(\frac{3.4}{3.4}\right)^{2\times2}}}{\sqrt{1 + \left(\frac{16 - 3.4}{3.4}\right)^{2\times2}}} = \frac{1.4142}{13.7699} = 10.26\%$$

In comparison with the result in Example 2.4, increasing the sampling rate can reduce the aliasing level.

The following example shows how to choose the order of the anti-aliasing filter.

EXAMPLE 2.6

Given the DSP system shown in Figure 2.16, where a sampling rate of 40,000 Hz is used, the anti-aliasing filter is the Butterworth lowpass filter with a cutoff frequency 8 kHz, and the percentage of aliasing level at the cutoff frequency is required to be less than 1%, determine the order of the anti-aliasing lowpass filter.

Solution:
Using $f_s = 40,000$, $f_c = 8,000$, and $f_a = 8,000$ Hz, we start at order 1 and increase the filter order until the requirement is met.

$$n = 1, \quad \text{aliasing level } \% = \frac{\sqrt{1 + \left(\frac{8}{8}\right)^{2\times1}}}{\sqrt{1 + \left(\frac{40 - 8}{8}\right)^{2\times1}}} = \frac{1.4142}{\sqrt{1 + (4)^2}} = 34.30\%$$

$$n = 2, \quad \text{aliasing level } \% = \frac{1.4142}{\sqrt{1 + (4)^4}} = 8.82\%$$

$$n = 3, \quad \text{aliasing level } \% = \frac{1.4142}{\sqrt{1 + (4)^6}} = 2.21\%$$

$$n = 4, \quad \text{aliasing level } \% = \frac{1.4142}{\sqrt{1 + (4)^8}} = 0.55\% < 1\%$$

To satisfy the 1% aliasing level requirement, we choose $n = 4$.

2.2.2 Practical Considerations for Signal Reconstruction: Anti-Image Filter and Equalizer

The analog signal recovery for a practical DSP system is illustrated in Figure 2.19.

As shown in Figure 2.19, the DAC unit converts the processed digital signal $y(n)$ to a sampled signal $y_s(t)$, and then the hold circuit produces the sample-and-hold voltage $y_H(t)$. The transfer function of the hold circuit can be derived as

$$H_h(s) = \frac{1 - e^{-sT}}{sT} \tag{2.12}$$

We can obtain the frequency response of the DAC with the hold circuit by substituting $s = j\omega$ in Equation (2.12). It follows that

$$H_h(\omega) = e^{-j\omega T/2} \frac{\sin(\omega T/2)}{\omega T/2} \tag{2.13}$$

The magnitude and phase responses are given by

$$|H_h(\omega)| = \left| \frac{\sin(\omega T/2)}{\omega T/2} \right| = \left| \frac{\sin(x)}{x} \right| \tag{2.14}$$

$$\angle H_h(\omega) = -\omega T/2 \tag{2.15}$$

where $x = \omega T/2$. In terms of Hz, we have

$$|H_h(f)| = \left| \frac{\sin(\pi f T)}{\pi f T} \right| \tag{2.16}$$

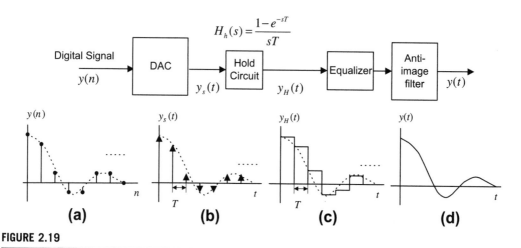

FIGURE 2.19

Signal notations at the practical reconstruction stage. (a) Processed digital signal. (b) Recovered ideal sampled signal. (c) Recovered sample-and-hold voltage. (d) Recovered analog signal.

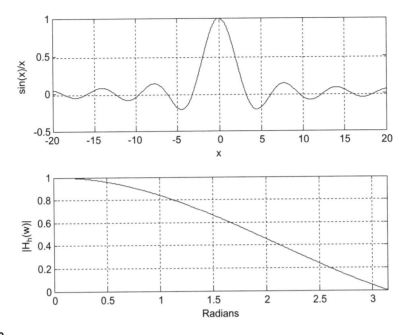

FIGURE 2.20

Sample-and-hold lowpass filtering effect.

$$\angle H_h(f) = -\pi f T \qquad (2.17)$$

The plot of the magnitude effect is shown in Figure 2.20.

The magnitude frequency response acts like lowpass filtering and shapes the sampled signal spectrum of $Y_s(f)$. This shaping effect distorts the sampled signal spectrum $Y_s(f)$ in the desired frequency band, as illustrated in Figure 2.21. On the other hand, the spectral images are attenuated

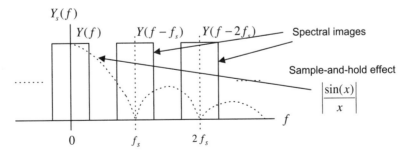

FIGURE 2.21

Sample-and-hold effect and distortion.

due to the lowpass effect of $\sin(x)/x$. This sample-and-hold effect can help us design the anti-image filter.

As shown in Figure 2.21, the percentage of distortion in the desired frequency band is given by

$$\text{distortion \%} = (1 - |H_h(f)|) \times 100\%$$

$$= \left(1 - \left|\frac{\sin(\pi fT)}{\pi fT}\right|\right) \times 100\% \tag{2.18}$$

EXAMPLE 2.7

Given a DSP system with a sampling rate of 8,000 Hz and a hold circuit used after DAC, determine

a. the percentage of distortion at a frequency of 3,400 Hz;
b. the percentage of distortion at a frequency of 1,000 Hz.

Solution:
a. Since $fT = 3,400 \times 1/8,000 = 0.425$,

$$\text{distortion \%} = \left(1 - \left|\frac{\sin(0.425\pi)}{0.425\pi}\right|\right) \times 100\% = 27.17\%$$

b. Since $fT = 1,000 \times 1/8,000 = 0.125$,

$$\text{distortion \%} = \left(1 - \left|\frac{\sin(0.125\pi)}{0.125\pi}\right|\right) \times 100\% = 2.55\%$$

To overcome the sample-and-hold effect, the following methods can be applied.

1. We can compensate the sample-and-hold shaping effect using an equalizer whose magnitude response is opposite to the shape of the hold circuit magnitude frequency response, which is shown as the solid line in Figure 2.22.

2. We can increase the sampling rate using oversampling and interpolation methods when a higher sampling rate is available at the DAC. Using the interpolation will increase the sampling rate without affecting the signal bandwidth, so that the baseband spectrum and its images are separated further apart and a lower-order anti-aliasing filter can be used. This subject will be discussed in Chapter 12.

3. We can change the DAC configuration and perform digital pre-equalization using a flexible digital filter whose magnitude frequency response is against the spectral shape effect due to the hold circuit. Figure 2.23 shows a possible implementation. In this way, the spectral shape effect can be balanced before the sampled signal passes through the hold circuit. Finally, the anti-image filter will remove the rest of images and recover the desired analog signal.

The following practical example will illustrate the design of an anti-image filter using a higher sampling rate while making use of the sample-and-hold effect.

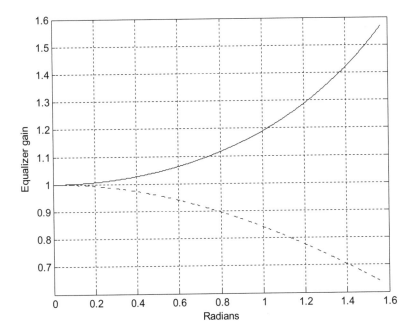

FIGURE 2.22

Ideal equalizer magnitude frequency response to overcome the distortion introduced by the sample-and-hold process.

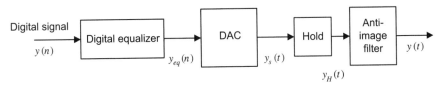

FIGURE 2.23

Possible implementation using a digital equalizer.

EXAMPLE 2.8

Determine the cutoff frequency and the order for the anti-image filter given a DSP system with a sampling rate of 16,000 Hz and specifications for the anti-image filter as shown in Figure 2.24.

Design requirements:

- Maximum allowable gain variation from 0 to 3,000 Hz = 2 dB
- 33 dB rejection at a frequency of 13,000 Hz
- Butterworth filter is assumed for the anti-image filter.

Solution:

We first determine the spectral shaping effects at $f = 3,000$ Hz and $f = 13,000$ Hz; that is,

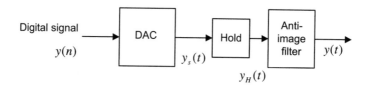

FIGURE 2.24

DSP recover system for Example 2.8.

$$f = 3,000 \text{ Hz}, \ fT = 3,000 \times 1/16,000 = 0.1785$$

$$\text{gain} = \left| \frac{\sin(0.1875\pi)}{0.1875\pi} \right| = 0.9484 = -0.46 \text{ dB}$$

and

$$f = 13,000 \text{ Hz}, \ fT = 13,000 \times 1/16,000 = 0.8125$$

$$\text{gain} = \left| \frac{\sin(0.8125\pi)}{0.8125\pi} \right| = 0.2177 \approx -13 \text{ dB}$$

This gain would help the attenuation requirement.
  Hence, the design requirements for the anti-image filter are

- Butterworth lowpass filter
- Maximum allowable gain variation from 0 to 3,000 Hz = 2−0.46 = 1.54 dB
- 33−13 = 20 dB rejection at frequency 13,000 Hz.

We set up equations using log operations of the Butterworth magnitude function as

$$20 \log(1 + (3,000/f_c)^{2n})^{1/2} \leq 1.54$$

$$20 \log(1 + (13,000/f_c)^{2n})^{1/2} \geq 20$$

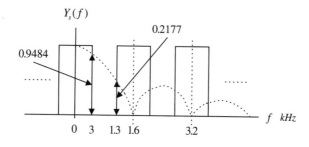

FIGURE 2.25

Spectral shaping by the sample-and-hold effect in Example 2.8.

From these two equations, we have to satisfy

$$(3,000/f_c)^{2n} = 10^{0.154} - 1$$

$$(13,000/f_c)^{2n} = 10^2 - 1$$

Taking the ratio of these two equations yields

$$\left(\frac{13,000}{3,000}\right)^{2n} = \frac{10^2 - 1}{10^{0.154} - 1}$$

Then

$$n = \frac{1}{2}\log((10^2 - 1)/(10^{0.154} - 1))/\log(13,000/3,000) = 1.86 \approx 2$$

Finally, the cutoff frequency can be computed as

$$f_c = \frac{13,000}{(10^2 - 1)^{1/(2n)}} = \frac{13,000}{(10^2 - 1)^{1/4}} = 4,121.30 \text{ Hz}$$

$$f_c = \frac{3,000}{(10^{0.154} - 1)^{1/(2n)}} = \frac{3,000}{(10^{0.154} - 1)^{1/4}} = 3,714.23 \text{ Hz}$$

We choose the smaller one, that is,

$$f_c = 3,714.23 \text{ Hz}$$

With the filter order and cutoff frequency, we can realize the anti-image (reconstruction) filter using the second-order unit gain Sallen-Key lowpass filter described in Figure 2.17.

Note that the specifications for anti-aliasing filter designs are similar to anti-image (reconstruction) filters, except for their stopband edges. The anti-aliasing filter is designed to block the frequency components beyond the folding frequency before the ADC operation, while the reconstruction filter is designed to block the frequency components beginning at the lower edge of the first image after the DAC.

2.3 ANALOG-TO-DIGITAL CONVERSION, DIGITAL-TO-ANALOG CONVERSION, AND QUANTIZATION

During the ADC process, amplitudes of the analog signal to be converted have infinite precision. The continuous amplitude must be converted to digital data with finite precision, which is called *quantization*. Figure 2.26 shows quantization as a part of ADC.

There are several ways to implement ADC. The most common ones are

- Flash ADC
- Successive approximation ADC
- Sigma-delta ADC.

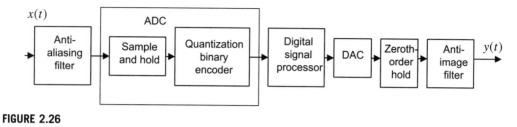

FIGURE 2.26

A block diagram for a DSP system.

In this chapter, we will focus on a simple 2-bit flash ADC unit, described in Figure 2.27, for illustrative purposes. Sigma-delta ADC will be studied in Chapter 12.

As shown in Figure 2.27, the 2-bit flash ADC unit consists of a serial reference voltage created by the equal value resistors, a set of comparators, and logic units. As an example, the reference voltages in the figure are 1.25 volts, 2.5 volts, 3.75 volts, and 5 volts, respectively. If an analog sample-and-hold voltage is $V_{in} = 3$ volts, then the lower two comparators will each output logic 1. Through the logic units, only the line labeled 10 is actively high, and the rest of lines are actively low. Hence, the encoding logic circuit outputs a 2-bit binary code of 10.

Flash ADC offers the advantage of high conversion speed, since all bits are acquired at the same time. Figure 2.28 illustrates a simple 2-bit DAC unit using an R-2R ladder. The DAC contains the R-2R ladder circuit, a set of single-throw switches, an adder, and a phase shifter. If a bit is logic 0, the switch connects a 2R resistor to ground. If a bit is logic 1, the corresponding 2R resistor is connected to the branch to the input of the operational amplifier (adder). When the operational amplifier operates in a linear range, the negative input is virtually equal to the positive input. The adder adds all the currents

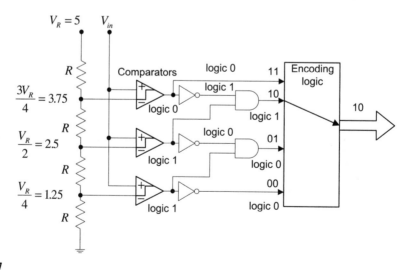

FIGURE 2.27

An example of a 2-bit flash ADC.

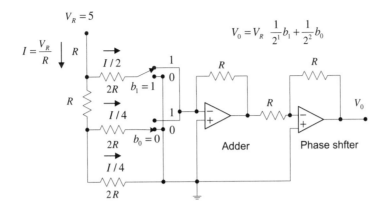

FIGURE 2.28

R-2R ladder DAC.

from all branches. The feedback resistor R in the adder provides overall amplification. The ladder network is equivalent to two 2R resistors in parallel. The entire network has a total current of

$$I = \frac{V_R}{R}$$

using Ohm's law, where V_R is the reference voltage, chosen to be 5 volts for our example. Hence, half of the total current flows into the b_1 branch, while the other half flows into the rest of the network. The halving process repeats for each branch successively to the lower bit branches to get lower bit weights. The second operational amplifier acts like a phase shifter to cancel the negative sign of the adder output. Using the basic electric circuit principle, we can determine the DAC output voltage as

$$V_0 = V_R \left(\frac{1}{2^1} b_1 + \frac{1}{2^2} b_0 \right)$$

where b_1 and b_0 are bits in the 2-bit binary code, with b_0 as the least significant bit (LSB). In Figure 2.28, where we set $V_R = 5$ and $b_1 b_0 = 10$, the ADC output is expected to be

$$V_0 = 5 \times \left(\frac{1}{2^1} \times 1 + \frac{1}{2^2} \times 0 \right) = 2.5 \text{ volts}$$

As we can see, the recovered voltage of $V_0 = 2.5$ volts introduces voltage error as compared with $V_{in} = 3$ volts, discussed in the ADC stage. This is due to the fact that in the flash ADC unit, we use only four (i.e., finite) voltage levels to represent continuous (infinitely possible) analog voltage values. This is called *quantization error*, obtained by subtracting the original analog voltage from the recovered analog voltage. For our example, the quantization error is

$$V_0 - V_{in} = 2.5 - 3 = -0.5 \text{ volts}$$

Next, we focus on quantization development. The process of converting analog voltage with infinite precision to finite precision is called the *quantization process*. For example, if the digital processor has only a 3-bit word, the amplitudes can be converted into eight different levels.

A *unipolar quantizer* deals with analog signals ranging from 0 volt to a positive reference voltage, and a *bipolar quantizer* deals with analog signals ranging from a negative reference to a positive reference. The notations and general rules for quantization are as follows:

$$\Delta = \frac{(x_{max} - x_{min})}{L} \tag{2.19}$$

$$L = 2^m \tag{2.20}$$

$$i = round\left(\frac{x - x_{min}}{\Delta}\right) \tag{2.21}$$

$$x_q = x_{min} + i\Delta \quad i = 0, 1, \cdots, L - 1 \tag{2.22}$$

where x_{max} and x_{min} are the maximum value and minimum values, respectively, of the analog input signal x. The symbol L denotes the number of quantization levels, which is determined by Equation (2.20), where m is the number of bits used in ADC. The symbol Δ is the step size of the quantizer or the ADC resolution. Finally, x_q indicates the quantization level, and i is an index corresponding to the binary code.

Figure 2.29 depicts a 3-bit unipolar quantizer and corresponding binary codes. From Figure 2.29, we see that $x_{min} = 0$, $x_{max} = 8\Delta$, and $m = 3$. Applying Equation (2.22) gives each quantization level as follows: $x_q = 0 + i\Delta$, $i = 0, 1, \cdots, L - 1$, where $L = 2^3 = 8$ and i is the integer corresponding to the 3-bit binary code. Table 2.1 details quantization for each input signal subrange.

Similarly, a 3-bit bipolar quantizer and binary codes are shown in Figure 2.30, where we have $x_{min} = -4\Delta$, $x_{max} = 4\Delta$, and $m = 3$. The corresponding quantization table is given in Table 2.2.

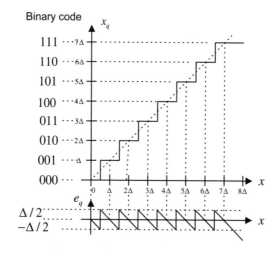

FIGURE 2.29

Characteristics of the unipolar quantizer.

Table 2.1 Quantization Table for the 3-Bit Unipolar Quantizer (step size $= \Delta = (x_{max} - x_{min})/2^3$, $x_{max} =$ maximum voltage, and $x_{min} = 0$)

Binary Code	Quantization Level x_q (V)	Input Signal Subrange (V)
0 0 0	0	$0 \le x < 0.5\Delta$
0 0 1	Δ	$0.5\Delta \le x < 1.5\Delta$
0 1 0	2Δ	$1.5\Delta \le x < 2.5\Delta$
0 1 1	3Δ	$2.5\Delta \le x < 3.5\Delta$
1 0 0	4Δ	$3.5\Delta \le x < 4.5\Delta$
1 0 1	5Δ	$4.5\Delta \le x < 5.5\Delta$
1 1 0	6Δ	$5.5\Delta \le x < 6.5\Delta$
1 1 1	7Δ	$6.5\Delta \le x < 7.5\Delta$

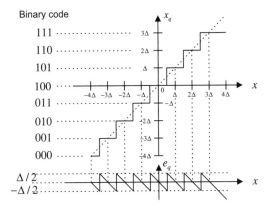

FIGURE 2.30

Characteristics for the bipolar quantizer.

EXAMPLE 2.9

Assuming that a 3-bit ADC channel accepts analog input ranging from 0 to 5 volts, determine

a. the number of quantization levels;
b. the step size of the quantizer or resolution;
c. the quantization level when the analog voltage is 3.2 volts;
d. the binary code produced by the ADC.

Solution:
Since the range is from 0 to 5 volts and a 3-bit ADC is used, we have

$$x_{min} = 0 \text{ volt}, \ x_{max} = 5 \text{ volts}, \quad \text{and} \quad m = 3 \text{ bits}$$

Table 2.2 Quantization Table for the 3-Bit Bipolar Quantizer (step size $= \Delta = (x_{max} - x_{min})/2^3$, $x_{max} =$ maximum voltage, and $x_{min} = -x_{max}$)

Binary Code	Quantization Level x_q (V)	Input Signal Subrange (V)
0 0 0	-4Δ	$-4\Delta \leq x < -3.5\Delta$
0 0 1	-3Δ	$-3.5\Delta \leq x < -2.5\Delta$
0 1 0	-2Δ	$-2.5\Delta \leq x < -1.5\Delta$
0 1 1	$-\Delta$	$-1.5 \leq x < -0.5\Delta$
1 0 0	0	$-0.5\Delta \leq x < 0.5\Delta$
1 0 1	Δ	$0.5\Delta \leq x < 1.5\Delta$
1 1 0	2Δ	$1.5\Delta \leq x < 2.5\Delta$
1 1 1	3Δ	$2.5\Delta \leq x < 3.5\Delta$

a. Using Equation (2.20), we get the number of quantization levels as

$$L = 2^m = 2^3 = 8$$

b. Applying Equation (2.19) yields

$$\Delta = \frac{5 - 0}{8} = 0.625 \text{ volt}$$

c. When $x = 3.2\dfrac{\Delta}{0.625} = 5.12\Delta$, from Equation (2.21) we get

$$i = round\left(\frac{x - x_{min}}{\Delta}\right) = round(5.12) = 5$$

From Equation (2.22), we determine the quantization level as

$$x_q = 0 + 5\Delta = 5 \times 0.625 = 3.125 \text{ volts}$$

d. The binary code is determined as 101, either from Figure 2.29 or Table 2.1.

After quantizing the input signal x, the ADC produces binary codes, as illustrated in Figure 2.31. The DAC process is shown in Figure 2.32. As shown in the figure, the DAC unit takes the binary codes from the DS processor. Then it converts the binary code using the zero-order hold circuit to reproduce the sample-and-hold signal. Assuming that the spectrum distortion due to sample-and-hold effect can be ignored for our illustration, the recovered sample-and-hold signal is further processed using the anti-image filter. Finally, the analog signal is produced.

When the DAC outputs the analog amplitude x_q with finite precision, it introduces quantization error defined as

$$e_q = x_q - x \tag{2.23}$$

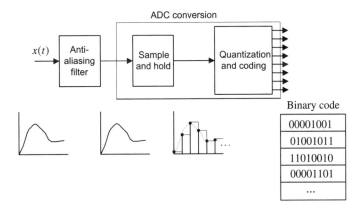

FIGURE 2.31

Typical ADC process.

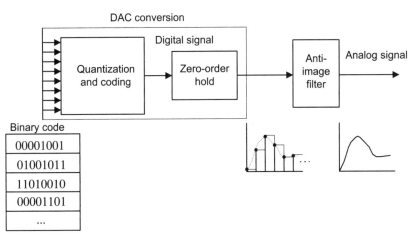

FIGURE 2.32

Typical DAC process.

The quantization error as shown in Figure 2.29 is bounded by half of the step size, that is,

$$-\frac{\Delta}{2} \le e_q \le \frac{\Delta}{2} \tag{2.24}$$

where Δ is the quantization step size, or the ADC resolution. We also refer to Δ as V_{min} (minimum detectable voltage) or the LSB value of the ADC.

EXAMPLE 2.10

Using Example 2.9, determine the quantization error when the analog input is 3.2 volts.

Solution:

Using Equation (2.23), we obtain

$$e_q = x_q - x = 3.125 - 3.2 = -0.075 \text{ volt}$$

Note that the quantization error is less than the half of the step size, that is,

$$|e_q| = 0.075 < \Delta/2 = 0.3125 \text{ volt}$$

In practice, we can empirically confirm that the quantization error appears in uniform distribution when the step size is much smaller than the dynamic range of the signal samples and we have a sufficiently large number of samples. Based on the theory of probability and random variables, the power of quantization noise is related to the quantization step and given by

$$E(e_q^2) = \frac{\Delta^2}{12} \tag{2.25}$$

where $E()$ is the expectation operator, which actually averages the squared values of the quantization error (the reader can get more information from the texts by Roddy and Coolen (1997); Tomasi (2004); and Stearns and Hush (1990)). The ratio of signal power to quantization noise power (SNR) can be expressed as

$$SNR = \frac{E(x^2)}{E(e_q^2)} \tag{2.26}$$

If we express the SNR in terms of decibels (dB), we have

$$SNR_{dB} = 10 \cdot \log_{10}(SNR) \text{ dB} \tag{2.27}$$

Substituting Equation (2.25) and $E(x^2) = x^2_{rms}$ into Equation (2.27), we achieve

$$SNR_{dB} = 10.79 + 20 \cdot \log_{10}\left(\frac{x_{rms}}{\Delta}\right) \tag{2.28}$$

where x_{rms} is the RMS (root mean squared) value of the signal to be quantized x.

Practically, the SNR can be calculated using the following formula:

$$SNR = \frac{\frac{1}{N}\sum_{n=0}^{N-1} x^2(n)}{\frac{1}{N}\sum_{n=0}^{N-1} e_q^2(n)} = \frac{\sum_{n=0}^{N-1} x^2(n)}{\sum_{n=0}^{N-1} e_q^2(n)} \tag{2.29}$$

where $x(n)$ is the nth sample amplitude and $e_q(n)$ the quantization error from quantizing $x(n)$.

EXAMPLE 2.11

If the analog signal to be quantized is a sinusoidal waveform, that is,

$$x(t) = A\sin(2\pi \times 1,000t)$$

and if the bipolar quantizer uses m bits, determine the SNR in terms of m bits.

Solution:
Since $x_{rms} = 0.707A$ and $\Delta = 2A/2^m$, substituting x_{rms} and Δ into Equation (2.28) leads to

$$SNR_{dB} = 10.79 + 20 \cdot \log_{10}\left(\frac{0.707A}{2A/2^m}\right)$$
$$= 10.79 + 20 \cdot \log_{10}(0.707/2) + 20m \cdot \log_{10}2$$

After simplifying the numerical values, we get

$$SNR_{dB} = 1.76 + 6.02m\,\text{dB} \qquad (2.30)$$

EXAMPLE 2.12

For a speech signal, if a ratio of the RMS value over the absolute maximum value of the analog signal (Roddy and Coolen, 1997) is given, that is, $\left(\frac{x_{rms}}{|x|_{max}}\right)$, and the ADC quantizer uses m bits, determine the SNR in terms of m bits.

Solution:
Since

$$\Delta = \frac{x_{max} - x_{min}}{L} = \frac{2|x|_{max}}{2^m}$$

substituting Δ in Equation (2.28) achieves

$$SNR_{dB} = 10.79 + 20 \cdot \log_{10}\left(\frac{x_{rms}}{2|x|_{max}/2^m}\right)$$
$$= 10.79 + 20 \cdot \log_{10}\left(\frac{x_{rms}}{|x|_{max}}\right) + 20m\log_{10}2 - 20\log_{10}2$$

Thus, after numerical simplification, we have

$$SNR_{dB} = 4.77 + 20 \cdot \log_{10}\left(\frac{x_{rms}}{|x|_{max}}\right) + 6.02m \qquad (2.31)$$

From Examples 2.11 and 2.12, we observed that increasing 1 bit of the ADC quantizer can improve SNR due to quantization by 6 dB.

EXAMPLE 2.13

Given a sinusoidal waveform with a frequency of 100 Hz,

$$x(t) = 4.5 \cdot \sin(2\pi \times 100t)$$

sampled at 8,000 Hz,

a. write a MATLAB program to quantize $x(t)$ using 4 bits to obtain and plot the quantized signal x_q, assuming the signal range is between -5 and 5 volts;
b. calculate the SNR due to quantization.

Solution:
a. Program 2.1. MATLAB program for Example 2.13.

```
%Example 2.13
clear all; close all
disp('Generate 0.02-second sine wave of 100 Hz and Vp=5');
fs=8000;                              % Sampling rate
T=1/fs;                               % Sampling interval
t=0:T:0.02;                           % Duration of 0.02 second
sig = 4.5*sin(2*pi*100*t);            % Generate sinusoids
bits = input('input number of bits =>');
lg = length(sig);                     % Length of signal vector sig
for x=1:lg
  [Index(x) pq] = biquant(bits, -5,5, sig(x));   % Output quantized index
```

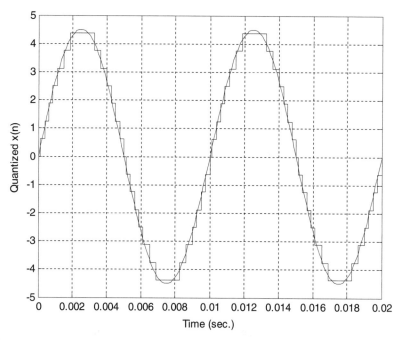

FIGURE 2.33

Comparison of the quantized signal and the original signal.

```
end
    % transmitted
    % received
    for x=1:lg
      qsig(x) = biqtdec(bits, -5,5, Index(x));        % Recover the quantized value
    end
      qerr = qsig-sig;                     % Calculate quantized error
      stairs(t,qsig); hold                 % Plot signal in staircase style
      plot(t,sig); grid;                   % Plot signal
      xlabel('Time (sec.)'); ylabel('Quantized x(n)')
      disp('Signal to noise ratio due to quantization noise')
      snr(sig,qsig);
```

b. Theoretically, applying Equation (2.30) gives

$$SNR_{dB} = 1.76 + 6.02 \cdot 4 = 25.84 \text{ dB}$$

Practically, using Equation (2.29), the simulated result is obtained as

$$SNR_{dB} = 25.78 \text{ dB}$$

It is clear from this example that the ratios of signal power to noise power due to quantization achieved from theory and from simulation are very close. Next, we look at an example for quantizing a speech signal.

EXAMPLE 2.14

Given the speech signal sampled at 8,000 Hz in the file *we.dat*,

a. write a MATLAB program to quantize $x(t)$ using 4-bit quantizers to obtain the quantized signal x_q, assuming the signal range is from -5 to 5 volts;

b. plot the original speech, quantized speech, and quantization error, respectively;

c. calculate the SNR due to quantization using the MATLAB program.

Solution:

a. Program 2.2 MATLAB program for Example 2.14.

```
%Example 2.14
clear all; close all
disp('load speech: We');
load we.dat                % Load speech data at the current folder
sig = we;                  % Provided by the instructor
fs=8000;                   % Sampling rate
lg=length(sig);            % Length of signal vector
T=1/fs;                    % Sampling period
t=[0:1:lg-1]*T;            % Time instants in seconds
sig=4.5*sig/max(abs(sig)); % Normalizes speech in the range from -4.5 to 4.5
Xmax = max(abs(sig));      % Maximum amplitude
Xrms = sqrt( sum(sig .*
sig) / length(sig))        % RMS value
disp('Xrms/Xmax')
k=Xrms/Xmax
```

```
disp('20*log10(k)=>');
k = 20*log10(k)
bits = input('input number of bits =>');
lg = length(sig);
for x=1:lg
  [Index(x) pq] = biquant(bits, -5,5, sig(x)); % Output quantized index
end
% transmitted
% received
for x=1:lg
  qsig(x) = biqtdec(bits, -5,5, Index(x));      % Recover the quantized value
end
  qerr = sig-qsig;                             % Calculate the quantized error
subplot(3,1,1);plot(t,sig);
ylabel('Original speech');title('we.dat: we');
subplot(3,1,2);stairs(t, qsig);grid
ylabel('Quantized speech')
subplot(3,1,3);stairs(t, qerr);grid
ylabel('Quantized error')
xlabel('Time (sec.)');axis([0 0.25 -1 1]);
disp('signal to noise ratio due to quantization noise')
snr(sig,qsig);                                 % Signal to ratio in dB:
sig = original signal vector,

                                               % qsig =quantized signal vector
```

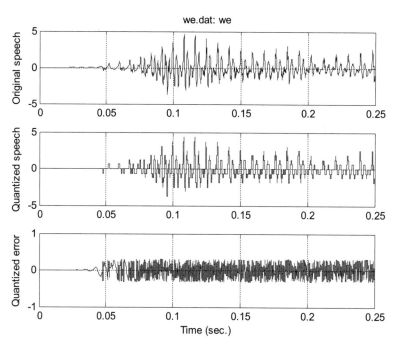

FIGURE 2.34

Original speech, quantized speech using the 4-bit bipolar quantizer, and quantization error.

b. In Figure 2.34, the top plot shows the speech wave to be quantized, while the middle plot displays the quantized speech signal using 4 bits. The bottom plot shows the quantization error. It also shows that the absolute value of quantization error is uniformly distributed in a range between -0.3125 and 0.3125.

c. From the MATLAB program, we have $\frac{X_{rms}}{|x|_{max}} = 0.203$. Theoretically, from Equation (2.31), it follows that

$$SNR_{dB} = 4.77 + 20\log_{10}\left(\frac{X_{rms}}{|x|_{max}}\right) + 6.02 \cdot 4$$

$$= 4.77 + 20\log_{10}(0.203) + 6.02 \cdot 4 = 15dB$$

On the other hand, the simulated result using Equation (2.29) gives

$$SNR_{dB} = 15.01 \text{ dB}$$

Results for SNRs from Equations (2.31) and (2.29) are very close in this example.

2.4 SUMMARY

1. Analog signal is sampled at a fixed time interval so the ADC will convert the sampled voltage level to the digital value; this is called the sampling process.
2. The fixed time interval between two samples is the sampling period, and the reciprocal of the sampling period is the sampling rate. Half of the sampling rate is the folding frequency (Nyquist limit).
3. The sampling theorem condition that the sampling rate must be larger than twice the highest frequency of the sampled analog signal must be met in order for the analog signal to be recovered.
4. The sampled spectrum is explained using the following well-known formula:

$$X_s(f) = \cdots + \frac{1}{T}X(f + f_s) + \frac{1}{T}X(f) + \frac{1}{T}X(f - f_s) + \cdots$$

That is, the sampled signal spectrum is a scaled and shifted version of its analog signal spectrum and its replicas centered at the frequencies that are multiples of the sampling rate.
5. The analog anti-aliasing lowpass filter is used before ADC to remove frequency components higher than the folding frequency to avoid aliasing.
6. The reconstruction (analog lowpass) filter is adopted after DAC to remove the spectral images that exist in the sample-and-hold signal and obtain the smoothed analog signal. The sample-and-hold DAC effect may distort the baseband spectrum, but it also reduces image spectrum.
7. Quantization occurs when the ADC unit converts the analog signal amplitude with infinite precision to digital data with finite precision (a finite number of codes).
8. When the DAC unit converts a digital code to a voltage level, quantization error occurs. The quantization error is bounded by half of the quantization step size (ADC resolution), which is a ratio of the full range of the signal over the number of quantization levels (number of codes).
9. The performance of the quantizer in terms of the signal to quantization noise ratio (SNR), in dB, is related to the number of bits in ADC. Increasing each ADC code by 1 bit will improve SNR by 6 dB due to quantization.

2.5 MATLAB PROGRAMS

Program 2.3. MATLAB function for uniform quantization encoding.

```
function [ I, pq]= biquant(NoBits,Xmin,Xmax,value)
% function pq = biquant(NoBits, Xmin, Xmax, value)
% This routine is created for simulation of the uniform quantizer.
%
% NoBits: number of bits used in quantization
% Xmax: overload value
% Xmin: minimum value
% value: input to be quantized
% pq: output of quantized value
% I: coded integer index
 L=2^NoBits;
 delta=(Xmax-Xmin)/L;
 I=round((value-Xmin)/delta);
 if ( I==L)
  I=I-1;
 end
 if I<0
  I=0;
 end
 pq=Xmin+I*delta;
```

Program 2.4. MATLAB function for uniform quantization decoding.

```
function pq = biqtdec(NoBits,Xmin,Xmax,I)
% function pq = biqtdec(NoBits,Xmin, Xmax, I)
% This routine recovers the quantized value.
%
% NoBits: number of bits used in quantization
% Xmax: overload value
% Xmin: minimum value
% pq: output of quantized value
% I: coded integer index
 L=2^NoBits;
 delta=(Xmax-Xmin)/L;
 pq=Xmin+I*delta;
```

Program 2.5. MATLAB function for calculation of signal to quantization noise ratio.

```
function snr = calcsnr(speech, qspeech)
% function snr = calcsnr(speech, qspeech)
% This routine was created to calculate SNR.
%
% speech: original speech waveform
% qspeech: quantized speech
% snr: output SNR in dB
%
 qerr = speech-qspeech;
 snr = 10*log10(sum(speech.*speech)/sum(qerr.*qerr))
```

2.6 PROBLEMS

2.1. Given an analog signal

$$x(t) = 5\cos(2\pi \cdot 1,500t), \text{ for } t \geq 0$$

sampled at a rate of 8,000 Hz,

a. sketch the spectrum of the original signal;

b. sketch the spectrum of the sampled signal from 0 kHz up to 20 kHz.

2.2. Given an analog signal

$$x(t) = 5\cos(2\pi \cdot 2,500t) + 2\cos(2\pi \cdot 3,200t), \text{ for } t \geq 0$$

sampled at a rate of 8,000 Hz,

a. sketch the spectrum of the sampled signal up to 20 kHz;

b. sketch the recovered analog signal spectrum if an ideal lowpass filter with a cutoff frequency of 4 kHz is used to filter the sampled signal in order to recover the original signal.

2.3. Given an analog signal

$$x(t) = 3\cos(2\pi \cdot 1,500t) + 2\cos(2\pi \cdot 2,200t), \text{ for } t \geq 0$$

sampled at a rate of 8,000 Hz,

a. sketch the spectrum of the sampled signal up to 20 kHz;

b. sketch the recovered analog signal spectrum if an ideal lowpass filter with a cutoff frequency of 4 kHz is used to filter the sampled signal in order to recover the original signal.

2.4. Given an analog signal

$$x(t) = 3\cos(2\pi \cdot 1,500t) + 2\cos(2\pi \cdot 4,200t), \text{ for } t \geq 0$$

sampled at a rate of 8,000 Hz,

a. sketch the spectrum of the sampled signal up to 20 kHz;

b. sketch the recovered analog signal spectrum if an ideal lowpass filter with a cutoff frequency of 4 kHz is used to filter the sampled signal in order to recover the original signal.

2.5. Given an analog signal

$$x(t) = 5\cos(2\pi \cdot 2,500t) + 2\cos(2\pi \cdot 4,500t), \text{ for } t \geq 0$$

sampled at a rate of 8,000 Hz,

a. sketch the spectrum of the sampled signal up to 20 kHz;

b. sketch the recovered analog signal spectrum if an ideal lowpass filter with a cutoff frequency of 4 kHz is used to filter the sampled signal in order to recover the original signal;

c. determine the frequency/frequencies of aliasing noise.

2.6. Assuming a continuous signal is given as

$$x(t) = 10\cos(2\pi \cdot 5, 500t) + 5\sin(2\pi \cdot 7, 500t), \text{ for } t \geq 0$$

sampled at a rate of 8,000 Hz,

a. sketch the spectrum of the sampled signal up to 20 kHz;

b. sketch the recovered analog signal spectrum if an ideal lowpass filter with a cutoff frequency of 4 kHz is used to filter the sampled signal in order to recover the original signal;

c. determine the frequency/frequencies of aliasing noise.

2.7. Assuming a continuous signal is given as

$$x(t) = 8\cos(2\pi \cdot 5, 000t) + 5\sin(2\pi \cdot 7, 000t), \text{ for } t \geq 0$$

sampled at a rate of 8,000 Hz,

a. sketch the spectrum of the sampled signal up to 20 kHz;

b. sketch the recovered analog signal spectrum if an ideal lowpass filter with a cutoff frequency of 4 kHz is used to filter the sampled signal in order to recover the original signal;

c. determine the frequency/frequencies of aliasing noise.

2.8. Assuming a continuous signal is given as

$$x(t) = 10\cos(2\pi \cdot 5, 000t) + 5\sin(2\pi \cdot 7, 500t), \text{ for } t \geq 0$$

sampled at a rate of 8,000 Hz,

a. sketch the spectrum of the sampled signal up to 20 kHz;

b. sketch the recovered analog signal spectrum if an ideal lowpass filter with a cutoff frequency of 4 kHz is used to filter the sampled signal in order to recover the original signal;

c. determine the frequency/frequencies of aliasing noise.

2.9. Given a Butterworth type second-order anti-aliasing lowpass filter (Figure 2.35), determine the values of circuit elements if we want the filter to have a cutoff frequency of 1,000 Hz.

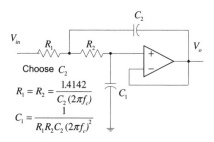

FIGURE 2.35

Filter circuit in Problem 2.9.

2.10. From Problem 2.9, determine the percentage of aliasing level at the frequency of 500 Hz, assuming that the sampling rate is 4,000 Hz.

2.11. Given a Butterworth type second-order anti-aliasing lowpass filter (Figure 2.36), determine the values of circuit elements if we want the filter to have a cutoff frequency of 800 Hz.

2.12. From Problem 2.11, determine the percentage of aliasing level at the frequency of 400 Hz, assuming that the sampling rate is 4,000 Hz.

2.13. Given a DSP system in which a sampling rate of 8,000 Hz is used and the anti-aliasing filter is a second-order Butterworth lowpass filter with a cutoff frequency of 3.2 kHz, determine

a. the percentage of aliasing level at the cutoff frequency;

b. the percentage of aliasing level at the frequency of 1,000 Hz.

2.14. Given a DSP system in which a sampling rate of 8,000 Hz is used and the anti-aliasing filter is a Butterworth lowpass filter with a cutoff frequency 3.2 kHz, determine the order of the Butterworth lowpass filter required to make the percentage of aliasing level at the cutoff frequency less than 10%.

2.15. Given a DSP system in which a sampling rate of 8,000 Hz is used and the anti-aliasing filter is a second-order Butterworth lowpass filter with a cutoff frequency of 3.1 kHz, determine

a. the percentage of aliasing level at the cutoff frequency;

b. the percentage of aliasing level at a frequency of 900 Hz.

2.16. Given a DSP system in which a sampling rate of 8,000 Hz is used and the anti-aliasing filter is a Butterworth lowpass filter with a cutoff frequency 3.1 kHz, determine the order of the Butterworth lowpass filter required to make the percentage of aliasing level at the cutoff frequency less than 10%.

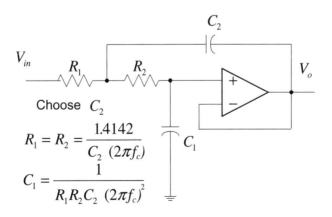

FIGURE 2.36

Filter circuit in Problem 2.11.

2.17. Given a DSP system with a sampling rate of 8,000 Hz and assuming that the hold circuit is used after DAC, determine

a. the percentage of distortion at a frequency of 3,200 Hz;

b. the percentage of distortion at a frequency of 1,500 Hz.

2.18. A DSP system (Figure 2.37) is given with the following specifications:

Design requirements:

- Sampling rate 20,000 Hz

- Maximum allowable gain variation from 0 to 4,000 Hz = 2 dB

- 40 dB rejection at a frequency of 16,000 Hz

- Butterworth filter.

Determine the cutoff frequency and order for the anti-image filter.

2.19. Given a DSP system with a sampling rate of 8,000 Hz and assuming that the hold circuit is used after DAC, determine

a. the percentage of distortion at a frequency of 3,000 Hz;

b. the percentage of distortion at a frequency of 1,600 Hz.

2.20. A DSP system (Figure 2.38) is given with the following specifications:

Design requirements:

- Sampling rate 22,000 Hz

- Maximum allowable gain variation from 0 to 4,000 Hz = 2 dB

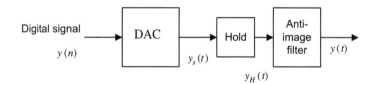

FIGURE 2.37

Analog signal reconstruction in Problem 2.18.

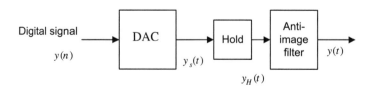

FIGURE 2.38

Analog signal reconstruction in Problem 2.20.

- 40 dB rejection at the frequency of 18,000 Hz
- Butterworth filter.

Determine the cutoff frequency and order for the anti-image filter.

2.21. Given the 2-bit flash ADC unit with an analog sample-and-hold voltage of 2 volts shown in Figure 2.39, determine the output bits.

2.22. Given the R-2R DAC unit with a 2-bit value defined as $b_1 b_0 = 01$ shown in Figure 2.40, determine the converted voltage.

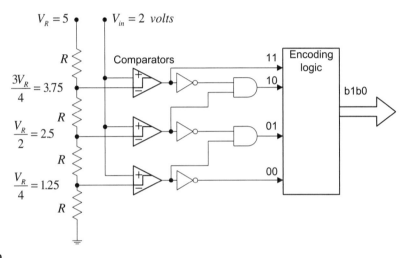

FIGURE 2.39

2-bit flash ADC in Problem 2.21.

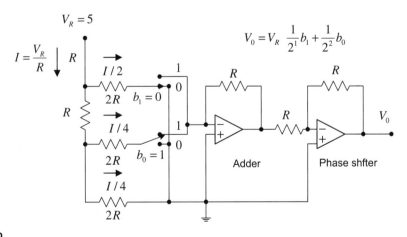

FIGURE 2.40

2-bit R-2R DAC in Problem 2.22.

2.23. Given the 2-bit flash ADC unit with an analog sample-and-hold voltage of 3.5 volts shown in Figure 2.39, determine the output bits.

2.24. Given the R-2R DAC unit with 2-bit values defined as $b_1 b_0 = 11$ and $b_1 b_0 = 10$ and shown in Figure 2.40, determine the converted voltages.

2.25. Assuming that a 4-bit ADC channel accepts analog input ranging from 0 to 5 volts, determine the following:

    a. number of quantization levels;

    b. step size of the quantizer or resolution;

    c. quantization level when the analog voltage is 3.2 volts;

    d. binary code produced by the ADC;

    e. quantization error.

2.26. Assuming that a 5-bit ADC channel accepts analog input ranging from 0 to 4 volts, determine the following:

    a. number of quantization levels;

    b. step size of the quantizer or resolution;

    c. quantization level when the analog voltage is 1.2 volts;

    d. binary code produced by the ADC;

    e. quantization error.

2.27. Assuming that a 3-bit ADC channel accepts analog input ranging from −2.5 to 2.5 volts, determine the following:

    a. number of quantization levels;

    b. step size of the quantizer or resolution;

    c. quantization level when the analog voltage is −1.2 volts;

    d. binary code produced by the ADC;

    e. quantization error.

2.28. Assuming that a 8-bit ADC channel accepts analog input ranging from −2.5 to 2.5 volts, determine the following:

    a. number of quantization levels;

    b. step size of the quantizer or resolution;

    c. quantization level when the analog voltage is 1.5 volts;

    d. binary code produced by the ADC;

    e. quantization error.

2.29. If the analog signal to be quantized is a sinusoidal waveform, that is,

$$x(t) = 9.5\sin(2,000 \times \pi t)$$

and if the bipolar quantizer uses 6 bits, determine

a. the number of quantization levels;

b. the quantization step size or resolution, Δ, assuming the signal range is from -10 to 10 volts;

c. the signal power to quantization noise power ratio.

2.30. For a speech signal, if the ratio of the RMS value over the absolute maximum value of the signal is given, that is, $\left(\dfrac{x_{rms}}{|x|_{max}}\right) = 0.25$, and the ADC bipolar quantizer uses 6 bits, determine

a. the number of quantization levels;

b. the quantization step size or resolution, Δ, if the signal range is 5 volts;

c. the signal power to quantization noise power ratio.

2.6.1 Computer Problems with MATLAB

Use the MATLAB programs in Section 2.5 to solve the following problems.

2.31. Given a sinusoidal waveform of 100 Hz,

$$x(t) = 4.5\sin(2\pi \times 100t)$$

sample it at 8,000 samples per second and

a. write a MATLAB program to quantize $x(t)$ using a 6-bit bipolar quantizer to obtain the quantized signal x_q, assuming that the signal range is from -5 to 5 volts;

b. plot the original signal and quantized signal;

c. calculate the SNR due to quantization using the MATLAB program.

2.32. Given a signal waveform,

$$x(t) = 3.25\sin(2\pi \times 50t) + 1.25\cos(2\pi \times 100t + \pi/4)$$

sample it at 8,000 samples per second and

a. write a MATLAB program to quantize $x(t)$ using a 6-bit bipolar quantizer to obtain the quantized signal x_q, assuming that the signal range is from -5 to 5 volts;

b. plot the original signal and quantized signal;

c. calculate the SNR due to quantization using the MATLAB program.

2.33. Given a speech signal sampled at 8,000 Hz, as shown in Example 2.14,

a. write a MATLAB program to quantize $x(t)$ using a 6-bit bipolar quantizer to obtain the quantized signal x_q, assuming that the signal range is from -5 to 5 volts;

b. plot the original speech waveform, quantized speech, and quantization error;

c. calculate the SNR due to quantization using the MATLAB program.

2.6.2 **MATLAB Projects**

2.34. Performance evaluation of speech quantization:

Given an original speech segment "speech.dat" sampled at 8,000 Hz with each sample encoded in 16 bits, use Programs 2.3 to 2.5 and modify Program 2.2 to quantize the speech segment using 3 to 15 bits, respectively. The SNR in dB must be measured for each quantization. The MATLAB function: "sound(x/max(abs(x)),fs)" can be used to evaluate sound quality, where "x" is the speech segment while "fs" is the sampling rate of 8,000 Hz. In this project, create a plot of the measured SNR (dB) versus the number of bits and discuss the effect on the sound quality. For comparisons, plot the original speech and the quantized one using 3 bits, 8 bits, and 15 bits.

2.35. Performance evaluation of seismic data quantization:

The seismic signal, a measurement of the acceleration of ground motion, is required for applications in the study of geophysics. The seismic signal ("seismic.dat" provided by the US Geological Survey, Albuquerque Seismological Laboratory) has a sampling rate of 15 Hz with 6,700 data samples, and each sample is encoded using 32 bits. Quantizing each 32-bit sample down to the lower number of bits per sample can reduce the memory storage requirement with the trade-off of reduced signal quality. Use Programs 2.3 to 2.5 and modify Program 2.2 to quantize the seismic data using 13, 15, 17, ..., 31 bits. The SNR in dB must be measured for each quantization. Create a plot of the measured SNR (dB) versus the number of bits. For comparison, plot the seismic data and the quantized one using 13 bits, 18 bits, 25 bits, and 31 bits.

Digital Signals and Systems

3

CHAPTER OUTLINE

3.1 Digital Signals 57
    3.1.1 Common Digital Sequences 58
    3.1.2 Generation of Digital Signals 61
3.2 Linear Time-Invariant, Causal Systems 63
    3.2.1 Linearity 63
    3.2.2 Time Invariance 65
    3.2.3 Causality 66
3.3 Difference Equations and Impulse Responses 67
    3.3.1 Format of the Difference Equation 67
    3.3.2 System Representation Using Its Impulse Response 68
3.4 Bounded-In and Bounded-Out Stability 71
3.5 Digital Convolution 72
3.6 Summary 79

OBJECTIVES:

This chapter introduces notations for digital signals and special digital sequences that are widely used in this book. The chapter continues to study some properties of linear systems such as time invariance, BIBO (bounded-in and bounded-out) stability, causality, impulse response, difference equations, and digital convolution.

3.1 DIGITAL SIGNALS

In our daily lives, analog signals appear in forms such as speech, audio, seismic, biomedical, and communications signals. To process an analog signal using a digital signal processor, the analog signal must be converted in to a digital signal, that is, analog-to-digital conversion (ADC) must take place, as discussed in Chapter 2. Then the digital signal is processed via digital signal processing (DSP) algorithm(s).

A typical digital signal $x(n)$ is shown in Figure 3.1, where both the time and the amplitude of the digital signal are discrete. Notice that the amplitudes of the digital signal samples are given and sketched only at their corresponding time indices, where $x(n)$ represents the amplitude of the nth sample and n is the time index or sample number. From Figure 3.1, we learn that

    $x(0)$: zeroth sample amplitude at sample number $n = 0$,
    $x(1)$: first sample amplitude at sample number $n = 1,$

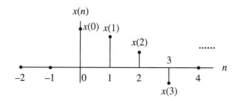

FIGURE 3.1

Digital signal notation.

$x(2)$: second sample amplitude at sample number $n = 2$,
$x(3)$: third sample amplitude at sample number $n = 3$, and so on.

Furthermore, Figure 3.2 illustrates the digital samples whose amplitudes are the discrete encoded values represented in the digital signal (DS) processor. Precision of the data is based on the number of bits used in the DSP system. The encoded data format can be either an integer if a fixed-point DS processor is used or a floating-point number if a floating-point DP processor is used. As shown in Figure 3.2 for the floating-point DS processor, we can identify the first five sample amplitudes at their time indices as follows:

$x(0) = 2.25$
$x(1) = 2.0$
$x(2) = 1.0$
$x(3) = -1.0$
$x(4) = 0.0$
......

Again, note that each sample amplitude is plotted using a vertical bar with a solid dot. This notation is well accepted in DSP literature.

3.1.1 Common Digital Sequences

Let us study some special digital sequences that are widely used. We define and plot each of them as follows:

Unit-impulse sequence (digital unit-impulse function):

$$\delta(n) = \begin{cases} 1 & n = 0 \\ 0 & n \neq 0 \end{cases} \tag{3.1}$$

FIGURE 3.2

Plot of the digital signal samples.

The plot of the unit-impulse function is given in Figure 3.3. The unit-impulse function has unit amplitude at only $n = 0$ and zero amplitude at other time indices.

Unit-step sequence (digital unit-step function):

$$u(n) = \begin{cases} 1 & n \geq 0 \\ 0 & n < 0 \end{cases} \tag{3.2}$$

The plot is given in Figure 3.4. The unit-step function has unit amplitude at $n = 0$ and at all the positive time indices, and zero amplitude at all negative time indices.

The shifted unit-impulse and unit-step sequences are displayed in Figure 3.5. As shown in the figure, the shifted unit-impulse function $\delta(n - 2)$ is obtained by shifting the unit-impulse function $\delta(n)$ to the right by two samples, and the shifted unit-step function $u(n - 2)$ is achieved by shifting the unit-step function $u(n)$ to the right by two samples; similarly, $\delta(n + 2)$ and $u(n + 2)$ are acquired by shifting $\delta(n)$ and $u(n)$ two samples to the left, respectively.

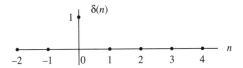

FIGURE 3.3

Unit-impulse sequence.

FIGURE 3.4

Unit-step sequence.

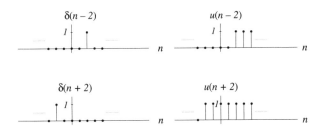

FIGURE 3.5

Shifted unit-impulse and unit-step sequences.

Sinusoidal and exponential sequences are depicted in Figures 3.6 and 3.7, respectively. For the sinusoidal sequence $x(n) = A\cos(0.125\pi n)u(n)$, and $A = 10$, we can calculate the digital values for the first eight samples and list their values in Table 3.1.

For the exponential sequence $x(n) = A(0.75)^n u(n)$, the calculated digital values for the first eight samples with $A = 10$ are listed in Table 3.2.

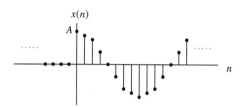

FIGURE 3.6

Plot of samples of the sinusoidal function.

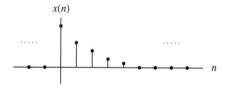

FIGURE 3.7

Plot of samples of the exponential function.

Table 3.1 Sample Values Calculated from the Sinusoidal Function

n	$x(n) = 10\cos(0.125\pi n)u(n)$
0	10.0000
1	9.2388
2	7.0711
3	3.8628
4	0.0000
5	−3.8628
6	−7.0711
7	−9.2388

Table 3.2 Sample Values Calculated from the Exponential Function	
n	$10(0.75)^n u(n)$
0	10.0000
1	7.5000
2	5.6250
3	4.2188
4	3.1641
5	2.3730
6	1.7798
7	1.3348

EXAMPLE 3.1

Sketch the following sequence

$$x(n) = \delta(n+1) + 0.5\delta(n-1) + 2\delta(n-2)$$

Solution:

According to the shift operation, $\delta(n+1)$ is obtained by shifting $\delta(n)$ to the left by one sample, while $\delta(n-1)$ and $\delta(n-2)$ are yielded by shifting $\delta(n)$ to right by one sample and two samples, respectively. Using the amplitude of each impulse function, we obtain the sketch in Figure 3.8.

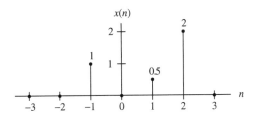

FIGURE 3.8

Plot of digital sequence in Example 3.1.

3.1.2 Generation of Digital Signals

Given the sampling rate of a DSP system to sample the analytical function of an analog signal, the corresponding digital function or digital sequence (assuming its sampled amplitudes are encoded to have finite precision) can be found. The digital sequence is often used to

1. Calculate the encoded sample amplitude for a given sample number n.
2. Generate the sampled sequence for simulation.

The procedure to develop the digital sequence from its analog signal function is as follows. Assuming that an analog signal $x(t)$ is uniformly sampled at the time interval of $\Delta t = T$, where T is the sampling period, the corresponding digital function (sequence) $x(n)$ gives the *instant encoded values* of the analog signal $x(t)$ at all the time instants $t = n\Delta t = nT$ and can be achieved by substituting time $t = nT$ into the analog signal $x(t)$, that is,

$$x(n) = x(t)|_{t=nT} = x(nT) \tag{3.3}$$

Also notice that for sampling the unit-step function $u(t)$, we have

$$u(t)|_{t=nT} = u(nT) = u(n) \tag{3.4}$$

The following example will demonstrate the use of Equations (3.3) and (3.4).

EXAMPLE 3.2

Assume we have a DSP system with a sampling time interval of 125 microseconds.

a. Convert each of following analog signals $x(t)$ to a digital signal $x(n)$:
  1. $x(t) = 10e^{-5,000t}u(t)$
  2. $x(t) = 10\sin(2,000\pi t)u(t)$
b. Determine and plot the sample values from each obtained digital function.

Solution:
a. Since $T = 0.000125$ seconds in Equation (3.3), substituting $t = nT = n \times 0.000125 = 0.000125\,n$ into the analog signal $x(t)$ expressed in (1) leads to the digital sequence

  1. $x(n) = x(nT) = 10e^{-5,000 \times 0.000125n}u(nT) = 10e^{-0.625n}u(n)$

Similarly, the digital sequence for (2) is achieved as follows:

  2. $x(n) = x(nT) = 10\sin(2,000\pi \times 0.000125n)u(nT) = 10\sin(0.25\pi n)u(n)$

b. The first five sample values for (1) are calculated and plotted in Figure 3.9.
  $x(0) = 10e^{-0.625 \times 0}u(0) = 10.0$
  $x(1) = 10e^{-0.625 \times 1}u(1) = 5.3526$

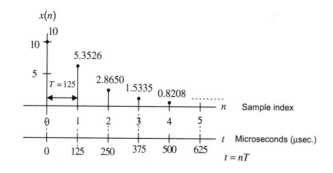

FIGURE 3.9

Plot of the digital sequence for (1) in Example 3.2.

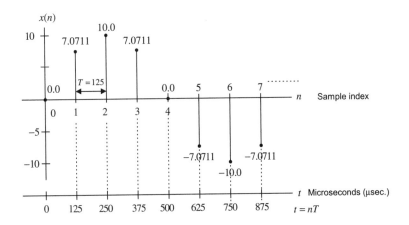

FIGURE 3.10

Plot of the digital sequence for (2) in Example 3.2.

$$x(2) = 10e^{-0.625\times2}u(2) = 2.8650$$
$$x(3) = 10e^{-0.625\times3}u(3) = 1.5335$$
$$x(4) = 10e^{-0.625\times4}u(4) = 0.8208$$

The first eight amplitudes for (2) are computed and sketched in Figure 3.10.

$$x(0) = 10\sin(0.25\pi \times 0)u(0) = 0$$
$$x(1) = 10\sin(0.25\pi \times 1)u(1) = 7.0711$$
$$x(2) = 10\sin(0.25\pi \times 2)u(2) = 10.0$$
$$x(3) = 10\sin(0.25\pi \times 3)u(3) = 7.0711$$
$$x(4) = 10\sin(0.25\pi \times 4)u(4) = 0.0$$
$$x(5) = 10\sin(0.25\pi \times 5)u(5) = -7.0711$$
$$x(6) = 10\sin(0.25\pi \times 6)u(6) = -10.0$$
$$x(7) = 10\sin(0.25\pi \times 7)u(7) = -7.0711$$

3.2 LINEAR TIME-INVARIANT, CAUSAL SYSTEMS

In this section, we study linear time-invariant causal systems and focus on properties such as linearity, time-invariance, and causality.

3.2.1 Linearity

A linear system is illustrated in Figure 3.11, where $y_1(n)$ is the system output using an input $x_1(n)$, and $y_2(n)$ the system output with an input $x_2(n)$.

Figure 3.11 illustrates that the system output due to the weighted sum inputs $\alpha x_1(n) + \beta x_2(n)$ is equal to the same weighted sum of the individual outputs obtained from their corresponding inputs, that is,

$$y(n) = \alpha y_1(n) + \beta y_2(n) \tag{3.5}$$

where α and β are constants.

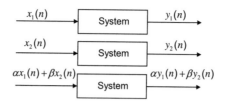

FIGURE 3.11

Digital linear system.

For example, assuming a digital amplifier is represented by $y(n) = 10x(n)$, the input is multiplied by 10 to generate the output. The inputs $x_1(n) = u(n)$ and $x_2(n) = \delta(n)$ generate the outputs

$$y_1(n) = 10u(n), \text{ and } y_2(n) = 10\delta(n), \text{ respectively}$$

If, as described in Figure 3.11, we apply the combined input $x(n)$ to the system, where the first input multiplied by a constant 2 while the second input multiplied by a constant 4,

$$x(n) = 2x_1(n) + 4x_2(n) = 2u(n) + 4\delta(n)$$

then the system output due to the combined input is obtained as

$$y(n) = 10x(n) = 10(2u(n) + 4\delta(n)) = 20u(n) + 40\delta(n) \tag{3.6}$$

If we verify the weighted sum of the individual outputs, we see that

$$2y_1(n) + 4y_2(n) = 20u(n) + 40\delta(n) \tag{3.7}$$

Comparing Equations (3.6) and (3.7) verifies that

$$y(n) = 2y_1(n) + 4y_2(n) \tag{3.8}$$

Hence, the system $y(n) = 10x(n)$ is a linear system. The linearity means that the system obeys the superposition principle, as shown in Equation (3.8). Let us verify a system whose output is a square of its input,

$$y(n) = x^2(n)$$

Applying the inputs $x_1(n) = u(n)$ and $x_2(n) = \delta(n)$ to the system leads to

$$y_1(n) = u^2(n) = u(n), \text{ and } y_2(n) = \delta^2(n) = \delta(n)$$

It is very easy to verify that $u^2(n) = u(n)$ and $\delta^2(n) = \delta(n)$.

We can determine the system output using a combined input, which is the weighed sum of the individual inputs with constants 2 and 4, respectively. Using algebra, we see that

$$
\begin{aligned}
y(n) = x^2(n) &= (4x_1(n) + 2x_2(n))^2 \\
&= (4u(n) + 2\delta(n))^2 = 16u^2(n) + 16u(n)\delta(n) + 4\delta^2(n) \\
&= 16u(n) + 20\delta(n)
\end{aligned}
\tag{3.9}
$$

Note that we use the fact that $u(n)\delta(n) = \delta(n)$, which can be easily verified.

Again, we express the weighted sum of the two individual outputs with the same constants 2 and 4 as

$$
4y_1(n) + 2y_2(n) = 4u(n) + 2\delta(n)
\tag{3.10}
$$

It is obvious that

$$
y(n) \neq 4y_1(n) + 2y_2(n)
\tag{3.11}
$$

Hence, the system is a nonlinear system, since the linear property, superposition, does not hold, as shown in Equation (3.11).

3.2.2 Time Invariance

A time-invariant system is illustrated in Figure 3.12, where $y_1(n)$ is the system output for the input $x_1(n)$. Let $x_2(n) = x_1(n - n_0)$ be the shifted version of $x_1(n)$ by n_0 samples. The output $y_2(n)$ obtained with the shifted input $x_2(n) = x_1(n - n_0)$ is equivalent to the output $y_2(n)$ acquired by shifting $y_1(n)$ by n_0 samples, $y_2(n) = y_1(n - n_0)$.

This can simply be viewed as the following:

If the system is time invariant and $y_1(n)$ is the system output due to the input $x_1(n)$, then the shifted system input $x_1(n - n_0)$ will produce a shifted system output $y_1(n - n_0)$ by the same amount of time n_0.

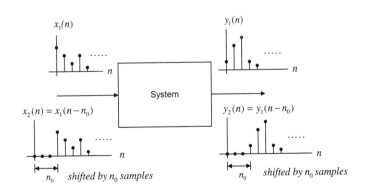

FIGURE 3.12

Illustration of the linear time-invariant digital system.

EXAMPLE 3.3

Determine whether the linear systems

a. $y(n) = 2x(n-5)$
b. $y(n) = 2x(3n)$

are time invariant.

Solution:

a. Let the input and output be $x_1(n)$ and $y_1(n)$, respectively; then the system output is $y_1(n) = 2x_1(n-5)$. Again, let $x_2(n) = x_1(n-n_0)$ be the shifted input and $y_2(n)$ be the output due to the shifted input. We determine the system output using the shifted input as

$$y_2(n) = 2x_2(n-5) = 2x_1(n-n_0-5)$$

Meanwhile, shifting $y_1(n) = 2x_1(n-5)$ by n_0 samples leads to

$$y_1(n-n_0) = 2x_1(n-5-n_0)$$

We can verify that $y_2(n) = y_1(n-n_0)$. Thus the shifted input of n_0 samples causes the system output to be shifted by the same n_0 samples. The system is thus time invariant.

b. Let the input and output be $x_1(n)$ and $y_1(n)$, respectively; then the system output is $y_1(n) = 2x_1(3n)$. Again, let the input and output be $x_2(n)$ and $y_2(n)$, where $x_2(n) = x_1(n-n_0)$, a shifted version, and the corresponding output is $y_2(n)$. We get the output due to the shifted input $x_2(n) = x_1(n-n_0)$ and note that $x_2(3n) = x_1(3n-n_0)$:

$$y_2(n) = 2x_2(3n) = 2x_1(3n-n_0)$$

On the other hand, if we shift $y_1(n)$ by n_0 samples, and replace n in $y_1(n) = 2x_1(3n)$ by $n-n_0$, we obtain

$$y_1(n-n_0) = 2x_1(3(n-n_0)) = 2x_1(3n-3n_0)$$

Clearly, we know that $y_2(n) \neq y_1(n-n_0)$. Since the system output $y_2(n)$ using the input shifted by n_0 samples is not equal to the system output $y_1(n)$ shifted by the same n_0 samples, the system is not time invariant.

3.2.3 Causality

A causal system is the one in which the output $y(n)$ at time n depends only on the current input $x(n)$ at time n, and its past input sample values such as $x(n-1)$, $x(n-2)$,.... Otherwise, if a system output depends on future input values such as $x(n+1)$, $x(n+2)$, ..., the system is noncausal. The noncausal system cannot be realized in real time.

EXAMPLE 3.4

Determine whether the systems

a. $y(n) = 0.5x(n) + 2.5x(n-2)$, for $n \geq 0$
b. $y(n) = 0.25x(n-1) + 0.5x(n+1) - 0.4y(n-1)$, for $n \geq 0$

are causal.

Solution:

a. Since for $n \geq 0$, the output $y(n)$ depends on the current input $x(n)$ and its past value $x(n-2)$, the system is causal.

b. Since for $n \geq 0$, the output $y(n)$ depends on the current input $x(n)$ and its future value $x(n+1)$, the system is noncausal.

3.3 DIFFERENCE EQUATIONS AND IMPULSE RESPONSES

Now we study the difference equation and its impulse response.

3.3.1 Format of the Difference Equation

A causal, linear, time-invariant system can be described by a difference equation having the following general form:

$$y(n) + a_1 y(n-1) + \cdots + a_N y(n-N) = b_0 x(n) + b_1 x(n-1) + \cdots + b_M x(n-M) \quad (3.12)$$

where $a_1, \ldots, a_N$, and $b_0, b_1, \ldots, b_M$ are the coefficients of the difference equation. Equation (3.12) can also be written as

$$y(n) = -a_1 y(n-1) - \cdots - a_N y(n-N) + b_0 x(n) + b_1 x(n-1) + \cdots + b_M x(n-M) \quad (3.13)$$

or

$$y(n) = -\sum_{i=1}^{N} a_i y(n-i) + \sum_{j=0}^{M} b_j x(n-j) \quad (3.14)$$

Notice that $y(n)$ is the current output, which depends on the past output samples $y(n-1), \ldots, y(n-N)$, the current input sample $x(n)$, and the past input samples, $x(n-1), \ldots, x(n-N)$.

We will examine the specific difference equations in the following examples.

EXAMPLE 3.5

Given the difference equation

$$y(n) = 0.25 y(n-1) + x(n)$$

identify the nonzero system coefficients.

Solution:

Comparison with Equation (3.13) leads to

$$b_0 = 1$$

$$-a_1 = 0.25, \quad \text{that is,} \quad a_1 = -0.25$$

EXAMPLE 3.6

Given a linear system described by the difference equation

$$y(n) = x(n) + 0.5 x(n-1)$$

determine the nonzero system coefficients.

Solution:

By comparing Equation (3.13), we have

$$b_0 = 1 \quad \text{and} \quad b_1 = 0.5$$

3.3.2 System Representation Using Its Impulse Response

A linear time-invariant system can be completely described by its unit-impulse response, which is defined as the system response due to the impulse input $\delta(n)$ with zero initial conditions, depicted in Figure 3.13.

With the obtained unit-impulse response $h(n)$, we can represent the linear time-invariant system as shown in Figure 3.14.

FIGURE 3.13

Unit-impulse response of a linear time-invariant system.

FIGURE 3.14

Representation of a linear time-invariant system using the impulse response.

EXAMPLE 3.7

Assume we have a linear time-invariant system

$$y(n) = 0.5x(n) + 0.25x(n-1)$$

with an initial condition $x(-1) = 0$.

a. Determine the unit-impulse response $h(n)$.
b. Draw the system block diagram.
c. Write the output using the obtained impulse response.

Solution:

a. According to Figure 3.13, let $x(n) = \delta(n)$, then

$$h(n) = y(n) = 0.5x(n) + 0.25x(n-1) = 0.5\delta(n) + 0.25\delta(n-1)$$

Thus, for this particular linear system, we have

$$h(n) = \begin{cases} 0.5 & n = 0 \\ 0.25 & n = 1 \\ 0 & elsewhere \end{cases}$$

b. The block diagram of the linear time-invariant system is shown in Figure 3.15.

c. The system output can be rewritten as

$$y(n) = h(0)x(n) + h(1)x(n-1)$$

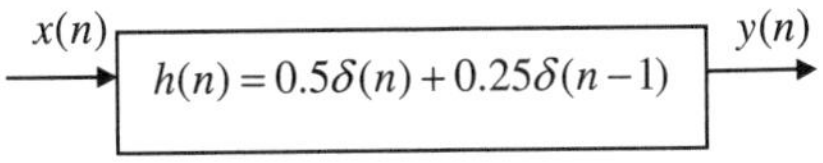

FIGURE 3.15

The system block diagram for Example 3.7.

From the result Example 3.7, it is noted that the difference equation does not have the past output terms, $y(n-1), ..., y(n-N)$, that is, the corresponding coefficients $a_1, ..., a_N$, are zeros, and the impulse response $h(n)$ has a finite number of terms. We call this system a *finite impulse response* (FIR) system. In general, Equation (3.12) contains the past output terms and the resulting impulse response $h(n)$ has an infinite number of terms. We can express the output sequence of a linear time-invariant system using its impulse response and inputs as

$$y(n) = \cdots + h(-1)x(n+1) + h(0)x(n) + h(1)x(n-1) + h(2)x(n-2) + \cdots \qquad (3.15)$$

Equation (3.15) is called the *digital convolution sum*, which will be explored in a later section. We can verify Equation (3.15) by substituting the impulse sequence $x(n) = \delta(n)$ to get the impulse response

$$h(n) = \cdots + h(-1)\delta(n+1) + h(0)\delta(n) + h(1)\delta(n-1) + h(2)\delta(n-2) + \cdots$$

where ... $h(-1), h(0), h(1), h(2)$ ... are the amplitudes of the impulse response at the corresponding time indices. Now let us look at another example.

EXAMPLE 3.8

Consider the difference equation

$$y(n) = 0.25y(n-1) + x(n) \quad \text{for} \quad n \geq 0 \quad \text{and} \quad y(-1) = 0$$

a. Determine the unit-impulse response $h(n)$.

b. Draw the system block diagram.

c. Write the output using the obtained impulse response.

d. For a step input $x(n) = u(n)$, verify and compare the output responses for the first three output samples using the difference equation and digital convolution sum (Equation (3.15)).

Solution:

a. Let $x(n) = \delta(n)$, then

$$h(n) = 0.25h(n-1) + \delta(n)$$

To solve for $h(n)$, we evaluate

$h(0) = 0.25h(-1) + \delta(0) = 0.25 \times 0 + 1 = 1$

$h(1) = 0.25h(0) + \delta(1) = 0.25 \times 1 + 0 = 0.25$

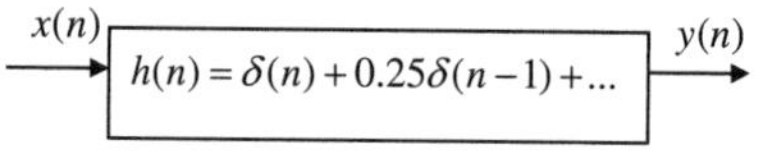

FIGURE 3.16

The system block diagram for Example 3.8.

$$h(2) = 0.25h(1) + \delta(2) = 0.25 \times 0.5 + 0 = 0.0625$$

....

With the calculated results, we can predict the impulse response as

$$h(n) = (0.25)^n u(n) = \delta(n) + 0.25\delta(n-1) + 0.0625\delta(n-2) + \cdots$$

b. The system block diagram is given in Figure 3.16.
c. The output sequence is a sum of infinite terms expressed as

$$y(n) = h(0)x(n) + h(1)x(n-1) + h(2)x(n-2) + \cdots$$

$$= x(n) + 0.25x(n-1) + 0.0625x(n-2) + \cdots$$

d. From the difference equation and using the zero initial condition, we have

$y(n) = 0.25y(n-1) + x(n)$ for $n \geq 0$ and $y(-1) = 0$
$n = 0,\ y(0) = 0.25y(-1) + x(0) = u(0) = 1$
$n = 1,\ y(1) = 0.25y(0) + x(1) = 0.25 \times u(0) + u(1) = 1.25$
$n = 2,\ y(2) = 0.25y(1) + x(2) = 0.25 \times 1.25 + u(2) = 1.3125$

....

Applying the convolution sum in Equation (3.15) yields

$$y(n) = x(n) + 0.25x(n-1) + 0.0625x(n-2) + \cdots$$

$n = 0,$
$$y(0) = x(0) + 0.25x(-1) + 0.0625x(-2) + \cdots$$
$$= u(0) + 0.25 \times u(-1) + 0.125 \times u(-2) + \cdots = 1$$

$n = 1,$
$$y(1) = x(1) + 0.25x(0) + 0.0625x(-1) + \cdots$$
$$= u(1) + 0.25 \times u(0) + 0.125 \times u(-1) + \cdots = 1.25$$

$n = 2,$
$$y(2) = x(2) + 0.25x(1) + 0.0625x(0) + \cdots$$
$$= u(2) + 0.25 \times u(1) + 0.0625 \times u(0) + \cdots = 1.3125$$

...

Comparing the results, we verify that a linear time-invariant system can be represented by the convolution sum using its impulse response and input sequence. Note that we verify only the causal system for simplicity, and the principle works for both causal and noncausal systems.

Notice that this impulse response $h(n)$ contains an infinite number of terms in its duration due to the past output term $y(n-1)$. Such a system as described in the preceding example is called an *infinite impulse response* (IIR) system, which will be studied in later chapters.

3.4 BOUNDED-IN AND BOUNDED-OUT STABILITY

We are interested in designing and implementing stable linear systems. A stable system is one for which every bounded input produces a bounded output (BIBO). There are many other stability definitions. To find the stability criterion, consider the linear time-invariant representation with all the inputs reaching the maximum value M for the worst case. Equation (3.15) becomes

$$y(n) = M(\cdots + h(-1) + h(0) + h(1) + h(2) + \cdots) \qquad (3.16)$$

Using the absolute values of the impulse response leads to

$$y(n) < M(\cdots + |h(-1)| + |h(0)| + |h(1)| + |h(2)| + \cdots) \qquad (3.17)$$

If the absolute sum in Equation (3.17) is a finite number, the product of the absolute sum and the maximum input value is therefore a finite number. Hence, we have a bounded input and bounded output. In terms of the impulse response, a linear system is stable if the sum of its absolute impulse response coefficients is a finite number. We can apply Equation (3.18) to determine whether a linear time-invariant system is stable or not stable, that is,

$$S = \sum_{k=-\infty}^{\infty} |h(k)| = \cdots + |h(-1)| + |h(0)| + |h(1)| + \cdots < \infty \qquad (3.18)$$

Figure 3.17 describes a linear stable system, where the impulse response decreases to zero in a finite amount of time so that the summation of its absolute impulse response coefficients is guaranteed to be finite.

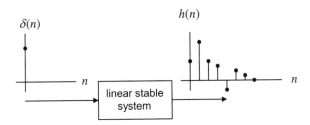

FIGURE 3.17

Illustration of stability of the digital linear system.

EXAMPLE 3.9

Given the linear system in Example 3.8,

$$y(n) = 0.25y(n-1) + x(n) \quad \text{for} \quad n \geq 0 \quad \text{and} \quad y(-1) = 0$$

which is described by the unit-impulse response

$$h(n) = (0.25)^n u(n)$$

determine whether this system is stable.

Solution:

Using Equation (3.18), we have

$$S = \sum_{k=-\infty}^{\infty} |h(k)| = \sum_{k=-\infty}^{\infty} \left| (0.25)^k u(k) \right|$$

Applying the definition of the unit-step function $u(k) = 1$ for $k \geq 0$, we have

$$S = \sum_{k=0}^{\infty} (0.25)^k = 1 + 0.25 + 0.25^2 + \cdots$$

Using the formula for a sum of the geometric series (see Appendix F),

$$\sum_{k=0}^{\infty} a^k = \frac{1}{1-a},$$

where $a = 0.25 < 1$, we conclude

$$S = 1 + 0.25 + 0.25^2 + \cdots = \frac{1}{1-0.25} = \frac{4}{3} < \infty$$

Since the summation is a finite number, the linear system is stable.

3.5 DIGITAL CONVOLUTION

Digital convolution plays an important role in digital filtering. As we verified in the last section, a linear time-invariant system can be represented using a digital convolution sum. Given a linear time-invariant system, we can determine its unit-impulse response $h(n)$, which relates the system input and output. To find the output sequence $y(n)$ for any input sequence $x(n)$, we write the digital convolution shown in Equation (3.15) as

$$
\begin{aligned}
y(n) &= \sum_{k=-\infty}^{\infty} h(k)x(n-k) \\
&= \cdots + h(-1)x(n+1) + h(0)x(n) + h(1)x(n-1) + h(2)x(n-2) + \cdots
\end{aligned}
\tag{3.19}
$$

The sequences $h(k)$ and $x(k)$ in Equation (3.19) are interchangeable. Hence, we have an alternative form:

$$
\begin{aligned}
y(n) &= \sum_{k=-\infty}^{\infty} x(k)h(n-k) \\
&= \cdots + x(-1)h(n+1) + x(0)h(n) + x(1)h(n-1) + x(2)h(n-2) + \cdots
\end{aligned}
\tag{3.20}
$$

Using conventional notation, we express the digital convolution as

$$y(n) = h(n) * x(n) \tag{3.21}$$

Note that for a causal system, which implies an impulse response of

$$h(n) = 0 \quad \text{for} \quad n < 0 \tag{3.16}$$

the lower limit of the convolution sum begins at 0 instead of ∞, that is

$$y(n) = \sum_{k=0}^{\infty} h(k)x(n-k) = \sum_{k=0}^{\infty} x(k)h(n-k) \tag{3.22}$$

We will focus on evaluating the convolution sum based on Equation (3.20). Let us examine first a few outputs from Equation (3.20):

$$
\begin{aligned}
y(0) &= \sum_{k=-\infty}^{\infty} x(k)h(-k) = \cdots + x(-1)h(1) + x(0)h(0) + x(1)h(-1) + x(2)h(-2) + \cdots \\
y(1) &= \sum_{k=-\infty}^{\infty} x(k)h(1-k) = \cdots + x(-1)h(2) + x(0)h(1) + x(1)h(0) + x(2)h(-1) + \cdots \\
y(2) &= \sum_{k=-\infty}^{\infty} x(k)h(2-k) = \cdots + x(-1)h(3) + x(0)h(2) + x(1)h(1) + x(2)h(0) + \cdots
\end{aligned}
$$

.....

We see that the convolution sum requires the sequence $h(n)$ to be reversed and shifted. The graphical, formula, and table methods for evaluating the digital convolution will be discussed via several examples. To evaluate the convolution sum graphically, we need to apply the reversed sequence and shifted sequence. The reversed sequence is defined as follows: if $h(n)$ is the given sequence, $h(-n)$ is the reversed sequence. The reversed sequence is a mirror image of the original sequence, assuming the vertical axis as the mirror. Let us study the reversed sequence and shifted sequence via the following example.

EXAMPLE 3.10

Consider a sequence

$$
h(k) = \begin{cases} 3, & k = 0, 1 \\ 1, & k = 2, 3 \\ 0 & elsewhere \end{cases}
$$

where k is the time index or sample number.

a. Sketch the sequence $h(k)$ and reversed sequence $h(-k)$.
b. Sketch the shifted sequences $h(-k + 3)$ and $h(-k - 2)$.

Solution:
a. Since $h(k)$ is defined, we plot it in Figure 3.18. Next, we need to find the reversed sequence $h(-k)$. We examine the following:

$k > 0, h(-k) = 0$
$k = 0, h(-0) = h(0) = 3$
$k = -1, h(-k) = h(-(-1)) = h(1) = 3$
$k = -2, h(-k) = h(-(-2)) = h(2) = 1$
$k = -3, h(-k) = h(-(-3)) = h(3) = 1$

One can verify that $k \le -4$, $h(-k) = 0$. Then the reversed sequence $h(-k)$ is shown as the second plot in Figure 3.18.

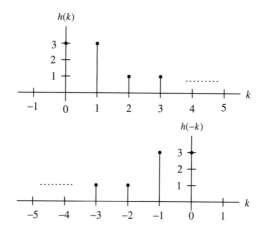

FIGURE 3.18

Plots of the digital sequence and its reversed sequence in Example 3.10.

As shown in the sketches, $h(-k)$ is just a mirror image of the original sequence $h(k)$.

b. Based on the definition of the original sequence, we know that $h(0) = h(1) = 3$, $h(2) = h(3) = 1$, and the others are zeros. The time indices correspond to the following:

$-k + 3 = 0, k = 3$
$-k + 3 = 1, k = 2$
$-k + 3 = 2, k = 1$
$-k + 3 = 3, k = 0$

Thus we can sketch $h(-k + 3)$ as shown in Figure 3.19.
    Similarly, $h(-k - 2)$ is shown in Figure 3.20.

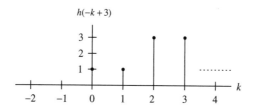

FIGURE 3.19

Plot of the sequence $h(-k + 3)$ in Example 3.10.

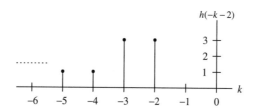

FIGURE 3.20

Plot of the sequence $h(-k - 2)$ in Example 3.10.

We can get $h(-k+3)$ by shifting $h(-k)$ to the right by three samples, and we can obtain $h(-k-2)$ by shifting $h(-k)$ to the left by two samples.

In summary, given $h(-k)$, we can obtain $h(n-k)$ by shifting $h(-k)$ n samples to the right or the left, depending on whether n is positive or negative.

Once we understand the shifted sequence and reversed sequence, we can perform digital convolution of the two sequences $h(k)$ and $x(k)$, defined in Equation (3.20) graphically. From that equation, we see that each convolution value $y(n)$ is the sum of the products of two sequences $x(k)$ and $h(n-k)$, the latter of which is the shifted version of the reversed sequence $h(-k)$ by $|n|$ samples. Hence, we can summarize the graphical convolution procedure in Table 3.3.

We illustrate the digital convolution sum in the following example.

Table 3.3 Digital Convolution Using the Graphical Method

Step 1. Obtain the reversed sequence $h(-k)$.
Step 2. Shift $h(-k)$ by $|n|$ samples to get $h(n-k)$. If $n \geq 0$, $h(-k)$ will be shifted to right by n samples; but if $n < 0$, $h(-k)$ will be shifted to the left by $|n|$ samples.
Step 3. Perform the convolution sum, which is the sum of the products of two sequences $x(k)$ and $h(n-k)$, to get $y(n)$.
Step 4. Repeat Steps 1 to 3 for the next convolution value $y(n)$.

EXAMPLE 3.11

Using the sequences defined in Figure 3.21, evaluate the digital convolution

$$y(n) = \sum_{k=-\infty}^{\infty} x(k)h(n-k)$$

a. By the graphical method.
b. By applying the formula directly.

Solution:
a. To obtain $y(0)$, we need the reversed sequence $h(-k)$; and to obtain $y(1)$, we need the reversed sequence $h(1-k)$, and so on. Using the technique we have discussed, sequences $h(-k)$, $h(-k+1)$, $h(-k+2)$, $h(-k+3)$, and $h(-k+4)$ are obtained and plotted in Figure 3.22.

Again, using the information in Figures 3.21 and 3.22, we can compute the convolution sum as

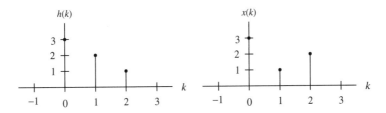

FIGURE 3.21

Plots of digital input sequence and impulse sequence in Example 3.11.

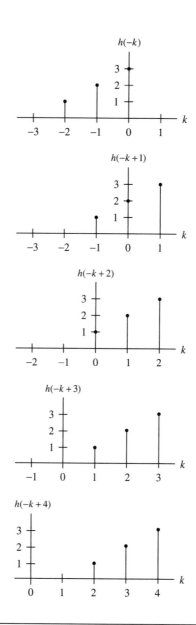

FIGURE 3.22

Illustration of convolution of two sequences $x(k)$ and $h(k)$ in Example 3.11.

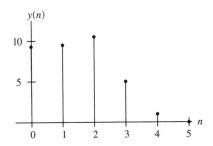

FIGURE 3.23

Plot of the convolution sum in Example 3.11.

sum of product of $x(k)$ and $h(-k)$: $y(0) = 3 \times 3 = 9$
sum of product of $x(k)$ and $h(1 - k)$: $y(1) = 1 \times 3 + 3 \times 2 = 9$
sum of product of $x(k)$ and $h(2 - k)$: $y(2) = 2 \times 3 + 1 \times 2 + 3 \times 1 = 11$
sum of product of $x(k)$ and $h(3 - k)$: $y(3) = 2 \times 2 + 1 \times 1 = 5$
sum of product of $x(k)$ and $h(4 - k)$: $y(4) = 2 \times 1 = 2$
sum of product of $x(k)$ and $h(5 - k)$: $y(n) = 0$ for $n > 4$, since sequences $x(k)$ and $h(n - k)$ do not overlap.

Finally, we sketch the output sequence $y(n)$ in Figure 3.23.
b. Applying Equation (3.20) with zero initial conditions leads to

$y(n) = x(0)h(n) + x(1)h(n - 1) + x(2)h(n - 2)$
$n = 0, y(0) = x(0)h(0) + x(1)h(-1) + x(2)h(-2) = 3 \times 3 + 1 \times 0 + 2 \times 0 = 9$
$n = 1, y(1) = x(0)h(1) + x(1)h(0) + x(2)h(-1) = 3 \times 2 + 1 \times 3 + 2 \times 0 = 9$
$n = 2, y(2) = x(0)h(2) + x(1)h(1) + x(2)h(0) = 3 \times 1 + 1 \times 2 + 2 \times 3 = 11$
$n = 3, y(3) = x(0)h(3) + x(1)h(2) + x(2)h(1) = 3 \times 0 + 1 \times 1 + 2 \times 2 = 5$
$n = 4, y(4) = x(0)h(4) + x(1)h(3) + x(2)h(2) = 3 \times 0 + 1 \times 0 + 2 \times 1 = 2$
$n \geq 5, y(n) = x(0)h(n) + x(1)h(n - 1) + x(2)h(n - 2) = 3 \times 0 + 1 \times 0 + 2 \times 0 = 0$

In simple cases such as this example, it is not necessary to use the graphical or formula methods. We can compute the convolution by treating the input sequence and impulse response as number sequences and sliding the reversed impulse response past the input sequence, cross-multiplying, and summing the nonzero overlap terms at each step. The procedure and calculated results are listed in Table 3.4.

Table 3.4 Convolution Sum Using the Table Method

k:	−2	−1	0	1	2	3	4	5	
$x(k)$:			3	1	2				
$h(-k)$:	1	2	3						$y(0) = 3 \times 3 = 9$
$h(1 - k)$		1	2	3					$y(1) = 3 \times 2 + 1 \times 3 = 9$
$h(2 - k)$			1	2	3				$y(2) = 3 \times 1 + 1 \times 2 + 2 \times 3 = 11$
$h(3 - k)$				1	2	3			$y(3) = 1 \times 1 + 2 \times 2 = 5$
$h(4 - k)$					1	2	3		$y(4) = 2 \times 1 = 2$
$h(5 - k)$						1	2	3	$y(5) = 0$ (no overlap)

We can see that the calculated results using all the methods are consistent. The steps using the table method are summarized in Table 3.5.

Table 3.5 Digital Convolution Steps via the Table Method

Step 1. List the index k covering a sufficient range.
Step 2. List the input $x(k)$.
Step 3. Obtain the reversed sequence $h(-k)$, and align the rightmost element of $h(n-k)$ to the leftmost element of $x(k)$.
Step 4. Cross-multiply and sum the nonzero overlap terms to produce $y(n)$.
Step 5. Slide $h(n-k)$ to the right by one position.
Step 6. Repeat Step 4; stop if all the output values are zero or if required.

EXAMPLE 3.12

Convolve the following two rectangular sequences using the table method:

$$x(n) = \begin{cases} 1 & n = 0,1,2 \\ 0 & \text{otherwise} \end{cases} \quad \text{and} \quad h(n) = \begin{cases} 0 & n = 0 \\ 1 & n = 1,2 \\ 0 & \text{otherwise} \end{cases}$$

Solution:
Using Table 3.5 as a guide, we list the operations and calculations in Table 3.6. Note that the output should show the trapezoidal shape.

Table 3.6 Convolution Sum in Example 3.12

k:	-2	-1	0	1	2	3	4	5	
$x(k)$:			1	1	1				
$h(-k)$:	1	1	0						$y(0) = 0$ (no overlap)
$h(1-k)$		1	1	0					$y(1) = 1 \times 1 = 1$
$h(2-k)$			1	1	0				$y(2) = 1 \times 1 + 1 \times 1 = 2$
$h(3-k)$				1	1	0			$y(3) = 1 \times 1 + 1 \times 1 = 2$
$h(4-k)$					1	1	0		$y(4) = 1 \times 1 = 1$
$h(n-k)$						1	1	0	$y(n) = 0, n \geq 5$ (no overlap)
									Stop

Let us examine convolving a finite long sequence with an infinite long sequence.

EXAMPLE 3.13

A system representation using the unit-impulse response for the linear system

$$y(n) = 0.25y(n-1) + x(n) \quad \text{for} \quad n \geq 0 \quad \text{and} \quad y(-1) = 0$$

was determined in Example 3.8 as

$$y(n) = \sum_{k=-\infty}^{\infty} x(k)h(n-k)$$

where $h(n) = (0.25)^n u(n)$. For a step input $x(n) = u(n)$, determine the output response for the first three output samples using the table method.

Solution:
Using Table 3.5 as a guide, we list the operations and calculations in Table 3.7. As expected, the output values are the same as those obtained in Example 3.8.

Table 3.7 Convolution Sum in Example 3.13.

k :	-2	-1	0	1	2	3	...	
$x(k)$:			1	1	1	1	...	
$h(-k)$:	0.0625	0.25	1					$y(0) = 1 \times 1 = 1$
$h(1-k)$		0.0625	0.25	1				$y(1) = 1 \times 0.25 + 1 \times 1 = 1.25$
$h(2-k)$			0.0625	0.25	1			$y(2) = 1 \times 0.0625 + 1 \times 0.25 +$
								$1 \times 1 = 1.3125$
								Stop as required

3.6 SUMMARY

1. Digital signal samples are sketched using their encoded amplitude versus sample numbers with vertical bars topped by solid circles located at their sampling instants, respectively. The impulse sequence, unit-step sequence, and their shifted versions are sketched in this notation.

2. The analog signal function can be sampled to its digital (discrete-time) version by substituting time $t = nT$ into the analog function, that is,

   $$x(n) = x(t)|_{t=nT} = x(nT)$$

   The digital function values can be calculated for the given time index (sample number).

3. The DSP system we wish to design must be a linear, time-invariant, causal system. Linearity means that the superposition principle exists. Time invariance requires that the shifted input generate the corresponding shifted output in the same amount of time. Causality indicates that the system output depends on only its current input sample and past input sample(s).

4. The difference equation describing a linear, time-invariant system has a format such that the current output depends on the current input, past input(s), and past output(s) in general.

5. The unit-impulse response can be used to fully describe a linear, time-invariant system. Given the impulse response, the system output is the sum of the products of the impulse response coefficients and corresponding input samples, called the digital convolution sum.
6. BIBO is a type of stability in which a bounded input will produce a bounded output. A BIBO system requires that the sum of the absolute impulse response coefficients be a finite number.
7. The digital convolution sum, which represents a DSP system, is evaluated in three ways: the graphical method, evaluation of the formula, and the table method. The table method is found to be most effective.

3.7 PROBLEMS

3.1. Sketch each of the following special digital sequences:

    a. $5\delta(n)$

    b. $-2\delta(n-5)$

    c. $-5u(n)$

    d. $5u(n-2)$

3.2. Calculate the first eight sample values and sketch each of the following sequences:

    a. $x(n) = 0.5^n u(n)$

    b. $x(n) = 5\sin(0.2\pi n)u(n)$

    c. $x(n) = 5\cos(0.1\pi n + 30^0)u(n)$

    d. $x(n) = 5(0.75)^n \sin(0.1\pi n)u(n)$

3.3. Sketch each of the following special digital sequences:

    a. $8\delta(n)$

    b. $-3.5\delta(n-4)$

    c. $4.5u(n)$

    d. $-6u(n-3)$

3.4. Calculate the first eight sample values and sketch each of the following sequences:

    a. $x(n) = 0.25^n u(n)$

    b. $x(n) = 3\sin(0.4\pi n)u(n)$

    c. $x(n) = 6\cos(0.2\pi n + 30^0)u(n)$

    d. $x(n) = 4(0.5)^n \sin(0.1\pi n)u(n)$

3.5. Sketch the following sequences:

    a. $x(n) = 3\delta(n+2) - 0.5\delta(n) + 5\delta(n-1) - 4\delta(n-5)$

    b. $x(n) = \delta(n+1) - 2\delta(n-1) + 5u(n-4)$

3.6. Given the digital signals $x(n)$ in Figures 3.24 and 3.25, write an expression for each digital signal using the unit-impulse sequence and its shifted sequences.

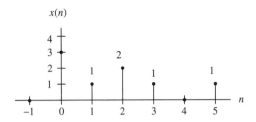

FIGURE 3.24

The first digital signal in Problem 3.6.

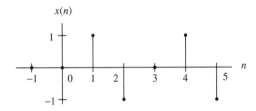

FIGURE 3.25

The second digital signal in Problem 3.6.

3.7. Sketch the following sequences:

a. $x(n) = 2\delta(n+3) - 0.5\delta(n+1) - 5\delta(n-2) - 4\delta(n-5)$

b. $x(n) = 2\delta(n+2) - 2\delta(n+1) + 5u(n-3)$

3.8. Given the digital signals $x(n)$ in Figures 3.26 and 3.27, write an expression for each digital signal using the unit-impulse sequence and its shifted sequences.

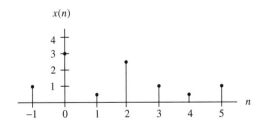

FIGURE 3.26

The first digital signal in Problem 3.8.

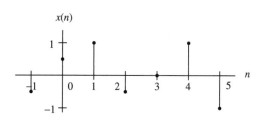

FIGURE 3.27

The second digital signal in Problem 3.8.

3.9. Assume that a DS processor with a sampling time interval of 0.01 second converts the following analog signals $x(t)$ to a digital signal $x(n)$; determine the digital sequence for each of the analog signals.

a. $x(t) = e^{-50t}u(t)$

b. $x(t) = 5\sin(20\pi t)u(t)$

c. $x(t) = 10\cos(40\pi t + 30^0)u(t)$

d. $x(t) = 10e^{-100t}\sin(15\pi t)u(t)$

3.10. Determine which of the following systems is a linear system.

a. $y(n) = 5x(n) + 2x^2(n)$

b. $y(n) = x(n-1) + 4x(n)$

c. $y(n) = 4x^3(n-1) - 2x(n)$

3.11. Assume that a DS processor with a sampling time interval of 0.005 second converts each of the following analog signals $x(t)$ to a digital signal $x(n)$; determine the digital sequence for each of the analog signals.

a. $x(t) = e^{-100t}u(t)$

b. $x(t) = 4\sin(60\pi t)u(t)$

c. $x(t) = 7.5\cos(20\pi t + 60^0)u(t)$

d. $x(t) = 20e^{-200t}\sin(60\pi t)u(t)$

3.12. Determine which of the following systems is a linear system.

a. $y(n) = 4x(n) + 8x^3(n)$

b. $y(n) = x(n-3) + 3x(n)$

c. $y(n) = 5x^2(n-1) - 3x(n)$

3.13. Determine which of the following linear systems is time invariant.

a. $y(n) = -5x(n-10)$

b. $y(n) = 4x(n^2)$

3.14. Determine which of the following linear systems is causal.

    a. $y(n) = 0.5x(n) + 100x(n - 2) - 20x(n - 10)$

    b. $y(n) = x(n + 4) + 0.5x(n) - 2x(n - 2)$

3.15. Determine the causality for each of the following linear systems.

    a. $y(n) = 0.5x(n) + 20x(n - 2) - 0.1y(n - 1)$

    b. $y(n) = x(n + 2) - 0.4y(n - 1)$

    c. $y(n) = x(n - 1) + 0.5y(n + 2)$

3.16. Find the unit-impulse response for each of the following linear systems.

    a. $y(n) = 0.5x(n) - 0.5x(n - 2)$; for $n \geq 0$, $x(-2) = 0$, $x(-1) = 0$

    b. $y(n) = 0.75y(n - 1) + x(n)$; for $n \geq 0$, $y(-1) = 0$

    c. $y(n) = -0.8y(n - 1) + x(n - 1)$; for $n \geq 0$, $x(-1) = 0$, $y(-1) = 0$

3.17. Determine the causality for each of the following linear systems.

    a. $y(n) = 5x(n) + 10x(n - 4) - 0.1y(n - 5)$

    b. $y(n) = 2x(n + 2) - 0.2y(n - 2)$

    c. $y(n) = 0.1x(n + 1) + 0.5y(n + 2)$

3.18. Find the unit-impulse response for each of the following linear systems.

    a. $y(n) = 0.2x(n) - 0.3x(n - 2)$; for $n \geq 0$, $x(-2) = 0$, $x(-1) = 0$

    b. $y(n) = 0.5y(n - 1) + 0.5x(n)$; for $n \geq 0$, $y(-1) = 0$

    c. $y(n) = -0.6y(n - 1) - x(n - 1)$; for $n \geq 0$, $x(-1) = 0$, $y(-1) = 0$

3.19. For each of the following linear systems, find the unit-impulse response, and draw the block diagram.

    a. $y(n) = 5x(n - 10)$

    b. $y(n) = x(n) + 0.5x(n - 1)$

3.20. Determine the stability of the following linear system.

$$y(n) = 0.5x(n) + 100x(n - 2) - 20x(n - 10)$$

3.21. For each of the following linear systems, find the unit-impulse response, and draw the block diagram.

    a. $y(n) = 2.5x(n - 5)$

    b. $y(n) = 2x(n) + 1.2x(n - 1)$

3.22. Determine the stability for the following linear system.

$$y(n) = 5x(n) + 30x(n - 3) - 10x(n - 20)$$

3.23. Determine the stability for each of the following linear systems.

    a. $y(n) = \sum_{k=0}^{\infty} 0.75^k x(n - k)$

    b. $y(n) = \sum_{k=0}^{\infty} 2^k x(n - k)$

3.24. Determine the stability for each of the following linear systems.

a. $y(n) = \sum_{k=0}^{\infty}(-1.5)^k x(n-k)$

b. $y(n) = \sum_{k=0}^{\infty}(-0.5)^k x(n-k)$

3.25. Given the sequence

$$h(k) = \begin{cases} 2, & k = 0,1,2 \\ 1, & k = 3,4 \\ 0 & elsewhere, \end{cases}$$

where k is the time index or sample number,

a. sketch the sequence $h(k)$ and the reverse sequence $h(-k)$;

b. sketch the shifted sequences $h(-k+2)$ and $h(-k-3)$.

3.26. Given the sequence

$$h(k) = \begin{cases} -1 & k = 0,1 \\ 2 & k = 2,3 \\ -2 & k = 4 \\ 0 & elsewhere \end{cases}$$

where k is the time index or sample number,

a. sketch the sequence $h(k)$ and the reverse sequence $h(-k)$;

b. sketch the shifted sequences $h(-k+1)$ and $h(-k-2)$.

3.27. Using the sequence definitions

$$h(k) = \begin{cases} 2, & k = 0,1,2 \\ 1, & k = 3,4 \\ 0 & elsewhere \end{cases} \quad and\ x(k) = \begin{cases} 2, & k = 0 \\ 1, & k = 1,2 \\ 0 & elsewhere \end{cases}$$

evaluate the digital convolution

$$y(n) = \sum_{k=-\infty}^{\infty} x(k)h(n-k)$$

a. using the graphical method;

b. using the table method;

c. applying the convolution formula directly.

3.28. Using the sequence definitions

$$x(k) = \begin{cases} -2, & k = 0,1,2 \\ 1, & k = 3,4 \\ 0 & elsewhere \end{cases} \quad and\ h(k) = \begin{cases} 2, & k = 0 \\ -1, & k = 1,2 \\ 0 & elsewhere \end{cases}$$

evaluate the digital convolution

$$y(n) = \sum_{k=-\infty}^{\infty} x(k)h(n-k)$$

a. using the graphical method;

b. using the table method;

c. applying the convolution formula directly.

3.29. Convolve the two rectangular sequences

$$x(n) = \begin{cases} 1 & n = 0,1 \\ 0 & otherwise \end{cases} \quad and \quad h(n) = \begin{cases} 0 & n = 0 \\ 1 & n = 1,2 \\ 0 & otherwise \end{cases}$$

using the table method.

Discrete Fourier Transform and Signal Spectrum

4

CHAPTER OUTLINE

4.1 **Discrete Fourier Transform** 87
    4.1.1 Fourier Series Coefficients of Periodic Digital Signals 88
    4.1.2 Discrete Fourier Transform Formulas..91
4.2 **Amplitude Spectrum and Power Spectrum** 97
4.3 **Spectral Estimation Using Window Functions** 107
4.4 **Application to Signal Spectral Estimation** 116
4.5 **Fast Fourier Transform** 123
    4.5.1 Decimation-in-Frequency Method 123
    4.5.2 Decimation-in-Time Method 128
4.6 **Summary** 132

OBJECTIVES:

This chapter investigates discrete Fourier transform (DFT) and fast Fourier transform (FFT) and their properties; introduces the DFT/FFT algorithms that compute the signal amplitude spectrum and power spectrum; and uses the window function to reduce spectral leakage. Finally, the chapter describes the FFT algorithm and shows how to apply FFT to estimate a speech spectrum.

4.1 DISCRETE FOURIER TRANSFORM

In the time domain, representation of digital signals describes the signal amplitude versus the sampling time instant or the sample number. However, in some applications, signal frequency content is very useful in ways other than as digital signal samples. The representation of the digital signal in terms of its frequency component in a frequency domain, that is, the signal spectrum, needs to be developed. As an example, Figure 4.1 illustrates the time domain representation of a 1,000-Hz sinusoid with 32 samples at a sampling rate of 8,000 Hz; the bottom plot shows the signal spectrum (frequency domain representation), where we can clearly observe that the amplitude peak is located at the frequency of 1,000 Hz in the calculated spectrum. Hence, the spectral plot better displays the frequency information of a digital signal.

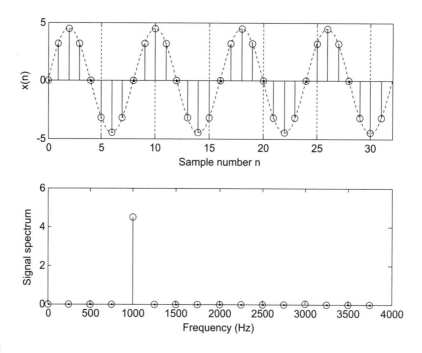

FIGURE 4.1

Example of the digital signal and its amplitude spectrum.

The algorithm transforming the time domain signal samples to the frequency domain components is known as the *discrete Fourier transform*, or DFT. The DFT also establishes a relationship between the time domain representation and the frequency domain representation. Therefore, we can apply the DFT to perform frequency analysis of a time domain sequence. In addition, the DFT is widely used in many other areas, including spectral analysis, acoustics, imaging/video, audio, instrumentation, and communications systems.

To be able to develop the DFT and understand how to use it, we first study the spectrum of periodic digital signals using the Fourier series. (There is a detailed discussion of the Fourier series in Appendix B.)

4.1.1 Fourier Series Coefficients of Periodic Digital Signals

Let us look at a process in which we want to estimate the spectrum of a periodic digital signal $x(n)$ sampled at a rate of f_s Hz with the fundamental period $T_0 = NT$, as shown in Figure 4.2, where there are N samples within the duration of the fundamental period and $T = 1/f_s$ is the sampling period. For the time being, we assume that the periodic digital signal is band limited such that all harmonic frequencies are less than the folding frequency $f_s/2$ so that aliasing does not occur.

According to Fourier series analysis (Appendix B), the coefficients of the Fourier series expansion of the periodic signal $x(t)$ in a complex form are

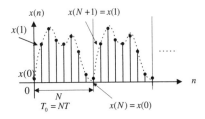

FIGURE 4.2

Periodic digital signal.

$$c_k = \frac{1}{T_0} \int_{T_0} x(t) e^{-jk\omega_0 t} dt \quad -\infty < k < \infty \qquad (4.1)$$

where k is the number of harmonics corresponding to the harmonic frequency of kf_0 and $\omega_0 = 2\pi/T_0$ and $f_0 = 1/T_0$ are the fundamental frequency in radians per second and the fundamental frequency in Hz, respectively. To apply Equation (4.1), we substitute $T_0 = NT$, $\omega_0 = 2\pi/T_0$ and approximate the integration over one period using a summation by substituting $dt = T$ and $t = nT$. We obtain

$$c_k = \frac{1}{N} \sum_{n=0}^{N-1} x(n) e^{-j\frac{2\pi kn}{N}}, \quad -\infty < k < \infty \qquad (4.2)$$

Since the coefficients c_k are obtained from the Fourier series expansion in the complex form, the resultant spectrum c_k will have two sides. There is an important feature of Equation (4.2) in which the Fourier series coefficient c_k is periodic of N. We can verify this as follows:

$$c_{k+N} = \frac{1}{N} \sum_{n=0}^{N-1} x(n) e^{-j\frac{2\pi(k+N)n}{N}} = \frac{1}{N} \sum_{n=0}^{N-1} x(n) e^{-j\frac{2\pi kn}{N}} e^{-j2\pi n} \qquad (4.3)$$

Since $e^{-j2\pi n} = \cos(2\pi n) - j\sin(2\pi n) = 1$, it follows that

$$c_{k+N} = c_k \qquad (4.4)$$

Therefore, the two-side line amplitude spectrum $|c_k|$ is periodic, as shown in Figure 4.3.

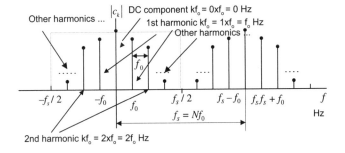

FIGURE 4.3

Amplitude spectrum of the periodic digital signal.

We note the following points:

a. As displayed in Figure 4.3, only the line spectral portion between the frequency $-f_s/2$ and frequency $f_s/2$ (folding frequency) represents frequency information of the periodic signal.

b. Notice that the spectral portion from $f_s/2$ to f_s is a copy of the spectrum in the negative frequency range from $-f_s/2$ to 0 Hz due to the spectrum being periodic for every Nf_0 Hz. Again, the amplitude spectral components indexed from $f_s/2$ to f_s can be folded at the folding frequency $f_s/2$ to match the amplitude spectral components indexed from 0 to $f_s/2$ in terms of $f_s - f$ Hz, where f is in the range from $f_s/2$ to f_s. For convenience, we compute the spectrum over the range from 0 to f_s Hz with nonnegative indices, that is,

$$c_k = \frac{1}{N} \sum_{n=0}^{N-1} x(n)e^{-j\frac{2\pi kn}{N}}, \quad k = 0, 1, \cdots, N-1 \tag{4.5}$$

We can apply Equation (4.4) to find the negative indexed spectral values if they are required.

c. For the kth harmonic, the frequency is

$$f = kf_0 \text{ Hz} \tag{4.6}$$

The frequency spacing between the consecutive spectral lines, called the frequency resolution, is f_0 Hz.

EXAMPLE 4.1

The periodic signal

$$x(t) = \sin(2\pi t)$$

is sampled using the sampling rate $f_s = 4$ Hz.

a. Compute the spectrum c_k using the samples in one period.
b. Plot the two-sided amplitude spectrum $|c_k|$ over the range from -2 to 2 Hz.

Solution:

a. From the analog signal, we can determine the fundamental frequency $\omega_0 = 2\pi$ radians per second and $f_0 = \frac{\omega_0}{2\pi} = \frac{2\pi}{2\pi} = 1$ Hz, and the fundamental period $T_0 = 1$ second.

Since using the sampling interval $T = 1/f_s = 0.25$ second, we get the sampled signal as

$$x(n) = x(nT) = \sin(2\pi nT) = \sin(0.5\pi n)$$

and plot the first eight samples as shown in Figure 4.4.

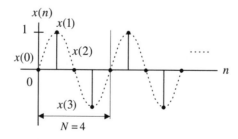

FIGURE 4.4

Periodic digital signal.

Choosing the duration of one period, $N = 4$, we have the following sample values:

$$x(0) = 0; \; x(1) = 1; \; x(2) = 0; \; \text{and } x(3) = -1$$

Using Equation (4.5),

$$c_0 = \frac{1}{4} \sum_{n=0}^{3} x(n) = \frac{1}{4}\left(x(0) + x(1) + x(2) + x(3)\right) = \frac{1}{4}(0 + 1 + 0 - 1) = 0$$

$$c_1 = \frac{1}{4} \sum_{n=0}^{3} x(n)e^{-j2\pi \times 1n/4} = \frac{1}{4}\left(x(0) + x(1)e^{-j\pi/2} + x(2)e^{-j\pi} + x(3)e^{-j3\pi/2}\right)$$

$$= \frac{1}{4}\left(x(0) - jx(1) - x(2) + jx(3)\right) = 0 - j(1) - 0 + j(-1) = -j0.5$$

Similarly, we get

$$c_2 = \frac{1}{4} \sum_{k=0}^{3} x(n)e^{-j2\pi \times 2n/4} = 0, \; \text{and } c_3 = \frac{1}{4} \sum_{n=0}^{3} x(k)e^{-j2\pi \times 3n/4} = j0.5$$

Using periodicity, it follows that

$$c_{-1} = c_3 = j0.5, \; \text{and } c_{-2} = c_2 = 0$$

b. The amplitude spectrum for the digital signal is sketched in Figure 4.5.

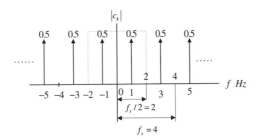

FIGURE 4.5

Two-sided spectrum for the periodic digital signal in Example 4.1.

As we know, the spectrum in the range of -2 to 2 Hz presents the information of the sinusoid with a frequency of 1 Hz and a peak value of $2|.c_1|. = 1$, which is obtained from converting two sides to one side by doubling the two-sided spectral value. Note that we do not double the direct-current (DC) component, that is, c_0.

4.1.2 Discrete Fourier Transform Formulas

Now let us concentrate on development of the DFT. Figure 4.6 shows one way to obtain the DFT formula.

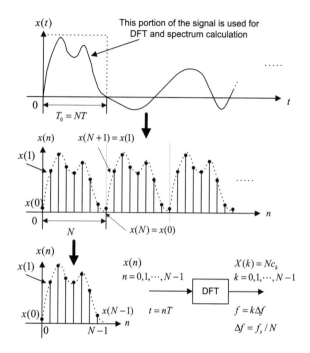

FIGURE 4.6

Development of DFT formula.

First, we assume that the process acquires data samples from digitizing the relevant continuous signal for T_0 seconds. Next, we assume that a periodic signal $x(n)$ is obtained by cascading the acquired N data samples with the duration of T_0 repetitively. Note that we assume continuity between the N data sample frames. This is not true in practice. We will tackle this problem in Section 4.3. Finally, we determine the Fourier series coefficients using one-period N data samples and Equation (4.5). Then we multiply the Fourier series coefficients by a factor of N to obtain

$$X(k) = Nc_k = \sum_{n=0}^{N-1} x(n)e^{-j\frac{2\pi kn}{N}}, \quad k = 0, 1, \cdots, N-1$$

where $X(k)$ constitutes the DFT coefficients. Notice that the factor of N is a constant and does not affect the relative magnitudes of the DFT coefficients $X(k)$. As shown in the last plot, applying DFT with N data samples of $x(n)$ sampled at a sampling rate of f_s (sampling period is $T = 1/f_s$) produces N complex DFT coefficients $X(k)$. The index n is the time index representing the sample number of the digital sequence, whereas k is the frequency index indicating each calculated DFT coefficient, and can be further mapped to the corresponding signal frequency in terms of Hz.

Now let us conclude the DFT definition. Given a sequence $x(n), 0 \leq n \leq N - 1$, its DFT is defined as

$$X(k) = \sum_{n=0}^{N-1} x(n)e^{-j2\pi kn/N} = \sum_{n=0}^{N-1} x(n)W_N^{kn}, \quad \text{for } k = 0, 1, \cdots, N - 1 \qquad (4.7)$$

Equation (4.7) can be expanded as

$$X(k) = x(0)W_N^{k0} + x(1)W_N^{k1} + x(2)W_N^{k2} + \cdots + x(N-1)W_N^{k(N-1)}, \quad \text{for } k = 0, 1, \cdots, N-1 \quad (4.8)$$

where the factor W_N (called the twiddle factor in some textbooks) is defined as

$$W_N = e^{-j2\pi/N} = \cos\left(\frac{2\pi}{N}\right) - j\sin\left(\frac{2\pi}{N}\right) \quad (4.9)$$

The inverse of the DFT is given by

$$x(n) = \frac{1}{N}\sum_{k=0}^{N-1} X(k)e^{j2\pi kn/N} = \frac{1}{N}\sum_{k=0}^{N-1} X(k)W_N^{-kn}, \quad \text{for } n = 0, 1, \cdots, N-1 \quad (4.10)$$

Proof can be found in Ahmed and Nataranjan (1983); Proakis and Manolakis (1996); Oppenheim, Schafer, and Buck (1997); and Stearns and Hush (1990).

Similar to Equation (4.7), the expansion of Equation (4.10) leads to

$$x(n) = \frac{1}{N}\left(X(0)W_N^{-0n} + X(1)W_N^{-1n} + X(2)W_N^{-2n} + \cdots + X(N-1)W_N^{-(N-1)n}\right),$$
$$\text{for } n = 0, 1, \cdots, N-1 \quad (4.11)$$

As shown in Figure 4.6, in time domain we use the sample number or time index n for indexing the digital sample sequence $x(n)$. However, in the frequency domain, we use index k for indexing N calculated DFT coefficients $X(k)$. We also refer to k as the frequency bin number in Equations (4.7) and (4.8).
We can use MATLAB functions **fft()** and **ifft()** to compute the DFT coefficients and the inverse DFT with the syntax listed in Table 4.1.

Table 4.1 MATLAB FFT Functions

X = fft(x)	% Calculate DFT coefficients
x = ifft(X)	% Inverse of DFT
x = input vector	
X = DFT coefficient vector	

The following examples serve to illustrate the application of DFT and the inverse DFT.

EXAMPLE 4.2

Given a sequence $x(n)$ for $0 \leq n \leq 3$, where $x(0) = 1$, $x(1) = 2$, $x(2) = 3$, and $x(3) = 4$, evaluate its DFT $X(k)$.

Solution:
Since $N = 4$ and $W_4 = e^{-j\frac{\pi}{2}}$, using Equation (4.7) we have a simplified formula,

$$X(k) = \sum_{n=0}^{3} x(n)W_4^{kn} = \sum_{n=0}^{3} x(n)e^{-j\frac{\pi kn}{2}}$$

Thus, for $k = 0$

$$X(0) = \sum_{n=0}^{3} x(n)e^{-j0} = x(0)e^{-j0} + x(1)e^{-j0} + x(2)e^{-j0} + x(3)e^{-j0}$$
$$= x(0) + x(1) + x(2) + x(3)$$
$$= 1 + 2 + 3 + 4 = 10$$

for $k = 1$

$$X(1) = \sum_{n=0}^{3} x(n)e^{-j\frac{\pi n}{2}} = x(0)e^{-j0} + x(1)e^{-j\frac{\pi}{2}} + x(2)e^{-j\pi} + x(3)e^{-j\frac{3\pi}{2}}$$
$$= x(0) - jx(1) - x(2) + jx(3)$$
$$= 1 - j2 - 3 + j4 = -2 + j2$$

for $k = 2$

$$X(2) = \sum_{n=0}^{3} x(n)e^{-j\pi n} = x(0)e^{-j0} + x(1)e^{-j\pi} + x(2)e^{-j2\pi} + x(3)e^{-j3\pi}$$
$$= x(0) - x(1) + x(2) - x(3)$$
$$= 1 - 2 + 3 - 4 = -2$$

and for $k = 3$

$$X(3) = \sum_{n=0}^{3} x(n)e^{-j\frac{3\pi n}{2}} = x(0)e^{-j0} + x(1)e^{-j\frac{3\pi}{2}} + x(2)e^{-j3\pi} + x(3)e^{-j\frac{9\pi}{2}}$$
$$= x(0) + jx(1) - x(2) - jx(3)$$
$$= 1 + j2 - 3 - j4 = -2 - j2$$

Let us verify the result using the MATLAB function **fft()**:

```
>> X = fft([1 2 3 4 ])
X = 10.0000 -2.0000 + 2.0000i -2.0000 -2.0000 -2.0000i
```

EXAMPLE 4.3

Using the DFT coefficients $X(k)$ for $0 \le k \le 3$ computed in Example 4.2, evaluate the inverse DFT to determine the time domain sequence $x(n)$.

Solution:
Since $N = 4$ and $W_4^{-1} = e^{j\frac{\pi}{2}}$, using Equation (4.10) we achieve a simplified formula,

$$x(n) = \frac{1}{4} \sum_{k=0}^{3} X(k) W_4^{-nk} = \frac{1}{4} \sum_{k=0}^{3} X(k) e^{j\frac{\pi kn}{2}}$$

Then for $n = 0$

$$x(0) = \frac{1}{4} \sum_{k=0}^{3} X(k) e^{j0} = \frac{1}{4}\left(X(0)e^{j0} + X(1)e^{j0} + X(2)e^{j0} + X(3)e^{j0}\right)$$
$$= \frac{1}{4}(10 + (-2 + j2) - 2 + (-2 - j2)) = 1$$

for $n = 1$

$$x(1) = \frac{1}{4} \sum_{k=0}^{3} X(k)e^{j\frac{k\pi}{2}} = \frac{1}{4}\left(X(0)e^{j0} + X(1)e^{j\frac{\pi}{2}} + X(2)e^{j\pi} + X(3)e^{j\frac{3\pi}{2}}\right)$$

$$= \frac{1}{4}(X(0) + jX(1) - X(2) - jX(3))$$

$$= \frac{1}{4}(10 + j(-2 + j2) - (-2) - j(-2 - j2)) = 2$$

for $n = 2$

$$x(2) = \frac{1}{4} \sum_{k=0}^{3} X(k)e^{jk\pi} = \frac{1}{4}\left(X(0)e^{j0} + X(1)e^{j\pi} + X(2)e^{j2\pi} + X(3)e^{j3\pi}\right)$$

$$= \frac{1}{4}(X(0) - X(1) + X(2) - X(3))$$

$$= \frac{1}{4}(10 - (-2 + j2) + (-2) - (-2 - j2)) = 3$$

and for $n = 3$

$$x(3) = \frac{1}{4} \sum_{k=0}^{3} X(k)e^{j\frac{k\pi 3}{2}} = \frac{1}{4}\left(X(0)e^{j0} + X(1)e^{j\frac{3\pi}{2}} + X(2)e^{j3\pi} + X(3)e^{j\frac{9\pi}{2}}\right)$$

$$= \frac{1}{4}(X(0) - jX(1) - X(2) + jX(3))$$

$$= \frac{1}{4}(10 - j(-2 + j2) - (-2) + j(-2 - j2)) = 4$$

This example actually verifies the inverse DFT. Applying the MATLAB function **ifft()** we obtain

```
>> x = ifft([10 -2+2j -2 -2 -2j])
x = 1 2 3 4
```

Now we explore the relationship between the frequency bin k and its associated frequency. Omitting the proof, the calculated N DFT coefficients $X(k)$ represent the frequency components ranging from 0 Hz (or radians/second) to f_s Hz (or ω_s radians/second), hence we can map the frequency bin k to its corresponding frequency as follows:

$$\omega = \frac{k\omega_s}{N} \text{ (radians per second)} \tag{4.12}$$

or in terms of Hz,

$$f = \frac{kf_s}{N} \text{ (Hz)} \tag{4.13}$$

where $\omega_s = 2\pi f_s$.

We can define the frequency resolution as the frequency step between two consecutive DFT coefficients to measure how fine the frequency domain presentation is and obtain

$$\Delta\omega = \frac{\omega_s}{N} \text{ (radians per second)} \tag{4.14}$$

or in terms of Hz, it follows that

$$\Delta f = \frac{f_s}{N} \text{ (Hz)} \tag{4.15}$$

Let us study the following example.

EXAMPLE 4.4

In Example 4.2, given a sequence $x(n)$ for $0 \le n \le 3$, where $x(0) = 1$, $x(1) = 2$, $x(2) = 3$, and $x(3) = 4$, we computed 4 DFT coefficients $X(k)$ for $0 \le k \le 3$ as $X(0) = 10$, $X(1) = -2 + j2$, $X(2) = -2$, and $X(3) = -2 - j2$. If the sampling rate is 10 Hz,

a. determine the sampling period, time index, and sampling time instant for a digital sample $x(3)$ in the time domain;

b. determine the frequency resolution, frequency bin, and mapped frequencies for the DFT coefficients $X(1)$ and $X(3)$ in the frequency domain.

Solution:

a. In the time domain, the sampling period is calculated as

$$T = 1/f_s = 1/10 = 0.1 \text{ second}$$

For $x(3)$, the time index is $n = 3$ and the sampling time instant is determined by

$$t = nT = 3 \cdot 0.1 = 0.3 \text{ second}$$

b. In the frequency domain, since the total number of DFT coefficients is four, the frequency resolution is determined by

$$\Delta f = \frac{f_s}{N} = \frac{10}{4} = 2.5 \text{ Hz}$$

The frequency bin for $X(1)$ should be $k = 1$ and its corresponding frequency is determined by

$$f = \frac{kf_s}{N} = \frac{1 \times 10}{4} = 2.5 \text{ Hz}$$

Similarly, for $X(3)$ and $k = 3$,

$$f = \frac{kf_s}{N} = \frac{3 \times 10}{4} = 7.5 \text{ Hz}$$

Note that from Equation (4.4), $k = 3$ is equivalent to $k - N = 3 - 4 = -1$, and $f = 7.5$ Hz is also equivalent to the frequency $f = (-1 \times 10)/4 = -2.5$ Hz, which corresponds to the negative side spectrum. The amplitude spectrum at 7.5 Hz after folding should match the one at $f_s - f = 10.0 - 7.5 = 2.5$ Hz. We will apply these developed notations in the next section for amplitude and power spectral estimation.

4.2 AMPLITUDE SPECTRUM AND POWER SPECTRUM

One DFT application is transformation of a finite-length digital signal $x(n)$ into the spectrum in the frequency domain. Figure 4.7 demonstrates such an application, where A_k and P_k are the computed amplitude spectrum and the power spectrum, respectively, using the DFT coefficients $X(k)$.

First, we obtain the digital sequence $x(n)$ by sampling the analog signal $x(t)$ and truncating the sampled signal with a data window of length $T_0 = NT$, where T is the sampling period and N the number of data points. The time for the data window is

$$T_0 = NT \qquad (4.16)$$

For the truncated sequence $x(n)$ with a range of $n = 0, 1, 2, \cdots, N - 1$, we get

$$x(0), \ x(1), \ x(2), \ \ldots, \ x(N - 1) \qquad (4.17)$$

Next, we apply the DFT to the obtained sequence, $x(n)$, to get the N DFT coefficients

$$X(k) = \sum_{n=0}^{N-1} x(n) W_N^{nk}, \quad \text{for } k = 0, 1, 2, \cdots, N - 1 \qquad (4.18)$$

Since each calculated DFT coefficient is a complex number, it is not convenient to plot it versus its frequency index. Hence, after evaluating Equation (4.18), the magnitude and phase of each DFT coefficient (we refer to them as the amplitude spectrum and phase spectrum, respectively) can be determined and plotted versus its frequency index. We define the amplitude spectrum as

$$A_k = \frac{1}{N}|X(k)| = \frac{1}{N}\sqrt{(\text{Real}[X(k)])^2 + (\text{Imag}[X(k)])^2}, \quad k = 0, 1, 2, \cdots, N - 1 \qquad (4.19)$$

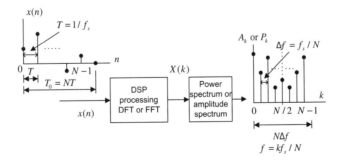

FIGURE 4.7

Applications of DFT/FFT.

We can modify the amplitude spectrum to a one-side amplitude spectrum by doubling the amplitudes in Equation (4.19), keeping the original DC term at $k = 0$. Thus we have

$$\bar{A}_k = \begin{cases} \dfrac{1}{N}|X(0)|, & k = 0 \\[2mm] \dfrac{2}{N}|X(k)|, & k = 1, \cdots, N/2 \end{cases} \qquad (4.20)$$

We can also map the frequency bin k to its corresponding frequency as

$$f = \frac{kf_s}{N} \qquad (4.21)$$

Correspondingly, the phase spectrum is given by

$$\phi_k = \tan^{-1}\left(\frac{\text{Imag}[X(k)]}{\text{Real}[X(k)]}\right), \qquad k = 0, 1, 2, \cdots, N-1 \qquad (4.22)$$

Besides the amplitude spectrum, the power spectrum is also used. The DFT power spectrum is defined as

$$P_k = \frac{1}{N^2}|X(k)|^2 = \frac{1}{N^2}\left\{(\text{Real}[X(k)])^2 + (\text{Imag}[X(k)])^2\right\}, \qquad k = 0, 1, 2, \cdots, N-1 \qquad (4.23)$$

Similarly, for a one-sided power spectrum, we get

$$\bar{P}_k = \begin{cases} \dfrac{1}{N^2}|X(0)|^2 & k = 0 \\[2mm] \dfrac{2}{N^2}|X(k)|^2 & k = 0, 1, \cdots, N/2 \end{cases} \qquad (4.24)$$

and

$$f = \frac{kf_s}{N} \qquad (4.25)$$

Again, notice that the frequency resolution, which denotes the frequency spacing between DFT coefficients in the frequency domain, is defined as

$$\Delta f = \frac{f_s}{N} \ (\text{Hz}) \qquad (4.26)$$

It follows that better frequency resolution can be achieved by using a longer data sequence.

EXAMPLE 4.5

Consider the sequence in Figure 4.8. Assuming that $f_s = 100$ Hz, compute the amplitude spectrum, phase spectrum, and power spectrum.

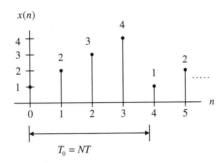

FIGURE 4.8

Sampled values in Example 4.5.

Solution:

Since $N = 4$, and using the DFT shown in Example 4.1, we find the DFT coefficients to be

$X(0) = 10$
$X(1) = -2 + j2$
$X(2) = -2$
$X(3) = -2 - j2$

The amplitude spectrum, phase spectrum, and power density spectrum are computed as follows:
for $k = 0$, $f = k \cdot f_s / N = 0 \times 100/4 = 0$ Hz,

$$A_0 = \frac{1}{4}|X(0)| = 2.5, \quad \phi_0 = \tan^{-1}\left(\frac{\text{Imag}[X(0)]}{\text{Real}([X(0)]}\right) = 0^0, \quad P_0 = \frac{1}{4^2}|X(0)|^2 = 6.25$$

for $k = 1$, $f = 1 \times 100/4 = 25$ Hz,

$$A_1 = \frac{1}{4}|X(1)| = 0.7071, \quad \phi_1 = \tan^{-1}\left(\frac{\text{Imag}[X(1)]}{\text{Real}[X(1)]}\right) = 135^0, \quad P_1 = \frac{1}{4^2}|X(1)|^2 = 0.5000$$

for $k = 2$, $f = 2 \times 100/4 = 50$ Hz,

$$A_2 = \frac{1}{4}|X(2)| = 0.5, \quad \phi_2 = \tan^{-1}\left(\frac{\text{Imag}[X(2)]}{\text{Real}[X(2)]}\right) = 180^0, \quad P_2 = \frac{1}{4^2}|X(2)|^2 = 0.2500$$

Similarly,
for $k = 3$, $f = 3 \times 100/4 = 75$ Hz,

$$A_3 = \frac{1}{4}|X(3)| = 0.7071, \quad \phi_3 = \tan^{-1}\left(\frac{\text{Imag}[X(3)]}{\text{Real}[X(3)]}\right) = -135^0, \quad P_3 = \frac{1}{4^2}|X(3)|^2 = 0.5000.$$

Thus, the sketches for the amplitude spectrum, phase spectrum, and power spectrum are given in Figures 4.9A and B.

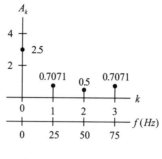

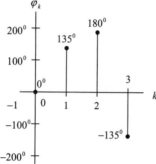

FIGURE 4.9A

Amplitude spectrum and phase spectrum in Example 4.5.

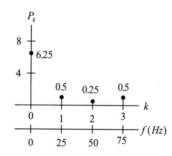

FIGURE 4.9B

Power density spectrum in Example 4.5.

Note that the folding frequency in this example is 50 Hz and the amplitude and power spectrum values at 75 Hz are each image counterparts (corresponding negative-indexed frequency components), respectively. Thus values at 0, 25, 50 Hz correspond to the positive-indexed frequency components.

We can easily find the one-sided amplitude spectrum and one-sided power spectrum as

$$\bar{A}_0 = 2.5, \ \bar{A}_1 = 1.4141, \ \bar{A}_2 = 1 \text{ and}$$

$$\bar{P}_0 = 6.25, \ \bar{P}_1 = 2, \ \bar{P}_2 = 1$$

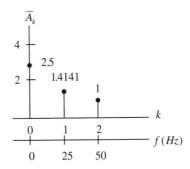

FIGURE 4.10

One-sided amplitude spectrum in Example 4.5.

We plot the one-sided amplitude spectrum for comparison in Figure 4.10.

Note that in the one-sided amplitude spectrum, the negative-indexed frequency components are added back to the corresponding positive-indexed frequency components; thus each amplitude value other than the DC term is doubled. It represents the frequency components up to the folding frequency.

EXAMPLE 4.6

Consider a digital sequence sampled at the rate of 10 kHz. If we use 1,024 data points and apply the 1,024-point DFT to compute the spectrum,

a. determine the frequency resolution;
b. determine the highest frequency in the spectrum.

Solution:

a. $\Delta f = \dfrac{f_s}{N} = \dfrac{10000}{1024} = 9.776$ Hz

b. The highest frequency is the folding frequency, given by

$$f_{max} = \frac{N}{2}\Delta f = \frac{f_s}{2}$$

$$= 512 \cdot 9.776 = 5000 \text{ Hz}.$$

As shown in Figure 4.7, the DFT coefficients may be computed via a *fast Fourier transform* (FFT) algorithm. The FFT is a very efficient algorithm for computing DFT coefficients. The FFT algorithm requires a time domain sequence $x(n)$ where the number of data points is equal to a power of 2; that is, 2^m samples, where m is a positive integer. For example, the number of samples in $x(n)$ can be $N = 2, 4, 8, 16$, etc.

When using the FFT algorithm to compute DFT coefficients, where the length of the available data is not equal to a power of 2 (as required by the FFT), we can pad the data sequence with zeros to create a new sequence with a larger number of samples, $\overline{N} = 2^m > N$. The modified data sequence for applying FFT, therefore, is

$$\overline{x}(n) = \begin{cases} x(n) & 0 \leq n \leq N-1 \\ 0 & N \leq n \leq \overline{N}-1 \end{cases} \tag{4.27}$$

It is very important to note that the signal spectra obtained via zero-padding the data sequence in Equation (4.27) do not add any new information and do not contain more accurate signal spectral presentation. In this situation, the frequency spacing is reduced due to more DFT points, and the achieved spectrum is an interpolated version with "better display." We illustrate the zero-padding effect via the following example instead of theoretical analysis. A theoretical discussion of zero padding in FFT can be found in Proakis and Manolakis (1996).

Figure 4.11(a) shows the 12 data samples from an analog signal containing frequencies of 10 Hz and 25 Hz at a sampling rate of 100 Hz, and the amplitude spectrum obtained by applying the DFT. Figure 4.11(b) displays the signal samples with padding of four zeros to the original data to make up a data sequence of 16 samples, along with the amplitude spectrum calculated by FFT. The data sequence padded with 20 zeros and its calculated amplitude spectrum using FFT are shown in Figure 4.11(c). It is evident that increasing the data length via zero padding to compute the signal spectrum does not add basic information and does not change the spectral shape but gives the

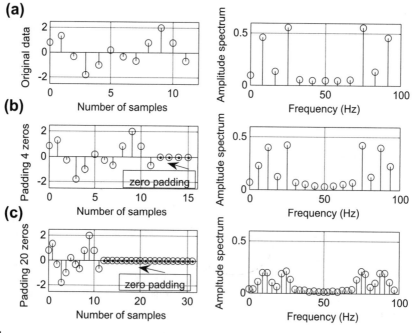

FIGURE 4.11

Zero-padding effect by using FFT.

"interpolated spectrum" with reduced frequency spacing. We can get a better view of the two spectral peaks described in this case.

The only way to obtain the detailed signal spectrum with a fine frequency resolution is to apply more available data samples, that is, a longer sequence of data. Here, we choose to pad the least number of zeros to satisfy the minimum FFT computational requirement. Let us look at another example.

EXAMPLE 4.7

We use the DFT to compute the amplitude spectrum of a sampled data sequence with a sampling rate $f_s = 10\,kHz$. Given a requirement that the frequency resolution be less than 0.5 Hz, determine the number of data points by using the FFT algorithm, assuming that the data samples are available.

Solution:

$$\Delta f = 0.5 \ Hz$$

$$N = \frac{f_s}{\Delta f} = \frac{10,000}{0.5} = 20,000$$

Since we use the FFT to compute the spectrum, the number of data points must be a power of 2, that is,

$$N = 2^{15} = 32,768$$

The resulting frequency resolution can be recalculated as

$$\Delta f = \frac{f_s}{N} = \frac{10,000}{32,768} = 0.31 \ Hz.$$

Next, we study a MATLAB example.

EXAMPLE 4.8

Consider the sinusoid

$$x(n) = 2 \cdot \sin\left(2,000\pi \frac{n}{8,000}\right)$$

obtained by sampling the analog signal

$$x(t) = 2 \cdot \sin(2,000\pi t)$$

with a sampling rate of $f_s = 8,000$ Hz,

a. Use the MATLAB DFT to compute the signal spectrum where the frequency resolution is equal to or less than 8 Hz.

b. Use the MATALB FFT and zero padding to compute the signal spectrum, assuming that the data samples in (a) are available.

Solution:

a. The number of data points is

$$N = \frac{f_s}{\Delta f} = \frac{8,000}{8} = 1,000$$

There is no zero padding needed if we use the DFT formula. The detailed implementation is given in Program 4.1. The first and second plots in Figure 4.12 show the two-sided amplitude and power spectra, respectively, using the DFT, where each frequency counterpart at 7,000 Hz appears. The third and fourth plots are the one-sided amplitude and power spectra, where the true frequency contents are displayed from 0 Hz to the Nyquist frequency of 4 kHz (folding frequency).

b. If the FFT is used, the number of data points must be a power of 2. Hence we choose

$$N = 2^{10} = 1,024$$

Assuming there are only 1,000 data samples available in (a), we need to pad 24 zeros to the original 1,000 data samples before applying the FFT algorithm, as required. Thus the calculated frequency resolution is $\Delta f = f_s/N = 8,000/1,024 = 7.8125$ Hz. Note that this is an interpolated frequency resolution by using zero padding. The zero padding actually interpolates a signal spectrum and carries no additional frequency information. Figure 4.13 shows the spectral plots using FFT. The detailed implementation is given in Program 4.1.

Program 4.1. MATLAB program for Example 4.8.

```
% Example 4.8
close all;clear all
% Generate the sine wave sequence
fs=8000;                                % Sampling rate
N=1000;                                 % Number of data points
x=2*sin(2000*pi*[0:1:N-1]/fs);
% Apply the DFT algorithm
figure(1)
xf=abs(fft(x))/N;                       % Compute the amplitude spectrum
P=xf.*xf;                               % Compute power spectrum
f=[0:1:N-1]*fs/N;                       % Map the frequency bin to frequency (Hz)
subplot(2,1,1); plot(f,xf);grid
xlabel('Frequency (Hz)'); ylabel('Amplitude spectrum (DFT)');
subplot(2,1,2);plot(f,P);grid
xlabel('Frequency (Hz)'); ylabel('Power spectrum (DFT)');
figure(2)
% Convert it to one side spectrum
xf(2:N)=2*xf(2:N);                      % Get the single-side spectrum
P=xf.*xf;                               % Calculate the power spectrum
f=[0:1:N/2]*fs/N                        % Frequencies up to the folding frequency
subplot(2,1,1); plot(f,xf(1:N/2+1));grid
xlabel('Frequency (Hz)'); ylabel('Amplitude spectrum (DFT)');
subplot(2,1,2);plot(f,P(1:N/2+1));grid
xlabel('Frequency (Hz)'); ylabel('Power spectrum (DFT)');
figure (3)
% Zero padding to the length of 1024
x=[x,zeros(1,23)];
N=length(x);
xf=abs(fft(x))/N;                       % Compute amplitude spectrum with zero padding
P=xf.*xf;                               % Compute power spectrum
f=[0:1:N-1]*fs/N;                       % Map frequency bin to frequency (Hz)
subplot(2,1,1); plot(f,xf);grid
```

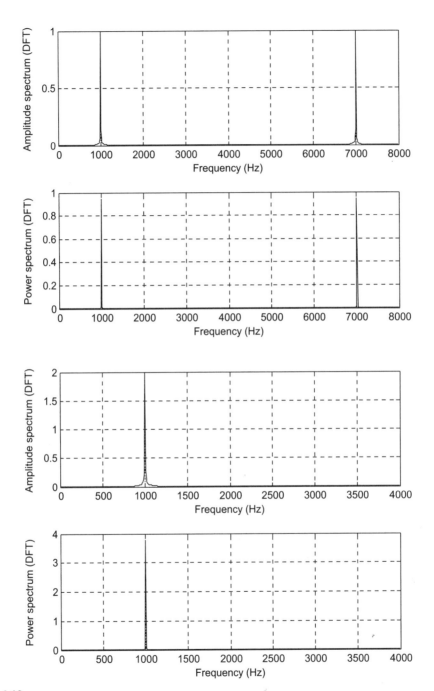

FIGURE 4.12

Amplitude spectrum and power spectrum using DFT for Example 4.8.

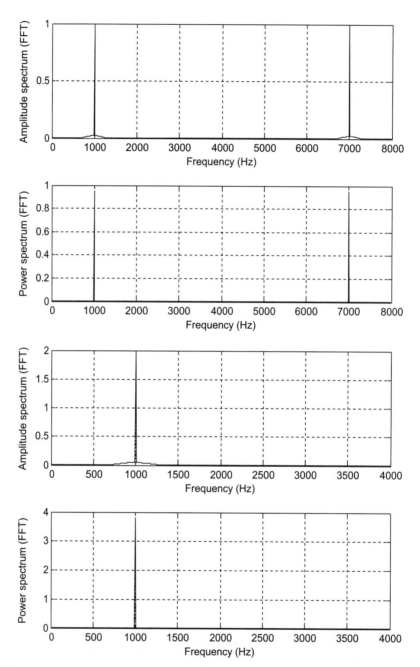

FIGURE 4.13

Amplitude spectrum and power spectrum using FFT for Example 4.8.

```
xlabel('Frequency (Hz)'); ylabel('Amplitude spectrum (FFT)');
subplot(2,1,2);plot(f,P);grid
xlabel('Frequency (Hz)'); ylabel('Power spectrum (FFT)');
figure(4)
% Convert it to one side spectrum
xf(2:N)=2*xf(2:N);
P=xf.*xf;
f=[0:1:N/2]*fs/N;
subplot(2,1,1); plot(f,xf(1:N/2+1));grid
xlabel('Frequency (Hz)'); ylabel('Amplitude spectrum (FFT)');
subplot(2,1,2);plot(f,P(1:N/2+1));grid
xlabel('Frequency (Hz)'); ylabel('Power spectrum (FFT)');
```

4.3 SPECTRAL ESTIMATION USING WINDOW FUNCTIONS

When we apply DFT to the sampled data in the previous section, we theoretically imply the following assumptions: first, that the sampled data are periodic (repeat themselves), and second, that the sampled data are continuous and band limited to the folding frequency. The second assumption is often violated, and the discontinuity produces undesired harmonic frequencies. Consider a pure 1-Hz sine wave with 32 samples shown in Figure 4.14.

As shown in the figure, if we use a window size of $N = 16$ samples, which is a multiple of the two waveform cycles, the second window has continuity with the first. However, when the window size is

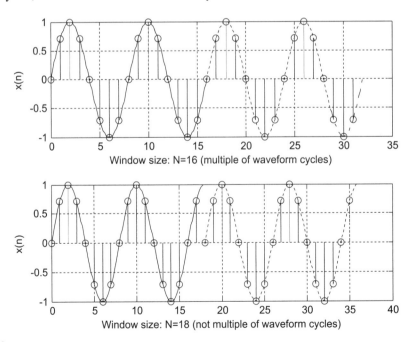

FIGURE 4.14

Sampling a 1-Hz sine wave using (top) 16 samples per cycle and (bottom) 18 samples per cycle.

chosen to be 18 samples, which is not multiple of the waveform cycles (2.25 cycles), there is a discontinuity in the second window. It is this discontinuity that produces harmonic frequencies that are not present in the original signal. Figure 4.15 shows the spectral plots for both cases using the DFT/FFT directly.

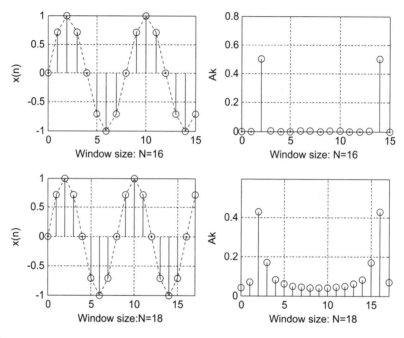

FIGURE 4.15

Signal samples and spectra without spectral leakage and with spectral leakage.

The first spectral plot contains a single frequency, as we expected, while the second spectrum has the expected frequency component plus many harmonics, which do not exist in the original signal. We called such an effect *spectral leakage*. The amount of spectral leakage shown in the second plot is due to amplitude discontinuity in time domain. The bigger the discontinuity, the more the leakage. To reduce the effect of spectral leakage, a window function can be used whose amplitude tapers smoothly and gradually toward zero at both ends. Applying the window function $w(n)$ to a data sequence $x(n)$ to obtain the windowed sequence $x_w(n)$ is illustrated in Figure 4.16 using Equation (4.28):

$$x_w(n) = x(n)w(n), \quad \text{for } n = 0, 1, \cdots, N-1 \tag{4.28}$$

The top plot is the data sequence $x(n)$, and the middle plot is the window function $w(n)$. The bottom plot in Figure 4.16 shows that the windowed sequence $x_w(n)$ is tapped down by a window function to zero at both ends such that the discontinuity is dramatically reduced.

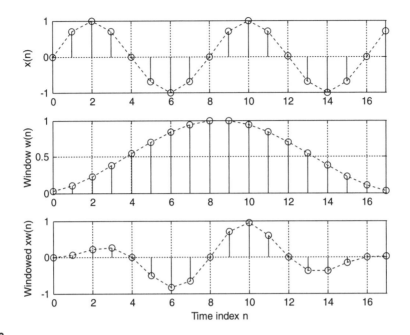

FIGURE 4.16

Illustration of the window operation.

EXAMPLE 4.9

In Figure 4.16, given

- $x(2) = 1$ and $w(2) = 0.2265$
- $x(5) = -0.7071$ and $w(5) = 0.7008$

calculate the windowed sequence data points $x_w(2)$ and $x_w(5)$.

Solution:

Applying the window function operation leads to

$x_w(2) = x(2) \times w(2) = 1 \times 0.2265 = 0.2265$ and
$x_w(5) = x(5) \times w(5) = -0.7071 \times 0.7008 = -0.4956$

which agree with the values shown in the bottom plot in Figure 4.16.

Using the window function shown in Example 4.9, the spectral plot is reproduced. As a result, the spectral leakage is greatly reduced, as shown in Figure 4.17.

The common window functions are listed as follows.

The rectangular window (no window function):

$$w_R(n) = 1, \, 0 \le n \le N - 1 \qquad (4.29)$$

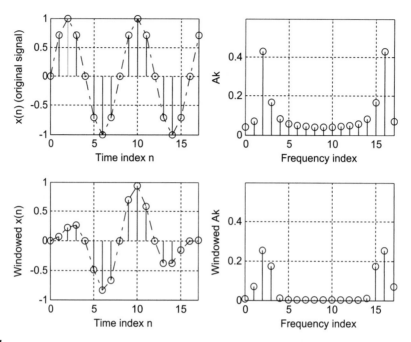

FIGURE 4.17

Comparison of spectra calculated without using a window function and using a window function to reduce spectral leakage.

The triangular window:

$$w_{tri}(n) = 1 - \frac{|2n - N + 1|}{N - 1}, \quad 0 \leq n \leq N - 1 \tag{4.30}$$

The Hamming window:

$$w_{hm}(n) = 0.54 - 0.46\cos\left(\frac{2\pi n}{N - 1}\right), \quad 0 \leq n \leq N - 1 \tag{4.31}$$

The Hanning window:

$$w_{hn}(n) = 0.5 - 0.5\cos\left(\frac{2\pi n}{N - 1}\right), \quad 0 \leq n \leq N - 1 \tag{4.32}$$

Plots for each window function for a size of 20 samples are shown in Figure 4.18.
The following example details each step for computing the spectral information using the window functions.

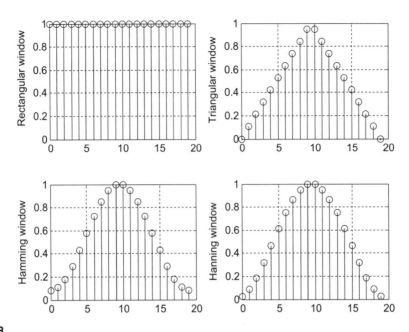

FIGURE 4.18

Plots of window sequences.

EXAMPLE 4.10

Considering the sequence $x(0) = 1$, $x(1) = 2$, $x(2) = 3$, $x(3) = 4$, and given $f_s = 100$ Hz, $T = 0.01$ seconds, compute the amplitude spectrum, phase spectrum, and power spectrum

a. using the triangle window function;
b. using the Hamming window function.

Solution:
a. Since $N = 4$, from the triangular window function, we have

$$w_{tri}(0) = 1 - \frac{|2 \times 0 - 4 + 1|}{4 - 1} = 0$$

$$w_{tri}(1) = 1 - \frac{|2 \times 1 - 4 + 1|}{4 - 1} = 0.6667$$

Similarly, $w_{tri}(2) = 0.6667$, $w_{tri}(3) = 0$. Next, the windowed sequence is computed as

$x_w(0) = x(0) \times w_{tri}(0) = 1 \times 0 = 0$
$x_w(1) = x(1) \times w_{tri}(1) = 2 \times 0.6667 = 1.3334$
$x_w(2) = x(2) \times w_{tri}(2) = 3 \times 0.6667 = 2$
$x_w(3) = x(3) \times w_{tri}(3) = 4 \times 0 = 0$

Applying DFT Equation (4.8) to $x_w(n)$ for $k = 0, 1, 2, 3$, respectively,

$$X(k) = x_w(0)W_4^{k \times 0} + x(1)W_4^{k \times 1} + x(2)W_4^{k \times 2} + x(3)W_4^{k \times 3}$$

We obtain the following results:

$X(0) = 3.3334$
$X(1) = -2 - j1.3334$
$X(2) = 0.6666$
$X(3) = -2 + j1.3334$

$$\Delta f = \frac{1}{NT} = \frac{1}{4 \cdot 0.01} = 25 \text{ Hz}$$

Applying Equations (4.19), (4.22), and (4.23) leads to

$$A_0 = \frac{1}{4}|X(0)| = 0.8334, \quad \phi_0 = \tan^{-1}\left(\frac{0}{3.3334}\right) = 0^0, \quad P_0 = \frac{1}{4^2}|X(0)|^2 = 0.6954$$

$$A_1 = \frac{1}{4}|X(1)| = 0.6009, \quad \phi_1 = \tan^{-1}\left(\frac{-1.3334}{-2}\right) = -146.31^0, \quad P_1 = \frac{1}{4^2}|X(1)|^2$$
$$= 0.3611$$

$$A_2 = \frac{1}{4}|X(2)| = 0.1667, \quad \phi_2 = \tan^{-1}\left(\frac{0}{0.6666}\right) = 0^0, \quad P_1 = \frac{1}{4^2}|X(2)|^2 = 0.0278$$

Similarly,

$$A_3 = \frac{1}{4}|X(3)| = 0.6009, \quad \phi_3 = \tan^{-1}\left(\frac{1.3334}{-2}\right) = 146.31^0, \quad P_3 = \frac{1}{4^2}|X(3)|^2 = 0.3611$$

b. Since $N = 4$, from the Hamming window function, we have

$$w_{hm}(0) = 0.54 - 0.46 \cos\left(\frac{2\pi \times 0}{4 - 1}\right) = 0.08$$

$$w_{hm}(n) = 0.54 - 0.46 \cos\left(\frac{2\pi \times 1}{4 - 1}\right) = 0.77$$

Similarly, $w_{hm}(2) = 0.77$, $w_{hm}(3) = 0.08$. Next, the windowed sequence is computed as

$x_w(0) = x(0) \times w_{hm}(0) = 1 \times 0.08 = 0.08$
$x_w(1) = x(1) \times w_{hm}(1) = 2 \times 0.77 = 1.54$
$x_w(2) = x(2) \times w_{hm}(2) = 3 \times 0.77 = 2.31$
$x_w(0) = x(3) \times w_{hm}(3) = 4 \times 0.08 = 0.32$

Applying DFT Equation (4.8) to $x_w(n)$ for $k = 0, 1, 2, 3$, respectively,

$$X(k) = x_w(0)W_4^{k \times 0} + x(1)W_4^{k \times 1} + x(2)W_4^{k \times 2} + x(3)W_4^{k \times 3}$$

We obtain the following:

$X(0) = 4.25$
$X(1) = -2.23 - j1.22$
$X(2) = 0.53$
$X(3) = -2.23 + j1.22$

$$\Delta f = \frac{1}{NT} = \frac{1}{4 \cdot 0.01} = 25 \text{ Hz}$$

Applying Equations (4.19), (4.22), and (4.23), we achieve

$$A_0 = \frac{1}{4}|X(0)| = 1.0625, \quad \phi_0 = \tan^{-1}\left(\frac{0}{4.25}\right) = 0^0, \quad P_0 = \frac{1}{4^2}|X(0)|^2 = 1.1289$$

$$A_1 = \frac{1}{4}|X(1)| = 0.6355, \quad \phi_1 = \tan^{-1}\left(\frac{-1.22}{-2.23}\right) = -151.32^0, \quad P_1 = \frac{1}{4^2}|X(1)|^2 = 0.4308$$

$$A_2 = \frac{1}{4}|X(2)| = 0.1325, \quad \phi_2 = \tan^{-1}\left(\frac{0}{0.53}\right) = 0^0, \quad P_2 = \frac{1}{4^2}|X(2)|^2 = 0.0176$$

Similarly,

$$A_3 = \frac{1}{4}|X(3)| = 0.6355, \quad \phi_3 = \tan^{-1}\left(\frac{1.22}{-2.23}\right) = 151.32^0, \quad P_3 = \frac{1}{4^2}|X(3)|^2 = 0.4308$$

EXAMPLE 4.11

Given the sinusoid

$$x(n) = 2 \cdot \sin\left(2,000\pi \frac{n}{8,000}\right)$$

obtained using a sampling rate of $f_s = 8,000$ Hz, use the DFT to compute the spectrum with the following specifications:

a. Compute the spectrum of a triangular window function with window size = 50.
b. Compute the spectrum of a Hamming window function with window size = 100.
c. Compute the spectrum of a Hanning window function with window size = 150 and a one-sided spectrum.

Solution:
The MATLAB program is listed in Program 4.2, and results are plotted in Figures 4.19 to 4.21. As compared with the no-window (rectangular window) case, all three windows are able to effectively reduce the spectral leakage, as shown in the figures.
Program 4.2. MATLAB program for Example 4.11.

```
%Example 4.11
close all;clear all
% Generate the sine wave sequence
fs=8000; T=1/fs;                    % Sampling rate and sampling period
x=2*sin(2000*pi*[0:1:50]*T);        % Generate 51 2000-Hz samples.
% Apply the FFT algorithm
N=length(x);
```

```
index_t=[0:1:N-1];
f=[0:1:N-1]*8000/N;            % Map frequency bin to frequency (Hz)
xf=abs(fft(x))/N;             % Calculate amplitude spectrum
figure(1)
%Using Bartlett window
x_b=x.*bartlett(N)';          % Apply triangular window function
xf_b=abs(fft(x_b))/N;         % Calculate amplitude spectrum
subplot(2,2,1);plot(index_t,x);grid
xlabel('Time index n'); ylabel('x(n)');
subplot(2,2,3); plot(index_t,x_b);grid
xlabel('Time index n'); ylabel('Triangular windowed x(n)');
subplot(2,2,2);plot(f,xf);grid;axis([0 8000 0 1]);
xlabel('Frequency (Hz)'); ylabel('Ak (no window)');
subplot(2,2,4); plot(f,xf_b);grid; axis([0 8000 0 1]);
xlabel('Frequency (Hz)'); ylabel('Triangular windowed Ak');
figure(2)
% Generate the sine wave sequence
x=2*sin(2000*pi*[0:1:100]*T); % Generate 101 2000-Hz samples.
% Apply the FFT algorithm
N=length(x);
index_t=[0:1:N-1];
f=[0:1:N-1]*fs/N;
xf=abs(fft(x))/N;
% Using Hamming window
x_hm=x.*hamming(N)';                  % Apply Hamming window function
xf_hm=abs(fft(x_hm))/N;               % Calculate amplitude spectrum
subplot(2,2,1);plot(index_t,x);grid
xlabel('Time index n'); ylabel('x(n)');
subplot(2,2,3); plot(index_t,x_hm);grid
xlabel('Time index n'); ylabel('Hamming windowed x(n)');
subplot(2,2,2);plot(f,xf);grid;axis([0 fs 0 1]);
xlabel('Frequency (Hz)'); ylabel('Ak (no window)');
subplot(2,2,4); plot(f,xf_hm);grid;axis([0 fs 0 1]);
xlabel('Frequency (Hz)'); ylabel('Hamming windowed Ak');
figure(3)
% Generate the sine wave sequence
x=2*sin(2000*pi*[0:1:150]*T);  % Generate 151 2-kHz samples
% Apply the FFT algorithm
N=length(x);
index_t=[0:1:N-1];
f=[0:1:N-1]*fs/N;
xf=2*abs(fft(x))/N;xf(1)=xf(1)/2;   % Single-sided spectrum
%Using Hanning window
x_hn=x.*hanning(N)';
xf_hn=2*abs(fft(x_hn))/N;xf_hn(1)=xf_hn(1)/2;   % Single-sided spectrum
subplot(2,2,1);plot(index_t,x);grid
xlabel('Time index n'); ylabel('x(n)');
subplot(2,2,3); plot(index_t,x_hn);grid
xlabel('Time index n'); ylabel('Hanning windowed x(n)');
subplot(2,2,2);plot(f(1:(N-1)/2),xf(1:(N-1)/2));grid;axis([0 fs/2 0 1]);
xlabel('Frequency (Hz)'); ylabel('Ak (no window)');
subplot(2,2,4); plot(f(1:(N-1)/2),xf_hn(1:(N-1)/2));grid;axis([0 fs/2 0 1]);
xlabel('Frequency (Hz)'); ylabel('Hanning windowed Ak');
```

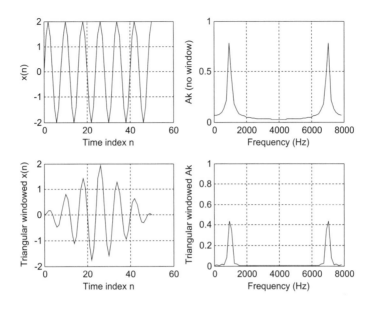

FIGURE 4.19

Comparison of a spectrum without using a window function and a spectrum using a triangular window with 50 samples in Example 4.11.

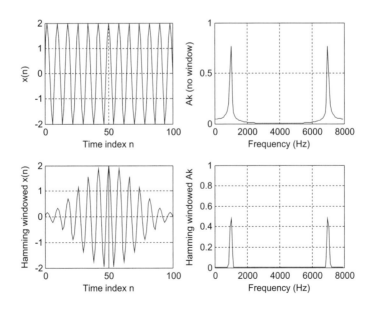

FIGURE 4.20

Comparison of a spectrum without using a window function and a spectrum using a Hamming window with 100 samples in Example 4.11.

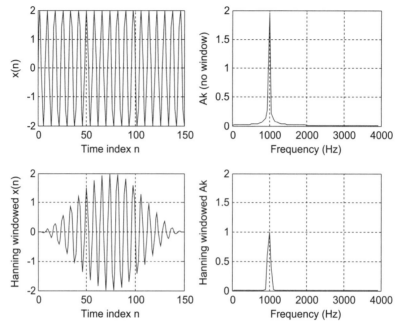

FIGURE 4.21

Comparison of a one-sided spectrum without using the window function and a one-sided spectrum using a Hanning window with 150 samples in Example 4.11.

4.4 APPLICATION TO SIGNAL SPECTRAL ESTIMATION

The following plots compare amplitude spectra for speech data (we.dat) with 2,001 samples and a sampling rate of 8,000 Hz using the rectangular window (no window) function and the Hamming window function. As demonstrated in Figure 4.22 (two-sided spectrum) and Figure 4.23 (one-sided spectrum), there is little difference between the amplitude spectrum using the Hamming window function and the spectrum without using the window function. This is due to the fact that when the data length of the sequence (e.g., 2,001 samples) increases, the frequency resolution will be improved and the spectral leakage will become less significant. However, when data length is short, the reduction in spectral leakage using a window function will be more prominent.

Next, we compute the one-sided spectrum for 32-bit seismic data sampled at 15 Hz (provided by the US Geological Survey, Albuquerque Seismological Laboratory) with 6,700 data samples. The computed spectral plots without using a window function and using the Hamming window are displayed in Figure 4.24. We can see that most of seismic signal components are below 3 Hz.

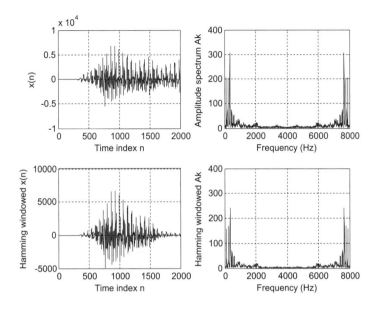

FIGURE 4.22

Comparison of a spectrum without using a window function and a spectrum using the Hamming window for speech data.

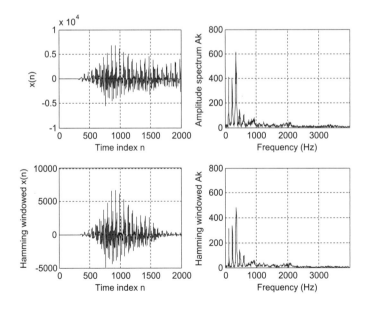

FIGURE 4.23

Comparison of a one-sided spectrum without using a window function and a one-sided spectrum using the Hamming window for speech data.

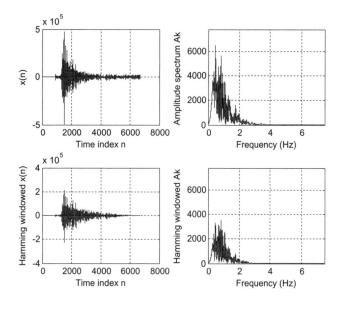

FIGURE 4.24

Comparison of a one-sided spectrum without using a window function and a one-sided spectrum using the Hamming window for seismic data.

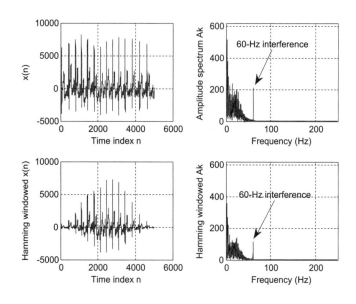

FIGURE 4.25

Comparison of a one-sided spectrum without using a window function and a one-sided spectrum using the Hamming window for electrocardiogram data.

We also compute the one-sided spectrum for a standard electrocardiogram (ECG) signal from the MIT–BIH (Massachusetts Institute of Technology–Beth Israel Hospital) Database. The ECG signal contains frequency components ranging from 0.05 to 100 Hz sampled at 500 Hz. As shown in Figure 4.25, there is a spike located at 60 Hz. This is due to the 60-Hz power line interference when the ECG is acquired via the ADC acquisition process. This 60-Hz interference can be removed by using a digital notch filter, which will be studied in Chapter 8.

Figure 4.26 shows a vibration signal and its spectrum. The vibration signal is captured using an accelerometer sensor attached to a simple supported beam while an impulse force is applied to a location that is close to the middle of the beam. The sampling rate is 1 kHz. As shown in Figure 4.26, four dominant modes (natural frequencies corresponding to locations of spectral peaks) can be easily identified from the displayed spectrum.

We now present another practical example for vibration signature analysis of a defective gear tooth, described in Section 1.3.5. Figure 4.27 shows a gearbox containing two straight bevel gears with a transmission ratio of 1.5:1 and the number of teeth on the pinion and gear are 18 and 27. The vibration data is collected by an accelerometer installed on the top of the gearbox. The data acquisition system uses a sampling rate of 12.8 kHz. The meshing frequency is determined as $f_m = f_i(\text{RPM}) \times 18/60 = 300$ Hz, where the input shaft frequency is $f_i = 1000$ RPM $= 16.67$ Hz. Figures 4.28–4.31 show the baseline vibration signal and spectrum for a gearbox in good condition,

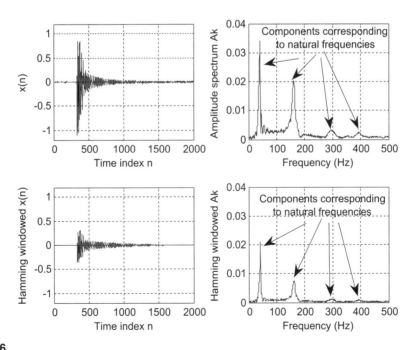

FIGURE 4.26

Comparison of a one-sided spectrum without using a window function and a one-sided spectrum using the Hamming window for vibration signal.

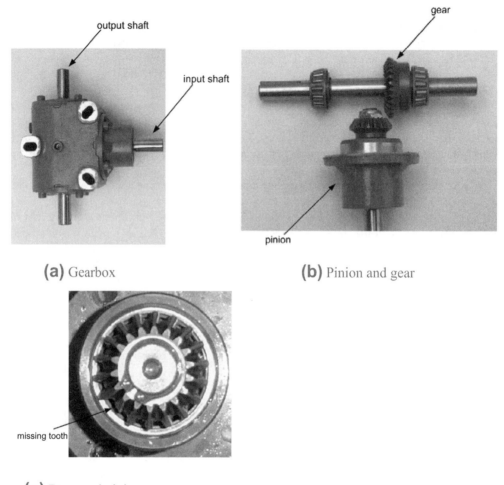

(a) Gearbox   **(b)** Pinion and gear

(c) Damaged pinion

FIGURE 4.27

Vibration signature analysis of a gearbox.

(Courtesy of SpectaQuest, Inc.)

along with the vibration signals and spectra for three different damage severity levels (there are five levels classified by SpectraQuest, Inc; the spectrums shown are for severity level 1 [lightly chipped]; severity level 4 [heavily chipped]; and severity level 5 [missing tooth]). As we can see, the baseline spectrum contains the meshing frequency component of 300 Hz and a sideband frequency component of 283.33 Hz (300–16.67). We can observe that the sidebands ($f_m \pm f_i$, $f_m \pm 2f_i$ ...) become more dominant when the severity level increases. Hence, the spectral information is very useful for monitoring the health condition of the gearbox.

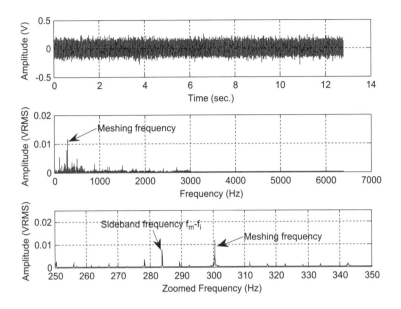

FIGURE 4.28

Vibration signal and spectrum from the good condition gearbox.

(Data provided by SpectaQuest, Inc.)

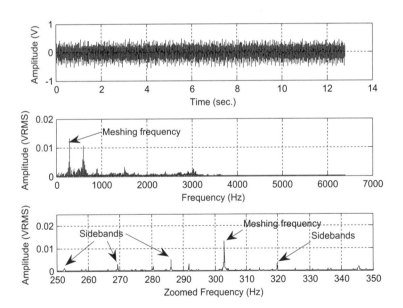

FIGURE 4.29

Vibration signal and spectrum for damage severity level 1.

(Data provided by SpectaQuest, Inc.)

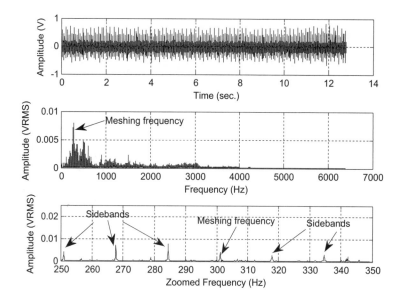

FIGURE 4.30

Vibration signal and spectrum for damage severity level 4.

(Data provided by SpectaQuest, Inc.)

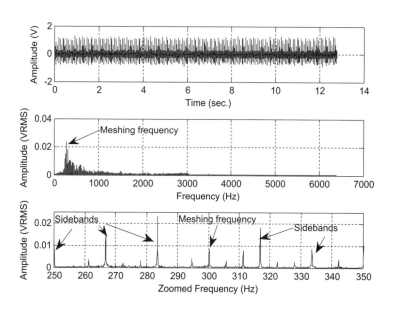

FIGURE 4.31

Vibration signal and spectrum for damage severity level 5.

(Data provided by SpectaQuest, Inc.)

4.5 FAST FOURIER TRANSFORM

Now we study FFT in detail. FFT is a very efficient algorithm in computing DFT coefficients and can reduce a very large amount of computational complexity (multiplications). Without loss of generality, we consider the digital sequence $x(n)$ consisting of 2^m samples, where m is a positive integer, that is, the number of samples of the digital sequence $x(n)$ is a power of 2, $N = 2, 4, 8, 16$, etc. If $x(n)$ does not contain 2^m samples, then we simply append it with zeros until the number of the appended sequence is a power of 2.

In this section, we focus on two formats. One is called the decimation-in-frequency algorithm, while the other is the decimation-in-time algorithm. They are referred to as the *radix-2* FFT algorithms. Other types of FFT algorithms are the radix-4 and the split radix and their advantages can be explored in more detail in other texts (see Proakis and Manolakis, 1996).

4.5.1 Decimation-in-Frequency Method

We begin with the definition of DFT studied in the opening section in this chapter:

$$X(k) = \sum_{n=0}^{N-1} x(n) W_N^{kn} \quad \text{for} \quad k = 0, 1, \cdots, N-1 \tag{4.33}$$

where $W_N = e^{-j\frac{2\pi}{N}}$ is the twiddle factor, and $N = 2, 4, 8, 16, \cdots$. Equation (4.33) can be expanded as

$$X(k) = x(0) + x(1) W_N^k + \cdots + x(N-1) W_N^{k(N-1)} \tag{4.34}$$

Again, if we split Equation (4.34) into

$$X(k) = x(0) + x(1) W_N^k + \cdots + x\left(\frac{N}{2} - 1\right) W_N^{k(N/2-1)}$$

$$+ x\left(\frac{N}{2}\right) W^{kN/2} + \cdots + x(N-1) W_N^{k(N-1)} \tag{4.35}$$

then we can rewrite it as a sum of the following two parts:

$$X(k) = \sum_{n=0}^{(N/2)-1} x(n) W_N^{kn} + \sum_{n=N/2}^{N-1} x(n) W_N^{kn} \tag{4.36}$$

Modifying the second term in Equation (4.36) yields

$$X(k) = \sum_{n=0}^{(N/2)-1} x(n) W_N^{kn} + W_N^{(N/2)k} \sum_{n=0}^{(N/2)-1} x\left(n + \frac{N}{2}\right) W_N^{kn} \tag{4.37}$$

Recall $W_N^{N/2} = e^{-j\frac{2\pi(N/2)}{N}} = e^{-j\pi} = -1$; then we have

$$X(k) = \sum_{n=0}^{(N/2)-1} \left(x(n) + (-1)^k x\left(n + \frac{N}{2} \right) \right) W_N^{kn} \tag{4.38}$$

Now letting $k = 2m$ be an even number we obtain

$$X(2m) = \sum_{n=0}^{(N/2)-1} \left(x(n) + x\left(n + \frac{N}{2} \right) \right) W_N^{2mn} \tag{4.39}$$

while substituting $k = 2m + 1$ (an odd number) yields

$$X(2m+1) = \sum_{n=0}^{(N/2)-1} \left(x(n) - x\left(n + \frac{N}{2} \right) \right) W_N^n W_N^{2mn} \tag{4.40}$$

Using the fact that $W_N^2 = e^{-j\frac{2\pi\times2}{N}} = e^{-j\frac{2\pi}{(N/2)}} = W_{N/2}$, it follows that

$$X(2m) = \sum_{n=0}^{(N/2)-1} a(n) W_{N/2}^{mn} = DFT\{a(n) \text{ with} (N/2) \text{points}\} \tag{4.41}$$

$$X(2m+1) = \sum_{n=0}^{(N/2)-1} b(n) W_N^n W_{N/2}^{mn} = DFT\{b(n) W_N^n \text{ with } (N/2) \text{points}\} \tag{4.42}$$

where $a(n)$ and $b(n)$ are introduced and expressed as

$$a(n) = x(n) + x\left(n + \frac{N}{2} \right), \quad \text{for } n = 0, 1 \cdots, \frac{N}{2} - 1 \tag{4.43a}$$

$$b(n) = x(n) - x\left(n + \frac{N}{2} \right), \quad \text{for } n = 0, 1, \cdots, \frac{N}{2} - 1 \tag{4.43b}$$

Equations (4.33), (4.41), and (4.42) can be summarized as

$$DFT\{x(n) \text{ with } N \text{ points}\} = \begin{cases} DFT\{a(n) \text{ with } (N/2) \text{ points}\} \\ DFT\{b(n) W_N^n \text{ with } (N/2) \text{ points}\} \end{cases} \tag{4.44}$$

The computation process is illustrated in Figure 4.32. As shown in this figure, there are three graphical operations, which are illustrated Figure 4.33.

If we continue the process described by Figure 4.32, we obtain the block diagrams shown in Figures 4.34 and 4.35.

Figure 4.35 illustrates the FFT computation for the eight-point DFT, where there are 12 complex multiplications. This is a big saving as compared with the eight-point DFT with 64 complex

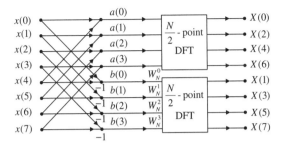

FIGURE 4.32

The first iteration of the eight-point FFT.

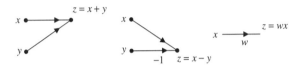

FIGURE 4.33

Definitions of the graphical operations.

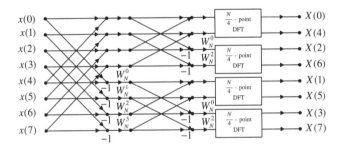

FIGURE 4.34

The second iteration of the eight-point FFT.

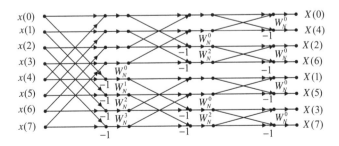

FIGURE 4.35

Block diagram for the eight-point FFT (total 12 multiplications).

multiplications. For a data length of N, the number of complex multiplications for DFT and FFT, respectively, are determined by

$$\text{Complex multiplications of DFT } = N^2, \text{ and}$$

$$\text{Complex multiplications of FFT } = \frac{N}{2}\log_2(N)$$

To see the effectiveness of FFT, let us consider a sequence with 1,024 data points. Applying DFT will require $1,024 \times 1,024 = 1,048,576$ complex multiplications; however, applying FFT will require only $(1024/2)\log_2(1,024) = 5,120$ complex multiplications. Next, the index (bin number) of the eight-point DFT coefficient $X(k)$ becomes 0, 4, 2, 6, 1, 5, 3, and 7, respectively, which is not the natural order. This can be fixed by index matching. The index matching between the input sequence and output frequency bin number by applying reversal bits is described in Table 4.2.

Figure 4.36 explains the bit reversal process. First, the input data with indices 0, 1, 2, 3, 4, 5, 6, 7 are split into two parts. The first half contains even indices—0, 2, 4, 6—while the second half contains odd indices. The first half with indices 0, 2, 4, 6 at the first iteration continues to be split into even indices 0, 4 and odd indices 2, 6 as shown in the second iteration. The second half with indices 1, 3, 5, 7 at the first iteration is split to even indices 1, 5 and odd indices 3, 7 in the second iteration. The splitting process continues to the end at the third iteration. The bit patterns of the output data indices are just the respective reversed bit patterns of the input data indices.

Although Figure 4.36 illustrates the case of an eight-point FFT, this bit reversal process works as long as N is a power of 2.

The inverse FFT is defined as

$$x(n) = \frac{1}{N}\sum_{k=0}^{N-1}X(k)W_N^{-kn} = \frac{1}{N}\sum_{k=0}^{N-1}X(k)\tilde{W}_N^{kn}, \quad \text{for } k = 0, 1, \cdots, N - 1 \qquad (4.45)$$

Table 4.2 Index Mapping for Fast Fourier Transform

Input Data	Index Bits	Reversal Bits	Output Data
$x(0)$	000	000	$X(0)$
$x(1)$	001	100	$X(4)$
$x(2)$	010	010	$X(2)$
$x(3)$	011	110	$X(6)$
$x(4)$	100	001	$X(1)$
$x(5)$	101	101	$X(5)$
$x(6)$	110	011	$X(3)$
$x(7)$	111	111	$X(7)$

Binary index	1st split	2nd split	3rd split	Bit reversal
000	0	0	0	0	000
001	1	2	4	4	100
010	2	4	2	2	010
011	3	6	6	6	011
100	4	1	1	1	001
101	5	3	5	5	101
110	6	5	3	3	011
111	7	7	7	7	111

FIGURE 4.36

Bit reversal process in FFT.

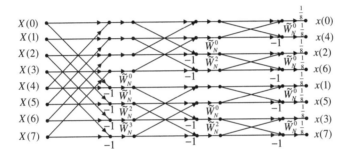

FIGURE 4.37

Block diagram for the inverse of eight-point FFT.

Comparing Equation (4.45) with Equation (4.33), we notice the difference as follows: the twiddle factor W_N is changed to $\tilde{W}_N = W_N^{-1}$, and the sum is multiplied by a factor of $1/N$. Hence, by modifying the FFT block diagram as shown in Figure 4.35, we achieve the inverse FFT block diagram shown in Figure 4.37.

EXAMPLE 4.12

Given a sequence $x(n)$ for $0 \le n \le 3$, where $x(0) = 1$, $x(1) = 2$, $x(2) = 3$, and $x(3) = 4$,

a. evaluate its DFT $X(k)$ using the decimation-in-frequency FFT method;
b. determine the number of complex multiplications.

Solution:
a. Using the FFT block diagram in Figure 4.35, the result is shown in Figure 4.38.
b. From Figure 4.38, the number of complex multiplications is four, which can also be determined by

$$\frac{N}{2}\log_2(N) = \frac{4}{2}\log_2(4) = 4$$

Bit index
00 $x(0) = 1$
01 $x(1) = 2$
10 $x(2) = 3$
11 $x(3) = 4$

4
6
-2 $W_4^0 = 1$
-1
-2 $W_4^1 = -j$
-1

10
$W_4^0 = 1$ -2
-1 $-2 + j2$
$W_4^0 = 1$ $-2 - j2$
-1

Bit reversal
$X(0)$ 00
$X(2)$ 10
$X(1)$ 01
$X(3)$ 11

FIGURE 4.38

Four-point FFT block diagram in Example 4.12.

EXAMPLE 4.13

Given the DFT sequence $X(k)$ for $0 \le k \le 3$ computed in Example 4.12, evaluate its inverse DFT $x(n)$ using the decimation-in-frequency FFT method.

Solution:
Using the inverse FFT block diagram in Figure 4.37, we have the result shown in Figure 4.39.

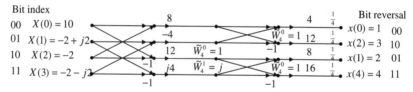

Bit index
00 $X(0) = 10$
01 $X(1) = -2 + j2$
10 $X(2) = -2$
11 $X(3) = -2 - j2$

8
-4
12 $\tilde{W}_4^0 = 1$
-1
$j4$ $\tilde{W}_4^1 = j$
-1

4 $\frac{1}{4}$
$\tilde{W}_4^0 = 1$ 12 $\frac{1}{4}$
-1 8 $\frac{1}{4}$
$\tilde{W}_4^0 = 1$ 16 $\frac{1}{4}$
-1

Bit reversal
$x(0) = 1$ 00
$x(2) = 3$ 10
$x(1) = 2$ 01
$x(4) = 4$ 11

FIGURE 4.39

Four-point inverse FFT block diagram in Example 4.13.

4.5.2 Decimation-in-Time Method

In this method, we split the input sequence $x(n)$ into the even indexed $x(2m)$ and $x(2m+1)$, each with N data points. Then Equation (4.33) becomes

$$X(k) = \sum_{m=0}^{(N/2)-1} x(2m)W_N^{2mk} + \sum_{m=0}^{(N/2)-1} x(2m+1)W_N^k W_N^{2mk}, \quad \text{for } k = 0, 1, \cdots, N-1 \qquad (4.46)$$

Using the relation $W_N^2 = W_{N/2}$, it follows that

$$X(k) = \sum_{m=0}^{(N/2)-1} x(2m)W_{N/2}^{mk} + W_N^k \sum_{m=0}^{(N/2)-1} x(2m+1)W_{N/2}^{mk}, \quad \text{for } k = 0, 1, \cdots, N-1 \quad (4.47)$$

Define new functions as

$$G(k) = \sum_{m=0}^{(N/2)-1} x(2m)W_{N/2}^{mk} = DFT\{x(2m) \text{ with } (N/2) \text{ points}\} \quad (4.48)$$

$$H(k) = \sum_{m=0}^{(N/2)-1} x(2m+1)W_{N/2}^{mk} = DFT\{x(2m+1) \text{ with } (N/2) \text{ points}\} \quad (4.49)$$

Note that

$$G(k) = G\left(k + \frac{N}{2}\right), \quad \text{for } k = 0, 1, \cdots, \frac{N}{2} - 1 \quad (4.50)$$

$$H(k) = H\left(k + \frac{N}{2}\right), \quad \text{for } k = 0, 1, \cdots, \frac{N}{2} - 1 \quad (4.51)$$

Substituting Equations (4.50) and (4.51) into Equation (4.47) yields the first half frequency bins

$$X(k) = G(k) + W_N^k H(k), \quad \text{for } k = 0, 1, \cdots, \frac{N}{2} - 1 \quad (4.52)$$

Considering Equations (4.50) and (4.51) and the fact that

$$W_N^{(N/2+k)} = -W_N^k \quad (4.53)$$

the second half of frequency bins can be computed as follows:

$$X\left(\frac{N}{2} + k\right) = G(k) - W_N^k H(k), \quad \text{for } k = 0, 1, \cdots, \frac{N}{2} - 1 \quad (4.54)$$

If we perform backward iterations, we can obtain the FFT algorithm. The procedure using Equations (4.52) and (4.54) is illustrated in Figure 4.40, the block diagram for the eight-point FFT algorithm. From a further computation, we obtain Figure 4.41. Finally, after three recursions, we end up with the block diagram in Figure 4.42.

The index for each input sequence element can be achieved by bit reversal of the frequency index in sequential order. Similar to the decimation-in-frequency method, after we change W_N to $\tilde{W}_N$ in Figure 4.42 and multiply the output sequence by a factor of $1/N$, we derive the inverse FFT block diagram for the eight-point inverse FFT in Figure 4.43.

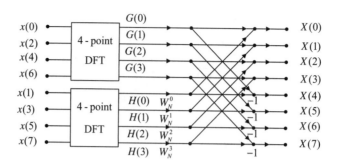

FIGURE 4.40

The first iteration.

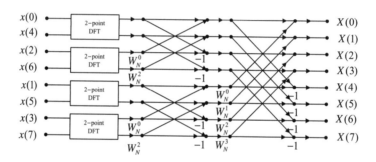

FIGURE 4.41

The second iteration.

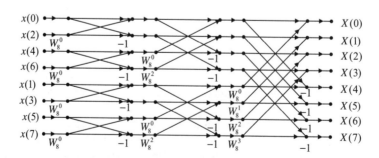

FIGURE 4.42

The eight-point FFT algorithm using decimation-in-time (12 complex multiplications).

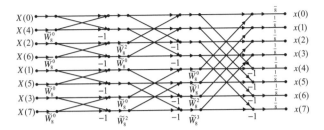

FIGURE 4.43

The eight-point IFFT using decimation-in-time.

EXAMPLE 4.14

Given a sequence $x(n)$ for $0 \le n \le 3$, where $x(0) = 1$, $x(1) = 2$, $x(2) = 3$, and $x(3) = 4$, evaluate its DFT $X(k)$ using the decimation-in-time FFT method.

Solution:
Using the block diagram in Figure 4.42 leads to the result shown in Figure 4.44.

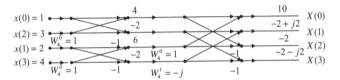

FIGURE 4.44

The four-point FFT using decimation-in-time.

EXAMPLE 4.15

Given the DFT sequence $X(k)$ for $0 \le k \le 3$ computed in Example 4.14, evaluate its inverse DFT $x(n)$ using the decimation-in-time FFT method.

Solution:
Using the block diagram in Figure 4.43 yields Figure 4.45.

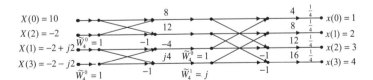

FIGURE 4.45

The four-point IFFT using decimation-in-time.

4.6 SUMMARY

1. The Fourier series coefficients for a periodic digital signal can be used to develop the DFT.
2. The DFT transforms a time sequence to the complex DFT coefficients, while the inverse DFT transforms DFT coefficients back to the time sequence.
3. The *frequency bin number* is the same as the frequency index. *Frequency resolution* is the frequency spacing between two consecutive frequency indices (two consecutive spectrum components).
4. The DFT coefficients for a given digital signal are applied to compute the amplitude spectrum, power spectrum, or phase spectrum.
5. The spectrum calculated from all the DFT coefficients represents the signal frequency range from 0 Hz to the sampling rate. The spectrum beyond the folding frequency is equivalent to the negative-indexed spectrum from the negative folding frequency to 0 Hz. This two-sided spectrum can be converted into a single-sided spectrum by doubling alternation-current (AC) components from 0 Hz to the folding frequency and retaining the DC component as is.
6. To reduce the burden of computing DFT coefficients, the FFT algorithm is used, which requires the data length to be a power of 2. Sometimes zero padding is employed to make up the data length. The zero padding actually interpolates the spectrum and does not carry any new information about the signal; even the calculated frequency resolution is smaller due to the zero-padded longer length.
7. Applying a window function to the data sequence before DFT reduces the spectral leakage due to abrupt truncation of the data sequence when performing spectral calculation for a short sequence.
8. Two radix-2 FFT algorithms—decimation-in-frequency and decimation-in-time—are developed via graphical illustrations.

4.7 PROBLEMS

4.1. Given a sequence $x(n)$ for $0 \leq n \leq 3$, where $x(0) = 1, x(1) = 1, x(2) = -1$, and $x(3) = 0$, compute its DFT $X(k)$.

4.2. Given a sequence $x(n)$ for $0 \leq n \leq 3$, where $x(0) = 4, x(1) = 3, x(2) = 2$, and $x(3) = 1$, evaluate its DFT $X(k)$.

4.3. Given a sequence $x(n)$ for $0 \leq n \leq 3$, where $x(0) = 0.2, x(1) = 0.2, x(2) = -0.2$, and $x(3) = 0$, compute its DFT $X(k)$.

4.4. Given a sequence $x(n)$ for $0 \leq n \leq 3$, where $x(0) = 0.8, x(1) = 0.6, x(2) = 0.4$, and $x(3) = 0.2$, evaluate its DFT $X(k)$.

4.5. Given the DFT sequence $X(k)$ for $0 \leq k \leq 3$ obtained in Problem 4.2, evaluate its inverse DFT $x(n)$.

4.6. Given a sequence $x(n)$, where $x(0) = 4, x(1) = 3, x(2) = 2$, and $x(3) = 1$ with two additional zero-padded data points $x(4) = 0$ and $x(5) = 0$, evaluate its DFT $X(k)$.

4.7. Given the DFT sequence $X(k)$ for $0 \leq k \leq 3$ obtained in Problem 4.4, evaluate its inverse DFT $x(n)$.

4.8. Given a sequence $x(n)$, where $x(0) = 0.8$, $x(1) = 0.6$, $x(2) = 0.4$, and $x(3) = 0.2$ with two additional zero-padded data points $x(4) = 0$ and $x(5) = 0$, evaluate its DFT $X(k)$.

4.9. Using the DFT sequence $X(k)$ for $0 \leq k \leq 5$ computed in Problem 4.6, evaluate the inverse DFT for $x(0)$ and $x(4)$.

4.10. Consider a digital sequence sampled at the rate of 20,000 Hz. If we use the 8,000-point DFT to compute the spectrum, determine

a. the frequency resolution;

b. the folding frequency in the spectrum.

4.11. Using the DFT sequence $X(k)$ for $0 \leq k \leq 5$ computed in Problem 4.8, evaluate the inverse DFT for $x(0)$ and $x(4)$.

4.12. Consider a digital sequence sampled at the rate of 16,000 Hz. If we use the 4,000-point DFT to compute the spectrum, determine

a. the frequency resolution;

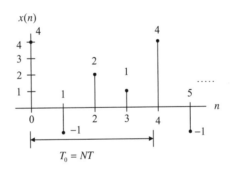

FIGURE 4.46

Data sequence for Problem 4.14.

b. the folding frequency in the spectrum.

4.13. We use the DFT to compute the amplitude spectrum of a sampled data sequence with a sampling rate $f_s = 2,000$ Hz. It requires the frequency resolution to be less than 0.5 Hz. Determine the number of data points used by the FFT algorithm and actual frequency resolution in Hz, assuming that the data samples are available for selecting the number of data points.

4.14. Given the sequence in Figure 4.46 and assuming $f_s = 100$ Hz, compute the amplitude spectrum, phase spectrum, and power spectrum.

4.15. Compute the following window functions for a size of eight:

a. Hamming window function;

b. Hanning window function.

4.16. Consider the following data sequence of length six:

$$x(0) = 0, \ x(1) = 1, \ x(2) = 0, \ x(3) = -1, \ x(4) = 0, \ x(5) = 1$$

Compute the windowed sequence $x_w(n)$ using the

a. triangular window function;

b. Hamming window function;

c. Hanning window function.

4.17. Compute the following window functions for a size of 10:

a. Hamming window function;

b. Hanning window function.

4.18. Consider the following data sequence of length six:

$$x(0) = 0, \ x(1) = 0.2, \ x(2) = 0, \ x(3) = -0.2, \ x(4) = 0, \ x(5) = 0.2$$

Compute the windowed sequence $x_w(n)$ using the

a. triangular window function;

b. Hamming window function;

c. Hanning window function.

4.19. Given the sequence in Figure 4.47 where $f_s = 100$ Hz and $T = 0.01$ sec., compute the amplitude spectrum, phase spectrum, and power spectrum using the

a. triangular window;

b. Hamming window;

c. Hanning window.

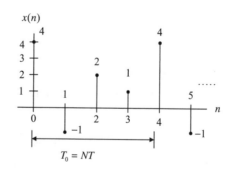

FIGURE 4.47

Data sequence for Problem 4.19.

4.20. Given the sinusoid

$$x(n) = 2 \cdot \sin\left(2{,}000 \cdot 2\pi \cdot \frac{n}{8{,}000}\right)$$

obtained using a sampling rate of $f_s = 8,000$ Hz, we apply the DFT to compute the amplitude spectrum.

   a. Determine the frequency resolution when the data length is 100 samples. Without using the window function, is there any spectral leakage in the computed spectrum? Explain.

   b. Determine the frequency resolution when the data length is 73 samples. Without using the window function, is there any spectral leakage in the computed spectrum? Explain.

4.21. Given a sequence $x(n)$ for $0 \le n \le 3$, where $x(0) = 4$, $x(1) = 3$, $x(2) = 2$, and $x(3) = 1$, evaluate its DFT $X(k)$ using the decimation-in-frequency FFT method, and determine the number of complex multiplications.

4.22. Given the DFT sequence $X(k)$ for $0 \le k \le 3$ obtained in Problem 4.21, evaluate its inverse DFT $x(n)$ using the decimation-in-frequency FFT method.

4.23. Given a sequence $x(n)$ for $0 \le n \le 3$, where $x(0) = 0.8$, $x(1) = 0.6$, $x(2) = 0.4$, and $x(3) = 0.2$, evaluate its DFT $X(k)$ using the decimation-in-frequency FFT method, and determine the number of complex multiplications.

4.24. Given the DFT sequence $X(k)$ for $0 \le k \le 3$ obtained in Problem 4.23, evaluate its inverse DFT $x(n)$ using the decimation-in-frequency FFT method.

4.25. Given a sequence $x(n)$ for $0 \le n \le 3$, where $x(0) = 4$, $x(1) = 3$, $x(2) = 2$, and $x(3) = 1$, evaluate its DFT $X(k)$ using the decimation-in-time FFT method, and determine the number of complex multiplications.

4.26. Given the DFT sequence $X(k)$ for $0 \le k \le 3$ computed in Problem 4.25, evaluate its inverse DFT $x(n)$ using the decimation-in-time FFT method.

4.27. Given a sequence $x(n)$ for $0 \le n \le 3$, where $x(0) = 0.8$, $x(1) = 0.4$, $x(2) = -0.4$, and $x(3) = -0.2$, evaluate its DFT $X(k)$ using the decimation-in-time FFT method, and determine the number of complex multiplications.

4.28. Given the DFT sequence $X(k)$ for $0 \le k \le 3$ computed in Problem 4.27, evaluate its inverse DFT $x(n)$ using the decimation-in-time FFT method.

4.7.1 Computer Problems with MATLAB
Use MATLAB to solve Problems 4.29 and 4.30.

4.29. Consider three sinusoids with the following amplitudes and phases:

$$x_1(t) = 5\cos(2\pi(500)t)$$
$$x_2(t) = 5\cos(2\pi(1200)t + 0.25\pi)$$
$$x_3(t) = 5\cos(2\pi(1800)t + 0.5\pi)$$

   a. Create a MATLAB program to sample each sinusoid and generate a sum of three sinusoids, that is, $x(n) = x_1(n) + x_2(n) + x_3(n)$, using a sampling rate of 8,000 Hz. Plot $x(n)$ over a range of 0.1 seconds.

   b. Use the MATLAB function fft() to compute DFT coefficients, and plot and examine the spectrum of the signal $x(n)$.

4.30. Consider the sum of sinusoids in Problem 4.29.

   a. Generate the sum of sinusoids for 240 samples using a sampling rate of 8,000 Hz.

   b. Write a MATLAB program to compute and plot the amplitude spectrum of the signal $x(n)$ with the FFT using each of the following window functions:

   (1) Rectangular window (no window);

   (2) Triangular window;

   (3) Hamming window.

   c. Examine the effect of spectral leakage for each window use in (b).

4.7.2 MATLAB Projects

4.31. Signal spectral analysis:

Given the four practical signals below, compute their one-sided spectra and create their time-domain plots and spectral plots, respectively:

   a. Speech signal ("speech.dat"), sampling rate = 8,000 Hz. From the spectral plot, identify the first 5 formants.

   b. ECG signal ("ecg.dat"), sampling rate = 500 Hz. From the spectral plot, identify the 60 Hz-interference component.

   c. Seismic data ("seismic.dat"), sampling rate =15 Hz. From the spectral plot, determine the dominant frequency component.

   d. Vibration signal of the acceleration response from a simple supported beam ("vbrdata.dat"), sampling rate =1,000 Hz. From the spectral plot, determine four dominant frequencies (modes).

4.32. Vibration signature analysis:

The acceleration signals measured from a gearbox can be used to monitor the condition of the gears inside the gearbox. The early diagnosis of any gear issues can prevent the future catastrophic failure of the system. Assume the following measurements and specifications (courtesy of SpectraQuest, Inc.):

   a. The input shaft has a speed of 1,000 RPM and meshing frequency is approximately 300 Hz.

   b. Data specifications:

Sampling rate = 12.8 kHz
v0.dat: healthy condition
v1.dat: damage severity level 1 (lightly chipped gear)
v2.dat: damage severity level 2 (moderately chipped gear)
v3.dat: damage severity level 3 (chipped gear)
v4.dat: damage severity level 4 (heavily chipped gear)
v5.dat: damage severity level 5 (missing tooth)
Investigate the spectrum for each measurement and identify sidebands. For each measurement, determine the ratio of the largest sideband amplitude over the amplitude of meshing frequency. Investigate the relation between the computed ratio values and the damage severity.

The z-Transform

CHAPTER OUTLINE

5.1 Definition 137
5.2 Properties of the z-Transform 140
5.3 Inverse z-Transform 144
    5.3.1 Partial Fraction Expansion Using MATLAB 150
5.4 Solution of Difference Equations Using the z-Transform 152
5.5 Summary 156

OBJECTIVES

This chapter introduces the z-transform and its properties; illustrates how to determine the inverse z-transform using partial fraction expansion; and applies the z-transform to solve linear difference equations.

5.1 DEFINITION

The *z-transform* is a very important tool in describing and analyzing digital systems. It also supports the techniques for digital filter design and frequency analysis of digital signals. We begin with the definition of the z-transform.

The z-transform of a causal sequence $x(n)$, designated by $X(z)$ or $Z(x(n))$, is defined as

$$X(z) = Z(x(n)) = \sum_{n=0}^{\infty} x(n)z^{-n}$$

$$= x(0)z^{-0} + x(1)z^{-1} + x(2)z^{-2} + \cdots$$

(5.1)

where z is the complex variable. Here, the summation taken from $n = 0$ to $n = \infty$ is according to the fact that for most situations, the digital signal $x(n)$ is the causal sequence, that is, $x(n) = 0$ for $n < 0$. Thus, the definition in Equation (5.1) is referred to as a *one-sided z-transform* or a *unilateral transform*. In Equation (5.1), all the values of z that make the summation exist form a *region of convergence* in the z-transform domain, while all other values of z outside the region of convergence will cause the summation to diverge. The region of convergence is defined based on the particular sequence $x(n)$ being applied. Note that we deal with the unilateral z-transform in this book, and hence when performing inverse z-transform (which we shall study later), we are restricted to the causal sequence. Now let us study the following typical examples.

EXAMPLE 5.1

Given the sequence

$$x(n) = u(n)$$

find the z-transform of $x(n)$.

Solution:

From the definition of Equation (5.1), the z-transform is given by

$$X(z) = \sum_{n=0}^{\infty} u(n)z^{-n} = \sum_{n=0}^{\infty} \left(z^{-1}\right)^n = 1 + \left(z^{-1}\right) + \left(z^{-1}\right)^2 + \cdots$$

This is an infinite geometric series that converges to

$$X(z) = \frac{z}{z-1}$$

with a condition $|z^{-1}|<1$. Note that for an infinite geometric series, we have $1 + r + r^2 + \cdots = \frac{1}{1-r}$ when $|r|<1$. The region of convergence for all values of z is given as $|z| > 1$.

EXAMPLE 5.2

Consider the exponential sequence

$$x(n) = a^n u(n)$$

and find the z-transform of the sequence $x(n)$.

Solution:

From the definition of the z-transform in Equation (5.1), it follows that

$$X(z) = \sum_{n=0}^{\infty} a^n u(n)z^{-n} = \sum_{n=0}^{\infty} \left(az^{-1}\right)^n = 1 + \left(az^{-1}\right) + \left(az^{-1}\right)^2 + \cdots$$

Since this is a geometric series that will converge for $|az^{-1}|<1$, it is further expressed as

$$X(z) = \frac{z}{z-a}, \quad \text{for } |z| > |a|$$

The z-transforms for common sequences are summarized in Table 5.1. Example 5.3 illustrates how to find the z-transform using Table 5.1.

Table 5.1 Table of z-Transform Pairs

Line No.	$x(n), n \geq 0$	z-Transform $X(z)$	Region of Convergence
1	$x(n)$	$\sum_{n=0}^{\infty} x(n) z^{-n}$	
2	$\delta(n)$	1	$|z| > 0$
3	$au(n)$	$\dfrac{az}{z-1}$	$|z| > 1$
4	$nu(n)$	$\dfrac{z}{(z-1)^2}$	$|z| > 1$
5	$n^2 u(n)$	$\dfrac{z(z+1)}{(z-1)^3}$	$|z| > 1$
6	$a^n u(n)$	$\dfrac{z}{z-a}$	$|z| > |a|$
7	$e^{-na} u(n)$	$\dfrac{z}{(z-e^{-a})}$	$|z| > e^{-a}$
8	$na^n u(n)$	$\dfrac{az}{(z-a)^2}$	$|z| > |a|$
9	$\sin(an)u(n)$	$\dfrac{z \sin(a)}{z^2 - 2z \cos(a) + 1}$	$|z| > 1$
10	$\cos(an)u(n)$	$\dfrac{z[z - \cos(a)]}{z^2 - 2z \cos(a) + 1}$	$|z| > 1$
11	$a^n \sin(bn)u(n)$	$\dfrac{[a \sin(b)]z}{z^2 - [2a \cos(b)]z + a^2}$	$|z| > |a|$
12	$a^n \cos(bn)u(n)$	$\dfrac{z[z - a \cos(b)]}{z^2 - [2a \cos(b)]z + a^{-2}}$	$|z| > |a|$
13	$e^{-an} \sin(bn)u(n)$	$\dfrac{[e^{-a} \sin(b)]z}{z^2 - [2e^{-a} \cos(b)]z + e^{-2a}}$	$|z| > e^{-a}$
14	$e^{-an} \cos(bn)u(n)$	$\dfrac{z[z - e^{-a} \cos(b)]}{z^2 - [2e^{-a} \cos(b)]z + e^{-2a}}$	$|z| > e^{-a}$
15	$2|A||P|^n \cos(n\theta + \varphi)u(n)$ where P and A are complex constants defined by $P = |P| \angle \theta$, $A = |A| \angle \varphi$	$\dfrac{Az}{z-P} + \dfrac{A^*z}{z-P^*}$	

EXAMPLE 5.3

Find the z-transform for each of the following sequences:

a. $x(n) = 10u(n)$
b. $x(n) = 10\sin(0.25\pi n)u(n)$

c. $x(n) = (0.5)^n u(n)$
d. $x(n) = (0.5)^n \sin(0.25\pi n) u(n)$
e. $x(n) = e^{-0.1n} \cos(0.25\pi n) u(n)$

Solution:

a. From Line 3 in Table 5.1, we get

$$X(z) = Z(10u(n)) = \frac{10z}{z-1}$$

b. Line 9 in Table 5.1 leads to

$$X(z) = 10Z(\sin(0.2\pi n)u(n))$$

$$= \frac{10\sin(0.25\pi)z}{z^2 - 2z\cos(0.25\pi) + 1} = \frac{7.07z}{z^2 - 1.414z + 1}$$

c. From Line 6 in Table 5.1, we obtain

$$X(z) = Z((0.5)^n u(n)) = \frac{z}{z - 0.5}$$

d. From Line 11 in Table 5.1, it follows that

$$X(z) = Z((0.5)^n \sin(0.25\pi n) u(n)) = \frac{0.5 \times \sin(0.25\pi)z}{z^2 - 2 \times 0.5 \cos(0.25\pi)z + 0.5^2}$$

$$= \frac{0.3536z}{z^2 - 1.4142z + 0.25}$$

e. From Line 14 in Table 5.1, it follows that

$$X(z) = Z(e^{-0.1n} \cos(0.25\pi n) u(n)) = \frac{z(z - e^{-0.1}\cos(0.25\pi))}{z^2 - 2e^{-0.1}\cos(0.25\pi)z + e^{-0.2}}$$

$$= \frac{z(z - 0.6397)}{z^2 - 1.2794z + 0.8187}$$

5.2 PROPERTIES OF THE Z-TRANSFORM

In this section, we study some important properties of the z-transform. These properties are widely used in deriving the z-transfer functions of difference equations and solving the system output responses of linear digital systems with constant system coefficients, which will be discussed in the next chapter.

Linearity: The z-transform is a linear transformation, which implies

$$Z(ax_1(n) + bx_2(n)) = aZ(x_1(n)) + bZ(x_2(n)) \tag{5.2}$$

where $x_1(n)$ and $x_2(n)$ denote the sampled sequences, while a and b are the arbitrary constants.

EXAMPLE 5.4

Find the z-transform of the sequence defined by

$$x(n) = u(n) - (0.5)^n u(n)$$

Solution:

Applying the linearity of the z-transform discussed above, we have

$$X(z) = Z(x(n)) = Z(u(n)) - Z(0.5^n(n))$$

Using Table 5.1 yields

$$Z(u(n)) = \frac{z}{z-1}$$

and

$$Z(0.5^n u(n)) = \frac{z}{z - 0.5}$$

Substituting these results in $X(z)$ leads to the final solution,

$$X(z) = \frac{z}{z-1} - \frac{z}{z - 0.5}$$

Shift theorem: Given $X(z)$, the z-transform of a sequence $x(n)$, the z-transform of $x(n - m)$, the time-shifted sequence, is given by

$$Z(x(n - m)) = z^{-m}X(z) \tag{5.3}$$

Note that if $m \geq 0$, then $x(n - m)$ is obtained by right shifting $x(n)$ by m samples. Since the shift theorem plays a very important role in developing the transfer function from a difference equation, we verify the shift theorem for the causal sequence. Note that the shift theorem also works for the noncausal sequence.

Verification: Applying the z-transform to the shifted causal signal $x(n - m)$ leads to

$$Z(x(n - m)) = \sum_{n=0}^{\infty} x(n - m)z^{-n}$$

$$= x(-m)z^{-0} + \cdots + x(-1)z^{-(m-1)} + x(0)z^{-m} + x(1)z^{-m-1} + \cdots$$

Since $x(n)$ is assumed to be a causal sequence, this means that

$$x(-m) = x(-m+1) = \cdots = x(-1) = 0$$

Then we achieve

$$Z(x(n - m)) = x(0)z^{-m} + x(1)z^{-m-1} + x(2)z^{-m-2} + \cdots \tag{5.4}$$

Factoring z^{-m} from Equation (5.4) and applying the definition of z-transform of $X(z)$, we get

$$Z(x(n-m)) = z^{-m}(x(0) + x(1)z^{-1} + x(2)z^{-2} + ...) = z^{-m}X(z)$$

EXAMPLE 5.5

Determine the z-transform of the following sequence:

$$y(n) = (0.5)^{(n-5)} \cdot u(n-5)$$

where $u(n-5) = 1$ for $n \geq 5$ and $u(n-5) = 0$ for $n < 5$.

Solution:
We first use the shift theorem to obtain

$$Y(z) = Z\left[(0.5)^{n-5}u(n-5)\right] = z^{-5}Z\left[(0.5)^n u(n)\right]$$

Using Table 5.1 leads to

$$Y(z) = z^{-5} \cdot \frac{z}{z - 0.5} = \frac{z^{-4}}{z - 0.5}$$

Convolution: Given two sequences $x_1(n)$ and $x_2(n)$, their convolution can be determined as follows:

$$x(n) = x_1(n) * x_2(n) = \sum_{k=0}^{\infty} x_1(n-k)x_2(k) \tag{5.5}$$

where $*$ designates the linear convolution. In the z-transform domain, we have

$$X(z) = X_1(z)X_2(z) \tag{5.6}$$

Here, $X(z) = Z(x(n))$, $X_1(z) = Z(x_1(n))$, and $X_2(z) = Z(x_2(n))$.

EXAMPLE 5.6

Verify Equation (5.6) using causal sequences $x_1(n)$ and $x_2(n)$.

Solution:
Taking the z-transform of Equation (5.5) leads to

$$X(z) = \sum_{n=0}^{\infty} x(n)z^{-n} = \sum_{n=0}^{\infty}\sum_{k=0}^{\infty} x_1(n-k)x_2(k)z^{-n}$$

This expression can be further modified to

$$X(z) = \sum_{n=0}^{\infty}\sum_{k=0}^{\infty} x_2(k)z^{-k}x_1(n-k)z^{-(n-k)}$$

Now interchanging the order of the previous summation gives

$$X(z) = \sum_{k=0}^{\infty} x_2(k)z^{-k} \sum_{n=0}^{\infty} x_1(n-k)z^{-(n-k)}$$

Now, let $m = n - k$:

$$X(z) = \sum_{k=0}^{\infty} x_2(k)z^{-k} \sum_{m=0}^{\infty} x_1(m)z^{-m}$$

By the definition of Equation (5.1), it follows that

$$X(z) = X_2(z)X_1(z) = X_1(z)X_2(z)$$

EXAMPLE 5.7

Consider two sequences,

$$x_1(n) = 3\delta(n) + 2\delta(n-1)$$

$$x_2(n) = 2\delta(n) - \delta(n-1)$$

a. Find the z-transform of the convolution:

$$X(z) = Z(x_1(n) * x_2(n))$$

b. Determine the convolution sum using the z-transform:

$$x(n) = x_1(n) * x_2(n) = \sum_{k=0}^{\infty} x_1(k)x_2(n-k)$$

Solution:

a. Applying the z-transform for $x_1(n)$ and $x_2(n)$, respectively, it follows that

$$X_1(z) = 3 + 2z^{-1}$$

$$X_2(z) = 2 - z^{-1}$$

Using the convolution property, we have

$$X(z) = X_1(z)X_2(z) = (3 + 2z^{-1})(2 - z^{-1})$$
$$= 6 + z^{-1} - 2z^{-2}$$

b. Applying the inverse z-transform and using the shift theorem and Line 1 of Table 5.1 leads to

$$x(n) = Z^{-1}(6 + z^{-1} - 2z^{-2}) = 6\delta(n) + \delta(n-1) - 2\delta(n-2)$$

Table 5.2 z-Transform Properties

Property	Time Domain	z-Transform
Linearity	$ax_1(n) + bx_2(n)$	$aZ(x_1(n)) + bZ(x_2(n))$
Shift theorem	$x(n-m)$	$z^{-m}X(z)$
Linear convolution	$x_1(n) * x_2(n) = \sum_{k=0}^{\infty} x_1(n-k)x_2(k)$	$X_1(z)X_2(z)$

The properties of the z-transform discussed in this section are listed in Table 5.2.

5.3 INVERSE Z-TRANSFORM

The z-transform of the sequence $x(n)$ and the inverse z-transform for the function $X(z)$ are defined as, respectively

$$X(z) = Z(x(n)) \tag{5.7}$$

and

$$x(n) = Z^{-1}(X(z)) \tag{5.8}$$

where $Z(\ )$ is the z-transform operator, and $Z^{-1}(\ )$ is the inverse z-transform operator. The inverse of the z-transform may be obtained by at least three methods:

1. partial fraction expansion and lookup table;
2. power series expansion;
3. residue method.

The first method is widely utilized, and it is assumed that the reader is well familiar with the partial fraction expansion method in learning Laplace transform. Therefore, we concentrate on the first method in this book. As for the power series expansion and residue methods, the interested reader is referred to the textbook by Oppenheim and Schafer (1975). The key idea of the partial fraction expansion is that if $X(z)$ is a proper rational function of z, we can expand it to a sum of the first-order factors or higher-order factors using the partial fraction expansion that can be inverted by inspecting the z-transform table. The partial fraction expansion method is illustrated via the following examples. (For simple z-transform functions, we can directly find the inverse z-transform using Table 5.1.)

EXAMPLE 5.8

Find the inverse z-transform for each of the following functions:

a. $X(z) = 2 + \dfrac{4z}{z-1} - \dfrac{z}{z-0.5}$

b. $X(z) = \dfrac{5z}{(z-1)^2} - \dfrac{2z}{(z-0.5)^2}$

c. $X(z) = \dfrac{10z}{z^2 - z + 1}$

d. $X(z) = \dfrac{z^{-4}}{z - 1} + z^{-6} + \dfrac{z^{-3}}{z + 0.5}$

Solution:

a. $x(n) = 2Z^{-1}(1) + 4Z^{-1}\left(\dfrac{z}{z - 1}\right) - Z^{-1}\left(\dfrac{z}{z - 0.5}\right)$

From Table 5.1, we have

$$x(n) = 2\delta(n) + 4u(n) - (0.5)^n u(n)$$

b. $x(n) = Z^{-1}\left(\dfrac{5z}{(z - 1)^2}\right) - Z^{-1}\left(\dfrac{2z}{(z - 0.5)^2}\right) = 5Z^{-1}\left(\dfrac{z}{(z - 1)^2}\right) - \dfrac{2}{0.5}Z^{-1}\left(\dfrac{0.5z}{(z - 0.5)^2}\right)$

Then

$$x(n) = 5nu(n) - 4n(0.5)^n u(n)$$

c. $X(z) = \dfrac{10z}{z^2 - z + 1} = \left(\dfrac{10}{\sin(a)}\right)\dfrac{\sin(a)z}{z^2 - 2z\cos(a) + 1}$

By coefficient matching, we have

$$-2\cos(a) = -1$$

Hence, $\cos(a) = 0.5$, and $a = 60^0$. Substituting $a = 60^0$ into the sine function leads to

$$\sin\left(a\right) = \sin\left(60^0\right) = 0.866$$

Finally, we have

$$x(n) = \dfrac{10}{\sin(a)}Z^{-1}\left(\dfrac{\sin(a)z}{z^2 - 2z\cos\left(a\right) + 1}\right) = \dfrac{10}{0.866}\sin(n{\cdot}60^0) = 11.547\sin(n{\cdot}60^0)$$

d.

$$x(n) = Z^{-1}\left(z^{-5}\dfrac{z}{z - 1}\right) + Z^{-1}\left(z^{-6}{\cdot}1\right) + Z^{-1}\left(z^{-4}\dfrac{z}{z + 0.5}\right)$$

Using Table 5.1 and the shift property, we get

$$x(n) = u(n - 5) + \delta(n - 6) + (-0.5)^{n-4}u(n - 4)$$

Now, we are ready to deal with the inverse z-transform using the partial fraction expansion and lookup table. The general procedure is as follows:

1. Eliminate the negative powers of z for the z-transform function $X(z)$.
2. Determine the rational function $X(z)/z$ (assuming it is proper), and apply the partial fraction expansion to the determined rational function $X(z)/z$ using the formula in Table 5.3.
3. Multiply the expanded function $X(z)/z$ by z on both sides of the equation to obtain $X(z)$.
4. Apply the inverse z-transform using Table 5.1.

Table 5.3 Partial Fraction(s) and Formulas for Constant(s)

Partial fraction with the first-order real pole: $\dfrac{R}{z-p}$ $R = (z-p)\dfrac{X(z)}{z}\Big|_{z=p}$

Partial fraction with the first-order complex poles: $\dfrac{Az}{(z-P)} + \dfrac{A^*z}{(z-P^*)}$ $A = (z-P)\dfrac{X(z)}{z}\Big|_{z=P}$

$P^* = $ complex conjugate of P $\ A^* = $ complex conjugate of A

Partial fraction with mth-order real poles:

$\dfrac{R_m}{(z-p)} + \dfrac{R_{m-1}}{(z-p)^2} + \cdots + \dfrac{R_1}{(z-p)^m}$ $R_k = \dfrac{1}{(k-1)!}\dfrac{d^{k-1}}{dz^{k-1}}\left((z-p)^m\dfrac{X(z)}{z}\right)\Big|_{z=p}$

The partial fraction format and the formulas for calculating the constants are listed in Table 5.3. Example 5.9 considers the situation of the z-transform function having first-order poles.

EXAMPLE 5.9

Find the inverse of the following z-transform:

$$X(z) = \frac{1}{(1 - z^{-1})(1 - 0.5z^{-1})}$$

Solution:

Eliminating the negative power of z by multiplying the numerator and denominator by z^2 yields

$$X(z) = \frac{z^2}{z^2(1 - z^{-1})(1 - 0.5z^{-1})}$$

$$= \frac{z^2}{(z-1)(z-0.5)}$$

Dividing both sides by z leads to

$$\frac{X(z)}{z} = \frac{z}{(z-1)(z-0.5)}$$

Again, we write

$$\frac{X(z)}{z} = \frac{A}{(z-1)} + \frac{B}{(z-0.5)}$$

where A and B are constants found using the formula in Table 5.3, that is,

$$A = (z-1)\frac{X(z)}{z}\Big|_{z=1} = \frac{z}{(z-0.5)}\Big|_{z=1} = 2$$

$$B = (z - 0.5)\frac{X(z)}{z}\bigg|_{z=0.5} = \frac{z}{(z-1)}\bigg|_{z=0.5} = -1$$

Thus

$$\frac{X(z)}{z} = \frac{2}{(z-1)} + \frac{-1}{(z-0.5)}$$

Multiplying z on both sides gives

$$X(z) = \frac{2z}{(z-1)} + \frac{-z}{(z-0.5)}$$

Using Table 5.1 with the z-transform pairs, it follows that

$$x(n) = 2u(n) - (0.5)^n u(n)$$

Tabulating this solution in terms of integer values of n, we obtain the results in Table 5.4.

Table 5.4 Determined Sequence in Example 5.9

n	0	1	2	3	4	...	∞
$x(n)$	1.0	1.5	1.75	1.875	1.9375	...	2.0

The following example considers the case where $X(z)$ has first-order complex poles.

EXAMPLE 5.10

Find $y(n)$ if

$$Y(z) = \frac{z^2(z+1)}{(z-1)(z^2 - z + 0.5)}$$

Solution:
Dividing $Y(z)$ by z, we have

$$\frac{Y(z)}{z} = \frac{z(z+1)}{(z-1)(z^2 - z + 0.5)}$$

Applying the partial fraction expansion leads to

$$\frac{Y(z)}{z} = \frac{B}{z-1} + \frac{A}{(z-0.5 - j0.5)} + \frac{A^*}{(z-0.5 + j0.5)}$$

We first find B :

$$B = (z-1)\frac{Y(z)}{z}\bigg|_{z=1} = \frac{z(z+1)}{(z^2-z+0.5)}\bigg|_{z=1} = \frac{1\times(1+1)}{(1^2-1+0.5)} = 4$$

Notice that A and A^* are a complex conjugate pair. We determine A as follows:

$$A = (z-0.5-j0.5)\frac{Y(z)}{z}\bigg|_{z=0.5+j0.5} = \frac{z(z+1)}{(z-1)(z-0.5+j0.5)}\bigg|_{z=0.5+j0.5}$$

$$= \frac{(0.5+j0.5)(0.5+j0.5+1)}{(0.5+j0.5-1)(0.5+j0.5-0.5+j0.5)} = \frac{(0.5+j0.5)(1.5+j0.5)}{(-0.5+j0.5)j}$$

Using the polar form, we get

$$A = \frac{(0.707\angle45^0)(1.58114\angle18.43^0)}{(0.707\angle135^0)(1\angle90^0)} = 1.58114\angle-161.57^0$$

$$A^* = \overline{A} = 1.58114\angle161.57^0$$

Assume that a first-order complex pole takes the form

$$P = 0.5+0.5j = |P|\angle\theta = 0.707\angle45^0 \text{ and } P^* = |P|\angle-\theta = 0.707\angle-45^0$$

We have

$$Y(z) = \frac{4z}{z-1} + \frac{Az}{(z-P)} + \frac{A^*z}{(z-P^*)}$$

Applying the inverse z-transform from Line 15 in Table 5.1 leads to

$$y(n) = 4Z^{-1}\left(\frac{z}{z-1}\right) + Z^{-1}\left(\frac{Az}{(z-P)} + \frac{A^*z}{(z-P^*)}\right)$$

Using the previous formula, the inversion and subsequent simplification yield

$$y(n) = 4u(n) + 2|A|(|P|)^n\cos(n\theta+\varphi)u(n)$$
$$= 4u(n) + 3.1623(0.7071)^n\cos(45^0n-161.57^0)u(n)$$

The situation dealing with real repeated poles is presented next.

EXAMPLE 5.11

Find $x(n)$ if

$$X(z) = \frac{z^2}{\left(z-1\right)\left(z-0.5\right)^2}$$

Solution:
Dividing both sides of the previous z-transform by z yields

$$\frac{X(z)}{z} = \frac{z}{(z-1)(z-0.5)^2} = \frac{A}{z-1} + \frac{B}{z-0.5} + \frac{C}{(z-0.5)^2}$$

where

$$A = (z-1)\frac{X(z)}{z}\bigg|_{z=1} = \frac{z}{(z-0.5)^2}\bigg|_{z=1} = 4$$

Using the formulas for mth-order real poles in Table 5.3, where $m = 2$ and $p = 0.5$, to determine B and C yields

$$B = R_2 = \frac{1}{(2-1)!}\frac{d}{dz}\left\{(z-0.5)^2\frac{X(z)}{z}\right\}_{z=0.5}$$

$$= \frac{d}{dz}\left(\frac{z}{z-1}\right)\bigg|_{z=0.5} = \frac{-1}{(z-1)^2}\bigg|_{z=0.5} = -4$$

$$C = R_1 = \frac{1}{(1-1)!}\frac{d^0}{dz^0}\left\{(z-0.5)^2\frac{X(z)}{z}\right\}_{z=0.5}$$

$$= \frac{z}{z-1}\bigg|_{z=0.5} = -1$$

Then

$$X(z) = \frac{4z}{z-1} + \frac{-4z}{z-0.5} + \frac{-1z}{(z-0.5)^2} \tag{5.9}$$

The inverse z-transform for each term on the right-hand side of Equation (5.9) can be obtained using the result listed in Table 5.1, that is,

$$Z^{-1}\left\{\frac{z}{z-1}\right\} = u(n)$$

$$Z^{-1}\left\{\frac{z}{z-0.5}\right\} = (0.5)^n u(n)$$

$$Z^{-1}\left\{\frac{z}{(z-0.5)^2}\right\} = 2n(0.5)^n u(n)$$

From these results, it follows that

$$x(n) = 4u(n) - 4(0.5)^n u(n) - 2n(0.5)^n u(n)$$

5.3.1 Partial Fraction Expansion Using MATLAB

The MATLAB function **residue()** can be applied to perform the partial fraction expansion of a z-transform function $X(z)/z$. The syntax is given as

$$[\mathbf{R}, \mathbf{P}, \mathbf{K}] = \mathbf{residue}(\mathbf{B}, \mathbf{A})$$

Here, B and A are the vectors consisting of coefficients for the numerator and denominator polynomials, $B(z)$ and $A(z)$, respectively. Notice that $B(z)$ and $A(z)$ are the polynomials with increasing positive powers of z.

$$\frac{B(z)}{A(z)} = \frac{b_0 z^M + b_1 z^{M-1} + b_2 z^{M-2} + \cdots + b_M}{z^N + a_1 z^{N-1} + a_2 z^{-2} + \cdots + a_N}$$

The function returns the residues in vector R, corresponding poles in vector P, and polynomial coefficients (if any) in vector K. The expansion format is shown as

$$\frac{B(z)}{A(z)} = \frac{r_1}{z - p_1} + \frac{r_2}{z - p_2} + \cdots + k_0 + k_1 z^{-1} + \cdots$$

For a pole p_j of multiplicity m, the partial fraction includes the following terms:

$$\frac{B(z)}{A(z)} = \cdots + \frac{r_j}{z - p_j} + \frac{r_{j+1}}{(z - p_j)^2} + \cdots + \frac{r_{j+m}}{(z - p_j)^m} + \cdots + k_0 + k_1 z^{-1} + \cdots$$

EXAMPLE 5.12

Find the partial expansion for each of the following z-transform functions:

a. $X(z) = \dfrac{1}{(1 - z^{-1})(1 - 0.5z^{-1})}$

b. $Y(z) = \dfrac{z^2(z+1)}{(z-1)(z^2 - z + 0.5)}$

c. $X(z) = \dfrac{z^2}{(z-1)(z-0.5)^2}$

Solution:

a. From MATLAB, we can show the denominator polynomial as
```
» conv([1 −1],[1 −0.5])
D =
1.0000 −1.5000 0.5000
```

This leads to

$$X(z) = \frac{1}{(1 - z^{-1})(1 - 0.5z^{-1})} = \frac{1}{1 - 1.5z^{-1} + 0.5^{-2}} = \frac{z^2}{z^2 - 1.5z + 0.5}$$

and

$$\frac{X(z)}{z} = \frac{z}{z^2 - 1.5z + 0.5}$$

From MATLAB, we have
» **[R,P,K]=residue([1 0],[1 −1.5 0.5])**
R =
2
−1
P =
1.0000
0.5000
K =
[]
»

Then the expansion is written as

$$X(z) = \frac{2z}{z-1} - \frac{z}{z-0.5}$$

b. From the MATLAB entry
» **N=conv([1 0 0],[1 1])**
N =
1 1 0 0
» **D=conv([1 −1], [1 −1 0.5])**
D =
1.0000 −2.0000 1.5000 −0.5000

we get

$$Y(z) = \frac{z^2(z+1)}{(z-1)(z^2 - z + 0.5)} = \frac{z^3 + z^2}{z^3 - 2z^2 + 1.5z - 0.5}$$

and

$$\frac{Y(z)}{z} = \frac{z^2 + z}{z^3 - 2z^2 + 1.5z - 0.5}$$

Using the MATLAB residue function yields
» **[R,P,K]=residue([1 1 0],[1 −2 1.5 −0.5])**
R =
4.0000
−1.5000 − 0.5000i
−1.5000 + 0.5000i
P =
1.0000
0.5000 + 0.5000i
0.5000 − 0.5000i
K =
[]
»

Then the expansion is shown below

$$X(z) = \frac{Bz}{z - p_1} + \frac{Az}{z - p} + \frac{A^*z}{z - p^*}$$

where $B = 4$, $p_1 = 1$,

$$A = -1.5 - 0.5j, \quad p = 0.5 + 0.5j,$$

$$A^* = -1.5 + 0.5j, \quad \text{and } p = 0.5 - 0.5j$$

c. Similarly, if we use
```
» D=conv(conv([1 −1], [1 −0.5]),[1 −0.5])
D =
1.0000 −2.0000 1.2500 −0.2500
```

then

$$X(z) = \frac{z^2}{(z-1)(z-0.5)^2} = \frac{z^2}{z^3 - 2z^2 + 1.25z - 0.25}$$

and we yield

$$\frac{X(z)}{z} = \frac{z}{z^3 - 2z^2 + 1.25z - 0.25}$$

From MATLAB, we obtain
```
» [R,P,K]=residue([1 0],[1 −2 1.25 −0.25])
R =
4.0000
−4.0000
−1.0000
P =
1.0000
0.5000
0.5000
K =
[]
»
```

Using the previous results leads to

$$X(z) = \frac{4z}{z-1} - \frac{4z}{z-0.5} - \frac{z}{(z-0.5)^2}$$

5.4 SOLUTION OF DIFFERENCE EQUATIONS USING THE Z-TRANSFORM

To solve a difference equation with initial conditions, we have to deal with time-shifted sequences such as $y(n-1)$, $y(n-2)$,..., $y(n-m)$, and so on. Let us examine the z-transform of these terms. Using the definition of the z-transform, we have

$$Z(y(n-1)) = \sum_{n=0}^{\infty} y(n-1)z^{-n}$$
$$= y(-1) + y(0)z^{-1} + y(1)z^{-2} + \cdots$$
$$= y(-1) + z^{-1}(y(0) + y(1)z^{-1} + y(2)z^{-2} + \cdots)$$

It holds that

$$Z(y(n-1)) = y(-1) + z^{-1}Y(z) \tag{5.10}$$

Similarly, we have

$$Z(y(n-2)) = \sum_{n=0}^{\infty} y(n-2)z^{-n}$$
$$= y(-2) + y(-1)z^{-1} + y(0)z^{-2} + y(1)z^{-3} + \cdots$$
$$= y(-2) + y(-1)z^{-1} + z^{-2}(y(0) + y(1)z^{-1} + y(2)z^{-2} + \cdots)$$

$$Z(y(n-2)) = y(-2) + y(-1)z^{-1} + z^{-2}Y(z) \tag{5.11}$$

$$Z(y(n-m)) = y(-m) + y(-m+1)z^{-1} + \cdots + y(-1)z^{-(m-1)} + z^{-m}Y(z) \tag{5.12}$$

where $y(-m), y(-m+1), \ldots, y(-1)$ are the initial conditions. If all initial conditions are considered to be zero, that is,

$$y(-m) = y(-m+1) = \cdots y(-1) = 0 \tag{5.13}$$

then Equation (5.12) becomes

$$Z(y(n-m)) = z^{-m}Y(z) \tag{5.14}$$

which is the same as the shift theorem in Equation (5.3).
The following two examples serve as illustrations of applying the z-transform to find the solutions of difference equations. The procedure is as follows:

1. Apply the z-transform to the difference equation.
2. Substitute the initial conditions.
3. Solve for the difference equation in the z-transform domain.
4. Find the solution in the time domain by applying the inverse z-transform.

EXAMPLE 5.13

A digital signal processing (DSP) system is described by the difference equation

$$y(n) - 0.5y(n-1) = 5(0.2)^n u(n)$$

Determine the solution when the initial condition is given by $y(-1) = 1$.

Solution:

Applying the z-transform on both sides of the difference equation and using Equation (5.12), we have

$$Y(z) - 0.5(y(-1) + z^{-1}Y(z)) = 5Z(0.2^n u(n))$$

Substituting the initial condition and $Z(0.2^n u(n)) = Z(0.2^n u(n)) = z/(z - 0.2)$, we achieve

$$Y(z) - 0.5(1 + z^{-1}Y(z)) = 5z/(z - 0.2)$$

Simplification leads to

$$Y(z) - 0.5z^{-1}Y(z) = 0.5 + 5z/(z - 0.2)$$

Factoring out $Y(z)$ and combining the right-hand side of the equation, it follows that

$$Y(z)(1 - 0.5z^{-1}) = (5.5z - 0.1)/(z - 0.2)$$

Then we obtain

$$Y(z) = \frac{(5.5z - 0.1)}{(1 - 0.5z^{-1})(z - 0.2)} = \frac{z(5.5z - 0.1)}{(z - 0.5)(z - 0.2)}$$

Using the partial fraction expansion method leads to

$$\frac{Y(z)}{z} = \frac{5.5z - 0.1}{(z - 0.5)(z - 0.2)} = \frac{A}{z - 0.5} + \frac{B}{z - 0.2}$$

where

$$A = (z - 0.5)\frac{Y(z)}{z}\bigg|_{z=0.5} = \frac{5.5z - 0.1}{z - 0.2}\bigg|_{z=0.5} = \frac{5.5 \times 0.5 - 0.1}{0.5 - 0.2} = 8.8333$$

$$B = (z - 0.2)\frac{Y(z)}{z}\bigg|_{z=0.2} = \frac{5.5z - 0.1}{z - 0.5}\bigg|_{z=0.2} = \frac{5.5 \times 0.2 - 0.1}{0.2 - 0.5} = -3.3333$$

Thus

$$Y(z) = \frac{8.8333z}{(z - 0.5)} + \frac{-3.3333z}{(z - 0.2)}$$

which gives the solution as

$$y(n) = 8.3333(0.5)^n u(n) - 3.3333(0.2)^n u(n)$$

EXAMPLE 5.14

A relaxed (zero initial conditions) DSP system is described by a difference equation

$$y(n) + 0.1y(n - 1) - 0.2y(n - 2) = x(n) + x(n - 1)$$

a. Determine the impulse response $y(n)$ due to the impulse sequence $x(n) = \delta(n)$.

b. Determine the system response $y(n)$ due to the unit step function excitation, where $u(n) = 1$ for $n \geq 0$.

Solution:

a. Applying the z-transform on both sides of the difference equation and using Equation (5.3) or Equation (5.14), we obtain

$$Y(z) + 0.1Y(z)z^{-1} - 0.2Y(z)z^{-2} = X(z) + X(z)z^{-1} \tag{5.15}$$

Factoring out $Y(z)$ on the left side and substituting $X(z) = Z(\delta(n)) = 1$ to the right side in the Equation (5.15) we get

$$Y(z)(1 + 0.1z^{-1} - 0.2z^{-2}) = 1(1 + z^{-1})$$

Then $Y(z)$ can be expressed as

$$Y(z) = \frac{1 + z^{-1}}{1 + 0.1z^{-1} - 0.2z^{-2}}$$

To obtain the impulse response, which is the inverse z-transform of the transfer function, we multiply the numerator and denominator by z^2.
Thus

$$Y(z) = \frac{z^2 + z}{z^2 + 0.1z - 0.2} = \frac{z(z+1)}{(z-0.4)(z+0.5)}$$

Using the partial fraction expansion method leads to

$$\frac{Y(z)}{z} = \frac{z+1}{(z-0.4)(z+0.5)} = \frac{A}{z-0.4} + \frac{B}{z+0.5}$$

where

$$A = (z-0.4)\frac{Y(z)}{z}\bigg|_{z=0.4} = \frac{z+1}{z+0.5}\bigg|_{z=0.4} = \frac{0.4+1}{0.4+0.5} = 1.5556$$

$$B = (z+0.5)\frac{Y(z)}{z}\bigg|_{z=-0.5} = \frac{z+1}{z-0.4}\bigg|_{z=-0.5} = \frac{-0.5+1}{-0.5-0.4} = -0.5556$$

Thus

$$Y(z) = \frac{1.5556z}{(z-0.4)} + \frac{-0.5556z}{(z+0.5)}$$

which gives the impulse response

$$y(n) = 1.5556(0.4)^n u(n) - 0.5556(-0.5)^n u(n)$$

b. To obtain the response due to a unit step function, the input sequence is set to be

$$x(n) = u(n)$$

and the corresponding z-transform is given by

$$X(z) = \frac{z}{z-1}$$

Notice that

$$Y(z) + 0.1Y(z)z^{-1} - 0.2Y(z)z^{-2} = X(z) + X(z)z^{-1}$$

Then the z-transform of the output sequence $y(n)$ can be obtained as

$$Y(z) = \left(\frac{z}{z-1}\right)\left(\frac{1+z^{-1}}{1+0.1z^{-1}-0.2z^{-2}}\right) = \frac{z^2(z+1)}{(z-1)(z-0.4)(z+0.5)}$$

Using the partial fraction expansion method as before gives

$$Y(z) = \frac{2.2222z}{z-1} + \frac{-1.0370z}{z-0.4} + \frac{-0.1852z}{z+0.5}$$

and the system response is found by using Table 5.1:

$$y(n) = 2.2222u(n) - 1.0370(0.4)^n u(n) - 0.1852(-0.5)^n u(n)$$

5.5 SUMMARY

1. The one-sided (unilateral) z-transform, which can be used to transform the causal sequence to the z-transform domain, was defined.
2. The lookup table of the z-transform determines the z-transform for a simple causal sequence, or the causal sequence from a simple z-transform function.
3. The important properties of the z-transform, such as linearity, the shift theorem, and convolution were introduced. The shift theorem can be used to solve a difference equation. The z-transform of a digital convolution of two digital sequences is equal to the product of their z-transforms.
4. Methods to determine the inverse of the z-transform, such as partial fraction expansion, invert the complicated z-transform function, which can have first-order real poles, multiple-order real poles, and first-order complex poles assuming that the z-transform function is proper. The MATLAB techniques to determine the inverse were introduced.
5. The z-transform can be applied to solve linear difference equations with nonzero initial conditions and zero initial conditions.

5.6 PROBLEMS

5.1. Find the z-transform for each of the following sequences:

  a. $x(n) = 4u(n)$
  b. $x(n) = (-0.7)^n u(n)$
  c. $x(n) = 4e^{-2n}u(n)$
  d. $x(n) = 4(0.8)^n \cos(0.1\pi n)u(n)$
  e. $x(n) = 4e^{-3n}\sin(0.1\pi n)u(n)$

5.2. Using the properties of the z-transform, find the z-transform for each of the following

    a. $x(n) = u(n) + (0.5)^n u(n)$

    b. $x(n) = e^{-3(n-4)}\cos(0.1\pi(n-4))u(n-4)$, where $u(n-4) = 1$ for $n \geq 4$ while
    $u(n-4) = 0$ for $n < 4$

5.3. Find the z-transform for each of the following sequences:

    a. $x(n) = 3u(n-4)$

    b. $x(n) = 2(-0.2)^n u(n)$

    c. $x(n) = 5e^{-2(n-3)}u(n-3)$

    d. $x(n) = 6(0.6)^n \cos(0.2\pi n)u(n)$

    e. $x(n) = 4e^{-3(n-1)}\sin(0.2\pi(n-1))u(n-1).$

5.4. Using the properties of the z-transform, find the z-transform for each of the following sequences:

    a. $x(n) = -2u(n) - (0.75)^n u(n)$

    b. $x(n) = e^{-2(n-3)}\sin(0.2\pi(n-3))u(n-3)$, where $u(n-3) = 1$ for $n \geq 3$ while
    $u(n-3) = 0$ for $n < 3$

5.5. Given two sequences

$$x_1(n) = 5\delta(n) - 2\delta(n-2) \text{ and } x_2(n) = 3\delta(n-3)$$

    a. determine the z-transform of the convolution of the two sequences using the convolution property of z-transform

$$X(z) = X_1(z)X_2(z)$$

    b. determine the convolution by the inverse z-transform

$$x(n) = Z^{-1}(X_1(z)X_2(z))$$

    from the result in (a).

5.6. Using Table 5.1 and the z-transform properties, find the inverse z-transform for each of the following functions:

    a. $X(z) = 4 - \dfrac{10z}{z-1} - \dfrac{z}{z+0.5}$

    b. $X(z) = \dfrac{-5z}{(z-1)} + \dfrac{10z}{(z-1)^2} + \dfrac{2z}{(z-0.8)^2}$

    c. $X(z) = \dfrac{z}{z^2 + 1.2z + 1}$

    d. $X(z) = \dfrac{4z^{-4}}{z-1} + \dfrac{z^{-1}}{(z-1)^2} + z^{-8} + \dfrac{z^{-5}}{z-0.5}$

5.7. Given two sequences

$$x_1(n) = -2\delta(n) + 5\delta(n-2) \text{ and } x_2(n) = 4\delta(n-4)$$

a. determine the z-transform of convolution of the two sequences using the convolution property of z-transform

$$X(z) = X_1(z)X_2(z)$$

b. determine the convolution by the inverse z-transform

$$x(n) = Z^{-1}(X_1(z)X_2(z))$$

from the result in (a).

5.8. Using Table 5.1 and z-transform properties, find the inverse z-transform for each of the following functions:

a. $X(z) = 5 - \dfrac{7z}{z+1} - \dfrac{3z}{z-0.5}$

b. $X(z) = \dfrac{-3z}{(z-0.5)} + \dfrac{8z}{(z-0.8)} + \dfrac{2z}{(z-0.8)^2}$

c. $X(z) = \dfrac{3z}{z^2 + 1.414z + 1}$

d. $X(z) = \dfrac{5z^{-5}}{z-1} - \dfrac{z^{-2}}{(z-1)^2} + z^{-10} + \dfrac{z^{-3}}{z-0.75}$

5.9. Using the partial fraction expansion method, find the inverse of the following z-transforms:

a. $X(z) = \dfrac{1}{z^2 - 0.3z - 0.24}$

b. $X(z) = \dfrac{z}{(z-0.2)(z+0.4)}$

c. $X(z) = \dfrac{z}{(z+0.2)(z^2 - z + 0.5)}$

d. $X(z) = \dfrac{z(z+0.5)}{(z-0.1)^2(z-0.6)}$

5.10. A system is described by the difference equation

$$y(n) + 0.5y(n-1) = 2(0.8)^n u(n)$$

Determine the solution when the initial condition is $y(-1) = 2$.

5.11. Using the partial fraction expansion method, find the inverse of the following z-transforms:

a. $X(z) = \dfrac{1}{z^2 + 0.2z + 0.2}$

b. $X(z) = \dfrac{z}{(z + 0.3)(z - 0.5)}$

c. $X(z) = \dfrac{5z}{(z - 0.75)(z^2 - z + 0.5)}$

d. $X(z) = \dfrac{2z(z - 0.4)}{(z - 0.2)^2(z + 0.8)}$

5.12. A system is described by the difference equation

$$y(n) + 0.2y(n - 1) = 4(0.3)^n u(n)$$

Determine the solution when the initial condition is $y(-1) = 1$.

5.13. A system is described by the difference equation

$$y(n) - 0.5y(n - 1) + 0.06y(n - 2) = (0.4)^{n-1} u(n - 1)$$

Determine the solution when the initial conditions are $y(-1) = 1$, and $y(-2) = 2$.

5.14. Given the following difference equation with the input–output relationship of a certain initially relaxed system (all initial conditions are zero),

$$y(n) - 0.7y(n - 1) + 0.1y(n - 2) = x(n) + x(n - 1)$$

a. find the impulse response sequence $y(n)$ due to the impulse sequence $\delta(n)$;

b. find the output response of the system when the unit step function $u(n)$ is applied.

5.15. A system is described by the difference equation

$$y(n) - 0.6y(n - 1) + 0.08y(n - 2) = (0.5)^{n-1} u(n - 1)$$

Determine the solution when the initial conditions are $y(-1) = 2$, and $y(-2) = 1$.

5.16. Given the following difference equation with the input–output relationship of a certain initially relaxed system (all initial conditions are zero),

$$y(n) - 0.6y(n - 1) + 0.25y(n - 2) = x(n) + x(n - 1)$$

a. find the impulse response sequence $y(n)$ due to the impulse sequence $\delta(n)$;

b. find the output response of the system when the unit step function $u(n)$ is applied.

5.17. Given the following difference equation with the input–output relationship of a certain initially relaxed DSP system (all initial conditions are zero),

$$y(n) - 0.4y(n - 1) + 0.29y(n - 2) = x(n) + 0.5x(n - 1)$$

a. find the impulse response sequence $y(n)$ due to the impulse sequence $\delta(n)$;

b. find the output response of the system when the unit step function $u(n)$ is applied.

5.18. Given the following difference equation with the input–output relationship of a certain initially relaxed DSP system (all initial conditions are zero),

$$y(n) - 0.2y(n-1) + 0.17y(n-2) = x(n) + 0.3x(n-1)$$

a. find the impulse response sequence $y(n)$ due to the impulse sequence $\delta(n)$;

b. find the output response of the system when the unit step function $u(n)$ is applied.

Digital Signal Processing Systems, Basic Filtering Types, and Digital Filter Realizations

6

CHAPTER OUTLINE

6.1 The Difference Equation and Digital Filtering 161
6.2 Difference Equation and Transfer Function 166
    6.2.1 Impulse Response, Step Response, and System Response 169
6.3 The z-Plane Pole-Zero Plot and Stability... 172
6.4 Digital Filter Frequency Response 178
6.5 Basic Types of Filtering 186
6.6 Realization of Digital Filters.. 192
    6.6.1 Direct-Form I Realization 193
    6.6.2 Direct-Form II Realization 193
    6.6.3 Cascade (Series) Realization 195
    6.6.4 Parallel Realization..196
6.7 Application: Signal Enhancement and Filtering... 199
    6.7.1 Pre-Emphasis of Speech 200
    6.7.2 Bandpass Filtering of Speech 203
    6.7.3 Enhancement of ECG Signal Using Notch Filtering..205
6.8 Summary 206

OBJECTIVES:

This chapter illustrates digital filtering operations for a given input sequence; derives transfer functions from the difference equations; analyzes the stability of the linear systems using the z-plane pole-zero plot; and calculates the frequency responses of digital filters. Then the chapter further investigates realizations of the digital filters, and examines spectral effects by filtering speech data using the digital filters.

6.1 THE DIFFERENCE EQUATION AND DIGITAL FILTERING

In this chapter, we begin with developing the filtering concept of digital signal processing (DSP) systems. With the knowledge acquired in Chapter 5, dealing with the z-transform, we will learn how to describe and analyze linear time-invariant systems. We also will become familiar with digital filtering types and their realization structures.

161

A DSP system (digital filter) is described in Figure 6.1.

Let $x(n)$ and $y(n)$ be a DSP system's input and output, respectively. We can express the relationship between the input and the output of a DSP system by the following *difference equation*:

$$y(n) = b_0 x(n) + b_1 x(n-1) + \cdots + b_M x(n-M)$$
$$- a_1 y(n-1) - \cdots - a_N y(n-N) \tag{6.1}$$

where b_i, $0 \le i \le M$ and a_j, $1 \le j \le N$, represent the coefficients of the system and n is the time index. Equation (6.1) can also be written as

$$y(n) = \sum_{i=0}^{M} b_i x(n-i) - \sum_{j=1}^{N} a_j y(n-j) \tag{6.2}$$

From Equations (6.1) and (6.2), we observe that the DSP system output is the weighted summation of the current input value $x(n)$, past values $x(n-1)$, ... , $x(n-M)$, and the past output sequence $y(n-1)$, ... , $y(n-N)$. The system can be verified as linear, time-invariant, and causal. If the initial conditions are specified, we can compute system output (time response) $y(n)$ recursively. This process is referred to as *digital filtering*. We will illustrate filtering operations in Examples 6.1 and 6.2.

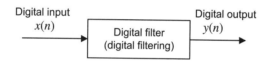

FIGURE 6.1

DSP system with input and output.

EXAMPLE 6.1

Compute the system output

$$y(n) = 0.5y(n-2) + x(n-1)$$

for the first four samples using the following initial conditions:

a. initial conditions $y(-2) = 1$, $y(-1) = 0$, $x(-1) = -1$, and input $x(n) = (0.5)^n u(n)$
b. zero initial conditions $y(-2) = 0$, $y(-1) = 0$, $x(-1) = 0$, and input $x(n) = (0.5)^n u(n)$

Solution:

According to Equation (6.1), we identify the system coefficients as

$$N = 2, M = 1, a_1 = 0, a_2 = -0.5, b_0 = 0, \quad \text{and} \quad b_1 = 1$$

a. Setting $n = 0$, and using initial conditions, we obtain the input and output as

$$x(0) = (0.5)^0 u(0) = 1$$
$$y(0) = 0.5y(-2) + x(-1) = 0.5 \cdot 1 + (-1) = -0.5$$

Setting $n = 1$, and using the initial condition $y(-1) = 0$, we achieve

$$x(1) = (0.5)^1 u(1) = 0.5$$
$$y(1) = 0.5y(-1) + x(0) = 0.5 \cdot 0 + 1 = 1.0$$

Similarly, using the past output $y(0) = -0.5$, we get

$x(2) = (0.5)^2 u(2) = 0.25$
$y(2) = 0.5y(0) + x(1) = 0.5 \cdot (-0.5) + 0.5 = 0.25$

and with $y(1) = 1.0$, we yield

$x(3) = (0.5)^3 u(3) = 0.125$
$y(3) = 0.5y(1) + x(2) = 0.5 \cdot 1 + 0.25 = 0.75$

....
Clearly, $y(n)$ could be recursively computed for $n > 3$.
b. Setting $n = 0$, we obtain

$x(0) = (0.5)^0 u(0) = 1$
$y(0) = 0.5y(-2) + x(-1) = 0 \cdot 1 + 0 = 0$

Setting $n = 1$, we achieve

$x(1) = (0.5)^1 u(1) = 0.5$
$y(1) = 0.5y(-1) + x(0) = 0 \cdot 0 + 1 = 1$

Similarly, with the past output $y(0) = 0$, we determine

$x(2) = (0.5)^2 u(2) = 0.25$
$y(2) = 0.5y(0) + x(1) = 0.5 \cdot 0 + 0.5 = 0.5$

and with $y(1) = 1$, we obtain

$x(3) = (0.5)^3 u(3) = 0.125$
$y(3) = 0.5y(1) + x(2) = 0.5 \cdot 1 + 0.25 = 0.75$

....
Clearly, $y(n)$ could be recursively computed for $n > 3$

EXAMPLE 6.2

Given the DSP system
$$y(n) = 2x(n) - 4x(n-1) - 0.5y(n-1) - y(n-2)$$

with initial conditions $y(-2) = 1$, $y(-1) = 0$, $x(-1) = -1$, and the input $x(n) = (0.8)^n u(n)$, compute the system response $y(n)$ for 20 samples using MATLAB.

Solution:
Program 6.1 lists the MATLAB program for computing the system response $y(n)$. The top plot in Figure 6.2 shows the input sequence. The middle plot displays the filtered output using the initial conditions, and the bottom plot shows the filtered output for zero initial conditions. As we can see, the system outputs are different at the beginning, but they approach the same value later.
Program 6.1. MATLAB program for Example 6.2.

```
% Example 6.2
% Compute y(n)=2x(n)-4x(n-1)-0.5y(n-1)-0.5y(n-2)
% Nonzero initial conditions:
% y(-2)=1, y(-1)=0, x(-1) =-1, and x(n)=(0.8)^n*u(n)
%
y = zeros(1,20);      % Set up a vector to store y(n)
y = [ 1 0 y];         % Add initial condition of y(-2) and y(-1)
n=0:1:19;             % Compute time indexes
x=(0.8).^n;           % Compute 20 input samples of x(n)
```

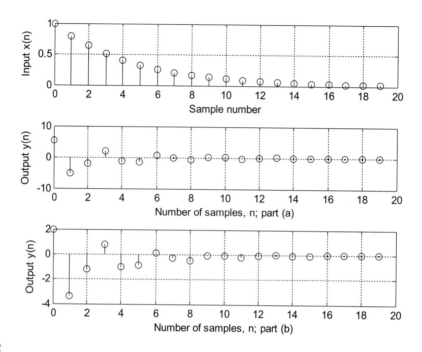

FIGURE 6.2

Plots of the input and system outputs $y(n)$ for Example 6.2.

```
x = [ 0 -1 x];                                    % Add initial condition of
                                                      x(-2)=0 and x(-1)=1
for n=1:20
  y(n+2)= 2*x(n+2)-4*x(n+1)-0.5*y(n+1)-0.5*y(n); % Compute 20 outputs of y(n)
end
n=0:1:19;
subplot(3,1,1);stem(n,x(3:22));grid;ylabel('Input x(n)');xlabel('Sample number');
Subplot(3,1,2); stem(n,y(3:22)),grid;
xlabel('Number of samples, n; part (a)'); ylabel('Output y(n)');
y(3:22)     %Output y(n)
% Zero initial conditions:
% y(-2)=0, y(-1)=0, x(-1) =0, and x(n)=1/(n+1)
%
y = zeros(1,20);      % Set up a vector to store y(n)
y = [ 0 0 y];         % Add zero initial conditions for y(-2) and y(-1)
n=0:1:19;             % Compute time indexes
x=(0.8).^n;           % Compute 20 input samples of x(n)
x = [ 0 0 x];         % Add zero initial conditions for x(-2) and x(-1)
for n=1:20
y(n+2)= 2*x(n+2)-4*x(n+1)-0.5*y(n+1)-0.5*y(n); % Compute 20 outputs of y(n)
end
n=0:1:19
subplot(3,1,3); stem(n,y(3:22)),grid;
   xlabel('Number of samples, n; part (b)'); ylabel('Output y(n)');
   y(3:22)              %Output y(n)
```

The MATLAB function **filter()**, developed using a direct-form II realization (which will be discussed in a later section), can be used to operate digital filtering, and the syntax is

Zi=filtic(B, A, Yi, Xi)
y=filter(B, A, x, Zi)

where B and A are vectors for the coefficients b_j and a_j whose formats are

$$A = \begin{bmatrix} 1 & a_1 & a_2 & \cdots & a_N \end{bmatrix} \quad \text{and} \quad B = \begin{bmatrix} b_0 & b_1 & b_2 & \cdots & b_M \end{bmatrix}$$

and x and y are the input data vector and output data vector, respectively.

Note that the filter function **filtic()** is a MATLAB function which is used to obtain initial states from initial conditions in the difference equation. The initial states are required by the MATLAB filter function **filter()** since it is implemented in a direct-form II. Hence, Z_i contains initial states required for operating MATLAB function **filter()**, that is,

$$Z_i = \begin{bmatrix} w(-1) & w(-2) & \cdots & \end{bmatrix}$$

which can be recovered by another MATLAB function **filtic()**. X_i and Y_i are initial conditions with the length of the greater of M or N, given by

$$X_i = \begin{bmatrix} x(-1) & x(-2) & \cdots & \end{bmatrix} \quad \text{and} \quad Y_i = \begin{bmatrix} y(-1) & y(-2) & \cdots & \end{bmatrix}$$

For zero initial conditions in particular, the syntax is reduced to

y=filter(B, A, x)

Let us verify the filter operation results in Example 6.1 using the MATLAB functions. The MATLAB codes and results for Example 6.1(a) with the nonzero initial conditions are listed as

```
» B = [0 1]; A = [1 0 −0.5];
» x = [1 0.5 0.25 0.125];
» Xi = [−1 0];Yi=[0 1];
» Zi = filtic(B, A, Yi, Xi);
» y = filter(B, A, x, Zi)
y =
 − 0.5000 1.0000 0.2500 0.7500
»
```

For the case of zero initial conditions in Example 6.1(b), the MATLAB codes and results are

```
» B = [0 1]; A = [1 0 −0.5];
» x = [1 0.5 0.25 0.125];
» y = filter(B, A, x)
y =
0 1.0000 0.5000 0.7500
»
```

As we expected, the filter outputs match the ones in Example 6.1.

6.2 DIFFERENCE EQUATION AND TRANSFER FUNCTION

To proceed in this section, Equation (6.1) is rewritten as

$$y(n) = b_0 x(n) + b_1 x(n-1) + \cdots + b_M x(n-M)$$
$$- a_1 y(n-1) - \cdots - a_N y(n-N)$$

With an assumption that all initial conditions of this system are zero, and with $X(z)$ and $Y(z)$ denoting the z-transforms of $x(n)$ and $y(n)$, respectively, taking the z-transform of Equation (6.1) yields

$$Y(z) = b_0 X(z) + b_1 X(z)z^{-1} + \cdots + b_M X(z)z^{-M}$$
$$- a_1 Y(z)z^{-1} - \cdots - a_N Y(z)z^{-N} \tag{6.3}$$

Rearranging Equation (6.3), we obtain

$$H(z) = \frac{Y(z)}{X(z)} = \frac{b_0 + b_1 z^{-1} + \cdots + b_M z^{-M}}{1 + a_1 z^{-1} + \cdots + a_N z^{-N}} = \frac{B(z)}{A(z)} \tag{6.4}$$

where $H(z)$ is defined as the transfer function with its numerator and denominator polynomials defined below:

$$B(z) = b_0 + b_1 z^{-1} + \cdots + b_M z^{-M} \tag{6.5}$$

$$A(z) = 1 + a_1 z^{-1} + \cdots + a_N z^{-N} \tag{6.6}$$

Clearly the z-transfer function is defined as

$$\text{ratio} = \frac{\text{z-transform of the output}}{\text{z-transform of the input}}$$

In DSP applications, given the difference equation, we can develop the z-transfer function and represent the digital filter in the z-domain as shown in Figure 6.3. Then the stability and frequency response can be examined based on the developed transfer function.

FIGURE 6.3

Digital filter transfer function.

EXAMPLE 6.3

A DSP system is described by the following difference equation:

$$y(n) = x(n) - x(n-2) - 1.3y(n-1) - 0.36y(n-2)$$

Find the transfer function $H(z)$, the denominator polynomial $A(z)$, and the numerator polynomial $B(z)$.

Solution:

Taking the z-transform on both sides of the previous difference equation, we obtain

$$Y(z) = X(z) - X(z)z^{-2} - 1.3Y(z)z^{-1} - 0.36Y(z)z^{-2}$$

Moving the last two terms to the left side of the difference equation and factoring $Y(z)$ on the left side and $X(z)$ on the right side, we obtain

$$Y(z)(1 + 1.3z^{-1} + 0.36z^{-2}) = (1 - z^{-2})X(z)$$

Therefore, the transfer function, which is the ratio of $Y(z)$ over $X(z)$, can be found to be

$$H(z) = \frac{Y(z)}{X(z)} = \frac{1 - z^{-2}}{1 + 1.3z^{-1} + 0.36z^{-2}}$$

From the derived transfer function $H(z)$, we can obtain the denominator polynomial and numerator polynomial as

$$A(z) = 1 + 1.3z^{-1} + 0.36z^{-2}$$

and

$$B(z) = 1 - z^{-2}$$

The difference equation and its transfer function, as well as the stability issue of the linear time-invariant system, will be discussed in the following sections

EXAMPLE 6.4

A digital system is described by the following difference equation:

$$y(n) = x(n) - 0.5x(n-1) + 0.36x(n-2)$$

Find the transfer function $H(z)$, the denominator polynomial $A(z)$, and the numerator polynomial $B(z)$.

Solution:

Taking the z-transform on both sides of the previous difference equation, we obtain

$$Y(z) = X(z) - 0.5X(z)z^{-2} + 0.36X(z)z^{-2}$$

Therefore, the transfer function, that is the ratio of $Y(z)$ to $X(z)$, can be found as

$$H(z) = \frac{Y(z)}{X(z)} = 1 - 0.5z^{-1} + 0.36z^{-2}$$

From the derived transfer function $H(z)$, it follows that

$$A(z) = 1$$

$$B(z) = 1 - 0.5z^{-1} + 0.36z^{-2}$$

In DSP applications, the given transfer function of a digital system can converted into a difference equation for DSP implementation. The following example illustrates the procedure.

EXAMPLE 6.5

Convert each of the following transfer functions into its difference equation.

a. $H(z) = \dfrac{z^2 - 1}{z^2 + 1.3z + 0.36}$

b. $H(z) = \dfrac{z^2 - 0.5z + 0.36}{z^2}$

Solution:

a. Dividing the numerator and denominator by z^2 to obtain the transfer function whose numerator and denominator polynomials have the negative power of z, it follows that

$$H(z) = \frac{(z^2 - 1)/z^2}{(z^2 + 1.3z + 0.36)/z^2} = \frac{1 - z^{-2}}{1 + 1.3z^{-1} + 0.36z^{-2}}$$

We write the transfer function using the ratio of $Y(z)$ to $X(z)$:

$$\frac{Y(z)}{X(z)} = \frac{1 - z^{-2}}{1 + 1.3z^{-1} + 0.36z^{-2}}$$

Then we have

$$Y(z)(1 + 1.3z^{-1} + 0.36z^{-2}) = X(z)(1 - z^{-2})$$

By distributing $Y(z)$ and $X(z)$, we yield

$$Y(z) + 1.3z^{-1}Y(z) + 0.36z^{-2}Y(z) = X(z) - z^{-2}X(z)$$

Applying the inverse z-transform and using the shift property in Equation (5.3) of Chapter 5, we get

$$y(n) + 1.3y(n - 1) + 0.36y(n - 2) = x(n) - x(n - 2)$$

Writing the output $y(n)$ in terms of inputs and past outputs leads to

$$y(n) = x(n) - x(n - 2) - 1.3y(n - 1) - 0.36y(n - 2)$$

b. Similarly, dividing the numerator and denominator by z^2, we obtain

$$H(z) = \frac{Y(z)}{X(z)} = \frac{(z^2 - 0.5z + 0.36)/z^2}{z^2/z^2} = 1 - 0.5z^{-1} + 0.36z^{-2}$$

Thus

$$Y(z) = X(z)(1 - 0.5z^{-1} + 0.36z^{-2})$$

By distributing $X(z)$, we yield

$$Y(z) = X(z) - 0.5z^{-1}X(z) + 0.36z^{-2}X(z)$$

Applying the inverse z-transform with using the shift property in Equation (5.3), we obtain

$$y(n) = x(n) - 0.5x(n - 1) + 0.36x(n - 2)$$

The transfer function $H(z)$ can be factored into the *pole-zero form*:

$$H(z) = \frac{b_0(z - z_1)(z - z_2)\cdots(z - z_M)}{(z - p_1)(z - p_2)\cdots(z - p_N)} \qquad (6.7)$$

where the zeros z_i can be found by solving roots of the numerator polynomial, while the poles p_i can be solved for the roots of the denominator polynomial.

EXAMPLE 6.6

Consider the following transfer functions:

$$H(z) = \frac{1 - z^{-2}}{1 + 1.3z^{-1} + 0.36z^{-2}}$$

Convert it into the pole-zero form.

Solution:
We first multiply the numerator and denominator polynomials by z^2 to achieve the advanced form in which both numerator and denominator polynomials have positive powers of z, that is,

$$H(z) = \frac{(1 - z^{-2})z^2}{(1 + 1.3z^{-1} + 0.36z^{-2})z^2} = \frac{z^2 - 1}{z^2 + 1.3z + 0.36}$$

Letting $z^2 - 1 = 0$, we get $z = 1$ and $z = -1$. Setting $z^2 + 1.3z + 0.36 = 0$ leads to $z = -0.4$ and $z = -0.9$. We then can write numerator and denominator polynomials in the factored form to obtain the pole-zero form:

$$H(z) = \frac{(z - 1)(z + 1)}{(z + 0.4)(z + 0.9)}$$

6.2.1 Impulse Response, Step Response, and System Response

The impulse response $h(n)$ of the DSP system $H(z)$ can be obtained by solving its difference equation using a unit impulse input $\delta(n)$. With the help of the z-transform and noticing that $X(z) = Z\{\delta(n)\}1$, we yield

$$h(n) = Z^{-1}\{H(z)X(z)\} = Z^{-1}\{H(z)\} \qquad (6.8)$$

Similarly, for a step input, we can determine step response assuming zero initial conditions. Letting

$$X(z) = Z[u(n)] = \frac{z}{z - 1}$$

the step response can be found as

$$y(n) = Z^{-1}\left\{H(z)\frac{z}{z - 1}\right\} \qquad (6.9)$$

Furthermore, the z-transform of the general system response is given by

$$Y(z) = H(z)X(z) \qquad (6.10)$$

If we know the transfer function $H(z)$ and z-transform of the input $X(z)$, we are able to determine the system response $y(n)$ by finding the inverse z-transform of the output $Y(z)$:

$$y(n) = Z^{-1}\{Y(z)\} \tag{6.11}$$

EXAMPLE 6.7

Given a transfer function depicting a DSP system

$$H(z) = \frac{z+1}{z-0.5}$$

determine

a. the impulse response $h(n)$,
b. step response $y(n)$, and
c. system response $y(n)$ if the input is given as $x(n) = (0.25)^n u(n)$.

Solution:

a. The transfer function can be rewritten as

$$\frac{H(z)}{z} = \frac{z+1}{z(z-0.5)} = \frac{A}{z} + \frac{B}{z-0.5}$$

where

$$A = \frac{z+1}{(z-0.5)}\Big|_{z=0} = -2 \text{ and } B = \frac{z+1}{z}\Big|_{z=0.5} = 3$$

Thus we have

$$\frac{H(z)}{z} = \frac{-2}{z} + \frac{3}{z-0.5}$$

and

$$H(z) = \left(-\frac{2}{z} + \frac{3}{z-0.5}\right)z = -2 + \frac{3z}{z-0.5}$$

By taking the inverse z-transform as shown in Equation (6.8), we yield the impulse response

$$h(n) = -2\delta(n) + 3(0.5)^n u(n)$$

b. For the step input $x(n) = u(n)$ and its z-transform $X(z) = \frac{z}{z-1}$, we can determine the z-transform of the step response as

$$Y(z) = H(z)X(z) = \frac{z+1}{z-0.5}\frac{z}{z-1}$$

Applying the partial fraction expansion leads to

$$\frac{Y(z)}{z} = \frac{z+1}{(z-0.5)(z-1)} = \frac{A}{z-0.5} + \frac{B}{z-1}$$

where

$$A = \frac{z+1}{z-1}\bigg|_{z=0.5} = -3 \quad \text{and} \quad B = \frac{z+1}{z-0.5}\bigg|_{z=1} = 4$$

The z-transform step response is therefore

$$Y(z) = \frac{-3z}{z-0.5} + \frac{4z}{z-1}$$

Applying the inverse z-transform table yields the step response as

$$y(n) = -3(0.5)^n u(n) + 4u(n)$$

c. To determine the system output response, we first find the z-transform of the input $x(n)$,

$$X(z) = Z\{(0.25)^n u(n)\} = \frac{z}{z-0.25}$$

Then $Y(z)$ can be obtained via Equation (6.10), that is,

$$Y(z) = H(z)X(z) = \frac{z+1}{z-0.5} \cdot \frac{z}{z-0.25} = \frac{z(z+1)}{(z-0.5)(z-0.25)}$$

Using the partial fraction expansion, we have

$$\frac{Y(z)}{z} = \frac{(z+1)}{(z-0.5)(z-0.25)} = \left(\frac{A}{z-0.5} + \frac{B}{z-0.25}\right)$$

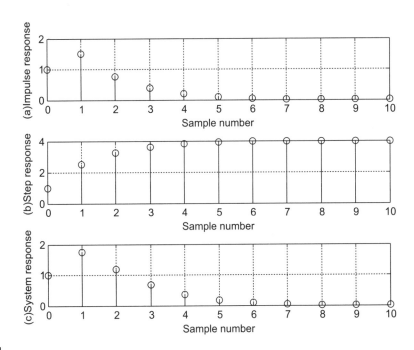

FIGURE 6.4

Impulse, step, and system response in Example 6.7.

$$Y(z) = \left(\frac{6z}{z - 0.5} + \frac{-5z}{z - 0.25} \right)$$

Using Equation (6.11) and Table 5.1 in Chapter 5, we finally yield

$$y(n) = Z^{-1}\{Y(z)\} = 6(0.5)^n u(n) - 5(0.25)^n u(n)$$

The impulse response for (a), step response for (b), and system response for (c) are each plotted in Figure 6.4.

6.3 THE Z-PLANE POLE-ZERO PLOT AND STABILITY

A very useful tool to analyze digital systems is the z-plane pole-zero plot. This graphical technique allows us to investigate characteristics of the digital system shown in Figure 6.1, including the system stability. In general, a digital transfer function can be written in the pole-zero form as shown in Equation (6.7), and we can plot the poles and zeros on the z-plane. The z-plane is depicted in Figure 6.5 and has the following features:

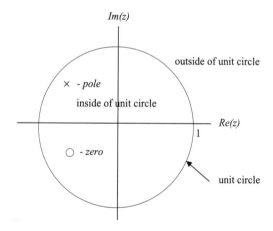

FIGURE 6.5

z-plane and pole-zero plot.

1. The horizontal axis is the real part of the variable z, and the vertical axis represents the imaginary part of the variable z.
2. The z-plane is divided into two parts by a unit circle.
3. Each pole is marked on z-plane using the cross symbol x, while each zero is plotted using the small circle symbol o.

Let's investigate the z-plane pole-zero plot of a digital filter system via the following example.

EXAMPLE 6.8

Given the digital transfer function

$$H(z) = \frac{z^{-1} - 0.5z^{-2}}{1 + 1.2z^{-1} + 0.45z^{-2}}$$

plot poles and zeros.

Solution:

Converting the transfer function to its advanced form by multiplying the numerator and denominator by z^2, it follows that

$$H(z) = \frac{(z^{-1} - 0.5z^{-2})z^2}{(1 + 1.2z^{-1} + 0.45z^{-2})z^2} = \frac{z - 0.5}{z^2 + 1.2z + 0.45}$$

By setting $z^2 + 1.2z + 0.45 = 0$ and $z - 0.5 = 0$, we obtain two poles

$p_1 = -0.6 + j0.3$
$p_2 = p_1^* = -0.6 - j0.3$

and a zero $z_1 = 0.5$, which are plotted on the z-plane shown in Figure 6.6. According to the form of Equation (6.7), we also yield the pole-zero form as

$$H(z) = \frac{z^{-1} - 0.5z^{-2}}{1 + 1.2z^{-1} + 0.45z^{-2}} = \frac{(z - 0.5)}{(z + 0.6 - j0.3)(z + 0.6 + j0.3)}$$

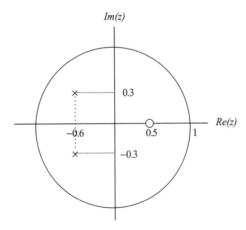

FIGURE 6.6

The z-plane pole-zero plot of Example 6.8.

With zeros and poles plotted on the z-plane, we are able to study system stability. We first establish the relationship between the s-plane in the Laplace domain and the z-plane in the z-transform domain, as illustrated in Figure 6.7.

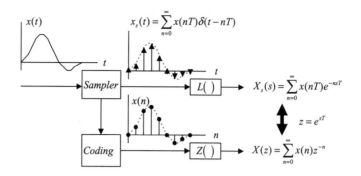

FIGURE 6.7

Relationship between Laplace transform and z-transform.

As shown in Figure 6.7, the sampled signal, which is not quantized, with a sampling period of T is written as

$$x_s(t) = \sum_{n=0}^{\infty} x(nT)\delta(t - nT) = x(0)\delta(t) + x(T)\delta(t - T) + x(2T)\delta(t - 2T) + \cdots \qquad (6.12)$$

Taking the Laplace transform and using the Laplace shift property as

$$L(\delta(t - nT)) = e^{-nTs} \qquad (6.13)$$

leads to

$$X_s(s) = \sum_{n=0}^{\infty} x(nT)e^{-nTs} = x(0)e^{-0 \times Ts} + x(T)e^{-Ts} + x(2T)e^{-2Ts} + \cdots \qquad (6.14)$$

Compare Equation (6.14) with the definition of a one-sided z-transform of the data sequence $x(n)$ from analog-to-digital conversion (ADC):

$$X(z) = Z(x(n)) = \sum_{n=0}^{\infty} x(n)z^{-n} = x(0)z^{-0} + x(1)z^{-1} + x(2)z^{-2} + \cdots \qquad (6.15)$$

Clearly, we see the relationship of the sampled system in the Laplace domain and its digital system in the z-transform domain by the following mapping:

$$z = e^{sT} \qquad (6.16)$$

Substituting $s = -\alpha \pm j\omega$ into Equation (6.16), it follows that $z = e^{-\alpha T \pm j\omega T}$. In the polar form, we have

$$z = e^{-\alpha T} \angle \pm \omega T \qquad (6.17)$$

Equations (6.16) and (6.17) give the following important conclusions.

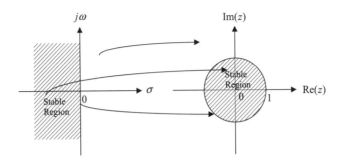

FIGURE 6.8

Mapping between s-plane and z-plane.

If $\alpha > 0$, this means $|z| = e^{-\alpha T} < 1$. Then the left-hand side half plane (LHHP) of the s-plane is mapped to the inside of the unit circle of the z-plane. When $\alpha = 0$, this causes $|z| = e^{-\alpha T} = 1$. Thus the $j\omega$ axis of the s-plane is mapped on the unit circle of the z-plane, as shown in Figure 6.8. Obviously, the right-hand half plane (RHHP) of the s-plane is mapped to the outside of the unit cycle in the z-plane. A stable system means that for a given bounded input, the system output must be bounded. Similar to the analog system, the digital system requires that all poles plotted on the z-plane must be inside the unit circle. We summarize the rules for determining the stability of a DSP system as follows:

1. If the outmost pole(s) of the z-transfer function $H(z)$ describing the DSP system is (are) inside the unit circle on the z-plane pole-zero plot, then the system is stable.
2. If the outmost pole(s) of the z-transfer function $H(z)$ is (are) outside the unit circle on the z-plane pole-zero plot, the system is unstable.
3. If the outmost pole(s) is (are) first-order pole(s) of the z-transfer function $H(z)$ and on the unit circle on the z-plane pole-zero plot, then the system is marginally stable.
4. If the outmost pole(s) is (are) multiple-order pole(s) of the z-transfer function $H(z)$ and on the unit circle on the z-plane pole-zero plot, then the system is unstable.
5. The zeros do not affect the system stability.

Notice that the following facts apply to a stable system (bounded-in/bounded-out [BIBO] stability discussed in Chapter 3):

1. If the input to the system is bounded, then the output of the system will also be bounded, or the impulse response of the system will go to zero in a finite number of steps.
2. An unstable system is one where the output of the system will grow without bound due to any bounded input, initial condition, or noise, or the impulse response will grow without bound.
3. The impulse response of a marginally stable system stays at a constant level or oscillates between two finite values.

Examples illustrating these rules are shown in Figure 6.9.

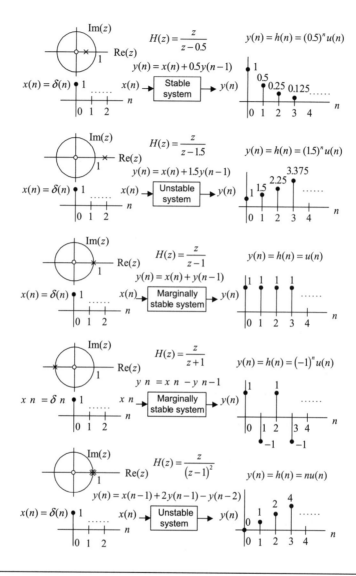

FIGURE 6.9

Stability illustrations.

EXAMPLE 6.9

The following transfer functions describe digital systems.

a. $H(z) = \dfrac{z + 0.5}{(z - 0.5)(z^2 + z + 0.5)}$

b. $H(z) = \dfrac{z^2 + 0.25}{(z - 0.5)(z^2 + 3z + 2.5)}$

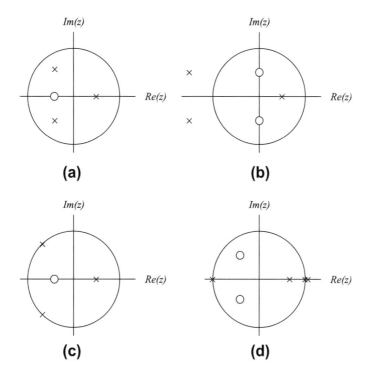

FIGURE 6.10

Pole-zero plots for Example 6.9.

c. $H(z) = \dfrac{z + 0.5}{(z - 0.5)(z^2 + 1.4141z + 1)}$

d. $H(z) = \dfrac{z^2 + z + 0.5}{(z - 1)^2(z + 1)(z - 0.6)}$

For each, sketch the z-plane pole-zero plot and determine the stability status for the digital system.

Solution:

a. A zero is located at $z = -0.5$.

Poles: $z = 0.5$, $|z| = 0.5 < 1$; $z = -0.5 \pm j0.5$,

$|z| = \sqrt{(-0.5)^2 + (\pm 0.5)^2} = 0.707 < 1$.

The plot of poles and a zero is shown in Figure 6.10. Since the outmost poles are inside the unit circle, the system is stable.

b. Zeros are $z = \pm j0.5$.

Poles: $z = 0.5$, $|z| = 0.5 < 1$; $z = -1.5 \pm j0.5$

$|z| = \sqrt{(1.5)^2 + (\pm 0.5)^2} = 1.5811 > 1$.

The plot of poles and zeros is shown in Figure 6.10. Since we have two poles at $z = -1.5 \pm j0.5$ that are outside the unit circle, the system is unstable.

c. A zero is located at $z = -0.5$.

Poles: $z = 0.5$, $|z| = 0.5 < 1$; $z = -0.707 \pm j0.707$,

$$|z| = \sqrt{(0.707)^2 + (\pm 0.707)^2} = 1.$$

The zero and poles are plotted in Figure 6.10. Since the outmost poles are first order at $z = -0.707 \pm j0.707$ and are on the unit circle, the system is marginally stable.

d. Zeros are $z = -0.5 \pm j0.5$.

Poles: $z = 1$, $|z| = 1$; $z = 1$, $|z| = 1$; $z = -1$, $|z| = 1$;, $z = 0.6$
$|z| = 0.6 < 1$.

The zeros and poles are plotted in Figure 6.10. Since the outmost pole is a multiple order (second order) pole at $z = 1$ and is on the unit circle, the system is unstable.

6.4 DIGITAL FILTER FREQUENCY RESPONSE

From the Laplace transfer function, we can achieve the analog filter steady-state frequency response $H(j\omega)$ by substituting $s = j\omega$ into the transfer function $H(s)$. That is,

$$H(s)|_{s=j\omega} = H(j\omega)$$

Then we can study the magnitude frequency response $|H(j\omega)|$ and phase response $\angle H(j\omega)$. Similarly, in a DSP system, using the mapping in Equation (6.16), we substitute $z = e^{sT}|_{s=j\omega} = e^{j\omega T}$ into the z-transfer function $H(z)$ to acquire the digital frequency response, which is converted into the magnitude frequency response $|H(e^{j\omega T})|$ and phase response $\angle |H(e^{j\omega T})|$. That is,

$$H(z)|_{z=e^{j\omega T}} = H(e^{j\omega T}) = |H(e^{j\omega T})|\angle H(e^{j\omega T}) \tag{6.18}$$

Let us introduce a normalized digital frequency in radians in the digital domain:

$$\Omega = \omega T \tag{6.19}$$

Then the digital frequency response in Equation (6.18) becomes

$$H(e^{j\Omega}) = H(z)|_{z=e^{j\Omega}} = |H(e^{j\Omega})|\angle H(e^{j\Omega}) \tag{6.20}$$

The formal derivation for Equation (6.20) can be found in Appendix D.

Now we verify the frequency response via the following simple digital filter. Consider a digital filter with a sinusoidal input of amplitude K (Figure 6.11).

We can determine the system output $y(n)$, which consists of the transient response $y_{tr}(n)$ and the steady-state response $y_{ss}(n)$. We find the z-transform output as

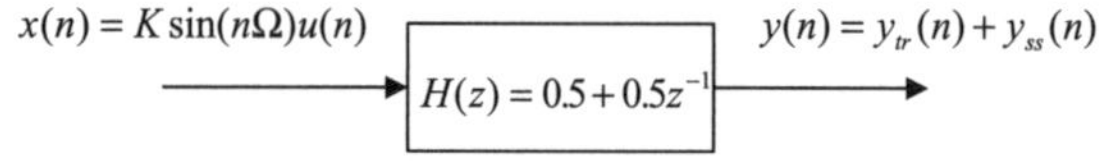

$$x(n) = K\sin(n\Omega)u(n) \quad \boxed{H(z) = 0.5 + 0.5z^{-1}} \quad y(n) = y_{tr}(n) + y_{ss}(n)$$

FIGURE 6.11

System transient and steady-state frequency responses.

$$Y(z) = \left(\frac{0.5z + 0.5}{z}\right)\frac{Kz\sin\Omega}{z^2 - 2z\cos\Omega + 1} \tag{6.21}$$

To perform the inverse z-transform to find the system output, we further rewrite Equation (6.21) as

$$\frac{Y(z)}{z} = \left(\frac{0.5z + 0.5}{z}\right)\frac{K\sin\Omega}{(z - e^{j\Omega})(z - e^{-j\Omega})} = \frac{A}{z} + \frac{B}{z - e^{j\Omega}} + \frac{B^*}{z - e^{-j\Omega}}$$

where A, B and the complex conjugate B^* are the constants for the partial fractions. Applying the partial fraction expansion leads to

$$A = 0.5K\sin\Omega$$

$$B = \left.\frac{0.5z + 0.5}{z}\right|_{z=e^{j\Omega}}\frac{K}{2j} = |H(e^{j\Omega})|e^{j\angle H(e^{j\Omega})}\frac{K}{2j}$$

Notice that the first part of constant B is a complex function, which is obtained by substituting $z = e^{j\Omega}$ into the filter z-transfer function. We can also express the complex function in terms of the polar form:

$$\left.\frac{0.5z + 0.5}{z}\right|_{z=e^{j\Omega}} = 0.5 + 0.5z^{-1}\big|_{z=e^{j\Omega}} = H(z)|_{z=e^{j\Omega}} = H(e^{j\Omega}) = |H(e^{j\Omega})|e^{j\angle H(e^{j\Omega})}$$

where $H(e^{j\Omega}) = 0.5 + 0.5e^{-j\Omega}$. We call this complex function the steady-state frequency response. Based on the complex conjugate property, we get another residue as

$$B^* = |H(e^{j\Omega})|e^{-j\angle H(e^{j\Omega})}\frac{K}{-j2}$$

The z-transform system output is then given by

$$Y(z) = A + \frac{Bz}{z - e^{j\Omega}} + \frac{B^*z}{z - e^{-j\Omega}}$$

Taking the inverse z-transform, we achieve the following system transient and steady-state responses:

$$y(n) = \underbrace{0.5K\sin\Omega\delta(n)}_{y_{tr}(n)} + \underbrace{|H(e^{j\Omega})|e^{j\angle H(e^{j\Omega})}\frac{K}{j2}e^{jn\Omega}u(n) + |H(e^{j\Omega})|e^{-j\angle H(e^{j\Omega})}\frac{K}{-j2}e^{-jn\Omega}u(n)}_{y_{ss}(n)}$$

Simplifying the response yields the form

$$y(n) = 0.5K\sin\Omega\delta(n) + |H(e^{j\Omega})|K\frac{e^{jn\Omega+j\angle H(e^{j\Omega})}u(n) - e^{-jn\Omega-j\angle H(e^{j\Omega})}u(n)}{j2}$$

We can further combine the last term using Euler's formula to express the system response as

$$y(n) = \underbrace{0.5K\sin\Omega\delta(n)}_{y_{tr}(n)\text{ will decay to zero after the first sample}} + \underbrace{|H(e^{j\Omega})|K\sin(n\Omega + \angle H(e^{j\Omega}))u(n)}_{y_{ss}(n)}$$

Finally, the steady-state response is identified as

$$y_{ss}(n) = K|H(e^{j\Omega})| \sin(n\Omega + \angle H(e^{j\Omega}))u(n)$$

For this particular filter, the transient response exists for only the first sample in the system response. By substituting $n = 0$ into $y(n)$ and after simplifying algebra, we achieve the response for the first output sample:

$$y(0) = y_{tr}(0) + y_{ss}(0) = 0.5K \sin(\Omega) - 0.5K \sin(\Omega) = 0$$

Note that the first output sample of the transient response cancels the first output sample of the steady-state response, so the combined first output sample has a value of zero for this particular filter. The system response reaches the steady-state response after the first output sample. At this point, we can conclude that

$$\text{Steady-state magnitude frequency response} = \frac{\text{Peak amplitude of steady state response at } \Omega}{\text{Peak amplitude of sinusoidal input at } \Omega}$$

$$= \frac{|H(e^{j\Omega})|K}{K} = |H(e^{j\Omega})|$$

$$\text{Steady-state phase frequency response} = \text{Phase difference} = \angle H(e^{j\Omega})$$

Figure 6.12 shows the system response with sinusoidal inputs at $\Omega = 0.25\pi$, $\Omega = 0.5\pi$, and $\Omega = 0.75\pi$, respectively.

Next, we examine the properties of the filter frequency response $H(e^{j\Omega})$. From Euler's identity and the trigonometric identity, we know that

$$e^{j(\Omega+k2\pi)} = \cos(\Omega + k2\pi) + j\sin(\Omega + k2\pi)$$
$$= \cos\Omega + j\sin\Omega = e^{j\Omega}$$

where k is an integer taking values of $k = 0, \pm1, \pm2, \cdots$. Then the frequency response has the following property (assuming all input sequences are real):

1. Periodicity
   a. Frequency response: $H(e^{j\Omega}) = H(e^{j(\Omega+k2\pi)})$
   b. Magnitude frequency response: $|H(e^{j\Omega})| = |H(e^{j(\Omega+k2\pi)})|$
   c. Phase response: $\angle H(e^{j\Omega}) = \angle H(e^{j\Omega+k2\pi})$

The second property is given without proof (see proof in Appendix D):

2. Symmetry
   a. Magnitude frequency response: $|H(e^{-j\Omega})| = |H(e^{j\Omega})|$
   b. Phase response: $\angle H(e^{-j\Omega}) = -\angle H(e^{j\Omega})$

Since the maximum frequency in a DSP system is the folding frequency, $f_s/2$, where $f_s = 1/T$, and T designates the sampling period, the corresponding maximum normalized frequency of the system frequency can be calculated as

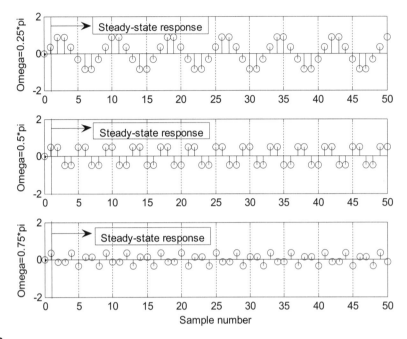

FIGURE 6.12

The digital filter responses to different input sinusoids.

$$\Omega = \omega T = 2\pi \frac{f_s}{2} \times T = \pi \text{ radians} \tag{6.22}$$

The frequency response $H(e^{j\Omega})$ for $|\Omega| \rangle \pi$ consists of the image replicas of $H(e^{j\Omega})$ for $|\Omega| \leq \pi$ and will be removed via the reconstruction filter later. Hence, we need to evaluate $H(e^{j\Omega})$ for only the positive normalized frequency range from $\Omega = 0$ to $\Omega = \pi$ radians. The frequency, in Hz, can be determined by

$$f = \frac{\Omega}{2\pi} f_s \tag{6.23}$$

The magnitude frequency response is often expressed in decibels, defined as

$$\left| H(e^{j\Omega}) \right|_{dB} = 20 \log_{10}(\left| H(e^{j\Omega}) \right|) \tag{6.24}$$

The DSP system stability, magnitude response, and phase response are investigated via the following examples.

EXAMPLE 6.10

Given the digital system

$$y(n) = 0.5x(n) + 0.5x(n-1)$$

with a sampling rate of 8,000 Hz, determine the frequency response.

Solution:

Taking the z-transform on both sides on the difference equation leads to

$$Y(z) = 0.5X(z) + 0.5z^{-1}X(z)$$

Then the transfer function describing the system is easily found to be

$$H(z) = \frac{Y(z)}{X(z)} = 0.5 + 0.5z^{-1}$$

Substituting $z = e^{j\Omega}$, we obtain the frequency response as

$$H(e^{j\Omega}) = 0.5 + 0.5e^{-j\Omega}$$
$$= 0.5 + 0.5\cos(\Omega) - j0.5\sin(\Omega).$$

Therefore, the magnitude frequency response and phase response are given by

$$\left|H(e^{j\Omega})\right| = \sqrt{(0.5 + 0.5\cos(\Omega))^2 + (0.5\sin(\Omega))^2}$$

and

$$\angle H(e^{j\Omega}) = \tan^{-1}\left(\frac{-0.5\sin(\Omega)}{0.5 + 0.5\cos(\Omega)}\right)$$

Several points for the magnitude response and phase response are calculated and shown in Table 6.1. According to the data, we plot the frequency response and phase response of the DSP system in Figure 6.13.

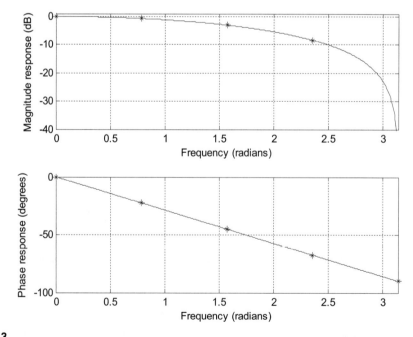

FIGURE 6.13

Frequency responses of the digital filter in Example 6.10.

Table 6.1 Frequency Response Calculations for Example 6.10

Ω (radians)	$f = \dfrac{\Omega}{2\pi} f_s$ (Hz)	$\left\|H(e^{j\Omega})\right\|$	$\left\|H(e^{j\Omega})\right\|_{dB}$	$\angle H(e^{j\Omega})$
0	0	1.000	0 dB	0^0
0.25π	1,000	0.924	-0.687 dB	-22.5^0
0.50π	2,000	0.707	-3.012 dB	-45.00^0
0.75π	3,000	0.383	-8.336 dB	-67.50^0
1.00π	4,000	0.000	$-\infty$	-90^0

It is observed that when the frequency increases, the magnitude response decreases. The DSP system acts like a digital lowpass filter, and its phase response is linear.

We can also verify the periodicity for $\left|H(e^{j\Omega})\right|$ and $\angle H(e^{j\Omega})$ when $\Omega = 0.25\pi + 2\pi$:

$$\left|H(e^{j(0.25\pi+2\pi)})\right| = \sqrt{(0.5 + 0.5\cos(0.25\pi + 2\pi))^2 + (0.5\sin(0.25\pi + 2\pi))^2}$$
$$= 0.924 = \left|H(e^{j0.25\pi})\right|$$

$$\angle H(e^{j(0.25\pi+2\pi)}) = \tan^{-1}\left(\frac{-0.5\sin(0.25\pi + 2\pi)}{0.5 + 0.5\cos(0.25\pi + 2\pi)}\right) = -22.5^0 = \angle H(e^{j0.25\pi}).$$

For $\Omega = -0.25\pi$, we can verify the symmetry property as

$$\left|H(e^{-j0.25\pi})\right| = \sqrt{(0.5 + 0.5\cos(-0.25\pi))^2 + (0.5\sin(-0.25\pi))^2}$$
$$= 0.924 = \left|H(e^{j0.25\pi})\right|$$

$$\angle H(e^{-j0.25\pi}) = \tan^{-1}\left(\frac{-0.5\sin(-0.25\pi)}{0.5 + 0.5\cos(-0.25\pi)}\right) = 22.5^0 = -\angle H(e^{j0.25\pi})$$

The properties can be observed in Figure 6.14, where the frequency range is chosen from $\Omega = -2\pi$ to $\Omega = 4\pi$ radians. As shown in the figure, the magnitude and phase responses are periodic with a period of 2π. For a period between $\Omega = -\pi$ to $\Omega = \pi$, the magnitude responses for the portion $\Omega = -\pi$ to $\Omega = 0$ and the portion $\Omega = 0$ to $\Omega = \pi$ are the same, while the phase responses are opposite. Since the magnitude and phase responses calculated for the range from $\Omega = 0$ to $\Omega = \pi$ are sufficient to present frequency response information, this range is only required for generating the frequency response plots.

Again, note that the phase plot shows a sawtooth shape instead of a linear straight line for this particular filter. This is due to the phase wrapping at $\Omega = 2\pi$ radians since $e^{j(\Omega+k2\pi)} = e^{j\Omega}$ is used in the calculation. However, the phase plot shows that the phase is linear in the range from $\Omega = 0$ to $\Omega = \pi$ radians.

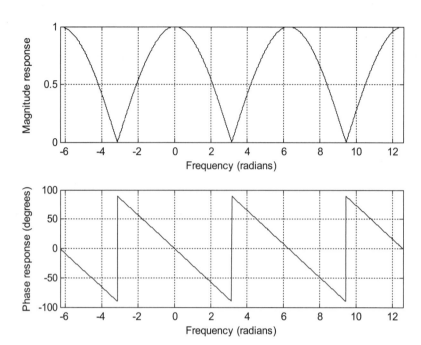

FIGURE 6.14

Periodicity of the magnitude response and phase response in Example 6.10.

EXAMPLE 6.11

Given a digital system

$$y(n) = x(n) - 0.5y(n-1)$$

with a sampling rate of 8,000 Hz, determine the frequency response.

Solution:

Taking the z-transform on both sides of the difference equation leads to

$$Y(z) = X(z) - 0.5z^{-1}Y(z)$$

Then the transfer function describing the system is easily found to be

$$H(z) = \frac{Y(z)}{X(z)} = \frac{1}{1 + 0.5z^{-1}} = \frac{z}{z + 0.5}$$

Substituting $z = e^{j\Omega}$, we have the frequency response as

$$H(e^{j\Omega}) = \frac{1}{1 + 0.5e^{-j\Omega}}$$

$$= \frac{1}{1 + 0.5\cos(\Omega) - j0.5\sin(\Omega)}$$

Table 6.2 Frequency Response Calculations in Example 6.11

Ω (radians)	$f = \dfrac{\Omega}{2\pi} f_s$ (Hz)	$\left\|H(e^{j\Omega})\right\|$	$\left\|H(e^{j\Omega})\right\|_{dB}$	$\angle H(e^{j\Omega})$
0	0	0.670	−3.479 dB	0^0
0.25π	1,000	0.715	−2.914 dB	14.64^0
0.50π	2,000	0.894	−0.973 dB	26.57^0
0.75π	3,000	1.357	2.652 dB	28.68^0
1.00π	4,000	2.000	6.021 dB	0^0

Therefore, the magnitude frequency response and phase response are given by

$$\left|H\left(e^{j\Omega}\right)\right| = \frac{1}{\sqrt{(1+0.5\cos(\Omega))^2+(0.5\sin(\Omega))^2}}$$

and

$$\angle H(e^{j\Omega}) = -\tan^{-1}\left(\frac{-0.5\sin(\Omega)}{1+0.5\cos(\Omega)}\right)$$

Several points for the magnitude response and phase response are calculated and shown in Table 6.2. The magnitude response and phase response of the DSP system are roughly plotted in Figure 6.15 in accordance with the obtained data.

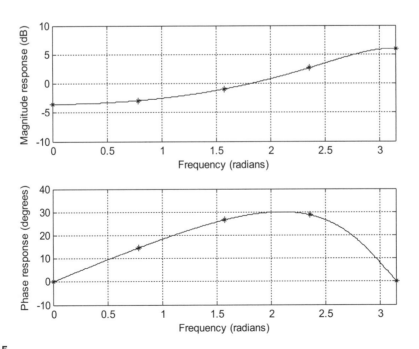

FIGURE 6.15

Frequency responses of the digital filter in Example 6.11.

From Table 6.2 and Figure 6.15, we can see that when the frequency increases, the magnitude response increases. The DSP system actually performs digital highpass filtering.

Notice that if all the coefficients a_i for $i = 0, 1, \cdots, M$ in Equation (6.1) are zeros, Equation (6.2) is reduced to

$$
\begin{aligned}
y(n) &= \sum_{i=0}^{M} b_i x(n - i) \\
&= b_0 x(n) + b_1 x(n - 1) + \cdots + b_K x(n - M)
\end{aligned}
\tag{6.25}
$$

Notice that b_i is the ith impulse response coefficient. Also, since M is a finite positive integer, b_i in this particular case is a finite set, $H(z) = B(z)$; note that the denominator $A(z) = 1$. Such systems are called *finite impulse response* (FIR) systems. If not all a_i in Equation (6.1) are zeros, the impulse response $h(i)$ would consist of an infinite number of coefficients. Such systems are called *infinite impulse response* (IIR) systems. The z-transform of the IIR $h(i)$, in general, is given by

$$
H(z) = \frac{B(z)}{A(z)}
$$

where $A(z) \neq 1$.

6.5 BASIC TYPES OF FILTERING

The basic filter types can be classified into four categories: *lowpass, highpass, ... bandpass,* and *bandstop.* Each of them finds a specific application in digital signal processing. One of the objectives in applications may involve the design of digital filters. In general, the filter is designed based on the specifications primarily for the passband, stopband, and transition band of the filter frequency response. The filter passband is the frequency range with the amplitude gain of the filter response being approximately unity. The filter stopband is defined as the frequency range over which the filter magnitude response is attenuated to eliminate the input signal whose frequency components are within that range. The transition band denotes the frequency range between the passband and stopband.

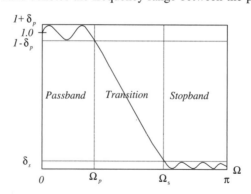

FIGURE 6.16

Magnitude response of the normalized lowpass filter.

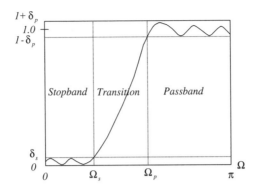

FIGURE 6.17

Magnitude response of the normalized highpass filter.

The design specifications of the lowpass filter are illustrated in Figure 6.16, where the low-frequency components are passed through the filter while the high-frequency components are attenuated. As shown in Figure 6.16, Ω_p and Ω_s are the passband cutoff frequency and the stopband cutoff frequency, respectively; δ_p is the design parameter to specify the ripple (fluctuation) of the frequency response in the passband; and δ_s specifies the ripple of the frequency response in the stopband.

The highpass filter keeps high-frequency components and rejects low-frequency components. The magnitude frequency response for the highpass filter is demonstrated in Figure 6.17.

The bandpass filter attenuates both low- and high-frequency components while keeping the middle-frequency components, as shown in Figure 6.18.

As illustrated in Figure 6.18, Ω_{pL} and Ω_{sL} are the lower passband cutoff frequency and lower stopband cutoff frequency, respectively. Ω_{pH} and Ω_{sH} are the upper passband cutoff frequency and upper stopband cutoff frequency, respectively. δ_p is the design parameter to specify the ripple of the frequency response in the passband, while δ_s specifies the ripple of the frequency response in the stopband.

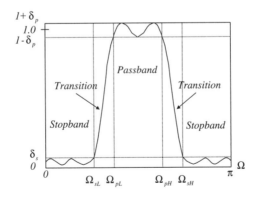

FIGURE 6.18

Magnitude response of the normalized bandpass filter.

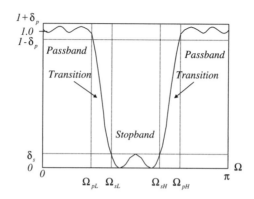

FIGURE 6.19

Magnitude of the normalized bandstop filter.

Finally, the bandstop (band reject or notch) filter shown in Figure 6.19 rejects the middle-frequency components and accepts both the low- and the high-frequency components.

As a matter of fact, all kinds of digital filters are implemented using FIR or IIR systems. Furthermore, the FIR and IIR systems can each be realized by various filter configurations, such as direct forms, cascade forms, and parallel forms. Such topics will be included in the next section.

Given a transfer function, the MATLAB function **freqz()** can be used to determine the frequency response. The syntax is given by

[h, w] = freqz(B, A, N)

where the parameters are defined as follows:

h = an output vector containing frequency response
w = an output vector containing normalized frequency values distributed in the range from 0 to π radians
B = an input vector for numerator coefficients
A = an input vector for denominator coefficients
N = the number of normalized frequency points used for calculating the frequency response

Let's consider Example 6.12.

EXAMPLE 6.12

Consider the following digital transfer function:

a. $H(z) = \dfrac{z}{z - 0.5}$

b. $H(z) = 1 - 0.5z^{-1}$

c. $H(z) = \dfrac{0.5z^2 - 0.32}{z^2 - 0.5z + 0.25}$

d. $H(z) = \dfrac{1 - 0.9z^{-1} + 0.81z^{-2}}{1 - 0.6z^{-1} + 0.36z^{-2}}$

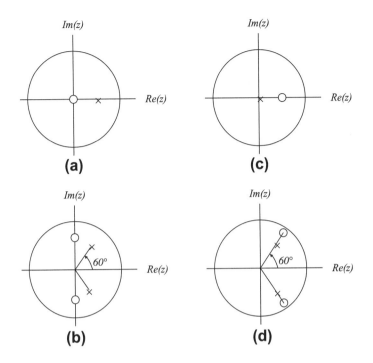

FIGURE 6.20

Pole-zero plots of Example 6.12.

1. Plot the poles and zeros on the z-plane.
2. Use the MATLAB function **freqz()** to plot the magnitude frequency response and phase response for each transfer function.
3. Identify the corresponding filter type (e.g., lowpass, highpass, bandpass, or bandstop).

Solution:
1. The pole-zero plot for each transfer function is demonstrated in Figure 6.20. The transfer functions of (a) and (c) need to be converted into the standard form (delay form) required by the MATLAB function **freqz()**, in which both numerator and denominator polynomials have negative powers of z. Hence, we obtain

$$H(z) = \frac{z}{z - 0.5} = \frac{1}{1 - 0.5z^{-1}}$$

$$H(z) = \frac{0.5z^2 - 0.32}{z^2 - 0.5z + 0.25} = \frac{0.5 - 0.32z^{-2}}{1 - 0.5z^{-1} + 0.25z^{-2}}$$

while the transfer functions of (b) and (d) are already in their standard forms (delay forms).
2. The MATLAB program for plotting the magnitude frequency response and the phase response for each case is listed in Program 6.2.
3. From the plots in Figures 6.21A–6.21D of magnitude frequency responses for all cases, we can conclude that case (a) is a low pass filter, (b) is a high pass filter, (c) is a bandpass filter, and (d) is a bandstop (band reject) filter.

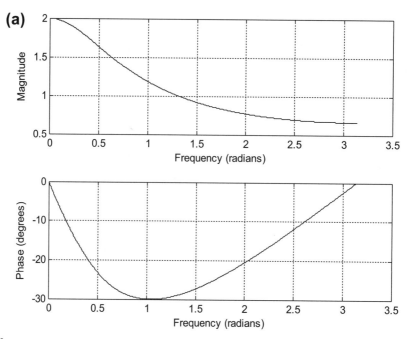

(a)

FIGURE 6.21A

Plots of frequency responses for Example 6.12 for (a).

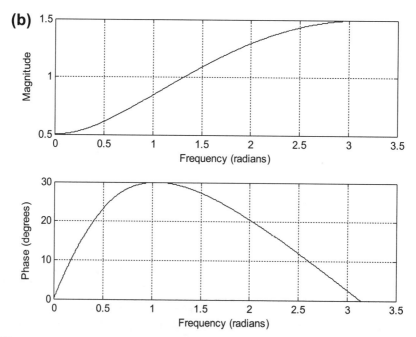

(b)

FIGURE 6.21B

Plots of frequency responses for Example 6.12 for (b).

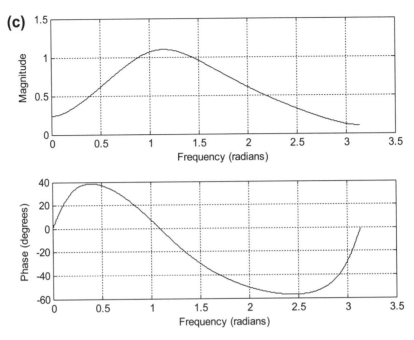

FIGURE 6.21C

Plots of frequency responses for Example 6.12 for (c).

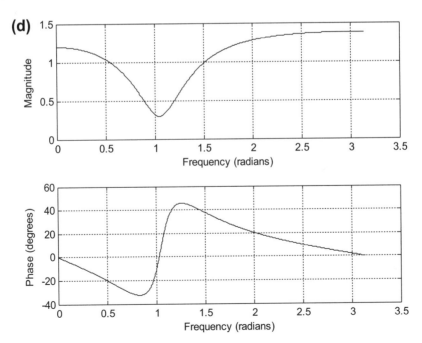

FIGURE 6.21D

Plots of frequency responses for Example 6.12 for (d).

Program 6.2. MATLAB program for Example 6.12.

```
% Example 6.12
% Plot the frequency response and phase response
% Case a
  figure (1)
  [h w] = freqz([1],[1 -0.5],1024);  % Calculate frequency response
  phi=180*unwrap(angle(h))/pi;
  subplot(2,1,1), plot(w,abs(h)),grid; xlabel('Frequency (radians)'),
ylabel('Magnitude')
subplot(2,1,2), plot(w,phi),grid; xlabel('Frequency (radians)'), ylabel('Phase
(degrees)')
  % Case b
  figure (2)
  [h w] = freqz([1 -0.5],[1],1024);  % Calculate frequency response
  phi=180*unwrap(angle(h))/pi;
  subplot(2,1,1), plot(w,abs(h)),grid;xlabel('Frequency (radians)'),
ylabel('Magnitude')
  subplot(2,1,2), plot(w,phi),grid; xlabel('Frequency (radians)'), ylabel('Phase
(degrees)')
  % Case c
  figure (3)
  [h w] = freqz([0.5 0 -0.32],[1 -0.5 0.25],1024); % Calculate frequency response
  phi=180*unwrap(angle(h))/pi;
subplot(2,1,1), plot(w,abs(h)),grid;
xlabel('Frequency (radians)'),ylabel('Magnitude')
subplot(2,1,2), plot(w,phi),grid;
xlabel('Frequency (radians)'), ylabel('Phase (degrees)')
  % Case d
  figure (4)
  [h w] = freqz([1 -0.9 0.81], [1 -0.6 0.36],1024); % Calculate frequency response
  phi=180*unwrap(angle(h))/pi;
  subplot(2,1,1), plot(w,abs(h)),grid; xlabel('Frequency (radians)'),
ylabel('Magnitude')
  subplot(2,1,2), plot(w,phi),grid; xlabel('Frequency (radians)'), ylabel('Phase
(degrees)')
  %
```

6.6 REALIZATION OF DIGITAL FILTERS

In this section, basic realization methods for digital filters are discussed. Digital filters described by the transfer function $H(z)$ may be generally realized into the following forms:

- Direct-form I
- Direct-form II
- Cascade
- Parallel

(The reader can explore various lattice realizations in the textbook by Stearns and Hush [1990].)

6.6.1 Direct-Form I Realization

As we know, a digital filter transfer function, $H(z)$, is given by

$$H(z) = \frac{B(z)}{A(z)} = \frac{b_0 + b_1 z^{-1} + \cdots + b_M z^{-M}}{1 + a_1 z^{-1} + \cdots + a_N z^{-N}} \qquad (6.26)$$

Let $x(n)$ and $y(n)$ be the digital filter input and output, respectively. We can express the relationship in z-transform domain as

$$Y(z) = H(z)X(z) \qquad (6.27)$$

where $X(z)$ and $Y(z)$ are the z-transforms of $x(n)$ and $y(n)$, respectively. If we substitute Equation (6.26) into $H(z)$ in Equation (6.27), we have

$$Y(z) = \left(\frac{b_0 + b_1 z^{-1} + \cdots + b_M z^{-M}}{1 + a_1 z^{-1} + \cdots + a_N z^{-N}} \right) X(z) \qquad (6.28)$$

Taking the inverse of the z-transform of Equation (6.28), we yield the relationship between input $x(n)$ and output $y(n)$ in the time domain, as follows:

$$y(n) = b_0 x(n) + b_1 x(n-1) + \cdots + b_M x(n-M)$$
$$-a_1 y(n-1) - a_2 y(n-2) - \cdots - a_N y(n-N) \qquad (6.29)$$

This difference equation thus can be implemented by the direct-form I realization shown in Figure 6.22(a). Figure 6.22(b) illustrates the realization of the second-order IIR filter ($M = N = 2$). Note that the notation used in Figures 6.22(a) and (b) are defined in Figure 22(c) and will be applied for discussion of other realizations.

Notice that any of the a_j and b_i can be zero, thus not all the paths need to exist for realization.

6.6.2 Direct-Form II Realization

Considering Equations (6.26) and (6.27) with $N = M$, we can express

$$Y(z) = H(z)X(z) = \frac{B(z)}{A(z)}X(z) = B(z)\left(\frac{X(z)}{A(z)} \right)$$

$$= (b_0 + b_1 z^{-1} + \cdots + b_M z^{-M}) \underbrace{\left(\frac{X(z)}{1 + a_1 z^{-1} + \cdots + a_M z^{-M}} \right)}_{W(z)} \qquad (6.30)$$

(a)

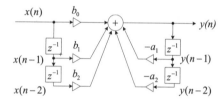

(b)

(c)

$b_i \cdot x(n)$

*This denotes that the output is the product
of the weight b_i and input $x(n)$; that is, $b_i \cdot x(n)$*

$x(n)$ $x(n-1)$

z^{-1}

*This denotes a unit delay element, which
implies that the output of this stage is $x(n-1)$*

FIGURE 6.22

(a) Direct-form I realization; (b) direct-form I realization with $M = 2$; (c) notation.

Also, if we define a new z-transform function as

$$W(z) = \frac{X(z)}{1 + a_1 z^{-1} + \cdots + a_M z^{-M}} \tag{6.31}$$

we have

$$Y(z) = (b_0 + b_1 z^{-1} + \cdots + b_M z^{-M})W(z) \tag{6.32}$$

The corresponding difference equations for Equations (6.31) and (6.32), respectively, become

$$w(n) = x(n) - a_1 w(n-1) - a_2 w(n-2) - \cdots - a_M w(n-M) \tag{6.33}$$

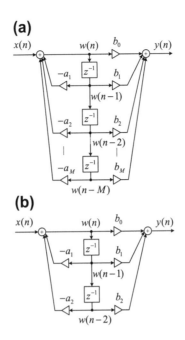

(a)

(b)

FIGURE 6.23

(a) Direct-form II realization; (b) direct-form II realization with $M = 2$.

and

$$y(n) = b_0 w(n) + b_1 w(n-1) + \ldots + b_M w(n-M) \qquad (6.34)$$

Realization of Equations (6.33) and (6.34) produces another direct-form II realization, which is demonstrated in Figure 6.23(a). Again, the corresponding realization of the second-order IIR filter is described in Figure 6.23(b). Note that in Figure 6.23(a), the variables $w(n)$, $w(n-1)$, $w(n-2),\ldots$, $w(n-M)$ are different from the filter inputs $x(n-1)$, $x(n-2)$, ..., $x(n-M)$.

6.6.3 Cascade (Series) Realization

An alternate way to filter realization is to cascade the factorized $H(z)$ in the following form:

$$H(z) = H_1(z) \cdot H_2(z) \cdots H_k(z) \qquad (6.35)$$

where $H_k(z)$ is chosen to be the first- or second-order transfer function (section), which is defined by

$$H_k(z) = \frac{b_{k0} + b_{k1}z^{-1}}{1 + a_{k1}z^{-1}} \qquad (6.36)$$

or

$$H_k(z) = \frac{b_{k0} + b_{k1}z^{-1} + b_{k2}z^{-2}}{1 + a_{k1}z^{-1} + a_{k2}z^{-2}} \qquad (6.37)$$

respectively. The block diagram of the cascade, or series, realization is depicted in Figure 6.24.

$$x(n) \longrightarrow \boxed{H_1(z)} \longrightarrow \boxed{H_2(z)} \longrightarrow \cdots \longrightarrow \boxed{H_k(z)} \longrightarrow y(n)$$

FIGURE 6.24

Cascade realization.

6.6.4 Parallel Realization

Now we convert $H(z)$ into the form

$$H(z) = H_1(z) + H_2(z) + \cdots + H_k(z) \tag{6.38}$$

where $H_k(z)$ is defined as the first- or second-order transfer function (section) given by

$$H_k(z) = \frac{b_{k0}}{1 + a_{k1}z^{-1}} \tag{6.39}$$

or

$$H_k(z) = \frac{b_{k0} + b_{k1}z^{-1}}{1 + a_{k1}z^{-1} + a_{k2}z^{-2}} \tag{6.40}$$

respectively. The resulting parallel realization is illustrated in the block diagram in Figure 6.25.

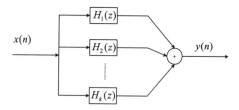

FIGURE 6.25

Parallel realization.

EXAMPLE 6.13

Given a second-order transfer function

$$H(z) = \frac{0.5(1 \quad z^{-2})}{1 + 1.3z^{-1} + 0.36z^{-2}}$$

perform the filter realizations and write difference equations using the following realizations:

a. direct-form I and direct-form II;
b. cascade form via first-order sections;
c. parallel form via first-order sections.

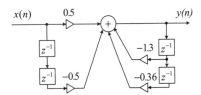

FIGURE 6.26

Direct-form I realization for Example 6.13.

Solution:

a. To perform the filter realizations using direct-form I and direct-form II, we rewrite the given second-order transfer function as

$$H(z) = \frac{0.5 - 0.5z^{-2}}{1 + 1.3z^{-1} + 0.36z^{-2}}$$

and identify that

$$a_1 = 1.3,\ a_2 = 0.36,\ b_0 = 0.5,\ b_1 = 0,\ \text{and}\ b_2 = -0.5$$

Based on the realizations in Figure 6.22, we sketch the direct-form I realization in Figure 6.26. The difference equation for the direct-form I realization is given by

$$y(n) = 0.5x(n) - 0.5x(n-2) - 1.3y(n-1) - 0.36y(n-2)$$

Using the direct-form II realization shown in Figure 6.23, we present the realization in Figure 6.27. The difference equations for the direct-form II realization are expressed as

$$w(n) = x(n) - 1.3w(n-1) - 0.36w(n-2)$$

$$y(n) = 0.5w(n) - 0.5w(n-2)$$

b. To achieve the cascade (series) form realization, we factor $H(z)$ into two first-order sections to yield

$$H(z) = \frac{0.5(1 - z^{-2})}{1 + 1.3z^{-1} + 0.36z^{-2}} = \frac{0.5 - 0.5z^{-1}}{1 + 0.4z^{-1}} \frac{1 + z^{-1}}{1 + 0.9z^{-1}}$$

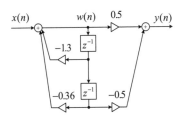

FIGURE 6.27

Direct-form II realization for Example 6.13.

where $H_1(z)$ and $H_2(z)$ are chosen to be

$$H_1(z) = \frac{0.5 - 0.5z^{-1}}{1 + 0.4z^{-1}}$$

$$H_2(z) = \frac{1 + z^{-1}}{1 + 0.9z^{-1}}$$

Notice that the obtained $H_1(z)$ and $H_2(z)$ are not the unique selections for realization. For example, there is another way of choosing $H_1(z) = \dfrac{0.5 - 0.5z^{-1}}{1 + 0.9z^{-1}}$ and $H_2(z) = \dfrac{1 + z^{-1}}{1 + 0.4z^{-1}}$ to yield the same $H(z)$. Using the $H_1(z)$ and $H_2(z)$ we have obtained, and with the direct-form II realization, we achieve the cascade form depicted in Figure 6.28.

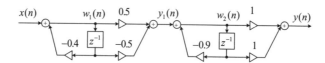

FIGURE 6.28

Cascade realization for Example 6.13.

The difference equations for the direct-form II realization have two cascaded sections, expressed as
Section 1:

$$w_1(n) = x(n) - 0.4w(n-1)$$

$$y_1(n) = 0.5w_1(n) - 0.5w_1(n-1)$$

Section 2:

$$w_2(n) = y_1(n) - 0.9w_2(n-1)$$

$$y(n) = w_2(n) + w_2(n-1)$$

c. In order to yield the parallel form of realization, we need to make use of the partial fraction expansion, and we first let

$$\frac{H(z)}{z} = \frac{0.5(z^2 - 1)}{z(z + 0.4)(z + 0.9)} = \frac{A}{z} + \frac{B}{z + 0.4} + \frac{C}{z + 0.9}$$

where

$$A = z\left(\frac{0.5(z^2 - 1)}{z(z + 0.4)(z + 0.9)}\right)\bigg|_{z=0} = \frac{0.5(z^2 - 1)}{(z + 0.4)(z + 0.9)}\bigg|_{z=0} = -1.39$$

$$B = (z + 0.4)\left(\frac{0.5(z^2 - 1)}{z(z + 0.4)(z + 0.9)}\right)\bigg|_{z=-0.4} = \frac{0.5(z^2 - 1)}{z(z + 0.9)}\bigg|_{z=-0.4} = 2.1$$

$$C = \left(z + 0.9\right)\left(\frac{0.5(z^2 - 1)}{z(z + 0.4)(z + 0.9)}\right)\bigg|_{z=-0.9} = \frac{0.5(z^2 - 1)}{z(z + 0.4)}\bigg|_{z=-0.9} = -0.21$$

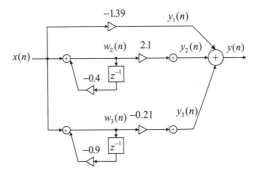

FIGURE 6.29

Parallel realization for Example 6.13.

Therefore

$$H(z) = -1.39 + \frac{2.1z}{z+0.4} + \frac{-0.21z}{z+0.9} = -1.39 + \frac{2.1}{1+0.4z^{-1}} + \frac{-0.21}{1+0.9z^{-1}}$$

Again, using the direct-form II realization for each section, we obtain the parallel realization in Figure 6.29. The difference equations for the direct-form II realization have three parallel sections, expressed as

$$y_1(n) = -1.39x(n)$$

$$w_2(n) = x(n) - 0.4w_2(n-1)$$

$$y_2(n) = 2.1w_2(n)$$

$$w_3(n) = x(n) - 0.9w_3(n-1)$$

$$y_3(n) = -0.21w_3(n)$$

$$y(n) = y_1(n) + y_2(n) + y_3(n)$$

In practice, the second-order filter module with the direct-form I or direct-form II realization is used. The high-order filter can be factored in the cascade form with the first- or second-order sections. In cases where the first order-filter is required, we can still modify the second-order filter module by setting the corresponding filter coefficients to be zero.

6.7 APPLICATION: SIGNAL ENHANCEMENT AND FILTERING

This section investigates applications of signal enhancement using a pre-emphasis filter and speech filtering using a bandpass filter. Enhancement also includes biomedical signals such as electrocardiogram (ECG) signals.

6.7.1 Pre-Emphasis of Speech

A speech signal may have frequency components that fall off at high frequencies. In some applications such as speech coding, to avoid overlooking the high frequencies, the high frequency components are compensated using pre-emphasis filtering. A simple digital filter used for such compensation is given as

$$y(n) = x(n) - \alpha x(n-1) \qquad (6.41)$$

where α is the positive parameter to control the degree of pre-emphasis filtering and usually is chosen to be less than 1. The filter described in Equation (6.41) is essentially a highpass filter. Applying z-transform on both sides of Equation (6.41) and solving for the transfer function, we have

$$H(z) = 1 - \alpha z^{-1} \qquad (6.42)$$

The magnitude and phase responses adopting the pre-emphasis parameter $\alpha = 0.9$ and the sampling rate $f_s = 8{,}000$ Hz are plotted in Figure 6.30A using MATLAB.

Figure 6.30B compares the original speech waveform and the pre-emphasized speech using the filter in Equation (6.42). Again, we apply the fast Fourier transform (FFT) to estimate the spectrum of

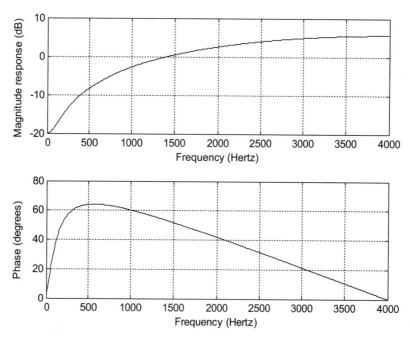

FIGURE 6.30A

Frequency responses of the pre-emphasis filter.

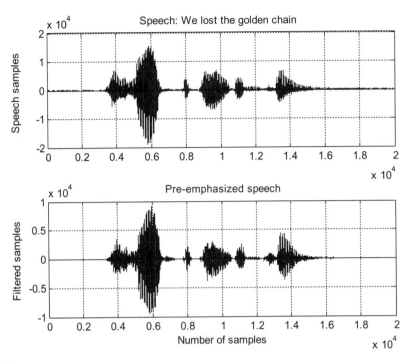

FIGURE 6.30B

Original speech and pre-emphasized speech waveforms.

the original speech and the spectrum of the pre-emphasized speech. The plots are displayed in Figure 6.31.

From Figure 6.31, we can conclude that the filter does its job to boost the high-frequency components and attenuate the low-frequency components. We can also try this filter with different values of α to examine the degree of the pre-emphasis filtering of the digitally recorded speech. The MATLAB list is in Program 6.3.

Program 6.3. MATLAB program for pre-emphasis of speech.

```
% MATLAB program for Figures 6.30 and 6.31
close all;clear all
fs=8000;                        % Sampling rate
alpha =0.9;                     % Degree of pre-emphasis
figure(1);
freqz([1 -alpha],1,512,fs);     % Calculate and display frequency response
load speech.dat
figure(2);
y=filter([1 -alpha],1,speech);  % Filtering speech
subplot(2,1,1),plot(speech);grid;
ylabel('Speech samples')
```

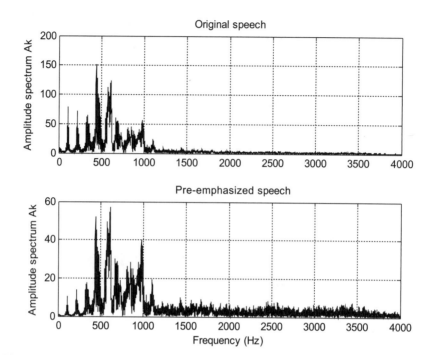

FIGURE 6.31

Amplitude spectral plots for the original speech and pre-emphasized speech.

```
title('Speech: We lost the golden chain.')
subplot(2,1,2),plot(y);grid
ylabel('Filtered samples')
xlabel('Number of samples');
title('Preemphasized speech.')
figure(3);
N=length(speech);                      % Length of speech
Axk=abs(fft(speech.*hamming(N)'))/N;   % Two-sided spectrum of speech
Ayk=abs(fft(y.*hamming(N)'))/N;        % Two-sided spectrum of pre-emphasized speech
f=[0:N/2]*fs/N;
Axk(2:N)=2*Axk(2:N);                   % Get one-sided spectrum of speech
Ayk(2:N)=2*Ayk(2:N);                   % Get one-sided spectrum of filtered speech
subplot(2,1,1),plot(f,Axk(1:N/2+1));grid
ylabel('Amplitude spectrum Ak')
title('Original speech');
subplot(2,1,2),plot(f,Ayk(1:N/2+1));grid
ylabel('Amplitude spectrum Ak')
xlabel('Frequency (Hz)');
title('Preemphasized speech');
%
```

6.7.2 **Bandpass Filtering of Speech**

Bandpass filtering plays an important role in DSP applications. It can be used to pass the signals according to the specified frequency passband and reject the frequency other than the passband specification. Then the filtered signal can be further used for the signal feature extraction. Filtering can also be applied to perform applications such as noise reduction, frequency boosting, digital audio equalizing, and digital crossover, among others.

Let us consider the following digital fourth-order bandpass Butterworth filter with a lower cutoff frequency of 1,000 Hz, an upper cutoff frequency of 1,400 Hz (that is, the bandwidth is 400 Hz), and a sampling rate of 8,000 Hz:

$$H(z) = \frac{0.0201 - 0.0402z^{-2} + 0.0201z^{-4}}{1 - 2.1192z^{-1} + 2.6952z^{-2} - 1.6924z^{-3} + 0.6414z^{-4}} \tag{6.43}$$

Converting the z-transfer function into the DSP difference equation yields

$$y(n) = 0.0201x(n) - 0.0402x(n-2) + 0.0201x(n-4)$$
$$+2.1192y(n-1) - 2.6952y(n-2) + 1.6924y(n-3) - 0.6414y(n-4) \tag{6.44}$$

The filter frequency responses are computed and plotted in Figure 6.32A with MATLAB. Figure 6.32B shows the original speech and filtered speech, while Figure 6.32C displays the spectral plots for the original speech and filtered speech.

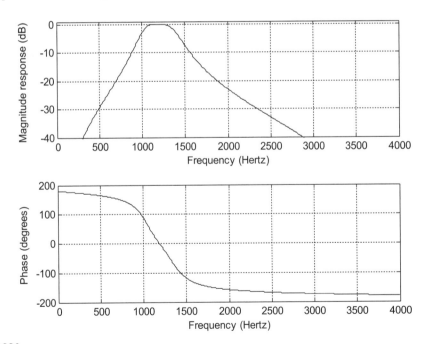

FIGURE 6.32A

Frequency responses of the designed bandpass filter.

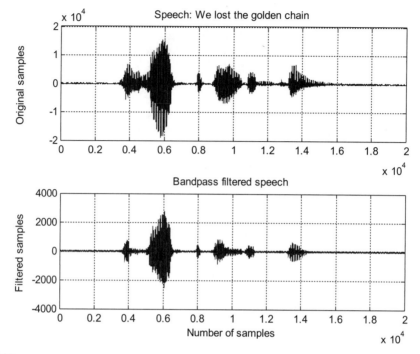

FIGURE 6.32B

Plots of the original speech and filtered speech.

As shown in Figure 6.32C, the designed bandpass filter significantly reduces low-frequency components, which are less than 1,000 Hz, and the high-frequency components above 1,400 Hz, while letting the signals with the frequencies ranging from 1,000 Hz to 1,400 Hz pass through the filter. Similarly, we can design and implement other types of filters, such as lowpass, highpass, bandpass, and band reject (bandstop) to filter the signals and examine the performance of their designs. MATLAB implementation details are given in Program 6.4.

Program 6.4. MATLAB program for bandpass filtering of speech.

```
fs=8000;                             % Sampling rate
freqz([0.0201 0.00 -0.0402 0 0.0201],[1 -2.1192 2.6952 -1.6924 0.6414],512,fs);
axis([0 fs/2 -40 1]);                % Frequency response of bandpass filter
figure
load speech.dat
y=filter([0.0201 0.00 -0.0402 0.0201],[1 -2.1192 2.6952 -1.6924 0.6414],speech);
subplot(2,1,1),plot(speech); grid;   % Filtering speech
ylabel('Origibal Samples')
title('Speech: We lost the golden chain.')
subplot(2,1,2),plot(y);grid
xlabel('Number of Samples');ylabel('Filtered Samples')
```

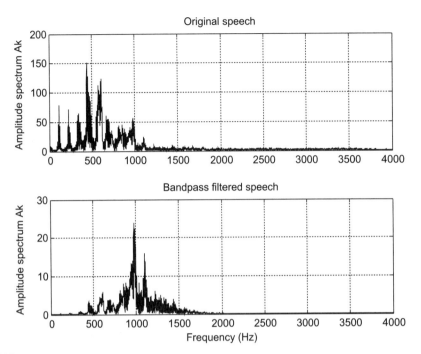

FIGURE 6.32C

Amplitude spectra of the original speech and bandpass filtered speech.

```
title('Bandpass filtered speech.')
figure
N=length(speech);
Axk=abs(fft(speech.*hamming(N)'))/N;        % One-sided spectrum of speech
Ayk=abs(fft(y.*hamming(N)'))/N;             % One-sided spectrum of filtered speech
f=[0:N/2]*fs/N;
Axk(2:N)=2*Axk(2:N);Ayk(2:N)=2*Ayk(2:N);    % One-sided spectra
subplot(2,1,1),plot(f,Axk(1:N/2+1));grid
ylabel('Amplitude spectrum Ak')
title('Original speech');
subplot(2,1,2),plot(f,Ayk(1:N/2+1));grid
ylabel('Amplitude spectrum Ak');xlabel('Frequency (Hz)');
title('Bandpass filtered speech');
```

6.7.3 Enhancement of ECG Signal Using Notch Filtering

A notch filter is a bandstop filter with a very narrow bandwidth. It can be applied to enhance an ECG signal that is corrupted during the data acquisition stage, where the signal is exposed to 60-Hz interference induced from the power line. Let us consider the following digital second-order notch

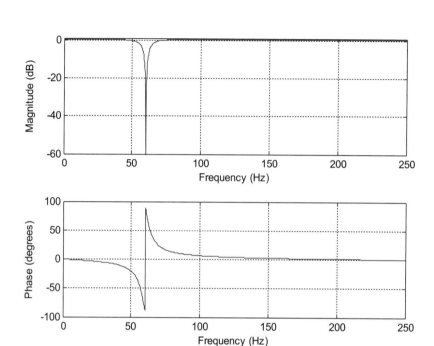

FIGURE 6.33

Notch filter frequency responses.

filter with a notch frequency of 60 Hz where the digital system has a sampling frequency of 500 Hz. We obtain a notch filter (details can be found in Chapter 8) as follows:

$$H(z) = \frac{1 - 1.4579z^{-1} + z^{-2}}{1 - 1.3850z^{-1} + 0.9025z^{-2}} \tag{6.45}$$

The DSP difference equation is expressed as

$$y(n) = x(n) - 1.4579x(n-1) + 1.3850y(n-1) - 0.9025y(n-2) \tag{6.46}$$

The frequency responses are computed and plotted in Figure 6.33. Comparisons of the raw ECG signal corrupted by 60-Hz interference with the enhanced ECG signal for both the time domain and frequency domain are displayed in Figures 6.34 and 6.35, respectively. As we can see, the notch filter completely removes the 60-Hz interference.

6.8 SUMMARY

1. The digital filter (DSP system) is represented by a difference equation, which is linear and time invariant.
2. The filter output depends on the filter current input, past input(s), and past output(s) in general. Given arbitrary inputs and nonzero or zero initial conditions, operating the difference equation can generate the filter output recursively.

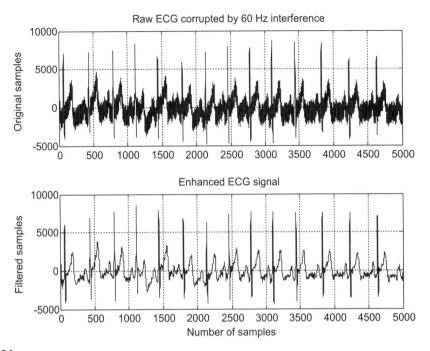

FIGURE 6.34

The corrupted ECG signal and the enhanced ECG signal.

3. System responses such as the impulse response and step response can be determined analytically using the z-transform.
4. The transfer function can be obtained by applying z-transform to the difference equation to determine the ratio of the output z-transform over the input z-transform. A digital filter (DSP system) can be represented by its transfer function.
5. System stability can be studied using a very useful tool, a z-plane pole-zero plot.
6. The frequency response of the DSP system was developed and illustrated to investigate the magnitude and phase responses. In addition, the FIR (finite impulse response) and IIR (infinite impulse response) systems were defined.
7. Digital filters and their specifications, such as lowpass, highpass, bandpass, and bandstop, were reviewed.
8. A digital filter can be realized using standard realization methods such as direct-form I; direct-form II; cascade, or series form; and parallel form.
9. Digital processing of speech using the pre-emphasis filter and bandpass filter was investigated to study spectral effects of the processed digital speech. The pre-emphasis filter boosts the high-frequency components, while bandpass filtering keeps the midband frequency components and rejects other lower- and upper-band frequency components.

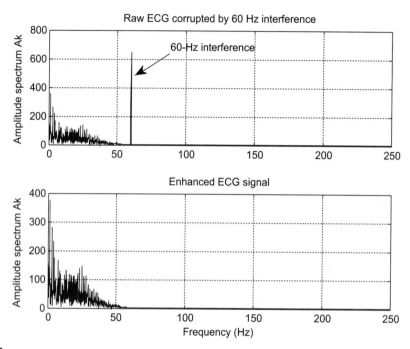

FIGURE 6.35

The corrupted ECG signal spectrum and the enhanced ECG signal spectrum.

6.9 PROBLEMS

6.1. Given the difference equation

$$y(n) = x(n) - 0.5y(n-1)$$

a. calculate the system response $y(n)$ for $n = 0, 1, ..., 4$ with the input $x(n) = (0.5)^n u(n)$ and initial condition $y(-1) = 1$;

b. calculate the system response $y(n)$ for $n = 0, 1, ..., 4$ with the input $x(n) = (0.5)^n u(n)$ and zero initial condition $y(-1) = 0$

6.2. Given the difference equation

$$y(n) = 0.5x(n-1) + 0.6y(n-1)$$

a. calculate the system response $y(n)$ for $n = 0, 1, ..., 4$ with the input $x(n) = (0.5)^n u(n)$ and initial conditions $x(-1) = -1$, and $y(-1) = 1$;

b. calculate the system response $y(n)$ for $n = 0, 1, ..., 4$ with the input $x(n) = (0.5)^n u(n)$ and zero initial conditions $x(-1) = 0$, and $y(-1) = 0$.

6.3. Given the difference equation

$$y(n) = x(n-1) - 0.75y(n-1) - 0.125y(n-2)$$

a. calculate the system response $y(n)$ for $n = 0, 1, ..., 4$ with the input $x(n) = (0.5)^n u(n)$ and initial conditions: $x(-1) = -1$, $y(-2) = 2$, and $y(-1) = 1$;

b. calculate the system response $y(n)$ for $n = 0, 1, ..., 4$ with the input $x(n) = (0.5)^n u(n)$ and zero initial conditions: $x(-1) = 0$, $y(-2) = 0$, and $y(-1) = 0$.

6.4. Given the difference equation

$$y(n) = 0.5x(n) + 0.5x(n-1)$$

a. find the $H(z)$;

b. determine the impulse response $y(n)$ if the input $x(n) = 4\delta(n)$;

c. determine the step response $y(n)$ if the input $x(n) = 10u(n)$.

6.5. Given the difference equation,

$$y(n) = x(n) - 0.5y(n-1)$$

a. find the $H(z)$;

b. determine the impulse response $y(n)$ if the input $x(n) = \delta(n)$;

c. determine the step response $y(n)$ if the input $x(n) = u(n)$.

6.6. A digital system is described by the following difference equation:

$$y(n) = x(n) - 0.25x(n-2) - 1.1y(n-1) - 0.28y(n-2)$$

Find the transfer function $H(z)$, the denominator polynomial $A(z)$, and the numerator polynomial $B(z)$.

6.7. A digital system is described by the following difference equation:

$$y(n) = 0.5x(n) + 0.5x(n-1) - 0.6y(n-2)$$

Find the transfer function $H(z)$, the denominator polynomial $A(z)$, and the numerator polynomial $B(z)$.

6.8. A digital system is described by the following difference equation:

$$y(n) = 0.25x(n-2) + 0.5y(n-1) - 0.2y(n-2)$$

Find the transfer function $H(z)$, the denominator polynomial $A(z)$, and the numerator polynomial $B(z)$.

6.9. A digital system is described by the following difference equation:

$$y(n) = x(n) - 0.3x(n-1) + 0.28x(n-2)$$

Find the transfer function $H(z)$, the denominator polynomial $A(z)$, and the numerator polynomial $B(z)$.

6.10. Convert each of the following transfer functions into difference equations:

a. $H(z) = 0.5 + 0.5z^{-1}$

b. $H(z) = \dfrac{1}{1 - 0.3z^{-1}}$

6.11. Convert each of the following transfer functions into difference equations:

a. $H(z) = 0.1 + 0.2z^{-1} + 0.3z^{-2}$

b. $H(z) = \dfrac{0.5 - 0.5z^{-2}}{1 - 0.3z^{-1} + 0.8z^{-2}}$

6.12. Convert each of the following transfer functions into difference equations:

a. $H(z) = \dfrac{z^2 - 0.25}{z^2 + 1.1z + 0.18}$

b. $H(z) = \dfrac{z^2 - 0.1z + 0.3}{z^3}$

6.13. Convert each of the following transfer functions into pole-zero form:

a. $H(z) = \dfrac{z^2 + 2z + 1}{z^2 + 5z + 6}$

b $H(z) = \dfrac{1 - 0.16z^{-2}}{1 + 0.7z^{-1} + 0.1z^{-2}}$

c $H(z) = \dfrac{z^2 + 4z + 5}{z^3 + 2z^2 + 6z}$

6.14. A transfer function depicting a discrete-time system is given by

$$H(z) = \frac{10(z + 1)}{(z + 0.75)}$$

a. Determine the impulse response $h(n)$ and step response.

b. Determine the system response $y(n)$ if the input is $x(n) = (0.25)^n u(n)$.

6.15. Given each of the following transfer functions that describe digital systems, sketch the z-plane pole-zero plot and determine the stability for each digital system.

a. $H(z) = \dfrac{z - 0.5}{(z + 0.25)(z^2 + z + 0.8)}$

b. $H(z) = \dfrac{z^2 + 0.25}{(z - 0.5)(z^2 + 4z + 7)}$

c. $H(z) = \dfrac{z + 0.95}{(z + 0.2)(z^2 + 1.414z + 1)}$

d. $H(z) = \dfrac{z^2 + z + 0.25}{(z - 1)(z + 1)^2(z - 0.36)}$

6.16. Given the digital system
$$y(n) = 0.5x(n) + 0.5x(n-2)$$
with a sampling rate of 8,000 Hz,

a. determine the frequency response;

b. calculate and plot the magnitude and phase frequency responses;

c. determine the filter type based the magnitude frequency response.

6.17. Given the digital system,
$$y(n) = 0.5x(n-1) + 0.5x(n-2)$$
with a sampling rate of 8,000 Hz,

a. determine the frequency response;

b. calculate and plot the magnitude and phase frequency responses;

c. determine the filter type based the magnitude frequency response.

6.18. For the digital system
$$y(n) = 0.5x(n) + 0.5y(n-1)$$

with a sampling rate of 8,000 Hz,

a. determine the frequency response;

b. calculate and plot the magnitude and phase frequency responses;

c. determine the filter type based the magnitude frequency response.

6.19. For the digital system
$$y(n) = x(n) - 0.5y(n-2)$$

with a sampling rate of 8,000 Hz,

a. determine the frequency response;

b. calculate and plot the magnitude and phase frequency responses;

c. determine the filter type based the magnitude frequency response.

6.20. Given the difference equation
$$y(n) = x(n) - 2 \cdot \cos{(\alpha)}x(n-1) + x(n-2) + 2\gamma \cdot \cos{(\alpha)} - \gamma^2$$
where $\gamma = 0.8$ and $\alpha = 60^0$,

a. find the transfer function $H(z)$;

b. plot the poles and zeros on the z-plan with the unit circle;

c. determine the stability of the system from the pole-zero plot;

d. calculate the amplitude (magnitude) response of $H(z)$;

e. calculate the phase response of $H(z)$.

6.21. For the difference equations

    a. $y(n) = 0.5x(n) + 0.5x(n-1)$

    b $y(n) = 0.5x(n) - 0.5x(n-1)$

    c. $y(n) = 0.5x(n) + 0.5x(n-2)$

    d. $y(n) = 0.5x(n) - 0.5x(n-2)$

    1. find $H(z)$;

    2. calculate the magnitude response;

    3. specify the filtering type based on the calculated magnitude response.

6.22. Given an IIR system expressed as

$$y(n) = 0.5x(n) + 0.2y(n-1), \quad y(-1) = 0$$

    a. find $H(z)$;

    b. find the system response $y(n)$ due to the input $x(n) = (0.5)^n u(n)$.

6.23. Given the IIR system

$$y(n) = 0.5x(n) - 0.7y(n-1) - 0.1y(n-2)$$

    with zero initial conditions,

    a. find $H(z)$;

    b. find the unit step response.

6.24. Given the first-order IIR system

$$H(z) = \frac{1 + 2z^{-1}}{1 - 0.5z^{-1}}$$

    realize $H(z)$ and develop the difference equations using the following forms:

    a. direct-form I;

    b. direct-form II.

6.25. Given the filter

$$H(z) = \frac{1 - 0.9z^{-1} - 0.1z^{-2}}{1 + 0.3z^{-1} - 0.04z^{-2}}$$

    realize $H(z)$ and develop difference equations using the following forms:

    a. direct-form I;

    b. direct-form II;

c. cascade (series) form via the first-order sections;

d. parallel form via the first-order sections.

6.26. Given the pre-emphasis filters:

$$H(z) = 1 - 0.5z^{-1}$$

$$H(z) = 1 - 0.7z^{-1}$$

$$H(z) = 1 - 0.9z^{-1}$$

a. write the difference equation for each;

b. determine which emphasizes high frequency components most.

6.9.1 MATLAB Problems

6.27. Given a filter

$$H(z) = \frac{1 + 2z^{-1} + z^{-2}}{1 - 0.5z^{-1} + 0.25z^{-2}}$$

use MATLAB to plot

a. its magnitude frequency response;

b. its phase response.

6.28. Given a difference equation

$$y(n) = x(n-1) - 0.75y(n-1) - 0.125y(n-2)$$

a. use the MATLAB functions **filter()** and **filtic()** to calculate the system response $y(n)$ for $n = 0, 1, ..., 4$ with the input of $x(n) = (0.5)^n u(n)$ and initial conditions $x(-1) = -1$, $y(-2) = 2$, and $y(-1) = 1$;

b. use the MATLAB function **filter()** to calculate the system response $y(n)$ for $n = 0, 1, ..., 4$ with the input of $x(n) = (0.5)^n u(n)$ and zero initial conditions $x(-1) = 0$, $y(-2) = 0$, and $y(-1) = 0$.

6.29. Given a filter

$$H(z) = \frac{1 - z^{-1} + z^{-2}}{1 - 0.9z^{-1} + 0.81z^{-2}}$$

a. plot the magnitude frequency response and phase response using MATLAB;

b. specify the type of filtering;

c. find the difference equation;

d. perform filtering, that is, calculate $y(n)$ for the first 1,000 samples for each of the following inputs and plot the filter outputs using MATLAB, assuming that all initial conditions are zeros and the sampling rate is 8,000 Hz:

$$x(n) = \cos\left(\pi \cdot 10^3 \frac{n}{8,000}\right)$$

$$x(n) = \cos\left(\frac{8}{3}\pi \cdot 10^3 \frac{n}{8,000}\right)$$

$$x(n) = \cos\left(6\pi \cdot 10^3 \frac{n}{8,000}\right)$$

e. repeat (d) using the MATLAB function **filter()**.

6.30. Repeat (d) in Problem 6.29 using direct-form II structure.

6.9.2 MATLAB Projects

6.31. Sound effects of pre-emphasis filtering:

A pre-emphasis filter is shown in Figure 6.36 with a selective parameter $0 \leq \alpha < 1$, which controls the degree of pre-emphasis filtering. Assuming the system has a sampling rate of 8,000 Hz, plot the frequency responses for $\alpha = 0, \alpha = 0.4, \alpha = 0.8, \alpha = 0.95, \alpha = 0.99$, respectively. For each case, apply the pre-emphasis filter to the given speech ("speech.dat") and discuss the sound effects.

6.32. Echo generation (sound regeneration):

Echo is the repetition of sound due to sound wave reflection from the objects. It can easily be generated using an FIR filter such as that in Figure 6.37, where $|\alpha| < 1$ is an attenuation

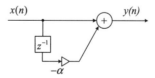

FIGURE 6.36

A pre-emphasis filter.

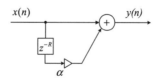

FIGURE 6.37

A single echo generator using an FIR filter.

factor and R the delay of the echo. The echo signal is generated by the sum of a delayed version of sound with the attenuation of α and the nondelayed version.

However, a single echo generator may not be useful, so a multiple-echo generator using an IIR filter is usually applied, as shown in Figure 6.38.

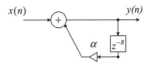

FIGURE 6.38

A multiple-echo generator using an IIR filter.

a. Assuming the system has a sampling rate of 8,000 Hz, plot the IIR filter frequency responses for the following cases: $\alpha = 0.5$ and $R = 1$; $\alpha = 0.6$ and $R = 4$; $\alpha = 0.7$ and $R = 10$, and characterize the frequency responses.

b. Implement the multiple-echo generator using the following code:

$$y = \text{filter}([1], [1 \ \text{zeros}(1, R\text{-}1) \ \text{alpha}], x)$$

Following that, evaluate the sound effects of the speech file ("speech.dat") for the following cases: $\alpha = 0.5$ and $R = 500$ (62.5 ms); $\alpha = 0.7$ and $R = 1000$ (125 ms); $\alpha = 0.5$, $R = 2000$ (250 ms); and $\alpha = 0.5$, $R = 4000$ (500 ms).

Finite Impulse Response Filter Design

CHAPTER OUTLINE

7.1 Finite Impulse Response Filter Format 217
7.2 Fourier Transform Design 219
7.3 Window Method 230
7.4 Applications: Noise Reduction and Two-Band Digital Crossover 253
    7.4.1 Noise Reduction 253
    7.4.2 Speech Noise Reduction 256
    7.4.3 Noise Reduction in Vibration Signals 257
    7.4.4 Two-Band Digital Crossover 258
7.5 Frequency Sampling Design Method 262
7.6 Optimal Design Method 269
7.7 Realization Structures of Finite Impulse Response Filters 280
    7.7.1 Transversal Form 280
    7.7.2 Linear Phase Form 281
7.8 Coefficient Accuracy Effects on Finite Impulse Response Filters 282
7.9 Summary of FIR Design Procedures and Selection of FIR Filter Design Methods in Practice 285
7.10 Summary 288
7.11 MATLAB Programs 288

OBJECTIVES:

This chapter introduces principles of the finite impulse response (FIR) filter design and investigates design methods such as the Fourier transform method, window method, frequency sampling method, and optimal design method. Then the chapter illustrates how to apply the designed FIR filters to solve real-world problems such as noise reduction and digital crossover for audio applications.

7.1 FINITE IMPULSE RESPONSE FILTER FORMAT

In this chapter, we describe techniques for designing *finite impulse response* (FIR) filters. An FIR filter is completely specified by the following input–output relationship:

$$
\begin{aligned}
y(n) &= \sum_{i=0}^{K} b_i x(n-i) \\
&= b_0 x(n) + b_1 x(n-1) + b_2 x(n-2) + \cdots + b_K x(n-K)
\end{aligned}
\tag{7.1}
$$

Digital Signal Processing.
Copyright © 2013 Elsevier Inc. All rights reserved.

where b_i represents FIR filter coefficients and $K + 1$ denotes the FIR filter length. Applying the z-transform on both sides on Equation (7.1) leads to

$$Y(z) = b_0 X(z) + b_1 z^{-1} X(z) + \cdots + b_K z^{-K} X(z) \qquad (7.2)$$

Factoring out $X(z)$ on the right-hand side of Equation (7.2) and then dividing $X(z)$ on both sides, we have the transfer function, which depicts the FIR filter, as

$$H(z) = \frac{Y(z)}{X(z)} = b_0 + b_1 z^{-1} + \cdots + b_K z^{-K} \qquad (7.3)$$

The following example serves to illustrate the notations used in Equations (7.1) and (7.3) numerically.

EXAMPLE 7.1

Given the FIR filter

$$y(n) = 0.1x(n) + 0.25x(n-1) + 0.2x(n-2)$$

determine the transfer function, filter length, nonzero coefficients, and impulse response.

Solution:
Applying the z-transform on both sides of the difference equation yields

$$Y(z) = 0.1X(z) + 0.25X(z)z^{-1} + 0.2X(z)z^{-2}$$

Then the transfer function is found to be

$$H(z) = \frac{Y(z)}{X(z)} = 0.1 + 0.25z^{-1} + 0.2z^{-2}$$

The filter length is $K + 1 = 3$, and the identified coefficients are

$$b_0 = 0.1, \ b_1 = 0.25 \quad \text{and} \quad b_2 = 0.2$$

Taking the inverse z-transform of the transfer function, we have

$$h(n) = 0.1\delta(n) + 0.25\delta(n-1) + 0.2\delta(n-2)$$

This FIR filter impulse response has only three terms.

The foregoing example is to help us understand the FIR filter format. We can conclude the following:

1. The transfer function in Equation (7.3) has a constant term, all the other terms have negative powers of z, and all the poles are at the origin on the z-plane. Hence, the stability of the filter is guaranteed. Its impulse response has only a finite number of terms.

2. The FIR filter operations involve only multiplying the filter inputs by their corresponding coefficients and accumulating them; the implementation of this filter type in real time is straightforward.

From the FIR filter format, the design objective is to obtain b_i coefficients for the FIR filter such that the magnitude frequency response of the FIR filter $H(z)$ will approximate the desired magnitude frequency response, such as that of a lowpass, highpass, bandpass, or bandstop filter. The following sections will introduce design methods to calculate the FIR filter coefficients.

7.2 FOURIER TRANSFORM DESIGN

We begin with an ideal lowpass filter with a normalized cutoff frequency Ω_c, whose magnitude frequency response in terms of the normalized digital frequency Ω is plotted in Figure 7.1 and is characterized by

$$H(e^{j\Omega}) = \begin{cases} 1, & 0 \leq |\Omega| \leq \Omega_c \\ 0, & \Omega_c \leq |\Omega| \leq \pi \end{cases} \tag{7.4}$$

Since the frequency response is periodic with a period of $\Omega = 2\pi$ radians, as we discussed in Chapter 6, we can extend the frequency response of the ideal filter $H(e^{j\Omega})$, as shown in Figure 7.2.

The periodic frequency response can be approximated using a complex Fourier series expansion (see Appendix B) in terms of the normalized digital frequency Ω, that is,

$$H(e^{j\Omega}) = \sum_{n=-\infty}^{\infty} c_n e^{-j\omega_0 n\Omega} \tag{7.5}$$

and the Fourier coefficients are given by

$$c_n = \frac{1}{2\pi} \int_{-\pi}^{\pi} H(e^{j\Omega}) e^{j\omega_0 n\Omega} d\Omega \quad \text{for} \quad -\infty < n < \infty \tag{7.6}$$

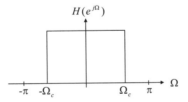

FIGURE 7.1

Frequency response of an ideal lowpass filter.

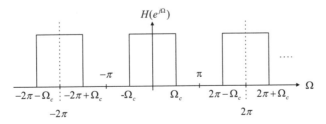

FIGURE 7.2

Periodicity of the ideal lowpass frequency response.

Notice that we obtain Equations (7.5) and (7.6) simply by treating the Fourier series expansion in the time domain with the time variable t replaced by the normalized digital frequency variable Ω. The fundamental frequency is easily found to be

$$\omega_0 = 2\pi/(period\ of\ waveform) = 2\pi/2\pi = 1 \qquad (7.7)$$

Substituting $\omega_0 = 1$ into Equation (7.6) and introducing $h(n) = c_n$, called the desired impulse response of the ideal filter, we obtain the Fourier transform design as

$$h(n) = \frac{1}{2\pi} \int_{-\pi}^{\pi} H(e^{j\Omega}) e^{j\Omega n} d\Omega \qquad for \quad -\infty < n < \infty \qquad (7.8)$$

Now, let us look at the possible z-transfer function. If we substitute $e^{j\Omega} = z$ and $\omega_0 = 1$ back to Equation (7.5), we yield a z-transfer function in the following format:

$$H(z) = \sum_{n=-\infty}^{\infty} h(n) z^{-n}$$
$$\cdots + h(-2)z^2 + h(-1)z^1 + h(0) + h(1)z^{-1} + h(2)z^{-2} + \cdots \qquad (7.9)$$

This is a noncausal FIR filter. We will deal with this later in this section. Using the Fourier transform design shown in Equation (7.8), the desired impulse response approximation of the ideal lowpass filter is solved as

$$For\ n = 0 \quad h(n) = \frac{1}{2\pi} \int_{-\pi}^{\pi} H(e^{j\Omega}) e^{j\Omega \times 0} d\Omega$$

$$= \frac{1}{2\pi} \int_{-\Omega_c}^{\Omega_c} 1 d\Omega = \frac{\Omega_c}{\pi}$$

$$For\ n \neq 0 \quad h(n) = \frac{1}{2\pi} \int_{-\pi}^{\pi} H(e^{j\Omega}) e^{j\Omega n} d\Omega = \frac{1}{2\pi} \int_{-\Omega_c}^{\Omega_c} e^{j\Omega n} d\Omega$$

$$= \left. \frac{e^{j n \Omega}}{2\pi j n} \right|_{-\Omega_c}^{\Omega_c} = \frac{1}{\pi n} \frac{e^{j n \Omega_c} - e^{-j n \Omega_c}}{2j} = \frac{\sin(\Omega_c n)}{\pi n} \qquad (7.10)$$

The desired impulse response $h(n)$ is plotted versus the sample number n in Figure 7.3.

Theoretically, $h(n)$ in Equation (7.10) exists for $-\infty < n < \infty$ and is symmetrical about $n = 0$; that is, $h(n) = h(-n)$. The amplitude of the impulse response sequence $h(n)$ becomes smaller when n increases in both directions. The FIR filter design must first be completed by truncating the infinite-length sequence $h(n)$ to achieve the $2M + 1$ dominant coefficients using the coefficient symmetry, that is,

$$H(z) = h(M)z^M + \cdots + h(1)z^1 + h(0) + h(1)z^{-1} + \cdots + h(M)z^{-M}$$

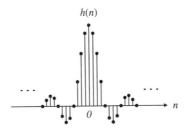

FIGURE 7.3

Impulse response of an ideal digital lowpass filter.

The obtained filter is a noncausal z-transfer function of the FIR filter, since the filter transfer function contains terms with the positive powers of z, which in turn means that the filter output depends on the future filter inputs. To remedy the noncausal z-transfer function, we delay the truncated impulse response $h(n)$ by M samples to yield the following causal FIR filter:

$$H(z) = b_0 + b_1 z^{-1} + \cdots + b_{2M}(2M)z^{-2M} \tag{7.11}$$

Table 7.1 Summary of Ideal Impulse Responses for Standard FIR Filters

Filter	Type Ideal Impulse Response $h(n)$ (noncausal FIR coefficients)
Lowpass:	$h(n) = \begin{cases} \dfrac{\Omega_c}{\pi} & \text{for } n = 0 \\ \dfrac{\sin(\Omega_c n)}{n\pi} & \text{for } n \neq 0 \end{cases} \quad -M \leq n \leq M$
Highpass:	$h(n) = \begin{cases} \dfrac{\pi - \Omega_c}{\pi} & \text{for } n = 0 \\ -\dfrac{\sin(\Omega_c n)}{n\pi} & \text{for } n \neq 0 \end{cases} \quad -M \leq n \leq M$
Bandpass:	$h(n) = \begin{cases} \dfrac{\Omega_H - \Omega_L}{\pi} & \text{for } n = 0 \\ \dfrac{\sin(\Omega_H n)}{n\pi} - \dfrac{\sin(\Omega_L n)}{n\pi} & \text{for } n \neq 0 \end{cases} \quad -M \leq n \leq M$
Bandstop:	$h(n) = \begin{cases} \dfrac{\pi - \Omega_H + \Omega_L}{\pi} & \text{for } n = 0 \\ -\dfrac{\sin(\Omega_H n)}{n\pi} + \dfrac{\sin(\Omega_L n)}{n\pi} & \text{for } n \neq 0 \end{cases} \quad -M \leq n \leq M$

Causal FIR filter coefficients: shifting $h(n)$ to the right by M samples.
Transfer function:
$H(z) = b_0 + b_1 z^{-1} + b_2 z^{-2} + \cdots b_{2M} z^{-2M}$
where $b_n = h(n - M), n = 0, 1, \cdots, 2M.$

where the delay operation is given by

$$b_n = h(n - M) \quad \text{for} \quad n = 0, 1, \cdots, 2M \tag{7.12}$$

Similarly, we can obtain the design equations for other types of FIR filters, such as highpass, bandpass, and bandstop, using their ideal frequency responses and Equation (7.8). The derivations are omitted here. Table 7.1 gives a summary of all the formulas for FIR filter coefficient calculations.

The following example illustrates the coefficient calculation for the lowpass FIR filter.

EXAMPLE 7.2

a. Calculate the filter coefficients for a 3-tap FIR lowpass filter with a cutoff frequency of 800 Hz and a sampling rate of 8,000 Hz using the Fourier transform method.
b. Determine the transfer function and difference equation of the designed FIR system.
c. Compute and plot the magnitude frequency response for $\Omega = 0, \pi/4, \pi/2, 3\pi/4,$ and π radians.

Solution:

a. Calculating the normalized cutoff frequency leads to

$$\Omega_c = 2\pi f_c T_s = 2\pi \times 800/8,000 = 0.2\pi \text{ radians}$$

Since $2M + 1 = 3$ in this case, using the equation in Table 7.1 results in

$$h(0) = \frac{\Omega_c}{\pi} \quad \text{for} \quad n = 0$$

$$h(n) = \frac{\sin(\Omega_c n)}{n\pi} = \frac{\sin(0.2\pi n)}{n\pi} \quad \text{for} \quad n \neq 1$$

The computed filter coefficients via the previous expression are listed as:

$$h(0) = \frac{0.2\pi}{\pi} = 0.2$$

$$h(1) = \frac{\sin[0.2\pi \times 1]}{1 \times \pi} = 0.1871$$

Using the symmetry leads to

$$h(-1) = h(1) = 0.1871$$

Thus delaying $h(n)$ by $M = 1$ sample using Equation (7.12) gives

$$b_0 = h(0 - 1) = h(-1) = 0.1871$$
$$b_1 = h(1 - 1) = h(0) = 0.2$$
$$b_2 = h(2 - 1) = h(1) = 0.1871$$

b. The transfer function is achieved as

$$H(z) = 0.1871 + 0.2z^{-1} + 0.1871z^{-2}$$

Using the technique described in Chapter 6, we have

$$\frac{Y(z)}{X(z)} = H(z) = 0.1871 + 0.2z^{-1} + 0.1871z^{-2}$$

Table 7.2 Frequency Response Calculation in Example 7.2

Ω radians	$f = \Omega f_s / (2\pi)$ Hz	$0.2 + 0.3742 \cos \Omega$	$\left|H(e^{j\Omega})\right|$	$\left|H(e^{j\Omega})\right|_{dB}$	$\angle H(e^{j\Omega})$ degree
0	0	0.5742	0.5742	−4.82	0
$\pi/4$	1000	0.4646	0.4646	−6.66	−45
$\pi/2$	2000	0.2	0.2	−14.0	−90
$3\pi/4$	3000	−0.0646	0.0646	−23.8	45
π	4000	−0.1742	0.1742	−15.2	0

Multiplying $X(z)$ leads to

$$Y(z) = 0.1871X(z) + 0.2z^{-1}X(z) + 0.1871z^{-2}X(z)$$

Applying the inverse z-transform on both sides, the difference equation is yielded as

$$y(n) = 0.1871x(n) + 0.2x(n-1) + 0.1871x(n-2)$$

c. The magnitude frequency response and phase response can be obtained using the technique introduced in Chapter 6. Substituting $z = e^{j\Omega}$ into $H(z)$, it follows that

$$H(e^{j\Omega}) = 0.1871 + 0.2e^{-j\Omega} + 0.1871e^{-j2\Omega}$$

Factoring term $e^{-j\Omega}$ and using the Euler formula $e^{jx} + e^{-jx} = 2\cos(x)$, we achieve

$$H(e^{j\Omega}) = e^{-j\Omega}(0.1871e^{j\Omega} + 0.2 + 0.1871e^{-j\Omega})$$
$$= e^{-j\Omega}(0.2 + 0.3742 \cos(\Omega))$$

Then the magnitude frequency response and phase response are found to be

$$\left|H(e^{j\Omega})\right| = |0.2 + 0.3472 \cos \Omega|$$

and

$$\angle H(e^{j\Omega}) = \begin{cases} -\Omega & \text{if} \quad 0.2 + 0.3472 \cos \Omega > 0 \\ -\Omega + \pi & \text{if} \quad 0.2 + 0.3472 \cos \Omega < 0 \end{cases}$$

Details of the magnitude calculations for several typical normalized frequencies are listed in Table 7.2.

Due to the symmetry of the coefficients, the obtained FIR filter has a linear phase response as shown in Figure 7.4. The sawtooth shape is produced by the contribution of the negative sign of the real magnitude term $0.2 + 0.3742 \cos \Omega$ in the 3-tap filter frequency response, that is,

$$H(e^{j\Omega}) = e^{-j\Omega}(0.2 + 0.3742 \cos \Omega)$$

In general, the FIR filter with symmetric coefficients has a linear phase response (linear function of Ω) as follows:

$$\angle H(e^{j\Omega}) = -M\Omega + \text{possible phase of } 180° \tag{7.13}$$

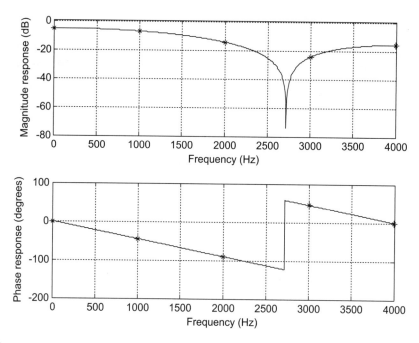

FIGURE 7.4

Magnitude frequency response in Example 7.2.

Next, we see that the 3-tap FIR filter does not give an acceptable magnitude frequency response.

To explore this response further, Figure 7.5 displays the magnitude and phase responses of 3-tap ($M = 1$) and 17-tap ($M = 8$) FIR lowpass filters with a normalized cutoff frequency $\Omega_c = 0.2\pi$ radians. The calculated coefficients for the 17-tap FIR lowpass filter are listed in Table 7.3.

We can make the following observations at this point:

1. The oscillations (ripples) exhibited in the passband (main lobe) and stopband (side lobes) of the magnitude frequency response constitute the *Gibbs effect*. The Gibbs oscillatory behavior originates from the abrupt truncation of the infinite impulse response in Equation (7.11). To remedy this problem, window functions will be used and will be discussed in the next section.

2. Using a larger number of the filter coefficients will produce the sharp roll-off characteristic of the transition band but may cause increased time delay and increase computational complexity for implementing the designed FIR filter.

3. The phase response is linear in the passband. This is consistent with Equation (7.13), which means that all frequency components of the filter input within the passband are subjected to the same time delay at the filter output. This is a requirement for applications in audio and speech filtering, where phase distortion needs to be avoided. Note that we impose a linear phase requirement, that is, the FIR coefficients are symmetric about the middle coefficient, and the FIR filter order is an odd number. If the design methods cannot produce the symmetric coefficients or generate anti-symmetric coefficients (Proakis and Manolakis, 1996), the resultant FIR filter does not have the

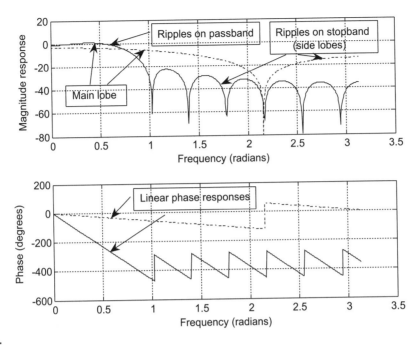

FIGURE 7.5

Magnitude and phase frequency responses of the lowpass FIR filters with 3 coefficients (dash-dotted line) and 17 coefficients (solid line).

linear phase property. (Linear phase even-order FIR filters and FIR filters using the anti-symmetry of coefficients are discussed in Proakis and Manolakis [1996].)

To further probe the linear phase property, we consider a sinusoidal sequence $x(n) = A \sin(n\Omega)$ as the FIR filter input, with the output expected to be

$$y(n) = A|H|\sin(n\Omega + \phi)$$

where $\phi = -M\Omega$. Substituting $\phi = -M\Omega$ into $y(n)$ leads to

$$y(n) = A|H|\sin[\Omega(n - M)]$$

Table 7.3 17-Tap FIR Lowpass Filter Coefficients in Example 7.2 ($M = 8$)

$b_0 = b_{16} = -0.0378$	$b_1 = b_{15} = -0.0432$
$b_2 = b_{14} = -0.0312$	$b_3 = b_{13} = 0.0000$
$b_4 = b_{12} = 0.0468$	$b_5 = b_{11} = 0.1009$
$b_6 = b_{10} = 0.1514$	$b_7 = b_9 = 0.1871 b_8 = 0.2000$

This clearly indicates that within the passband, all frequency components passing through the FIR filter will have the same constant delay at the output, which equals M samples. Hence, phase distortion is avoided.

Figure 7.6 verifies the linear phase property using an FIR filter with 17 taps. Two sinusoids of the normalized digital frequencies 0.05π and 0.15π radians, respectively, are used as inputs. These two input signals are within the passband, so their magnitudes are not changed. As shown in Figure 7.6, beginning at the ninth sample the output matches the input, which is delayed by eight samples for each case.

What would happen if the filter phase were nonlinear? This can be illustrated using the following combined sinusoids as the filter input:

$$x(n) = x_1(n) + x_2(n) = \sin(0.05\pi n)u(n) - \frac{1}{3}\sin(0.15\pi n)u(n)$$

The original $x(n)$ is the top plot shown in Figure 7.7. If the linear phase response of a filter is considered, such as $\phi = -M\Omega_0$, where $M = 8$ in our illustration, we have the filtered output as

$$y_1(n) = \sin[0.05\pi(n-8)] - \frac{1}{3}\sin[0.15\pi(n-8)]$$

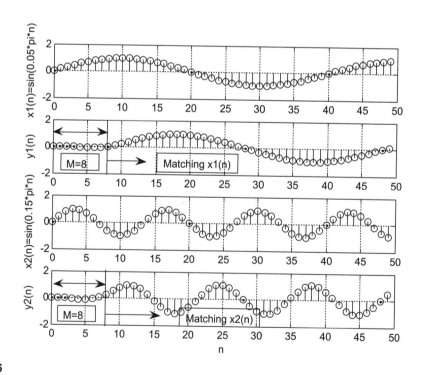

FIGURE 7.6

Illustration of FIR filter linear phase property (constant delay of eight samples).

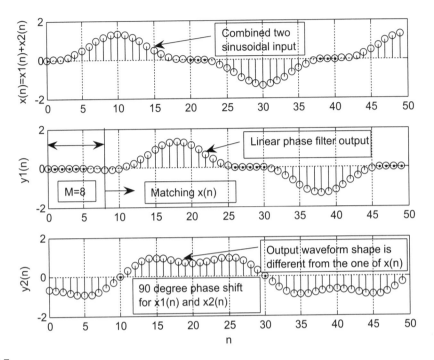

FIGURE 7.7

Comparison of linear and nonlinear phase responses.

The linear phase effect is shown in the middle plot of Figure 7.7. We see that $y_1(n)$ is the eight-sample delayed version of $x(n)$. However, considering a unit gain filter with a phase delay of 90 degrees for all the frequency components, we obtain the filtered output as

$$y_2(n) = \sin(0.05\pi n - \pi/2) - \frac{1}{3}\sin(0.15\pi n - \pi/2)$$

where the first term has a phase shift of 10 samples (see $\sin[0.05\pi(n-10)]$), while the second term has a phase shift of $10/3$ samples $\left(\text{see } \frac{1}{3}\sin\left[0.15\pi\left(n - \frac{10}{3}\right)\right]\right)$. Certainly, we do not have the linear phase feature. The signal $y_2(n)$ plotted in Figure 7.7 shows that the waveform shape is different from that of the original signal $x(n)$, and hence has significant phase distortion. This phase distortion is audible for audio applications and can be avoided by using an FIR filter, which has the linear phase feature.

We now have finished discussing the coefficient calculation for the FIR lowpass filter, which has a good linear phase property. To explain the calculation of filter coefficients for the other types of filters and examine the Gibbs effect, we look at another simple example.

EXAMPLE 7.3

a. Calculate the filter coefficients for a 5-tap FIR bandpass filter with a lower cutoff frequency of 2,000 Hz and an upper cutoff frequency of 2,400 Hz and a sampling rate of 8,000 Hz.
b. Determine the transfer function and plot the frequency responses with MATLAB.

Solution:

a. Calculating the normalized cutoff frequencies leads to

$$\Omega_L = 2\pi f_L/f_s = 2\pi \times 2,000/8,000 = 0.5\pi \text{ radians}$$

$$\Omega_H = 2\pi f_H/f_s = 2\pi \times 2,400/8,000 = 0.6\pi \text{ radians}$$

Since $2M + 1 = 5$ in this case, using the equation in Table 7.1 yields

$$h(n) = \begin{cases} \dfrac{\Omega_H - \Omega_L}{\pi} & n = 0 \\ \dfrac{\sin(\Omega_H n)}{n\pi} - \dfrac{\sin(\Omega_L n)}{n\pi} & n \neq 0 \quad -2 \le n \le 2 \end{cases} \tag{7.14}$$

Calculations for noncausal FIR coefficients are listed as

$$h(0) = \frac{\Omega_H - \Omega_L}{\pi} = \frac{0.6\pi - 0.5\pi}{\pi} = 0.1$$

The other computed filter coefficients via Equation (7.14) are

$$h(1) = \frac{\sin[0.6\pi \times 1]}{1 \times \pi} - \frac{\sin[0.5\pi \times 1]}{1 \times \pi} = -0.01558$$

$$h(2) = \frac{\sin[0.6\pi \times 2]}{2 \times \pi} - \frac{\sin[0.5\pi \times 2]}{2 \times \pi} = -0.09355$$

Using symmetry leads to

$$h(-1) = h(1) = -0.01558$$

$$h(-2) = h(2) = -0.09355$$

Thus, delaying $h(n)$ by $M = 2$ samples gives

$$b_0 = b_4 = -0.09355$$

$$b_1 = b_3 = -0.01558, \quad \text{and} \quad b_2 = 0.1$$

b. The transfer function is achieved as

$$H(z) = -0.09355 - 0.01558z^{-1} + 0.1z^{-2} - 0.01558z^{-3} - 0.09355z^{-4}$$

To complete Example 7.3, the magnitude frequency response plotted in terms of $\left|H(e^{j\Omega})\right|_{dB} = 20\log_{10}\left|H(e^{j\Omega})\right|$ using MATLAB Program 7.1 is displayed in Figure 7.8.

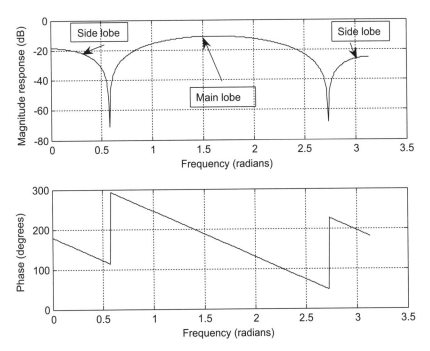

FIGURE 7.8

Frequency responses for Example 7.3.

Program 7.1. MATLAB program for Example 7.3.

```
% Example 7.3
% MATLAB program to plot frequency response
%
  [hz,w]=freqz([-0.09355 -0.01558 0.1 -0.01558 -0.09355], [1], 512);
  phi=180*unwrap(angle(hz))/pi;
  subplot(2,1,1), plot(w,20*log10(abs(hz))),grid;
  xlabel('Frequency (radians)');
  ylabel('Magnitude Response (dB)')
  subplot(2,1,2), plot(w, phi); grid;
  xlabel('Frequency (radians)');
  ylabel('Phase (degrees)');
```

To summarize Example 7.3, the magnitude frequency response demonstrates the Gibbs oscillatory behavior existing in the passband and stopband. The peak of the main lobe in the passband is dropped from 0 dB to approximately −10 dB, while for the stopband, the lower side lobe in the magnitude response plot swings approximately between −18 dB and −70 dB, and the upper side lobe swings between −25 dB and −68 dB. As we have pointed out, this is due to the abrupt truncation of the infinite impulse sequence $h(n)$. The oscillations can be reduced by increasing the number of coefficients and using a window function, which will be studied next.

7.3 WINDOW METHOD

In this section, the *window method* (Fourier transform design with window functions) is developed to remedy the undesirable Gibbs oscillations in the passband and stopband of the designed FIR filter. Recall that the Gibbs oscillations originate from the abrupt truncation of the infinite-length coefficient sequence. Then it is natural to seek a window function, which is symmetrical and can gradually weight the designed FIR coefficients down to zeros at both ends for the range $-M \le n \le M$. Applying the window sequence to the filter coefficients gives

$$h_w(n) = h(n) \cdot w(n)$$

where $w(n)$ designates the window function. Common window functions used in the FIR filter design are as follows:

1. Rectangular window:

$$w_{rec}(n) = 1, \ -M \le n \le M \tag{7.15}$$

2. Triangular (Bartlett) window:

$$w_{tri}(n) = 1 - \frac{|n|}{M}, \ -M \le n \le M \tag{7.16}$$

3. Hanning window:

$$w_{han}(n) = 0.5 + 0.5 \ \cos\left(\frac{n\pi}{M}\right), -M \le n \le M \tag{7.17}$$

4. Hamming window:

$$w_{ham}(n) = 0.54 + 0.46 \ \cos\left(\frac{n\pi}{M}\right), -M \le n \le M \tag{7.18}$$

5. Blackman window:

$$w_{black}(n) = 0.42 + 0.5 \ \cos\left(\frac{n\pi}{M}\right) + 0.08 \ \cos\left(\frac{2n\pi}{M}\right), \ -M \le n \le M \tag{7.19}$$

In addition, there is another popular window function, called the Kaiser window (detailed information can be found in Oppenheim, Shaffer, and Buck [1999]). As we expected, the rectangular window function has a constant value of 1 within the window, and hence only does truncation. For comparison, shapes of the other window functions from Equations (7.16) to (7.19) are plotted in Figure 7.9 for the case of $2M + 1 = 81$.

We apply the Hamming window function in Example 7.4.

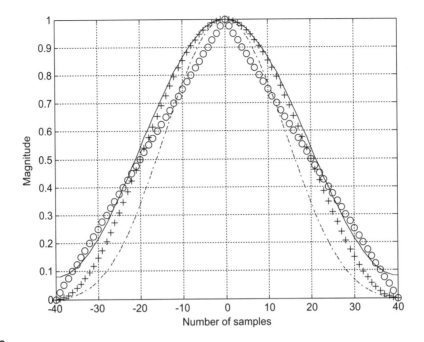

FIGURE 7.9

Shapes of window functions for the case of $2M + 1 = 81$. "o" line, triangular window; "+" line, Hanning window; solid line, Hamming window; dashed line, Blackman window.

EXAMPLE 7.4

Given the calculated filter coefficients

$h(0) = 0.25,\ h(-1) = h(1) = 0.22508,\ h(-2) = h(2) = 0.15915,\ h(-3) = h(3) = 0.07503$

a. apply the hamming window function to obtain windowed coefficients $h_w(n)$;
b. plot the impulse response $h(n)$ and windowed impulse response $h_w(n)$.

Solution:
a. Since $M = 3$, applying Equation (7.18) leads to the window sequence

$$w_{ham}(-3) = 0.54 + 0.46 \cos \left(\frac{-3 \times \pi}{3} \right) = 0.08$$

$$w_{ham}(-2) = 0.54 + 0.46 \cos \left(\frac{-2 \times \pi}{3} \right) = 0.31$$

$$w_{ham}(-1) = 0.54 + 0.46 \cos \left(\frac{-1 \times \pi}{3} \right) = 0.77$$

$$w_{ham}(0) = 0.54 + 0.46 \cos \left(\frac{0 \times \pi}{3} \right) = 1$$

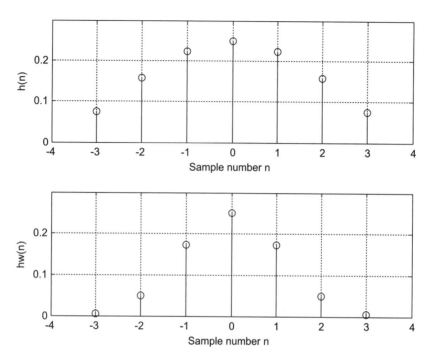

FIGURE 7.10

Plots of FIR noncausal coefficients and windowed FIR coefficients in Example 7.4.

$$w_{ham}(1) = 0.54 + 0.46 \cos \left(\frac{1 \times \pi}{3} \right) = 0.77$$

$$w_{ham}(2) = 0.54 + 0.46 \cos \left(\frac{2 \times \pi}{3} \right) = 0.31$$

$$w_{ham}(3) = 0.54 + 0.46 \cos \left(\frac{3 \times \pi}{3} \right) = 0.08$$

Applying the Hamming window function and its symmetric property to the filter coefficients, we get

$$h_w(0) = h(0) \cdot w_{ham}(0) = 0.25 \times 1 = 0.25$$

$$h_w(1) = h(1) \cdot w_{ham}(1) = 0.22508 \times 0.77 = 0.17331 = h_w(-1)$$

$$h_w(2) = h(2) \cdot w_{ham}(2) = 0.15915 \times 0.31 = 0.04934 = h_w(-2)$$

$$h_w(3) = h(3) \cdot w_{ham}(3) = 0.07503 \times 0.08 = 0.00600 = h_w(-3)$$

b. Noncausal impulse responses $h(n)$ and $h_w(n)$ are plotted in Figure 7.10.

We observe that the Hamming window does its job and weights the FIR filter coefficients to zero gradually at both ends. Hence, we can expect a reduced Gibbs effect in the magnitude frequency response.

Now lowpass FIR filter design via the window method can be achieved. The design procedure includes three steps. The first step is to obtain the truncated impulse response $h(n)$, where $-M \leq n \leq M$; then we multiply the obtained sequence $h(n)$ by the selected window data sequence to yield the windowed noncausal FIR filter coefficients $h_w(n)$; the final step is to delay the windowed noncausal sequence $h_w(n)$ by M samples to obtain the causal FIR filter coefficients, $b_n = h_w(n - M)$. The design procedure of the FIR filter via windowing is summarized as follows:

1. Obtain the FIR filter coefficients $h(n)$ via the Fourier transform method (Table 7.1).
2. Multiply the generated FIR filter coefficients by the selected window sequence

$$h_w(n) = h(n)w(n), \quad n = -M, \cdots 0, 1, \cdots, M \qquad (7.20)$$

where $w(n)$ is chosen to be one of the window functions listed in Equations (7.15) to (7.19).
3. Delay the windowed impulse sequence $h_w(n)$ by M samples to get the windowed FIR filter coefficients

$$b_n = h_w(n - M), \quad \text{for} \quad n = 0, 1, \cdots, 2M \qquad (7.21)$$

Let us study the following design examples.

EXAMPLE 7.5

a. Design a 3-tap FIR lowpass filter with a cutoff frequency of 800 Hz and a sampling rate of 8,000 Hz using the Hamming window function.
b. Determine the transfer function and difference equation of the designed FIR system.
c. Compute and plot the magnitude frequency response for $\Omega = 0, \pi/4, \pi/2, 3\pi/4,$ and π radians.

Solution:
a. The normalized cutoff frequency is calculated as
$$\Omega_c = 2\pi f_c T_s = 2\pi \times 800/8,000 = 0.2\pi \text{ radians}$$

Since $2M + 1 = 3$ in this case, FIR coefficients obtained by using the equation in Table 7.1 are listed as
$$h(0) = 0.2 \quad \text{and} \quad h(-1) = h(1) = 0.1871$$

(see Example 7.2). Applying the Hamming window function defined in Equation (7.18), we have
$$w_{ham}(0) = 0.54 + 0.46 \cos\left(\frac{0\pi}{1}\right) = 1$$

$$w_{ham}(1) = 0.54 + 0.46 \cos\left(\frac{1 \times \pi}{1}\right) = 0.08$$

Using the symmetry of the window function gives
$$w_{ham}(-1) = w_{ham}(1) = 0.08$$

The windowed impulse response is calculated as
$$h_w(0) = h(0)w_{ham}(0) = 0.2 \times 1 = 0.2$$

$$h_w(1) = h(1)w_{ham}(1) = 0.1871 \times 0.08 = 0.01497$$

$$h_w(-1) = h(-1)w_{ham}(-1) = 0.1871 \times 0.08 = 0.01497$$

Thus delaying $h_w(n)$ by $M = 1$ sample gives

$$b_0 = b_2 = 0.01496 \quad \text{and} \quad b_1 = 0.2$$

b. The transfer function is

$$H(z) = 0.01497 + 0.2z^{-1} + 0.01497z^{-2}$$

Using the technique described in Chapter 6, we have

$$\frac{Y(z)}{X(z)} = H(z) = 0.01497 + 0.2z^{-1} + 0.01497z^{-2}$$

Multiplying X(z) leads to

$$Y(z) = 0.01497X(z) + 0.2z^{-1}X(z) + 0.01497z^{-2}X(z)$$

Applying the inverse z-transform on both sides, the difference equation is obtained as

$$y(n) = 0.01497x(n) + 0.2x(n-1) + 0.01497x(n-2)$$

c. The magnitude frequency response and phase response can be obtained using the technique introduced in Chapter 6. Substituting $z = e^{j\Omega}$ into $H(z)$, it follows that

$$H(e^{j\Omega}) = 0.01497 + 0.2e^{-j\Omega} + 0.01497e^{-j2\Omega}$$
$$= e^{-j\Omega}(0.01497e^{j\Omega} + 0.2 + 0.01497e^{-j\Omega})$$

Using Euler's formula leads to

$$H(e^{j\Omega}) = e^{-j\Omega}(0.2 + 0.02994 \cos \Omega)$$

Then the magnitude frequency response and phase response are found to be

$$\left| H(e^{j\Omega}) \right| = |0.2 + 0.2994 \cos \Omega|$$

Table 7.4 Frequency Response Calculation in Example 7.5

Ω radians	$f = \Omega f_s/(2\pi)$ Hz	$0.2 + 0.02994 \cos \Omega$	$\left\| H(e^{j\Omega}) \right\|$	$\left\| H(e^{j\Omega}) \right\|_{dB}$ dB	$\angle H(e^{j\Omega})$ degrees
0	0	0.2299	0.2299	−12.77	0
$\pi/4$	1,000	0.1564	0.2212	−13.11	−45
$\pi/2$	2,000	0.2000	0.2000	−13.98	−90
$3\pi/4$	3,000	0.1788	0.1788	−14.95	−135
π	4,000	0.1701	0.1701	−15.39	−180

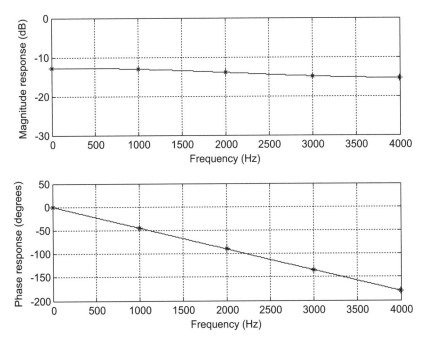

FIGURE 7.11

The frequency responses in Example 7.5.

and

$$\angle H(e^{j\Omega}) = \begin{cases} -\Omega & \text{if} \quad 0.2 + 0.02994 \cos \Omega > 0 \\ -\Omega + \pi & \text{if} \quad 0.2 + 0.02994 \cos \Omega < 0 \end{cases}$$

The calculation details of the magnitude response for several normalized values are listed in Table 7.4. Figure 7.11 shows the plots of the frequency responses.

EXAMPLE 7.6

a. Design a 5-tap FIR band reject (bandstop) filter with a lower cutoff frequency of 2,000 Hz, an upper cutoff frequency of 2,400 Hz, and a sampling rate of 8,000 Hz using the Hamming window method.

b. Determine the transfer function.

Solution:

a. Calculating the normalized cutoff frequencies leads to

$$\Omega_L = 2\pi f_L T = 2\pi \times 2,000/8,000 = 0.5\pi \text{ radians}$$

$$\Omega_H = 2\pi f_H T = 2\pi \times 2,400/8,000 = 0.6\pi \text{ radians}$$

Since $2M + 1 = 5$ in this case, using the equation in Table 7.1 yields

$$h(n) = \begin{cases} \dfrac{\pi - \Omega_H + \Omega_L}{\pi} & n = 0 \\[2ex] -\dfrac{\sin(\Omega_H n)}{n\pi} + \dfrac{\sin(\Omega_L n)}{n\pi} & n \neq 0 \quad -2 \le n \le 2 \end{cases}$$

When $n = 0$, we have

$$h(0) = \frac{\pi - \Omega_H + \Omega_L}{\pi} = \frac{\pi - 0.6\pi + 0.5\pi}{\pi} = 0.9$$

The other computed filter coefficients for the previous expression are listed below:

$$h(1) = \frac{\sin[0.5\pi \times 1]}{1 \times \pi} - \frac{\sin[0.6\pi \times 1]}{1 \times \pi} = 0.01558$$

$$h(2) = \frac{\sin[0.5\pi \times 2]}{2 \times \pi} - \frac{\sin[0.6\pi \times 2]}{2 \times \pi} = 0.09355$$

Using symmetry leads to

$$h(-1) = h(1) = 0.01558$$

$$h(-2) = h(2) = 0.09355$$

Applying the Hamming window function in Equation (7.18), we have

$$w_{ham}(0) = 0.54 + 0.46 \cos\left(\frac{0 \times \pi}{2}\right) = 1.0$$

$$w_{ham}(1) = 0.54 + 0.46 \cos\left(\frac{1 \times \pi}{2}\right) = 0.54$$

$$w_{ham}(2) = 0.54 + 0.46 \cos\left(\frac{2 \times \pi}{2}\right) = 0.08$$

Using the symmetry of the window function gives

$$w_{ham}(-1) = w_{ham}(1) = 0.54$$

$$w_{ham}(-2) = w_{ham}(2) = 0.08$$

The windowed impulse response is calculated as

$$h_w(0) = h(0)w_{ham}(0) = 0.9 \times 1 = 0.9$$

$$h_w(1) = h(1)w_{ham}(1) = 0.01558 \times 0.54 = 0.00841$$

$$h_w(2) = h(2)w_{ham}(2) = 0.09355 \times 0.08 = 0.00748$$

$$h_w(-1) = h(-1)w_{ham}(-1) = 0.00841$$

$$h_w(-2) = h(-2)w_{ham}(-2) = 0.00748$$

Thus, delaying $h_w(n)$ by $M = 2$ samples gives
$$b_0 = b_4 = 0.00748, \quad b_1 = b_3 = 0.00841, \quad \text{and} \quad b_2 = 0.9$$

b. The transfer function is
$$H(z) = 0.00748 + 0.00841z^{-1} + 0.9z^{-2} + 0.00841z^{-3} + 0.00748z^{-4}$$

The following design examples are demonstrated using MATLAB programs. The MATLAB function **firwd(N, Ftype, WnL, WnH, Wtype)** is listed in the "MATLAB Programs" section at the end of this chapter. Table 7.5 lists comments to show how the function is used.

Table 7.5 Illustration of the MATLAB Function for FIR Filter Design Using Window Methods

```
function B=firwd(N,Ftype,WnL,WnH,Wtype)
% B = firwd(N,Ftype,WnL,WnH,Wtype)
% FIR filter design using the window function method.
% Input parameters:
% N: the number of the FIR filter taps.
% Note: It must be odd number.
% Ftype: the filter type
%      1. Lowpass filter
%      2. Highpass filter
%      3. Bandpass filter
%      4. Band reject (Bandstop) filter
% WnL: lower cutoff frequency in radians. Set WnL=0 for the highpass filter.
% WnH: upper cutoff frequency in radians. Set WnH=0 for the lowpass filter.
% Wtypw: window function type
%      1. Rectangular window
%      2. Triangular window
%      3. Hanning window
%      4. Hamming window
%      5. Blackman window
```

EXAMPLE 7.7

a. Design a lowpass FIR filter with 25 taps using the MATLAB program listed in the "MATLAB Programs" section at the end of this chapter. The cutoff frequency of the filter is 2,000 Hz, assuming a sampling frequency of 8,000 Hz. The rectangular window and Hamming window functions are used for each design.
b. Plot the frequency responses along with those obtained using the rectangular window and Hamming window for comparison.
c. List the FIR filter coefficients for each window design method.

Solution:
a. With a given sampling rate of 8,000 Hz, the normalized cutoff frequency can be found as

$$\Omega_c = \frac{2,000 \times 2\pi}{8,000} = 0.5\pi \text{ radians}$$

Now we are ready to design FIR filters via the MATLAB program. The function **firwd(N, Ftype, WnL, WnH, Wtype)** listed in the "MATLAB Programs" section at the end of this chapter, has five input parameters, which are described as follows:

- "N" is the number of specified filter coefficients (the number of filter taps).
- "Ftype" denotes the filter type, that is, input "1" for the lowpass filter design, input "2" for the highpass filter design, input "3" for the bandpass filter design, and input "4" for the band reject filter design.

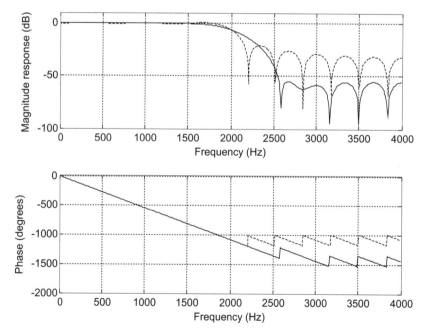

FIGURE 7.12

Frequency responses using the rectangular and Hamming windows.

- "WnL" and "WnH" are the lower and upper cutoff frequency inputs, respectively. Note that WnH = 0 when specifying WnL for the lowpass filter design, while WnL = 0 when specifying WnH for the highpass filter design.
- "Wtype" specifies the window data sequence to be used in the design, that is, input "1" for the rectangular window, input "2" for the triangular window, input "3" for the Hanning window, input "4" for the Hamming window, and input "5" for the Blackman window.

b. The following program (Program 7.2) is used to generate FIR filter coefficients using the rectangular window. Its frequency responses will be plotted together with the results of the FIR filter design obtained using the Hamming window, as shown in Program 7.3.

Program 7.2. MATLAB program for Example 7.7.

```
% Example 7.7
% MATLAB program to generate FIR coefficients
% using the rectangular window.
%
N=25; Ftype=1; WnL=0.5*pi; WnH=0; Wtype=1;
B=firwd(N,Ftype,WnL,WnH,Wtype);
```

Program 7.3. MATLAB program for Example 7.7.

```
%Figure 7.12
% MATLAB program to create Figure 7.12
%
N=25; Ftype=1; WnL=0.5*pi; WnH=0; Wtype=1;fs=8000;
%design using the rectangular window;
Brec=firwd(N,Ftype,WnL,WnH,Wtype);
N=25; Ftype=1; WnL=0.5*pi; WnH=0; Wtype=4;
%design using the Hamming window;
Bham=firwd(N,Ftype,WnL,WnH,Wtype);
[hrec,f]=freqz(Brec,1,512,fs);
[hham,f]=freqz(Bham,1,512,fs);
prec=180*unwrap(angle(hrec))/pi;
pham=180*unwrap(angle(hham))/pi;
subplot(2,1,1);
```

Table 7.6 FIR Filter Coefficients in Example 7.7 (rectangular and Hamming windows)

B: FIR Filter Coefficients (Rectangular Window)	Bham: FIR Filter Coefficients (Hamming Window)
$b_0 = b_{24} = 0.000000$	$b_0 = b_{24} = 0.000000$
$b_1 = b_{23} = -0.028937$	$b_1 = b_{23} = -0.002769$
$b_2 = b_{22} = 0.000000$	$b_2 = b_{22} = 0.000000$
$b_3 = b_{21} = 0.035368$	$b_3 = b_{21} = 0.007595$
$b_4 = b_{20} = 0.000000$	$b_4 = b_{20} = 0.000000$
$b_5 = b_{19} = -0.045473$	$b_5 = b_{19} = -0.019142$
$b_6 = b_{18} = 0.000000$	$b_6 = b_{18} = 0.000000$
$b_7 = b_{17} = 0.063662$	$b_7 = b_{17} = 0.041957$
$b_8 = b_{16} = 0.000000$	$b_8 = b_{16} = 0.000000$
$b_9 = b_{15} = -0.106103$	$b_9 = b_{15} = -0.091808$
$b_{10} = b_{14} = 0.000000$	$b_{10} = b_{14} = 0.000000$
$b_{11} = b_{13} = 0.318310$	$b_{11} = b_{13} = 0.313321$
$b_{12} = 0.500000$	$b_{12} = 0.500000$

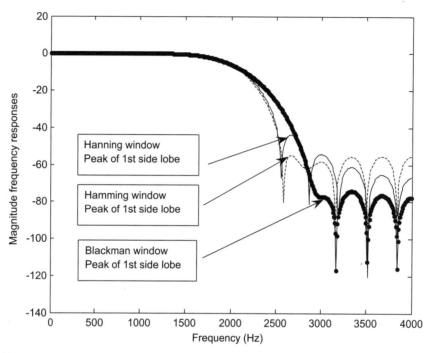

FIGURE 7.13

Comparisons of magnitude frequency responses for the Hanning, Hamming, and Blackman windows.

```
plot(f,20*log10(abs(hrec)),'-.',f,20*log10(abs(hham)));grid
axis([0 4000 -100 10]);
xlabel('Frequency (Hz)'); ylabel('Magnitude Response (dB)');
subplot(2,1,2);
plot(f,prec,'-.',f,pham);grid
xlabel('Frequency (Hz)'); ylabel('Phase (degrees)');
```

For comparison, the frequency responses achieved from the rectangular window and the Hamming window are plotted in Figure 7.12, where the dash-dotted line indicates the frequency response via the rectangular window, and the solid line indicates the frequency response via the Hamming window.

c. The FIR filter coefficients for both methods are listed in Table 7.6.

For comparison with other window functions, Figure 7.13 shows the magnitude frequency responses using the Hanning, Hamming, and Blackman windows, with 25 taps and a cutoff frequency of 2,000 Hz. The Blackman window offers the lowest side lobe, but with an increased width of the main lobe. The Hamming window and Hanning have a similar narrow width of the main lobe, but the Hamming window accommodates a lower side lobe than the Hanning window. Next, we will study how to choose a window in practice.

Applying the window to remedy the Gibbs effect will change the characteristics of the magnitude frequency response of the FIR filter, as the width of the main lobe becomes wider and more attenuation of the side lobes occurs.

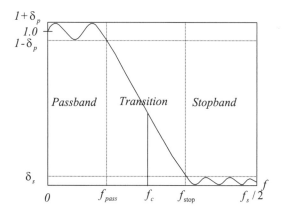

FIGURE 7.14

Lowpass filter frequency domain specifications.

Next, we illustrate the design for customer specifications in practice. Given the required stopband attenuation and passband ripple specifications shown in Figure 7.14, where the lowpass filter specifications are given for illustrative purposes, the appropriate window can be selected based on the performance of the window functions listed in Table 7.7. For example, the Hamming window offers a passband ripple of 0.0194 dB and stopband attenuation of 53 dB. With the selected Hamming window and the normalized transition band defined in Table 7.7,

$$\Delta f = \left| f_{stop} - f_{pass} \right| \big/ f_s \qquad (7.22)$$

TABLE 7.7 FIR Filter Length Estimation Using Window Functions
(normalized transition width $\Delta f = \left| f_{stop} - f_{pass} \right| / f_s$)

Window Type	Window Function $w(n), -M \leq n \leq M$	Window Length N	Passband Ripple (dB)	Stopband Attenuation (dB)
Rectangular	1	$N = 0.9/\Delta f$	0.7416	21
Hanning	$0.5 + 0.5\cos\left(\dfrac{\pi n}{M}\right)$	$N = 3.1/\Delta f$	0.0546	44
Hamming	$0.54 + 0.46\cos\left(\dfrac{\pi n}{M}\right)$	$N = 3.3/\Delta f$	0.0194	53
Blackman	$0.42 + 0.5\cos\left(\dfrac{n\pi}{M}\right)$ $+0.08\cos\left(\dfrac{2n\pi}{M}\right)$	$N = 5.5/\Delta f$	0.0017	74

the filter length using the Hamming window can be determined by

$$N = \frac{3.3}{\Delta f} \tag{7.23}$$

Note that the passband ripple is defined as

$$\delta_p dB = 20 \cdot \log_{10}(1 + \delta_p) \tag{7.24}$$

while the stopband attenuation is defined as

$$\delta_s dB = -20 \log_{10}(\delta_s) \tag{7.25}$$

The cutoff frequency used for the design will be chosen at the middle of the transition band, as illustrated for the lowpass filter case shown in Figure 7.14.

As a rule of thumb, the cutoff frequency used for design is determined by

$$f_c = (f_{pass} + f_{stop})/2 \tag{7.26}$$

Note that Equation (7.23) and formulas for other window lengths in Table 7.7 are empirically derived based on the normalized spectral transition width of each window function. The spectrum of each window function appears to be shaped like the lowpass filter magnitude frequency response with ripples in the passband and side lobes in the stopband. The passband frequency edge of the spectrum is the frequency where the magnitude just begins to drop below the passband ripple and where the stop frequency edge is at the peak of the first side lobe in the spectrum. With the passband ripple and stopband attenuation specified for a particular window, the normalized transition width of the window is in inverse proportion to the window length N multiplied by a constant. For example, the normalized spectral transition Δf for the Hamming window is $3.3/N$. Hence, matching the FIR filter transition width with the transition width of the window spectrum gives the filter length estimation listed in Table 7.7.

The following examples illustrate the determination of each filter length and cutoff frequency/frequencies for the design of lowpass, highpass, bandpass, and bandstop filters. Application of each designed filter to the processing of speech data is included, along with an illustration of filtering effects in both the time domain and frequency domain.

EXAMPLE 7.8

A lowpass FIR filter has the following specifications:

Passband 0–1,850 Hz
Stopband 2,150–4,000 Hz
Stopband attenuation 20 dB
Passband ripple 1 dB
Sampling rate 8,000 Hz

Determine the FIR filter length and the cutoff frequency to be used in the design equation.

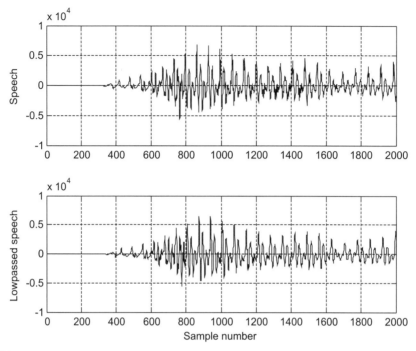

FIGURE 7.15A

Original speech and processed speech using the lowpass filter.

Solution:

The normalized transition band as defined in Equation (7.22) and Table 7.7 is given by

$$\Delta f = |2,150 - 1,850|/8,000 = 0.0375$$

Again, based on Table 7.7, selecting the rectangular window will result in a passband ripple of 0.74 dB and stopband attenuation of 21 dB. Thus, this window selection would satisfy the design requirement for a passband ripple of 1 dB and stopband attenuation of 20 dB. Next, we determine the length of the filter as

$$N = 0.9/\Delta f = 0.9/0.0375 = 24$$

We choose the odd number $N = 25$. The cutoff frequency is determined by $(1,850 + 2,150)/2 = 2,000$ Hz. Such a filter has been designed in Example 7.7, its filter coefficients are listed in Table 7.6, and its frequency responses can be found in Figure 7.12 (dashed lines).

Now we look at the time domain and frequency domain results from filtering a speech signal by using the lowpass filter we have just designed. Figure 7.15A shows the original speech and lowpass filtered speech. The spectral comparison is given in Figure 7.15B, where, as we can see, the frequency components beyond 2 kHz are filtered. The lowpass filtered speech would sound muffled.

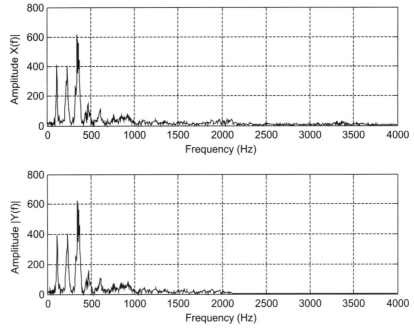

FIGURE 7.15B

Spectral plots of the original speech and processed speech by the lowpass filter.

We will continue to illustrate the determination of the filter length and cutoff frequency for other types of filters via the following examples.

EXAMPLE 7.9

Design a highpass FIR filter with the following specifications:

Stopband 0–1,500 Hz
Passband 2,500–4,000 Hz
Stopband attenuation 40 dB
Passband ripple 0.1 dB
Sampling rate 8,000 Hz

Solution:
Based on the specification, the Hanning window will do the job since it has a passband ripple of 0.0546 dB and stopband attenuation of 44 dB.
    Then

$$\Delta f = |1,500 - 2,500|/8,000 = 0.125$$

Table 7.8 FIR Filter Coefficients in Example 7.9 (Hanning window)

Bhan: FIR Filter Coefficients (Hanning Window)

$b_0 = b_{24} = 0.000000$	$b_1 = b_{23} = 0.000493$
$b_2 = b_{22} = 0.000000$	$b_3 = b_{21} = -0.005179$
$b_4 = b_{20} = 0.000000$	$b_5 = b_{19} = 0.016852$
$b_6 = b_{18} = 0.000000$	$b_7 = b_{17} = -0.040069$
$b_8 = b_{16} = 0.0000000$	$b_9 = b_{15} = 0.090565$
$b_{10} = b_{14} = 0.000000$	$b_{11} = b_{13} = -0.312887$
$b_{12} = 0.500000$	

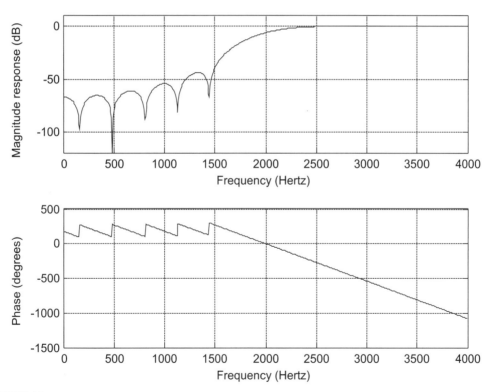

FIGURE 7.16

Frequency responses of the designed highpass filter using the Hanning window.

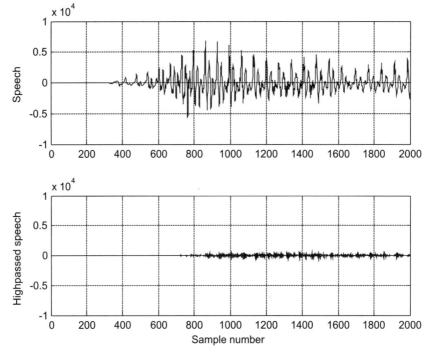

FIGURE 7.17A

Original speech and processed speech using the highpass filter.

$$N = 3.1/\Delta f = 24.2 \text{ (choose } N = 25)$$

Hence, we choose 25 filter coefficients using the Hanning window method. The cutoff frequency is $(1,500 + 2,500)/2 = 2,000$ Hz. The normalized cutoff frequency can be easily found as

$$\Omega_c = \frac{2,000 \times 2\pi}{8,000} = 0.5\pi \text{ radians}$$

Notice that $2M + 1 = 25$. The application program and design results are listed in Program 7.4 and Table 7.8.

Program 7.4. MATLAB program for Example 7.9

```
%Figure 7.16(Example 7.9)
% MATLAB program to create Figure 7.16
%
N=25; Ftype=2; WnL=0; WnH=0.5*pi; Wtype=3;fs=8000;
Bhan=firwd(N,Ftype,WnL,WnH,Wtype);
freqz(Bhan,1,512,fs);
axis([0 fs/2 -120 10]);
```

The corresponding frequency responses of the designed highpass FIR filter are displayed in Figure 7.16.

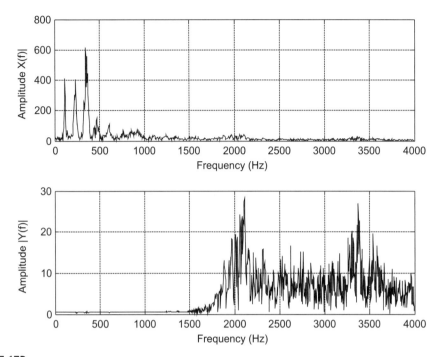

FIGURE 7.17B

Spectral comparison of the original speech and processed speech using the highpass filter.

Comparisons are given in Figure 7.17A, where the original speech and processed speech using the highpass filter are plotted, respectively. The high-frequency components of speech generally contain a small amount of energy. Figure 7.17B displays the spectral plots, where clearly the frequency components lower than 1.5 kHz are filtered. The processed speech would sound crisp.

EXAMPLE 7.10

Design a bandpass FIR filter with the following specifications:

Lower stopband 0–500 Hz
Passband 1,600–2,300 Hz
Upper stopband 3,500–4,000 Hz
Stopband attenuation 50 dB
Passband ripple 0.05 dB
Sampling rate 8,000 Hz

Solution:

$$\Delta f_1 = |1,600 - 500|/8,000 = 0.1375 \quad \text{and} \quad \Delta f_2 = |3,500 - 2,300|/8,000 = 0.15$$

$$N_1 = 3.3/0.1375 = 24 \quad \text{and} \quad N_2 = 3.3/0.15 = 22$$

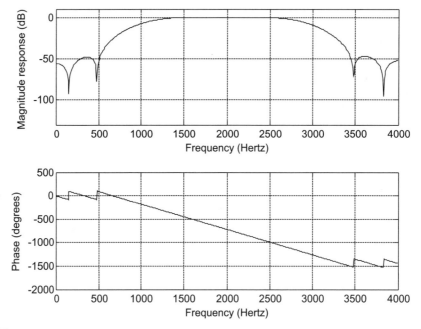

FIGURE 7.18

Frequency responses of the designed bandpass filter using the Hamming window.

We select $N = 25$ for the number of filter coefficients and using the Hamming window method, we next obtain

$$f_1 = (1,600 + 500)/2 = 1,050 \text{ Hz} \quad \text{and} \quad f_2 = (3,500 + 2,300)/2 = 2,900 \text{ Hz}$$

The normalized lower and upper cutoff frequencies are calculated as

$$\Omega_L = \frac{1,050 \times 2\pi}{8,000} = 0.2625\pi \text{ radians}$$

$$\Omega_H = \frac{2,900 \times 2\pi}{8,000} = 0.725\pi \text{ radians}$$

and $N = 2M + 1 = 25$. The design results are shown using the MATLAB program in Program 7.5.

Program 7.5. MATLAB program for Example 7.10.

```
%Figure 7.18(Example 7.10)
% MATLAB program to create Figure 7.18
%
N=25; Ftype=3; WnL=0.2625*pi; WnH=0.725*pi; Wtype=4;fs=8000;
Bham=firwd(N,Ftype,WnL,WnH,Wtype);
freqz(Bham,1,512,fs);
axis([0 fs/2 -130 10]);
```

Figure 7.18 depicts the frequency responses of the designed bandpass FIR filter. Table 7.9 lists the designed FIR filter coefficients.

Table 7.9 FIR Filter Coefficients in Example 7.10 (Hamming Window)	
Bham: FIR Filter Coefficients (Hamming Window)	
$b_0 = b_{24} = 0.002680$	$b_1 = b_{23} = -0.001175$
$b_2 = b_{22} = -0.007353$	$b_3 = b_{21} = 0.000674$
$b_4 = b_{20} = -0.011063$	$b_5 = b_{19} = 0.004884$
$b_6 = b_{18} = 0.053382$	$b_7 = b_{17} = -0.003877$
$b_8 = b_{16} = 0.028520$	$b_9 = b_{15} = -0.008868$
$b_{10} = b_{14} = -0.296394$	$b_{11} = b_{13} = 0.008172$
$b_{12} = 0.462500$	

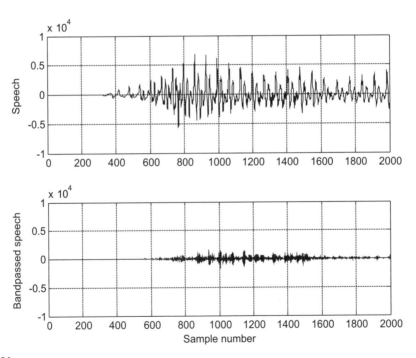

FIGURE 7.19A

Original speech and processed speech using the bandpass filter.

For comparison, the original speech and bandpass filtered speech are plotted in Figure 7.19A, where the bandpass frequency components contain a small portion of speech energy. Figure 7.19B shows a comparison indicating that low and high frequencies are removed by the bandpass filter.

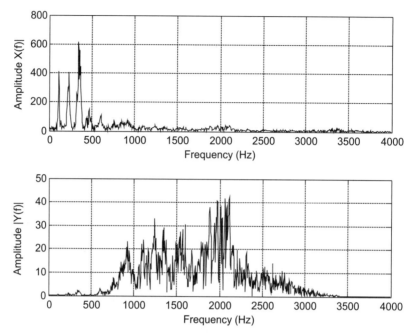

FIGURE 7.19B

Spectral comparison of the original speech and processed speech using the bandpass filter.

EXAMPLE 7.11

Design a bandstop FIR filter with the following specifications:

Lower cutoff frequency 1,250 Hz
Lower transition width 1,500 Hz
Upper cutoff frequency 2,850 Hz
Upper transition width 1,300 Hz
Stopband attenuation 60 dB
Passband ripple 0.02 dB
Sampling rate 8,000 Hz

Solution:
We can directly compute the normalized transition width:

$$\Delta f_1 = 1,500/8,000 = 0.1875 \quad \text{and} \quad \Delta f_2 = 1,300/8,000 = 0.1625$$

The filter lengths are determined using the Blackman windows as

$$N_1 = 5.5/0.1875 = 29.33, \quad \text{and} \quad N_2 = 5.5/0.1625 = 33.8$$

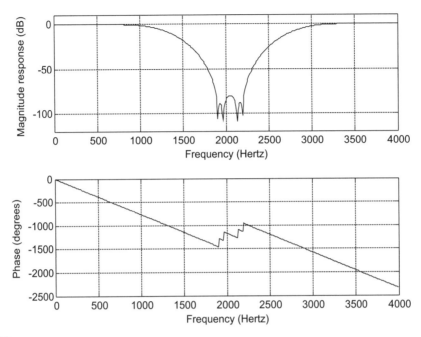

FIGURE 7.20

Frequency responses of the designed bandstop filter using the Blackman window.

We choose $N = 35$, an odd number. The normalized lower and upper cutoff frequencies are calculated as

$$\Omega_L = \frac{2\pi \times 1,250}{8,000} = 0.3125\pi \text{ radians}$$

$$\Omega_H = \frac{2\pi \times 2,850}{8,000} = 0.7125\pi \text{ radians}$$

and $N = 2M + 1 = 35$. Using MATLAB, the design results are demonstrated in Program 7.6.

Program 7.6. MATLAB program for Example 7.11.

```
%Figure 7.20 (Example 7.11)
% MATLAB program to create Figure 7.20
%
N=35; Ftype=4; WnL=0.3125*pi; WnH=0.7125*pi; Wtype=5;fs=8000;
Bblack=firwd(N,Ftype,WnL,WnH,Wtype);
freqz(Bblack,1,512,fs);
axis([0 fs/2 -120 10]);
```

Figure 7.20 shows the plot of the frequency responses of the designed bandstop filter. The designed filter coefficients are listed in Table 7.10.

Comparisons of filtering effects are illustrated in Figures 7.21A and 7.21B. In Figure 7.21A, the original speech and speech processed by the bandstop filter are plotted. The processed speech contains most of the energy of the original speech because most of the energy of the speech signal exists in the low-frequency band. Figure 7.21B verifies the filtering frequency effects. The requency components ranging from 2,000 Hz to 2,200 Hz have been completely removed.

Table 7.10 FIR Filter Coefficients in Example 7.11 (Blackman Window)

Black: FIR Filter Coefficients (Blackman Window)	
$b_0 = b_{34} = 0.000000$	$b_1 = b_{33} = 0.000059$
$b_2 = b_{32} = 0.000000$	$b_3 = b_{31} = 0.000696$
$b_4 = b_{30} = 0.001317$	$b_5 = b_{29} = -0.004351$
$b_6 = b_{28} = -0.002121$	$b_7 = b_{27} = 0.000000$
$b_8 = b_{26} = -0.004249$	$b_9 = b_{25} = 0.027891$
$b_{10} = b_{24} = 0.011476$	$b_{11} = b_{23} = -0.036062$
$b_{12} = b_{22} = 0.000000$	$b_{13} = b_{21} = -0.073630$
$b_{14} = b_{20} = -0.020893$	$b_{15} = b_{19} = 0.285306$
$b_{16} = b_{18} = 0.014486$	$b_{17} = 0.600000$

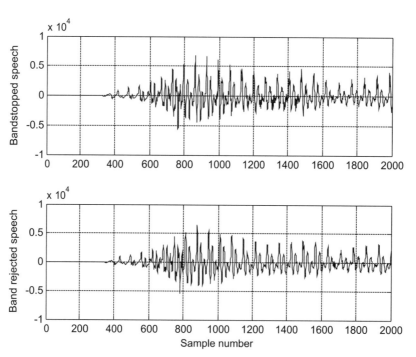

FIGURE 7.21A

Original speech and processed speech using the bandstop filter.

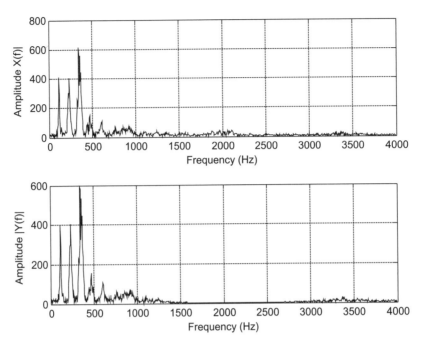

FIGURE 7.21B

Spectral comparison of the original speech and processed speech using the bandstop filter.

7.4 APPLICATIONS: NOISE REDUCTION AND TWO-BAND DIGITAL CROSSOVER

In this section, we will investigate noise reduction and digital crossover design using FIR filters.

7.4.1 Noise Reduction

One of the key digital signal processing (DSP) applications is noise reduction. In this application, a digital FIR filter removes noise in a signal that is contaminated by noise existing in a broad frequency range. For example, such noise often appears during the data acquisition process. In real-world applications, the desired signal usually occupies a certain frequency range. We can design a digital filter to remove frequency components other than the desired frequency range.

In a data acquisition system, we record a 500 Hz sine wave at a sampling rate of 8,000 samples per second. The signal is corrupted by broadband noise $v(n)$:

$$x(n) = 1.4141 \cdot \sin(2\pi \cdot 500n/8,000) + v(n)$$

The 500 Hz signal with noise and its spectrum are plotted in Figure 7.22, from which it is obvious that the digital sine wave contains noise. The spectrum is also displayed to give a better understanding of the noise frequency level. We can see that noise is broadband, existing from 0 Hz to the folding frequency of 4,000 Hz. Assuming that the desired signal has a frequency range of only 0 to 800 Hz, we can filter noise from 800 Hz and beyond. A lowpass filter would complete such a task. Then we develop the filter specifications:

Passband frequency range: 0 Hz to 800 Hz with the passband ripple less than 0.02 dB
Stopband frequency range: 1 kHz to 4 kHz with 50 dB attenuation

As we will see, lowpass filtering will remove noise in the range 1,000 Hz to 4,000 Hz , and hence the signal to noise power ratio will be improved.

Based on the specifications, we design an FIR filter with the Hamming window, a cutoff frequency of 900 Hz, and an estimated filter length of 133 taps. The enhanced signal is depicted in Figure 7.23, where the clean signal can be observed. The amplitude spectrum for the enhanced signal is also plotted. As shown in the spectral plot, the noise level is negligible between 1 and 4 kHz. Notice that since we use the higher-order FIR filter, the signal experiences a linear phase delay of 66 samples, as is expected. We also see some transient response effects. However, the transient response effects will end after the first 132 samples due to the length of the FIR filter. MATLAB implementation is given in Program 7.7.

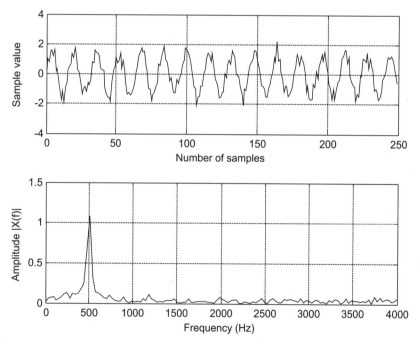

FIGURE 7.22

Signal with noise and its spectrum.

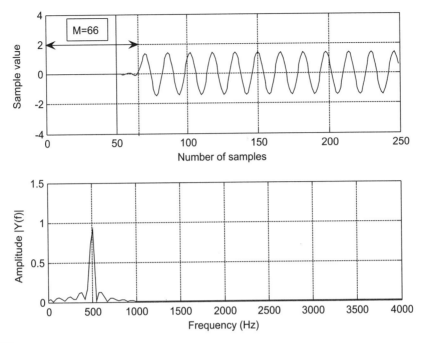

FIGURE 7.23

The clean signal and spectrum with noise removed.

Program 7.7. MATLAB program for the application of noise filtering.

```
close all; clear all
fs=8000;                              % Sampling rate
T=1/fs;                               % Sampling period
v=sqrt(0.1)*randn(1,250);             % Generate Gaussian random noise
n=0:1:249;                            % Indexes
x=sqrt(2)*sin(2*pi*500*n*T)+v;        % Generate 500-Hz plus noise
subplot(2,1,1);plot(n,x);
xlabel('Number of samples');ylabel('Sample value');grid;
N=length(x);
f=[0:N/2]*fs/N;
Axk=2*abs(fft(x))/N;Axk(1)=Axk(1)/2;  % Calculate single side spectrum for x(n)
subplot(2,1,2); plot(f,Axk(1:N/2+1));
xlabel('Frequency (Hz)'); ylabel('Amplitude |X(f)| ');grid;
figure
Wnc=2*pi*900/fs;                      % Determine the normalized digital cutoff
frequency
B=firwd(133,1,Wnc,0,4);               % Design FIR filter
y=filter(B,1,x);                      % Perform digital filtering
Ayk=2*abs(fft(y))/N;Ayk(1)=Ayk(1)/2;  % Single-side spectrum of the filtered data
subplot(2,1,1); plot(n,y);
```

```
xlabel('Number of samples');ylabel('Sample value');grid;
subplot(2,1,2);plot(f,Ayk(1:N/2+1)); axis([0 fs/2 0 1.5]);
xlabel('Frequency (Hz)'); ylabel('Amplitude |Y(f)| ');grid;
```

7.4.2 Speech Noise Reduction

In a speech recording system, we digitally record speech in a noisy environment at a sampling rate of 8,000 Hz. Assuming the recorded speech contains information within 1,800 Hz, we can design a lowpass filter to remove the noise between 1,800 Hz and the Nyquist limit (the folding frequency of 4,000 Hz). The filter specifications are listed below:

> Filter type: lowpass FIR filter
> Passband frequency range: 0–1,800 Hz
> Passband ripple: 0.02 dB
> Stopband frequency range: 2,000–4,000 Hz
> Stopband attenuation: 50 dB

According to these specifications, we can determine the following parameters for filter design:

> Window type = Hamming window
> Number of filter taps = 133
> Lowpass cutoff frequency = 1,900 Hz

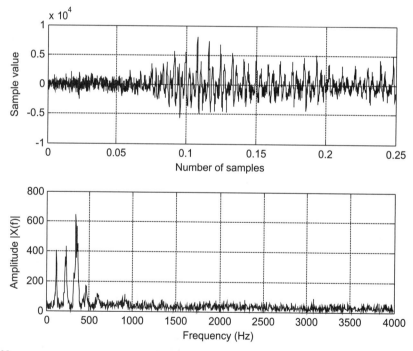

FIGURE 7.24A

Noisy speech and its spectrum.

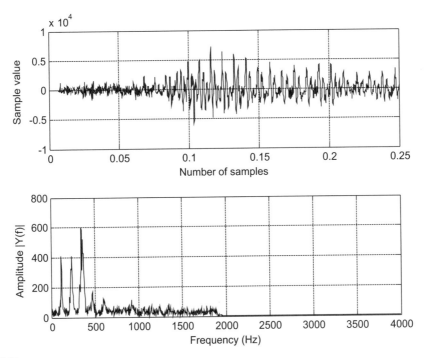

FIGURE 7.24B

Enhanced speech and its spectrum.

Figure 7.24A shows the plots of the recorded noisy speech and its spectrum. As we can see in the noisy spectrum, the noise level is high and broadband. After applying the designed lowpass filter, we plot the filtered speech and its spectrum shown in Figure 7.24B, where the clean speech is clearly identified, while the spectrum shows that the noise components above 2 kHz have been completely removed.

7.4.3 Noise Reduction in Vibration Signals

In a data acquisition system for vibration analysis, we captured a vibration signal using an accelerometer sensor in the noisy environment. The sampling rate is 1,000 Hz. The captured signal is significantly corrupted by a broadband noise. Vibration analysis requires the first dominant frequency component in the range 35 to 50 Hz to be retrieved. We list the filter specifications below:

    Filter type = bandpass FIR filter
    Passband frequency range = 35–50 Hz
    Passband ripple = 0.02 dB
    Stopband frequency ranges = 0–15 and 70–500 Hz
    Stopband attenuation: 50 dB

According to these specifications, we can determine the following parameters for the filter design:

Window type = Hamming window
Number of filter taps = 167
Low cutoff frequency = 25 Hz
High cutoff frequency = 65 Hz

Figure 7.25 displays the plots of the recorded noisy vibration signal and its spectrum. Figure 7.26 shows the retrieved vibration signal with noise reduction by a bandpass filter.

7.4.4 Two-Band Digital Crossover

In audio systems, there is often a situation where the application requires the entire audible range of frequencies, but this is beyond the capability of any single speaker driver. So, we combine several drivers, such as the speaker cone and horns, each covering a different frequency range, to reproduce the full audio frequency range.

A typical two-band digital crossover can be designed as shown in Figure 7.27. There are two speaker drivers. The woofer responds to low frequencies, and the tweeter responds to high frequencies. The incoming digital audio signal is split into two bands by using a lowpass filter and a highpass filter in parallel. We then amplify the separated audio signals and send them to their respective corresponding speaker drivers. Hence, the objective is to design the lowpass filter and the highpass filter so

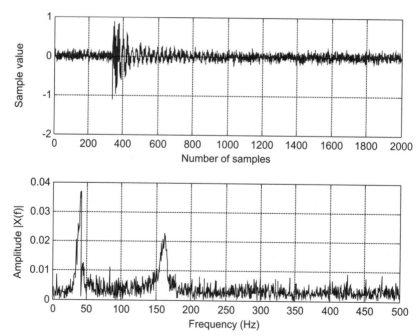

FIGURE 7.25

Noisy vibration signal and its spectrum.

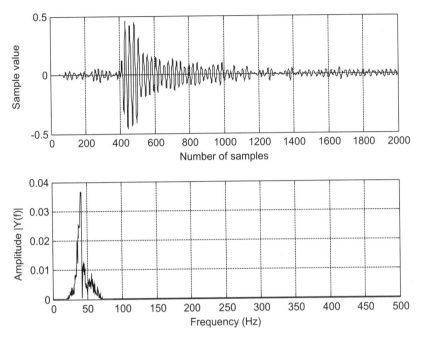

FIGURE 7.26

Retrieved vibration signal and its spectrum.

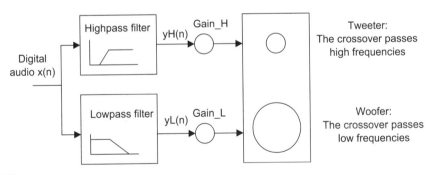

FIGURE 7.27

Two-band digital crossover.

that their combined frequency response is flat, while keeping transitions as sharp as possible to prevent audio signal distortion in the transition frequency range. Although traditional crossover systems are designed using active circuits (analog systems) or passive circuits, the digital crossover system provides a cost-effective solution with programmability, flexibility, and high quality.

A crossover system has the following specifications:

Sampling rate = 44,100 Hz

Crossover frequency = 1,000 Hz (cutoff frequency)

Transition band = 600 Hz to 1,400 Hz

Lowpass filter = passband frequency range from 0 to 600 Hz with a ripple of 0.02 dB and stopband edge at 1,400 Hz with an attenuation of 50 dB

Highpass filter = passband frequency range from 1.4 to 44.1 kHz with ripple of 0.02 dB and stopband edge at 600 Hz with an attenuation of 50 dB

In the design of this crossover system, one possibility is to use an FIR filter, since it provides a linear phase for the audio system. However, an infinite impulse response (IIR) filter (which will be discussed in the next chapter) is a possible alternative. Based on the transition band of 800 Hz and the passband ripple and stopband attenuation requirements, the Hamming window is chosen for both lowpass and highpass filters. We can determine the number of filter taps as 183, each with a cutoff frequency of 1,000 Hz.

The frequency responses for the designed lowpass and highpass filters are given in Figure 7.28A; the lowpass filter, highpass filter, and combined responses appear in Figure 7.28B. As we can see, the crossover frequency for both filters is at 1,000 Hz, and the combined frequency response is perfectly flat. The impulse responses (filter coefficients) for the lowpass and highpass filters are plotted in Figure 7.28C.

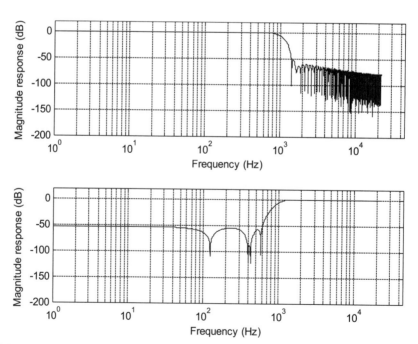

FIGURE 7.28A

Magnitude frequency responses for lowpass filter and highpass filter.

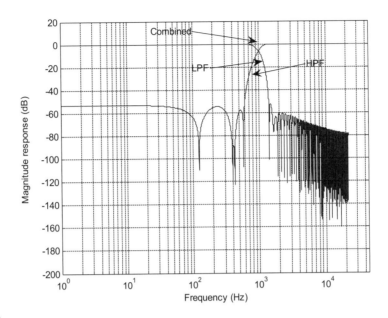

FIGURE 7.28B

Magnitude frequency responses for both the lowpass and highpass filters, and the combined magnitude frequency response for the digital audio crossover system.

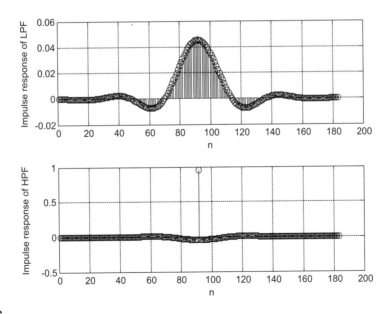

FIGURE 7.28C

Impulse responses of both the FIR lowpass filter and the FIR highpass filter for the digital audio crossover system.

7.5 FREQUENCY SAMPLING DESIGN METHOD

In addition to methods of Fourier transform design and Fourier transform with windowing discussed in the previous section, *frequency sampling* is another alternative. The key feature of frequency sampling is that the filter coefficients can be calculated based on the specified magnitudes of the desired filter frequency response uniformly in the frequency domain. Hence, it has design flexibility.

To begin development, we let $h(n)$, for $n = 0, 1, \cdots, N - 1$, be the causal impulse response (FIR filter coefficients) that approximates the FIR filter, and we let $H(k)$, for $k = 0, 1, \cdots, N - 1$, represent the corresponding discrete Fourier transform (DFT) coefficients. We obtain $H(k)$ by sampling the desired frequency filter response $H(k) = H(e^{j\Omega_k})$ at equally spaced instants in frequency domain, as shown in Figure 7.29.

Then, according to the definition of the inverse DFT (IDFT), we can calculate the FIR coefficients:

$$h(n) = \frac{1}{N} \sum_{k=0}^{N-1} H(k) W_N^{-kn}, \quad \text{for} \quad n = 0, 1, \cdots, N - 1 \tag{7.27}$$

where

$$W_N = e^{-j\frac{2\pi}{N}} = \cos\left(\frac{2\pi}{N}\right) - j\sin\left(\frac{2\pi}{N}\right)$$

We assume that the FIR filter has linear phase and the number of taps is $N = 2M + 1$. Equation (7.27) can be significantly simplified as

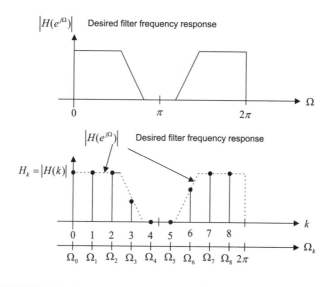

FIGURE 7.29

Desired filter frequency response and sampled frequency response.

$$h(n) = \frac{1}{2M+1}\left\{ H_0 + 2\sum_{k=1}^{M} H_k \cos\left(\frac{2\pi k(n-M)}{2M+1}\right)\right\}, \quad \text{for} \quad n = 0, 1, \cdots, 2M \qquad (7.28)$$

where H_k, for $k = 0, 1, \cdots, 2M$, represents the magnitude values specifying the desired filter frequency response sampled at $\Omega_k = \frac{2\pi k}{(2M+1)}$. The derivation is detailed in Appendix E. The design procedure is therefore simply summarized as follows:

1. Given the filter length of $2M + 1$, specify the magnitude frequency response for the normalized frequency range from 0 to π:

$$H_k \text{ at } \Omega_k = \frac{2\pi k}{(2M+1)} \quad \text{for} \quad k = 0, 1, \cdots, M \qquad (7.29)$$

2. Calculate the FIR filter coefficients:

$$b_n = h(n) = \frac{1}{2M+1}\left\{ H_0 + 2\sum_{k=1}^{M} H_k \cos\left(\frac{2\pi k(n-M)}{2M+1}\right)\right\} \quad \text{for} \quad n = 0, 1, \cdots, M \qquad (7.30)$$

3. Use symmetry (linear phase requirement) to determine the rest of coefficients:

$$h(n) = h(2M - n) \quad \text{for} \quad n = M+1, \cdots, 2M \qquad (7.31)$$

Example 7.12 illustrates the design procedure.

EXAMPLE 7.12

Design a linear phase lowpass FIR filter with 7 taps and a cutoff frequency of $\Omega_c = 0.3\pi$ radians using the frequency sampling method.

Solution:

Since $N = 2M + 1 = 7$ and $M = 3$, the sampled frequencies are given by

$$\Omega_k = \frac{2\pi}{7}k \text{ radians}, \quad k = 0, 1, 2, 3$$

Next we specify the magnitude values H_k at the specified frequencies as follows:

$$\text{for } \Omega_0 = 0 \text{ radians}, \quad H_0 = 1.0$$

$$\text{for } \Omega_1 = \frac{2}{7}\pi \text{ radians}, \quad H_1 = 1.0$$

$$\text{for } \Omega_2 = \frac{4}{7}\pi \text{ radians}, \quad H_2 = 0.0$$

$$\text{for } \Omega_3 = \frac{6}{7}\pi \text{ radians}, \quad H_3 = 0.0$$

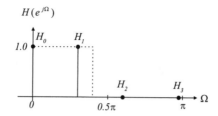

FIGURE 7.30

Sampled values of the frequency response in Example 7.12.

Figure 7.30 shows the specifications.
   Using Equation (7.30), we achieve

$$h(n) = \frac{1}{7}\left\{1 + 2\sum_{k=1}^{3} H_k \cos\left[2\pi k(n-3)/7\right]\right\}, \quad n = 0, 1, \cdots, 3.$$

$$= \frac{1}{7}\{1 + 2\cos\left[2\pi(n-3)/7\right]\}$$

Thus, computing the FIR filter coefficients yields

$$h(0) = \frac{1}{7}\{1 + 2\cos\left(-6\pi/7\right)\} = -0.11456$$

$$h(1) = \frac{1}{7}\{1 + 2\cos\left(-4\pi/7\right)\} = 0.07928$$

$$h(2) = \frac{1}{7}\{1 + 2\cos\left(-2\pi/7\right)\} = 0.32100$$

$$h(3) = \frac{1}{7}\{1 + 2\cos\left(-0 \times \pi/7\right)\} = 0.42857$$

By symmetry, we obtain the rest of the coefficients as follows:

$$h(4) = h(2) = 0.32100$$
$$h(5) = h(1) = 0.07928$$
$$h(6) = h(0) = -0.11456$$

The following two examples are devoted to illustrating the FIR filter design using the frequency sampling method. A MATLAB program, **firfs(N, Hk),** is provided in the "MATLAB Programs" section at the end of this chapter (see its usage in Table 7.11) to implement the design in Equation (7.30) with the input parameters of $N = 2M + 1$

Table 7.11 Illustrative Usage for MATLAB Function **firfs(N, Hk)**

function B=firfs(N,Hk)
% B=firls(N,Hk)
% FIR filter design using the frequency sampling method.
% Input parameters:
% N: the number of filter coefficients.
%   note: N must be odd number.
% Hk: sampled frequency response for k=0,1,2,...,M=(N-1)/2.
% Output:
% B: FIR filter coefficients.

(number of taps) and a vector **Hk** containing the specified magnitude values $H_k, k = 0, 1, \cdots, M$. Finally, the MATLAB function will return the calculated FIR filter coefficients.

EXAMPLE 7.13

a. Design a linear phase lowpass FIR filter with 25 coefficients using the frequency sampling method. Let the cutoff frequency be 2,000 Hz and assume a sampling frequency of 8,000 Hz.
b. Plot the frequency responses.
c. List the FIR filter coefficients.

Solution:
a. The normalized cutoff frequency for the lowpass filter is $\Omega_c = \omega T = 2\pi 2000/8000 = 0.5\pi$ radians, $N = 2M + 1 = 25$, and the specified values of the sampled frequency response are chosen to be

$$H_k = [1\ 1\ 1\ 1\ 1\ 1\ 1\ 0\ 0\ 0\ 0\ 0\ 0]$$

MATLAB Program 7.8 produces the design results.

Program 7.8. MATLAB program for Example 7.13.

```
%Figure 7.31(Example 7.13)
% MATLAB program to create Figure 7.31
fs=8000;  % Sampling frequency
H1=[1 1 1 1 1 1 1 0 0 0 0 0 0];   % Magnitude specifications
B1=firfs(25,H1);                  % Design filter
[h1,f]=freqz(B1,1,512,fs);        % Calculate magnitude frequency response
H2=[1 1 1 1 1 1 1 0.5 0 0 0 0 0]; % Magnitude specifications
B2=firfs(25,H2);                  % Frequency response
[h2,f]=freqz(B2,1,512,fs);        % Calculate magnitude frequency response
p1=180*unwrap(angle(h1))/pi;
p2=180*unwrap(angle(h2))/pi
subplot(2,1,1); plot(f,20*log10(abs(h1)),'-.',f,20*log10(abs(h2)));grid
axis([0 fs/2 -80 10]);
xlabel('Frequency (Hz)'); ylabel('Magnitude Response (dB)');
subplot(2,1,2); plot(f,p1,'-.',f,p2);grid
xlabel('Frequency (Hz)'); ylabel('Phase (degrees)');
```

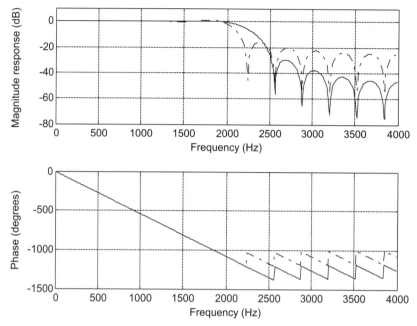

FIGURE 7.31

Frequency responses using the frequency sampling method in Example 7.13.

Table 7.12 FIR Filter Coefficients in Example 7.13 (frequency sampling method)

B1: FIR Filter Coefficients	B2: FIR Filter Coefficients
$b_0 = b_{24} = 0.027436$	$b_0 = b_{24} = 0.001939$
$b_1 = b_{23} = -0.031376$	$b_1 = b_{23} = 0.003676$
$b_2 = b_{22} = -0.024721$	$b_2 = b_{22} = -0.012361$
$b_3 = b_{21} = 0.037326$	$b_3 = b_{21} = -0.002359$
$b_4 = b_{20} = 0.022823$	$b_4 = b_{20} = 0.025335$
$b_5 = b_{19} = -0.046973$	$b_5 = b_{19} = -0.008229$
$b_6 = b_{18} = -0.021511$	$b_6 = b_{18} = -0.038542$
$b_7 = b_{17} = 0.064721$	$b_7 = b_{17} = 0.032361$
$b_8 = b_{16} = 0.020649$	$b_8 = b_{16} = 0.049808$
$b_9 = b_{15} = -0.106734$	$b_9 = b_{15} = -0.085301$
$b_{10} = b_{14} = -0.020159$	$b_{10} = b_{14} = -0.057350$
$b_{11} = b_{13} = 0.318519$	$b_{11} = b_{13} = 0.311024$
$b_{12} = 0.520000$	$b_{12} = 0.560000$

b. The magnitude frequency response plotted using the dash-dotted line is displayed in Figure 7.31, where it is observed that oscillations occur in the passband and stopband of the designed FIR filter. This is due to the abrupt change of the specification in the transition band (between the passband and the stopband). To reduce this ripple effect, the modified specification with a smooth transition band, H_k, $k = 0, 1, \cdots, 13$, is used:

$$H_k = [1\ 1\ 1\ 1\ 1\ 1\ 1\ 0.5\ 0\ 0\ 0\ 0\ 0]$$

The improved magnitude frequency response is shown in Figure 7.31 via the solid line.
c. The calculated FIR coefficients for both filters are listed in Table 7.12.

EXAMPLE 7.14

a. Design a linear phase bandpass FIR filter with 25 coefficients using the frequency sampling method. Let the lower and upper cutoff frequencies be 1,000 Hz and 3,000 Hz, respectively, and assume a sampling frequency of 8,000 Hz.
b. List the FIR filter coefficients.
c. Plot the frequency responses.

Solution:
a. First we calculate the normalized lower and upper cutoff frequencies for the bandpass filter; that is, $\Omega_L = 2\pi \times 1,000/8,000 = 0.25\pi$ radians and $\Omega_H = 2\pi \times 3,000/8,000 = 0.75\pi$ radians, respectively. The sampled values of the bandpass frequency response are specified by the following vector:

$$H_k = [0\ 0\ 0\ 0\ 1\ 1\ 1\ 1\ 1\ 0\ 0\ 0\ 0]$$

For comparison, a second specification of H_k with a smooth transition band is used:

$$H_k = [0\ 0\ 0\ 0.5\ 1\ 1\ 1\ 1\ 1\ 0.5\ 0\ 0\ 0]$$

b. The MATLAB list is shown in Program 7.9. The generated FIR coefficients are listed in Table 7.13.

Program 7.9 MATLAB program for Example 7.14.

```
% Figure 7.32 (Example 7.14)
% MATLAB program to create Figure 7.32
%
fs=8000;
H1=[0 0 0 0 1 1 1 1 1 0 0 0 0];        % Magnitude specifications
B1=firfs(25,H1);                        % Design filter
[h1,w]=freqz(B1,1,512);                 % Calculate magnitude frequency response
H2=[0 0 0 0.5 1 1 1 1 1 0.5 0 0 0];    % Magnitude spectrum
B2=firfs(25,H2);                        % Design filter
[h2,w]=freqz(B2,1,512);                 % Calculate magnitude frequency response
p1=180*unwrap(angle(h1)')/pi;
p2=180*unwrap(angle(h2)')/pi
subplot(2,1,1); plot(f,20*log10(abs(h1)),'-.',f,20*log10(abs(h2)));grid
axis([0 fs/2 -100 10]);
xlabel('Frequency (Hz)'); ylabel('Magnitude Response (dB)');
subplot(2,1,2); plot(f,p1,'-.',f,p2);grid
xlabel('Frequency (Hz)'); ylabel('Phase (degrees)');
```

Table 7.13 FIR Filter Coefficients in Example 7.14 (frequency sampling method)

B1: FIR Filter Coefficients	B2: FIR Filter Coefficients
$b_0 = b_{24} = 0.055573$	$b_0 = b_{24} = 0.001351$
$b_1 = b_{23} = -0.030514$	$b_1 = b_{23} = -0.008802$
$b_2 = b_{22} = 0.000000$	$b_2 = b_{22} = -0.020000$
$b_3 = b_{21} = -0.027846$	$b_3 = b_{21} = 0.009718$
$b_4 = b_{20} = -0.078966$	$b_4 = b_{20} = -0.011064$
$b_5 = b_{19} = 0.042044$	$b_5 = b_{19} = 0.023792$
$b_6 = b_{18} = 0.063868$	$b_6 = b_{18} = 0.077806$
$b_7 = b_{17} = 0.000000$	$b_7 = b_{17} = -0.020000$
$b_8 = b_{16} = 0.094541$	$b_8 = b_{16} = 0.017665$
$b_9 = b_{15} = -0.038728$	$b_9 = b_{15} = -0.029173$
$b_{10} = b_{14} = -0.303529$	$b_{10} = b_{14} = -0.308513$
$b_{11} = b_{13} = 0.023558$	$b_{11} = b_{13} = 0.027220$
$b_{12} = 0.400000$	$b_{12} = 0.480000$

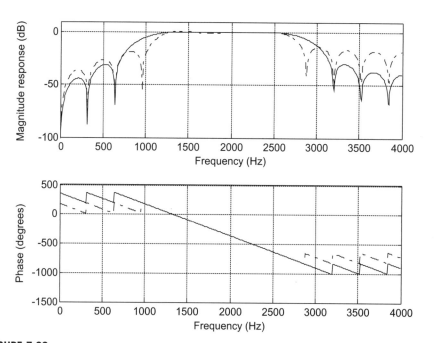

FIGURE 7.32

Frequency responses using the frequency sampling method in Example 7.14.

c. Similar to the preceding example, Figure 7.32 shows the frequency responses. Focusing on the magnitude frequency responses depicted in Figure 7.32, the dash-dotted line indicates the magnitude frequency response obtained without specifying the smooth transition band, while the solid line indicates the magnitude frequency response achieved with the specification of the smooth transition band, resulting in the reduced ripple effect.

Observations can be made from examining Examples 7.13 and 7.14. First, the oscillations (Gibbs behavior) in the passband and stopband can be reduced at the expense of increasing the width of the main lobe. Second, we can modify the specification of the magnitude frequency response with a smooth transition band to reduce the oscillations and thus improve the performance of the FIR filter. Third, the magnitude values H_k, $k = 0, 1, \cdots, M$ in general can be arbitrarily specified. This indicates that the frequency sampling method is more flexible and can be used to design the FIR filter with an arbitrary specification of the magnitude frequency response.

7.6 OPTIMAL DESIGN METHOD

This section introduces the Parks–McClellan algorithm, which is one of the most popular optimal design method used in the industry due to its efficiency and flexibility. The FIR filter design using the Parks–McClellan algorithm is developed based on the idea of minimizing the maximum approximation error between a Chebyshev polynomial and the desired filter magnitude frequency response. The details of this design development are beyond the scope of this text and can be found in Ambardar (1999) and Porat (1997). We will outline the design criteria and notation and then focus on the design procedure.

Given an ideal magnitude response $H_d(e^{j\omega T})$, the approximation error $E(\omega)$ is defined as

$$E(\omega) = W(\omega)[H(e^{j\omega T}) - H_d(e^{j\omega T})] \tag{7.32}$$

where $H(e^{j\omega T})$ is the frequency response of the linear phase FIR filter to be designed, and $W(\omega)$ is the weight function for emphasizing certain frequency bands over others during the optimization process. The goal is to minimize the error over the set of FIR coefficients:

$$\min(\max|E(\omega)|) \tag{7.33}$$

With the help of the Remez exchange algorithm, which is also beyond the scope of this book, we can obtain the best FIR filter whose magnitude response has an equiripple approximation to the ideal magnitude response. The achieved filters are optimal in the sense that the algorithms minimize the maximum error between the desired frequency response and actual frequency response. These are often called *minimax filters*.

Next, we establish the notation that will be used in the design procedure. Figure 7.33 shows the characteristics of the FIR filter designed by the Parks–McClellan and Remez exchange algorithms. As illustrated in the top graph of Figure 7.33, the passband frequency response and stopband frequency response have equiripples. δ_p is used to specify the magnitude ripple in the passband, while δ_s

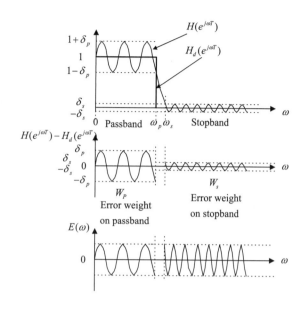

FIGURE 7.33

(Top) Magnitude frequency response of an ideal lowpass filter and a typical lowpass filter designed using the Parks–McClellan algorithm. (Middle) Error between the ideal and practical responses. (Bottom) Weighted error between the ideal and practical responses.

specifies the stopband magnitude attenuation. In terms of dB value specification, we have $\delta_p dB = 20 \times \log_{10}(1 + \delta_p)$ and $\delta_s dB = 20 \times \log_{10}\delta_s$.

The middle graph in Figure 7.33 describes the error between the ideal frequency response and the actual frequency response. In general, the error magnitudes in the passband and stopband are different. This makes optimization unbalanced, since the optimization process involves an entire band. When the error magnitude in a band dominates the other(s), the optimization process may deemphasize the contribution due to a small magnitude error. To make the error magnitudes balanced, a weight function can be introduced. The idea is to weight a band with a bigger magnitude error with a small weight factor and to weight a band with a smaller magnitude error with a big weight factor. We use a weight factor W_p for weighting the passband error and W_s for weighting the stopband error. The bottom graph in Figure 7.33 shows the weighted error, and clearly, the error magnitudes on both bands are at the same level. Selection of the weighting factors is further illustrated in the following design procedure.

Optimal FIR Filter Design Procedure for the Parks–McClellan Algorithm

1. Specify the band edge frequencies such as passband and stopband frequencies, passband ripple, stopband attenuation, filter order, and sampling frequency of the DSP system.
2. Normalize band edge frequencies to the Nyquist limit (folding frequency $= f_s/2$) and specify the ideal magnitudes.

3. Calculate the absolute values of the passband ripple and stopband attenuation if they are given in terms if dB values:

$$\delta_p = 10^{\left(\frac{\delta_p dB}{20}\right)} - 1 \tag{7.34}$$

$$\delta_s = 10^{\left(\frac{\delta_s dB}{20}\right)} \tag{7.35}$$

Then calculate the ratio and put it into fraction form:

$$\frac{\delta_p}{\delta_s} = fraction\ form = \frac{numerator}{denominator} = \frac{W_s}{W_p} \tag{7.36}$$

Next, set the error weight factors for passband and stopband, respectively:

$$W_s = numerator$$

$$W_p = denominator \tag{7.37}$$

4. Apply the Remez algorithm to calculate filter coefficients.
5. If the specifications are not met, then increase the filter order and repeat steps 1 to 4.

The following two examples are given to illustrate the design procedure.

EXAMPLE 7.15

Design a lowpass filter with the following specifications:

DSP system sampling rate = 8,000 Hz
Passband = 0–800 Hz
Stopband = 1,000–4,000 Hz
Passband ripple = 1 dB
Stopband attenuation = 40 dB
Filter order = 53

Solution:
From the specifications, we have two bands: a lowpass band and a stopband. We perform normalization and specify ideal magnitudes as follows:

Folding frequency: $f_s/2 = 8,000/2 = 4,000$ Hz
For 0 Hz: $0/4,000 = 0$ magnitude: 1
For 800 Hz: $800/4,000 = 0.2$ magnitude: 1
For 1,000 Hz: $1,000/4,000 = 0.25$ magnitude: 0
For 4,000 Hz: $4,000/4,000 = 1$ magnitude: 0

Next, let us determine the weights:

$$\delta_p = 10^{\left(\frac{1}{20}\right)} - 1 = 0.1220$$

$$\delta_s = 10^{\left(\frac{-40}{20}\right)} = 0.01$$

Then, applying Equation (7.36) gives

$$\frac{\delta_p}{\delta_s} = 12.2 \approx \frac{12}{1} = \frac{W_s}{W_p}$$

Hence, we have

$$W_s = 12 \text{ and } W_p = 1$$

We apply the **remez()** routine provided by MATLAB in Program 7.10. The filter coefficients are listed in Table 7.14. Figure 7.34 shows the frequency responses.

Program 7.10. MATLAB program for Example 7.15.

```
%Figure 7.34 (Example 7.15)
% MATLAB program to create Figure 7.34
%
fs=8000;
f=[ 0 0.2 0.25 1]; % Edge frequencies
m=[ 1 1 0 0] ; % Ideal magnitudes
w=[ 1 12 ]; % Error weight factors
b=remez(53,f,m,w); % (53+1)Parks-McClellan algorithm and Remez exchange
format long
freqz(b,1,512,fs) % Plot the frequency response
axis([0 fs/2 -80 10]);
```

Clearly, the stopband attenuation is satisfied. We plot the details for the filter passband in Figure 7.35.

Table 7.14 FIR Filter Coefficients in Example 7.15	
B: FIR Filter Coefficients (optimal design method)	
$b_0 = b_{53} = -0.006075$	$b_1 = b_{52} = -0.00197$
$b_2 = b_{51} = 0.001277$	$b_3 = b_{50} = 0.006937$
$b_4 = b_{49} = 0.013488$	$b_5 = b_{48} = 0.018457$
$b_6 = b_{47} = 0.019347$	$b_7 = b_{46} = 0.014812$
$b_8 = b_{45} = 0.005568$	$b_9 = b_{44} = -0.005438$
$b_{10} = b_{43} = -0.013893$	$b_{11} = b_{42} = -0.015887$
$b_{12} = b_{41} = -0.009723$	$b_{13} = b_{40} = 0.002789$
$b_{14} = b_{39} = 0.016564$	$b_{15} = b_{38} = 0.024947$
$b_{16} = b_{37} = 0.022523$	$b_{17} = b_{36} = 0.007886$
$b_{18} = b_{35} = -0.014825$	$b_{19} = b_{34} = -0.036522$
$b_{20} = b_{33} = -0.045964$	$b_{21} = b_{32} = -0.033866$
$b_{22} = b_{31} = 0.003120$	$b_{23} = b_{30} = 0.060244$
$b_{24} = b_{29} = 0.125252$	$b_{25} = b_{28} = 0.181826$
$b_{26} = b_{27} = 0.214670$	

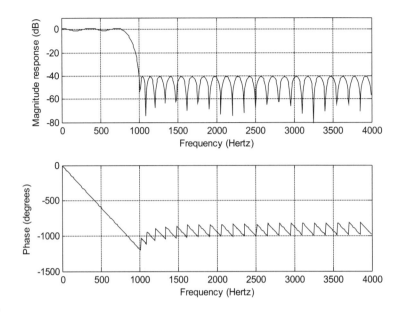

FIGURE 7.34

Frequency and phase responses for Example 7.15.

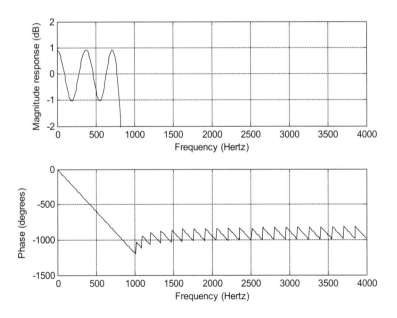

FIGURE 7.35

Frequency response details for passband in Example 7.15.

As shown in Figure 7.35, the ripples in the passband are between -1 and 1 dB. Hence, all the specifications are met. Note that if the specification is not satisfied, we will increase the order until the stopband attenuation and passband ripple are met.

The next example illustrates bandpass filter design.

EXAMPLE 7.16

Design a bandpass filter with the following specifications:

DSP system sampling rate = 8,000 Hz
Passband = 1,000–1,600 Hz
Stopband = 0–600 Hz and 2,000–4,000 Hz
Passband ripple = 1 dB
Stopband attenuation = 30 dB
Filter order = 25

Solution:

From the specifications, we have three bands: a passband, a lower stopband, and an upper stopband. We perform normalization and specify ideal magnitudes as follows:

Folding frequency: $f_s/2 = 8,000/2 = 4,000$ Hz
For 0 Hz: $0/4,000 = 0$ magnitude: 0
For 600 Hz: $600/4,000 = 0.15$ magnitude: 0
For 1,000 Hz: $1,000/4,000 = 0.25$ magnitude: 1
For 1,600 Hz: $1,600/4,000 = 0.4$ magnitude: 1
For 2,000 Hz: $2,000/4,000 = 0.5$ magnitude: 0
For 4,000 Hz: $4,000/4,000 = 1$ magnitude: 0

Next, let us determine the weights:

$$\delta_p = 10^{\left(\frac{1}{20}\right)} - 1 = 0.1220$$

$$\delta_s = 10^{\left(\frac{-30}{20}\right)} = 0.0316$$

Then applying Equation (7.36), we get

$$\frac{\delta_p}{\delta_s} = 3.86 \approx \frac{39}{10} = \frac{W_s}{W_p}$$

Hence, we have

$$W_s = 39 \text{ and } W_p = 10$$

We apply the **Remez()** routine provided by MATLAB and check performance in Program 7.11. Table 7.15 lists the filter coefficients. The frequency responses are depicted in Figure 7.36.

Program 7.11. MATLAB program for Example 7.16.

```
%Figure 7.36(Example 7.16)
% MATLAB program to create Figure 7.36
```

Table 7.15 FIR Filter Coefficients in Example 7.16

B: FIR Filter Coefficients (optimal design method)

$b_0 = b_{25} = -0.022715$	$b_1 = b_{24} = -0.012753$
$b_2 = b_{23} = 0.005310$	$b_3 = b_{22} = 0.009627$
$b_4 = b_{21} = -0.004246$	$b_5 = b_{20} = 0.006211$
$b_6 = b_{19} = 0.057515$	$b_7 = b_{18} = 0.076593$
$b_8 = b_{17} = -0.015655$	$b_9 = b_{16} = -0.156828$
$b_{10} = b_{15} = -0.170369$	$b_{11} = b_{14} = 0.009447$
$b_{12} = b_{13} = 0.211453$	

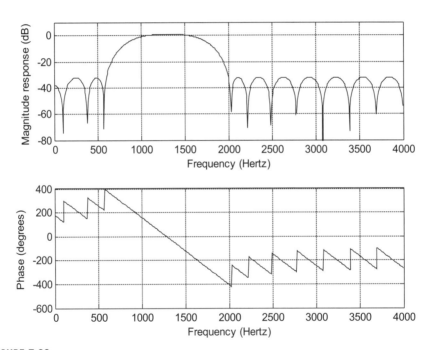

FIGURE 7.36

Frequency and phase responses for Example 7.16.

```
%
fs=8000;
f=[ 0 0.15 0.25 0.4 0.5 1];    % Edge frequencies
m=[ 0 0 1 1 0 0];              % Ideal magnitudes
w=[ 39 10 39 ];               % Error weight factors
format long
```

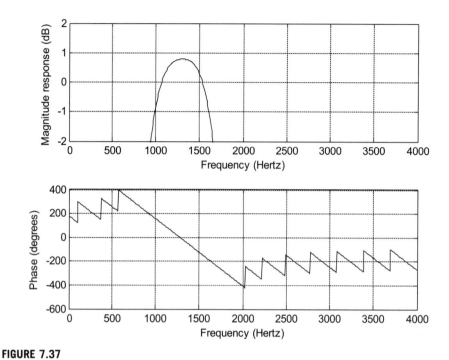

FIGURE 7.37

Frequency response details for passband in Example 7.16.

```
b=remez(25,f,m,w) % (25+1) taps Parks-McClellan algorithm and Remez exchange
freqz(b,1,512,fs); % Plot the frequency response
axis([0 fs/2 -80 10])
```

Clearly, the stopband attenuation is satisfied. We also check the details for the passband as shown in Figure 7.37.

As shown in Figure 7.37, the ripples in the passband between 1,000 Hz and 1,600 Hz are between -1 and 1 dB. Hence, all specifications are satisfied.

EXAMPLE 7.17

Now we show how the Remez exchange algorithm in Equation (7.32) is processed using a linear phase 3-tap FIR filter represented as follows:

$$H(z) = b_0 + b_1 z^{-1} + b_0 z^{-2}$$

The ideal frequency response specifications are shown in Figure 7.38(a), where the filter gain increases linearly from a gain of 0.5 at $\Omega = 0$ radians to a gain of 1 at $\Omega = \pi/4$ radians. The band between $\Omega = \pi/4$ radians and $\Omega = \pi/2$ radians is a transition band. Finally, the filter gain decreases linearly from the gain 0.75 at $\Omega = \pi/2$ radians to the gain of 0 at $\Omega = \pi$ radians.

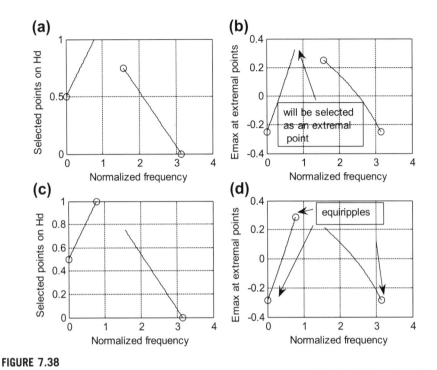

FIGURE 7.38

Determining the 3-tap FIR filter coefficients using the Remez algorithm in Example 7.17.

For simplicity, we set all the weight factors to 1, that is, $W(\Omega) = 1$. Equation (7.32) is simplified to

$$E(\Omega) = H(e^{j\Omega}) - H_d(e^{j\Omega})$$

Substituting $z = e^{j\Omega}$ into the transfer function $H(z)$ gives

$$H(e^{j\Omega}) = b_0 + b_1 e^{-j\Omega} + b_0 e^{-j2\Omega}$$

After simplification using Euler's identity $e^{j\Omega} + e^{-j\Omega} = 2\cos\Omega$, the filter frequency response is given by

$$H(e^{j\Omega}) = e^{j\Omega}(b_1 + 2b_0 \cos\Omega)$$

Disregarding the linear phase shift term $e^{j\Omega}$ for the time being, we have a Chebyshev real magnitude function (there are a few other types as well):

$$H(e^{j\Omega}) = b_1 + 2b_0 \cos\Omega$$

The *alternation theorem* (Ambardar, 1999; Porat, 1997) must be used. The alternation theorem states that given a Chebyshev polynomial $H(e^{j\Omega})$ to approximate the ideal magnitude response $H_d(e^{j\Omega})$, we can find at least $M + 2$ (where $M = 1$ for our case) frequencies $\Omega_0, \Omega_1, \ldots \Omega_{M+1}$, called the extremal frequencies, so that signs of the error at the extremal frequencies alternate and the absolute error value at each extremal point reaches the maximum absolute error, that is,

$$E(\Omega_k) = -E(\Omega_{k+1}) \quad \text{for} \quad \Omega_0, \Omega_1, \dots \Omega_{M+1}$$

and

$$|E(\Omega_k)| = E_{\max}$$

But the alternation theorem does not tell us how to execute the algorithm. The Remez exchange algorithm actually is employed to solve this problem. The equations and steps (Ambardar, 1999; Porat, 1997) are briefly summarized for our illustrative example:

1. Given an order of $N = 2M+1$, choose the initial extremal frequencies:
$$\Omega_0, \Omega_1, \dots, \Omega_{M+1} \text{ (can be uniformly distributed first)}$$
2. Solve the following equation to satisfy the alternate theorem:

$$-(-1)^k E = W(\Omega_k)(H_d(e^{j\Omega_k}) - H(e^{j\Omega_k})) \quad \text{for} \quad \Omega_0, \Omega_1, \dots, \Omega_{M+1}$$

Note that since $H(e^{j\Omega}) = b_1 + 2b_0 \cos \Omega$, for example, the solution will include solving for three unknowns: b_0, b_1, and $E_{\max}$.
3. Determine the extremal points including band edges (can be more than $M+2$ points), and retain $M+2$ extremal points with the largest error values $E_{\max}$.
4. Output the coefficients, if the extremal frequencies are not changed; otherwise, go to step 2 using the new set of extremal frequencies.

Now let us apply the Remez exchange algorithm.
First Iteration

1. We use uniformly distributed extremal points $\Omega_0 = 0$, $\Omega_1 = \pi/2$, $\Omega_2 = \pi$ whose ideal magnitudes are marked by the symbol "o" in Figure 7.38(a).
2. The alternation theorem requires $-(-1)^k E = H_d(e^{j\Omega}) - (b_1 + 2b_0 \cos \Omega)$.
Applying extremal points yields the following three simultaneous equations with three unknowns: b_0, b_1, and E:

$$\begin{cases} -E = 0.5 - b_1 - 2b_0 \\ E = 0.75 - b_1 \\ -E = 0 - b_1 + 2b_0 \end{cases}$$

We solve these three equations to get

$$b_0 = 0.125, \quad b_1 = 0.5, \quad E = 0.25, \quad H(e^{j\Omega}) = 0.5 + 0.25 \cos \Omega$$

3. We then determine the extremal points, including at the band edge, with their error values from Figure 7.38(b) using the following error function:

$$E(\Omega) = H_d(e^{j\Omega}) - 0.5 - 0.25 \cos \Omega$$

These extremal points are marked by symbol "o" and their error values are listed in Table 7.16.
4. Since the band edge $\Omega = \pi/4$ has an larger error than the others, it must be chosen as the extremal frequency. After deleting the extremal point at $\Omega = \pi/2$, a new set of extremal points are found according the largest error values as

$$\Omega_0 = 0$$

$$\Omega_1 = \pi/4$$

Table 7.16 Extremal Points and Band Edges with Their Error Values for the First Iteration.

Ω	0	$\pi/4$	$\pi/2$	π
E_{max}	-0.25	0.323	0.25	-0.25

Table 7.17 Error Values at Extremal Frequencies and Band Edge

Ω	0	$\pi/4$	$\pi/2$	π
E_{max}	-0.287	0.287	0.213	-0.287

$$\Omega_2 = \pi$$

The ideal magnitudes at these three extremal points are given in Figure 7.38(c), that is, 0.5, 1, 0. Now let us examine the second iteration.

Second Iteration

Applying the alternation theorem at the new set of extremal points, we have

$$\begin{cases} -E = 0.5 - b_1 - 2b_0 \\ E = 1 - b_1 - 1.4142b_0 \\ -E = 0 - b_1 + 2b_0 \end{cases}$$

Solving these three simultaneous equations leads to

$$b_0 = 0.125, \quad b_1 = 0.537, \quad E = 0.287, \quad \text{and} \quad H(e^{j\Omega}) = 0.537 + 0.25\cos\Omega$$

The error values at the extremal points and band edge are listed in Table 7.17 and shown in Figure 7.38(d), where the determined extremal points are marked by the symbol "o".

Since the extremal points have the same maximum error value of 0.287, they are $\Omega_0 = 0, \Omega_1 = \pi/4$, and $\Omega_2 = \pi$, which is unchanged. We stop the iteration and output the filter transfer function as

$$H(z) = 0.125 + 0.537z^{-1} + 0.125z^{-2}$$

As shown in Figure 7.37(d), we obtain the equiripples of error at the extemal points $\Omega_0 = 0$, $\Omega_1 = \pi/4$, and $\Omega_2 = \pi$; their signs are alternating, and the maximum absolute error of 0.287 is obtained at each point. It takes two iterations to determine the coefficients for this simplified example.

As we mentioned, the Parks–McClellan algorithm is one of the most popular filter design methods in industry due to its flexibility and performance. However, there are two disadvantages. The filter length has to be estimated by the empirical method. Once the frequency edges, magnitudes, and weighting factors are specified, the Remez exchange algorithm cannot control the actual ripple obtained from the design. We may often need to try a longer length of filter or different weight factors to remedy situations where the ripple is unacceptable.

7.7 REALIZATION STRUCTURES OF FINITE IMPULSE RESPONSE FILTERS

Using the direct-form I realization (discussed in Chapter 6), we will obtain a special realization form, called the *transversal form*. Using the linear phase property will produce a linear phase realization structure.

7.7.1 Transversal Form

Given the transfer function of the FIR filter in Equation (7.38),

$$H(z) = b_0 + b_1 z^{-1} + \cdots + b_K z^{-K} \tag{7.38}$$

we obtain the difference equation as

$$y(n) = b_0 x(n) + b_1 x(n-1) + b_2 x(n-2) + \cdots + b_K x(n-K)$$

Realization of such a transfer function is the transversal form, displayed in Figure 7.39.

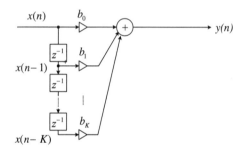

FIGURE 7.39

FIR filter realization (transversal form).

EXAMPLE 7.18

Given an FIR filter transfer function

$$H(z) = 1 + 1.2z^{-1} + 0.36z^{-2}$$

perform the FIR filter realization.

Solution:
From the transfer function, we can identify that

$$b_0 = 1, \ b_1 = 1.2, \text{ and } b_2 = 0.36$$

Using Figure 7.39, we find the FIR realization to be as displayed in Figure 7.40. We determine the DSP equation for implementation to be

$$y(n) = x(n) + 1.2x(n-1) + 0.36x(n-2)$$

Program 7.12 shows the MATLAB implementation.

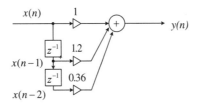

FIGURE 7.40

FIR filter realization for Example 7.18.

Program 7.12. MATLAB program for Example 7.18.

```
% Sample MATLAB code
sample =1:1:10;                 % Input test array
x=[ 0 0 0];                     % Input buffer [x(n) x(n-1) ...]
y=[0];                          %output buffer [y(n) y(n-1) ... ]
b=[1.0 1.2 0.36];               % FIR filter coefficients [b0 b1 ...]
KK=length(b);
for n=1:1:length(sample)        % Loop processing
for k=KK:-1:2                   % Shift input by one sample
x(k)=x(k-1);
end
x(1)=sample(n);                 % Get new sample
y(1)=0;                         % Perform FIR filtering
for k=1:1:KK
y(1)=y(1)+b(k)*x(k);
end
out(n)=y(1); %send filtered sample to the output array
end
out
```

7.7.2 Linear Phase Form

We illustrate the linear phase structure using the following simple example.

Consider the following transfer function with 5 taps obtained from the design:

$$H(z) = b_0 + b_1 z^{-1} + b_2 z^{-2} + b_1 z^{-3} + b_0 z^{-4} \qquad (7.39)$$

We can see that the coefficients are symmetrical and the difference equation is

$$y(n) = b_0 x(n) + b_1 x(n-1) + b_2 x(n-2) + b_1 x(n-3) + b_0 x(n-4)$$

This DSP equation can further be combined to yield

$$y(n) = b_0(x(n) + x(n-4)) + b_1(x(n-1) + x(n-3)) + b_2 x(n-2)$$

Then we obtain the realization structure in a linear phase form as shown in Figure 7.41.

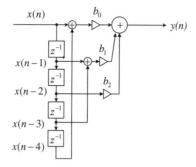

Linear phase FIR filter realization.

7.8 COEFFICIENT ACCURACY EFFECTS ON FINITE IMPULSE RESPONSE FILTERS

In practical applications, the filter coefficients achieved through high-level software such as MATLAB must be quantized using finite word length. This may have two effects. First, the locations of zeros are changed; second, due to the location change of zeros, the filter frequency response will change correspondingly. In practice, there are two types of digital signal (DS) processors: *fixed-point processors* and *floating-point processors*. The fixed-point DS processor uses integer arithmetic, and the floating-point processor employs floating-point arithmetic. Such effects of filter coefficient quantization will be covered in Chapter 9.

In this section, we will study the effects of FIR filter coefficient quantization in general, since during practical filter realization, obtaining filter coefficients with infinite precision is impossible. Filter coefficients are usually truncated or rounded off for the application. Assume that the FIR filter transfer function with infinite precision is given by

$$H(z) = \sum_{n=0}^{K} b_n z^{-n} = b_0 + b_1 z^{-1} + \cdots + b_{2M} z^{-K} \tag{7.40}$$

where each filter coefficient b_n has infinite precision. Now let the quantized FIR filter transfer function be

$$H^q(z) = \sum_{n=0}^{K} b_n^q z^{-n} = b_0^q + b_1^q z^{-1} + \cdots + b_K^q z^{-K} \tag{7.41}$$

where each filter coefficient b_n^q is quantized (rounded off) using the specified number of bits. Then the error of the magnitude frequency response can be bounded as

$$\left| H(e^{j\Omega}) - H^q(e^{j\Omega}) \right| = \sum_{n=0}^{K} \left| (b_n - b_n^q) e^{j\Omega} \right|$$

$$< \sum_{n=0}^{K} \left| b_n - b_n^q \right| \langle (K+1) \cdot 2^{-B} \tag{7.42}$$

where B is the number of bits used to encode each magnitude of the filter coefficient.

EXAMPLE 7.19

In Example 7.7, a lowpass FIR filter with 25 taps using a Hamming window was designed, and FIR filter coefficients are listed below for comparison in Table 7.18. One sign bit is used, and 7 bits are used for fractional parts, since all FIR filter coefficients are less than 1. We will multiply each filter coefficient by a scale factor of 2^7 and round off each scaled magnitude to an integer whose magnitude can be encoded using 7 bits. When the coefficient integer is scaled back, the coefficient with finite precision (quantized filter coefficient) using 8 bits, including the sign bit, will be achieved.

To understand quantization, we take a look at one of the infinite precision coefficients $Bham(3) = 0.00759455135346$, for illustration. The quantization using 7 magnitude bits is

$$0.00759455135346 \times 2^7 = 0.9721 = 1$$

Then the quantized filter coefficient is obtained as

$$BhamQ(3) = 1/2^7 = 0.0078125$$

Since the poles for both FIR filters always reside at the origin, we need to examine only their zeros. The z-plane zero plots for both FIR filters are shown in Figure 7.42A, where the circles are zeros from the FIR filter with infinite precision, while the crosses are zeros from the FIR filter with the quantized coefficients.

Most importantly, Figure 7.42B shows the difference of the frequency responses for both filters obtained using Program 7.13. In the figure, the solid line represents the frequency response with infinite filter coefficient precision, and the dot-dashed line indicates the frequency response with finite filter coefficients. It is observed that the stopband performance is degraded due to the filter coefficient quantization. The degradation in the passband is not severe.

TABLE 7.18 FIR Filter Coefficients and Their Quantized Filter Coefficients in Example 7.19 (Hamming window)

Bham: FIR Filter Coefficients	BhamQ: FIR Filter Coefficients
$b_0 = b_{24} = 0.00000000000000$	$b_0 = b_{24} = 0.0000000$
$b_1 = b_{23} = -0.00276854711076$	$b_1 = b_{23} = -0.0000000$
$b_2 = b_{22} = 0.00000000000000$	$b_2 = b_{22} = 0.0000000$
$b_3 = b_{21} = 0.00759455135346$	$b_3 = b_{21} = 0.0078125$
$b_4 = b_{20} = 0.00000000000000$	$b_4 = b_{20} = 0.0000000$
$b_5 = b_{19} = -0.01914148493949$	$b_5 = b_{19} = -0.0156250$
$b_6 = b_{18} = 0.00000000000000$	$b_6 = b_{18} = 0.0000000$
$b_7 = b_{17} = 0.04195685650042$	$b_7 = b_{17} = 0.0390625$
$b_8 = b_{16} = 0.00000000000000$	$b_8 = b_{16} = 0.0000000$
$b_9 = b_{15} = -0.09180790496577$	$b_9 = b_{15} = -0.0859375$
$b_{10} = b_{14} = 0.00000000000000$	$b_{10} = b_{14} = 0.0000000$
$b_{11} = b_{13} = 0.31332065886015$	$b_{11} = b_{13} = 0.3125000$
$b_{12} = 0.50000000000000$	$b_{12} = 0.5000000$

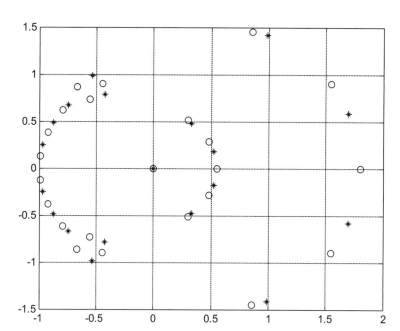

FIGURE 7.42A

The z-plane zero plots for both FIR filters. The circles are zeros for infinite precision; the crosses are zeros for round-off coefficients.

Program 7.13. MATLAB program for Example 7.19.

```
fs=8000;
[hham,f]=freqz(Bham,1,512,fs);
[hhamQ,f]=freqz(BhamQ,1,512,fs);
p=180*unwrap(angle(hham))/pi;
pQ=180*unwrap(angle(hhamQ))/pi
subplot(2,1,1); plot(f,20*log10(abs(hham)),f,20*log10(abs(hhamQ)),':');grid
axis([0 4000 -100 10]);
xlabel('Frequency (Hz)'); ylabel('Magnitude Response (dB)');
subplot(2,1,2); plot(f,p,f,pQ,':');grid
```

Using Equation (7.42), the error of the magnitude frequency response due to quantization is bounded by

$$\left|H(e^{j\Omega}) - H^q(e^{j\Omega})\right| \langle 25/256 = 0.0977$$

This can be easily verified at the stopband of the magnitude frequency response for the worst condition as follows:

$$\left|H(e^{j\Omega}) - H^q(e^{j\Omega})\right| = \left|10^{-100/20} - 10^{-30/20}\right| = 0.032 < 0.0977$$

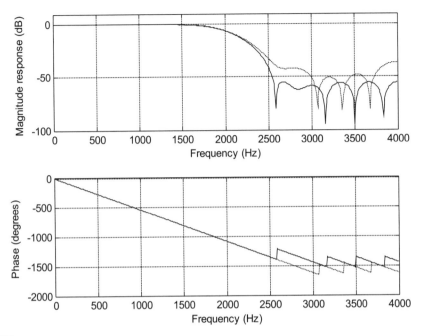

FIGURE 7.42B

Frequency responses. The solid line indicates the FIR filter with infinite precision; the dashed line indicates the FIR filter with the round-off coefficients.

In practical situations, a similar procedure can be used to analyze the effects of filter coefficient quantization to make sure that the designed filter meets the requirements.

7.9 SUMMARY OF FIR DESIGN PROCEDURES AND SELECTION OF FIR FILTER DESIGN METHODS IN PRACTICE

In this section, we first summarize the design procedures of the window design, frequency sampling design, and optimal design methods, and then discuss the selection of the particular filter for typical applications.

The window method (Fourier transform design using windows):

1. Given the filter frequency specifications, determine the filter order (odd number used in this book) and the cutoff frequency/frequencies using Table 7.7 and Equation (7.26).
2. Compute the impulse sequence $h(n)$ via the Fourier transform method using the appropriate equations (in Table 7.1).
3. Multiply the generated FIR filter coefficients $h(n)$ in step 2 by the selected window sequence using Equation (7.20) to obtain the windowed impulse sequence $h_w(n)$.

4. Delay the windowed impulse sequence $h_w(n)$ by M samples to get the causal windowed FIR filter coefficients $b_n = h_w(n - M)$ using Equation (7.21).
5. Output the transfer function and plot the frequency responses.
6. If the frequency specifications are satisfied, output the difference equation. If the frequency specifications are not satisfied, increase the filter order and repeat beginning with step 2.

The frequency sampling method:

1. Given the filter frequency specifications, choose the filter order (odd number used in the book), and specify the equally spaced magnitudes of the frequency response for the normalized frequency range from 0 to π using Equation (7.29).
2. Calculate FIR filter coefficients using Equation (7.30).
3. Use the symmetry in Equation (7.31) and the linear phase requirement to determine the rest of the coefficients.
4. Output the transfer function and plot the frequency responses.
5. If the frequency specifications are satisfied, output the difference equation. If the frequency specifications are not satisfied, increase the filter order and repeat beginning with step 2.

The optimal design method (Parks–McClellan algorithm):

1. Given the band edge frequencies, choose the filter order , normalize each band edge frequency to the Nyquist limit (folding frequency $= f_s/2$), and specify the ideal magnitudes.
2. Calculate the absolute values of the passband ripple and stopband attenuation, if they are given in terms of dB values, using Equations (7.34) and (7.35).
3. Determine the error weight factors for the passband and stopband, respectively, using Equations (7.36) and (7.37).
4. Apply the Remez algorithm to calculate filter coefficients.

Table 7.19 Comparisons of Three Design Methods

Design Method	Window	Frequency Sampling	Optimal
Filter type	1. Lowpass, highpass, bandpass, bandstop. 2. Formulas are not valid for arbitrary frequency selectivity.	1. Any type of filter 2. The formula is valid for arbitrary frequency selectivity.	1. Any type of filter 2. Valid for arbitrary frequency selectivity
Linear phase	Yes	Yes	Yes
Ripple and stopband specifications	Used for determining the filter order and cutoff frequency/-cies	Need to be checked after each design trial	Used in the algorithm; need to be checked after each design trial
Algorithm complexity for coefficients	Moderate: 1. Impulse sequence calculation 2. Window function weighting	Simple: Single equation	Complicated: 1. Parks–McClellan algorithm 2. Remez exchange algorithm
Minimal design tool	Calculator	Calculator	Software

5. Output the transfer function and check the frequency responses.
6. If the frequency specifications are satisfied, output the difference equation. If the frequency specifications are not satisfied, increase the filter order and repeat beginning with step 4.

Table 7.19 shows the comparisons for the window, frequency sampling, and optimal methods. The table can be used as a selection guide for each design method in this book.

Example 7.20 describes the possible selection of the design method by a DSP engineer to solve a real-world problem.

EXAMPLE 7.20

Determine the appropriate FIR filter design method for each of the following DSP applications.

a. A DSP engineer implements a digital two-band crossover system as described in Section 7.4.4 in this book. He selects the FIR filters to satisfy the following specifications:
Sampling rate = 44,100 Hz
Crossover frequency = 1,000 Hz (cutoff frequency)
Transition band = 600 Hz to 1,400 Hz
Lowpass filter = passband frequency range from 0 to 600 Hz with a ripple of 0.02 dB and stopband edge at 1,400 Hz with an attenuation of 50 dB.
Highpass filter = passband frequency range from 1.4 to 44.1 kHz with a ripple of 0.02 dB and stopband edge at 600 Hz with an attenuation of 50 dB.

The engineer does not have the software routine for the Remez algorithm.
b. An audio engineer tries to equalize a speech signal sampled at 8,000 Hz using a linear phase FIR filter based on the magnitude specifications in Figure 7.43. The engineer does not have the software routine for the Remez algorithm.

Solution:
a. The window design method is the first choice, since this formula is expressed in terms of the cutoff frequency (crossover frequency), the filter order is based on the transient band, and the filter types are standard lowpass and highpass. The ripple and stopband specifications can be satisfied by selecting the Hamming window. The optimal design method will also do the job if the **remez()** algorithm is available. But there exists a challenge to satisfy the combined unity gains at the crossover frequency of 1,000 Hz.
b. Since the magnitude frequency response is not a standard filter type such as lowpass, highpass, bandpass, or bandstop, and the **remez()** algorithm is not available, the first choice should be the frequency sampling method.

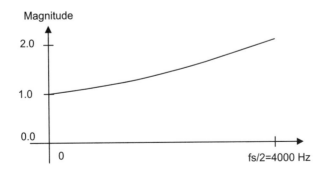

FIGURE 7.43

Magnitude frequency response in Example 7.20(b).

7.10 SUMMARY

1. The Fourier transform method is used to compute noncausal FIR filter coefficients, including those of lowpass, highpass, bandpass, and bandstop filters.

2. Converting noncausal FIR filter coefficients to causal FIR filter coefficients only introduces linear phase, which is a good property for audio applications. The linear phase filter output has the same amount of delay for all the input signals whose frequency components are within the passband.

3. The causal FIR filter using the Fourier transform method generates ripple oscillations (Gibbs effect) in the passband and stopband in its filter magnitude frequency response due to abrupt truncation of the FIR filter coefficient sequence.

4. To reduce the oscillation effect, the window method is introduced to tap down the coefficient values towards both ends. A substantial improvement of the magnitude frequency response is achieved.

5. Real-life DSP applications such as noise reduction systems and two-band digital audio crossover systems were investigated.

6. Frequency sampling design is feasible for an FIR filter with an arbitrary magnitude response specification.

7. An optimal design method, the Parks–McClellan algorithm using the Remez exchange algorithm, offers flexibility for filter specifications. The Remez exchange algorithm was explained using a simplified example.

8. Realization structures of FIR filters have special forms, such as the transversal form and the linear phase form.

9. The effect of quantizing FIR filter coefficients for implementation changes the zero locations of the FIR filter. More effects on the stopband in the magnitude and phase responses are observed.

10. Guidelines for selecting an appropriate design method in practice were summarized with consideration of the filter type, linear phase, ripple and stopband specifications, algorithm complexity, and design tools.

7.11 MATLAB PROGRAMS

Program 7.14 enables one to design FIR filters via the window method using window functions such as the rectangular window, triangular window, Hanning window, Hamming window, and Blackman window. Filter types of the design include lowpass, highpass, bandpass, and bandstop.

Program 7.14. MATLAB function for FIR filter design using the window method.

```
function B=firwd(N,Ftype,WnL,WnH,Wtype)
% B = firwd(N,Ftype,WnL,WnH,Wtype)
% FIR filter design using the window function method.
% Input parameters:
% N: the number of the FIR filter taps.
%   Note: It must be odd number.
```

```
% Ftype: the filter type
%       1. Lowpass filter
%       2. Highpass filter
%       3. Bandpass filter
%       4. Bandstop filter
% WnL: lower cutoff frequency in radians. Set WnL=0 for the highpass filter.
% WnH: upper cutoff frequency in radians. Set WnL=0 for the lowpass filter.
% Wtypw: window function type
%       1. Rectangular window
%       2. Triangular window
%       3. Hanning window
%       4. Hamming window
%       5. Balckman window
% Output:
% B: FIR filter coefficients.
      M=(N-1)/2;
      hH=sin(WnH*[-M:1:-1])./([-M:1:-1]*pi);
      hH(M+1)=WnH/pi;
      hH(M+2:1:N)=hH(M:-1:1);
      hL=sin(WnL*[-M:1:-1])./([-M:1:-1]*pi);
      hL(M+1)=WnL/pi;
      hL(M+2:1:N)=hL(M:-1:1);
      if Ftype == 1
        h(1:N)=hL(1:N);
      end
      if Ftype == 2
        h(1:N)=-hH(1:N);
        h(M+1)=1+h(M+1);
end
if Ftype ==3
  h(1:N)=hH(1:N)-hL(1:N);
end
if Ftype == 4
  h(1:N)=hL(1:N)-hH(1:N);
  h(M+1)=1+h(M+1);
end
% Window functions
      if Wtype ==1
      w(1:N)=ones(1,N);
  end
  if Wtype ==2
  w=1-abs([-M:1:M])/M;
  end
  if Wtype ==3
  w= 0.5+0.5*cos([-M:1:M]*pi/M);
  end
  if Wtype ==4
  w=0.54+0.46*cos([-M:1:M]*pi/M);
```

```
end
if Wtype ==5
w=0.42+0.5*cos([-M:1:M]*pi/M)+0.08*cos(2*[-M:1:M]*pi/M);
end
B=h .* w
```

Program 7.15 enables one to design FIR filters using the frequency sampling method. Note that values of the frequency response, which correspond to the equally spaced DFT frequency components, must be specified for design. Besides the lowpass, highpass, bandpass and bandstop filter designs, the method can be used to design FIR filters with an arbitrarily specified magnitude frequency response.

Program 7.15. MATLAB function for FIR filter design using the frequency sampling method.

```
function B=firfs(N,Hk)
% B=firls(N,Hk)
% FIR filter design using the frequency sampling method.
% Input parameters:
% N: the number of filter coefficients.
%   note: N must be odd number.
% Hk: sampled frequency reponse for k=0,1,2,...,M=(N-1)/2.
% Output:
% B: FIR filter coefficients.
     M=(N-1)/2;
     for n=1:1:N
     B(n)=(1/N)*(Hk(1)+...
     2*sum(Hk(2:1:M+1)...
     .*cos(2*pi*([1:1:M])*(n-1-M)/N)));
end
```

7.12 PROBLEMS

7.1. Design a 3-tap FIR lowpass filter with a cutoff frequency of 1,500 Hz and a sampling rate of 8,000 Hz using a

   a. rectangular window function

   b. Hamming window function

   Determine the transfer function and difference equation of the designed FIR system, and compute and plot the magnitude frequency response for $\Omega = 0, \pi/4, \pi/2, 3\pi/4,$ and π radians.

7.2. Design a 3-tap FIR highpass filter with a cutoff frequency of 1,600 Hz and a sampling rate of 8,000 Hz using a

   a. rectangular window function

   b. Hamming window function

   Determine the transfer function and difference equation of the designed FIR system, and compute and plot the magnitude frequency response for $\Omega = 0, \pi/4, \pi/2, 3\pi/4,$ and π radians.

7.3. Design a 5-tap FIR lowpass filter with a cutoff frequency of 100 Hz and a sampling rate of 1,000 Hz using a

a. rectangular window function

b. Hamming window function

Determine the transfer function and difference equation of the designed FIR system, and compute and plot the magnitude frequency response for $\Omega = 0, \pi/4, \pi/2, 3\pi/4,$ and π radians.

7.4. Design a 5-tap FIR highpass filter with a cutoff frequency of 250 Hz and a sampling rate of 1,000 Hz using a

a. rectangular window function

b. Hamming window function

Determine the transfer function and difference equation of the designed FIR system, and compute and plot the magnitude frequency response for $\Omega = 0, \pi/4, \pi/2, 3\pi/4,$ and π radians.

7.5. Design a 5-tap FIR bandpass filter with a lower cutoff frequency of 1,600 Hz, an upper cut-off frequency of 1,800 Hz and a sampling rate of 8,000 Hz using a

a. rectangular window function

b. Hamming window function

Determine the transfer function and difference equation of the designed FIR system, and compute and plot the magnitude frequency response for $\Omega = 0, \pi/4, \pi/2, 3\pi/4,$ and π radians.

7.6. Design a 5-tap FIR band reject filter with a lower cutoff frequency of 1,600 Hz, an upper cutoff frequency of 1,800 Hz, and a sampling rate of 8,000 Hz using a

a. rectangular window function

b. Hamming window function

Determine the transfer function and difference equation of the designed FIR system, and compute and plot the magnitude frequency response for $\Omega = 0, \pi/4, \pi/2, 3\pi/4,$ and π radians.

7.7. Consider an FIR lowpass filter design with the following specifications:

Passband $= 0–800$ Hz

Stopband $= 1,200–4,000$ Hz

Passband ripple $= 0.1$ dB

Stopband attenuation $= 40$ dB

Sampling rate $= 8,000$ Hz

Determine the following:

a. window method

b. length of the FIR filter

c. cutoff frequency for the design equation

7.8. Consider an FIR highpass filter design with the following specifications:

Stopband = 0–1,500 Hz

Passband = 2,000–4,000 Hz

Passband ripple = 0.02 dB

Stopband attenuation = 60 dB

Sampling rate = 8,000 Hz

Determine the following:

a. window method

b. length of the FIR filter

c. cutoff frequency for the design equation

7.9. Consider an FIR bandpass filter design with the following specifications:

Lower cutoff frequency = 1,500 Hz

Lower transition width = 600 Hz

Upper cutoff frequency = 2,300 Hz

Upper transition width = 600 Hz

Passband ripple = 0.1 dB

Stopband attenuation = 50 dB

Sampling rate: 8,000 Hz

Determine the following:

a. window method

b. length of the FIR filter

c. cutoff frequencies for the design equation

7.10. Consider an FIR bandstop filter design with the following specifications:

Lower passband = 0–1,200 Hz

Stopband = 1,600–2,000 Hz

Upper passband = 2,400–4,000 Hz

Passband ripple = 0.05 dB

Stopband attenuation = 60 dB

Sampling rate = 8,000 Hz

Determine the following:

a. window method

b. length of the FIR filter

c. cutoff frequencies for the design equation

7.11. Given an FIR system

$$H(z) = 0.25 - 0.5z^{-1} + 0.25z^{-2}$$

realize $H(z)$ using each of the following specified methods:

a. transversal form (write the difference equation for implementation)

b. linear phase form (write the difference equation for implementation)

7.12. Given an FIR filter transfer function

$$H(z) = 0.2 + 0.5z^{-1} - 0.3z^{-2} + 0.5z^{-3} + 0.2z^{-4}$$

perform the linear phase FIR filter realization, and write the difference equation for implementation.

7.13. Determine the transfer function for a 3-tap FIR lowpass filter with a cutoff frequency of 150 Hz and a sampling rate of 1,000 Hz using the frequency sampling method.

7.14. Determine the transfer function for a 3-tap FIR highpass filter with a cutoff frequency of 250 Hz and a sampling rate of 1,000 Hz using the frequency sampling method.

7.15. Determine the transfer function for a 5-tap FIR lowpass filter with a cutoff frequency of 2,000 Hz and a sampling rate of 8,000 Hz using the frequency sampling method.

7.16. Determine the transfer function for a 5-tap FIR highpass filter with a cutoff frequency of 3,000 Hz and a sampling rate of 8,000 Hz using the frequency sampling method.

7.17. Given the following specifications, determine the transfer function:

- 7-tap FIR bandpass filter
- lower cutoff frequency of 1,500 Hz and upper cutoff frequency of 3,000 Hz
- sampling rate of 8,000 Hz
- frequency sampling design method

7.18. Given the following specifications, determine the transfer function:

- 7-tap FIR bandstop filter
- lower cutoff frequency of 1,500 Hz and upper cutoff frequency of 3,000 Hz
- sampling rate of 8,000 Hz
- frequency sampling design method

7.19. A lowpass FIR filter to be designed has the following specifications:

Design method: Parks–McClellan algorithm

Sampling rate = 1,000 Hz

Passband = 0–200 Hz

Stopband = 300–500 Hz

Passband ripple = 1 dB

Stopband attenuation = 40 dB

Determine the error weights W_p and W_s for the passband and stopband in the Parks–McClellan algorithm.

7.20. A bandpass FIR filter to be designed has the following specifications:

Design method: Parks–McClellan algorithm

Sampling rate = 1,000 Hz

Passband = 200–250 Hz

Lower stopband = 0–150 Hz

Upper stopband = 300–500 Hz

Passband ripple = 1 dB

Stopband attenuation = 30 dB

Determine the error weights W_p and W_s for the passband and stopband in the Parks–McClellan algorithm.

7.21. A highpass FIR filter to be designed has the following specifications:

Design method: Parks–McClellan algorithm

Sampling rate = 1,000 Hz

Passband = 350–500 Hz

Stopband = 0–250 Hz

Passband ripple = 1 dB

Stopband attenuation = 60 dB

Determine the error weights W_p and W_s for the passband and stopband in the Parks–McClellan algorithm.

7.22. A bandstop FIR filter to be designed has the following specifications:

Design method: Parks–McClellan algorithm

Sampling rate = 1,000 Hz

Stopband = 250–350 Hz

Lower passband = 0–200 Hz

Upper passband = 400–500 Hz

Passband ripple = 1 dB

Stopband attenuation = 25 dB

Determine the error weights W_p and W_s for the passband and stopband in the Parks–McClellan algorithm.

7.23. In a speech recording system with a sampling rate of 10,000 Hz, the speech is corrupted by broadband random noise. To remove the random noise while preserve speech information, the following specifications are given:

Speech frequency range = 0–3,000 Hz

Stopband range = 4,000–5,000 Hz

Passband ripple = 0.1 dB

Stopband attenuation = 45 dB

FIR filter with Hamming window

Determine the FIR filter length (number of taps) and the cutoff frequency; use MATLAB to design the filter; and plot the frequency response.

7.24. Consider the speech equalizer shown in Figure 7.44 to compensate for midrange frequency loss of hearing that has the following specifications:

Sampling rate = 8,000 Hz

Bandpass FIR filter with Hamming window

Frequency range to be emphasized = 1,500–2,000 Hz

Lower stopband = 0–1,000 Hz

Upper stopband = 2,500–4,000 Hz

Passband ripple = 0.1 dB

Stopband attenuation = 45 dB

Determine the filter length and the lower and upper cutoff frequencies.

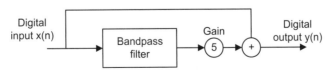

FIGURE 7.44

Speech equalizer in Problem 7.24.

7.25. A digital crossover can be designed as shown in Figure 7.45.

Consider the following audio specifications:

Sampling rate = 44,100 Hz

Crossover frequency = 2,000 Hz

Transition band range = 1,600 Hz

Passband ripple = 0.1 dB

Stopband attenuation = 50 dB

Filter type = FIR

Determine the following for each filter in Figure 7.45:

a. window function

b. filter length

c. cutoff frequency

Use MALAB to design both filters and plot frequency responses for both filters.

7.12.1 Computer Problems with MATLAB

Use the MATLAB programs provided in Section 7.11 to design the following FIR filters.

7.26. Design a 41-tap lowpass FIR filter whose cutoff frequency is 1,600 Hz using the following window functions. Assume that the sampling frequency is 8,000 Hz.

a. rectangular window function

b. triangular window function

c. Hanning window function

d. Hamming window function

e. Blackman window function

List the FIR filter coefficients and plot the frequency responses for each case.

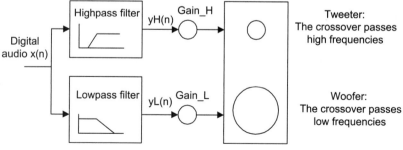

FIGURE 7.45

Two-band crossover in Problem 7.25.

7.27. Design a lowpass FIR filter whose cutoff frequency is 1,000 Hz using the Hamming window function for the following specified filter length. Assume that the sampling frequency is 8,000 Hz.

a. 21 filter coefficients

b. 31 filter coefficients

c. 41 filter coefficients

List the FIR filter coefficients for each design and compare the magnitude frequency responses.

7.28. Design a 31-tap highpass FIR filter whose cutoff frequency is 2,500 Hz using the following window functions. Assume that the sampling frequency is 8,000 Hz.

a. Hanning window function

b. Hamming window function

c. Blackman window function

List the FIR filter coefficients and plot the frequency responses for each design.

7.29. Design a 41-tap bandpass FIR filter with lower and upper cutoff frequencies of 2,500 Hz and 3,000 Hz, respectively, using the following window functions. Assume a sampling frequency of 8,000 Hz.

a. Hanning window function

b. Blackman window function.

List the FIR filter coefficients and plot the frequency responses for each design.

7.30. Design a 41-tap band reject FIR filter with cutoff frequencies of 2,500 Hz and 3,000 Hz, respectively, using the Hamming window function. Assume a sampling frequency of 8,000 Hz. List the FIR filter coefficients and plot the frequency responses.

7.31. Use the frequency sampling method to design a linear phase lowpass FIR filter with 17 coefficients. Let the cutoff frequency be 2,000 Hz and assume a sampling frequency of 8,000 Hz. List the FIR filter coefficients and plot the frequency responses.

7.32. Use the frequency sampling method to design a linear phase bandpass FIR filter with 21 coefficients. Let the lower and upper cutoff frequencies be 2,000 Hz and 2,500 Hz, respectively, and assume a sampling frequency of 8,000 Hz. List the FIR filter coefficients and plot the frequency responses.

7.33. Given an input data sequence

$$x(n) = 1.2 \cdot \sin(2\pi(1,000)n/8,000) - 1.5 \cdot \cos(2\pi(2,800)n/8,000)$$

with a sampling frequency of 8,000 Hz, use the designed FIR filter with a Hamming window in Problem 7.26 to filter 400 data points of $x(n)$, and plot the 400 samples of the input and output data.

7.34. Design a lowpass FIR filter with the following specifications:

Design method: Parks−McClellan algorithm

Sampling rate = 8,000 Hz

Passband = 0−1,200 Hz

Stopband = 1,500−4,000 Hz

Passband ripple = 1 dB

Stopband attenuation = 40 dB

List the filter coefficients and plot the frequency responses.

7.35. Design a bandpass FIR filter with the following specifications:

Design method: Parks−McClellan algorithm

Sampling rate = 8,000 Hz

Passband = 1,200−1,600 Hz

Lower stopband = 0−800 Hz

Upper stopband = 2,000−4,000 Hz

Passband ripple = 1 dB

Stopband attenuation = 40 dB

List the filter coefficients and plot the frequency responses.

7.12.2 MATLAB Projects

7.36. Speech enhancement:

Digitally recorded speech in a noisy environment can be enhanced using a lowpass filter if the recorded speech with a sampling rate of 8,000 Hz contains the desired frequency components lower than 1,600 Hz. Design a lowpass filter to remove the high frequency noise above 1,600 Hz with the following filter specifications: passband frequency range: 0 − 1,600 Hz; passband ripple: 0.02 dB; stopband frequency range: 1,800−4,000 Hz; stopband attenuation: 50 dB.

Use the designed lowpass filter to filter the noisy speech and adopt the following code to simulate the noisy speech:

```
load speech.dat
t=[0:length(speech)-1]*T;
th=mean(speech.*speech)/4; %Noise power =(1/4) speech power
noise=sqrt(th)*randn([1,length(speech)]); %Generate Gaussian noise
nspeech=speech+noise; % Generate noisy speech
```

In this project, plot the speech samples and spectra for both noisy speech and the enhanced speech and use the MATLAB **sound()** function to evaluate the sound quality. For example, to hear the noisy speech, use the following:

$$sound(nspeech/max(abs(nspeech)), 8000);$$

7.37. Digital crossover system:

Design a two-band digital crossover system with the following specifications:

Sampling rate = 44,100 Hz

Crossover frequency = 1,200 Hz (cutoff frequency)

Transition band = 800–1,600 Hz

Lowpass filter: passband frequency range from 0 to 800 Hz with a ripple of 0.02 dB and stopband edge at 1,400 Hz with the attenuation of 50 dB

Highpass filter: passband frequency range from 1.6 to 44.1 kHz with a ripple of 0.02 dB and stopband edge at 1,600 Hz with the attenuation of 50 dB

In this project, plot the magnitude frequency responses for both lowpass and highpass filters. Use the following MATLAB code to read stereo audio data ("No9seg.wav"):

$$[\text{x fs Nbits}] = \text{wavread}('\text{No9seg.wav}');$$

Process the given stereo audio segment. Listen to and describe the sound effects of the processed audio in the following sequences:

Channel 1: original, lowband, and highband

Channel 2: original, lowband, and highband

Stereo (both channels): original, lowband, and highband

Infinite Impulse Response Filter Design

CHAPTER OUTLINE

8.1 Infinite Impulse Response Filter Format 302
8.2 Bilinear Transformation Design Method... 303
  8.2.1 Analog Filters Using Lowpass Prototype Transformation 304
  8.2.2 Bilinear Transformation and Frequency Warping.......................................308
  8.2.3 Bilinear Transformation Design Procedure 314
8.3 Digital Butterworth and Chebyshev Filter Designs 318
  8.3.1 Lowpass Prototype Function and Its Order 318
  8.3.2 Lowpass and Highpass Filter Design Examples...322
  8.3.3 Bandpass and Bandstop Filter Design Examples..331
8.4 Higher-Order Infinite Impulse Response Filter Design Using the Cascade Method 338
8.5 Application: Digital Audio Equalizer 341
8.6 Impulse-Invariant Design Method 345
8.7 Pole-Zero Placement Method for Simple Infinite Impulse Response Filters 351
  8.7.1 Second-Order Bandpass Filter Design..352
  8.7.2 Second-Order Bandstop (Notch) Filter Design 354
  8.7.3 First-Order Lowpass Filter Design..355
  8.7.4 First-Order Highpass Filter Design..357
8.8 Realization Structures of Infinite Impulse Response Filters 358
  8.8.1 Realization of Infinite Impulse Response Filters in Direct-Form I and Direct-Form II 358
  8.8.2 Realization of Higher-Order Infinite Impulse Response Filters via the Cascade Form........361
8.9 Application: 60-Hz Hum Eliminator and Heart Rate Detection Using Electrocardiography 362
8.10 Coefficient Accuracy Effects on Infinite Impulse Response Filters.......................... 369
8.11 Application: Generation and Detection of DTMF Tones Using the Goertzel Algorithm 373
  8.11.1 Single-Tone Generator...374
  8.11.2 Dual-Tone Multifrequency Tone Generator 375
  8.11.3 Goertzel Algorithm..377
  8.11.4 Dual-Tone Multifrequency Tone Detection Using the Modified Goertzel Algorithm 383
8.12 Summary of Infinite Impulse Response (IIR) Design Procedures and Selection of the IIR Filter Design Methods in Practice 388
8.13 Summary 391

OBJECTIVES:

This chapter investigates a bilinear transformation method for infinite impulse response (IIR) filter design and develops a procedure to design digital Butterworth and Chebyshev filters. The chapter also investigates other IIR filter design methods, such as impulse-invariant design and pole-zero placement design. Finally, the chapter illustrates how to apply the designed IIR filters to solve real-world problems such as digital audio equalization, 60-Hz interference cancellation in audio and electrocardiography signals, dual-tone multifrequency tone generation, and detection using the Goertzel algorithm.

8.1 INFINITE IMPULSE RESPONSE FILTER FORMAT

In this chapter, we will study several methods for infinite impulse response (IIR) filter design. An IIR filter is described using the difference equation, as discussed in Chapter 6:

$$y(n) = b_0 x(n) + b_1 x(n-1) + \cdots + b_M x(n-M)$$
$$- a_1 y(n-1) - \cdots - a_N y(n-N)$$

Chapter 6 also gives the IIR filter transfer function as

$$H(z) = \frac{Y(z)}{X(z)} = \frac{b_0 + b_1 z^{-1} + \cdots + b_M z^{-M}}{1 + a_1 z^{-1} + \cdots + a_N z^{-N}}$$

where b_i and a_i are the $(M+1)$ numerator and N denominator coefficients, respectively. $Y(z)$ and $X(z)$ are the z-transform functions of the filter input $x(n)$ and filter output $y(n)$. To become familiar with the form of the IIR filter, let us look at the following example.

EXAMPLE 8.1

Given the IIR filter

$$y(n) = 0.2x(n) + 0.4x(n-1) + 0.5y(n-1)$$

determine the transfer function, nonzero coefficients, and impulse response.

Solution:

Applying the z-transform and solving for a ratio of the z-transform output over input, we have

$$H(z) = \frac{Y(z)}{X(z)} = \frac{0.2 + 0.4z^{-1}}{1 - 0.5z^{-1}}$$

We also identify the nonzero numerator coefficients and denominator coefficient as

$$b_0 = 0.2, \; b_1 = 0.4, \quad \text{and} \quad a_1 = -0.5$$

To determine the impulse response, we rewrite the transfer function as

$$H(z) = \frac{0.2}{1 - 0.5z^{-1}} + \frac{0.4z^{-1}}{1 - 0.5z^{-1}}$$

Using the inverse z-transform and shift theorem, we obtain the impulse response as

$$h(n) = 0.2(0.5)^n u(n) + 0.4(0.5)^{n-1} u(n-1)$$

The obtained impulse response has an infinite number of terms, where the first several terms are calculated as

$$h(0) = 0.2, \ h(1) = 0.7, \ h(2) = 0.25, \ ....$$

At this point, we can make following remarks:

1. The IIR filter output $y(n)$ depends not only on the current input $x(n)$ and past inputs $x(n-1)$, ..., but also on the past output(s) $y(n-1)$, ..., (recursive terms). Its transfer function is a ratio of the numerator polynomial over the denominator polynomial, and its impulse response has an infinite number of terms.

2. Since the transfer function has the denominator polynomial, the pole(s) of a designed IIR filter must be inside the unit circle on the z-plane to ensure its stability.

3. Compared with the finite impulse response (FIR) filter (see Chapter 7), the IIR filter offers a much smaller filter size. Hence, the filter operation requires a fewer number of computations, but the linear phase is not easily obtained. The IIR filter is preferred when a small filter size is called for but the application does not require a linear phase.

The objective of IIR filter design is to determine the filter numerator and denominator coefficients to satisfy filter specifications such as passband gain and stopband attenuation, as well as cutoff frequency/frequencies for the lowpass, highpass, bandpass, and bandstop filters.

We first focus on the bilinear transformation (BLT) design method. Then we introduce other design methods such as the impulse invariant design and the pole-zero placement design.

8.2 BILINEAR TRANSFORMATION DESIGN METHOD

Figure 8.1 illustrates a flow chart of the BLT design used in this book. The design procedure includes the following steps: (1) transforming digital filter specifications into analog filter specifications,

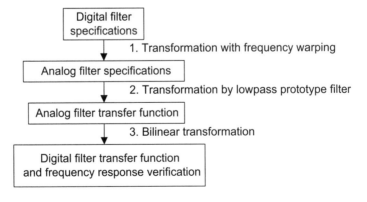

FIGURE 8.1

General procedure for IIR filter design using bilinear transformation.

(2) performing analog filter design, and (3) applying bilinear transformation (which will be introduced in the next section) and verifying the frequency response.

8.2.1 Analog Filters Using Lowpass Prototype Transformation

Before we begin to develop the BLT design, let us review analog filter design using *lowpass prototype transformation*. This method converts an analog lowpass filter with a cutoff frequency of 1 radians per second, called the lowpass prototype, into practical analog lowpass, highpass, banspass, and bandstop filters with specified frequencies.

Letting $H_P(s)$ be a transfer function of the lowpass prototype, the transformation of the lowpass prototype into a lowpass filter is given in Figure 8.2.

As shown in Figure 8.2, $H_{LP}(s)$ designates the analog lowpass filter with a cutoff frequency ω_c radians/second. The lowpass prototype to lowpass filter transformation substitutes s in the lowpass prototype function $H_P(s)$ with s/ω_c, where v is the normalized frequency of the lowpass prototype and ω_c is the cutoff frequency of the lowpass filter. Let us consider the following first-order lowpass prototype:

$$H_P(s) = \frac{1}{s+1} \tag{8.1}$$

Its frequency response is obtained by substituting $s = jv$ into Equation (8.1), that is,

$$H_P(jv) = \frac{1}{jv + 1}$$

and the magnitude gain is

$$|H_P(jv)| = \frac{1}{\sqrt{1 + v^2}} \tag{8.2}$$

We compute the gains at $v = 0$, $v = 1$, $v = 100$, $v = 10,000$ to obtain 1, $1/\sqrt{2}$, 0.0995, and 0.01, respectively. The cutoff frequency gain at $v = 1$ equals $1/\sqrt{2}$, which is equivalent to -3 dB, and the direct-current (DC) gain is 1. The gain approaches zero when the frequency goes to $v = +\infty$. This verifies that the lowpass prototype is a normalized lowpass filter with a normalized cutoff frequency of 1. Applying the prototype transformation $s = s/\omega_c$ in Figure 8.2, we get an analog lowpass filter with a cutoff frequency of ω_c:

$$H(s) = \frac{1}{s/\omega_c + 1} = \frac{\omega_c}{s + \omega_c} \tag{8.3}$$

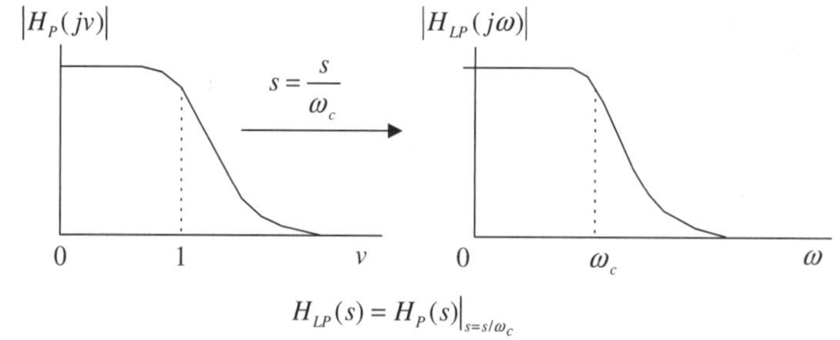

FIGURE 8.2

Analog lowpass prototype transformation into a lowpass filter.

We can obtain the analog frequency response by substituting $s = j\omega$ into Equation (8.3), that is,

$$H(j\omega) = \frac{1}{j\omega/\omega_c + 1}$$

The magnitude response is determined by

$$|H(j\omega)| = \frac{1}{\sqrt{1 + \left(\dfrac{\omega}{\omega_c}\right)^2}} \tag{8.4}$$

Similarly, we verify the gains at $\omega = 0, \omega = \omega_c, \omega = 100\omega_c, \omega = 10,000\omega_c$ to be $1, 1/\sqrt{2}, 0.0995$, and 0.01, respectively. The filter gain at the cutoff frequency ω_c equals $1/\sqrt{2}$, and the DC gain is 1. The gain approaches zero when $\omega = +\infty$. We notice that filter gains do not change but that the filter frequency is scaled up by a factor of ω_c. This verifies that the prototype transformation converts the lowpass prototype to the analog lowpass filter with the specified cut-off frequency of ω_c without an effect on the filter gain.

This first-order prototype function is used here for illustrative purposes. We will obtain general functions for Butterworth and Chebyshev lowpass prototypes in Section 8.3.

The highpass, bandpass, and bandstop filters using the specified lowpass prototype transformation can be easily verified. We review them in Figures 8.3, 8.4 and 8.5, respectively.

The transformation from the lowpass prototype to the highpass filter $H_{HP}(s)$ with a cutoff frequency ω_c radians/second is given in Figure 8.3, where $s = \omega_c/s$ in the lowpass prototype transformation.

The transformation of the lowpass prototype function to a bandpass filter with a center frequency ω_0, a lower cutoff frequency ω_l, and an upper cutoff frequency ω_h in the passband is depicted in Figure 8.4, where $s = (s^2 + \omega_0^2)/(sW)$ is substituted into the lowpass prototype.

As shown in Figure 8.4, ω_0 is the geometric center frequency, which is defined as $\omega_0 = \sqrt{\omega_l\omega_h}$ while the passband bandwidth is given by $W = \omega_h - \omega_l$. Similarly, the transformation from the lowpass prototype to a bandstop (band reject) filter is illustrated in Figure 8.5 with $s = sW/(s^2 + \omega_0^2)$ substituted into the lowpass prototype.

Finally, the lowpass prototype transformations are summarized in Table 8.1.

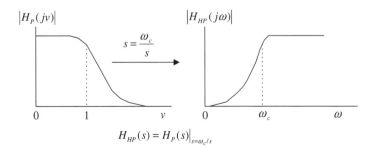

FIGURE 8.3

Analog lowpass prototype transformation to the highpass filter.

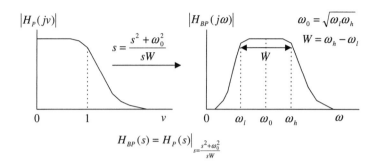

$$H_{BP}(s) = H_P(s)\Big|_{s=\frac{s^2+\omega_0^2}{sW}}$$

FIGURE 8.4

Analog lowpass prototype transformation to the bandpass filter.

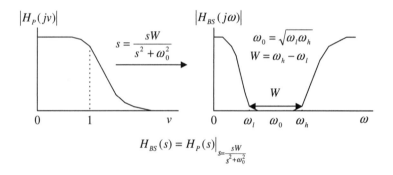

$$H_{BS}(s) = H_P(s)\Big|_{s=\frac{sW}{s^2+\omega_0^2}}$$

FIGURE 8.5

Analog lowpass prototype transformation to a bandstop filter.

Table 8.1 Analog Lowpass Prototype Transformations

Filter Type	Prototype Transformation
Lowpass	$\dfrac{s}{\omega_c}$, ω_c is the cutoff frequency
Highpass	$\dfrac{\omega_c}{s}$, ω_c is the cutoff frequency
Bandpass	$\dfrac{s^2+\omega_0^2}{sW}$, $\omega_0 = \sqrt{\omega_l\omega_h}$, $W = \omega_h - \omega_l$
Bandstop	$\dfrac{sW}{s^2+\omega_0^2}$, $\omega_0 = \sqrt{\omega_l\omega_h}$, $W = \omega_h - \omega_l$

The MATLAB function **freqs()** can be used to plot analog filter frequency responses for verification with the following syntax:

H = freqs(B, A, W)
B = the vector containing the numerator coefficients
A = the vector containing the denominator coefficients
W = the vector containing the specified analog frequency points (radians per second)
H = the vector containing the frequency response

The following example verifies the lowpass prototype transformation.

EXAMPLE 8.2

Given a lowpass prototype

$$H_P(s) = \frac{1}{s+1}$$

determine each of the following analog filters and plot their magnitude responses from 0 to 200 radians per second.

a. A highpass filter with a cutoff frequency of 40 radians per second.
b. A bandpass filter with a center frequency of 100 radians per second and bandwidth of 20 radians per second.

Solution:

a. Applying the lowpass prototype transformation by substituting $s = 40/s$ into the lowpass prototype, we obtain an analog highpass filter:

$$H_{HP}(s) = \frac{1}{\frac{40}{s}+1} = \frac{s}{s+40}$$

b. Similarly, substituting the lowpass-to-bandpass transformation $s = (s^2 + 100)/(20s)$ into the lowpass prototype leads to

$$H_{BP}(s) = \frac{1}{\frac{s^2+100}{20s}+1} = \frac{20s}{s^2+20s+100}$$

The program for plotting the magnitude responses for the highpass filter and bandpass filter is shown in Program 8.1, and Figure 8.6 displays the magnitude responses for the highpass filter and bandpass filter, respectively.
Program 8.1 MATLAB program in Example 8.2.

```
W=0:1:200;                                    % Analog frequency points for computing the
                                              % filter gains

Ha=freqs([1 0],[1 40],W);                     % Frequency response for the highpass filter
Hb=freqs([20 0],[1 20 100],W);               % Frequency response for the bandpass filter
subplot(2,1,1);plot(W,abs(Ha),'k'); grid      % The filter gain plot for highpass filter
xlabel('(a) Frequency (radians per second)')
ylabel('Absolute filter gain');
subplot(2,1,2);plot(W,abs(Hb),'k');grid       % The filter gain plot for bandpass filter
xlabel('(b) Frequency (radians per second)')
ylabel('Absolute filter gain');
```

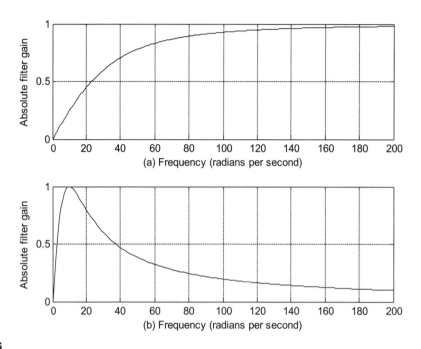

FIGURE 8.6

Magnitude responses for the analog highpass filter and bandpass filter in Example 8.2.

Figure 8.6 confirms the transformation of the lowpass prototype into a highpass filter and a bandpass filter, respectively. To obtain the transfer function of an analog filter, we always begin with the lowpass prototype and apply the corresponding lowpass prototype transformation. To transfer from a lowpass prototype to a bandpass or bandstop filter, the resultant order of the analog filter is twice that of the lowpass prototype order.

8.2.2 Bilinear Transformation and Frequency Warping

In this subsection, we develop the BLT, which converts an analog filter to a digital filter. We begin by finding the area under a curve using the integration of calculus and the numerical recursive method. The area under the curve is a common problem in early calculus courses. As shown in Figure 8.7, the area under the curve can be determined using the following integration:

$$y(t) = \int_0^t x(t)dt \tag{8.5}$$

where $y(t)$ (area under the curve) and $x(t)$ (curve function) are the output and input of the analog integrator, respectively, and t is the upper limit of the integration.

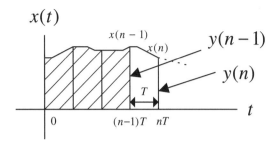

FIGURE 8.7

Digital integration method to calculate the area under the curve.

Applying the Laplace transform on Equation (8.5), we have

$$Y(s) = \frac{X(s)}{s} \tag{8.6}$$

and find that the Laplace transfer function is

$$G(s) = \frac{Y(s)}{X(s)} = \frac{1}{s} \tag{8.7}$$

Now we examine the numerical integration method shown in Figure 8.7 to approximate the integration of Equation (8.5) using the following difference equation:

$$y(n) = y(n-1) + \frac{x(n) + x(n-1)}{2} T \tag{8.8}$$

where T denotes the sampling period. $y(n) = y(nT)$ is the output sample that represents the whole area under the curve, while $y(n-1) = y(nT - T)$ is the previous output sample from the integrator indicating the previously computed area under the curve (the shaded area in Figure 8.7). Notice that $x(n) = x(nT)$ and $x(n-1) = x(nT - T)$, sample amplitudes from the curve, are the current input sample and the previous input sample in Equation (8.8). Applying the z-transform on both sides of Equation (8.8) leads to

$$Y(z) = z^{-1} Y(z) + \frac{T}{2} \left(X(z) + z^{-1} X(z) \right)$$

Solving for the ratio of $Y(z)/X(z)$, we obtain the z-transfer function as

$$H(z) = \frac{Y(z)}{X(z)} = \frac{T}{2} \frac{1 + z^{-1}}{1 - z^{-1}} \tag{8.9}$$

Next, comparing Equation (8.9) with Equation (8.7), it follows that

$$\frac{1}{s} = \frac{T}{2} \frac{1 + z^{-1}}{1 - z^{-1}} = \frac{T}{2} \frac{z + 1}{z - 1} \tag{8.10}$$

Solving for s in Equation (8.10) gives the bilinear transformation

$$s = \frac{2}{T} \frac{z - 1}{z + 1} \tag{8.11}$$

The BLT method is a mapping or transformation of points on the s-plane to the z-plane. Equation (8.11) can be alternatively written as

$$z = \frac{1 + sT/2}{1 - sT/2} \tag{8.12}$$

The general mapping properties are summarized as following:

1. The left-half s-plane is mapped onto the inside of the unit circle of the z-plane.
2. The right-half s-plane is mapped into the outside of the unit circle of the z-plane.
3. The positive $j\omega$ axis portion in the s-plane is mapped onto the positive half circle (the dashed line arrow in Figure 8.8) on the unit circle, while the negative $j\omega$ axis is mapped onto the negative half circle (the dotted line arrow in Figure 8.8) on the unit circle.

To verify these features, let us look at the following illustrative example.

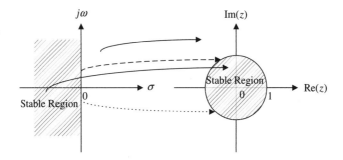

FIGURE 8.8

Mapping between the s-plane and the z-plane by the bilinear transformation.

EXAMPLE 8.3

Assume that $T = 2$ seconds in Equation (8.12), and that the following points are given:

1. $s = -1 + j$, on the left half of the s-plane
2. $s = 1 - j$, on the right half of the s-plane
3. $s = j$, on the positive $j\omega$ on the s-plane
4. $s = -j$, on the negative $j\omega$ on the s-plane

Convert each of the points in the s-plane to the z-plane, and verify mapping properties (1) to (3).

Solution:
Substituting $T = 2$ into Equation (8.12) leads to

$$z = \frac{1 + s}{1 - s}$$

We can carry out mapping for each point as follows:

1. $z = \dfrac{1 + (-1 + j)}{1 - (-1 + j)} = \dfrac{j}{2 - j} = \dfrac{1\angle 90^\circ}{\sqrt{5}\angle - 26.57^\circ} = 0.4472\angle 116.57^\circ,$

since $|z| = 0.4472 < 1$, which is inside the unit circle on the z-plane.

2. $z = \dfrac{1+(1-j)}{1-(1-j)} = \dfrac{2-j}{j} = \dfrac{\sqrt{5}\angle -26.57°}{1\angle 90°} = 2.2361\angle -116.57°,$

since $|z| = 2.2361 > 1$, which is outside the unit circle on the z-plane.

3. $z = \dfrac{1+j}{1-j} = \dfrac{\sqrt{2}\angle 45°}{\sqrt{2}\angle -45°} = 1\angle 90°,$

since $|z| = 1$ and $\theta = 90°$, which is on the positive half circle on the unit circle on the z-plane.

4. $z = \dfrac{1-j}{1-(-j)} = \dfrac{1-j}{1+j} = \dfrac{\sqrt{2}\angle -45°}{\sqrt{2}\angle 45°} = 1\angle -90°,$

since $|z| = 1$ and $\theta = -90°$, which is on the negative half circle on the unit circle on the z-plane.

As shown in Example (8.3), the BLT offers convertion of an analog transfer function to a digital transfer function. Example (8.4) shows how to perform the BLT.

EXAMPLE 8.4

Given an analog filter whose transfer function is

$$H(s) = \frac{10}{s+10}$$

convert it to the digital filter transfer function and difference equation, respectively, when the sampling period is given as $T = 0.01$ second.

Solution:
Applying the BLT, we have

$$H(z) = H(s)|_{s=\frac{2}{T}\frac{z-1}{z+1}} = \frac{10}{s+10}\Big|_{s=\frac{2}{T}\frac{z-1}{z+1}}$$

Substituting $T = 0.01$, it follows that

$$H(z) = \frac{10}{\frac{200(z-1)}{z+1}+10} = \frac{0.05}{\frac{z-1}{z+1}+0.05} = \frac{0.05(z+1)}{z-1+0.05(z+1)} = \frac{0.05z+0.05}{1.05z-0.95}.$$

Finally, we get

$$H(z) = \frac{(0.05z+0.05)/(1.05z)}{(1.05z-0.95)/(1.05z)} = \frac{0.0476+0.0476z^{-1}}{1-0.9048z^{-1}}.$$

Applying the technique from Chapter 6, we obtain the difference equation as

$$y(n) = 0.0476x(n) + 0.0476x(n-1) + 0.9048y(n-1).$$

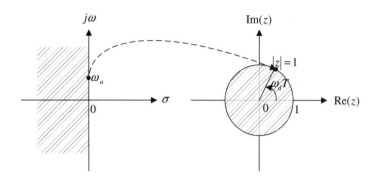

FIGURE 8.9

Frequency mapping from the analog domain to the digital domain.

Next, we examine frequency mapping between the s-plane and the z-plane. As illustrated in Figure 8.9, the analog frequency ω_a is marked on the $j\omega$ -axis on the s-plane, whereas ω_d is the digital frequency labeled on the unit circle in the z-plane.

We substitute $s = jw_a$ and $z = e^{j\omega T}$ into the bilinear transformation in Equation (8.11) to get

$$j\omega_a = \frac{2}{T} \frac{e^{j\omega_d T} - 1}{e^{j\omega_d T} + 1} \tag{8.13}$$

Simplifying Equation (8.13) leads to

$$\omega_a = \frac{2}{T} \tan\left(\frac{\omega_d T}{2}\right) \tag{8.14}$$

Equation (8.14) explores the relation between the analog frequency on the $j\omega$ axis and the corresponding digital frequency ω_d on the unit circle. We can also write its inverse as

$$\omega_d = \frac{2}{T} \tan^{-1}\left(\frac{\omega_a T}{2}\right) \tag{8.15}$$

The range of the digital frequency ω_d is from 0 radians per second to the folding frequency $\omega_s/2$ radians per second, where ω_s is the sampling frequency in terms of radians per second. We present a plot of Equation (8.14) in Figure 8.10.

From Figure 8.10, when the digital frequency range $0 \leq \omega_d \leq 0.25\omega_s$ is mapped to the analog frequency range $0 \leq \omega_a \leq 0.32\omega_s$, the transformation appears to be linear; however, when the digital frequency range $0.25\omega_s \leq \omega_d \leq 0.5\omega_s$ is mapped to the analog frequency range for $\omega_a > 0.32\omega_s$, the transformation is nonlinear. The analog frequency range for $\omega_a > 0.32\omega_s$ is compressed into the digital frequency range $0.25\omega_s \leq \omega_d \leq 0.5\omega_s$. This nonlinear frequency mapping effect is called *frequency warping*. We must incorporate frequency warping into IIR filter design.

The following example will illustrate the frequency warping effect in the BLT.

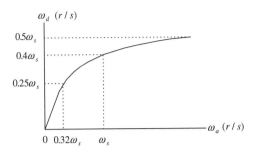

FIGURE 8.10

Frequency warping from bilinear transformation.

EXAMPLE 8.5

Assume the following analog frequencies:

$\omega_a = 10$ radians per second
$\omega_a = \omega_s/4 = 50\pi = 157$ radians per second
$\omega_a = \omega_s/2 = 100\pi = 314$ radians per second.

Find their digital frequencies using the BLT with a sampling period of 0.01 second, given the analog filter in Example 8.4 and the developed digital filter.

Solution:
From Equation (8.15), we can calculate digital frequency ω_d as follows:
When $\omega_a = 10$ radians per second and $T = 0.01$ second

$$\omega_d = \frac{2}{T}\tan^{-1}\left(\frac{\omega_a T}{2}\right) = \frac{2}{0.01}\tan^{-1}\left(\frac{10 \times 0.01}{2}\right) = 9.99 \text{ rad/sec}$$

which is close to the analog frequency of 10 radians per second. When $\omega_a = 157$ rad/sec and $T = 0.01$ second

$$\omega_d = \frac{2}{0.01}\tan^{-1}\left(\frac{157 \times 0.01}{2}\right) = 133.11 \text{ rad/sec}$$

which is somewhat off as compared with the desired value 157. When $\omega_a = 314$ rad/sec and $T = 0.01$ second,

$$\omega_d = \frac{2}{0.01}\tan^{-1}\left(\frac{314 \times 0.01}{2}\right) = 252.5 \text{ rad/sec}$$

which is significantly different than the digital folding frequency of 314 radians per second.

Figure 8.11 shows how to correct the frequency warping error. First, given the digital frequency specification, we prewarp the digital frequency specification to the analog frequency specification by Equation (8.14).

Second, we obtain the analog lowpass filter $H(s)$ using the prewarped analog frequency ω_a and the lowpass prototype. For the lowpass analog filter, we have

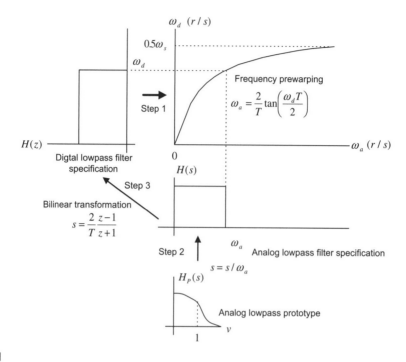

FIGURE 8.11

Graphical representation of IIR filter design using the bilinear transformation.

$$H(s) = H_P(s)|_{s=\frac{s}{\omega_a}} = H_P\left(\frac{s}{\omega_a}\right)$$ (8.16)

Finally, substituting the BLT Equation (8.11) into Equation (8.16) yields the digital filter:

$$H(z) = H(s)|_{s=\frac{2}{T}\frac{z-1}{z+1}}$$ (8.17)

This approach can be similarly extended to other types of filter design.

8.2.3 Bilinear Transformation Design Procedure

Now we can summarize the BLT design procedure.

1. Given the digital filter frequency specifications, prewarp the digital frequency specifications to the analog frequency specifications.

   For the lowpass filter and highpass filter:

$$\omega_a = \frac{2}{T} \tan\left(\frac{\omega_d T}{2}\right)$$ (8.18)

For the bandpass filter and bandstop filter :

$$\omega_{al} = \frac{2}{T} \tan\left(\frac{\omega_l T}{2}\right), \quad \omega_{ah} = \frac{2}{T} \tan\left(\frac{\omega_h T}{2}\right) \tag{8.19}$$

and $\omega_0 = \sqrt{\omega_{al}\omega_{ah}}$, $W = \omega_{ah} - \omega_{al}$

2. Perform the prototype transformation using the lowpass prototype $H_P(s)$.

$$\text{From lowpass to lowpass:} \quad H(s) = H_P(s)|_{s = \frac{s}{\omega_a}} \tag{8.20}$$

$$\text{From lowpass to highpass:} \quad H(s) = H_P(s)|_{s = \frac{\omega_a}{s}} \tag{8.21}$$

$$\text{From lowpass to bandpass:} \quad H(s) = H_P(s)|_{s = \frac{s^2 + \omega_0^2}{sW}} \tag{8.22}$$

$$\text{From lowpass to bandstop:} \quad H(s) = H_P(s)|_{s = \frac{sW}{s^2 + \omega_0^2}} \tag{8.23}$$

3. Substitute the BLT to obtain the digital filter

$$H(z) = H(s)|_{s = \frac{2}{T}\frac{z-1}{z+1}} \tag{8.24}$$

Table 8.2 lists MATLAB functions for the BLT design.

We illustrate the lowpass filter design procedure in Example (8.6). Other types of filters, such as highpass, bandpass, and bandstop, will be illustrated in the next section.

EXAMPLE 8.6

The normalized lowpass filter with a cutoff frequency of 1 rad/sec is given as

$$H_P(s) = \frac{1}{s+1}$$

a. Use the given $H_P(s)$ and the BLT to design a corresponding digital IIR lowpass filter with a cutoff frequency of 15 Hz and a sampling rate of 90 Hz.
b. Use MATLAB to plot the magnitude response and phase response of $H(z)$.

Solution:

a. First, we obtain the digital frequency as

$$\omega_d = 2\pi f = 2\pi(15) = 30\pi \text{ radians/second,} \quad \text{and} \quad T = 1/f_s = 1/90 \text{ sec}$$

We then follow the design procedure:

1. First calculate the prewarped analog frequency as

Table 8.2 MATLAB Functions for Bilinear Transformation Design

Lowpass to lowpass: $H(s) = H_P(s)\big|_{s=\frac{s}{\omega_a}}$

>>[B, A]=lp2lp(Bp, Ap, wa)

Lowpass to highpass: $H(s) = H_P(s)\big|_{s=\frac{\omega_a}{s}}$

>>[B, A]=lp2hp(Bp, Ap, wa)

Lowpass to bandpass: $H(s) = H_P(s)\big|_{s=\frac{s^2+\omega_0^2}{sW}}$

>>[B, A]=lp2bp(Bp, Ap, w0, W)

Lowpass to bandstop: $H(s) = H_P(s)\big|_{s=\frac{sW}{s^2+\omega_0^2}}$

>>[B, A]=lp2bs(Bp, Ap, w0, W)

Bilinear transformation to achieve the digital filter:

>>[b, a]=bilinear(B, A, fs)

Plot of the magnitude and phase frequency responses of the digital filter:

>>freqz(b, a, 512, fs)

Definitions of design parameters:

Bp = vector containing the numerator coefficients of the lowpass prototype

Ap = vector containing the denominator coefficients of the lowpass prototype

wa = cutoff frequency for the lowpass or highpass analog filter (rad/sec)

w0 = center frequency for the bandpass or bandstop analog filter (rad/sec)

W = bandwidth for the bandpass or bandstop analog filter (rad/sec)

B = vector containing the numerator coefficients of the analog filter

A = vector containing the denominator coefficients of the analog filter

b = vector containing the numerator coefficients of the digital filter

a = vector containing the denominator coefficients of the digital filter

fs = sampling rate (samples/sec)

$$\omega_a = \frac{2}{T}\tan\left(\frac{\omega_d T}{2}\right) = \frac{2}{1/90}\tan\left(\frac{30\pi/90}{2}\right)$$

that is, $\omega_a = 180 \times \tan(\pi/6) = 180 \times \tan(30°) = 103.92$ rad/sec.

2. Then perform the prototype transformation (lowpass to lowpass) as follows:

$$H(s) = H_P(s)\big|_{s=\frac{s}{\omega_a}} = \frac{1}{\frac{s}{\omega_a}+1} = \frac{\omega_a}{s+\omega_a}$$

This yields an analog filter:

$$H(s) = \frac{103.92}{s + 103.92}$$

3. Apply the BLT, which yields

$$H(z) = \frac{103.92}{s + 103.92}\bigg|_{s=\frac{2}{T}\frac{z-1}{z+1}}$$

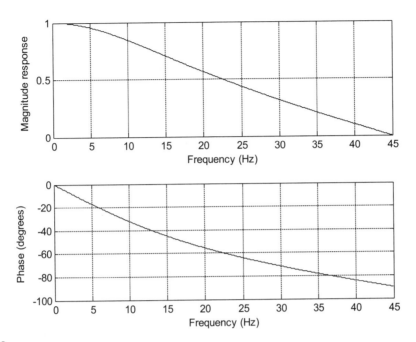

FIGURE 8.12

Frequency responses of the designed digital filter for Example 8.6.

We simplify the algebra by dividing both the numerator and the denominator by 180:

$$H(z) = \frac{103.92}{180 \times \dfrac{z-1}{z+1} + 103.92} = \frac{103.92/180}{\dfrac{z-1}{z+1} + 103.92/180} = \frac{0.5773}{\dfrac{z-1}{z+1} + 0.5773}$$

Then we multiply both numerator and denominator by $(z+1)$ to obtain

$$H(z) = \frac{0.5773(z+1)}{\left(\dfrac{z-1}{z+1} + 0.5773\right)(z+1)} = \frac{0.5773z + 0.5773}{(z-1) + 0.5773(z+1)} = \frac{0.5773z + 0.5773}{1.5773z - 0.4227}$$

Finally, we divide both numerator and denominator by $1.5773z$ to get the transfer function in the standard format:

$$H(z) = \frac{(0.5773z + 0.5773)/(1.5773z)}{(1.5773z - 0.4227)/(1.5773z)} = \frac{0.3660 + 0.3660z^{-1}}{1 - 0.2679z^{-1}}$$

b. The corresponding MATLAB design is listed in Program 8.2. Figure 8.12 shows the magnitude and phase frequency responses.

Program 8.2. MATLAB program for Example 8.6.

```
%Example 8.6
% Plot the magnitude and phase responses
fs=90;     % Sampling rate (Hz)
```

```
[B, A]=lp2lp([1],[1 1],103.92);
[b,a]=bilinear(B,A,fs)
% b = [0.3660 0.3660] numerator coefficients of the digital filter from MATLAB
% a = [1 -0.2679]    denominator coefficients of the digital filter from MATLAB
[hz, f]=freqz([0.3660 0.3660],[1 -0.2679],512,fs); % Frequency response
phi = 180*unwrap(angle(hz))/pi;
subplot(2,1,1), plot(f, abs(hz)),grid;
axis([0 fs/2 0 1]);
xlabel('Frequency (Hz)'); ylabel('Magnitude Response')
subplot(2,1,2), plot(f, phi); grid;
axis([0 fs/2 -100 0]);
xlabel('Frequency (Hz)'); ylabel('Phase (degrees)')
```

8.3 DIGITAL BUTTERWORTH AND CHEBYSHEV FILTER DESIGNS

In this section, we design various types of digital Butterworth and Chebyshev filters using the BLT design method developed in the previous section.

8.3.1 Lowpass Prototype Function and Its Order

As described in Section 8.2, BLT design requires obtaining the analog filter with prewarped frequency specifications. These analog filter design requirements include the ripple specification at the passband frequency edge, the attenuation specification at the stopband frequency edge, and the type of lowpass prototype (which we shall discuss) and its order.

Table 8.3 lists the Butterworth prototype functions with 3 dB passband ripple specification. Tables 8.4 and 8.5 contain the Chebyshev prototype functions (type I) with 1 dB and 0.5 dB passband ripple specifications, respectively. Other lowpass prototypes with different ripple specifications and orders can be computed using the methods described in Appendix C.

In this section, we will focus on the Chebyshev type I filter. The Chebyshev type II filter design can be found in Proakis and Manolakis (1996) and Porat (1997).

The magnitude response function of the Butterworth lowpass prototype with order n is shown in Figure 8.13, where the magnitude response $|H_P(v)|$ versus the normalized frequency v is given by

$$|H_P(v)| = \frac{1}{\sqrt{1 + \varepsilon^2 v^{2n}}} \tag{8.25}$$

With the given passband ripple A_p dB at the normalized passband frequency edge $v_p = 1$, and the stopband attenuation A_s dB at the normalized stopband frequency edge v_s, the following two equations must be satisfied to determine the prototype filter order:

$$A_P \, dB = -20 \cdot \log_{10}\left(\frac{1}{\sqrt{1 + \varepsilon^2}}\right) \tag{8.26}$$

Table 8.3 3-dB Butterworth Lowpass Prototype Transfer Functions ($\varepsilon = 1$)

n	$H_P(s)$
1	$\dfrac{1}{s+1}$
2	$\dfrac{1}{s^2 + 1.4142s + 1}$
3	$\dfrac{1}{s^3 + 2s^2 + 2s + 1}$
4	$\dfrac{1}{s^4 + 2.6131s^3 + 3.4142s^2 + 2.6131s + 1}$
5	$\dfrac{1}{s^5 + 3.2361s^4 + 5.2361s^3 + 5.2361s^2 + 3.2361s + 1}$
6	$\dfrac{1}{s^6 + 3.8637s^5 + 7.4641s^4 + 9.1416s^3 + 7.4641s^2 + 3.8637s + 1}$

$$A_s \, dB = -20 \cdot \log_{10} \left(\frac{1}{\sqrt{1 + \varepsilon^2 v_s^{2n}}} \right) \tag{8.27}$$

Solving Equations (8.26) and (8.27), we determine the lowpass prototype order as

$$\varepsilon^2 = 10^{0.1A_p} - 1 \tag{8.28}$$

Table 8.4 Chebyshev Lowpass Prototype Transfer Functions with 0.5 dB Ripple ($\varepsilon = 0.3493$)

n	$H_P(s)$
1	$\dfrac{2.8628}{s + 2.8628}$
2	$\dfrac{1.4314}{s^2 + 1.4256s + 1.5162}$
3	$\dfrac{0.7157}{s^3 + 1.2529s^2 + 1.5349s + 0.7157}$
4	$\dfrac{0.3579}{s^4 + 1.1974s^3 + 1.7169s^2 + 1.0255s + 0.3791}$
5	$\dfrac{0.1789}{s^5 + 1.1725s^4 + 1.9374s^3 + 1.3096s^2 + 0.7525s + 0.1789}$
6	$\dfrac{0.0895}{s^6 + 1.1592s^5 + 2.1718s^4 + 1.5898s^3 + 1.1719s^2 + 0.4324s + 0.0948}$

Table 8.5 Chebyshev Lowpass Prototype Transfer Functions with 1 dB Ripple ($\varepsilon = 0.5088$)

n	$H_P(s)$
1	$\dfrac{1.9652}{s + 1.9652}$
2	$\dfrac{0.9826}{s^2 + 1.0977s + 1.1025}$
3	$\dfrac{0.4913}{s^3 + 0.9883s^2 + 1.2384s + 0.4913}$
4	$\dfrac{0.2456}{s^4 + 0.9528s^3 + 1.4539s^2 + 0.7426s + 0.2756}$
5	$\dfrac{0.1228}{s^5 + 0.9368s^4 + 1.6888s^3 + 0.9744s^2 + 0.5805s + 0.1228}$
6	$\dfrac{0.0614}{s^6 + 0.9283s^5 + 1.9308s^4 + 1.20121s^3 + 0.9393s^2 + 0.3071s + 0.0689}$

$$n \geq \frac{\log_{10}\left(\dfrac{10^{0.1A_s} - 1}{\varepsilon^2}\right)}{[2 \cdot \log_{10}(v_s)]} \tag{8.29}$$

where ε is the absolute ripple specification.

The magnitude response function of a Chebyshev lowpass prototype with order n is shown in Figure 8.14, where the magnitude response $|H_P(v)|$ versus the normalized frequency v is given by

$$|H_P(v)| = \frac{1}{\sqrt{1 + \varepsilon^2 C_n^2(v)}} \tag{8.30}$$

where

$$C_n(v_s) = \cosh\left[n\cosh^{-1}(v_s)\right] \tag{8.31}$$

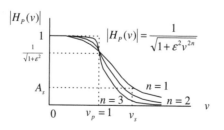

FIGURE 8.13

Normalized Butterworth magnitude response function.

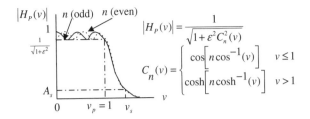

FIGURE 8.14

Normalized Chebyshev magnitude response function.

$$\cosh^{-1}(v_s) = \ln\left(v_s + \sqrt{v_s^2 - 1}\right) \tag{8.32}$$

As shown in Figure 8.14, the magnitude response for the Chebyshev lowpass prototype with an odd-numbered order begins with a filter DC gain of 1. In the case of a Chebyshev lowpass prototype with an even-numbered order, the magnitude starts at a filter DC gain of $1/\sqrt{1+\epsilon^2}$. For both cases, the filter gain at the normalized cutoff frequency $v_p = 1$ is $1/\sqrt{1+\epsilon^2}$.

Similarly, Equations (8.33) and (8.34) must be satisfied:

$$A_p \; dB = -20 \cdot \log_{10}\left(\frac{1}{\sqrt{1+\epsilon^2}}\right) \tag{8.33}$$

$$A_s \; dB = -20 \cdot \log_{10}\left(\frac{1}{\sqrt{1+\epsilon^2 C_n^2(v_s)}}\right) \tag{8.34}$$

The lowpass prototype order can be solved in Equation (8.35):

$$\epsilon^2 = 10^{0.1A_p} - 1 \tag{8.35a}$$

$$n \geq \frac{\cosh^{-1}\left[\left(\dfrac{10^{0.1A_s} - 1}{\epsilon^2}\right)^{0.5}\right]}{\cosh^{-1}(v_s)} \tag{8.35b}$$

where $\cosh^{-1}(x) = \ln(x + \sqrt{x^2 - 1})$, ϵ is the absolute ripple parameter.

The normalized stopband frequency v_s can be determined from the frequency specifications of the analog filter in Table 8.6. Then the order of the lowpass prototype can be determined by Equation (8.29) for the Butterworth function and Equation (8.35b) for the Chebyshev function. Figure 8.15 gives frequency edge notations for analog lowpass and bandpass filters. The notations for analog highpass and bandstop filters can be defined correspondingly.

Table 8.6 Conversion from Analog Filter Specifications to Lowpass Prototype Specifications

Analog Filter Specifications	Lowpass Prototype Specifications
Lowpass: ω_{ap}, ω_{as}	$v_p = 1$, $v_s = \omega_{as}/\omega_{ap}$
Highpass: ω_{ap}, ω_{as}	$v_p = 1$, $v_s = \omega_{ap}/\omega_{as}$
Bandpass: ω_{apl}, ω_{aph}, ω_{asl}, ω_{ash}, $\omega_0 = \sqrt{\omega_{apl}\omega_{aph}}$, $\omega_0 = \sqrt{\omega_{asl}\omega_{ash}}$	$v_p = 1$, $v_s = \dfrac{\omega_{ash} - \omega_{asl}}{\omega_{aph} - \omega_{apl}}$
Bandstop: ω_{apl}, ω_{aph}, ω_{asl}, ω_{ash}, $\omega_0 = \sqrt{\omega_{apl}\omega_{aph}}$, $\omega_0 = \sqrt{\omega_{asl}\omega_{ash}}$	$v_p = 1$, $v_s = \dfrac{\omega_{aph} - \omega_{apl}}{\omega_{ash} - \omega_{asl}}$

ω_{ap}, passband frequency edge; ω_{as}, stopband frequency edge;
ω_{apl}, lower cutoff frequency in passband; ω_{aph}, upper cutoff frequency in passband;
ω_{asl}, lower cutoff frequency in stopband; ω_{ash}, upper cutoff frequency in stopband;
ω_0, geometric center frequency.

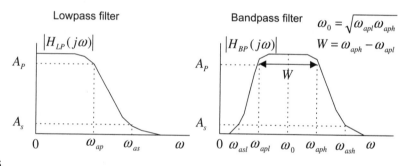

FIGURE 8.15

Specifications for analog lowpass and bandpass filters.

8.3.2 Lowpass and Highpass Filter Design Examples

The following examples illustrate various designs for the Butterworth and Chebyshev lowpass and highpass filters.

EXAMPLE 8.7

a. Design a digital lowpasss Butterworth filter with the following specifications:
   1. 3-dB attenuation at the passband frequency of 1.5 kHz
   2. 10-dB stopband attenuation at the frequency of 3 kHz
   3. Sampling frequency at 8,000 Hz
b. Use MATLAB to plot the magnitude and phase responses.

Solution:

a. First, we obtain the digital frequencies in radians per second:
$$\omega_{dp} = 2\pi f = 2\pi(1500) = 3,000\pi \text{ rad/sec}$$

$\omega_{ds} = 2\pi f = 2\pi(3,000) = 6,000\pi$ rad/sec
$T = 1/f_s = 1/8,000$ sec

We then follow the design procedure steps

1. We apply the warping equation as follows:

$$\omega_{ap} = \frac{2}{T}\tan\left(\frac{\omega_d T}{2}\right) = 16,000 \times \tan\left(\frac{3,000\pi/8,000}{2}\right) = 1.0691 \times 10^4 \text{ rad/sec.}$$

$$\omega_{as} = \frac{2}{T}\tan\left(\frac{\omega_d T}{2}\right) = 16,000 \times \tan\left(\frac{6,000\pi/8,000}{2}\right) = 3.8627 \times 10^4 \text{ rad/sec.}$$

We then find the lowpass prototype specifications using Table 8.6 as follows:

$$v_s = \omega_{as}/\omega_{ap} = 3.8627 \times 10^4 / \left(1.0691 \times 10^4\right) = 3.6130 \text{ rad/s and } A_s = 10 \text{ dB}$$

The filter order is computed as

$$\epsilon^2 = 10^{0.1\times3} - 1 = 1$$

$$n = \frac{\log_{10}\left(10^{0.1\times10} - 1\right)}{2 \cdot \log_{10}(3.6130)} = 0.8553$$

2. Rounding n up, we choose $n = 1$ for the lowpass prototype. From Table 8.3, we have

$$H_P(s) = \frac{1}{s+1}$$

Applying the prototype transformation (lowpass to lowpass) yields the analog filter

$$H(s) = H_P(s)\Big|_{\frac{s}{\omega_{ap}}} = \frac{1}{\dfrac{s}{\omega_{ap}}+1} = \frac{\omega_{ap}}{s+\omega_{ap}} = \frac{1.0691 \times 10^4}{s+1.0691 \times 10^4}$$

3. Finally, using the BLT, we have

$$H(z) = \frac{1.0691 \times 10^4}{s+1.0691 \times 10^4}\Bigg|_{s=16,000(z-1)/(z+1)}$$

Substituting the BLT leads to

$$H(z) = \frac{1.0691 \times 10^4}{\left(16,000\dfrac{z-1}{z+1}\right) + 1.0691 \times 10^4}$$

To simply the algebra, we both numerator and denominator by 16,000 to get

$$H(z) = \frac{0.6682}{\left(\dfrac{z-1}{z+1}\right) + 0.6682}$$

Then multiplying $(z + 1)$ on both numerator and denominator leads to

$$H(z) = \frac{0.6682(z + 1)}{(z - 1) + 0.6682(z + 1)} = \frac{0.6682z + 0.6682}{1.6682z - 0.3318}$$

Dividing both numerator and denominator by $(1.6682 \cdot z)$ leads to

$$H(z) = \frac{0.4006 + 0.4006z^{-1}}{1 - 0.1989z^{-1}}$$

b. Steps 2 and 3 can be carried out using MATLAB Program 8.3, as shown in the first three lines of the MATLAB code. Figure 8.16 describes the filter frequency responses.

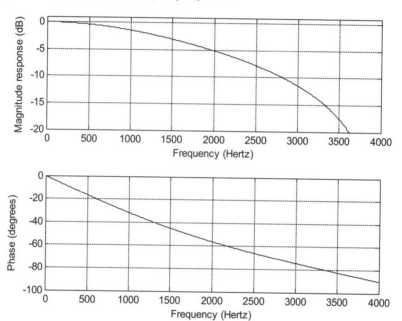

FIGURE 8.16

Frequency responses of the designed digital filter for Example 8.7.

Program 8.3. MATLAB program for Example 8.7.

```
% Example 8.7
% Design of the digital lowpass Butterworth filter
  format long
  fs=8000;                              % Sampling rate
  [B A]=lp2lp([1],[1 1], 1.0691*10^4)  % Complete step 2
  [b a]=bilinear(B,A,fs)                % Complete step 3
% Plot the magnitude and phase responses |H(z)|
% b=[0.4005 0.4005]; numerator coefficients from MATLAB
% a=[1 -0.1989]; denominator coefficients from MATLAB
  freqz(b,a,512,fs);
  axis([0 fs/2 -20 1])
```

EXAMPLE 8.8

a. Design a first-order highpass digital Chebyshev filter with a cutoff frequency of 3 kHz and 1 dB ripple on the passband using a sampling frequency of 8,000 Hz.
b. Use MATLAB to plot the magnitude and phase responses.

Solution:

a. First, we obtain the digital frequency in radians per second:

$$\omega_d = 2\pi f = 2\pi(3,000) = 6,000\pi \text{ rad/sec}, \quad \text{and} \quad T = 1/f_s = 1/8,000 \text{ sec}$$

Following the steps of the design procedure, we have

1.

$$\omega_a = \frac{2}{T} \tan\left(\frac{\omega_d T}{2}\right) = 16,000 \times \tan\left(\frac{6,000\pi/8,000}{2}\right) = 3.8627 \times 10^4 \text{ rad/sec}$$

2. Since the filter order is given as 1, we select the first-order lowpass prototype from Table 8.5:

$$H_P(s) = \frac{1.9625}{s + 1.9625}$$

Applying the prototype transformation (lowpass to highpass), we obtain

$$H(s) = H_P(s)|_{\frac{\omega_a}{s}} = \frac{1.9625}{\frac{\omega_a}{s} + 1.9625} = \frac{1.9625s}{1.9625s + 3.8627 \times 10^4}$$

Dividing both numerator and denominator by 1.9625 gives

$$H(s) = \frac{s}{s + 1.9683 \times 10^4}$$

3. Using the BLT, we have

$$H(z) = \frac{s}{s + 1.9683 \times 10^4}\bigg|_{s=16,000(z-1)/(z+1)}$$

The algebra work proceeds as follows:

$$H(z) = \frac{16,000\frac{z-1}{z+1}}{16,000\frac{z-1}{z+1} + 1.9683 \times 10^4}$$

Simplifying the transfer function yields

$$H(z) = \frac{0.4484 - 0.4484z^{-1}}{1 + 0.1032z^{-1}}$$

b. Steps 2 and 3 and frequency response plots shown in Figure 8.17 can be carried out using MATLAB Program 8.4.

Program 8.4. MATLAB program for Example 8.8.

```
% Example 8.8
% Design of the digital highpass Butterworth filter
```

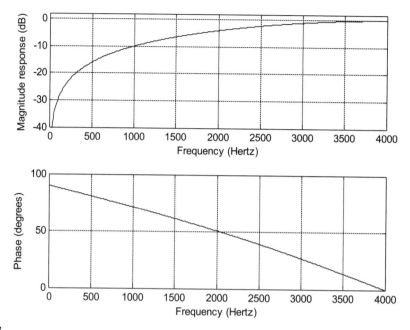

FIGURE 8.17

Frequency responses of the designed digital filter for Example 8.8.

```
format long
fs=8000;                                           % Sampling rate
[B A]=lp2hp([1.9625],[1 1.9625], 3.8627*10^4)      % Complete step 2
[b a]=bilinear(B,A,fs)                             % Complete step 3
% Plot the magnitude and phase responses |H(z)|
% b=[0.4484 -0.4484 ]; numerator coefficients from MATLAB
% a=[1 0.1032]; denominator coefficients from MATLAB
  freqz(b,a,512,fs);
  axis([0 fs/2 -40 2])
```

EXAMPLE 8.9

a. Design a second-order lowpass digital Butterworth filter with a cutoff frequency of 3.4 kHz at a sampling frequency of 8,000 Hz.

b. Use MATLAB to plot the magnitude and phase responses.

Solution:

a. First, we obtain the digital frequency in radians per second:

$$\omega_d = 2\pi f = 2\pi(3,400) = 6,800\pi \text{ rad/sec}, \quad \text{and} \quad T = 1/f_s = 1/8,000 \text{ sec}$$

Following the steps of the design procedure, we compute the prewarped analog frequencies as

1.

$$\omega_a = \frac{2}{T}\tan\left(\frac{\omega_d T}{2}\right) = 16,000 \times \tan\left(\frac{6,800\pi/8,000}{2}\right) = 6.6645 \times 10^4 \text{ rad/sec}$$

2. Since the order of 2 is given in the specification, we directly pick the second-order lowpass prototype from Table 8.3:

$$H_P(s) = \frac{1}{s^2 + 1.4142s + 1}$$

After applying the prototype transformation (lowpass to lowpass), we have

$$H(s) = H_P(s)|_{\frac{s}{\omega_a}} = \frac{4.4416 \times 10^9}{s^2 + 9.4249 \times 10^4 s + 4.4416 \times 10^9}$$

3. Carrying out the BLT yields

$$H(z) = \frac{4.4416 \times 10^9}{s^2 + 9.4249 \times 10^4 s + 4.4416 \times 10^9}\bigg|_{s=16,000(z-1)/(z+1)}$$

Algebra work yields the following:

$$H(z) = \frac{4.4416 \times 10^9}{\left(16,000\dfrac{z-1}{z+1}\right)^2 + 9.4249 \times 10^4\left(16,000\dfrac{z-1}{z+1}\right) + 4.4416 \times 10^9}$$

To simplify, we divide both numerator and denominator by $(16,000)^2$ to get

$$H(z) = \frac{17.35}{\left(\dfrac{z-1}{z+1}\right)^2 + 5.8906\left(\dfrac{z-1}{z+1}\right) + 17.35}$$

Then multiplying both numerator and denominator by $(z+1)^2$ leads to

$$H(z) = \frac{17.35(z+1)^2}{(z-1)^2 + 5.8906(z-1)(z+1) + 17.35(z+1)^2}$$

Using identities, we have

$$H(z) = \frac{17.35(z^2 + 2z + 1)}{(z^2 - 2z + 1) + 5.8906(z^2 - 1) + 17.35(z^2 + 2z + 1)} = \frac{17.35z^2 + 34.7z + 17.35}{24.2406z^2 + 32.7z + 12.4594}$$

Dividing both numerator and denominator by $(24.2406z^2)$ leads to

$$H(z) = \frac{0.7157 + 1.4314z^{-1} + 0.7151z^{-2}}{1 + 1.3490z^{-1} + 0.5140z^{-2}}$$

b. Steps 2 and 3 require a certain amount of algebra work and can be verified using MATLAB Program 8.5, as shown in the first three lines of the code. Figure 8.18 plots the filter magnitude and phase frequency responses.

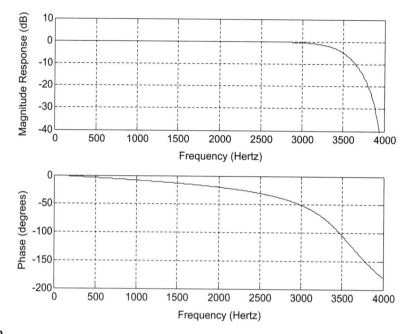

FIGURE 8.18

Frequency responses of the designed digital filter for Example 8.9.

Program 8.5. MATLAB program for Example 8.9.
```
% Example 8.9
% Design of the digital lowpass Butterworth filter
  format long
  fs=8000;                                        % Sampling rate
  [B A]=lp2lp([1],[1 1.4142 1], 6.6645*10^4)      % Complete step 2
  [b a]=bilinear(B,A,fs)                          % Complete step 3
% Plot the magnitude and phase responses |H(z)|
% b=[0.7157 1.4315 0.7157]; numerator coefficients from MATLAB
%a=[1 1.3490 0.5140]; denominator coefficients from MATLAB
  freqz(b,a,512,fs);
  axis([0 fs/2 -40 10])
```

EXAMPLE 8.10

a. Design a highpass digital Chebyshev filter with the following specifications:
   1. 0.5 dB ripple on the passband at a frequency of 3,000 Hz
   2. 25 dB attenuation at a frequency of 1,000 Hz
   3. Sampling frequency at 8,000 Hz.
b. Use MATLAB to plot the magnitude and phase responses.

Solution:

a. From the specifications, the digital frequencies are
$$\omega_{dp} = 2\pi f = 2\pi(3,000) = 6,000\pi \text{ rad/sec}$$
$$\omega_{ds} = 2\pi f = 2\pi(1,000) = 2,000\pi \text{ rad/sec}$$
and $T = 1/f_s = 1/8,000$ sec

Using the design procedure, it follows that

$$\omega_{ap} = \frac{2}{T}\tan\left(\frac{\omega_{dp}T}{2}\right) = 16,000 \times \tan\left(\frac{6,000\pi/8,000}{2}\right) = 3.8627 \times 10^4 \text{ rad/sec}$$

$$\omega_{as} = 16,000 \times \tan\left(\frac{\omega_{ds}T}{2}\right) = 16,000 \times \tan\left(\frac{2,000\pi/8,000}{2}\right) = 6.6274 \times 10^3 \text{ rad/sec}$$

We find the lowpass prototype specification as follows:

$$v_s = \omega_{ps}/\omega_{sp} = 3.8627 \times 10^4/6.6274 \times 10^3 = 5.8284 \text{ rad/s and } A_s = 25 \text{ dB}$$

Then the filter order is computed as

$$\varepsilon^2 = 10^{0.1\times0.5} - 1 = 0.1220$$

$$\left(10^{0.1\times25} - 1\right)/0.1220 = 2,583.8341$$

$$n = \frac{\cosh^{-1}\left[(2,583.8341)^{0.5}\right]}{\cosh^{-1}(5.8284)} = \frac{\ln\left(50.8314 + \sqrt{50.8314^2 - 1}\right)}{\ln\left(5.8284 + \sqrt{5.8284^2 - 1}\right)} = 1.8875$$

We select $n = 2$ for the lowpass prototype function. Following the steps of the design procedure, it follows that

1.
$$\omega_p = 3.8627 \times 10^4 \text{ rad/sec}$$

2. Performing the prototype transformation (lowpass to lowpass) using the prototype filter in Table 8.4, we have

$$H_P(s) = \frac{1.4314}{s^2 + 1.4256s + 1.5162} \text{ and}$$

$$H(s) = H_P(s)\big|_{\frac{s}{\omega_a}} = \frac{1.4314}{\left(\frac{\omega_p}{s}\right)^2 + 1.4256\left(\frac{\omega_p}{s}\right) + 1.5162} = \frac{0.9441s^2}{s^2 + 3.6319 \times 10^4 s + 9.8407 \times 10^8}$$

3. Applying the BLT, we convert the analog filter to the digital filter as follows:

$$H(z) = \frac{0.9441s^2}{s^2 + 3.6319 \times 10^4 s + 9.8407 \times 10^8}\bigg|_{s=16,000(z-1)/(z+1)}$$

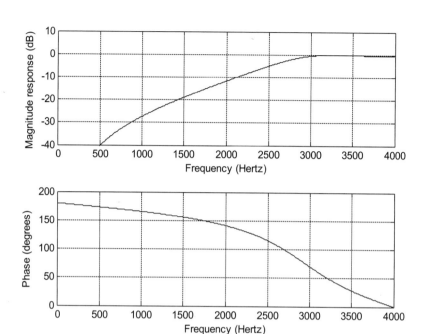

FIGURE 8.19

Frequency responses of the designed digital filter for Example 8.10.

After algebra simplification, it follows that

$$H(z) = \frac{0.1327 - 0.2654z^{-1} + 0.1327z^{-2}}{1 + 0.7996z^{-1} + 0.3618z^{-2}}$$

b. MATLAB Program 8.6 is listed for this example, and frequency responses are given in Figure 8.19.

Program 8.6. MATLAB program for Example 8.10.

```
% Example 8.10
% Design of the digital lowpass Chebyshev filter
format long
fs=8000;                                    % Sampling rate
% BLT design
[B A]=lp2hp([1.4314],[1 1.4256 1.5162], 3.8627*10^4) % Complete step 2
[b a]=bilinear(B,A,fs)                      % Complete step 3
% Plot the magnitude and phase responses |H(z)|
% b=[0.1327 -0.2654 0.1327]; numerator coefficients from MATLAB
% a=[1 0.7996 0.3618]; denominator coefficients from MATLAB
freqz(b,a,512,fs);
axis([0 fs/2 -40 10])
```

8.3.3 Bandpass and Bandstop Filter Design Examples

EXAMPLE 8.11

a. Design a second-order digital bandpass Butterworth filter with the following specifications:
   - upper cutoff frequency of 2.6 kHz
   - lower cutoff frequency of 2.4 kHz
   - sampling frequency of 8,000 Hz
b. Use MATLAB to plot the magnitude and phase responses.

Solution:

a. Let us find the digital frequencies in radians per second:

$\omega_h = 2\pi f_h = 2\pi(2600) = 5,200\pi$ rad/sec
$\omega_l = 2\pi f_l = 2\pi(2400) = 4,800\pi$ rad/sec, and $T = 1/f_s = 1/8,000$ sec

Following the steps of the design procedure, we have the following:

1.

$$\omega_{ah} = \frac{2}{T}\tan\left(\frac{\omega_h T}{2}\right) = 16,000 \times \tan\left(\frac{5,200\pi/8,000}{2}\right) = 2.6110 \times 10^4 \text{ rad/sec}$$

$$\omega_{al} = 16,000 \times \tan\left(\frac{\omega_l T}{2}\right) = 16,000 \times \tan(0.3\pi) = 2.2022 \times 10^4 \text{ rad/sec}$$

$$W = \omega_{ah} - \omega_{al} = 26,110 - 22,022 = 4,088 \text{ rad/sec}$$

$$\omega_0^2 = \omega_{ah} \times \omega_{al} = 5.7499 \times 10^8$$

2. We perform the prototype transformation (lowpass to bandpass) to obtain $H(s)$. From Table 8.3, we pick the lowpass prototype with order 1 to produce the bandpass filter with order 2:

$$H_P(s) = \frac{1}{s+1}$$

Applying the lowpass-to-bandpass transformation, it follows that

$$H(s) = H_P(s)|_{\frac{s^2+\omega_0^2}{sW}} = \frac{Ws}{s^2 + Ws + \omega_0^2} = \frac{4,088s}{s^2 + 4,088s + 5.7499 \times 10^8}$$

3. Hence we apply the BLT to yield

$$H(z) = \frac{4,088s}{s^2 + 4,088s + 5.7499 \times 10^8}\bigg|_{s=16,000(z-1)/(z+1)}$$

Via algebra work, we obtain the digital filter as

$$H(z) = \frac{0.0730 - 0.0730z^{-2}}{1 + 0.7117z^{-1} + 0.8541z^{-2}}$$

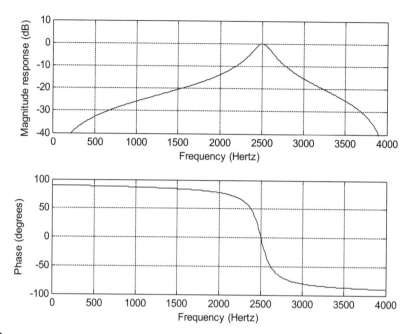

FIGURE 8.20

Frequency responses of the designed digital filter for Example 8.11.

b. MATLAB Program 8.7 is given for this example, and the corresponding frequency response plots are illustrated in Figure 8.20.

Program 8.7. MATLAB program for Example 8.11.

```
% Example 8.11
% Design of the digital bandpass Butterworth filter
  format long
  fs=8000;
  [B A]=lp2bp([1],[1 1],sqrt(5.7499*10^8),4088)    % Complete step 2
  [b a]=bilinear(B,A,fs)                           % Complete step 3
% Plot the magnitude and phase responses |H(z)|
% b=[0.0730 0 -0.0730]; numerator coefficients from MATLAB
% a=[1 0.7117 0.8541]; denominator coefficients form MATLAB
  freqz(b, a,512,fs);
  axis([0 fs/2 -40 10])
```

EXAMPLE 8.12

Now let us examine the bandstop Butterworth filter design.

a. Design a digital bandstop Butterworth filter with the following specifications:
   • Center frequency of 2.5 kHz
   • Passband width of 200 Hz and ripple of 3dB

- Stopband width of 50 Hz and attenuation of 10 dB
- Sampling frequency of 8,000 Hz

b. Use MATLAB to plot the magnitude and phase responses.

Solution:

a. The digital frequencies of the digital filter are

$\omega_h = 2\pi f_h = 2\pi(2,600) = 5,200\pi$ rad/sec
$\omega_l = 2\pi f_l = 2\pi(2,400) = 4,800\pi$ rad/sec
$\omega_{d0} = 2\pi f_0 = 2\pi(2,500) = 5,000\pi$ rad/sec, and $T = 1/f_s = 1/8,000$ sec

Applying the three steps of the IIR filter design approach, it follows that

1.

$$\omega_{ah} = \frac{2}{T}\tan\left(\frac{\omega_h T}{2}\right) = 16,000 \times \tan\left(\frac{5,200\pi/8,000}{2}\right) = 2.6110 \times 10^4 \text{ rad/sec}$$

$$\omega_{al} = 16,000 \times \tan\left(\frac{\omega_l T}{2}\right) = 16,000 \times \tan(0.3\pi) = 2.2022 \times 10^4 \text{ rad/sec}$$

$$\omega_0 = 16,000 \times \tan\left(\frac{\omega_{d0} T}{2}\right) = 16,000 \times \tan(0.3125\pi) = 2.3946 \times 10^4 \text{ rad/sec}$$

$$\omega_{sh} = \frac{2}{T}\tan\left(\frac{2,525 \times 2\pi/8,000}{2}\right) = 16,000 \times \tan(56.8125^0) = 2.4462 \times 10^4 \text{ rad/sec}$$

$$\omega_{sl} = 16,000 \times \tan\left(\frac{2,475 \times 2\pi/8,000}{2}\right) = 16,000 \times \tan(55.6875^0) = 2.3444 \times 10^4 \text{ rad/sec}$$

To adjust the unit passband gain at the center frequency of 2,500 Hz, we perform the following:

Fixing $\omega_{al} = 2.2022 \times 10^4$, we compute $\omega_{ah} = \omega_0^2/\omega_{al} = \dfrac{(2.3946 \times 10^4)^2}{2.2022 \times 10^4} = 2.6037 \times 10^4$

and the passband bandwidth: $W = \omega_{ah} - \omega_{al} = 4,015$

Fixing $\omega_{sl} = 2.3444 \times 10^4$, $\omega_{sh} = \omega_0^2/\omega_{sl} = \dfrac{(2.3946 \times 10^4)^2}{2.3444 \times 10^4} = 2.4459 \times 10^4$

and the stopband bandwidth: $W_s = \omega_{sh} - \omega_{sl} = 1,015$

Again, Fixing $\omega_{ah} = 2.6110 \times 10^4$, we got $\omega_{al} = \omega_0^2/\omega_{ah} = \dfrac{(2.3946 \times 10^4)^2}{2.6110 \times 10^4} = 2.1961 \times 10^4$

and the passband bandwidth: $W = \omega_{ah} - \omega_{al} = 4,149$

Fixing $\omega_{sh} = 2.4462 \times 10^4$, $\omega_{sl} = \omega_0^2/\omega_{sh} = \dfrac{(2.3946 \times 10^4)^2}{2.4462 \times 10^4} = 2.3441 \times 10^4$

and the stopband bandwidth: $W_s = \omega_{sh} - \omega_{sl} = 1,021$

For an aggressive bandstop design, we choose $\omega_{al} = 2.6110 \times 10^4$, $\omega_{ah} = 2.1961 \times 10^4$, $\omega_{sl} = 2.3441 \times 10^4$, $\omega_{sh} = 2.4462 \times 10^4$ and $\omega_0 = 2.3946 \times 10^4$ to satisfy a larger bandwidth.

Thus we develop the prototype specification:

$$v_s = (26,110 - 21,916)/(24,462 - 23,441) = 4.0177$$

$$n = \left(\frac{\log_{10}(10^{0.1 \times 10} - 1)}{2 \cdot \log_{10}(4.0177)}\right) = 0.7899, \text{ choose } n = 1$$

$$W = \omega_{ah} - \omega_{al} = 26,110 - 21,961 = 4,149 \text{ rad/sec}, \omega_0^2 = 5.7341 \times 10^8.$$

2. Then, carrying out the prototype transformation (lowpass to bandstop) using the first-order lowpass prototype filter given by

$$H_P(s) = \frac{1}{s+1}$$

it follows that

$$H(s) = H_P(s)\Big|_{\frac{sW}{s^2+\omega_0^2}} = \frac{(s^2 + \omega_0^2)}{s^2 + Ws + \omega_0^2}$$

Substituting the values of ω_0^2 and W yields

$$H(s) = \frac{s^2 + 5.7341 \times 10^8}{s^2 + 4,149s + 5.7341 \times 10^8}$$

3. Hence, applying the BLT leads to

$$H(z) = \frac{s^2 + 5.7341 \times 10^8}{s^2 + 4,149s + 5.73411 \times 10^8}\Big|_{s=16,000(z-1)/(z+1)}$$

After algebra, we get

$$H(z) = \frac{0.9259 + 0.7078z^{-1} + 0.9249z^{-2}}{1 + 0.7078z^{-1} + 0.8518z^{-2}}$$

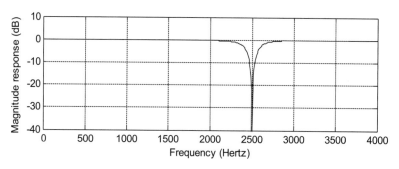

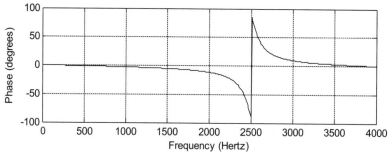

FIGURE 8.21

Frequency responses of the designed digital filter for Example 8.12.

b. MATLAB Program 8.8 includes the design steps. Figure 8.21 shows the filter frequency responses.

Program 8.8. MATLAB program for Example 8.12.

```
% Example 8.12
% Design of the digital bandstop Butterworth filter
  format long
  fs=8000;                                   % Sampling rate
  [B A]=lp2bs([1],[1 1],sqrt(5.7341*10^8),4149) % Complete step 2
  [b a]=bilinear(B,A,fs)                     % Complete step 3
% Plot the magnitude and phase responses |H(z)|
% b=[0.9259 0.7078 0.9259]; numerator coefficients from MATLAB
% a=[1 0.7078 0.8518]; denominator coefficients from MATLAB
  freqz(b,a,512,fs);
  axis([0 fs/2 -40 10])
```

EXAMPLE 8.13

a. Design a digital bandpass Chebyshev filter with the following specifications:
   - Center frequency of 2.5 kHz
   - Passband bandwidth of 200 Hz, 0.5 dB ripple on passband
   - Lower stop frequency of 1.5 kHz, upper stop frequency of 3.5 kHz
   - Stopband attenuation of 10 dB
   - Sampling frequency of 8,000 Hz
b. Use MATLAB to plot the magnitude and phase responses.

Solution:

a. The digital frequencies are given as
$$\omega_{dph} = 2\pi f_{dph} = 2\pi(2,600) = 5,200\pi \text{ rad/sec}$$
$$\omega_{dpl} = 2\pi f_{dpl} = 2\pi(2,400) = 4,800\pi \text{ rad/sec}$$
$$\omega_{d0} = 2\pi f_0 = 2\pi(2,500) = 5,000\pi \text{ rad/sec, and } T = 1/f_s = 1/8,000 \text{ sec}$$

Applying the frequency prewarping equation, it follows that

$$\omega_{aph} = \frac{2}{T}\tan\left(\frac{\omega_d T}{2}\right) = 16,000 \times \tan\left(\frac{5,200\pi/8,000}{2}\right) = 2.6110 \times 10^4 \text{ rad/sec}$$

$$\omega_{apl} = 16,000 \times \tan\left(\frac{\omega_{dpl} T}{2}\right) = 16,000 \times \tan(0.3\pi) = 2.2022 \times 10^4 \text{ rad/sec}$$

$$\omega_0 = 16,000 \times \tan\left(\frac{\omega_{d0} T}{2}\right) = 16,000 \times \tan(0.3125\pi) = 2.3946 \times 10^4 \text{ rad/sec}$$

$$\omega_{ash} = 16,000 \times \tan\left(\frac{3,500 \times 2\pi/8,000}{2}\right) = 16,000 \times \tan(78.75^0) = 8.0437 \times 10^4 \text{ rad/sec}$$

$$\omega_{asl} = 16,000 \times \tan\left(\frac{1,500 \times 2\pi/8,000}{2}\right) = 1.0691 \times 10^4 \text{ rad/sec}$$

Now, adjusting the unit gain for the center frequency of 2,500 Hz leads to the following:

Fixing $\omega_{apl} = 2.2022 \times 10^4$, we have $\omega_{aph} = \dfrac{\omega_0^2}{\omega_{apl}} = \dfrac{(2.3946 \times 10^4)^2}{2.2022 \times 10^4} = 2.6038 \times 10^4$

and the passband bandwidth: $W = \omega_{aph} - \omega_{apl} = 4,016$

Fixing $\omega_{asl} = 1.0691 \times 10^4$, $\omega_{ash} = \dfrac{\omega_0^2}{\omega_{asl}} = \dfrac{(2.3946 \times 10^4)^2}{2.10691 \times 10^4} = 5.3635 \times 10^4$

and the stopband bandwidth: $W_s = \omega_{ash} - \omega_{asl} = 42,944$

Again, fixing $\omega_{aph} = 2.6110 \times 10^4$, we have $\omega_{apl} = \dfrac{\omega_0^2}{\omega_{aph}} = \dfrac{(2.3946 \times 10^4)^2}{2.6110 \times 10^4} = 2.1961 \times 10^4$

and the passband bandwidth: $W = \omega_{aph} - \omega_{apl} = 4,149$

Fixing $\omega_{ash} = 8.0437 \times 10^4$, $\omega_{asl} = \dfrac{\omega_0^2}{\omega_{ash}} = \dfrac{(2.3946 \times 10^4)^2}{8.0437 \times 10^4} = 0.7137 \times 10^4$

and the stopband bandwidth: $W_s = \omega_{ash} - \omega_{asl} = 73,300$
For an aggressive bandpass design, we select $\omega_{apl} = 2.2022 \times 10^4$, $\omega_{aph} = 2.6038 \times 10^4$, $\omega_{asl} = 1.0691 \times 10^4$, $\omega_{ash} = 5.3635 \times 10^4$ and for a smaller bandwidth for passband.
Thus, we obtain the prototype specifications:

$$v_s = (53,635 - 10,691)/(26,110 - 21,961) = 10.6932$$

$$\epsilon^2 = 10^{0.1 \times 0.5} - 1 = 0.1220$$

$$\left(10^{0.1 \times 10} - 1\right)/0.1220 = 73.7705$$

$$n = \frac{\cosh^{-1}\left[(73.7705)^{0.5}\right]}{\cosh^{-1}(10.6932)} = \frac{\ln\left(8.5890 + \sqrt{8.5890^2 - 1}\right)}{\ln\left(10.6932 + \sqrt{10.6932^2 - 1}\right)} = 0.9280$$

Rounding up n leads to $n = 1$.
Next, we apply the design steps:

1.

$$\omega_{aph} = 2.6038 \times 10^4 \text{ rad/sec}, \ \omega_{apl} = 2.2022 \times 10^4 \text{ rad/sec}$$

$$W = 4,016 \text{ rad/sec}, \ \omega_0^2 = 5.7341 \times 10^8$$

2. Performing the prototype transformation (lowpass to bandpass), we obtain

$$H_P(s) = \frac{2.8628}{s + 2.8628}$$

and

$$H(s) = H_P(s)\Big|_{s = \frac{s^2 + \omega_0^2}{sW}} = \frac{2.8628Ws}{s^2 + 2.8628Ws + \omega_0^2} = \frac{1.1878 \times 10^4 s}{s^2 + 1.1878 \times 10^4 s + 5.7341 \times 10^8}$$

3. Applying the BLT, the analog filter is converted into a digital filter as follows:

$$H(z) = \frac{1.1878 \times 10^4 s}{s^2 + 1.1878 \times 10^4 s + 5.7341 \times 10^8}\Bigg|_{s=16,000(z-1)/(z+1)}$$

This is simplified and arranged as follows:

$$H(z) = \frac{0.1864 - 0.1864z^{-2}}{1 + 0.6227z^{-1} + 0.6272z^{-2}}$$

b. Program 8.9 lists the MATLAB details. Figure 8.22 displays the frequency responses.

Program 8.9. MATLAB program for Example 8.13.

```
% Example 8.13
% Design of the digital bandpass Chebyshev filter
 format long
 fs=8000;
 [B A]=lp2bp([2.8628],[1 2.8628],sqrt(5.7341*10^8),4149) % Complete step 2
 [b a]=bilinear(B,A,fs)                                  % Complete step 3
% Plot the magnitude and phase responses |H(z)|
% b=[0.1864 0.0 -0.1864]; numerator coefficients from MATLAB
% a=[1 0.6227 0.6272]; denominator coefficients from MATLAB
 freqz(b,a,512,fs);
 axis([0 fs/2 -40 10])
```

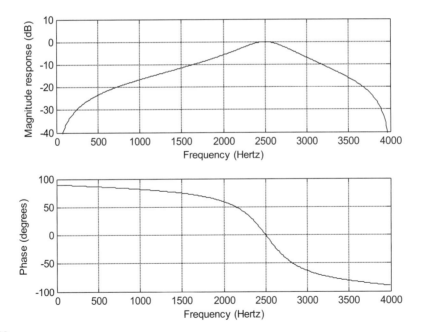

FIGURE 8.22

Frequency responses of the designed digital filter for Example 8.13.

8.4 HIGHER-ORDER INFINITE IMPULSE RESPONSE FILTER DESIGN USING THE CASCADE METHOD

For higher order IIR filter design, use of a cascade transfer function is preferred. The factored forms for the lowpass prototype transfer functions for Butterworth and Chebyshev filters are given in Tables 8.7, 8.8 and 8.9. A Butterworth filter design example will be provided and a similar procedure can be adopted for the Chebyshev filters.

Table 8.7 3-dB Butterworth Prototype Functions in Cascade Form

n	$H_P(s)$
3	$\dfrac{1}{(s+1)(s^2+s+1)}$
4	$\dfrac{1}{(s^2+0.7654s+1)(s^2+1.8478s+1)}$
5	$\dfrac{1}{(s+1)(s^2+0.6180s+1)(s^2+1.6180s+1)}$
6	$\dfrac{1}{(s^2+0.5176s+1)(s^2+1.4142s+1)(s^2+1.9319s+1)}$

Table 8.8 Chebyshev Prototype Functions in Cascade Form with 0.5-dB Ripple ($\varepsilon = 0.3493$)

n	$H_P(s)$ 0.5-dB Ripple ($\varepsilon = 0.3493$)
3	$\dfrac{0.7157}{(s+0.6265)(s^2+0.6265s+1.1425)}$
4	$\dfrac{0.3579}{(s^2+0.3507s+1.0635)(s^2+0.8467s+0.3564)}$
5	$\dfrac{0.1789}{(s+0.3623)(s^2+0.2239s+1.0358)(s^2+0.5862s+0.4768)}$
6	$\dfrac{0.0895}{(s^2+0.1553s+1.0230)(s^2+0.4243s+0.5900)(s^2+0.5796s+0.1570)}$

EXAMPLE 8.14

a. Design a fourth-order digital lowpass Butterworth filter with a cutoff frequency of 2.5 kHz at a sampling frequency of 8,000 Hz.
b. Use MATLAB to plot the magnitude and phase responses.

Table 8.9 Chebyshev Prototype Functions in Cascade Form with 1-dB Ripple $(\varepsilon = 0.5088)$

n	$H_P(s)$ 1-dB Ripple $(\varepsilon = 0.5088)$
3	$\dfrac{0.4913}{(s + 0.4942)(s^2 + 0.4942s + 0.9942)}$
4	$\dfrac{0.2456}{(s^2 + 0.2791s + 0.9865)(s^2 + 0.6737s + 0.2794)}$
5	$\dfrac{0.1228}{(s + 0.2895)(s^2 + 0.1789s + 0.9883)(s^2 + 0.4684s + 0.4293)}$
6	$\dfrac{0.0614}{(s^2 + 0.1244s + 0.9907)(s^2 + 0.3398s + 0.5577)(s^2 + 0.4641s + 0.1247)}$

Solution:

a. First, we obtain the digital frequency in radians per second:

$$\omega_d = 2\pi f = 2\pi(2,500) = 5,000\pi \text{ rad/sec, and } T = 1/f_s = 1/8,000 \text{ sec}$$

Following the design steps, we compute the specifications for the analog filter.

1.

$$\omega_a = \frac{2}{T} \tan\left(\frac{\omega_d T}{2}\right) = 16,000 \times \tan\left(\frac{5,000\pi/8,000}{2}\right) = 2.3946 \times 10^4 \text{ rad/sec}$$

2. From Table 8.7, we obtain the fourth-order factored prototype transfer function:

$$H_P(s) = \frac{1}{(s^2 + 0.7654s + 1)(s^2 + 1.8478s + 1)}$$

Applying the prototype transformation, we yield

$$H(s) = H_P(s)|_{\frac{s}{\omega_a}} = \frac{\omega_a^2 \times \omega_a^2}{(s^2 + 0.7654\omega_a s + \omega_a^2)(s^2 + 1.8478\omega_a s + \omega_a^2)}$$

Substituting $\omega_a = 2.3946 \times 10^4$ rad/sec yields

$$H(s) = \frac{(5.7340 \times 10^8) \times (5.7340 \times 10^8)}{(s^2 + 1.8328s + 5.7340 \times 10^8)(s^2 + 4.4247 \times 10^4 s + 5.7340 \times 10^8)}$$

3. Hence, after applying BLT, we have

$$H(z) = \frac{(5.7340 \times 10^8) \times (5.7340 \times 10^8)}{(s^2 + 1.8328s + 5.7340 \times 10^8)(s^2 + 4.4247 \times 10^4 s + 5.7340 \times 10^8)}\Bigg|_{s=16,000(z-1)/(z+1)}$$

After simplifying with algebra, we obtain the digital filter as

$$H(z) = \frac{0.5108 + 1.0215z^{-1} + 0.5108z^{-2}}{1 + 0.5654z^{-1} + 0.4776z^{-2}} \times \frac{0.3730 + 0.7460z^{-1} + 0.3730z^{-2}}{1 + 0.4129z^{-1} + 0.0790z^{-2}}$$

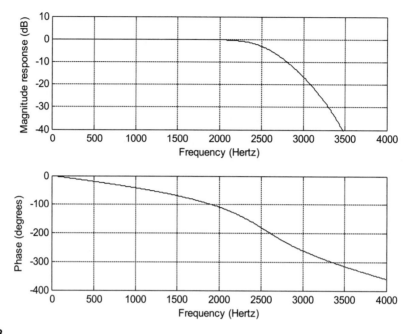

FIGURE 8.23

Frequency responses of the designed digital filter for Example 8.14.

b. A MATLAB program is preferable to carry out the algebra and is listed in Program 8.10. Figure 8.23 shows the filter magnitude and phase frequency responses.

Program 8.10. MATLAB program for Example 8.14.

```
% Example 8.14
% Design of the fourth-order digital lowpass Butterworth filter
% in the cascade form
format long
fs=8000;                                % Sampling rate
[B1 A1]=lp2lp([1],[1 0.7654 1], 2.3946*10^4) % Complete step 2
[b1 a1]=bilinear(B1,A1,fs) % complete step 3
[B2 A2]=lp2lp([1],[1 1.8478 1], 2.3946*10^4) % Complete step 2
[b2 a2]=bilinear(B2,A2,fs) % complete step 3
% Plot the magnitude and phase responses |H(z)|
% b1=[0.5108 1.0215 0.5108]; a1=[1 0.5654 0.4776]; coefficients from MATLAB
% b2=[0.3730 0.7460 0.3730]; a2=[1 0.4129 0.0790]; coefficients from MATLAB
 freqz(conv(b1,b2),conv(a1,a2),512,fs); % Combined filter responses
 axis([0 fs/2 -40 10]);
```

The higher-order bandpass, highpass and bandstop filters using the cascade form can be designed similarly.

8.5 APPLICATION: DIGITAL AUDIO EQUALIZER

In this section, the design of a digital audio equalizer is introduced. For an audio application such as the CD player, the digital audio equalizer is used to adjust the sound as one desires by changing filter gains for different audio frequency bands. Other applications include adjusting the sound source to take room acoustics into account, removing undesired noise, and boosting the desired signal in the specified passband. The simulation is based on the consumer digital audio processor—such as a CD player—handling the 16-bit digital samples with a sampling rate of 44.1 kHz and an audio signal bandwidth at 22.05 kHz. A block diagram of the digital audio equalizer is depicted in Figure 8.24.

A seven-band audio equalizer is adopted for discussion. The center frequencies are listed in Table 8.10. The 3-dB bandwidth for each bandpass filter is chosen to be 50% of the center frequency. As shown in Figure 8.24, g_0 through g_6 are the digital gains for each banspass filter output and can be adjusted to make sound effects, while $y_0(n)$ through $y_6(n)$ are the digital amplified bandpass filter outputs. Finally, the equalized signal is the sum of the amplified bandpass filter outputs and itself. By changing the digital gains of the equalizer, many sound effects can be produced.

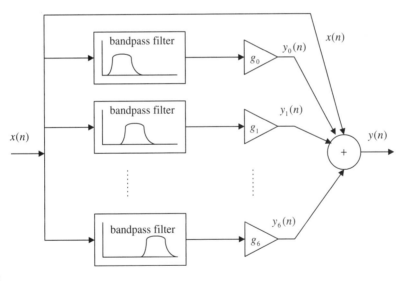

FIGURE 8.24

Simplified block diagram of the audio equalizer.

Table 8.10 Specifications for an Audio Equalizer

Center frequency (Hz)	100	200	400	1000	2500	6000	15000
Bandwidth (Hz)	50	100	200	500	1250	3000	7500

Table 8.11 Designed Filter Banks

Filter Banks	Coefficients for the Numerator	Coefficients for the Denominator
Bandpass filter 0	0.0031954934, 0, −0.0031954934	1, −1.9934066716 , 0.9936090132
Bandpass filter 1	0.0063708102, 0, −0.0063708102	1, −1.9864516324, 0.9872583796
Bandpass filter 2	0.0126623878, 0, −0.0126623878	1, −1.9714693192, 0.9746752244
Bandpass filter 3	0.0310900413, 0, −0.0310900413	1, −1.9181849043, 0.9378199174
Bandpass filter 4	0.0746111954, 0, −0.0746111954	1, −1.7346085867, 0.8507776092
Bandpass filter 5	0.1663862883, 0, −0.1663862884	1, −1.0942477187, 0.6672274233
Bandpass filter 6	0.3354404899, 0, −0.3354404899	1, 0.7131366534, 0.3291190202

To complete the design and simulation, second-order IIR bandpass Butterworth filters are chosen for the audio equalizer; the coefficients are determined using the BLT method, and are given in Table 8.11.

The magnitude response for each filter bank is plotted in Figure 8.25 for design verification. As shown in Figure 8.25, after careful examination, the magnitude response of each filter band meets the design specification. We will perform simulation next.

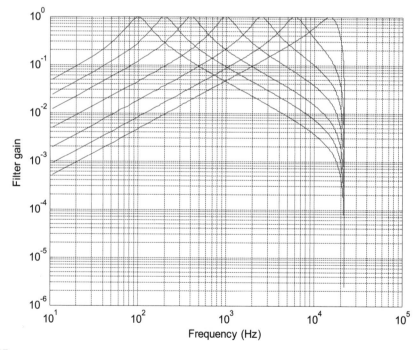

FIGURE 8.25

Magnitude frequency responses for the audio equalizer.

Simulation in the MATLAB environment is based on the following settings. The audio test signal with frequency components of 100 Hz, 200 Hz, 400 Hz, 1,000 Hz, 2,500 Hz, 6,000 Hz, and 15,000 Hz is generated from Equation (8.36):

$$
\begin{aligned}
x(n) = \; & \sin(200\pi n/44{,}100) + \sin(400\pi n/44{,}100 + \pi/14) \\
& + \sin(800\pi n/44{,}100 + \pi/7) + \sin(2{,}000\pi n/44{,}100 + 3\pi/14) \\
& + \sin(5{,}000\pi n/44{,}100 + 2\pi/7) + \sin(12{,}000\pi n/44{,}100 + 5\pi/14) \\
& + \sin(30{,}000\pi n/44{,}100 + 3\pi/7)
\end{aligned}
\tag{8.36}
$$

The gains set for the filter banks are

$$
g_0 = 10; \; g_1 = 10; \; g_2 = 0; \; g_3 = 0; \; g_4 = 0; \; g_5 = 10; \; g_6 = 10
$$

After simulation, we notice that the frequency components at 100 Hz, 200 Hz, 6,000 Hz, and 15,000 Hz will be boosted by $20 \cdot \log_{10} 10 = 20$ dB. The top plot in Figure 8.26 shows the spectrum for the audio test signal, while the bottom plot depicts the spectrum for the equalized audio test signal. As shown in the plots, before audio digital equalization, the spectral peaks at all bands are at the same level; after audio digital equalization, the frequency components at bank 0, bank 1, bank 5, and bank 6 are amplified. Therefore, as we expected, the operation of the digital equalizer boosts the low

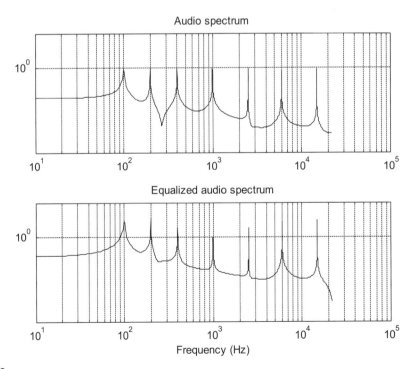

FIGURE 8.26

Audio spectrum and equalized audio spectrum.

frequency components and the high frequency components. The MATLAB list for the simulation is shown in Program 8.11.

Program 8.11. MATLAB program for the digital audio equalizer.

```
close all; clear all
% Filter Coefficients (Butterworth type designed using the BLT)
B0=[0.0031954934 0 -0.0031954934]; A0=[1.0000000000 -1.9934066716 0.9936090132];
B1=[0.0063708102 0 -0.0063708102]; A1=[1.0000000000 -1.9864516324 0.9872583796];
B2=[0.0126623878 0 -0.0126623878]; A2=[1.0000000000 -1.9714693192 0.9746752244];
B3=[0.0310900413 0 -0.0310900413]; A3=[ 1.0000000000 -1.9181849043 0.9378199174];
B4=[0.0746111954 0.000000000 -0.0746111954];
A4=[1.0000000000 -1.7346085867 0.8507776092];
B5=[0.1663862883 0.0000000000 -0.1663862884];
A5=[1.0000000000 -1.0942477187 0.6672274233];
B6=[0.3354404899 0.0000000000 -0.3354404899];
A6=[1.0000000000 0.7131366534  0.3291190202];
[h0,f]=freqz(B0,A0,2048,44100);
[h1,f]=freqz(B1,A1,2048,44100);
[h2,f]=freqz(B2,A2,2048,44100);
[h3,f]=freqz(B3,A3,2048,44100);
[h4,f]=freqz(B4,A4,2048,44100);
[h5,f]=freqz(B5,A5,2048,44100);
[h6,f]=freqz(B6,A6,2048,44100);
loglog(f,abs(h0),f,abs(h1), f,abs(h2), ...
f,abs(h3),f,abs(h4),f,abs(h5),f,abs(h6));
xlabel('Frequency (Hz)');
ylabel('Filter Gain');grid
axis([10 10^5 10^(-6) 1]);
figure(2)
g0=10;g1=10;g2=0;g3=0;g4=0;g5=10;g6=10;
p0=0;p1=pi/14;p2=2*p1;p3=3*p1;p4=4*p1;p5=5*p1;p6=6*p1;
n=0:1:20480;        % Indices of samples
fs=44100;           % Sampling rate
x=sin(2*pi*100*n/fs)+sin(2*pi*200*n/fs+p1)+...
  sin(2*pi*400*n/fs+p2)+sin(2*pi*1000*n/fs+p3)+...
  sin(2*pi*2500*n/fs+p4)+sin(2*pi*6000*n/fs+p5)+...
  sin(2*pi*15000*n/fs+p6); % Generate test audio signals
y0=filter(B0,A0,x);     % Bandpass filter 0
y1=filter(B1,A1,x);     % Bandpass filter 1
y2=filter(B2,A2,x);     % Bandpass filter 2
y3=filter(B3,A3,x);     % Bandpass filter 3
y4=filter(B4,A4,x);     % Bandpass filter 4
y5=filter(B5,A5,x);     % Bandpass filter 5
y6=filter(B6,A6,x);     % Bandpass filter 6
y=g0.*y0+g1.*y1+g2.*y2+g3.*y3+g4.*y4+g5.*y5+g6.*y6+x; % Equalizer output
N=length(x);
Axk=2*abs(fft(x))/N;Axk(1)=Axk(1)/2; % One-sided amplitude spectrum of the input
f=[0:N/2]*fs/N;
```

```
subplot(2,1,1);loglog(f,Axk(1:N/2+1));
title('Audio spectrum');
axis([10 100000 0.00001 100]);grid;
Ayk=2*abs(fft(y))/N; Ayk(1)=Ayk(1)/2; % One-sided amplitude
                                      % spectrum of the output
subplot(2,1,2);loglog(f,Ayk(1:N/2+1));
xlabel('Frequency (Hz)');
title('Equalized audio spectrum');
axis([10 100000 0.00001 100]);grid;
```

8.6 IMPULSE-INVARIANT DESIGN METHOD

We illustrate the concept of the impulse-invariant design in Figure 8.27. Given the transfer function of a designed analog filter, an analog impulse response can be easily found by the inverse Laplace transform of the transfer function. To replace the analog filter by the equivalent digital filter, we apply an approximation in the time domain in which the digital filter impulse response must be equivalent to the analog impulse response. Therefore, we can sample the analog impulse response to get the digital impulse response, and take the z-transform of the sampled analog impulse response to obtain the transfer function of the digital filter.

The analog impulse response can be achieved by taking the inverse Laplace transform of the analog filter $H(s)$, that is,

$$h(t) = L^{-1}(H(s)) \tag{8.37}$$

Now, if we sample the analog impulse response with a sampling interval of T and use T as a scale factor, it follows that

$$T \cdot h(n) = T \cdot h(t)|_{t=nT}, \; n \geq 0 \tag{8.38}$$

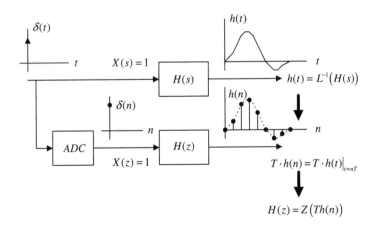

FIGURE 8.27

Impulse-invariant design method.

Taking the z-transform on both sides of Equation (8.38) yields the digital filter as

$$H(z) = Z[T \cdot h(n)] \tag{8.39}$$

The effect of the scale factor T in Equation (8.38) can be explained as follows. We approximate the area under the curve specified by the analog impulse function $h(t)$ using a digital sum given by

$$\text{area} = \int_0^\infty h(t)dt \approx T \cdot h(0) + T \cdot h(1) + T \cdot h(2) + \cdots \tag{8.40}$$

Note that the area under the curve indicates the DC gain of the analog filter while the digital sum in Equation (8.40) is the DC gain of the digital filter.

The rectangular approximation is used, since each sample amplitude is multiplied by the sampling interval T. Due to the interval size for approximation in practice, we cannot guarantee that the digital sum has exactly the same value as the one from the integration unless the sampling interval T in Equation (8.40) approaches zero. This means that the higher the sampling rate—that is, the smaller the sampling interval—the more accurately the digital filter gain matches the analog filter gain. Hence, in practice, we need to further apply gain scaling for adjustment if it is a requirement.

EXAMPLE 8.15

Consider the following Laplace transfer function:

$$H(s) = \frac{2}{s+2}$$

a. Determine $H(z)$ using the impulse-invariant method if the sampling rate $f_s = 10$ Hz.
b. Use MATLAB to plot the following:
   1. The magnitude response $|H(f)|$ and phase response $\varphi(f)$ with respect to $H(s)$ for the frequency range from 0 to $f_s/2$ Hz;
   2. The magnitude response $|H(e^{j\Omega})| = |H(e^{j2\pi fT})|$ and phase response $\varphi(f)$ with respect to $H(z)$ for the frequency range from 0 to $f_s/2$ Hz.

Solution:

a. Taking the inverse Laplace transform of the analog transfer function, the impulse response is found to be

$$h(t) = L^{-1}\left[\frac{2}{s+2}\right] = 2e^{-2t}u(t)$$

Sampling the impulse response $h(t)$ with $T = 1/f_s = 0.1$ second, we have

$$Th(n) = T2e^{-2nT}u(n) = 0.2e^{-0.2n}u(n)$$

Using the z-transform table in Chapter 5, we yield

$$Z\left[e^{-an}u(n)\right] = \frac{z}{z - e^{-a}}$$

And noting that $e^{-a} = e^{-0.2} = 0.8187$, the digital filter transfer function $H(z)$ is finally given by

$$H(z) = \frac{0.2z}{z - 0.8187} = \frac{0.2}{1 - 0.8187z^{-1}}$$

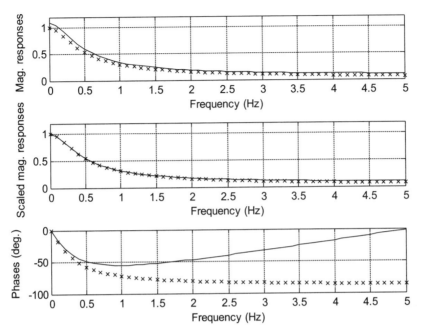

FIGURE 8.28

Frequency responses. The line of "x"s represent the frequency responses of the analog filter; the solid line represents the frequency responses of the designed digital filter.

b. The MATLAB code is listed in Program 8.12. The first and third plots in Figure 8.28 show comparisons of the magnitude and phase frequency responses. The shape of the magnitude response (first plot) closely matches that of the analog filter, while the phase response (third plot) differs from the analog phase response in this example.

Program 8.12. MATLAB program for Example 8.15.

```
% Example 8.15.
% Plot the magnitude responses |H(s)| and |H(z)|
% For the Laplace transfer function H(s)
  f=0:0.1:5;T=0.1;           % Frequency range and sampling interval
  w=2*pi*f;                  % Frequency range in rad/sec
  hs=freqs([2], [1 2],w);    % Analog frequency response
  phis=180*angle(hs)/pi;
% For the z-transfer function H(z)
  hz=freqz([0.2],[1 -0.8187],length(w)); % Digital frequency resoonse
  hz_scale=freqz([0.1813],[1 -0.8187],length(w)); % Scaled digital mag. response
  phiz=180*angle(hz)/pi;
% Plot magnitude and phase responses.
  subplot(3,1,1), plot(f,abs(hs),'kx',f, abs(hz),'k-'),grid; axis([0 5 0 1.2]);
  xlabel('Frequency (Hz)'); ylabel('Mag. Responses')
  subplot(3,1,2), plot(f,abs(hs),'kx',f, abs(hz_scale),'k-'),grid; axis([0 5
  0 1.2]);
```

```
xlabel('Frequency (Hz)'); ylabel('Scaled Mag. Responses')
subplot(3,1,3), plot(f,phis,'kx',f, phiz,'k-'); grid;
xlabel('Frequency (Hz)'); ylabel('Phases (deg.)');
```

The filter DC gain is given by

$$H(e^{j\Omega})\big|_{\Omega=0} = H(1) = 1.1031$$

We can further scale the filter to have a unit gain of

$$H(z) = \frac{1}{1.1031} \frac{0.2}{1 - 0.8187z^{-1}} = \frac{0.1813}{1 - 0.8187z^{-1}}$$

The scaled magnitude frequency response is shown in the middle plot along with that of analog filter in Figure 8.28, where the magnitudes are matched very well below 1.8 Hz.

Example 8.15 demonstrates the procedure for using the impulse-invariant design. The filter performance depends on the sampling interval (Lynn and Fuerst,1999). As shown in Figure 8.27, the analog impulse response $h(t)$ is not a band-limited signal whose frequency extends to infinity, which is certainly larger than the Nyquist limit (folding frequency); hence, sampling $h(t)$ could cause aliasing. Figure 8.29(a) shows the analog impulse response $Th(t)$ in Example 8.15 and its sampled version $Th(nT)$, where the sampling interval is 0.125 second. The analog filter and digital filter magnitude responses are plotted in Figure 8.29(b). The aliasing occurs because the impulse response contains frequency components beyond the Nyquist limit, that is, 4 Hz in this case. Furthermore, using the lower sampling rate of 8 Hz causes less accuracy in the digital filter magnitude response, so more aliasing develops.

Figure 8.29(c) shows the analog impulse response and its sampled version using a higher sampling rate of 16 Hz. Figure 8.29(d) displays the more accurate magnitude response of the digital filter. Hence, we can obtain a reduced aliasing level. Note that the aliasing cannot be avoided, due to sampling of the analog impulse response. The only way to reduce the aliasing is to use a higher sampling frequency or design a filter with a very low cutoff frequency to reduce the aliasing to a minimum level.

Investigation of the sampling interval effect leads us to the following conclusions. Note that the analog impulse response for an analog highpass filter or bandstop filter contains frequency up to infinity, which is larger than the Nyquist limit (folding frequency), even assuming that the sampling rate is much higher than the cutoff frequency of a highpass filter or the upper cutoff frequency of a bandstop filter. Hence, sampling the analog impulse response always produces aliasing. Without using an additional anti-aliasing filter, the impulse invariant method alone cannot be used for designing the highpass filter or bandstop filter.

Instead, in practice, we should apply the BLT design method. The impulse-invariant design method is only appropriate for designing a lowpass filter or bandpass filter with a sampling rate much larger than the lower cutoff frequency of the lowpass filter or the upper cutoff frequency of the bandpass filter.

Next, let us focus on second-order filter design via Example 8.16.

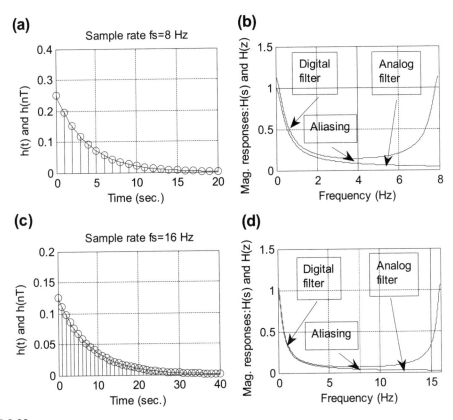

FIGURE 8.29

Sampling interval effect in the impulse invariant IIR filter design.

EXAMPLE 8.16

Consider the following Laplace transfer function:

$$H(s) = \frac{s}{s^2 + 2s + 5}$$

a. Determine $H(z)$ using the impulse-invariant method if the sampling rate $f_s = 10$ Hz.

b. Use MATLAB to plot the following:
   1. the magnitude response $|H(f)|$ and phase response $\phi(f)$ with respect to $H(s)$ for the frequency range from 0 to $f_s/2$ Hz;
   2. the magnitude response $|H(e^{j\Omega})| = |H(e^{j2\pi fT})|$ and phase response $\phi(f)$ with respect to $H(z)$ for the frequency range from 0 to $f_s/2$ Hz.

Solution:

a. Since $H(s)$ has complex poles located at $s = -1 \pm 2j$, we can write it in a quadratic form as

$$H(s) = \frac{s}{s^2 + 2s + 5} = \frac{s}{(s+1)^2 + 2^2}$$

We can further write the transfer function as

$$H(s) = \frac{(s+1)-1}{(s+1)^2+2^2} = \frac{(s+1)}{(s+1)^2+2^2} - 0.5 \times \frac{2}{(s+1)^2+2^2}$$

From the Laplace transform table (Appendix B), the analog impulse response can easily be found as

$$h(t) = e^{-t}\cos(2t)u(t) - 0.5e^{-t}\sin(2t)u(t)$$

Sampling the impulse response $h(t)$ using a sampling interval $T = 0.1$ and using the scale factor of $T = 0.1$, we have

$$Th(n) = Th(t)|_{t=nT} = 0.1e^{-0.1n}\cos(0.2n)u(n) - 0.05e^{-0.1n}\sin(0.2n)u(n)$$

Applying the z-transform (Chapter 5) leads to

$$H(z) = Z[0.1e^{-0.1n}\cos(0.2n)u(n) - 0.05e^{-0.1n}\sin(0.2n)u(n)]$$

$$= \frac{0.1z(z - e^{-0.1}\cos(0.2))}{z^2 - 2e^{-0.1}\cos(0.2)z + e^{-0.2}} - \frac{0.05e^{-0.1}\sin(0.2)z}{z^2 - 2e^{-0.1}\cos(0.2)z + e^{-0.2}}$$

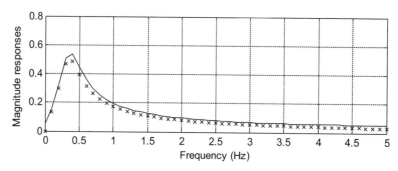

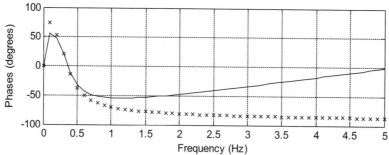

FIGURE 8.30

Frequency responses. The line of "x"s represents frequency responses of the analog filter; the solid line represents frequency responses of the designed digital filter.

After algebra simplification, we obtain the second-order digital filter as

$$H(z) = \frac{0.1 - 0.09767z^{-1}}{1 - 1.7735z^{-1} + 0.8187z^{-2}}$$

b. The magnitude and phase frequency responses are shown in Figure 8.30 and MATLAB Program 8.13 is given. The passband gain of the digital filter is higher than that of the analog filter, but their shapes are same.

Program 8.13. MATLAB program for Example 8.16.

```
% Example 8.16
% Plot the magnitude responses |H(s)| and |H(z)|
% For the Laplace transfer function H(s)
f=0:0.1:5;T=0.1; % Initialize analog frequency range in Hz and sampling interval
w=2*pi*f;    % Convert the frequency range to radians/second
hs=freqs([1 0], [1 2 5],w); % Calculate analog filter frequency responses
phis=180*angle(hs)/pi;
% For the z-transfer function H(z)
% Calculate digital filter frequency responses
hz=freqz([0.1 -0.09766],[1 -1.7735 0.8187],length(w));
phiz=180*angle(hz)/pi;
% Plot magnitude and phase responses
subplot(2,1,1), plot(f,abs(hs),'x',f, abs(hz),'-'),grid;
xlabel('Frequency (Hz)'); ylabel('Magnitude Responses')
subplot(2,1,2), plot(f,phis,'x',f, phiz,'-'); grid;
xlabel('Frequency (Hz)'); ylabel('Phases (degrees)')
```

8.7 POLE-ZERO PLACEMENT METHOD FOR SIMPLE INFINITE IMPULSE RESPONSE FILTERS

This section introduces a pole-zero placement method for a simple IIR filter design. Let us first examine the effects of the pole-zero placement on the magnitude response in the z-plane (Figure 8.31).

In the z-plane, when we place a pair of complex conjugate zeros at a given point on the unit circle with an angle θ (usually we do), we will have a numerator factor of $(z - e^{j\theta})(z - e^{-j\theta})$ in the transfer function. Its magnitude contribution to the frequency response at $z = e^{j\Omega}$ is $(e^{j\Omega} - e^{j\theta})(e^{j\Omega} - e^{-j\theta})$. When $\Omega = \theta$, the magnitude will reach zero, since the first factor $(e^{j\theta} - e^{j\theta}) = 0$ contributes zero magnitude. When a pair of complex conjugate poles are placed at a given point within the unit cycle, we have a denominator factor of $(z - re^{j\theta})(z - re^{-j\theta})$, where r is the radius chosen to be less than and close to 1 to place the poles inside the unit circle. The magnitude contribution to the frequency response at $\Omega = \theta$ will rise to a large magnitude, since the first factor $(e^{j\theta} - re^{j\theta}) = (1 - r)e^{j\theta}$ gives a small magnitude of $1 - r$ in the denominator. This small magnitude $(1-r)$ is the length between the pole location and the unit circle at the angle $\Omega = \theta$ as shown in Figure 8.31. Note that the magnitude of $e^{j\theta}$ is 1.

Therefore, we can reduce the magnitude response using zero placement, while we increase the magnitude response using pole placement. Placing a combination of poles and zeros will result in

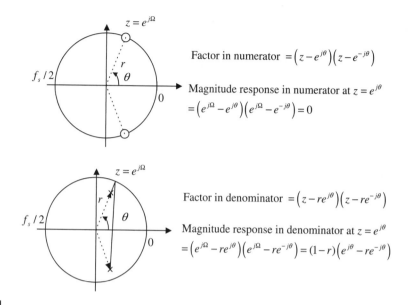

FIGURE 8.31

Effects of the pole-zero placement on the magnitude response.

different frequency responses, such as lowpass, highpass, bandpass, and bandstop. The method is intuitive and approximate. Furthermore, it is easy to compute filter coefficients for simple IIR filters. Here, we describe the design procedures for second-order bandpass and bandstop filters, as well first-order lowpass and highpass filters. (For details of derivations, readers are referred to Lynn and Fuerst [1999]). Practically, the pole-zero placement method delivers good performance when the bandpass and bandstop filters have very narrow bandwidth requirements and the lowpass and highpass filters have either very low cutoff frequency close to DC or very high cutoff frequency close to the folding frequency (Nyquist limit).

8.7.1 Second-Order Bandpass Filter Design

Typical pairs of poles and zeros for a bandpass filter are placed in Figure 8.32. Poles are complex conjugate, with the magnitude r controlling the bandwidth and the angle θ controlling the center frequency. The zeros are placed at $z = 1$ corresponding to DC, and $z = -1$, corresponding to the folding frequency.

The poles will increase the magnitude response at the center frequency while the zeros will cause zero gains at DC (zero frequency) and at the folding frequency.

The following equations give the bandpass filter design formulas using pole-zero placement:

$$r \approx 1 - (BW_{3dB}/f_s) \times \pi, \text{ good for } 0.9 \leq r < 1 \tag{8.41}$$

$$\theta = \left(\frac{f_0}{f_s}\right) \times 360° \tag{8.42}$$

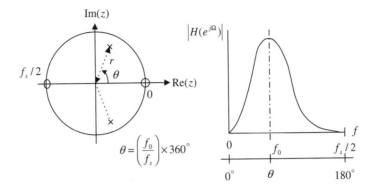

FIGURE 8.32

Pole-zero placement for a second-order narrow bandpass filter.

$$H(z) = \frac{K(z-1)(z+1)}{(z-re^{j\theta})(z-re^{-j\theta})} = \frac{K(z^2-1)}{(z^2 - 2rz\cos\theta + r^2)} \tag{8.43}$$

where K is a scale factor to adjust the bandpass filter so it has a unit passband gain given by

$$K = \frac{(1-r)\sqrt{1 - 2r\cos 2\theta + r^2}}{2|\sin\theta|} \tag{8.44}$$

EXAMPLE 8.17

A second-order bandpass filter is required to satisfy the following specifications:

- Sampling rate = 8,000 Hz
- 3 dB bandwidth: $BW = 200$ Hz
- Narrow passband centered at $f_0 = 1,000$ Hz
- Zero gain at 0 Hz and 4,000 Hz

Find the transfer function using the pole-zero placement method.

Solution:
First, we calculate the required magnitude of the poles

$$r = 1 - (200/8,000)\pi = 0.9215,$$

which is a good approximation. Use the center frequency to obtain the angle of the pole location:

$$\theta = \left(\frac{1,000}{8,000}\right) \times 360 = 45°$$

Compute the unit-gain scale factor as

$$K = \frac{(1 - 0.9215)\sqrt{1 - 2 \times 0.9215 \times \cos2 \times 45° + 0.9215^2}}{2|\sin 45°|} = 0.0755$$

Finally, the transfer function is given by

$$H(z) = \frac{0.0755(z^2 - 1)}{(z^2 - 2 \times 0.9215z\cos 45° + 0.9215^2)} = \frac{0.0755 - 0.0755z^{-2}}{1 - 1.3031z^{-1} + 0.8491z^{-2}}$$

8.7.2 Second-Order Bandstop (Notch) Filter Design

For this type of filter, the pole placement is the same as the bandpass filter (Figure 8.33). The zeros are placed on the unit circle with the same angles with respect to poles. This will improve passband performance. The magnitude and the angle of the complex conjugate poles determine the 3 dB bandwidth and center frequency, respectively.

Design formulas for bandstop filters are given in the following equations:

$$r \approx 1 - (BW_{3dB}/f_s) \times \pi, \text{ good for } 0.9 \le r < 1 \tag{8.45}$$

$$\theta = \left(\frac{f_0}{f_s}\right) \times 360° \tag{8.46}$$

$$H(z) = \frac{K(z - e^{j\theta})(z + e^{-j\theta})}{(z - re^{j\theta})(z - re^{-j\theta})} = \frac{K(z^2 - 2z\cos\theta + 1)}{(z^2 - 2rz\cos\theta + r^2)} \tag{8.47}$$

The scale factor to adjust the bandstop filter so it has a unit passband gain is given by

$$K = \frac{(1 - 2r\cos\theta + r^2)}{(2 - 2\cos\theta)} \tag{8.48}$$

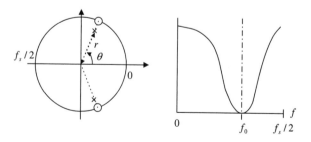

FIGURE 8.33

Pole-zero placement for a second-order notch filter.

EXAMPLE 8.18

A second-order notch filter is required to satisfy the following specifications:

- Sampling rate = 8,000 Hz
- 3 dB bandwidth: $BW = 100$ Hz
- Narrow passband centered at $f_0 = 1,500$ Hz

Find the transfer function using the pole-zero placement approach.

Solution:

We first calculate the required magnitude of the poles

$$r \approx 1 - (100/8,000) \times \pi = 0.9607$$

which is a good approximation. We use the center frequency to obtain the angle of the pole location:

$$\theta = \left(\frac{1,500}{8,000}\right) \times 360° = 67.5°$$

The unit-gain scale factor is calculated as

$$K = \frac{(1 - 2 \times 0.9607 \cos 67.5° + 0.9607^2)}{(2 - 2 \cos 67.5°)} = 0.9620$$

Finally, we obtain the transfer function:

$$H(z) = \frac{0.9620(z^2 - 2z \cos 67.5° + 1)}{(z^2 - 2 \times 0.9607 z \cos 67.5° + 0.9607^2)} = \frac{0.9620 - 0.7363z^{-1} + 0.9620z^{-2}}{1 - 0.7353z^{-1} + 0.9229}$$

8.7.3 First-Order Lowpass Filter Design

The first-order pole-zero placement can be utilized in two cases. The first situation is when the cutoff frequency is less than $f_s/4$. Then the pole-zero placement is shown in Figure 8.34.

As shown in Figure 8.34, the pole $z = \alpha$ is placed in the real axis. The zero is placed at $z = -1$ to ensure zero gain at the folding frequency (Nyquist limit). When the cutoff frequency is above $f_s/4$, the pole-zero placement is adopted as shown in Figure 8.35.

Design formulas for lowpass filters using the pole-zero placement are given in the following equations.

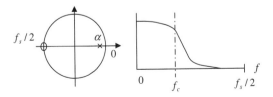

FIGURE 8.34

Pole-zero placement for the first-order lowpass filter with $f_c < f_s/4$.

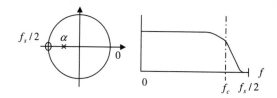

FIGURE 8.35

Pole-zero placement for the first-order lowpass filter with $f_c > f_s/4$.

When $f_c < f_s/4$,

$$\alpha \approx 1 - 2 \times (f_c/f_s) \times \pi, \quad \text{good for} \quad 0.9 \le r < 1 \tag{8.49}$$

When $f_c > f_s/4$,

$$\alpha \approx -(1 - \pi + 2 \times (f_c/f_s) \times \pi), \quad \text{good for} \quad -1 < r \le -0.9 \tag{8.50}$$

The transfer function is

$$H(z) = \frac{K(z+1)}{(z - \alpha)} \tag{8.51}$$

and the unit passband gain scale factor is given by

$$K = \frac{(1 - \alpha)}{2} \tag{8.52}$$

EXAMPLE 8.19

A first-order lowpass filter is required to satisfy the following specifications:

- Sampling rate = 8,000 Hz
- 3 dB cutoff frequency: $f_c = 100$ Hz
- Zero gain at 4,000 Hz

Find the transfer function using the pole-zero placement method.

Solution:

Since the cutoff frequency of 100 Hz is much less than $f_s/4 = 2,000$ Hz, we determine the pole as

$$\alpha \approx 1 - 2 \times (100/8,000) \times \pi = 0.9215$$

which is above 0.9. Hence, we have a good approximation. The unit gain scale factor is calculated by

$$K = \frac{(1 - 0.9215)}{2} = 0.03925$$

Last, we can develop the transfer function as

$$H(z) = \frac{0.03925(z+1)}{(z - 0.9215)} = \frac{0.03925 + 0.03925z^{-1}}{1 - 0.9215z^{-1}}$$

Note that we can also determine the unit-gain factor K by substituting $z = e^{j0} = 1$ in the transfer function $H(z) = \dfrac{(z+1)}{(z-\alpha)}$, then finding the DC gain. Set the scale factor to be a reciprocal of the DC gain. This can be easily done as follows:

$$DC\ gain = \left.\frac{z+1}{z-0.9215}\right|_{z=1} = \frac{1+1}{1-0.9215} = 25.4777$$

Hence, $K = 1/25.4777 = 0.03925$.

8.7.4 First-Order Highpass Filter Design

Similar to the lowpass filter design, the pole-zero placements for the first-order highpass filters in two cases are shown in Figure 8.36(a) and 8.36(b).

Formulas for designing highpass filters using the pole-zero placement are listed in the following equations:

When $f_c < f_s/4$,

$$\alpha \approx 1 - 2 \times (f_c/f_s) \times \pi, \quad \text{good for} \quad 0.9 \le r < 1 \tag{8.53}$$

When $f_c > f_s/4$,

$$\alpha \approx -(1 - \pi + 2 \times (f_c/f_s) \times \pi), \quad \text{good for} \quad -1 < r \le -0.9 \tag{8.54}$$

$$H(z) = \frac{K(z-1)}{(z-\alpha)} \tag{8.55}$$

$$K = \frac{(1+\alpha)}{2} \tag{8.56}$$

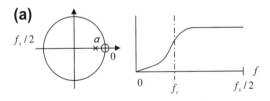

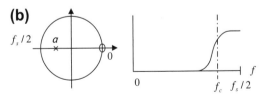

FIGURE 8.36

(a) Pole-zero placement for the first-order highpass filter with $f_c < f_s/4$. (b) Pole-zero placement for the first-order highpass filter with $f_c > f_s/4$.

EXAMPLE 8.20

A first-order highpass filter is required to satisfy the following specifications:

- Sampling rate = 8,000 Hz
- 3 dB cutoff frequency: $f_c = 3,800$ Hz
- Zero gain at 0 Hz

Find the transfer function using the pole-zero placement method.

Solution:

Since the cutoff frequency of 3,800 Hz is much larger than $f_s/4 = 2,000$ Hz, we determine the pole as

$$\alpha \approx -(1 - \pi + 2 \times (3,800/8,000) \times \pi) = -0.8429$$

The unit-gain scale factor and transfer function are obtained as

$$K = \frac{(1 - 0.8429)}{2} = 0.07854$$

$$H(z) = \frac{0.07854(z - 1)}{(z + 0.8429)} = \frac{0.07854 - 0.07854z^{-1}}{1 + 0.8429z^{-1}}$$

Note that we can also determine the unit-gain scale factor K by substituting $z = e^{j180^\circ} = -1$ to the transfer function $H(z) = \frac{(z-1)}{(z-\alpha)}$, finding a passband gain at the Nyquist limit $f_s/2 = 4,000$ Hz. We then set the scale factor to be a reciprocal of the passband gain. That is,

$$Passband\,gain = \frac{z-1}{z+0.8429}\bigg|_{z=1} = \frac{-1-1}{-1+0.8429} = 12.7307$$

Hence, $K = 1/12.7307 = 0.07854$.

8.8 REALIZATION STRUCTURES OF INFINITE IMPULSE RESPONSE FILTERS

In this section, we will realize the designed IIR filter using direct-form I as well as direct-form II. We will then realize a higher-order IIR filter using a cascade form.

8.8.1 Realization of Infinite Impulse Response Filters in Direct-Form I and Direct-Form II

EXAMPLE 8.21

Realize the first-order digital highpass Butterworth filter

$$H(z) = \frac{0.1936 - 0.1936z^{-1}}{1 + 0.6128z^{-1}}$$

using a direct-form I realization.

Solution:

From the transfer function, we can identify

$$b_0 = 0.1936, \ b_1 = -0.1936, \quad \text{and} \quad a_1 = 0.6128$$

Applying the direct-form I developed in Chapter 6 results in the diagram in Figure 8.37. The digital signal processing (DSP) equation for implementation is then given by

$$y(n) = -0.6128y(n-1) + 0.1936x(n) - 0.1936x(n-1)$$

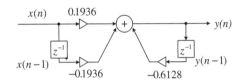

FIGURE 8.37

Realization of IIR filter in Example 8.21 in direct-form I.

Program 8.14 lists the MATLAB implementation.
Program 8.14. MATLAB program for Example 8.21.

```
% Sample MATLAB code
sample = 2:2:20; % Input test array
x=[ 0 0 ]; % Input buffer [x(n) x(n-1) ...]
y=[ 0 0]; % Output buffer [y(n) y(n-1) ... ]
b=[0.1936 -0.1936]; % Numerator coefficients [b0 b1 ... ]
a=[1 0.6128]; % Denominator coefficients [1 a0 a1 ...]
for n=1:1:length(sample) % Processing loop
  for k=2:-1:2
   x(k)=x(k-1); % Shift input by one sample
   y(k)=y(k-1); % Shift output by one sample
  end
  x(1)=sample(n);  % Get new sample
  y(1)=0;          % Digital filtering
  for k=1:1:2
   y(1)=y(1)+x(k)*b(k);
  end
  for k=2:2
   y(1)=y(1)-a(k)*y(k);
  end
  out(n)=y(1); % Output the filtered sample to output array
end
  out
```

EXAMPLE 8.22

Realize the following digital filter using direct-form II:

$$H(z) = \frac{0.7157 + 1.4314z^{-1} + 0.7151z^{-2}}{1 + 1.3490z^{-1} + 0.5140z^{-2}}$$

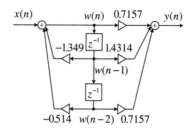

FIGURE 8.38

Realization of IIR filter in Example 8.22 in direct-form II.

Solution:

First, we can identify

$$b_0 = 0.7157, \ b_1 = 1.4314, \ b_2 = 0.7151$$

$$\text{and} \quad a_1 = 1.3490, \ a_2 = 0.5140$$

Applying the direct-form II realization developed in Chapter 6 leads to Figure 8.38. There are two difference equations required for implementation:

$$w(n) = x(n) - 1.3490w(n-1) - 0.5140w(n-2)$$

$$y(n) = 0.7157w(n) + 1.4314w(n-1) + 0.7157w(n-2)$$

The MATLAB implementation is listed in Program 8.15.

Program 8.15. MATLAB code for Example 8.22.

```
% Sample MATLAB code
sample =2:2:20;              % Input test array
x=[0];                       % Input buffer [x(n) ]
y=[0];                       % Output buffer [y(n)]
w=[0 0 0]; % Buffer for w(n) [w(n) w(n-1) ...]
b=[0.7157 1.4314 0.7157];    % Numerator coefficients [b0 b1 ...]
a=[1 1.3490 0.5140];         % Denominator coefficients [1 a1 a2 ...]
for n=1:1:length(sample)     % Processing loop
  for k=3:-1:2
  w(k)=w(k-1);               % Shift w(n) by one sample
  end
  x(1)=sample(n);            % Get new sample
  w(1)=x(1);                 % Perform IIR filtering
  for k=2:1:3
   w(1)=w(1)-a(k)*w(k);
  end
  y(1)=0;                    % Perform FIR filtering
  for k=1:1:3
   y(1)=y(1)+b(k)*w(k);
  end
  out(n)=y(1);               % Send the filtered sample to output array
end
  out
```

8.8.2 Realization of Higher-Order Infinite Impulse Response Filters via the Cascade Form

EXAMPLE 8.23

Given a fourth-order filter transfer function designed as

$$H(z) = \frac{0.5108z^2 + 1.0215z + 0.5108}{z^2 + 0.5654z + 0.4776} \times \frac{0.3730z^2 + 0.7460z + 0.3730}{z^2 + 0.4129z + 0.0790}$$

realize the digital filter using the cascade (series) form via second-order sections.

Solution:
Since the filter is designed using the cascade form, we have two sections of the second-order filters, whose transfer functions are

$$H_1(z) = \frac{0.5108z^2 + 1.0215z + 0.5108}{z^2 + 0.5654z + 0.4776} = \frac{0.5180 + 1.0215z^{-1} + 0.5108z^{-2}}{1 + 0.5654z^{-1} + 0.4776z^{-2}}$$

and

$$H_2(z) = \frac{0.3730z^2 + 0.7460z + 0.3730}{z^2 + 0.4129z + 0.0790} = \frac{0.3730 + 0.7460z^{-1} + 0.3730z^{-2}}{1 + 0.4129z^{-1} + 0.0790z^{-2}}$$

Each filter section is developed using the direct-form I realization, shown in Figure 8.39.

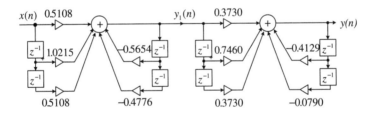

FIGURE 8.39

Cascade realization of IIR filter in Example 8.23 in direct-form I.

There are two sets of DSP equations for implementation of the first and second sections, respectively.
First section:

$$y_1(n) = -0.5654y_1(n-1) - 0.4776y_1(n-2)$$
$$+0.5108x(n) + 1.0215x(n-1) + 0.5108x(n-2)$$

Second section:

$$y(n) = -0.4129y(n-1) - 0.0790y(n-2)$$
$$+0.3730y_1(n) + 0.7460y_1(n-1) + 0.3730y_1(n-2)$$

Again, after we use the direct-form II for realizing each second-order filter, the realization shown in Figure 8.40 is developed.

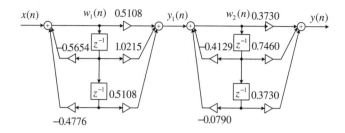

FIGURE 8.40

Cascade realization of IIR filter in Example 8.23 in direct-form II.

The difference equations for the implementation of the first section are

$$w_1(n) = x(n) - 0.5654w_1(n-1) - 0.4776w_1(n-2)$$

$$y_1(n) = 0.5108w_1(n) + 1.0215w_1(n-1) + 0.5108w_1(n-2)$$

The difference equations for the implementation of the second section are

$$w_2(n) = y_1(n) - 0.4129w_2(n-1) - 0.0790w_2(n-2)$$

$$y(n) = 0.3730w_2(n) + 0.7460w_2(n-1) + 0.3730w_2(n-2)$$

Note that for both direct-form I and direct-form II, the output from the first filter section becomes the input for the second filter section.

8.9 APPLICATION: 60-HZ HUM ELIMINATOR AND HEART RATE DETECTION USING ELECTROCARDIOGRAPHY

Hum noise created by poor power suppliers, transformers, or electromagnetic interference sourced by a main power supply is characterized by a frequency of 60 Hz and its harmonics. If this noise interferes with a desired audio or biomedical signal (e.g., in electrocardiography [ECG]), the desired signal could be corrupted. The corrupted signal is useless without signal processing. It is sufficient to eliminate the 60-Hz hum frequency with its second and third harmonics in most practical applications. We can complete this by cascading with notch filters having notch frequencies of 60 Hz, 120 Hz, and 180 Hz, respectively. Figure 8.41 depicts the functional block diagram.

Now let us apply the 60-Hz hum eliminator to an ECG recording system. ECG is a small electrical signal captured from an ECG sensor. The ECG signal is produced by the activity of the human heart, thus it can be used for heart rate detection, fetal monitoring, and diagnostic purposes. The single pulse of the ECG is depicted in Figure 8.42, which shows that the ECG signal is characterized by five peaks and valleys, labeled P, Q, R, S, and T. The highest positive wave is the R wave. Shortly before and after the R wave are negative waves called the Q wave and S wave. The P wave comes before the Q wave, while the T wave comes after the S wave. The Q, R, and S waves together are called the QRS complex.

The properties of the QRS complex, with its rate of occurrence and times, heights, and widths, provide information to cardiologists concerning various pathological conditions of the heart. The

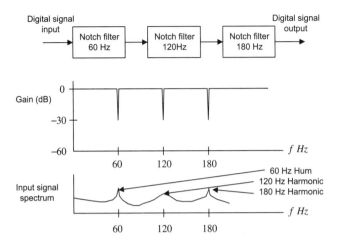

FIGURE 8.41

(Top) 60-Hz hum eliminator; (middle) the filter frequency response of the eliminator; (bottom) the input signal spectrum corrupted by the 60-Hz hum and its second and third harmonics.

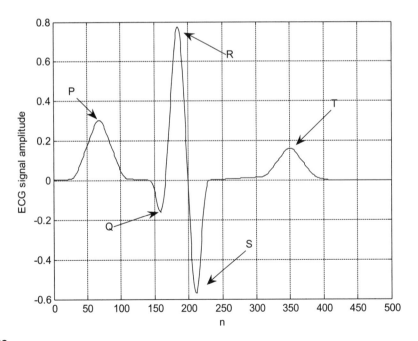

FIGURE 8.42

The characteristics of the ECG pulse.

reciprocal of the time period between R wave peaks (in milliseconds) multiplied by 60,000 gives instantaneous heart rate in beats per minute. On a modern ECG monitor, the acquired ECG signal is displayed for diagnostic purposes.

However, a major source of frequent interference is the electric-power system. Such interference appears on the recorded ECG data due to electrical-field coupling between the power lines and the electrocardiograph or patient, which is the cause of the electrical field surrounding power lines (mains). Another cause is magnetic induction in the power line, whereby current in the power line generates a magnetic field around the line. Sometimes, the harmonics of 60-Hz hum exist due to the nonlinear sensor and signal amplifier effects. If such interference is severe, the recorded ECG data becomes useless.

In this application, we focus on ECG enhancement for heart rate detection. To significantly reduce 60-Hz interference, we apply signal enhancement to the ECG recording system, as shown in Figure 8.43.

The 60-Hz eliminator removes the 60-Hz interference and has the capability to reduce its second harmonic of 120 Hz and third harmonic of 180 Hz.

The next objective is to detect the heart rate using the enhanced ECG signal. We need to remove DC drift and to filter muscle noise, which may occur at approximately 40 Hz or more. If we consider the lowest heart rate as 30 beats per minute, the corresponding frequency is $30/60 = 0.5$ Hz. Choosing a lower cutoff frequency of 0.25 Hz should be reasonable.

Thus, a bandpass filter with a passband from 0.25 Hz to 40 Hz (range from 0.67 Hz to 40 Hz, discussed in Webster [1998]), either FIR or IIR type, can be designed to reduce such effects. The resultant ECG signal is valid only for the detection of heart rate. Notice that the ECG signal after bandpass filtering with a passband from 0.25 Hz to 40 Hz is no longer valid for general ECG applications, since the original ECG signal occupies the frequency range from 0.01 Hz to 250 Hz (diagnostic-quality ECG), as discussed in Carr and Brown (2001) and Webster (1998). The enhanced ECG signal from the 60-Hz hum eliminator can serve for general ECG signal analysis (which is beyond the scope of this book). We summarize the design specifications for the heart rate detection application as follows:

System outputs:	Enhanced ECG signal with 60-Hz elimination
	Processed ECG signal for heart rate detection
60-Hz eliminator specifications:	
Harmonics to be removed:	60 Hz (fundamental)
	120 Hz (second harmonic)
	180 Hz (third harmonic)
3-dB bandwidth for each filter:	4 Hz
Sampling rate:	600 Hz
Notch filter type:	Second-order IIR

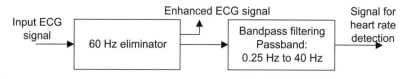

FIGURE 8.43

ECG signal enhancement system.

Design method:	Pole-zero placement
Bandpass filter specifications:	
Passband frequency range:	0.25–40 Hz
Passband ripple:	0.5 dB
Filter type:	Chebyshev fourth order
Design method:	Bilinear transformation method
DSP sampling rate:	600 Hz

Let us carry out the 60-Hz eliminator design and determine the transfer function and difference equation for each notch filter and bandpass filter. For the first section with the notch frequency of 60 Hz, applying Equations (8.45) to (8.48) leads to

$$r = 1 - (4/600) \times \pi = 0.9791$$

$$\theta = \left(\frac{60}{600} \right) \times 360° = 36°$$

We calculate $2 \cos (36°) = 1.6180$, $2r \cos (36°) = 1.5842$, and

$$K = \frac{(1 - 2r \cos \theta + r^2)}{(2 - 2 \cos \theta)} = 0.9803$$

Hence it follows that

$$H_1(z) = \frac{0.9803 - 1.5862z^{-1} + 0.9803z^{-2}}{1 - 1.5842z^{-1} + 0.9586z^{-2}}$$

$$y_1(n) = 0.9803x(n) - 1.5862x(n-1) + 0.9802x(n-2) + 1.5842y_1(n-1) - 0.9586y_1(n-2)$$

Similarly, we obtain the transfer functions and difference equations for the second section and third section as follows:

Second section:

$$H_2(z) = \frac{0.9794 - 0.6053z^{-1} + 0.9794z^{-2}}{1 - 0.6051z^{-1} + 0.9586z^{-2}}$$

$$y_2(n) = 0.9794y_1(n) - 0.6053y_1(n-1) + 0.9794y_1(n-2) + 0.6051y_2(n-1) - 0.9586y_2(n-2)$$

Third section:

$$H_3(z) = \frac{0.9793 + 0.6052z^{-1} + 0.9793z^{-2}}{1 + 0.6051z^{-1} + 0.9586z^{-2}}$$

$$y_3(n) = 0.9793y_2(n) + 0.6052y_2(n-1) + 0.9793y_2(n-2) - 0.6051y_3(n-1) - 0.9586y_3(n-2)$$

The cascaded frequency responses are plotted in Figure 8.44. As we can see, the rejection for each notch frequency is below 50 dB.

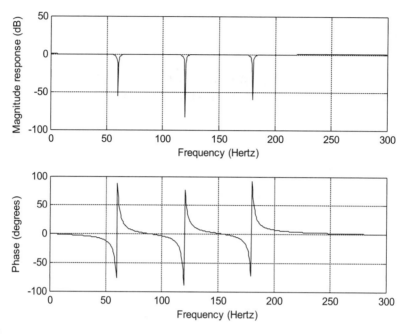

FIGURE 8.44

Frequency responses of three cascaded notch filters.

The second-stage design using the BLT gives the bandpass filter transfer function and difference equation:

$$H_4(z) = \frac{0.0464 - 0.0927z^{-2} + 0.0464z^{-4}}{1 - 3.3523z^{-1} + 4.2557z^{-2} - 2.4540z^{-3} + 0.5506z^{-4}}$$

$$y_4(n) = 0.046361y_3(n) - 0.092722y_3(n-2) + 0.046361y_3(n-4)$$
$$+03.352292y_4(n-1) - 4.255671y_4(n-2) + 2.453965y_4(n-3) - 0.550587y_4(n-4)$$

Figure 8.45 depicts the processed results at each stage. In Figure 8.45, plot (a) shows the initial corrupted ECG data, which includes 60-Hz interference and its 120 and 180 Hz harmonics, along with muscle noise. Plot (b) shows that the 60-Hz interference and its harmonics of 120 and 180 Hz have been removed. Finally, plot (c) displays the result after the bandpass filter. As we expected, the muscle noise has been removed; and the enhanced ECG signal is observed. A MATLAB simulation is provided in Program 8.16.

With the procssed ECG signal, a simple zero-cross algorithm can be designed to detect the heart rate. Based on plot (c) in Figure 8.45, we use a threshold value of 0.5 and continuously compare each of two consecutive samples with the threshold. If both results are opposite, then a zero crossing is detected. Each zero-crossing measure is given by

$$\text{zero crossing} = \frac{|cur\_sign - pre\_sign|}{2}$$

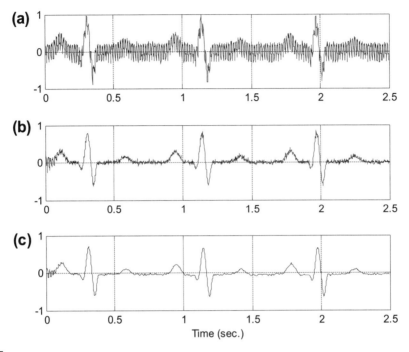

FIGURE 8.45

Results of ECG signal processing. (a) Initial corrupted ECG data; (b) ECG data enhanced by removing 60-Hz interference; (c) ECG data with DC blocking and noise removal for heart rate detection.

where *cur_sign* and *pre_sign* are determined based on the current input $x(n)$, the past input $x(n-1)$, and the threshold value, given as

$$\text{if} \quad x(n) \geq threshold \; cur\_sign = 1 \text{ else } cur\_sign = -1$$

$$\text{if} \quad x(n-1) \geq threshold \; pre\_sign = 1 \text{ else } pre\_sign = -1$$

Figure 8.46 summarizes the algorithm.

After detecting the total number of zero crossings, the number of peaks will be half the number of zero crossings. The heart rate in terms of pulses per minute can be determined by

$$\text{Heart rate} = \frac{60}{\left(\dfrac{Number\,of\,enhanced\,ECG\,data}{f_s}\right)} \times \left(\frac{zero-crossing\,number}{2}\right)$$

In our simulation, we have detected 6 zero-crossing points using 1,500 captured data points at a sampling rate of 600 samples per second. Hence,

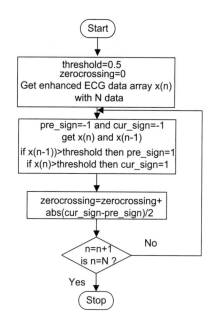

FIGURE 8.46

A simple zero-crossing algorithm.

$$\text{Heart rate} = \frac{60}{\left(\dfrac{1,500}{600}\right)} \times \left(\frac{6}{2}\right) = 72 \text{ pulses per minute}$$

The MATLAB implementation of the zero-crossing detection can be found in the last part in Program 8.16.

Program 8.16. MATLAB program for heart rate detection using an ECG signal.

```
load ecgbn.dat; % Load noisy ECG recording
b1=[0.9803 -1.5862 0.9803]; %Notch filter with a notch frequency of 60 Hz
a1=[1 -1.5842 0.9586];
b2=[0.9794 -0.6053 0.9794]; % Notch filter with a notch frequency 120 Hz
a2=[1 -0.6051 0.9586];
b3=[0.9793 0.6052 0.9793]; % Notch filter with a notch frequency of 180 Hz
a3=[1 0.6051 0.9586];
y1=filter(b1,a1,ecgbn); % First section filtering
y2=filter(b2,a2,y1);  % Second section filtering
y3=filter(b3,a3,y2);  % Third section filtering
%Bandpass filter
fs=600;    % Sampling rate
T=1/600;    % Sampling interval
% BLT design
wd1=2*pi*0.25;
```

```
wd2=2*pi*40;
wa1=(2/T)*tan(wd1*T/2);
wa2=(2/T)*tan(wd2*T/2);
[B,A]=lp2bp([1.4314], [1 1.4652 1.5162],sqrt(wa1*wa2),wa2-wa1);
[b,a]=bilinear(B,A,fs);
b =[ 0.046361 0 -0.092722 0 0.046361]
a =[1 -3.352292 4.255671 -2.453965 0.550587]
y4=filter(b,a,y3);        %Bandpass filtering
t=0:T:1499*T;        % Recover time
subplot(3,1,1);plot(t,ecgbn);grid;ylabel('(a)');
subplot(3,1,2);plot(t,y3);grid;ylabel('(b)');
subplot(3,1,3);plot(t,y4);grid;ylabel('(c)');
xlabel('Time (sec.)');
%Zero cross algorithm
zcross=0.0;threshold=0.5
for n=2:length(y4)
 pre_sign=-1;cur_sign=-1;
 if y4(n-1)>threshold
  pre_sign=1;
 end
 if y4(n)>threshold
  cur_sign=1;
 end
 zcross=zcross+abs(cur_sign-pre_sign)/2;
end
zcross                % Output the number of zero crossings
rate=60*zcross/(2*length(y4)/600)  % Output the heart rate
```

8.10 COEFFICIENT ACCURACY EFFECTS ON INFINITE IMPULSE RESPONSE FILTERS

In practical applications, the IIR filter coefficients with infinite precision may be quantized due to the finite word length. Quantization of infinite precision filter coefficients changes the locations of the zeros and poles of the designed filter transfer function, and thus changes the filter frequency responses. Since analysis of filter coefficient quantization for the IIR filter is very complicated and beyond the scope of this textbook, we pick only a couple of simple cases for discussion. Filter coefficient quantization for specific processors such as the fixed-point DSP processor and floating-point processor will be included in Chapter 9. To illustrate this effect, we look at the following first-order IIR filter transfer function with filter coefficients with infinite precision:

$$H(z) = \frac{b_0 + b_1 z^{-1}}{1 + a_1 z^{-1}} \tag{8.57}$$

After filter coefficient quantization, we have the quantized digital IIR filter transfer function:

$$H^q(z) = \frac{b_0^q + b_1^q z^{-1}}{1 + a_1^q z^{-1}} \tag{8.58}$$

Solving for the pole and zero, we get

$$z_1 = -\frac{b_1^q}{b_0^q} \tag{8.59}$$

$$p_1 = -a_1^q \tag{8.60}$$

Now considering a second-order IIR filter transfer function as

$$H(z) = \frac{b_0 + b_1 z^{-1} + b_2 z^{-2}}{1 + a_1 z^{-1} + a_2 z^{-2}} \tag{8.61}$$

and its quantized IIR filter transfer function

$$H^q(z) = \frac{b_0^q + b_1^q z^{-1} + b_2^q z^{-2}}{1 + a_1^q z^{-1} + a_2^q z^{-2}} \tag{8.62}$$

solving for poles and zeros yields

$$z_{1,2} = -0.5 \cdot \frac{b_1^q}{b_0^q} \pm j\left(\frac{b_2^q}{b_0^q} - 0.25 \cdot \left(\frac{b_1^q}{b_0^q}\right)^2\right)^{\frac{1}{2}} \tag{8.63}$$

$$p_{1,2} = -0.5 \cdot a_1^q \pm j\left(a_2^q - 0.25 \cdot \left(a_1^q\right)^2\right)^{\frac{1}{2}} \tag{8.64}$$

With Equations (8.59) and (8.60) for the first-order IIR filter, and Equations (8.63) and (8.64) for the second-order IIR filter, we can study the effects of location changes of the poles and zeros, and the frequency responses due to filter coefficient quantization.

EXAMPLE 8.24

Given the first-order IIR filter

$$H(z) = \frac{1.2341 + 0.2126 z^{-1}}{1 - 0.5126 z^{-1}}$$

and assuming that we use 1 sign bit and 6 bits for encoding the magnitude of the filter coefficients, find the quantized transfer function and pole-zero locations.

Solution:

Let us find the pole and zero for infinite precision filter coefficients. Solving $1.2341z + 0.2126 = 0$ leads to a zero location $z_1 = -0.17227$. Solving $z - 0.5126 = 0$ gives a pole location $p_1 = 0.5126$.

Now let us quantize the filter coefficients. Quantizing 1.2341 can be illustrated as

$$1.2341 \times 2^5 = 39.4912 = 39 \,(\text{rounded to integer})$$

Since the maximum magnitude of the filter coefficients is 1.2341, which is between 1 and 2, we scale all coefficient magnitudes by a factor of 2^5 and round off each value to an integer whose magnitude is encoded using 6 bits. As shown in the quantization, 6 bits are required to encode the integer 39. When the coefficient integer is scaled back by the same scale factor, the corresponding quantized coefficient with finite precision (7 bits, including the sign bit) is found to be

$$b_0^q = 39/2^5 = 1.21875$$

Following the same procedure, we can obtain

$$b_1^q = 0.1875$$

and

$$a_1^q = -0.5$$

Thus we achieve the quantized transfer function

$$H^q(z) = \frac{1.21875 + 0.1875z^{-1}}{1 - 0.5z^{-1}}$$

Solving for the pole and zero leads to

$$p_1 = 0.5$$

and

$$z_1 = -0.1538$$

It is clear that the pole and zero locations change after the filter coefficients are quantized. This effect can change the frequency response of the designed filter as well. In Example 8.25, we study quantization of the filter coefficients for the second-order IIR filter and examine the pole/zero location changes and magnitude/phase frequency responses.

EXAMPLE 8.25

A second-order digital lowpass Chebyshev filter with a cutoff frequency of 3.4 kHz and 0.5 dB ripple on passband at a sampling frequency of 8,000 Hz is designed. Assume that we use 1 sign bit and 7 bits for encoding the magnitude of each filter coefficient. The z-transfer function is given by

$$H(z) = \frac{0.7434 + 1.4865z^{-1} + 0.7434z^{-2}}{1 + 1.5149z^{-1} + 0.6346z^{-2}}$$

a. Find the quantized transfer function and pole and zero locations.
b. Plot the magnitude and phase responses, respectively.

Solution:

a. Since the maximum magnitude of the filter coefficients is between 1 and 2, the scale factor for quantization is chosen to be 2^6, so that the coefficient integer can be encoded using 7 bits.

After performing filter coefficient encoding, we have

$$H^q(z) = \frac{0.7500 + 1.484375z^{-1} + 0.7500z^{-2}}{1 + 1.515625z^{-1} + 0.640625z^{-2}}$$

For comparison, the uncoded zeros and encoded zeros of the transfer function $H(z)$ are

Uncoded zeros: $-1, -1$
Coded zeros: $-0.9896 + 0.1440i, -0.9896 -0.1440i$

Similarly, the uncoded poles and coded poles of the transfer function $H^q(z)$ are

Uncoded poles: $-0.7574 + 0.2467i, -0.7574 -0.2467i$
Coded poles: $-0.7578 + 0.2569i, -0.7578 -0.2569i$

b. The comparisons for the magnitude responses and phase responses are listed in Program 8.17 and plotted in Figure 8.47.

Program 8.17. MATLAB *m*-file for Example 8.25.

```
% Example 8.25
% Plot the magnitude and phase responses
 fs=8000;               % Sampling rate
 B=[0.7434 1.4868 0.7434];
 A=[1 1.5149 0.6346];
  [hz,f]=freqz(B,A,512,fs);   % Calculate reponses without coefficient quantization
 phi=180*unwrap(angle(hz))/pi;
 Bq=[0.750 1.4834375 0.75000];
```

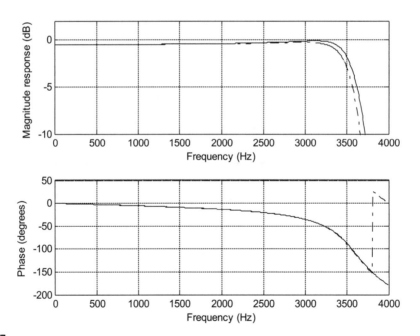

FIGURE 8.47

Frequency responses (dash-dotted line, quantized coefficients; solid line, unquantized coefficients).

```
Aq=[1 1.515625 0.640625];
[hzq,f]=freqz(Bq,Aq,512,fs); % Calculate responses with coefficient quantization
phiq=180*unwrap(angle(hzq))/pi;
subplot(2,1,1), plot(f,20*log10(abs(hz)),f,20*log10(abs(hzq)), '-.');grid;
axis([0 4000 -10 2])
xlabel('Frequency (Hz)');
ylabel('Magnitude Response (dB)');
subplot(2,1,2), plot(f, phi, f, phiq,'-.'); grid;
xlabel('Frequency (Hz)');
ylabel('Phase (degrees)');
```

From Figure 8.47, we observe that the quantization of IIR filter coefficients has more effect on magnitude response and less effect on phase response in the passband. In practice, one needs to verify this effect to make sure that the magnitude frequency response meets the filter specifications.

8.11 APPLICATION: GENERATION AND DETECTION OF DTMF TONES USING THE GOERTZEL ALGORITHM

In this section, we study an application of the digital filters to the generation and detection of dual-tone multifrequency (DTMF) signals used for telephone touch keypads. In our daily life, DTMF touch tones produced by telephone keypads on handsets are applied to dial telephone numbers routed to telephone companies, where the DTMF tones are digitized and processed and the detected dialed telephone digits are used for the telephone switching system to ring the party to be called. A telephone touch keypad is shown in Figure 8.48, where each key is represented by two tones with their specified frequencies. For example, if the key "7" is pressed, the DTMF signal with the designated frequencies of 852 Hz and 1,209 Hz is generated, which is sent to the central office at the telephone company for processing.

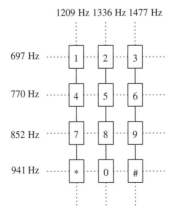

FIGURE 8.48

DTMF tone specifications.

At the central office, the received DTMF tones are detected through the digital filters and some logic operations are used to decode the dialed signal consisting of 852 Hz and 1,209 Hz as key "7". The frequencies defined for each key are in Figure 8.48.

8.11.1 Single-Tone Generator

Now, let us look at a digital tone generator whose transfer function is obtained from the z-transform function of a sinusoidal sequence $\sin(n\Omega_0)$ as

$$H(z) = \frac{z \sin \Omega_0}{z^2 - 2z \cos \Omega_0 + 1} = \frac{z^{-1} \sin \Omega_0}{1 - 2z^{-1} \cos \Omega_0 + z^{-2}} \tag{8.65}$$

where Ω_0 is the normalized digital frequency. Given the sampling rate of the DSP system and the frequency of the tone to be generated, we have the relationship

$$\Omega_0 = 2\pi f_0 / f_s \tag{8.66}$$

Applying the inverse z-transform to the transfer function leads to the difference equation

$$y(n) = \sin \Omega_0 x(n-1) + 2 \cos \Omega_0 y(n-1) - y(n-2) \tag{8.67}$$

since

$$Z^{-1}(H(z)) = Z^{-1}\left(\frac{z \sin \Omega_0}{z^2 - 2z \cos \Omega_0 + 1}\right) = \sin(\Omega_0 n) = \sin(2\pi f_0 n / f_s)$$

which is the impulse response. Hence, to generate a pure tone with an amplitude of A, an impulse function $x(n) = A\delta(n)$ must be used as the input to the digital filter, as illustrated in Figure 8.49.

Now, we illustrate implementation. Assuming that the sampling rate of the DSP system is 8,000 Hz, we need to generate a digital tone of 1 kHz. Then we compute

$$\Omega_0 = 2\pi \times 1,000/8,000 = \pi/4, \quad \sin \Omega_0 = 0.707107, \quad \text{and} \quad 2 \cos \Omega_0 = 1.414214$$

The required filter transfer function is determined as

$$H(z) = \frac{0.707107z^{-1}}{1 - 1.414214z^{-1} + z^{-2}}$$

The MATLAB simulation using the input $x(n) = \delta(n)$ is displayed in Figure 8.50, where the top plot is the generated tone of 1 kHz, and the bottom plot shows its spectrum. The corresponding MATLAB code is in Program 8.18.

$$x(n) = A\delta(n) \longrightarrow \boxed{H(z) = \frac{z^{-1} \sin \Omega_0}{1 - 2z^{-1} \cos \Omega_0 + z^{-2}}} \xrightarrow{\text{Tone}} y(n) = A \sin(2\pi f_0 n / f_s) u(n)$$

FIGURE 8.49

Single-tone generator.

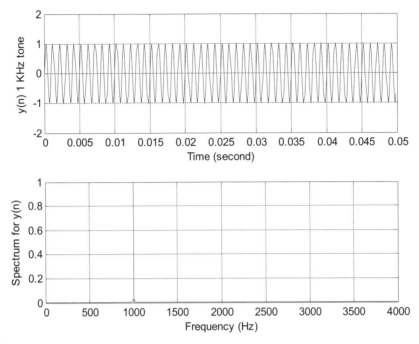

FIGURE 8.50

Plots of a generated single tone of 1,000 Hz and its spectrum.

Note that if we replace the filter $H(z)$ with the z-transform of other sequences such as a cosine function and use the impulse sequence as the filter input, the filter will generate the corresponding digital wave such as the digital cosine wave.

Program 8.18. MATLAB program for generating a sinusoid.

```
fs=8000;                                    % Sampling rate
t=0:1/fs:1;                                 % Time vector for 1 second
x=zeros(1,length(t));                       % Initialize input to be zero
x(1)=1;                                     % Set up impulse function
y=filter([0 0.707107],[1 -1.414214 1],x);  % Perform filtering
subplot(2,1,1);plot(t(1:400),y(1:400));grid
ylabel('y(n) 1 kHz tone'); xlabel('time (second)')
Ak=2*abs(fft(y))/length(y);Ak(1)=Ak(1)/2;  % One-sided amplitude spectrum
f=[0:1:(length(y)-1)/2]*fs/length(y);       % Indices to frequencies (Hz) for plot
subplot(2,1,2);plot(f,Ak(1:(length(y)+1)/2));grid
ylabel('Spectrum for y(n)'); xlabel('frequency (Hz)')
```

8.11.2 Dual-Tone Multifrequency Tone Generator

Now that the principle of a single-tone generator is illustrated, we can extend it to develop the DTMF tone generator using two digital filters in parallel. The DTMF tone generator for key "7" is depicted in Figure 8.51.

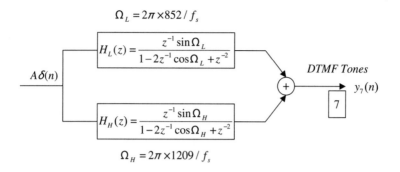

FIGURE 8.51

Digital DTMF tone generator for the keypad digit "7".

Here we generate the DTMF tone for key "7" for a duration of one second, assuming a sampling rate of 8,000 Hz. The generated tone and its spectrum are plotted in Figure 8.52 for verification, while the MATLAB implementation is given in Program 8.19.

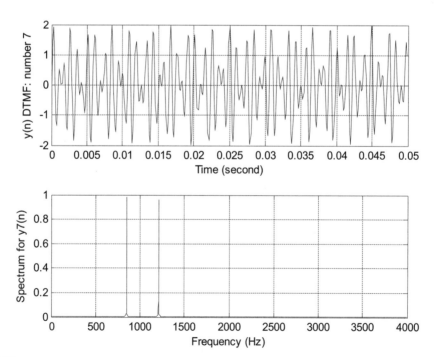

FIGURE 8.52

Plots of the generate DTMF tone of "7" and its spectrum.

Program 8.19. MATLAB program for DTMF tone generation.

```
close all; clear all
fs=8000;                                % Sampling rate
t=0:1/fs:1;                             % 1-second time vector
x=zeros(1,length(t));                   % Initialize input to be zero
x(1)=1;                                 % Set up impulse function
% Generate 852-Hz tone
y852=filter([0 sin(2*pi*852/fs)],[1 -2*cos(2*pi*852/fs) 1],x);
% Generate 1209-Hz tone
y1209=filter([0 sin(2*pi*1209/fs) ],[1 -2*cos(2*pi*1209/fs) 1],x); % Filtering
y7=y852+y1209;                          % Generate DTMF tone
subplot(2,1,1);plot(t(1:400),y7(1:400));grid
ylabel('y(n) DTMF: number 7');
xlabel('time (second)')
Ak=2*abs(fft(y7))/length(y7);Ak(1)=Ak(1)/2;  % One-sided amplitude spectrum
f=[0:1:(length(y7)-1)/2]*fs/length(y7);       % Map indices to frequencies (Hz) for plot
subplot(2,1,2);plot(f,Ak(1:(length(y7)+1)/2));grid
ylabel('Spectrum for y7(n)');
xlabel('frequency (Hz)');
```

8.11.3 **Goertzel Algorithm**

In practice, the DTMF tone detector is designed using the Goertzel algorithm. This is a special and powerful algorithm used for computing discrete Fourier tansform (DFT) coefficients and signal spectra using a digital filtering method. The modified Goertzel algorithm can be used for computing signal spectra without involving complex algebra like the DFT algorithm.

Specifically, the Goertzel algorithm is a filtering method for computing the DFT coefficient $X(k)$ at the specified frequency bin k with the given N digital data $x(0), x(1), \cdots, x(N-1)$. We can begin to illustrate the Goertzel algorithm using the second-order IIR digital Goertzel filter, whose transfer function is given by

$$H_k(z) = \frac{Y_k(z)}{X(z)} = \frac{1 - W_N^k z^{-1}}{1 - 2\cos\left(\dfrac{2\pi k}{N}\right)z^{-1} + z^{-2}} \tag{8.68}$$

with the input data $x(n)$ for $n = 0, 1, \cdots, N-1$, and the last element set to be $x(N) = 0$. Notice that $W_N^k = e^{-\frac{2\pi k}{N}}$. We will process the data sequence $N+1$ times to achieve the filter output as $y_k(n)$ for $n = 0, 1, \cdots, N$, where k is the frequency index (bin number) of interest. The DFT coefficient $X(k)$ is the last datum from the Goertzel filter, that is,

$$X(k) = y_k(N) \tag{8.69}$$

The implementation of the Goertzel filter is presented by direct-form II realization in Figure 8.53.

According to the direct-form II realization, we can write the Goertzel algorithm as follows:

$$x(N) = 0 \tag{8.70}$$

For $n = 0, 1, \cdots, N$

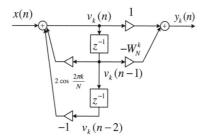

FIGURE 8.53

Second-order Goertzel IIR filter.

$$v_k(n) = 2\cos\left(\frac{2\pi k}{N}\right)v_k(n-1) - v_k(n-2) + x(n) \qquad (8.71)$$

$$y_k(n) = v_k(n) - W_N^k v_k(n-1) \qquad (8.72)$$

with initial conditions $v_k(-2) = 0$, $v_k(-1) = 0$

Then the DFT coefficient $X(k)$ is given as

$$X(k) = y_k(N) \qquad (8.73)$$

The squared magnitude $x(k)$ is computed as

$$|X(k)|^2 = v_k^2(N) + v_k^2(N-1) - 2\cos\left(\frac{2\pi k}{N}\right)v_k(N)v_k(N-1) \qquad (8.74)$$

We show the derivation of Equation (8.74) as follows. Note that Equation (8.72) involves complex algebra, since the equation contains only one complex number, a factor

$$W_N^k = e^{-j\frac{2\pi k}{N}} = \cos\left(\frac{2\pi k}{N}\right) - j\sin\left(\frac{2\pi k}{N}\right)$$

discussed in Chapter 4. If our objective is to compute the spectrum value, we can substitute $n = N$ into Equation (8.72) to obtain $X(k)$ and multiply $X(k)$ by its conjugate $X^*(k)$ to achieve the squared magnitude the DFT coefficient. It follows (Ifeachor and Jervis, 2002) that

$$|X(k)|^2 = X(k)X^*(k)$$

Since

$$X(k) = y_k(N) - W_N^k v_k(N-1)$$

$$X^*(k) = y_k(N) - W_N^{-k} v_k(N-1)$$

then

$$|X(k)|^2 = \left(y_k(N) - W_N^k y_k(N-1)\right)\left(y_k(N) - W_N^{-k} y_k(N-1)\right)$$
$$= y_k^2(N) + y_k^2(N-1) - \left(W_N^k + W_N^{-k}\right)y_k(N)y_k(N-1) \qquad (8.75)$$

Using Euler's identity yields

$$W_N^k + W_N^{-k} = e^{-j\frac{2\pi}{N}k} + e^{j\frac{2\pi}{N}k} = 2\cos\left(\frac{2\pi k}{N}\right) \qquad (8.76)$$

Substituting Equation (8.76) into Equation (8.75) leads to Equation (8.74).

We can see that the DSP equation for $v_k(k)$ and computation of the squared magnitude of the DFT coefficient $|X(k)|^2$ do not involve any complex algebra. Hence, we will use this advantage for later development. To illustrate the algorithm, let us consider Example 8.26.

EXAMPLE 8.26

Given a digital data sequence of length 4 as $x(0) = 1$, $x(1) = 2$, $x(2) = 3$, and $x(3) = 4$, use the Goertzel algorithm to compute DFT coefficient $X(1)$ and the corresponding spectral amplitude at the frequency bin $k = 1$.

Solution:
We have $k = 1$, $N = 4$, $x(0) = 1$, $x(1) = 2$, $x(2) = 3$, and $x(3) = 4$. Note that

$$2\cos\left(\frac{2\pi}{4}\right) = 0 \quad \text{and} \quad W_4^1 = e^{-j\frac{2\pi \times 1}{4}} = \cos\left(\frac{\pi}{2}\right) - j\sin\left(\frac{\pi}{2}\right) = -j$$

We first write the simplified difference equations:

$$x(4) = 0$$

For $n = 0, 1, \cdots, 4$

$$v_1(n) = -v_1(n-2) + x(n)$$

$$y_1(n) = v_1(n) + jv_1(n-1)$$

Then

$$X(1) = y_1(4)$$

$$|X(1)|^2 = v_1^2(4) + v_1^2(3)$$

The digital filter process is demonstrated in the following:

$$v_1(0) = -v_1(-2) + x(0) = 0 + 1 = 1$$

$$y_1(0) = v_1(0) + jv_1(-1) = 1 + j \times 0 = 1$$

$$v_1(1) = -v_1(-1) + x(1) = 0 + 2 = 2$$

$$y_1(1) = v_1(1) + jv_1(0) = 2 + j \times 1 = 2 + j$$

$$v_1(2) = -v_1(0) + x(2) = -1 + 3 = 2$$

$$y_1(2) = v_1(2) + jv_1(1) = 2 + j \times 2 = 2 + j2$$

$$v_1(3) = -v_1(1) + x(3) = -2 + 4 = 2$$

$$y_1(3) = v_1(3) + jv_1(2) = 2 + j \times 2 = 2 + j2$$

$$v_1(4) = -v_1(2) + x(4) = -2 + 0 = -2$$

$$y_1(4) = v_1(4) + jv_1(3) = -2 + j \times 2 = -2 + j2$$

Then the DFT coefficient and its squared magnitude are determined as

$$X(1) = y_1(4) = -2 + j2$$

$$|X(1)|^2 = v_1^2(4) + v_1^2(3) = (-2)^2 + (2)^2 = 8$$

Thus, the two-sided amplitude spectrum is computed as

$$A_1 = \frac{1}{4}\sqrt{\left(|X(1)|^2\right)} = 0.7071$$

and the corresponding single-sided amplitude spectrum is $A_1 = 2 \times 0.707 = 1.4141$.

From this simple illustrative example, we see that the Goertzel algorithm has the following advantages:

1. We can apply the algorithm for computing the DFT coefficient $X(k)$ for a specified frequency bin k; unlike the fast Fourier transform (FFT) algorithm, all the DFT coefficients are computed once it is applied.
2. If we want to compute the spectrum at frequency bin k, that is, $|X(k)|$, Equation (8.71) shows that we need to process $v_k(n)N + 1$ times and then compute $|X(k)|^2$. The operations avoid complex algebra.

If we use the modified Goertzel filter in Figure 8.54, then the corresponding transfer function is given by

$$G_k(z) = \frac{V_k(z)}{X(z)} = \frac{1}{1 - 2\cos\left(\dfrac{2\pi k}{N}\right)z^{-1} + z^{-2}} \tag{8.77}$$

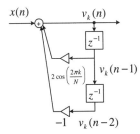

FIGURE 8.54

Modified second-order Goertzel IIR filter.

The modified Goertzel algorithm becomes the following:

$$x(N) = 0$$

For $n = 0, 1, \cdots, N$

$$v_k(n) = 2\cos\left(\frac{2\pi k}{N}\right)v_k(n-1) - v_k(n-2) + x(n)$$

with initial conditions $v_k(-2) = 0$ and $v_k(-1) = 0$
    Then the squared magnitude of the DFT coefficient is given by

$$|X(k)|^2 = v_k^2(N) + v_k^2(N-1) - 2\cos\left(\frac{2\pi k}{N}\right)v_k(N)v_k(N-1)$$

EXAMPLE 8.27

Given a digital data sequence of length 4 as $x(0) = 1$, $x(1) = 2$, $x(2) = 3$, and $x(3) = 4$, use the Goertzel algorithm to compute the spectral amplitude at the frequency bin $k = 0$.

Solution:

$$k = 0, \ N = 4, \ x(0) = 1, \ x(1) = 2, \ x(2) = 3, \quad \text{and} \quad x(3) = 4$$

Using the modified Goertzel algorithm and noting that $2 \cdot \cos\left(\frac{2\pi}{4} \times 0\right) = 2$, we obtain the simplified difference equations as follows:

$$x(4) = 0$$

For $n = 0, 1, \cdots, 4$

$$v_0(n) = 2v_0(n-1) - v_0(n-2) + x(n)$$

Then

$$|X(0)|^2 = v_0^2(4) + v_0^2(3) - 2v_0(4)v_0(3)$$

The digital filtering is performed as

$$v_0(0) = 2v_0(-1) - v_0(-2) + x(0) = 0 + 0 + 1 = 1$$

$$v_0(1) = 2v_0(0) - v_0(-1) + x(1) = 2 \times 1 + 0 + 2 = 4$$

$$v_0(2) = 2v_0(1) - v_0(0) + x(2) = 2 \times 4 - 1 + 3 = 10$$

$$v_0(3) = 2v_0(2) - v_0(1) + x(3) = 2 \times 10 - 4 + 4 = 20$$

$$v_0(4) = 2v_0(3) - v_0(2) + x(4) = 2 \times 20 - 10 + 0 = 30$$

Then the squared magnitude is determined by

$$|X(0)|^2 = v_0^2\left(4\right) + v_0^2\left(3\right) - 2v_0\left(4\right)v_0\left(3\right) = (30)^2 + (20)^2 - 2 \times 30 \times 20 = 100$$

Thus, the amplitude spectrum is computed as

$$A_0 = \frac{1}{4}\sqrt{\left(|X(0)|^2\right)} = 2.5$$

A MATLAB function for the Geortzel algorithm is shown in Program 8.20.

Program 8.20. MATLAB function for Geortzel Algorithm.

```
function [ Xk, Ak] = galg(x,k)
% Geortzel Algorithm
% [ Xk, Ak] = galg(x,k)
% x=input vetcor; k=frequency index
% Xk= kth DFT coeficient; Ak=magnitude of the kth DFT coefficient
N=length(x); x=[x 0];
vk=zeros(1,N+3);
for n=1:N+1
  vk(n+2)=2*cos(2*pi*k/N)*vk(n+1)-vk(n)+x(n);
end
Xk=vk(N+3)-exp(-2*pi*j*k/N)*vk(N+2);
Ak=vk(N+3)*vk(N+3)+vk(N+2)*vk(N+2)-2*cos(2*pi*k/N)*vk(N+3)*vk(N+2);
Ak=sqrt(Ak)/N;
end
```

EXAMPLE 8.28

Use Program 8.20 to verify the results in Examples 8.26 and 8.27.

Solution:

a. For Example 8.26, we obtain
>> x=[1 2 3 4]

```
x= 1 2 3 4
>> [X1, A1]=galg(x,1)
X1 = -2.0000 + 2.0000i
A1 =0.7071
```

b. For Example 8.27, we obtain

```
>> x=[1 2 3 4]
x= 1 2 3 4
>> [X0, A0]=galg(x,1)
X0 = 10
A0 = 2.5000
```

8.11.4 Dual-Tone Multifrequency Tone Detection Using the Modified Goertzel Algorithm

Based on the specified frequencies of each DTMF tone shown in Figure 8.48 and the modified Goertzel algorithm, we can develop the following design principles for DTMF tone detection.

1. When the digitized DTMF tone $x(n)$ is received, it has two nonzero frequency components from the following seven: 697, 770, 852, 941, 1,209, 1,336, and 1,477 Hz.
2. We can apply the modified Goertzel algorithm to compute seven spectral values, which correspond to the seven frequencies in (1). The single-sided amplitude spectrum is computed as

$$A_k = \frac{2}{N}\sqrt{|X(k)|^2} \tag{8.78}$$

3. Since the modified Goertzel algorithm is used, there is no complex algebra involved. Ideally, there are two nonzero spectral components. We will use these two nonzero spectral components to determine which key is pressed.
4. The frequency bin number (frequency index) can be determined based on the sampling rate f_s, and the data size of N via the following relation:

$$k = \frac{f}{f_s} \times N \text{ (round off to an integer)} \tag{8.79}$$

Given the key frequency specification in Table 8.12, we can determine the frequency bin k for each DTMF frequency with $f_s = 8,000$ Hz and $N = 205$.
The DTMF detector block diagram is shown in Figure 8.55.

5. The threshold value can be the sum of all seven spectral values divided by a factor of 4. Note that there are only two nonzero spectral values, hence the threshold value should ideally be half of the individual nonzero spectral value. If the spectrum value is larger than the threshold value, then the logic operation outputs logic 1; otherwise, it outputs logic 0. Finally, the logic operation at the last stage is to decode the key information based on the 7-bit binary pattern.

Table 8.12 DTMF Frequencies and Their Frequency Bins

DTMF Frequency (Hz)	Frequency Bin: $k = \dfrac{f}{f_s} \times N$
697	18
770	20
852	22
941	24
1209	31
1336	34
1477	38

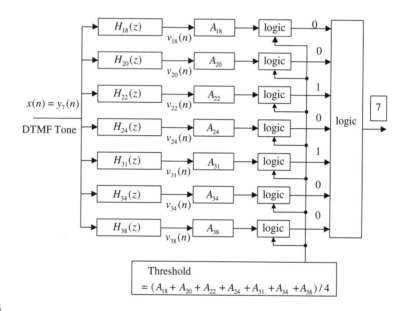

FIGURE 8.55

DTMF detector using the Goertzel algorithm.

EXAMPLE 8.29

Given a DSP system with $f_s = 8{,}000$ Hz and data size $N = 205$, seven Goertzel IIR filters are implemented for DTMF tone detection. Determine the following for the frequencies corresponding to key 7.

a. Frequency bin numbers
b. The Goertzel filter transfer functions and DSP equations
c. Equations for calculating amplitude spectral values

Solution:

For key 7, we have $f_L = 852$ Hz and $f_H = 1,209$ Hz.

a. Using Equation (8.79), we get

$$k_L = \frac{852}{8,000} \times 205 \approx 22 \quad \text{and} \quad k_H = \frac{1,209}{8,000} \times 205 \approx 31$$

b. Since $2\cos\left(\dfrac{2\pi \times 22}{205}\right) = 1.5623$, and $2\cos\left(\dfrac{2\pi \times 31}{205}\right) = 1.1631$, it follows that

$$H_{22}(z) = \frac{1}{1 - 1.5623z^{-1} + z^{-2}}$$

and

$$H_{31}(z) = \frac{1}{1 - 1.1631z^{-1} + z^{-2}}$$

The DSP equations are therefore given by

$$v_{22}(n) = 1.5623v_{22}(n-1) - v_{22}(n-2) + x(n) \text{ with } x(205) = 0, \text{ for } n = 0, 1, \cdots, 205$$

$$v_{31}(n) = 1.1631v_{31}(n-1) - v_{31}(n-2) + x(n) \text{ with } x(205) = 0, \text{ for } n = 0, 1, \cdots, 205$$

c. The amplitude spectral values are determined by

$$|X(22)|^2 = (v_{22}(205))^2 + (v_{22}(204))^2 - 1.5623(v_{22}(205)) \times (v_{22}(204))$$

$$A_{22} = \frac{2\sqrt{|X(22)|^2}}{205}$$

and

$$|X(31)|^2 = (v_{31}(205))^2 + (v_{31}(204))^2 - 1.1631(v_{31}(205)) \times (v_{31}(204))$$

$$A_{31} = \frac{2\sqrt{|X(31)|^2}}{205}$$

The MATLAB simulation for decoding key 7 is shown in Program 8.21. Figure 8.56(a) shows the frequency responses of the second-order Goertzel bandpass filters. The input is generated as shown in Figure 8.52. After filtering, the calculated spectral values and threshold value for decoding key 7 are displayed in Figure 8.56(b), where only two spectral values corresponding to the frequencies of 770 Hz and 1,209 Hz are above the threshold, and are encoded as logic 1. According to the key information in the Figure 8.55, the final logic operation decodes the key as 7.

The principle can easily be extended to transmit the ASCII (American Standard Code for Information Interchange) code or other types of code using the parallel Goertzel filter bank. If the calculated spectral value is larger than the threshold value, then the logic operation outputs logic 1; otherwise, it outputs logic 0. Finally, the logic operation at the last stage decodes the key information based on the 7-bit binary pattern.

Program 8.21. DTMF detection using the Goertzel algorithm.

```
close all;clear all;
% DTMF tone generator
N=205;
fs=8000; t=[0:1:N-1]/fs;           % Sampling rate and time vector
x=zeros(1,length(t));x(1)=1;       % Generate the impulse function
%Generation of tones
y697=filter([0 sin(2*pi*697/fs)],[1 -2*cos(2*pi*697/fs) 1],x);
y770=filter([0 sin(2*pi*770/fs)],[1 -2*cos(2*pi*770/fs) 1],x);
y852=filter([0 sin(2*pi*852/fs)],[1 -2*cos(2*pi*852/fs) 1],x);
y941=filter([0 sin(2*pi*941/fs)],[1 -2*cos(2*pi*941/fs) 1],x);
y1209=filter([0 sin(2*pi*1209/fs) ],[1 -2*cos(2*pi*1209/fs) 1],x);
y1336=filter([0 sin(2*pi*1336/fs)],[1 -2*cos(2*pi*1336/fs) 1],x);
y1477=filter([0 sin(2*pi*1477/fs)],[1 -2*cos(2*pi*1477/fs) 1],x);
key=input('input of the following keys: 1,2,3,4,5,6,7,8,9,*,0,# =>','s');
yDTMF=[];
if key=='1' yDTMF=y697+y1209; end
if key=='2' yDTMF=y697+y1336; end
if key=='3' yDTMF=y697+y1477; end
if key=='4' yDTMF=y770+y1209; end
```

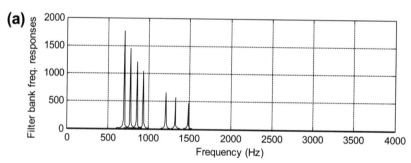

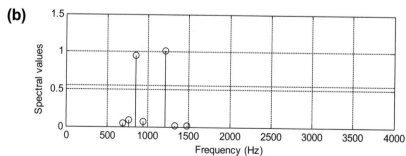

FIGURE 8.56

(a) Goertzel filter bank frequency responses; (b) display of spectral values and threshold for key 7.

```
if key=='5' yDTMF=y770+y1336; end
if key=='6' yDTMF=y770+y1477; end
if key=='7' yDTMF=y852+y1209; end
if key=='8' yDTMF=y852+y1336; end
if key=='9' yDTMF=y852+y1477; end
if key=='*' yDTMF=y941+y1209; end
if key=='0' yDTMF=y941+y1336; end
if key=='#' yDTMF=y941+y1477; end
if size(yDTMF)==0 disp('Invalid input key'); return; end
yDTMF=[yDTMF 0];     % DTMF signal appended with a zero
% DTMF detector (use Goertzel algorithm)
a697=[1 -2*cos(2*pi*18/N) 1];
a770=[1 -2*cos(2*pi*20/N) 1];
a852=[1 -2*cos(2*pi*22/N) 1];
a941=[1 -2*cos(2*pi*24/N) 1];
a1209=[1 -2*cos(2*pi*31/N) 1];
a1336=[1 -2*cos(2*pi*34/N) 1];
a1477=[1 -2*cos(2*pi*38/N) 1];
% Filter bank frequency responses
[w1, f]=freqz(1,a697,512,fs);
[w2, f]=freqz(1,a770,512,fs);
[w3, f]=freqz(1,a852,512,fs);
[w4, f]=freqz(1,a941,512,fs);
[w5, f]=freqz(1,a1209,512,fs);
[w6, f]=freqz(1,a1336,512,fs);
[w7, f]=freqz(1,a1477,512,fs);
subplot(2,1,1);plot(f,abs(w1),f,abs(w2),f,abs(w3), ...
f,abs(w4),f,abs(w5),f,abs(w6),f,abs(w7));grid
xlabel('Frequency (Hz)'); ylabel('(a) Filter bank freq. responses');
% Filter bank bandpass filtering
y697=filter(1,a697,yDTMF);
y770=filter(1,a770,yDTMF);
y852=filter(1,a852,yDTMF);
y941=filter(1,a941,yDTMF);
y1209=filter(1,a1209,yDTMF);
y1336=filter(1,a1336,yDTMF);
y1477=filter(1,a1477,yDTMF);
% Determine the absolute magnitude of DFT coefficents
m(1)=sqrt(y697(206)^2+y697(205)^2- ...
 2*cos(2*pi*18/205)*y697(206)*y697(205));
m(2)=sqrt(y770(206)^2+y770(205)^2- ...
 2*cos(2*pi*20/205)*y770(206)*y770(205));
m(3)=sqrt(y852(206)^2+y852(205)^2- ...
 2*cos(2*pi*22/205)*y852(206)*y852(205));
m(4)=sqrt(y941(206)^2+y941(205)^2- ...
 2*cos(2*pi*24/205)*y941(206)*y941(205));
m(5)=sqrt(y1209(206)^2+y1209(205)^2- ...
 2*cos(2*pi*31/205)*y1209(206)*y1209(205));
```

```
m(6)=sqrt(y1336(206)^2+y1336(205)^2- ...
 2*cos(2*pi*34/205)*y1336(206)*y1336(205));
m(7)=sqrt(y1477(206)^2+y1477(205)^2- ...
 2*cos(2*pi*38/205)*y1477(206)*y1477(205));
% Convert the magnitude of DFT coefficients to the single-side spectrum
m=2*m/205;
% Determine the threshold
th=sum(m)/4;
% Plot the DTMF spectrum with the threshold
f=[ 697 770 852 941 1209 1336 1477];
f1=[0 fs/2];
th=[ th th];
subplot(2,1,2);stem(f,m);grid;hold; plot(f1,th);
xlabel('Frequency (Hz)'); ylabel(' (b) Spectral values');
m=round(m); % Round to the binary pattern
if m== [ 1 0 0 0 1 0 0] disp('Detected Key 1'); end
if m== [ 1 0 0 0 0 1 0] disp('Detected Key 2'); end
if m== [ 1 0 0 0 0 0 1] disp('Detected Key 3'); end
if m== [ 0 1 0 0 1 0 0] disp('Detected Key 4'); end
if m== [ 0 1 0 0 0 1 0] disp('Detected Key 5'); end
if m== [ 0 1 0 0 0 0 1] disp('Detected Key 6'); end
if m== [ 0 0 1 0 1 0 0] disp('Detected Key 7'); end
if m== [ 0 0 1 0 0 1 0] disp('Detected Key 8'); end
if m== [ 0 0 1 0 0 0 1] disp('Detected Key 9'); end
if m== [ 0 0 0 1 1 0 0] disp('Detected Key *'); end
if m== [ 0 0 0 1 0 1 0] disp('Detected Key 0'); end
if m== [ 0 0 0 1 0 0 1] disp('Detected Key #'); end
```

8.12 SUMMARY OF INFINITE IMPULSE RESPONSE (IIR) DESIGN PROCEDURES AND SELECTION OF THE IIR FILTER DESIGN METHODS IN PRACTICE

In this section, we first summarize the design procedures of the BLT design, impulse-invariant design, and pole-zero placement design methods, and then discuss the selection of the particular filter for typical applications.

The BLT design method:

1. Given the digital filter frequency specifications, prewarp each digital frequency edge to the analog frequency edge using Equations (8.18) and (8.19).
2. Determine the prototype filter order using Equation (8.29) for the Butterworth filter or Equation (8.35b) for the Chebyshev filter, and perform lowpass prototype transformation using the lowpass prototype in Table 8.3 (Butterworth function) or Tables 8.4 and 8.5 (Chebyshev function) using Equations (8.20) to (8.23).
3. Apply the BLT to the analog filter using Equation (8.24) and output the transfer function.
4. Verify the frequency responses, and output the difference equation.

The impulse-invariant design method:

1. Given the lowpass or bandpass filter frequency specifications, perform analog filter design. For the highpass or bandstop filter design, quit this method and use the BLT.
   a. Determine the prototype filter order using Equation (8.29) for the Butterworth filter or Equation (8.35b) for the Chebyshev filter.
   b. Perform lowpass prototype transformation using the lowpass prototype in Table 8.3 (Butterworth function) or Tables 8.4 and 8.5 (Chebyshev functions) using Equations (8.20) to (8.23).
   c. Skip step 1 if the analog filter transfer function is given to begin with.
2. Determine the impulse response by applying the partial fraction expansion technique to the analog transfer function and inverse Laplace transform using Equation (8.37).
3. Sample the analog impulse response using Equation (8.38) and apply the z-transform to the digital impulse function to obtain the digital filter transfer function.
4. Verify the frequency response, and output the difference equation. If the frequency specifications are not net, quit the design method and use the BLT.

The pole-zero placement method:

1. Given the filter cutoff frequency specifications, determine the pole-zero locations using the corresponding equations:
   a. Second-order bandpass filter: Equations (8.41) and (8.42).
   b. Second-order notch filter: Equations (8.45) and (8.46).
   c. First-order lowpass filter: Equations (8.49) or (8.50).
   d. First-order highpass filter: Equations (8.53) or (8.54).
2. Apply the corresponding equation and scale factor to obtain the digital filter transfer function:
   a. Second-order bandpass filter: Equations (8.43) and (8.44).
   b. Second-order notch filter: Equations (8.47) and (8.48).
   c. First-order lowpass filter: Equations (8.51) and (8.52).
   d. First-order highpass filter: Equations (8.55) and (8.56).
3. Verify the frequency response, and output the difference equation. If the frequency specifications are not net, quit the design method and use BLT.

Table 8.13 compares the design parameters of the three design methods.

Performance comparisons using three methods are given in Figure 8.57, where the bandpass filter is designed using the following specifications:

Passband ripple $= -3$ dB
Center frequency $= 400$ Hz
Bandwidth $= 200$ Hz
Sampling rate $= 2,000$ Hz
Butterworth IIR filter $=$ second-order

As we expected, the BLT method satisfies the design requirement, and the pole-zero placement method has little performance degradation because $r = 1 - (f_0/f_s)\pi = 0.6858 < 0.9$, and this effect will also cause the center frequency to be shifted. For the bandpass filter designed using the impulse-invariant method, the gain at the center frequency is scaled to 1 for a frequency response shape

Table 8.13 Comparisons of Three IIR Design Methods

	Design Method		
	BLT	**Impulse Invariant**	**Pole-Zero Placement**
Filter type	Lowpass, highpass, bandpass, bandstop	Appropriate for lowpass and bandpass	Second-order for bandpass and band stop; first-order for lowpass and highpass
Linear phase	No	No	No
Ripple and stopband specifications	Used for determining the filter order	Used for determining the filter order	Not required; 3 dB on passband offered
Special requirement	None	Very high sampling relative to the cutoff frequency (LPF) or to upper cutoff frequency (BPF)	Narrow band for BPF or notch filter; lower cutoff frequency or higher cutoff frequency for LPF or HPF.
Algorithm complexity	High: Frequency prewarping, analog filter design, BLT	Moderate: Analog filter design determining digital impulse response; apply z-transform	Low: Design equations
Minimum design tool	Calculator, algebra	Calculator, algebra	Calculator

BLT = blinear transformation; LPF = lowpass filter; BPF = bandpass filter; HPF = highpass filter.

comparison. The performance of the impulse-invariant method is satisfactory in the passband. However, it has significant performance degradation in the stopband when compared with the other two methods. This is due to aliasing when sampling the analog impulse response in time domain.

Improvement in using the pole-zero placement and impulse-invariant methods can be achieved by using a very high sampling rate. Example 8.30 describes the possible selection of the design method by a DSP engineer to solve a real-world problem.

EXAMPLE 8.30

Determine an appropriate IIR filter design method for each of the following DSP applications. As described in a previous section, we apply a notch filter to remove the 60-Hz interference and cascade a bandpass filter to remove noise in an ECG signal for heart rate detection. The following specifications are required:

Notch filter:
  Harmonics to be removed = 60 Hz
  3dB bandwidth for the notch filter = 4 Hz
Bandpass filter:
  Passband frequency range = 0.25 to 40 Hz
  Passband ripple = 0.5 dB
  Sampling rate = 600 Hz

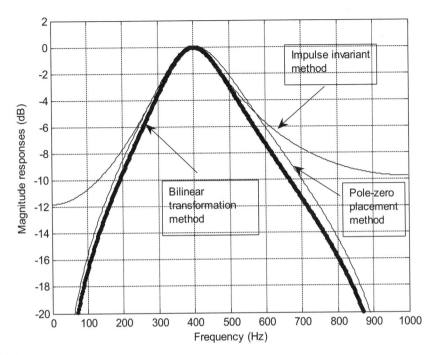

FIGURE 8.57

Performance comparisons for the BLT, pole-zero placement, and impulse-invariant methods.

The pole-zero placement method is the best choice, since the notch filter to be designed has a very narrow 3 dB bandwidth of 4 Hz. This simple design gives a quick solution. Since the bandpass filter requires a passband ripple of 0.5 dB from 0.25 to 40 Hz, the BLT can also be an appropriate choice. Even though the impulse-invariant method could work for this case, since the sampling rate of 600 Hz is much larger than 40 Hz, aliasing cannot be prevented completely. Hence, the BLT is a preferred design method for the bandpass filter.

8.13 SUMMARY

1. The BLT method is able to transform the transfer function of an anolog filter to the transfer function of the corresponding digital filter in general.
2. The BLT maps the left half of an s-plane to the inside unit circle of the z-plane. Stability of mapping is guaranteed.
3. The BLT causes analog frequency warping. The analog frequency range from 0 Hz to infinite is warped to a digital frequency range from 0 Hz to the folding frequency.
4. Given the digital frequency specifications, analog filter frequency specifications must be developed using the frequency warping equation before designing the corresponding analog filter and applying the BLT.

5. An analog filter transfer function can be obtained by using a lowpass prototype, which can be selected from the Butterworth and Chebyshev functions.

6. Higher-order IIR filters can be designed using a cascade form.

7. The impulse-invariant design method maps the analog impulse response to the digital equivalent impulse response. The method works for lowpass and bandpass filter design with a very high sampling rate. It is not appropriate for the highpass and bandstop filter design.

8. The pole-zero placement method can be applied for simple IIR filter designs such as second-order bandpass and bandstop filters with narrow band specifications, and first-order lowpass and highpass filters with the cutoff frequency close to either DC or the folding frequency.

9. Quantization of IIR filter coefficients has more effect on the magnitude frequency response than on the phase frequency response. It may cause the quantized IIR filter to be unstable.

10. A simple audio equalizer uses bandpass IIR filter banks to create sound effects.

11. The 60-Hz interference eliminator is designed to enhance biomedical ECG signals for heart rate detection. It can also be adapted for audio humming noise elimination.

12. A single tone or a DTMF tone can be generated using the IIR filter with the impulse sequence as the filter input.

13. The Goertzel algorithm is applied for DTMF tone detection. This is an important application in the telecommunication industry.

14. The procedures for the BLT, impulse-invariant, and pole-zero placement design methods were summarized, and their design feasibilities were compared, including the filter type, linear phase, ripple and stopband specifications, special requirements, algorithm comlexity, and design tool(s).

8.14 PROBLEMS

8.1. Given an analog filter with the transfer function

$$H(s) = \frac{1,000}{s + 1,000}$$

convert it to the digital filter transfer function and difference equation using the BLT if the DSP system has a sampling period of $T = 0.001$ second.

8.2. The lowpass filter with a cutoff frequency of 1 rad/sec is given as

$$H_P(s) = \frac{1}{s + 1}$$

a. Use $H_P(s)$ and the BLT to obtain a corresponding IIR digital lowpass filter with a cutoff frequency of 30 Hz, assuming a sampling rate of 200 Hz.

b. Use MATLAB to plot the magnitude and phase frequency responses of $H(z)$.

8.3. The normalized lowpass filter with a cutoff frequency of 1 rad/sec is given as

$$H_P(s) = \frac{1}{s + 1}$$

a. Use $H_P(s)$ and the BLT to obtain a corresponding IIR digital highpass filter with a cutoff frequency of 30 Hz, assuming a sampling rate of 200 Hz.

b. Use MATLAB to plot the magnitude and phase frequency responses of $H(z)$.

8.4. Consider the normalized lowpass filter with a cutoff frequency of 1 rad/sec:

$$H_P(s) = \frac{1}{s+1}$$

a. Use $H_P(s)$ and the BLT to design a corresponding IIR digital notch (bandstop) filter with a lower cutoff frequency of 20 Hz, an upper cutoff frequency of 40 Hz, and a sampling rate of 120 Hz.

b. Use MATLAB to plot the magnitude and phase frequency responses of $H(z)$.

8.5. Consider the following normalized lowpass filter with a cutoff frequency of 1 rad/sec:

$$H_P(s) = \frac{1}{s+1}$$

a. Use $H_P(s)$ and the BLT to design a corresponding IIR digital bandpass filter with a lower cutoff frequency of 15 Hz, an upper cutoff frequency of 25 Hz, and a sampling rate of 120 Hz.

b. Use MATLAB to plot the magnitude and phase frequency responses of $H(z)$.

8.6. Design a first-order digital lowpass Butterworth filter with a cutoff frequency of 1.5 kHz and a passband ripple of 3 dB at a sampling frequency of 8,000 Hz.

a. Determine the transfer function and difference equation.

b. Use MATLAB to plot the magnitude and phase frequency responses.

8.7. Design a second-order digital lowpass Butterworth filter with a cutoff frequency of 1.5 kHz and a passband ripple of 3 dB at a sampling frequency of 8,000 Hz.

a. Determine the transfer function and difference equation.

b. Use MATLAB to plot the magnitude and phase frequency responses.

8.8. Design a third-order digital highpass Butterworth filter with a cutoff frequency of 2 kHz and a passband ripple of 3dB at a sampling frequency of 8,000 Hz.

a. Determine the transfer function and difference equation.

b. Use MATLAB to plot the magnitude and phase frequency responses.

8.9. Design a second-order digital bandpass Butterworth filter with a lower cutoff frequency of 1.9 kHz, an upper cutoff frequency 2.1 kHz, and a passband ripple of 3dB at a sampling frequency of 8,000 Hz.

a. Determine the transfer function and difference equation.

b. Use MATLAB to plot the magnitude and phase frequency responses.

8.10. Design a second-order digital bandstop Butterworth filter with a center frequency of 1.8 kHz, a bandwidth of 200 Hz, and a passband ripple of 3dB at a sampling frequency of 8,000 Hz.

a. Determine the transfer function and difference equation.

b. Use MATLAB to plot the magnitude and phase frequency responses.

8.11. Design a first-order digital lowpass Chebyshev filter with a cutoff frequency of 1.5 kHz and a passband ripple of 1 dB at a sampling frequency of 8,000 Hz.

a. Determine the transfer function and difference equation.

b. Use MATLAB to plot the magnitude and phase frequency responses.

8.12. Design a second-order digital lowpass Chebyshev filter with a cutoff frequency of 1.5 kHz and a passband ripple of 0.5 dB at a sampling frequency of 8,000 Hz. Use MATLAB to plot the magnitude and phase frequency responses.

8.13. Design a third-order digital highpass Chebyshev filter with a cutoff frequency of 2 kHz and a passband ripple fo 1 dB at a sampling frequency of 8,000 Hz.

a. Determine the transfer function and difference equation.

b. Use MATLAB to plot the magnitude and phase frequency responses.

8.14. Design a second-order digital bandpass Chebyshev filter with the following specifications:

Center frequency of 1.5 kHz

Bandwidth of 200 Hz

0.5 dB passband ripple

Sampling frequency of 8,000 Hz

a. Determine the transfer function and difference equation.

b. Use MATLAB to plot the magnitude and phase frequency responses.

8.15. Design a second-order bandstop digital Chebyshev filter with the following specifications:

Center frequency of 2.5 kHz

Bandwidth of 200 Hz

1 dB stopband ripple

Sampling frequency of 8,000 Hz.

a. Determine the transfer function and difference equation.

b. Use MATLAB to plot the magnitude and phase frequency responses.

8.16. Design a fourth-order low pass digital Butterworth filter with a cutoff frequency of 2 kHz, and a passband ripple of 3 dB at a sampling frequency at 8,000 Hz.

a. Determine transfer function and difference equation;

b. Use MATLAB to plot the magnitude and phase frequency responses.

8.17. Design a fourth-order digital lowpass Chebyshev filter with a cutoff frequency of 1.5 kHz and a 0.5 dB passband ripple at a sampling frequency of 8,000 Hz.

    a. Determine the transfer function and difference equation.

    b. Use MATLAB to plot the magnitude and phase frequency responses.

8.18. Design a fourth-order digital bandpass Chebyshev filter with a center frequency of 1.5 kHz, a bandwidth of 200 Hz, and a 0.5 dB passband ripple at a sampling frequency of 8,000 Hz.

    a. Determine the transfer function and difference equation.

    b. Use MATLAB to plot the magnitude and phase frequency responses.

8.19. Consider the following Laplace transfer function:

$$H(s) = \frac{10}{s + 10}$$

    a. Determine $H(z)$ and the difference equation using the impulse-invariant method if the sampling rate $f_s = 10$ Hz.

    b. Use MATLAB to plot the magnitude frequency response $|H(f)|$ and the phase frequency response $\phi(f)$ with respect to $H(s)$ for the frequency range from 0 to $f_s/2$ Hz.

    c. Use MATLAB to plot the magnitude frequency response $|H(e^{j\Omega})| = |H(e^{j2\pi fT})|$ and the phase frequency response $\phi(f)$ with respect to $H(z)$ for the frequency range from 0 to $f_s/2$ Hz.

8.20. Consider the following Laplace transfer function:

$$H(s) = \frac{1}{s^2 + 3s + 2}$$

    a. Determine $H(z)$ and the difference equation using the impulse-invariant method if the sampling rate $f_s = 10$ Hz.

    b. Use MATLAB to plot the magnitude frequency response $|H(f)|$ and the phase frequency response $\phi(f)$ with respect to $H(s)$ for the frequency range from 0 to $f_s/2$ Hz.

    c. Use MATLAB to plot the magnitude frequency response $|H(e^{j\Omega})| = |H(e^{j2\pi fT})|$ and the phase frequency response $\varphi(f)$ with respect to $H(z)$ for the frequency range from 0 to $f_s/2$ Hz.

8.21. Consider the following Laplace transfer function:

$$H(s) = \frac{s}{s^2 + 4s + 5}$$

    a. Determine $H(z)$ and the difference equation using the impulse-invariant method if the sampling rate $f_s = 10$ Hz.

b. Use MATLAB to plot the magnitude frequency response $|H(f)|$ and the phase frequency response $\phi(f)$ with respect to $H(s)$ for the frequency range from 0 to $f_s/2$ Hz;

c. Use MATLAB to plot the magnitude frequency response $|H(e^{j\Omega})| = |H(e^{j2\pi fT})|$ and the phase frequency response $\phi(f)$ with respect to $H(z)$ for the frequency range from 0 to $f_s/2$ Hz.

8.22. A second-order bandpass filter is required to satisfy the following specifications:

Sampling rate = 8,000 Hz

3 dB bandwidth: $BW = 100$ Hz

Narrow passband centered at $f_0 = 2,000$ Hz

Zero gain at 0 Hz and 4,000 Hz

Find the transfer function and difference equation by the pole-zero placement method.

8.23. A second-order notch filter is required to satisfy the following specifications:

Sampling rate = 8,000 Hz

3 dB bandwidth: $BW = 200$ Hz

Narrow passband centered at $f_0 = 1,000$ Hz.

Find the transfer function and difference equation by the pole-zero placement method.

8.24. A first-order lowpass filter is required to satisfy the following specifications:

Sampling rate = 8,000 Hz

3 dB cutoff frequency: $f_c = 200$ Hz

Zero gain at 4,000 Hz

Find the transfer function and difference equation using the pole-zero placement method.

8.25. A first-order lowpass filter is required to satisfy the following specifications:

Sampling rate = 8,000 Hz

3 dB cutoff frequency: $f_c = 3,800$ Hz

Zero gain at 4,000 Hz

Find the transfer function and difference equation by the pole-zero placement method.

8.26. A first-order highpass filter is required to satisfy the following specifications:

Sampling rate = 8,000 Hz

3 dB cutoff frequency: $f_c = 3,850$ Hz

Zero gain at 0 Hz

Find the transfer function and difference equation by the pole-zero placement method.

8.27. A first-order highpass filter is required to satisfy the following specifications:

Sampling rate = 8,000 Hz

3 dB cutoff frequency: f_c = 100 Hz

Zero gain at 0 Hz

Find the transfer function and difference equation by the pole-zero placement method.

8.28. Given a filter transfer function

$$H(z) = \frac{0.3430z^2 + 0.6859z + 0.3430}{z^2 + 0.7075z + 0.7313}$$

a. realize the digital filter using direct-form I and using direct-form II;

b. determine the difference equations for each implementation.

8.29. Given a fourth-order filter transfer function,

$$H(z) = \frac{0.3430z^2 + 0.6859z + 0.3430}{z^2 + 0.7075z + 0.7313} \times \frac{0.4371z^2 + 0.8742z + 0.4371}{z^2 - 0.1316z + 0.1733}$$

a. realize the digital filter using the cascade (series) form via second-order sections using direct-form II;

b. determine the difference equations for implementation.

8.30. Given a DSP system with a sampling rate of 1,000 Hz, develop a 200 Hz single tone generator using the digital IIR filter by completing the following steps:

a. Determine the digital IIR filter transfer function.

b. Determine the DSP equation (difference equation).

8.31. Given a DSP system with a sampling rate of 8,000 Hz, develop a 250 Hz single tone generator using the digital IIR filter by completing the following steps:

a. Determine the digital IIR filter transfer function.

b. Determine the DSP equation (difference equation).

8.32. Given a DSP system with a sampling rate of 8,000 Hz, develop a DTMF tone generator for key 9 using the digital IIR filters by completing the following steps:

a. Determine the digital IIR filter transfer functions.

b. Determine the DSP equations (difference equation).

8.33. Given a DSP system with a sampling rate 8,000 Hz, develop a DTMF tone generator for key 3 using the digital IIR filters by completing the following steps:

a. Determine the digital IIR filter transfer functions.

b. Determine the DSP equations (difference equation).

8.34. Given $x(0) = 1$, $x(1) = 2$, $x(2) = 0$, $x(3) = -1$, use the Goertzel algorithm to compute the following DFT coefficients and their amplitude spectrum:

  a. $X(0)$

  b. $|X(0)|^2$

  c. A_0 (single side)

  d. $X(1)$

  e. $|X(1)|^2$

  f. A_1 (single side)

8.35. Repeat Problem 8.34 for $X(2)$ and $X(3)$.

8.36. Given a digital data sequence of length 4 as $x(0) = 4$, $x(1) = 3$, $x(2) = 2$, and $x(3) = 1$, use the modified Goertzel algorithm to compute the spectral amplitude at the frequency bin $k = 0$ and $k = 2$.

8.37. Repeat Problem 8.36 for $X(1)$ and $X(3)$.

Use MATLAB to solve Problems 8.38 to 8.50.

8.38. A speech sampled at 8,000 Hz is corrupted by a sine wave of 360 Hz. Design a notch filter to remove the noise with the following specifications:

Chebyshev notch filter

Center frequency: 360 Hz

Bandwidth: 60 Hz

Passband ripple: 0.5 dB

Stopband attenuation: 5 dB at 355 Hz and 365 Hz, respectively.

Determine the transfer function and difference equation.

8.39. In Problem 8.38, if the speech is corrupted by a sine wave of 360 Hz and its third harmonic, cascading two notch filters can be applied to remove noise signals. The possible specifications are given as

Chebyshev notch filter 1

Center frequency: 360 Hz

Bandwidth: 60 Hz

Passband ripple: 0.5 dB

Stopband attenuation: 5 dB at 355 Hz and 365 Hz, respectively

Chebyshev notch filter 2

Center frequency: 1,080 Hz

Bandwidth: 60 Hz

Passband ripple: 0.5 dB

Stopband attenuation: 5 dB at 1,075 Hz 1,085 Hz, respectively

Determine the transfer function and difference equation for each filter (Fig. 8.58).

8.40. In a speech recording system with a sampling frequency of 10,000 Hz, the speech is corrupted by random noise. To remove the random noise while preserving speech information, the following specifications are given:

Speech frequency range: 0–3,000 Hz

Stopband range: 4,000–5,000 Hz

Passband ripple: 3 dB

Stopband attenuation 25 dB

Butterworth IIR filter

Determine the filter order and transfer function.

8.41. In Problem 8.40, assume we instead use a Chebyshev IIR filter with the following specifications:

Speech frequency range: 0–3,000 Hz

Stopband range: 4,000–5,000 Hz

Passband ripple: 1 dB

Stopband attenuation: 35 dB

Chebyshev IIR filter

Determine the filter order and transfer function.

8.42. Consider a speech equalizer to compensate for midrange frequency loss of hearing (Figure 8.59).

The equalizer has the following specifications:

Sampling rate: 8,000 Hz

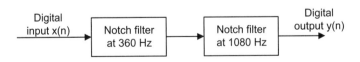

FIGURE 8.58

Cascaded notch filter in Problem 8.39.

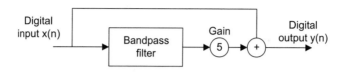

FIGURE 8.59

Speech equalizer in Problem 8.42.

Second-order bandpass IIR filter

Frequency range to be emphasized: 1,500–2,000 Hz

Passband ripple: 3 dB

Pole-zero placement design method

Determine the transfer function.

8.43. In Problem 8.42, assume we use an IIR filter with following specifications:

Sampling rate: 8,000 Hz

Butterworth IIR filter

Frequency range to be emphasized: 1,500–2,000 Hz

Lower stopband: 0–1,000 Hz

Upper stopband: 2,500–4,000 Hz

Passband ripple: 3 dB

Stopband attenuation: 20 dB

Determine the filter order and filter transfer function.

8.44. A digital crossover can be designed as shown in Figure 8.60.

Assume the following audio specifications:

Sampling rate: 44,100 Hz

Crossover frequency: 1,000 Hz

Highpass filter: third-order Butterworth type at a cutoff frequency of 1,000 Hz

Lowpass filter: third-order Butterworth type at a cutoff frequency of 1,000 Hz

Use the MATLAB BLT design method to determine

a. the transfer functions and difference equations for the highpass and lowpass filters;

b. frequency responses for the highpass filter and the lowpass filter;

c. combined frequency response for both filters.

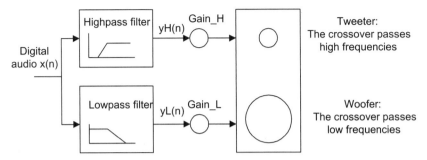

FIGURE 8.60

Two-band digital crossover system in Problem 8.44.

8.45. Given a DSP system with a sampling rate of 8,000 Hz, develop an 800 Hz single-tone generator using a digital IIR filter by completing the following steps:

a. Determine the digital IIR filter transfer function.

b. Determine the DSP equation (difference equation).

c. Write a MATLAB program using the MATLAB function **filter()** to generate and plot the 800-Hz tone for a duration of 0.01 second.

8.46. Given a DSP system with a sampling rate of 8,000 Hz, develop a DTMF tone generator for key "5" using digital IIR filters by completing the following steps:

a. Determine the digital IIR filter transfer functions.

b. Determine the DSP equations (difference equation).

c. Write a MATLAB program using the MATLAB function **filter()** to generate and plot the DTMF tone for key 5 for 205 samples.

8.47. Given $x(0) = 1$, $x(1) = 1$, $x(2) = 0$, $x(3) = -1$, use the Goertzel algorithm to compute the following DFT coefficients and their amplitude spectra:

a. $X(0)$

b. $|X(0)|^2$

c. A_0 (single sided)

d. $X(1)$

e. $|X(1)|^2$

f. A_1 (single sided)

8.48. Repeat Problem 8.47 for spectra: A_2, and A_3.

8.49. Given a DSP system with a sampling rate of 8,000 Hz and data size of 205 ($N = 205$), seven Goertzel IIR filters are implemented for DTMF tone detection. For the frequencies corresponding to key 5, determine

a. the modified Goertzel filter transfer functions;

b. the filter DSP equations for $v_k(n)$; and

c. the DSP equations for the squared magnitudes

$$|X(k)|^2 = |y_k(205)|^2$$

d. Using the data generated in Problem 8.46 (c), write a program using the MATLAB function **filter()** and Goertzel algorithm to detect the spectral values of the DTMF tone for key 5.

8.50. Given an input data sequence

$$x(n) = 1.2 \cdot \sin(2\pi(1,000)n/10,000)) - 1.5 \cdot \cos(2\pi(4,000)n/10,000)$$

assuming a sampling frequency of 10 kHz, implement the designed IIR filter in Problem 8.41 to filter 500 data points of $x(n)$ with the following specified method, and plot the 500 samples of the input and output data.

a. Direct-form I implementation

b. Direct-form II implementation

8.14.1 MATLAB Projects

8.51. The 60-Hz hum eliminator with harmonics and heart rate detection:

Given the recorded ECG data (ecgbn.dat) that is corrupted by 60-Hz interference with its harmonics and assuming a sampling rate of 600 Hz, plot the signal's spectrum and determine the harmonics. With the harmonic frequency information, design a notch filter to ehance the ECG signal. Then use the designed notch filter to process the given ECG signal and apply the zero-cross algorithm to determine the heart rate.

8.52. Digital speech and audio equalizer:

Design a seven-band audio equalizer using fourth-order bandpass filters. The center frequencies are listed below:

Center frequency (Hz)	160	320	640	1,280	2,560	5,120	10,240
Bandwidth (Hz)	80	160	320	640	1,280	2,560	5,120

In this project, use the designed equalizer to process stereo audio ("No9seg.wav"). Plot the magnitude response for each filter bank. Listen to and evaluate the processed audio with the following gain settings:

a. each filter bank gain = 0 (no equalization)

b. lowpass filtered

c. bandpass filtered

d. highpass filtered

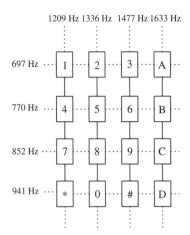

FIGURE 8.61

DTMF key frequencies.

8.53. DTMF tone generation and detection:

Implement the DTMF tone generation and detection according to Section 8.11 with the following specifications:

a. Input keys: 1, 2, 3, 4, 5, 6, 7, 8, 9, *, 0, #, A, B, C, D (key frequencies are given in Figure 8.61).

b. Sampling frequency is 8,000 Hz.

c. Program will respond to each input key with its DTMF tone and display the detected key.

Hardware and Software for Digital Signal Processors

CHAPTER OUTLINE

9.1 Digital Signal Processor Architecture 406
9.2 Digital Signal Processor Hardware Units 408
    9.2.1 Multiplier and Accumulator 408
    9.2.2 Shifters 409
    9.2.3 Address Generators 409
9.3 Digital Signal Processors and Manufacturers 411
9.4 Fixed-Point and Floating-Point Formats 411
    9.4.1 Fixed-Point Format 412
    9.4.2 Floating-Point Format 419
        Overflow... 422
        Underflow 423
    9.4.3 IEEE Floating-Point Formats 423
        Single Precision Format... 423
        Double Precision Format 425
    9.4.5 Fixed-Point Digital Signal Processors 426
    9.4.6 Floating-Point Processors 427
9.5 Finite Impulse Response and Infinite Impulse Response Filter Implementations in Fixed-Point
Systems 429
9.6 Digital Signal Processing Programming Examples 434
    9.6.1 Overview of TMS320C67x DSK 434
    9.6.2 Concept of Real-Time Processing...438
    9.6.3 Linear Buffering 440
        Finite Impulse Response Filtering 441
        Infinite Impulse Response Filtering 442
        Digital Oscillation with Infinite Impulse Response Filtering 442
    9.6.4 Sample C Programs 445
        Floating-Point Implementation Example 445
        Fixed-Point Implementation Example 445
9.7 Summary 448

OBJECTIVES:

This chapter introduces the basics of digital signal processors such as processor architectures and hardware units, investigates fixed-point and floating-point formats, and illustrates the implementation of digital filters in real time.

9.1 DIGITAL SIGNAL PROCESSOR ARCHITECTURE

Unlike microprocessors and microcontrollers, digital signal (DS) processors have special features that require operations such as fast Fourier transform (FFT), filtering, convolution and correlation, and real-time sample-based and block-based processing. Therefore, DS processors use a different dedicated hardware architecture.

We first compare the architecture of the general microprocessor with that of the DS processors. The design of general microprocessors and microcontrollers is based on the *Von Neumann architecture,* which was developed from a research paper written by John Von Neumann and others in 1946. Von Neumann suggested that computer instructions, as we shall discuss, be numerical codes instead of special wiring. Figure 9.1 shows the Von Neumann architecture.

As shown in Figure 9.1, a Von Neumann processor contains a single, shared memory for programs and data, a single bus for memory access, an arithmetic unit, and a program control unit. The processor proceeds in a serial fashion in terms of fetching and execution cycles. This means that the central processing unit (CPU) fetches an instruction from memory and decodes it to figure out what operation to do, then executes the instruction. The instruction (in machine code) has two parts: the *opcode* and the *operand.* The opcode specifies what the operation is, that is, tells the CPU what to do. The operand informs the CPU what data to operate on. These instructions will modify memory, or input and output (I/O). After an instruction is completed, the cycles will resume for the next instruction. One instruction or piece of data can be retrieved at a time. Since the processor proceeds in a serial fashion, it causes most units to stay in a wait state.

As noted, the Von Neumann architecture operates the cycles of fetching and execution by fetching an instruction from memory, decoding it via the program control unit, and finally executing instruction. When execution requires data movement—that is, data to be read from or written to memory—the next instruction will be fetched after the current instruction is completed. The Von Neumann-based processor has this bottleneck mainly due to the use of a single, shared memory for both program instructions and data. Increasing the speed of the bus, memory, and computational units can improve speed, but not significantly.

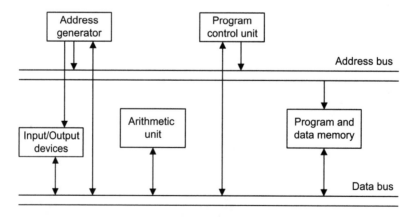

FIGURE 9.1

General microprocessor based on Von Neumann architecture.

To accelerate the execution speed of digital signal processing, DS processors are designed based on the *Harvard architecture*, which originated from the Mark 1 relay-based computers built by IBM in 1944 at Harvard University. This computer stored its instructions on punched tape and data using relay latches. Figure 9.2 shows today's Harvard architecture. As depicted, the DS processor has two separate memory spaces. One is dedicated for the program code, while the other is employed for data. Hence, to accommodate two memory spaces, two corresponding address buses and two data buses are used. In this way, the program memory and data memory have their own connections to the program memory bus and data memory bus, respectively. This means that the Harvard processor can fetch the program instruction and data in parallel at the same time, the former via the program memory bus and the latter via the data memory bus. There is an additional unit called a *multiplier and accumulator* (MAC), which is the dedicated hardware used for the digital filtering operation. The last additional unit, the shift unit, is used for the scaling operation for fixed-point implementation when the processor performs digital filtering.

Let us compare the executions of the two architectures. The Von Neumann architecture generally has the execution cycles described in Figure 9.3. The fetch cycle obtains the opcode from the memory, and the control unit will decode the instruction to determine the operation. Next is the execute cycle. Based the decoded information, execution will modify the content of the register or the memory. Once this is completed, the process will fetch the next instruction and continue. The processor operates one instruction at a time in a serial fashion.

To improve the speed of the processor operation, the Harvard architecture takes advantage of a common DS processor, in which one register holds the filter coefficient while the other register holds the data to be processed, as depicted in Figure 9.4.

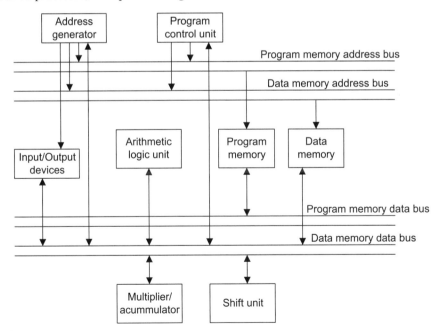

FIGURE 9.2

Digital signal processors based on the Harvard architecture.

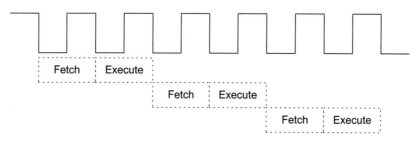

FIGURE 9.3

Execution cycle based on the Von Neumann architecture.

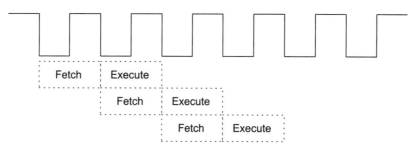

FIGURE 9.4

Execution cycle based on Harvard architecture.

As shown in Figure 9.4, the execute and fetch cycles are overlapped. We call this the *pipelining* operation. The DS processor performs one execution cycle while also fetching the next instruction to be executed. Hence, the processing speed is dramatically increased.

The Harvard architecture is preferred for all DS processors due to the requirements of most digital signal processing (DSP) algorithms, such as filtering, convolution, and FFT, which need repetitive arithmetic operations, including multiplications, additions, memory access, and heavy data flow through the CPU.

For the other applications, such as those dependent on simple microcontrollers with less of a timing requirement, the Von Neumann architecture may be a better choice, since it offers much less silica area and is thus less expensive.

9.2 DIGITAL SIGNAL PROCESSOR HARDWARE UNITS

In this section, we will briefly discuss special DS processor hardware units.

9.2.1 Multiplier and Accumulator

As compared with the general microprocessors based on the Von Newmann architecture, the DS processor uses the MAC, a special hardware unit for enhancing the speed of digital filtering. This is

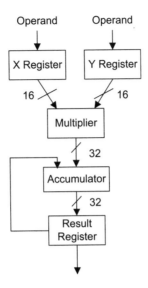

FIGURE 9.5

The multiplier and accumulator (MAC) dedicated to DSP.

dedicated hardware, and the corresponding instruction is generally referred to as a MAC operation. The basic structure of the MAC is shown in Figure 9.5.

As shown in Figure 9.5, in a typical hardware MAC, the multiplier has a pair of input registers, each holding the 16-bit input to the multiplier. The result of the multiplication is accumulated in the 32-bit accumulator unit. The result register holds the double precision data from the accumulator.

9.2.2 Shifters

In digital filtering, to prevent overflow, a scaling operation is required. A simple scaling-down operation shifts data to the right, while a scaling-up operation shifts data to the left. Shifting data to the right is the same as dividing the data by 2 and truncating the fraction part; shifting data to the left is equivalent to multiplying the data by 2. As an example for a 3-bit data word $011_2 = 3_{10}$, shifting 011 to the right gives $001_2 = 1$, that is, $3/2=1.5$, and truncating 1.5 results in 1. Shifting the same number to the left, we have $110_2 = 6_{10}$, that is, $3 \times 2 = 6$. The DS processor often shifts data by several bits for each data word. To speed up such operation, the special hardware shift unit is designed to accommodate the scaling operation, as depicted in Figure 9.2.

9.2.3 Address Generators

The DS processor generates the addresses for each datum on the data buffer to be processed. A special hardware unit for circular buffering is used (see the address generator in Figure 9.2). Figure 9.6 describes the basic mechanism of circular buffering for a buffer having eight data samples.

In circular buffering, a pointer is used and always points to the newest data sample, as shown in Figure 9.6. After the next sample is obtained from analog-to-digital conversion (ADC), the data will be

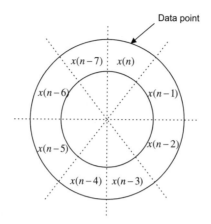

FIGURE 9.6

Illustration of circular buffering.

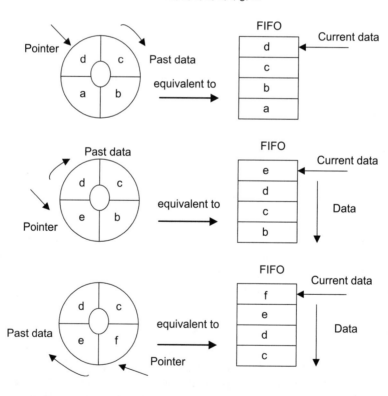

FIGURE 9.7

Circular buffer and equivalent FIFO.

placed at the location of $x(n - 7)$ and the oldest sample is pushed out. Thus, the location for $x(n - 7)$ becomes the location for the current sample. The original location for $x(n)$ becomes a location for the past sample of $x(n - 1)$. The process continues according to the mechanism just described. For each new data sample, only one location on the circular buffer needs to be updated.

The circular buffer acts like a first-in/first-out (FIFO) buffer, but each datum on the buffer does not have to be moved. Figure 9.7 gives a simple illustration of the 2-bit circular buffer. In the figure, there is data flow to the ADC ($a, b, c, d, e, f, g, \ldots$) and a circular buffer initially containing $a, b, c,$ and d. The pointer specifies the current data of d, and the equivalent FIFO buffer is shown on the right side with the current data of d at the top of the memory. When e comes in, as shown in the middle drawing in Figure 9.7, the circular buffer will change the pointer to the next position and update old a with a new datum e. It costs the pointer only one movement to update one datum in one step. However, on the right side, the FIFO has to move each of the other data down to let in the new datum e at the top. For this FIFO, it takes four data movements. In the bottom drawing in Figure 9.7, the incoming datum f for both the circular buffer and the FIFO buffer continues to confirm our observations.

Like finite impulse response (FIR) filtering, the data buffer size can reach several hundreds. Hence, using the circular buffer will significantly enhance the processing speed.

9.3 DIGITAL SIGNAL PROCESSORS AND MANUFACTURERS

DS processors are classified for general DSP and special DSP. The general-DSP processor is designed and optimized for applications such as digital filtering, correlation, convolution, and FFT. In addition to these applications, the special-DSP processor has features that are optimized for unique applications such as audio processing, compression, echo cancellation, and adaptive filtering. Here, we will focus on the general-DSP processor.

The major manufacturers in the DSP industry are Texas Instruments (TI), Analog Devices, and Motorola. TI and Analog Devices offer both fixed-point DSP families and floating-point DSP families, while Motorola offers fixed-point DSP families. We will concentrate on TI families, review their architectures, and study real-time implementation using the fixed- and floating-point formats.

9.4 FIXED-POINT AND FLOATING-POINT FORMATS

In order to process real-world data, we need to select an appropriate DS processor, as well as a DSP algorithm or algorithms for a certain application. Whether a DS processor uses a fixed- or floating-point method depends on how the processor's CPU performs arithmetic. A fixed-point DS processor represents data in *2's complement integer format* and manipulates data using integer arithmetic, while a floating-point processor represents number using a mantissa (fractional part) and an exponent in addition to the integer format and operates data using floating-point arithmetic (discussed in Section 9.4.2).

Since the fixed-point DS processor operates using the integer format, which represents only a very narrow dynamic range of the integer number, a problem such as overflow of data manipulation may occur. Hence, we need to spend much more coding effort to deal with such a problem. As we shall see, we may use floating-point DS processors, which offer a wider dynamic range of data, so that coding becomes much easier. However, the floating-point DS processor contains more hardware units to handle the integer arithmetic and the floating-point arithmetic; hence it is more expensive and slower than fixed-point processors in terms of instruction cycles. It is usually a choice for prototyping or proof-of-concept development.

When it is time to make the DSP an application-specific integrated circuit (ASIC), a chip designed for a particular application, a dedicated hand-coded fixed-point implementation is likely the best choice in terms of performance and small silica area.

The formats used by DSP implementation can be classified as fixed or floating point.

9.4.1 Fixed-Point Format

We begin with 2's complement representation. Considering a 3-bit 2's complement, we can represent all the decimal numbers shown in Table 9.1.

Table 9.1 A 3-Bit 2's Complement Number Representation

Decimal Number	Two's Complement
3	011
2	010
1	001
0	000
−1	111
−2	110
−3	101
−4	100

Let us review the 2's complement number system using Table 9.1. Converting a decimal number to its 2's complement form requires following steps:

1. Convert the magnitude in the decimal to its binary number using the required number of bits.
2. If the decimal number is positive, its binary number is its 2's complement representation; if the decimal number is negative, perform the 2's complement operation, where we negate the binary number by changing the logic 1s to logic 0s and logic 0s to logic 1s and then add a logic 1 to the data. For example, a decimal number of 3 is converted to its 3-bit 2's complement representation as 011; however, for converting a decimal number of −3, we first get a 3-bit binary number for the magnitude in decimal, that is, 011. Next, negating the binary number 011 yields the binary number 100. Finally, adding a binary logic 1 achieves the 3-bit 2's complement representation of −3, that is, $100 + 1 = 101$, as shown in Table 9.1.

As we see, a 3-bit 2's complement number system has a dynamic range from −4 to 3, which is very narrow. Since the basic DSP operations include multiplications and additions, results of operation can cause overflow problems. Let us examine multiplication in Example 9.1.

EXAMPLE 9.1

Given

a. $2 \times (-1)$
b. $2 \times (-3)$

operate each expression using 2's complement.

Solution:

a.
```
      0 1 0
    × 0 0 1
      0 1 0
      0 0 0
    + 0 0 0
    0 0 0 1 0
```

The 2's complement of 00010 = 11110. Removing two extended sign bits 1 gives 110. The answer is 110 (-2), which is within the system.

b.
```
      0 1 0
    × 0 1 1
      0 1 0
      0 1 0
    + 1 0 0
    0 0 1 1 0
```

The 2's complement of 00110 = 11010. Removing two extended sign bits leaves 010. Since the binary number 010 is 2, which is not (-6) as what we expect, overflow occurs; that is, the result of the multiplication (-6) is out of our dynamic range $(-4$ to 3$)$.

Let us design a system treating all the decimal values as fractional numbers, so that we obtain the fractional binary 2's complement system shown in Table 9.2.

To become familiar with the fractional binary 2's complement system, let us convert a positive fraction number $\frac{3}{4}$ and a negative fraction number $-\frac{1}{4}$ in decimals to their 2's complements. Since

$$\frac{3}{4} = 0 \times 2^0 + 1 \times 2^{-1} + 1 \times 2^{-2}$$

Table 9.2 A 3-Bit 2's Complement System Using Fractional Representation

Decimal Number	Decimal Fraction	Two's Complement
3	3/4	0.11
2	2/4	0.10
1	1/4	0.01
0	0	0.00
−1	−1/4	1.11
−2	−2/4	1.10
−3	−3/4	1.01
−4	−4/4 = −1	1.00

its 2's complement is 011. Note that we did not mark the binary point for clarity. Again, since

$$\frac{1}{4} = 0 \times 2^0 + 0 \times 2^{-1} + 1 \times 2^{-2}$$

its positive-number 2's complement is 001. For the negative number, applying the 2's complement to the binary number 001 leads to $110 + 1 = 111$, as we see in Table 9.2. By adding the binary points, we obtain 0.01 and 1.11, respectively.

Now let us focus on the fractional binary 2's complement system. The data are normalized to the fractional range from -1 to $1 - 2^{-2} = \frac{3}{4}$. When we carry out multiplications with two fractions, the result should be a fraction, so that multiplication overflow can be prevented. Let us verify the multiplication $(0.10) \times (1.01)$, which is the overflow case in Example 9.1. We first multiply two positive numbers:

$$
\begin{array}{r}
0.1\ 0 \\
\times\ 0.1\ 1 \\
\hline
0\ 1\ 0 \\
0\ 1\ 0 \\
+\ 1\ 0\ 0 \\
\hline
0.0\ 1\ 1\ 0
\end{array}
$$

The 2's complement of $0.0110 = 1.1010$.

The answer in decimal form should be

$$1.1010 = (-1) \times (0.0110)_2 = -(0 \times (2)^{-1} + 1 \times (2)^{-2} + 1 \times (2)^{-3} + 0 \times (2)^{-4}) = -\frac{3}{8}$$

This number is correct, as we can verify from Table 9.2, that is, $\left(\frac{2}{4} \times \left(-\frac{3}{4}\right)\right) = -\frac{3}{8}$.

If we truncate the last two least significant bits to keep the 3-bit binary number, we have an approximate answer:

$$1.10 = (-1) \times (0.01)_2 = -\left(0 \times (2)^{-1} + 1 \times (2)^{-2}\right) = -\frac{1}{2}$$

Truncation error occurs. The error should be bounded by $2^{-2} = \frac{1}{4}$. We can verify that

$$|-1/2 - (-3/8)| = 1/8 < 1/4$$

With such a scheme, we can avoid overflow due to multiplication but cannot prevent overflow due to addition. Consider the addition example

$$
\begin{array}{r}
0.1\ 1 \\
+\ 0.0\ 1 \\
\hline
1.0\ 0
\end{array}
$$

where the result 1.00 is a negative number. Adding two positive fractional numbers yields a negative number. Hence, overflow occurs.

We see that this signed fractional number scheme partially solves the overflow in multiplications. This fractional number format is called the signed Q-2 format, where there are 2 magnitude bits plus one sign bit. The overflow from addition will be tackled using a scaling method discussed in a later section.

Q-format number representation is the most common one used in fixed-point DSP implementation. It is defined in Figure 9.8.

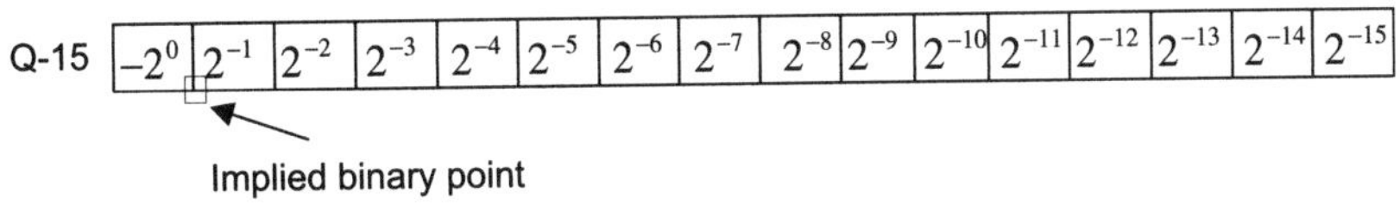

Implied binary point

FIGURE 9.8

Q-15 (fixed-point) format.

As indicated in Figure 9.8, Q-15 means that the data are in a sign magnitude form in which there are 15 bits for magnitude and one bit for sign. Note that after the sign bit, the dot shown in Figure 9.8 implies the binary point. The number is normalized to the fractional range from -1 to 1. The range is divided in to 2^{16} intervals, each with a size of 2^{-15}. The most negative number is -1, while the most positive number is $1 - 2^{-15}$. Any result from multiplication is within the fractional range of -1 to 1. Let us study the following examples to become familiar with Q-format number representation.

EXAMPLE 9.2

Find the signed Q-15 representation for the decimal number 0.560123.

Table 9.3 Conversion Process of Q-15 Representation

Number	Product	Carry
0.560123 × 2	1.120246	1 (MSB)
0.120246 × 2	0.240492	0
0.240492 × 2	0.480984	0
0.480984 × 2	0.961968	0
0.961968 × 2	1.923936	1
0.923936 × 2	1.847872	1
0.847872 × 2	1.695744	1
0.695744 × 2	1.391488	1
0.391488 × 2	0.782976	0
0.782976 × 2	1.565952	1
0.565952 × 2	1.131904	1
0.131904 × 2	0.263808	0
0.263808 × 2	0.527616	0
0.527616 × 2	1.055232	1
0.055232 × 2	0.110464	0 (LSB)

MSB, most significant bit; LSB, least-significant bit.

Solution:

The conversion process is illustrated using Table 9.3. For a positive fractional number, we multiply the number by 2 if the product is larger than 1, carry bit 1 as a most significant bit (MSB), and copy the fractional part to the next line for the next multiplication by 2; if the product is less than 1, we carry bit 0 to MSB. The procedure continues to collect all 15 magnitude bits.

We yield the Q-15 format representation as

$$0.100011110110010$$

Since we only use 16 bits to represent the number, we may lose accuracy after conversion. Like quantization, truncation error is introduced. However, this error should be less than the interval size, in this case, $2^{-15} = 0.0000305017$. We shall verify this in Example 9.5. An alternative method of conversion is to convert a fraction, let's say $\frac{3}{4}$ to Q-2 format, multiply it by 2^2, and then convert the truncated integer to its binary, that is,

$$(3/4) \times 2^2 = 3 = 011_2$$

In this way, it follows that

$$(0.560123) \times 2^{15} = 18,354$$

Converting 18,354 to its binary representation will achieve the same answer. The next example illustrates the signed Q-15 representation for a negative number.

EXAMPLE 9.3

Find the signed Q-15 representation for the decimal number −0.160123.

Solution:

Converting the Q-15 format for the corresponding positive number with the same magnitude using the procedure described in Example 9.2, we have

$$0.160123 = 0.001010001111110$$

Then after applying 2's complement, the Q-15 format becomes

$$-0.160123 = 1.110101110000010$$

Alternative method: Since $(-0.160123) \times 2^{15} = -5,246.9$, converting the truncated number −5,246 to its 16-bit 2's complement yields 1110101110000010.

EXAMPLE 9.4

Convert the Q-15 signed number 1.110101110000010 to the decimal number.

Solution:

Since the number is negative, applying the 2's complement yields
$$0.001010001111110$$

Then the decimal number is

$$-(2^{-3} + 2^{-5} + 2^{-9} + 2^{-10} + 2^{-11} + 2^{-12} + 2^{-13} + 2^{-14}) = -0.160095$$

EXAMPLE 9.5

Convert the Q-15 signed number 0.100011110110010 to the decimal number.

Solution:
The decimal number is

$$2^{-1} + 2^{-5} + 2^{-6} + 2^{-7} + 2^{-8} + 2^{-10} + 2^{-11} + 2^{-14} = 0.560120$$

As we know, the truncation error in Example 9.2 is less than $2^{-15} = 0.000030517$. We verify that the truncation error is bounded by

$$|0.560120 - 0.560123| = 0.000003 < 0.000030517$$

Note that the larger the number of bits used, the smaller the truncation error that may accompany it.
Examples 9.6 and 9.7 are devoted to illustrating data manipulations in the Q-15 format.

EXAMPLE 9.6

Add the two numbers in Examples 9.4 and 9.5 in Q-15 format.

Solution:
Binary addition is carried out as follows:

$$
\begin{array}{r}
1.110101110000010 \\
+\ 0.100011110110010 \\
\hline
10.011001100110100
\end{array}
$$

Then the result is

$$0.011001100110100$$

This number in decimal form is

$$2^{-2} + 2^{-3} + 2^{-6} + 2^{-7} + 2^{-10} + 2^{-11} + 2^{-13} = 0.400024$$

EXAMPLE 9.7

This is a simple illustration of fixed-point multiplication.
Determine the fixed-point multiplication of 0.25 and 0.5 in Q-3 fixed-point 2's complement format.

Solution:
Since $0.25 = 0.010$ and $0.5 = 0.100$, we carry out binary multiplication as follows:

$$
\begin{array}{r}
0.010 \\
\times\ 0.100 \\
\hline
0000 \\
0000 \\
0010 \\
+\ 0000 \\
\hline
0.001000
\end{array}
$$

Truncating the least significant bits to convert the result to Q-3 format, we have

$$0.010 \times 0.100 = 0.001$$

Note that $0.001 = 2^{-3} = 0.125$. We can also verify that $0.25 \times 0.5 = 0.125$.

The Q-format number representation is a better choice than the 2's complement integer representation, it can prevent multiplication overflow. But we need to be concerned with the following problems.

1. When converting a decimal number to its Q-N format, where N denotes the number of magnitude bits, we may lose accuracy due to the truncation error, which is bounded by the size of the interval, that is, 2^{-N}.

2. Addition and subtraction may cause overflow, where adding two positive numbers leads to a negative number, or adding two negative number yields a positive number; similarly, subtracting a positive number from a negative number gives a positive number, while subtracting a negative number from a positive number results in a negative number.

3. Multiplying two numbers in Q-15 format will lead to a Q-30 format, which has 31 bits in total. As in Example 9.7, the multiplication of Q-3 yields a Q-6 format, that is, 6 magnitude bits and a sign bit. In practice, it is common for a DS processor to hold the multiplication result using a double word size such as MAC operation, as shown in Figure 9.9 for multiplying two numbers in Q-15 format. In Q-30 format, there is one sign-extended bit. We may get rid of it by shifting left by one bit to obtain Q-31 format and maintaining the Q-31 format for each MAC operation.

   Sometimes, the number in Q-31 format needs to be converted to Q-15; for example, the 32-bit data in the accumulator needs to be sent for 16-bit digital-to-analog conversion (DAC), where the upper most-significant 16 bits in the Q-30 format must be used to maintain accuracy. We can shift the number in Q-30 to the right by 15 bits or shift the Q-31 number to the right by 16 bits. The useful result is stored in the lower 16-bit memory location. Note that after truncation, the maximum error is bounded by the interval size of 2^{-15}, which satisfies most applications. In using the Q-format in the fixed-point DS processor, it is costly to maintain the accuracy of data manipulation.

4. Underflow can happen when the result of multiplication is too small to be represented in the Q-format. As an example, in a Q-2 system shown in Table 9.2, multiplying 0.01×0.01 leads to 0.0001. To keep the result in Q-2, we truncate the last two bits of 0.0001 to achieve 0.00, which is zero. Hence, underflow occurs.

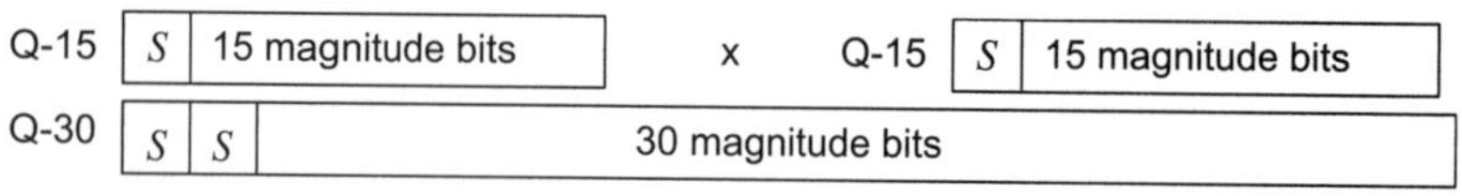

FIGURE 9.9

Sign bit extended Q-30 format.

9.4.2 Floating-Point Format

To increase the dynamic range of number representation, a floating-point format, which is similar to scientific notation, is used. The general format for floating-point number representation is given by

$$x = M \cdot 2^E \qquad (9.1)$$

where M is the mantissa, or fractional part in Q format, and E is the exponent. The mantissa and exponent are signed numbers. If we assign 12 bits for the mantissa and 4 bits for the exponent, the format looks like Figure 9.10.

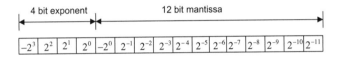

FIGURE 9.10

Floating-point format.

Since the 12-bit mantissa is limited to between -1 to $+1$, the number of bits assigned to the exponent controls the dynamic range. The bigger the number of bits designated to the exponent, the larger the dynamic range. The number of bits for the mantissa defines the interval in the normalized range; as shown in Figure 9.10, the interval size is 2^{-11} in the normalized range, which is smaller than the Q-15 format. However, when more mantissa bits are used, there will be a smaller interval size. Using the format in Figure 9.10, we can determine the most negative and most positive numbers as

$$\text{most negative number} = (1.00000000000)_2 \cdot 2^{0111_2} = (-1) \times 2^7 = -128.0$$

$$\text{most positive number} = (0.11111111111)_2 \cdot 2^{0111_2} = (1 - 2^{-11}) \times 2^7 = 127.9375$$

The smallest positive number is given by

$$\text{smallest positive number} = (0.00000000001)_2 \cdot 2^{1,000_2} = (2^{-11}) \times 2^{-8} = 2^{-19}$$

As we can see, the exponent acts like a scale factor to increase the dynamic range of the number representation. We study the floating-point format in the following example.

EXAMPLE 9.8

Convert each of the following decimal numbers to a floating-point number using the format specified in Figure 9.10.

a. 0.1601230
b. −20.430527

Solution:
a. We first scale the number 0.1601230 to $0.160123/2^{-2} = 0.640492$ with an exponent of −2 (other choices could be 0 or −1) to get $0.160123 = 0.640492 \times 2^{-2}$. Using 2's complement, we have $-2 = 1110$. Now we convert the value 0.640492 using the Q-11 format to get 010100011111. Cascading the exponent bits and the mantissa bits yields

$$1110010100011111$$

b. Since $-20.430527/2^5 = -0.638454$, we can convert it into the fractional part and exponent part as $-20.430527 = -0.638454 \times 2^5$. Note that this conversion is not particularly unique; the forms $-20.430527 = -0.319227 \times 2^6$ and $-20.430527 = -0.1596135 \times 2^7$ are still valid choices. Let us keep what we have now. Therefore, the exponent bits should be 0101.

Converting the number 0.638454 using the Q-11 format gives

$$010100011011$$

Using 2's complement, we obtain the representation for the decimal number -0.6438454 as

$$101011100101$$

Cascading the exponent bits and mantissa bits, we achieve

$$0101101011100100$$

Floating-point arithmetic is more complicated. We must obey the rules for manipulating two floating-point numbers. For arithmetic addition, with two floating point numbers given as

$$x_1 = M_1 2^{E_1}$$

$$x_2 = M_2 2^{E_2}$$

the floating-point sum is performed as follows:

$$x_1 + x_2 = \begin{cases} (M_1 + M_2 \times 2^{-(E_1-E_2)}) \times 2^{E_1}, & \text{if } E_1 \geq E_2 \\ (M_1 \times 2^{-(E_2-E_1)} + M_2) \times 2^{E_2} & \text{if } E_1 < E_2 \end{cases}$$

For multiplication, given two properly normalized floating-point numbers

$$x_1 = M_1 2^{E_1}$$

$$x_2 = M_2 2^{E_2}$$

where $0.5 \leq |.M_1|. < 1$ and $0.5 \leq |.M_2|. < 1$, the calculation can be performed as follows:

$$x_1 \times x_2 = (M_1 \times M_2) \times 2^{E_1+E_2} = M \times 2^E$$

That is, the mantissas are multiplied while the exponents are added:

$$M = M_1 \times M_2$$

$$E = E_1 + E_2$$

Examples 9.9 and 9.10 serve to illustrate manipulators.

EXAMPLE 9.9

Add the two floating point numbers obtained in Example 9.8:

$$1110\ 010100011111 = .640136718 \times 2^{-2}$$

$$0101\ 101011100101 = -0.638183593 \times 2^5$$

Solution:

Before addition, we change the first number so it has the same exponent as the second number, that is,

$$0101\ 000000001010 = 0.00500168 \times 2^5$$

Then we add the two mantissa numbers:

$$
\begin{array}{r}
0.0\,0\,0\,0\,0\,0\,0\,1\,0\,1\,0 \\
+\ 1.0\,1\,0\,1\,1\,1\,0\,0\,1\,0\,1 \\
\hline
1.0\,1\,0\,1\,1\,1\,0\,1\,1\,1\,1
\end{array}
$$

The floating number is

$$0101\ 101011101111$$

We can verify the result by the following:

$$0101\ 101011101111 = -(2^{-1} + 2^{-3} + 2^{-7} + 2^{-11}) \times 2^5 = -0.633300781 \times 2^5$$
$$= -20.265625$$

EXAMPLE 9.10

Multiply the two floating-point numbers obtained in Example 9.8:

$$1110\ 010100011111 = .640136718 \times 2^{-2}$$

$$0101\ 101011100101 = -0.638183593 \times 2^5$$

Solution:

From the results in Example 9.8, we have the bit patterns for these two numbers as

$$E_1 = 1110,\ E_2 = 0101,\ M_1 = 010100011111,\ M_2 = 101011100101$$

Adding two exponents in 2's complement form leads to

$$E = E_1 + E_2 = 1110 + 0101 = 0011$$

which is $+3$, as we expected, since in the decimal domain $(-2) + 5 = 3$. As previously shown when introducing the multiplication rule, when multiplying two mantissas, we need to apply their corresponding positive values. If the sign for the final value is negative, then we convert it to its 2's complement form. In our example, $M_1 = 010100011111$ is a positive mantissa. However, $M_2 = 101011100101$ is a negative mantissa, since the MSB is 1. To perform multiplication, we use 2's complement to convert M_2 to its positive value, 010100011011, and note that the multiplication result is negative. We multiply two positive mantissas and truncate the result to 12 bits to give

$$010100011111 \times 010100011011 = 001101010100$$

Now we need to add a negative sign to the multiplication result with the 2's complement operation. Taking the 2's complement, we have

$$M = 110010101100$$

Hence, the product is achieved by cascading the 4-bit exponent and 12-bit mantissa as

$$0011110010111100$$

Converting this number back to the decimal number, we verify the result to be

$$-0.408203125 \times 2^3 = -3.265625.$$

Next, we examine overflow and underflow in the floating-point number system.

Overflow

During an operation, overflow will occur when a number is too large to be represented in the floating-point number system. Adding two mantissa numbers may lead to a number larger than 1 or less than -1; and multiplying two numbers causes the addition of their two exponents so that the sum of the two exponents could overflow. Consider the following overflow cases.

Case 1. Add the following two floating-point numbers:

$$0111\ 011000000000 + 011101000000000$$

Note that the two exponents are the same and they are the biggest positive number in 4-bit 2's complement representation. We add two positive mantissa numbers as

$$
\begin{array}{r}
0.\,1\,1\,0\,0\,0\,0\,0\,0\,0\,0\,0 \\
+\quad 0.\,1\,0\,0\,0\,0\,0\,0\,0\,0\,0\,0 \\
\hline
1.\,0\,1\,0\,0\,0\,0\,0\,0\,0\,0\,0
\end{array}
$$

The result for adding mantissa numbers is negative. Hence the overflow occurs.

Case 2: Multiply the following two numbers:

$$0111\ 011000000000 \times 0111\ 01100000000$$

Adding the two positive exponents gives

$$0111 + 0111 = 1000 \text{ (negative, the overflow occurs)}$$

Multiplying the two mantissa numbers gives

$$0.11000000000 \times 0.1100000000 = 0.10010000000 \text{ (OK!)}$$

Underflow

As we discussed before, underflow will occur when a number is too small to be represented in the number system. Let us divide the following two floating-point numbers:

$$1001\ 001000000000 \div 0111\ 01000000000$$

First, subtracting the two exponents leads to

$$1001(\text{negative}) - 0111(\text{positive}) = 1001 + 1001 = 0010\ (\text{positive; the underflow occurs})$$

Then, dividing two mantissa numbers, it follows that

$$0.01000000000 \div 0.1000000000 = 0.10000000000\ (\text{OK!})$$

However, in this case, the expected resulting exponent is -14 in decimal, which is too small to be presented in the 4-bit 2's complement system. Hence the underflow occurs.

Now that we understand the basic principles of the floating-point formats, we can next examine two floating-point formats of the Institute of Electrical and Electronics Engineers (IEEE).

9.4.3 IEEE Floating-Point Formats

Single Precision Format

IEEE floating-point formats are widely used in many modern DS processors. There are two types of IEEE floating-point formats (IEEE 754 standard). One is the IEEE single precision format, and the other is the IEEE double precision format. The single precision format is described in Figure 9.11.

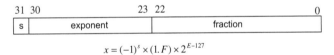

$$x = (-1)^s \times (1.F) \times 2^{E-127}$$

FIGURE 9.11

IEEE single precision floating-point format.

The format of IEEE single precision floating-point standard representation requires 23 fraction bits F, 8 exponent bits E, and 1 sign bit S, with a total of 32 bits for each word. F is the mantissa in 2's complement positive binary fraction represented from bit 0 to bit 22. The mantissa is within the normalized range limits between $+1$ and $+2$. The sign bit S is employed to indicate the sign of the number, where when $S = 1$ the number is negative, and when $S = 0$ the number is positive. The exponent E is in excess 127 form. The value of 127 is the offset from the 8-bit exponent range from 0 to 255, so that E-127 will have a range from -127 to $+128$. The formula shown in Figure 9.11 can be applied to convert the IEEE 754 standard (single precision) to the decimal number. The following simple examples also illustrate this conversion:

$$0\ 10000000\ 00000000000000000000000 = (-1)^0 \times (1.0_2) \times 2^{128-127} = 2.0$$

$$0\ 10000001\ 10100000000000000000000 = (-1)^0 \times (1.101_2) \times 2^{129-127} = 6.5$$

$$1\ 10000001\ 10100000000000000000000 = (-1)^1 \times (1.101_2) \times 2^{129-127} = -6.5$$

Let us look at Example 9.11 for more explanation.

EXAMPLE 9.11

Convert the following number in IEEE single precision format to decimal format:

$$110000000.010...0000$$

Solution:

From the bit pattern in Figure 9.11, we can identify the sign bit, exponent, and fractional as

$$s = 1,\ E = 2^7 = 128$$

$$1.F = 1.01_2 = (2)^0 + (2)^{-2} = 1.25$$

Then, applying the conversion formula leads to

$$x = (-1)^1(1.25) \times 2^{128-127} = -1.25 \times 2^1 = -2.5$$

In conclusion, the value x represented by the word can be determined based on the following rules, including all the exceptional cases:

- If $E = 255$ and F is nonzero, then $x = NaN$ ("Not a number").
- If $E = 255$, F is zero, and S is 1, then $x = -$Infinity.
- If $E = 255$, F is zero, and S is 0, then $x = +$Infinity.
- If $0 < E < 255$, then $x = (-1)^s \times (1.F) \times 2^{E-127}$, where $1.F$ represents the binary number created by prefixing F with an implicit leading 1 and a binary point.
- If $E = 0$ and F is nonzero, then $x = (-1)^s \times (0.F) \times 2^{-126}$. This is an "unnormalized" value.
- If $E = 0$, F is zero, and S is 1, then $x = -0$.
- If $E = 0$, F is zero, and S is 0, then $x = 0$.

Typical and exceptional examples are shown as follows:

$$000000000\ 00000000000000000000000 = 0$$
$$100000000\ 00000000000000000000000 = -0$$
$$011111111\ 00000000000000000000000 = \text{Infinity}$$
$$111111111\ 00000000000000000000000 = -\text{Infinity}$$
$$011111111\ 00000010000000000000000 = \text{NaN}$$
$$111111111\ 00100010001001010101010 = \text{NaN}$$
$$000000001\ 00000000000000000000000 = (-1)^0 \times (1.0_2) \times 2^{1-127} = 2^{-126}$$
$$000000000\ 10000000000000000000000 = (-1)^0 \times (0.1_2) \times 2^{0-126} = 2^{-127}$$

$$000000000\ 00000000000000000000001 =$$
$$(-1)^0 \times (0.00000000000000000000001_2) \times 2^{0-126} = 2^{-149} \text{ (smallest positive value)}$$

Double Precision Format

The IEEE double precision format is described in Figure 9.12.

The IEEE double precision floating-point standard representation requires a 64-bit word, which may be numbered from 0 to 63, left to right. The first bit is the sign bit S, the next eleven bits are the exponent bits E, and the final 52 bits are the fraction bits F. The IEEE floating-point format in double precision significantly increases the dynamic range of number representation since there are eleven exponent bits; the double-precision format also reduces the interval size in the mantissa normalized range of $+1$ to $+2$, since there are 52 mantissa bits as compare with the single precision case of 23 bits. Applying the conversion formula shown in Figure 9.12 is similar to the single precision case.

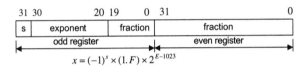

FIGURE 9.12

IEEE double precision floating-point format.

EXAMPLE 9.12

Convert the following number in IEEE double precision format to the decimal format:

$$001000...0.110...000$$

Solution:
Using the bit pattern in Figure 9.12, we have

$$s = 0,\ E = 2^9 = 512 \text{ and}$$

$$1.F = 1.11_2 = (2)^0 + (2)^{-1} + (2)^{-2} = 1.75$$

Then, applying the double precision formula yields

$$x = (-1)^0(1.75) \times 2^{512-1023} = 1.75 \times 2^{-511} = 2.6104 \times 10^{-154}$$

For the purpose of completeness, rules for determining the value x represented by the double-precision word are listed as follows:

- If $E = 2{,}047$ and F is nonzero, then $x = NaN$ ("Not a number").
- If $E = 2{,}047$, F is zero, and S is 1, then $x = -$Inifinity.
- If $E = 2{,}047$, F is zero, and S is 0, then $x = +$Inifinity.
- If $0 < E < 2{,}047$, then $x = (-1)^s \times (1.F) \times 2^{E-1{,}023}$, where "$1.F$" is intended to represent the binary number created by prefixing F with an implicit leading 1 and a binary point.

- If $E = 0$ and F is nonzero, then $x = (-1)^s \times (0.F) \times 2^{-1022}$. This is an "unnormalized" value.
- If $E = 0$, F is zero, and S is 1, then $x = -0$.
- If $E = 0$, F is zero, and S is 0, then $x = 0$.

9.4.5 Fixed-Point Digital Signal Processors

Analog Devices, Texas Instruments, and Motorola all manufacture fixed-point DS processors. Analog Devices offers a fixed-point DSP family such as the ADSP21xx. Texas Instruments provides various generations of fixed-point DS processors based on historical development, architecture features, and computational performance. Some of the most common ones are the TMS320C1x (first generation), TMS320C2x, TMS320C5x, and TMS320C62x. Motorola manufactures a variety of fixed-point processors, such as the DSP5600x family. The new families of fixed-point DS processors are expected

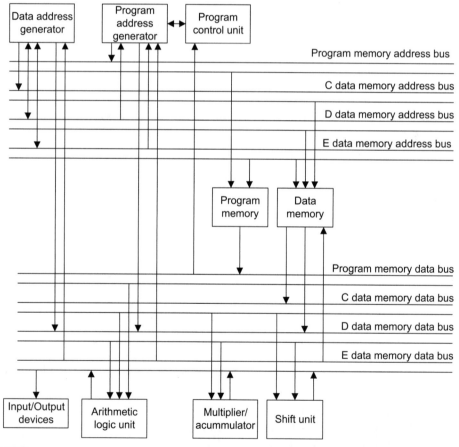

FIGURE 9.13

Basic architecture of the TMSC320C54x family.

to continue to grow. Since they share some basic common features such as program memory and data memory with associated address buses, arithmetic logic units (ALUs), program control units, MACs, shift units, and address generators, here we focus on an overview of the TMS320C54x processor. The typical TMS320C54x fixed-point DSP architecture appears in Figure 9.13.

The fixed-point TMS320C50 families supporting 16-bit data have on-chip program memory and data memory in various sizes and configurations. They include data RAM (random access memory) and program ROM (read-only memory) used for program code, instruction, and data. Four data buses and four address buses are accommodated to work with the data memory and program memory. The program memory address bus and program memory data bus are responsible for fetching program instructions. As shown in Figure 9.13, the C and D data memory address buses and the C and D data memory data buses deal with fetching data from the data memory while the E data memory address bus and E data memory data bus are dedicated to moving data into data memory. In addition, the E memory data bus can access the I/O devices.

Computational units consist of an ALU, a MAC, and a shift unit. For the TMS320C54x family, the ALU can fetch data from the C, D, and program memory data buses and access the E memory data bus. It has two independent 40-bit accumulators, which are able to operate 40-bit addition. The multiplier, which can fetch data from the C and D memory data buses and write data via the E data memory data bus, is capable of operating 17-bit × 17-bit multiplications. The 40-bit shifter has the same capability of bus access as the MAC, allowing all possible shifts for scaling and fractional arithmetic such as those we have discussed for the Q-format.

The program control unit fetches instructions via the program memory data bus. Again, in order to speed up memory access, there are two address generators available: one responsible for program addresses and one for data addresses.

Advanced Harvard architecture is employed, where several instructions operate at the same time for given a given single instruction cycle. Processing performance offers 40 MIPS (million instruction sets per second). To further explore this subject, the reader is referred to Dahnoun (2000), Embree (1995), Ifeachor and Jervis (2002), and Van der Vegte (2002), as well as the TI website (www.ti.com).

9.4.6 Floating-Point Processors

Floating-point DS processors perform DSP operations using floating-point arithmetic, as we discussed before. The advantages of using the floating-point processor include getting rid of finite word length effects such as overflows, round-off errors, truncation errors, and coefficient quantization error. Hence, in terms of coding, we do not need to scale input samples to avoid overflow, shift the accumulator result to fit the DAC word size, scale the filter coefficients, or apply Q-format arithmetic. A floating-point DS processor with high speed and calculation precision facilitates a friendly environment to develop and implement DSP algorithms.

Analog Devices provides floating-point DSP families such as ADSP210xx and TigerSHARC. Texas Instruments offers a wide range of the floating-point DSP families, in which the TMS320C3x is the first generation, followed by the TMSC320C4x and TMS320C67x families. Since the first generation of a floating-point DS processor is less complicated than later generations but still has the common basic features, we review the first-generation architecture first.

Figure 9.14 shows the typical architecture of Texas Instruments' TMS320C3x family of processors. We discuss some key features briefly. Further detail can be found in the TMS320C3x User's Guide

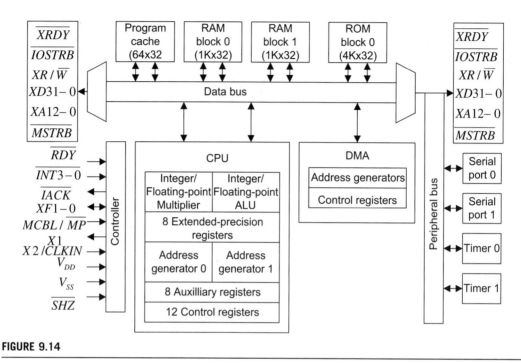

FIGURE 9.14

The typical TMS320C3x floating-point DS processor.

(Texas Instruments 1991), the TMS320C6x CPU and Instruction Set Reference Guide (Texas Instruments, 1998), and other studies (Dahnoun, 2000; Embree, 1995; Ifeachor and Jervis, 2002; Kehtaranavaz and Simsek, 2000; Sorensen and Chen, 1997; Van der Vegte, 2002). The TMS320C3x family consists of 32-bit single chip floating-point processors that support both integer and floating-point operations.

The processor has a large memory space and is equipped with dual-access on-chip memories. A program cache is employed to enhance the execution of commonly used codes. Similar to the fixed-point processor, it uses the Harvard architecture, where there are separate buses used for program and data so that instructions can be fetched at the same time that data are being accessed. There also exist memory buses and data buses for direct-memory access (DMA) for concurrent I/O and CPU operations, and peripheral access such as serial ports, I/O ports, memory expansion, and an external clock.

The C3x CPU contains the floating-point/integer multiplier; an ALU, which is capable of operating both integer and floating-point arithmetic; a 32-bit barrel shifter; internal buses; a CPU register file; and dedicated auxiliary register arithmetic units (ARAUs). The multiplier operates single-cycle multiplications on 24-bit integers and on 32-bit floating-point values. Using parallel instructions to perform a multiplication, an ALU will cost a single cycle, which means that a multiplication and an addition are equally fast. The ARAUs support addressing modes, in which some of them are specific to DSP such as circular buffering and bit-reversal addressing (digital filtering and FFT operations). The CPU register file offers 28 registers, which can be operated on by the multiplier and ALU. The special functions of the registers include eight-extended 40-bit precision registers for maintaining accuracy of the floating-point results. Eight auxiliary registers can be used for addressing and for integer arithmetic. These registers provide internal temporary storage of internal variables instead of external memory storage, to allow performance of arithmetic between registers. In this way, program efficiency is greatly increased.

The prominent feature of C3x is its floating-point capability, allowing operation of numbers with a very large dynamic range. It offers implementation of the DSP algorithm without worrying about problems such as overflows and coefficient quantization. Three floating-point formats are supported. A short 16-bit floating-point format has 4 exponent bits, 1 sign bit, and 11 mantissa bits. A 32-bit single precision format has 8 exponent bits, 1 sign bit, and 23 fraction bits. A 40-bit extended precision format contains 8 exponent bits, 1 sign bit, and 31 fraction bits. Although the formats are slightly different from the IEEE 754 standard, conversions are available between these formats.

The TMS320C30 offers high-speed performance with 60-nanosecond single-cycle instruction execution time, which is equivalent to 16.7 MIPS. For speech quality applications with an 8 kHz sampling rate, it can handle over 2,000 single-cycle instructions between two samples (125 microseconds). With instruction enhancements such as pipelines executing each instruction in a single cycle (four cycles required from fetch to execution by the instruction itself) and a multiple interrupt structure, this high-speed processor validates implementation of real-time applications in floating-point arithmetic.

9.5 FINITE IMPULSE RESPONSE AND INFINITE IMPULSE RESPONSE FILTER IMPLEMENTATIONS IN FIXED-POINT SYSTEMS

With knowledge of the IEEE format and of filter realization structures such as direct-form I, direct-form II, and parallel and cascade forms (Chapter 6), we can study digital filter implementation in the fixed-point processor. In the fixed-point system, where only integer arithmetic is used, we prefer input data, filter coefficients, and processed output data to be in the Q-format. In this way, we avoid overflow due to multiplication and can prevent overflow due to addition by scaling input data. When the filter coefficients are out of the Q-format range, coefficient scaling must be taken into account to maintain the Q-format. We develop FIR filter implementation in Q-format first, and then infinite impulse response (IIR) filter implementation next. In addition, we assume that with a given input range in Q-format, the filter output is always in Q-format even if the filter passband gain is larger than 1.

First, to avoid the overflow for an adder, we can scale the input down by a scale factor S, which can be safely determined by the following equation

$$S = I_{max} \cdot \sum_{k=0}^{\infty} |h(k)| = I_{max} \cdot (|h(0)| + |h(1)| + |h(2)| + \cdots) \qquad (9.2)$$

where $h(k)$ is the impulse response of the adder output and I_{max} the maximum amplitude of the input in Q-format. Note that this is not an optimal factor in terms of reduced signal-to-noise ratio. However, it shall prevent the overflow. Equation (9.2) means that the adder output can actually be expressed as a convolution output:

$$\text{adder output} = h(0)x(n) + h(1)x(n-1) + h(2)x(n-2) + \cdots$$

Assuming the worst conditions, that is, that all the inputs $x(n)$ reach a maximum value of I_{max} and all the impulse coefficients are positive, the sum of the adder gives the most conservative scale factor, as shown in Equation (9.2). Hence, scaling down the input by a factor of S will guarantee that the output of the adder is in Q-format.

When some of the FIR coefficients are larger than 1, which is beyond the range of Q-format representation, coefficient scaling is required. The idea is that scaling down the coefficients will make them less than 1, and later the filtered output will be scaled up by the same amount before it is sent to DAC. Figure 9.15 describes the modified implementation.

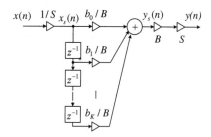

FIGURE 9.15

Direct-form I implementation of the FIR filter.

In the figure, the scale factor B makes the coefficients b_k/B convertible to the Q-format. The scale factors S and B are usually chosen to be a power of 2, so the simple shift operation can be used in the coding process. Let us implement an FIR filter containing filter coefficients larger than 1 in the fixed-point implementation.

EXAMPLE 9.13

Given the FIR filter

$$y(n) = 0.9x(n) + 3x(n-1) + 0.9x(n-2)$$

with a passband gain of 4, and assuming that the input range only occupies one quarter of the full range for a particular application, develop the DSP implementation equations in the Q-15 fixed-point system.

Solution:

The adder may cause overflow if the input data exist for one quarter of the full dynamic range. The scale factor is determined using the impulse response, which consists of the FIR filter coefficients, as discussed in Chapter 3.

$$S = \frac{1}{4}(|h(0)| + |h(1)| + |h(2)|) = \frac{1}{4}(0.9 + 3 + 0.9) = 1.2$$

Overflow may occur. Hence, we select $S = 2$ (a power of 2). We choose $B = 4$ to scale all the coefficients to be less than 1, so the Q-15 format can be used. According to Figure 9.15, the developed difference equations are given by

$$x_s(n) = \frac{x(n)}{2}$$

$$y_s(n) = 0.225x_s(n) + 0.75x_s(n-1) + 0.225x_s(n-2)$$

$$y(n) = 8y_s(n)$$

Next, the direct-form I implementation of the IIR filter is illustrated in Figure 9.16. As shown in the figure, the purpose of the scale factor C is to scale down the original filter coefficients to the Q-format. The factor C is usually chosen to be a power of 2 for using a simple shift operation in DSP.

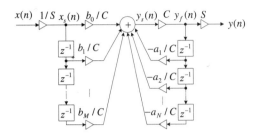

FIGURE 9.16

Direct-form I implementation of the IIR filter.

EXAMPLE 9.14

The IIR filter

$$y(n) = 2x(n) + 0.5y(n-1)$$

uses the direct-form I realization, and for a particular application, the maximum input is $I_{max} = 0.010....0_2 = 0.25$. Develop the DSP implementation equations in the Q-15 fixed-point system.

Solution:
This is an IIR filter whose transfer function is

$$H(z) = \frac{2}{1 - 0.5z^{-1}} = \frac{2z}{z - 0.5}$$

Applying the inverse z-transform, we obtain the impulse response

$$h(n) = 2 \times (0.5)^n u(n)$$

To prevent overflow in the adder, we can compute the S factor with the help of the Maclaurin series or approximate Equation (9.2) numerically. We get

$$S = 0.25 \times \left(2(0.5)^0 + 2(0.5)^1 + 2(0.5)^2 + \cdots \right) = \frac{0.25 \times 2 \times 1}{1 - 0.5} = 1$$

The MATLAB function **impz()** can also be applied to find the impulse response and the S factor:
```
>> h=impz(2,[1 -0.5]); % Find the impulse response
>> sf=0.25*sum(abs(h)) % Determine the sum of absolute values of h(k)
sf =1
```
Hence, we do not need to perform input scaling. However, we need scale down all the coefficients to use the Q-15 format. A factor of $C = 4$ is selected. From Figure 9.16, we get the difference equations as

$$x_s(n) = x(n)$$

$$y_s(n) = 0.5_s x(n) + 0.125 y_f(n-1)$$

$$y_f(n) = 4y_s(n)$$

$$y(n) = y_f(n)$$

We can develop these equations directly. First, we divide the original difference equation by a factor of 4 to scale down all the coefficients to be less than 1, that is,

$$\frac{1}{4}y_f(n) = \frac{1}{4} \times 2x_s(n) + \frac{1}{4} \times 0.5y_f(n-1)$$

and define a scaled output

$$y_s(n) = \frac{1}{4}y_f(n)$$

Finally, substituting $y_s(n)$ on the left side of the scaled equation and rescaling up the filter output as $y_f(n) = 4y_s(n)$, we have the same results as before.

The fixed-point implementation for direct-form II is more complicated. The developed direct-form II implementation of the IIR filter is illustrated in Figure 9.17.

As shown in the figure, two scale factors A and B are designated to scale denominator coefficients and numerator coefficients to their Q-format representations, respectively. Here S is a special factor to scale down the input sample so that the numerical overflow in the first sum in Figure 9.17 can be prevented. The difference equations are given in Chapter 6 and listed here:

$$w(n) = x(n) - a_1 w(n-1) - a_2 w(n-2) - \cdots - a_M w(n-M)$$

$$y(n) = b_0 w(n) + b_1 w(n-1) + \cdots + b_M w(n-M)$$

The first equation is scaled down by the factor A to ensure that all the denominator coefficients are less than 1, that is,

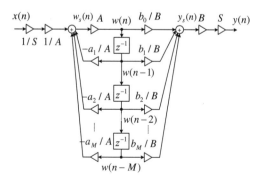

FIGURE 9.17

Direct-form II implementation of the IIR filter.

$$w_s(n) = \frac{1}{A}w(n) = \frac{1}{A}x(n) - \frac{1}{A}a_1 w(n-1) - \frac{1}{A}a_2 w(n-2) - \cdots - \frac{1}{A}a_M w(n-M)$$

$$w(n) = A \times w_s(n)$$

Similarly, scaling the second equation yields

$$y_s(n) = \frac{1}{B}y(n) = \frac{1}{B}b_0 w(n) + \frac{1}{B}b_1 w(n-1) + \cdots + \frac{1}{B}b_M w(n-M)$$

and

$$y(n) = B \times y_s(n)$$

To avoid the first adder overflow (first equation), the scale factor S can be safely determined by Equation (9.3):

$$S = I_{max}(|h(0)| + |h(1)| + |h(2)| + \cdots) \tag{9.3}$$

where $h(k)$ is the impulse response due to the denominator polynomial of the IIR filter, where the poles can cause a larger value to the first sum. Hence, h(k) is given by

$$h(n) = Z^{-1}\left(\frac{1}{1 + a_1 z^{-1} + \cdots + a z^{-M}}\right) \tag{9.4}$$

All the scale factors A, B, and S are usually chosen to be a power of 2, respectively, so that the shift operations can be used in the coding process. Example 9.15 serves as illustration.

EXAMPLE 9.15

Given the IIR filter

$$y(n) = 0.75x(n) + 1.49x(n-1) + 0.75x(n-2) - 1.52y(n-1) - 0.64y(n-2)$$

with a passband gain of 1 and a full range of input, use the direct-form II implementation to develop the DSP implementation equations in the Q-15 fixed-point system.

Solution:
The difference equations without scaling in the direct-form II implementation are given by

$$w(n) = x(n) - 1.52w(n-1) - 0.64w(n-2)$$

$$y(n) = 0.75w(n) + 1.49w(n-1) + 0.75w(n-2)$$

To prevent overflow in the first adder, we obtain the reciprocal of the denominator polynomial as

$$A(z) = \frac{1}{1 + 1.52z^{-1} + 0.64z^{-2}}$$

Using the MATLAB function **impz()** leads to:
```
>> h=impz(1,[1 1.52 0.64]);
>> sf=sum(abs(h))
sf = 10.4093
```

We choose the S factor as $S = 16$ and we choose $A = 2$ to scale down the denominator coefficients by half. Since the second adder output after scaling is

$$y_s(n) = \frac{0.75}{B}w(n) + \frac{1.49}{B}w(n-1) + \frac{0.75}{B}w(n-2)$$

to avoid second adder overflow we have to ensure that each coefficient is less than 1, along with the sum of the absolute values:

$$\frac{0.75}{B} + \frac{1.49}{B} + \frac{0.75}{B} < 1$$

Hence $B = 4$ is selected. We develop the DSP equations as

$$x_s(n) = x(n)/16$$

$$w_s(n) = 0.5x_s(n) - 0.76w(n-1) - 0.32w(n-2)$$

$$w(n) = 2w_s(n)$$

$$y_s(n) = 0.1875w(n) + 0.3725w(n-1) + 0.1875w(n-2)$$

$$y(n) = (B \times S)y_s(n) = 64y_s(n)$$

The implementation for cascading the second-order section filters can be found in Ifeachor and Jervis (2002).

A practical example will be presented in the next section. Note that if a floating-point DS processor is used, all the scaling concerns should be ignored, since the floating-point format offers a large dynamic range, so that overflow hardly ever happens.

9.6 DIGITAL SIGNAL PROCESSING PROGRAMMING EXAMPLES

In this section, we first review the TMS320C67x DSK (DSP Starter Kit), which offers floating-point and fixed-point arithmetic. We will then investigate real-time implementation of digital filters.

9.6.1 Overview of TMS320C67x DSK

In this section, a Texas Instruments TMS320C6713 DSK (DSP Starter Kit) shown in Figure 9.18 is chosen for demonstration. This DSK board has an approximate size of 5 x 8 inches, a clock rate of 225 MHz, and a 16-bit stereo codec TL V320AIC23 (AIC23), which deals with analog inputs and outputs. The onboard codec AIC23 applies sigma-delta technology for analog-to-digital conversion (ADC) and digital-to-analog conversion (DAC) functions. The codec runs using a 12 MHz system clock and the sampling rate can be selected from a range of 8 to 96 kHz for speech and audio processing. Other boards such as a Texas Instruments TMS320C6711 DSK can also be found in the references (Kehtaranavaz and Simsek, 2001; TMS320C6x CPU and Instruction Set Reference Guide, 1999). The

(a) TMS320C6713 DSK board

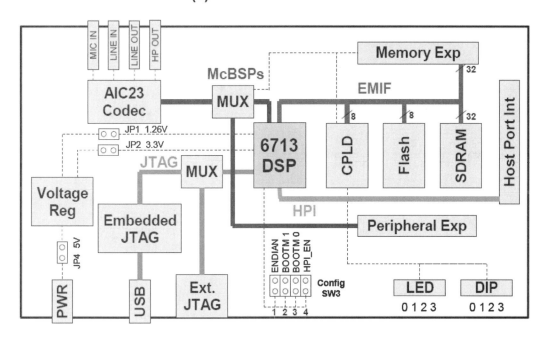

(b) TMS320C6713 DSK block diagram

FIGURE 9.18

(a) C6713 DSK board and (b) block diagram (courtesy of Texas Instruments).

on-board daughter card connections facilitate the external units for advanced applications such as external peripheral and external memory interfaces. The TMS320C6713 DSK board consists of 16 MB (megabytes) of synchronous dynamic RAM (SDRAM) and 512 kB (kilobytes) of flash memory. There are four onboard connections: MIC IN for microphone input; LINE IN for line input; LINE OUT for line output; and HEADPHONE for a headphone output (multiplexed with LINE OUT). The board components are tied by a CPLD (programmable logic device), which has a register based user interface. A user can configure the board by reading and writing to the CPLD registers. The DSK includes the four user DIP switches and four LEDs (light-emitting diodes) which provide a user a simple form of inputs and outputs. Both can be accessed by the CPLD registers. The onboard voltage regulators provide 1.26 V for the DSP core and 3.3 V for the memory and peripherals. The USB port provides the connection between the DSK board and the host computer, where the user program is developed, compiled, and downloaded to the DSK for real-time applications using the user-friendly software called Code Composer Studio, which we shall discuss later.

In general, the TMS320C67x operates at a high clock rate of 300 MHz. Combining this high speed and multiple units operating at the same time has pushed its performance up to 2,400 MIPS at 300 MHz. Using this number, the C67x can handle 0.3 MIPS between two speech samples at a sampling rate of 8 kHz, and can handle over 54,000 instructions between two audio samples with a sampling rate of 44.1 kHz. Hence, the C67x offers great flexibility for real-time applications with a high-level C language.

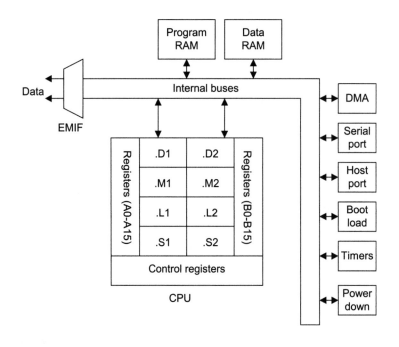

FIGURE 9.19

Block diagram of TMS320C67x floating-point DSP.

Figure 9.19 shows a C67x architecture overview, while Figure 9.20 displays a more detailed block diagram. The C67x contains three main parts, which are the CPU, the memory, and the peripherals. As shown in Figure 9.19, these three main parts are joined by an external memory interface (EMIF) interconnected by internal buses to facilitate interface with common memory devices, DMA, a serial port, and a host port interface (HPI).

Since this section is devoted to showing DSP coding examples, C67x key features and references are briefly listed here:

1. Architecture: The system uses Texas Instruments Veloci^{TM} architecture, which is an enhancement of the VLIW (very long instruction word architecture) (Dahnoun, 2000; Ifeachor and Jervis, 2002; Kehtaranavaz and Simsek, 2000).
2. CPU: As shown in Figure 9.20, the CPU has eight functional units divided into two sides A and B, each consisting of units .D, .M, .L, and .S. For each side, an .M unit is used for multiplication operation, an .L unit is used for logical and arithmetic operations, and a .D unit is used for loading/storing and arithmetic operations. Each side of the C67x CPU has sixteen 32-bit registers that the CPU must go through for interface. More detail can be found in Appendix D (Texas Instruments, 1991) as well as in Kehtaranavaz and Simsek (2000) and Texas Instruments (1998).

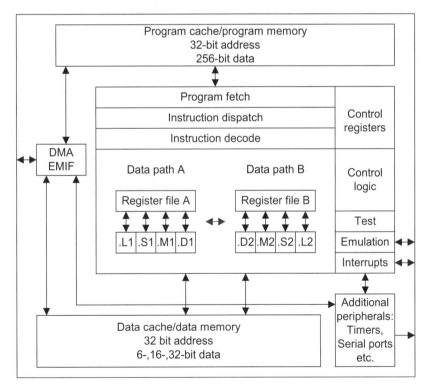

FIGURE 9.20

Registers of TMS320C67x floating-point DSP.

3. Memory and internal buses: Memory space is divided into internal program memory, internal data memory, and internal peripheral and external memory space. The internal buses include a 32-bit program address bus, a 256-bit program data bus to carry out eight 32-bit instructions (VLIW), two 32-bit data address buses, two 64-bit load data buses, two 64-bit store data buses, two 32-bit DMA buses, and two 32-bit DMA address buses responsible for reading and writing. There also is a 22-bit address bus and a 32-bit data bus for accessing off-chip or external memory.

4. Peripherals:
   a. EMIF, which provides the required timing for accessing external memory;
   b. DMA, which moves data from one memory location to another without interfering with CPU operations;
   c. Multichannel buffered serial port (McBSP) with a high-speed multichannel serial communication link;
   d. HPI, which lets a host access internal memory;
   e. Boot loader for loading code from off-chip memory or the HPI to internal memory;
   f. Timers (two 32-bit counters);
   g. Power-down units for saving power for periods when the CPU is inactive.

The software tool for the C67x is the Code Composer Studio (CCS) provided by TI. It allows the user to build and debug programs from a user-friendly graphical user interface (GUI) and extends the capabilities of code development tools to include real-time analysis. Installation, tutorial, coding, and debugging information can be found in the CCS Getting Started Guide (Texas Instruments, 2001) and in Kehtaranavaz and Simsek (2000).

Of particular note is the TMS320C6713 DSK with a clock rate of 225 MHz, which has the capability to fetch eight 32-bit instructions every 4.4 nanoseconds (1/225 MHz). The functional block diagram is shown in Figure 9.21. A detailed description can be found in Chassaing and Reay (2008).

9.6.2 Concept of Real-Time Processing

We illustrate real-time implementation in Figure 9.22, where the sampling rate is 8,000 samples per second; that is, the sampling period is $T = 1/f_s = 125$ microseconds, which is the time between two samples.

As shown in Figure 9.22, the required timing includes an input sample clock and an output sample clock. The input sample clock maintains the accuracy of the sampling time for each ADC operation, while the output sample clock keeps the accuracy of the time instant for each DAC operation. The time between the input sample clock n and output sample clock n consists of the ADC operation, algorithm processing, and the wait for the next ADC operation. The numbers of instructions for ADC and the DSP algorithm must be estimated and verified to ensure that all instructions have been completed before DAC begins. Similarly, the number of instructions for DAC must be verified so that DAC instructions will be finished between the output sample clock n and the next input sample clock $n + 1$. Timing usually is set up using the DSP interrupts (we will not pursue the interrupt setup here).

Next, we focus on the implementation of the DSP algorithm in a floating-point system for simplicity. A DSK setup example (Tan and Jiang, 2010) is depicted in Figure 9.23, while a skeleton code for verification of the input and output is depicted in Figure 9.24.

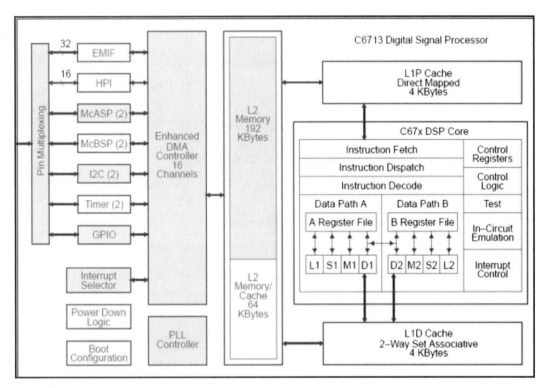

FIGURE 9.21

Functional block diagram and registers of TMS320C6713 (courtesy of Texas Instruments).

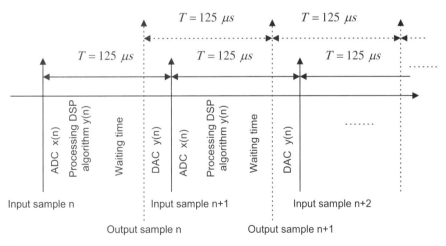

FIGURE 9.22

Concept of real-time processing.

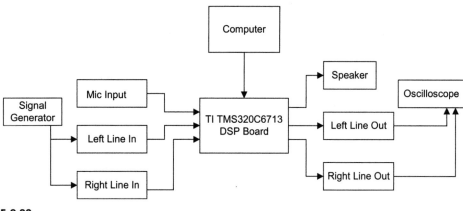

FIGURE 9.23

TMS320C6713 DSK setup example.

```
float x[1]={0.0};
float y[1]={0.0};
interrupt void c_int11()
{
    float lc; /*left channel input */
    float rc; /*right channel input */
    float lcnew; /*left channel output */
    float rcnew; /*right channel output */
    int i;
//Left channel and right channel inputs
    AIC23_data.combo=input_sample();
    lc=(float) (AIC23_data.channel[LEFT]);
    rc= (float) (AIC23_data.channel[RIGHT]);
// Insert DSP algorithm below
    x[0]=lc; /* Input from the left channel */
    y[0]=x[0];  /* simplest DSP equation */
// End of the DSP algorithm
    rcnew=y[0];
    rcnew=y[0];
    AIC23_data.channel[LEFT]=(short) lcnew;
    AIC23_data.channel[RIGHT]=(short) rcnew;
    output_sample(AIC23_data.combo);
}
```

FIGURE 9.24

Program segment for verifying input and output.

9.6.3 Linear Buffering

During DSP such as digital filtering, past inputs and past outputs are required to be buffered and updated for processing the next input sample. Let us first study the FIR filter implementation.

Finite Impulse Response Filtering

Consider implementation for the following 3-tap FIR filter:

$$y(n) = 0.5x(n) + 0.2x(n-1) + 0.5x(n-2)$$

The buffer requirements are shown in Figure 9.25. The coefficient buffer b[3] contains 3 FIR coefficients, and the coefficient buffer is fixed during the process. The input buffer x[3], which holds the

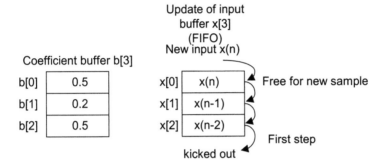

```
float x[3]={0.0, 0.0, 0.0};
float b[3]={0.5, 0.2, 0.5};
float y[1]={0.0};
interrupt void c_int11()
{
    float lc; /*left channel input */
    float rc; /*right channel input */
    float lcnew; /*left channel output */
    float rcnew; /*right channel output */
    int i;
//Left channel and right channel inputs
    AIC23_data.combo=input_sample();
    lc=(float) (AIC23_data.channel[LEFT]);
    rc= (float) (AIC23_data.channel[RIGHT]);
// Insert DSP algorithm below
    for(i=2; i>0; i--)   /* Update the input buffer x[3] */
    {  x[i]=x[i-1]; }
    x[0]= (float) lc;  /*Input from the left channel */
    y[0]=0;
    for(i=0; i<3; i++)
    {  y[0]=y[0]+b[i]*x[i]; } /* FIR filtering */
// End of the DSP algorithm
    rcnew=y[0];
    rcnew=y[0];
    AIC23_data.channel[LEFT]=(short) lcnew;
    AIC23_data.channel[RIGHT]=(short) rcnew;
    output_sample(AIC23_data.combo);
}
```

FIGURE 9.25

Example of FIR filtering with linear buffer update.

current and past inputs, must be updated. The FIFO update process is adopted here with the segment of code shown in Figure 9.25. For each input sample, we update the input buffer using FIFO, which begins at the end of the data buffer; the oldest sample is kicked out first from the buffer and updated with the value from the upper location. When the FIFO process is completed, the first memory location x[0] will be free to be used to store the current input sample. The segment of code in Figure 9.25 explains the implementation.

Note that in the code segment, $x[0]$ holds the current input sample $x(n)$, while $b[0]$ is the corresponding coefficient; $x[1]$ and $x[2]$ hold the past input samples $x(n-1)$ and $x(n-2)$, respectively; similarly, $b[1]$ and $b[2]$ are the corresponding coefficients.

Again, note that using the array and loop structures in the code segment is for simplicity in notation and the assumption is that the reader is not familiar with C pointers in the C language. The concern for simplicity has to do mainly with the DSP algorithm. More coding efficiency can be achieved using C pointers and a circular buffer. The DSP-oriented coding implementation can be found in Kehtarnavaz and Simsek (2000) and Chassaing and Reay (2008).

Infinite Impulse Response Filtering

Similarly, we can implement an IIR filter. It requires an input buffer, which holds the current and past inputs; an output buffer, which holds the past outputs; a numerator coefficient buffer; and a denominator coefficient buffer. Consider the following IIR filter for implementation:

$$y(n) = 0.5x(n) + 0.7x(n-1) - 0.5x(n-2) - 0.4y(n-1) + 0.6y(n-2)$$

We accommodate the numerator coefficient buffer b[3], the denominator coefficient buffer a[3], the input buffer x[3], and the output buffer y[3] shown in Figure 9.26. The buffer updates for input x[3] and output y[3] are FIFO. The implementation is illustrated in the segment of code listed in Figure 9.26.

Again, note that in the code segment, $x[0]$ holds the current input sample, while $y[0]$ holds the current processed output, which will be sent to the DAC unit for conversion. The coefficient $a[0]$ is never modified in the code. We keep that for a purpose of notation simplicity and consistency during the programming process.

Digital Oscillation with Infinite Impulse Response Filtering

The principle for generating digital oscillation is described in Chapter 8, where the input to the digital filter is the impulse sequence, and the transfer function is obtained by applying the z-transform of the digital sinusoid function. Applications can be found in dual-tone multifrequency (DTMF) tone generation, digital carrier generation for communications, and so on. Hence, we can modify the implementation of IIR filtering for tone generation with the input generated internally instead of using the ADC channel.

Let us generate an 800 Hz tone with a digital amplitude of 5,000. According to Section 8.11 "Application: Generation and Detection of DTMF Tones Using the Goertzel Algorithm" in Chapter 8, the transfer function, difference equation, and the impulse input sequence are found to be, respectively,

$$H(z) = \frac{0.587785z^{-1}}{1 - 1.618034z^{-1} + z^{-2}}$$

$$y(n) = 0.587785x(n-1) + 1.618034y(n-1) - y(n-2)$$

$$x(n) = 5000\delta(n)$$

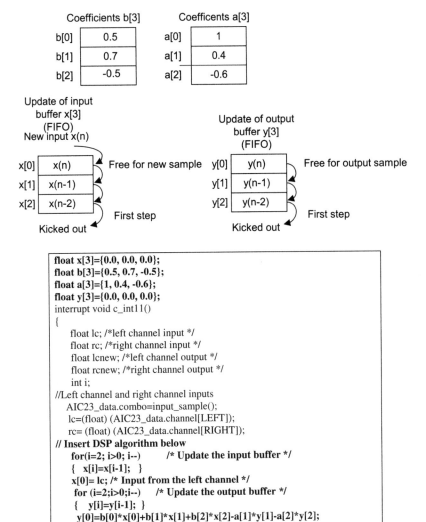

Coefficients b[3]

b[0]	0.5
b[1]	0.7
b[2]	-0.5

Coefficents a[3]

a[0]	1
a[1]	0.4
a[2]	-0.6

Update of input buffer x[3] (FIFO)
New input x(n)

x[0]	x(n)
x[1]	x(n-1)
x[2]	x(n-2)

Free for new sample

First step

Kicked out

Update of output buffer y[3] (FIFO)

y[0]	y(n)
y[1]	y(n-1)
y[2]	y(n-2)

Free for output sample

First step

Kicked out

```
float x[3]={0.0, 0.0, 0.0};
float b[3]={0.5, 0.7, -0.5};
float a[3]={1, 0.4, -0.6};
float y[3]={0.0, 0.0, 0.0};
interrupt void c_int11()
{
    float lc; /*left channel input */
    float rc; /*right channel input */
    float lcnew; /*left channel output */
    float rcnew; /*right channel output */
    int i;
//Left channel and right channel inputs
    AIC23_data.combo=input_sample();
    lc=(float) (AIC23_data.channel[LEFT]);
    rc= (float) (AIC23_data.channel[RIGHT]);
// Insert DSP algorithm below
    for(i=2; i>0; i--)       /* Update the input buffer */
    {   x[i]=x[i-1];  }
    x[0]= lc; /* Input from the left channel */
    for (i=2;i>0;i--)     /* Update the output buffer */
    {   y[i]=y[i-1];  }
    y[0]=b[0]*x[0]+b[1]*x[1]+b[2]*x[2]-a[1]*y[1]-a[2]*y[2];
// End of the DSP algorithm
    rcnew=y[0];
    rcnew=y[0];
    AIC23_data.channel[LEFT]=(short) lcnew;
    AIC23_data.channel[RIGHT]=(short) rcnew;
    output_sample(AIC23_data.combo);
}
```

FIGURE 9.26

Example of IIR filtering using linear buffer update.

We define the numerator coefficient buffer b[2], the denominator coefficient buffer a[3], the input buffer x[2], and the output buffer y[3] in Figure 9.27, which also shows the modified implementation for tone generation.

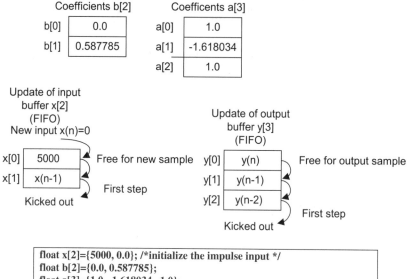

```
float x[2]={5000, 0.0}; /*initialize the impulse input */
float b[2]={0.0, 0.587785};
float a[3]={1.0, -1.618034,  1.0};
float y[3]={0.0};
interrupt void c_int11()
{    float lc; /*left channel input */
     float rc; /*right channel input */
     float lcnew; /*left channel output */
     float rcnew; /*right channel output */
     int i;
//Left channel and right channel inputs
   AIC23_data.combo=input_sample();
    lc=(float) (AIC23_data.channel[LEFT]);
    rc= (float) (AIC23_data.channel[RIGHT]);
// Insert DSP algorithm below
    y[0]=b[0]*x[0]+b[1]*x[1]+b[2]*x[2]-a[1]*y[1]-a[2]*y[2];
    for(i=2; i>0; i--)    /* Update the input buffer with zero input */
    {  x[i]=x[i-1];   }
    x[0]= 0;
    for (i=2;i>0;i--)    /* Update the output buffer */
    {   y[i]=y[i-1]; }
// End of the DSP algorithm
    rcnew=y[0];
    rcnew=y[0];
    AIC23_data.channel[LEFT]=(short) lcnew;
    AIC23_data.channel[RIGHT]=(short) rcnew;
    output_sample(AIC23_data.combo);
}
```

FIGURE 9.27

Example of IIR filtering using linear buffer update and the impulse sequence input.

Initially, we set $x[0] = 5,000$. Then it will be updated with $x[0] = 0$ for each current processed output sample $y[0]$.

9.6.4 Sample C Programs

Floating-Point Implementation Example

Real-time DSP implementation using a floating-point processor is easy to program. The overflow problem hardly ever occurs. Therefore, we do not need to consider scaling factors, as described in the last section. The code segment shown in Figure 9.28 demonstrates the simplicity of coding the floating-point IIR filter using the direct-form I structure.

Fixed-Point Implementation Example

When the execution time is critical, fixed-point implementation is preferred in a floating-point processor. We implement the following IIR filter with a unit passband gain in direct-form II:

```
float a[5]={1.00, -2.1192, 2.6952, -1.6924, 0.6414};
float b[5]={0.0201, 0.00, -0.0402, 0.00, 0.0201};
float x[5]={0.0, 0.0, 0.0, 0.0, 0.0};
float y[5]={0.0, 0.0, 0.0, 0.0, 0.0};

interrupt void c_int11()
{
    float lc; /*left channel input */
    float rc; /*right channel input */
    float lcnew; /*left channel output */
    float rcnew; /*right channel output */
    int i;
//Left channel and right channel inputs
    AIC23_data.combo=input_sample();
    lc=(float) (AIC23_data.channel[LEFT]);
    rc= (float) (AIC23_data.channel[RIGHT]);
// Insert DSP algorithm below
    for(i=4; i>0; i--)       /* Update the input buffer */
    {
        x[i]=x[i-1];
    }
    x[0]= lc; /* Input from the left channel */
    for (i=2;i>0;i--)     /* Update the output buffer */
    {
        y[i]=y[i-1];
    }
    y[0]=b[0]*x[0]+b[1]*x[1]+b[2]*x[2]+b[3]*x[3]+b[4]*x[4]-a[1]*y[1]-a[2]*y[2]-a[3]*y[3]-a[4]*y[4];
// End of the DSP algorithm
    rcnew=y[0];
    rcnew=y[0];
    AIC23_data.channel[LEFT]=(short) lcnew;
    AIC23_data.channel[RIGHT]=(short) rcnew;
    output_sample(AIC23_data.combo);
}
```

FIGURE 9.28

Sample C code for IIR filtering (floating-point implementation).

$$H(z) = \frac{0.0201 - 0.0402z^{-2} + 0.0201z^{-4}}{1 - 2.1192z^{-1} + 2.6952z^{-2} - 1.6924z^{-3} + 0.6414z^{-4}}$$

$$w(n) = x(n) + 2.1192w(n-1) - 2.6952w(n-2) + 1.6924w(n-3) - 0.6414w(n-4)$$

$$y(n) = 0.0201w(n) - 0.0402w(n-2) + 0.0201w(n-4)$$

Using MATLAB to calculate the scale factor S, it follows that

```
» h=impz([1],[1 -2.1192 2.6952 -1.6924 0.6414]);
» sf=sum(abs(h))
  sf=28.2196
```

Hence we choose $S = 32$. To scale the filter coefficients in the Q-15 format, we use the factors $A = 4$ and $B = 1$. Then the developed DSP equations are

$$x_s(n) = x(n)/32$$

$$w_s(n) = 0.25x_s(n) + 0.5298w_s(n-1) - 0.6738w_s(n-2) + 0.4231w_s(n-3) - 0.16035w_s(n-4)$$

$$w(n) = 4w_s(n)$$

$$y_s(n) = 0.0201w(n) - 0.0402w(n-2) + 0.0201w(n-4)$$

$$y(n) = 32y_s(n)$$

Using the method described in Section 9.5, we can convert filter coefficients into the Q-15 format; each coefficient is listed in Table 9.4.

Table 9.4 Filter Coefficients in Q-15 Format

IIR Filter	Filter Coefficients	Q-15 Format (Hex)
$-a_1$	0.5298	0 × 43D0
$-a_2$	−0.6738	0 × A9C1
$-a_3$	0.4230	0 × 3628
$-a_4$	−0.16035	0 × EB7A
b_0	0.0201	0 × 0293
b_1	0.0000	0 × 0000
b_2	−0.0402	0 × FADB
b_3	0.0000	0 × 000
b_4	0.0201	0 × 0293

```
/*float a[5]={1.00, -2.1192, 2.6952, -1.6924, 0.6414}; float b[5]={0.0201, 0.00, -0.0402, 0.00, 0.0201};*/
short a[5]={0x2000, 0x43D0, 0xA9C1, 0x3628, 0xEB7A}; /* coefficients in Q-15 format */
short b[5]={0x0293, 0x0000, 0xFADB, 0x0000, 0x0293};
int w[5]={0, 0, 0, 0, 0};
int sample;

interrupt void c_int11()
{
    float lc; /*left channel input */
    float rc; /*right channel input */
    float lcnew; /*left channel output */
    float rcnew; /*right channel output */
    int i, sum=0;
//Left channel and right channel inputs
    AIC23_data.combo=input_sample();
    lc= (float) (AIC23_data.channel[LEFT]);
    rc= (float) (AIC23_data.channel[RIGHT]);
// Insert DSP algorithm below
    sample = (int) lc; /*input sample from the left channel*/
    sample = (sample << 16); /* move to high 16 bits */
    sample = (sample>>5); /* scaled down by 32 to avoid overflow */
    for (i=4;i>0;i--)
    {
      w[i]=w[i-1];
    }
    sum= (sample >> 2); /* scaled down by 4 to use Q-15 */
    for (i=1;i<5;i++)
    {
      sum += (_mpyhl(w[i],a[i])) <<1;
    }
    sum = (sum <<2); /* scaled up by 4 */
    w[0]=sum;
    sum =0;
    for(i=0;i<5;i++)
    {
      sum += (_mpyhl(w[i],b[i]))<<1;
    }
    sum = (sum << 5); /* scaled up by 32 to get y(n) */
    sample= (sum>>16); /* move to low 16 bits */
// End of the DSP algorithm
    rcnew=sample;
    rcnew=sample;
    AIC23_data.channel[LEFT]=(short) lcnew;
    AIC23_data.channel[RIGHT]=(short) rcnew;
    output_sample(AIC23_data.combo);
}
```

FIGURE 9.29

Sample C code for IIR filtering (fixed point implementation).

```
short coefficient;     declaration of 16 bit signed integer
int sample, result;    declaration of 32 bit signed integer
MPYHL   assembly instruction (signed multiply high low 16 MSB x 16 LSB)
        result = (_mpyhl(sample,coefficient) ) <<1;
sample must be shifted left by 16 bits to be stored in the high 16 MSB.
coefficient is the 16 bit data to be stored in the low 16 LSB.
result is shifted left by one bit to get rid of the extended sign bit, and high 16
MSB's are designated for the processed data.
Final result will be shifted down to right by 16 bits before DAC conversion.
        sample = (result>>16);
```

FIGURE 9.30

Some coding notations for the Q-15 fixed-point implementation.

The code for the fixed-point implementation is displayed in Figure 9.29, and some coding notations are given in Figure 9.30.

Note that this chapter has provided only basic concepts and an introduction to real-time DSP implementation. The coding detail and real-time DSP applications will be treated in a separate DSP course, which deals with real-time implementations.

9.7 SUMMARY

1. The Von Neumann architecture consists of a single, shared memory for programs and data, a single bus for memory access, an arithmetic unit, and a program control unit. The Von Neumann processor operates fetching and execution cycles seriously.

2. The Harvard architecture has two separate memory spaces dedicated to program code and to data, respectively, two corresponding address buses, and two data buses for accessing two memory spaces. The Harvard processor offers fetching and executions in parallel.

3. The DSP special hardware units include a MAC dedicated to DSP filtering operations, a shifter unit for scaling, and address generators for circular buffering.

4. The fixed-point DS processor uses integer arithmetic. The data format Q-15 for the fixed-point system is preferred to avoid the overflows.

5. The floating-point processor uses floating-point arithmetic. The standard floating-point formats include the IEEE single precision and double precision formats.

6. The architectures and features of fixed-point processors and floating-point processors were briefly reviewed.

7. Implementing digital filters in the fixed-point DSP system requires scaling filter coefficients so that the filters are in Q-15 format, and input scaling for the adder so that overflow during MAC operations can be avoided.

8. The floating-point processor is easy to code using floating-point arithmetic and develops the prototype quickly. However, it is not efficient in terms of the number of instructions it has to complete compared with the fixed-point processor.

9. The fixed-point processor using fixed-point arithmetic takes much effort to code. But it offers the least number of the instructions for the CPU to execute.

9.8 PROBLEMS

9.1. Find the signed Q-15 representation for the decimal number 0.2560123.

9.2. Find the signed Q-15 representation for the decimal number −0.2160123.

9.3. Find the signed Q-15 representation for the decimal number −0.3567921.

9.4. Find the signed Q-15 representation for the decimal number 0.4798762.

9.5. Convert the Q-15 signed number = 1.010101110100010 to a decimal number.

9.6. Convert the Q-15 signed number = 0.001000111101110 to a decimal number.

9.7. Convert the Q-15 signed number = 0.110101000100010 to a decimal number.

9.8. Convert the Q-15 signed number = 1.101000100101111 to a decimal number.

9.9. Add the following two Q-15 numbers:

$$1.101010111000001 + 0.010001111011010$$

9.10. Add the following two Q-15 numbers:

$$0.001010101000001 + 0.010101111010010$$

9.11. Add the following two Q-15 numbers:

$$1.001010101000001 + 1.010101111010010$$

9.12. Add the following two Q-15 numbers:

$$0.001010101000001 + 1.010101111010010$$

9.13. Convert each of the following decimal numbers to a floating-point number using the format specified in Figure 9.10.

   a. 0.1101235
   b. −10.430527

9.14. Convert each of the following decimal numbers to a floating-point number using the format specified in Figure 9.10.

   a. 2.5568921
   b. −0.678903
   c. 0.0000000
   d. −1.0000000

9.15. Add the following floating-point numbers whose formats are defined in Figure 9.10, and determine the sum in decimal format:

$$1101\ 011100011011 + 0100\ 101111100101$$

9.16. Add the following floating-point numbers whose formats are defined in Figure 9.10, and determine the sum in decimal format:

$$0111\ 110100011011 + 0101\ 001000100101$$

9.17. Add the following floating-point numbers whose formats are defined in Figure 9.10, and determine the sum in decimal format:

$$0001\ 000000010011 + 0100\ 001000000101$$

9.18. Convert the following number in IEEE single precision format to the decimal format:

$$110100000.010...\ 0000$$

9.19. Convert the following number in IEEE single precision format to the decimal format:

$$010100100.101...\ 0000$$

9.20. Convert the following number in IEEE double precision format to the decimal format:

$$011000...0.1010...000$$

9.21. Convert the following number in IEEE double precision format to the decimal format:

$$011000...0.0110...0000$$

9.22. Given the FIR filter

$$y(n) = -0.2x(n) + 0.6x(n-1) + 0.2x(n-2)$$

with a passband gain of 1 and the input being a full range, develop the DSP implementation equations in the Q-15 fixed-point system.

9.23. Given the IIR filter

$$y(n) = 0.6x(n) + 0.3y(n-1)$$

with a passband gain of 1 and the input being a full range, use the direct-form I method to develop the DSP implementation equations in the Q-15 fixed-point system.

9.24. Repeat Problem 9.23 using the direct-form II method.

9.25. Given the FIR filter

$$y(n) = -0.36x(n) + 1.6x(n-1) + 0.36x(n-2)$$

with a passband gain of 2 and the input being half of the range, develop the DSP implementation equations in the Q-15 fixed-point system.

9.26. Given the IIR filter

$$y(n) = 1.35x(n) + 0.3y(n-1)$$

with a passband gain of 2, and the input being half of the range, use the direct-form I method to develop the DSP implementation equations in the Q-15 fixed-point system.

9.27. Repeat Problem 9.26 using the direct-form II method.

9.28. Given the IIR filter

$$y(n) = 0.72x(n) + 1.42x(n-2) + 0.72x(n-2) - 1.35y(n-1) - 0.5y(n-2)$$

with a passband gain of 1 and a full range of input, use the direct-form I to develop the DSP implementation equations in the Q-15 fixed-point system.

9.29. Repeat Problem 9.28 using the direct-form II method.

Adaptive Filters and Applications

CHAPTER OUTLINE

10.1 Introduction to Least Mean Square Adaptive Finite Impulse Response Filters 453
10.2 Basic Wiener Filter Theory and Least Mean Square Algorithm 457
10.3 Applications: Noise Cancellation, System Modeling, and Line Enhancement 462
    10.3.1 Noise Cancellation 462
    10.3.2 System Modeling 468
    10.3.3 Line Enhancement Using Linear Prediction 473
10.4 Other Application Examples 476
    10.4.1 Canceling Periodic Interferences Using Linear Prediction 476
    10.4.2 Electrocardiography Interference Cancellation 476
    10.4.3 Echo Cancellation in Long-Distance Telephone Circuits 479
10.5 Laboratory Examples Using the TMS320C6713 DSK 480
10.6 Summary 485

OBJECTIVES

This chapter introduces principles of adaptive filters and the adaptive least mean square algorithm and illustrates how to apply the adaptive filters to solve the real-world application problems such as adaptive noise cancellation, system modeling, adaptive line enhancement, and telephone echo cancellation.

10.1 INTRODUCTION TO LEAST MEAN SQUARE ADAPTIVE FINITE IMPULSE RESPONSE FILTERS

An *adaptive filter* is a digital filter that has self-adjusting characteristics. It is capable of adjusting its filter coefficients automatically to adapt the input signal via an adaptive algorithm. Adaptive filters play an important role in modern digital signal processing (DSP) products in areas such as telephone echo cancellation, noise cancellation, equalization of communications channels, biomedical signal enhancement, active noise control, and adaptive control systems. Adaptive filters work generally for adaptation of signal-changing environments, spectral overlap between noise and signal, and unknown or time-varying noise. For example, when the interference noise is strong and its spectrum overlaps that of the desired signal, removing the interference using a traditional filter such as a notch filter with fixed filter coefficients will fail to preserve the desired signal spectrum, as shown in Figure 10.1.

However, an adaptive filter will do the job. Note that adaptive filtering, with its applications, has existed for more than two decades in the research community and is still active there. This chapter can

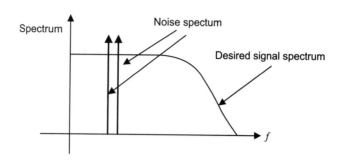

FIGURE 10.1

Spectrum illustration for using adaptive filters.

only introduce some fundaments of the subject, that is, adaptive finite impulse response (FIR) filters with a simple and popular least mean square (LMS) algorithm. Further exploration into adaptive infinite impulse response (IIR) filters, adaptive lattice filters, their associated algorithms and applications, and so on, can be found in comprehensive texts by Haykin (1991), Stearns (2003), and Widrow and Stearns (1985).

To understand the concept of adaptive filtering, we will first look at an illustrative example of the simplest noise canceller to see how it works before diving into detail. The block diagram for such a noise canceller is shown in Figure 10.2.

As shown in Figure 10.2, first, the DSP system consists of two analog-to-digital conversion (ADC) channels. The first microphone with ADC is used to capture the desired speech $s(n)$. However, due to a noisy environment, the signal is contaminated and the ADC channel produces a signal with the noise; that is, $d(n) = s(n) + n(n)$. The second microphone is placed where only noise is picked up and the second ADC channel captured noise $x(n)$, which is fed to the adaptive filter.

Note that the corrupting noise $n(n)$ in the first channel is uncorrelated to the desired signal $s(n)$, so that separation between them is possible. The noise signal $x(n)$ from the second channel is correlated

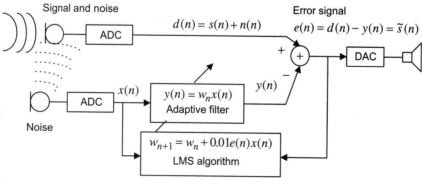

FIGURE 10.2

Simplest noise canceller using a one-tap adaptive filter.

to the corrupting noise $n(n)$ in the first channel, since both come from the same noise source. Similarly, the noise signal $x(n)$ is not correlated to the desired speech signal $s(n)$.

We assume that the corrupting noise in the first channel is a linear filtered version of the second-channel noise, since it has a different physical path from the second-channel noise, and the noise source is time varying, so that we can estimate the corrupting noise $n(n)$ using an adaptive filter. The adaptive filter contains a digital filter with adjustable coefficient(s) and the LMS algorithm to modify the value(s) of coefficient(s) for filtering each sample. The adaptive filter then produces an estimate of noise $y(n)$, which will be subtracted from the corrupted signal $d(n) = s(n) + n(n)$. When the noise estimate $y(n)$ equals or approximates the noise $n(n)$ in the corrupted signal, that is, $y(n) \approx n(n)$, the error signal $e(n) = s(n) + n(n) - y(n) \approx \tilde{s}(n)$ will approximate the clean speech signal $s(n)$. Hence, the noise is cancelled.

In our illustrative numerical example, the adaptive filter is set to be one-tap FIR filter to simplify numerical algebra. The filter adjustable coefficient w_n is adjusted based on the LMS algorithm (discussed later in detail) in the following:

$$w_{n+1} = w_n + 0.01 \cdot e(n) \cdot x(n)$$

where w_n is the coefficient used currently, while w_{n+1} is the coefficient obtained from the LMS algorithm and will be used for the next coming input sample. The value of 0.01 controls the speed of the coefficient change. To illustrate the concept of the adaptive filter in Figure 10.2, the LMS algorithm has the initial coefficient set to $w_0 = 0.3$ and leads to

$$y(n) = w_n x(n)$$
$$e(n) = d(n) - y(n)$$
$$w_{n+1} = w_n + 0.01 e(n) x(n)$$

The corrupted signal is generated by adding noise to a sine wave. The corrupted signal and noise reference are shown in Figure 10.3, and their first 16 values are listed in Table 10.1.

Let us perform adaptive filtering for several samples using the values for the corrupted signal and reference noise in Table 10.1. We see that

$$n = 0, \quad y(0) = w_0 x(0) = 0.3 \times (-0.5893) = -0.1768$$
$$e(0) = d(0) - y(0) = -0.2947 - (-0.1768) = -0.1179 = \tilde{s}(0)$$
$$w_1 = w_0 + 0.01 e(0) x(0) = 0.3 + 0.01 \times (-0.1179) \times (-0.5893) = 0.3007$$

$$n = 1, \quad y(1) = w_1 x(1) = 0.3007 \times 0.5893 = 0.1772$$
$$e(1) = d(1) - y(1) = 1.0017 - 0.1772 = 0.8245 = \tilde{s}(1)$$
$$w_2 = w_1 + 0.01 e(1) x(1) = 0.3007 + 0.01 \times 0.8245 \times 0.5893 = 0.3056$$

$$n = 2, \quad y(2) = w_2 x(2) = 0.3056 \times 3.1654 = 0.9673$$
$$e(2) = d(2) - y(2) = 2.5827 - 0.9673 = 1.6155 = \tilde{s}(2)$$
$$w_3 = w_2 + 0.01 e(2) x(2) = 0.3056 + 0.01 \times 1.6155 \times 3.1654 = 0.3567$$

$$n = 3, \quad \cdots$$

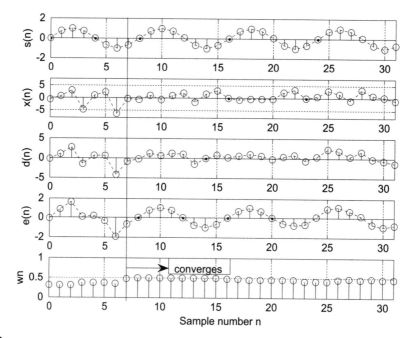

FIGURE 10.3

Original signal, reference noise, corrupted signal, enhanced signal, and adaptive coefficient in the noise cancellation.

TABLE 10.1 Adaptive Filtering Results for the Simplest Noise Canceller Example

n	d(n)	x(n)	$\tilde{s}(n) = e(n)$	Original s(n)	w_{n+1}
0	−0.2947	−0.5893	−0.1179	0	0.3000
1	1.0017	0.5893	0.8245	0.7071	0.3007
2	2.5827	3.1654	1.6155	1.0000	0.3056
3	−1.6019	−4.6179	0.0453	0.7071	0.3567
4	0.5622	1.1244	0.1635	0.0000	0.3546
5	0.4456	2.3054	−0.3761	−0.7071	0.3564
6	−4.2674	−6.5348	−1.9948	−1.0000	0.3478
7	−0.8418	−0.2694	−0.7130	−0.7071	0.4781
8	−0.3862	−0.7724	−0.0154	−0.0000	0.4800
9	1.2274	1.0406	0.7278	0.7071	0.4802
10	0.6021	−0.7958	0.9902	1.0000	0.4877
11	1.1647	0.9152	0.7255	0.7071	0.4799
12	0.9630	1.9260	0.0260	0.0000	0.4865
13	−1.5065	−1.5988	−0.7279	−0.7071	0.4870
14	−0.1329	1.7342	−0.9976	−1.0000	0.4986
15	0.8146	3.0434	−0.6503	−0.7071	0.4813

For comparison, results of the first 16 processed output samples, original samples, and filter coefficient values are also included in Table 10.1. Figure 10.3 also shows the original signal samples, reference noise samples, corrupted signal samples, enhanced signal samples, and filter coefficient values for each incoming sample, respectively.

As shown in Figure 10.3, after seven adaptations, the adaptive filter learns noise characteristics and cancels the noise in the corrupted signal. The adaptive coefficient is close to the optimal value of 0.5. The processed output is close to the original signal. The first 16 processed values for corrupted signal, reference noise, clean signal, original signal, and adaptive filter coefficient used at each step are listed in Table 10.1.

Clearly, the enhanced signal samples look much like the sinusoid input samples. Now our simplest one-tap adaptive filter works for this particular case. In general, an FIR filter with multiple taps is used and has the following format:

$$y(n) = \sum_{i=0}^{N-1} w_n(i)x(n-i) = w_n(0)x(n) + w_n(1)x(n-1) + \cdots + w_n(N-1)x(n-N+1) \quad (10.1)$$

The LMS algorithm for the adaptive FIR filter will be developed next.

10.2 BASIC WIENER FILTER THEORY AND LEAST MEAN SQUARE ALGORITHM

Many adaptive algorithms can be viewed as approximations of the discrete Wiener filter shown in Figure 10.4, where the Wiener filter output y(n) is a sum of its N weighted inputs, that is,

$$y(n) = w(0)x(n) + w(1)x(n-1) + \ldots + w(N-1)x(n-N+1).$$

The Wiener filter adjusts its weight(s) to produce a desired filter output y(n) which is close to the noise n(n) contained in the corrupted signal d(n). At the subtracted output, the noise n(n) is cancelled or attenuated. Hence, the output e(n) contains a clean signal.

Consider a single-weight case of $y(n) = wx(n)$, and note that the error signal e(n) is given by

$$e(n) = d(n) - wx(n) \quad (10.2)$$

Now let us determine the best weight w^*. Taking the square or enhanced the output error leads to

$$e^2(n) = (d(n) - wx(n))^2 = d^2(n) - 2d(n)wx(n) + w^2x^2(n) \quad (10.3)$$

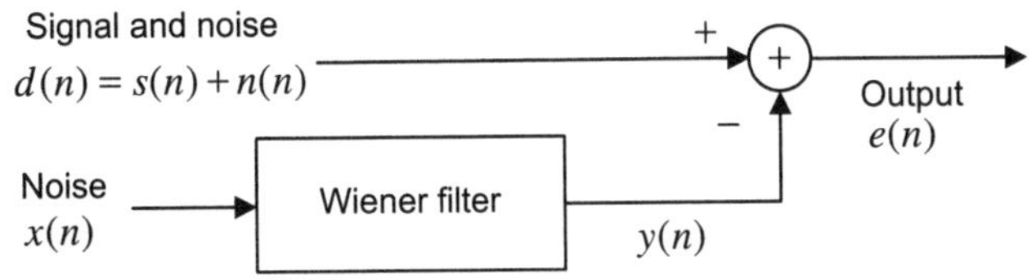

FIGURE 10.4

Wiener filter for noise cancellation.

Taking the statistical expectation of Equation (10.3), we have

$$E(e^2(n)) = E(d^2(n)) - 2wE(d(n)x(n)) + w^2 E(x^2(n)) \tag{10.4}$$

Using the notations in statistics, we define

$$
\begin{aligned}
J &= E(e^2(n)) = MSE = \textit{mean squared error} \\
\sigma^2 &= E(d^2(n)) = \textit{power of corrupted signal} \\
P &= E(d(n)x(n)) = \textit{cross-correlation between } d(n) \textit{ and } x(n) \\
R &= E(x^2(n)) = \textit{autocorrelation}
\end{aligned}
$$

We can view the statistical expectation as an average of the N signal terms, each being a product of two individual samples

$$E(e^2(n)) = \frac{e^2(0) + e^2(1) + \cdots + e^2(N-1)}{N}$$

or

$$E(d(n)x(n)) = \frac{d(0)x(0) + d(1)x(1) + \cdots + d(N-1)x(N-1)}{N}$$

for a sufficiently large sample number of N. We can write Equation (10.4) as

$$J = \sigma^2 - 2wP + w^2 R \tag{10.5}$$

Since σ^2, P, and R are constants, J is a quadratic function of w that may be plotted as shown in Figure 10.5.

The best weight (optimal) w^* is at the location where the minimum MSE J_{min} is achieved. To obtain w^*, taking a derivative of J and setting it to zero leads to

$$\frac{dJ}{dw} = -2P + 2wR = 0 \tag{10.6}$$

Solving Equation (10.6), we get the best weight solution as

$$w^* = R^{-1}P \tag{10.7}$$

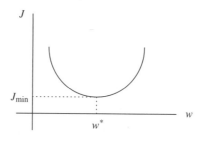

FIGURE 10.5

Mean square error quadratic function.

EXAMPLE 10.1

Consider the following quadratic MSE function for the Wiener filter:

$$J = 40 - 20w + 10w^2$$

Find the optimal solution for w^* to achieve the minimum MSE J_{min} and determine J_{min}.

Solution:

Taking a derivative of the MSE function and setting it to zero, we have

$$\frac{dJ}{dw} = -20 + 10 \times 2w = 0$$

Solving the equation leads to

$$w^* = 1$$

Finally, substituting $w^* = 1$ into the MSE function, we get the minimum J_{min} as

$$J_{min} = J|_{w=w^*} = 40 - 20w + 10w^2|_{w=1} = 40 - 20 \times 1 + 10 \times 1^2 = 30$$

Notice that a few points need to be clarified for Equation (10.7):

1. Optimal coefficient (s) can be different for every block of data, since the corrupted signal and reference signal are unknown. The autocorrelation and cross-correlation may vary.
2. If a larger number of coefficients (weights) are used, the inverse matrix of R^{-1} may require a larger number of computations and may become ill-conditioned. This will make real-time implementation impossible.
3. The optimal solution is based on the statistics, assuming that the size of the data block, N, is sufficient long. This will cause a long processing delay that will make real-time implementation impossible.

As we pointed out, solving the Wiener solution, Equation (10.7), requires a lot of computations, including matrix inversion for a general multiple-tap FIR filter. The well-known textbook authored by Widrow and Stearns (1985) described a powerful LMS algorithm by using the steepest descent algorithm to minimize the MSE sample by sample to locate the filter coefficient(s). We first study the steepest descent algorithm illustrated in the following:

$$w_{n+1} = w_n - \mu \frac{dJ}{dw} \tag{10.8}$$

where μ = constant controlling speed of convergence.

The illustration of the steepest decent algorithm for solving the optimal coefficient(s) is described in Figure 10.6.

As shown in the first plot in Figure 10.6, if $\frac{dJ}{dw} < 0$, notice that $-\mu \frac{dJ}{dw} > 0$. The new coefficient w_{n+1} will be increased to approach the optimal value w^* by Equation (10.8). On the other hand, if $\frac{dJ}{dw} > 0$, as shown in the second plot in Figure 10.6, we see that $-\mu \frac{dJ}{dw} < 0$. The new

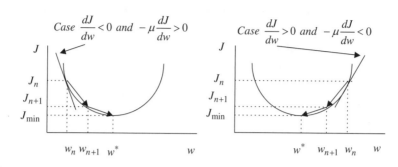

FIGURE 10.6

Illustration of the steepest descent algorithm.

coefficient w_{n+1} will be decreased to approach the optimal value w^*. When $\dfrac{dJ}{dw} = 0$, the best coefficient w_{n+1} is reached.

EXAMPLE 10.2

Consider the following quadratic MSE function for the Wiener filter:

$$J = 40 - 20w + 10w^2$$

Use the steepest decent method with an initial guess of $w_0 = 0$ and $\mu = 0.04$ to find the optimal solution for w^* and determine J_{min} by iterating three times.

Solution:

Taking a derivative of the MSE function, we have

$$\frac{dJ}{dw} = -20 + 10 \times 2w_n$$

When $n = 0$, we calculate

$$\mu \frac{dJ}{dw} = 0.04 \times (-20 + 10 \times 2w_0)\big|_{w_0=0} = -0.8$$

Applying the steepest decent algorithm, it follows that

$$w_1 = w_0 - \mu \frac{dJ}{dw} = 0 - (-0.8) = 0.8$$

Similarly for $n = 1$, we get

$$\mu \frac{dJ}{dw} = 0.04 \times (-20 + 10 \times 2w_1)\big|_{w_1=0.8} = -0.16$$

$$w_2 = w_1 - \mu \frac{dJ}{dw} = 0.8 - (-0.16) = 0.96$$

and for $n = 2$, it follows that

$$\mu \frac{dJ}{dw} = 0.04 \times (-20 + 10 \times 2w_2)\big|_{w_2=0.96} = -0.032$$

$$w_3 = w_2 - \mu \frac{dJ}{dw} = 0.96 - (-0.032) = 0.992$$

Finally, substituting $w^* \approx w_3 = 0.992$ into the MSE function, we get the minimum J_{min} as

$$J_{min} \approx 40 - 20w + 10w^2|_{w=0.992} = 40 - 20 \times 0.992 + 10 \times 0.992^2 = 30.0006$$

As we can see, after three iterations, the filter coefficient and minimum MSE values are very close to the theoretical values obtained in Example 10.1.

Application of the steepest descent algorithm still needs an estimation of the derivative of the MSE function that could include statistical calculation of a block of data. To change the algorithm to do sample-based processing, an LMS algorithm must be used. To develop the LMS algorithm in terms of sample-based processing, we take the statistical expectation out of J and then take the derivative to obtain an approximation of $\frac{dJ}{dw}$, that is,

$$J = e^2(n) = (d(n) - wx(n))^2 \tag{10.9}$$

$$\frac{dJ}{dw} = 2(d(n) - wx(n)) \frac{d(d(n) - wx(n))}{dw} = -2e(n)x(n) \tag{10.10}$$

Substituting $\frac{dJ}{dw}$ into the steepest descent algorithm in Equation (10.8), we achieve the LMS algorithm for updating a single-weight case as

$$w_{n+1} = w_n + 2\mu e(n)x(n) \tag{10.11}$$

where μ is the convergence parameter controlling speed of convergence. For example, let us choose $2\mu = 0.01$. In general, with an adaptive FIR filter of length N, we extend the single-tap LMS algorithm without going through derivation, as shown in the following equations:

$$y(n) = w_n(0)x(n) + w_n(1)x(n-1) + \cdots + w_n(N-1)x(n-N+1) \tag{10.12}$$

$$\begin{aligned} &\text{for } i = 0, \cdots, N-1 \\ &w_{n+1}(i) = w_n(i) + 2\mu e(n)x(n-i) \end{aligned} \tag{10.13}$$

The convergence factor is chosen to be

$$0 < \mu < \frac{1}{NP_x} \tag{10.14}$$

where P_x is the input signal power. In practice, if the ADC has 16-bit data, the maximum signal amplitude should be $A = 2^{15}$. Then the maximum input power must be less than

$$P_x < (2^{15})^2 = 2^{30}$$

Hence, we may make a selection of the convergence parameter as

$$\mu = \frac{1}{N \times 2^{30}} \approx \frac{9.3 \times 10^{-10}}{N} \tag{10.15}$$

We further neglect time index for $w_n(i)$ and use the notation $w(i) = w_n(i)$, since only the current updated coefficients are needed for next sample adaptation. We conclude the implementation of the LMS algorithm with the following steps:

1. Initialize $w(0)$, $w(1)$, ... $w(N-1)$ to arbitrary values.
2. Read $d(n)$, $x(n)$, and perform digital filtering:

$$y(n) = w(0)x(n) + w(1)x(n-1) + \cdots + w(N-1)x(n-N+1)$$

3. Compute the output error:

$$e(n) = d(n) - y(n)$$

4. Update each filter coefficient using the LMS algorithm:

$$\text{for } i = 0, \cdots, N-1$$
$$w(i) = w(i) + 2\mu e(n)x(n-i)$$

We will apply the adaptive filter to solve real-world problems in the next section.

10.3 APPLICATIONS: NOISE CANCELLATION, SYSTEM MODELING, AND LINE ENHANCEMENT

We now examine several applications of the LMS algorithm, such as noise cancellation, system modeling, and line enhancement via application examples. First, we begin with the noise cancellation problem to illustrate operations of the LMS adaptive FIR filter.

10.3.1 Noise Cancellation

The concept of noise cancellation was introduced in the previous section. Figure 10.7 shows the main concept.

The DSP system consists of two ADC channels. The first microphone with ADC captures the noisy speech, $d(n) = s(n) + n(n)$, which contains the clean speech $s(n)$ and noise $n(n)$ due to a noisy environment, while the second microphone with ADC resides where it picks up only the correlated noise and feeds the noise reference $x(n)$ to the adaptive filter. The adaptive filter uses the LMS algorithm to adjust its coefficients to produce the best estimate of noise $y(n) \approx n(n)$, which will be subtracted from the corrupted signal $d(n) = s(n) + n(n)$. The output of the error signal $e(n) = s(n) + n(n) - y(n) \approx \tilde{s}(n)$ is expected to be the best estimate of the clean speech signal. Through digital-to-analog conversion (DAC), the cleaned digital speech becomes analog voltage, which drives the speaker.

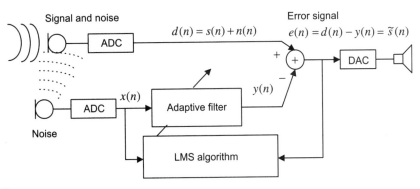

FIGURE 10.7

Simplest noise canceller using a one-tap adaptive filter.

We first study the noise cancellation problem using a simple two-tap adaptive filter via Example 10.3 and assumed data. The purpose of doing so is to become familiar with the setup and operations of the adaptive filter and LMS algorithm. The simulation for real adaptive noise cancellation follows.

EXAMPLE 10.3

Consider the DSP system for the noise cancellation application using an adaptive filter with two coefficients shown in Figure 10.8.

a. Set up the LMS algorithm for the adaptive filter.
b. Perform adaptive filtering to obtain outputs $e(n)$ for $n = 0, 1, 2$ given the following inputs and outputs:

$$x(0) = 1,\ x(1) = 1,\ x(2) = -1,\ d(0) = 2,\ d(1) = 1,\ d(2) = -2$$

The initial weights are $w(0) = w(1) = 0$, and the convergence factor is set to be $\mu = 0.1$.

Solution:

a. The adaptive LMS algorithm is set up as:
   Initialization: $w(0) = 0,\ w(1) = 0$
   Digital filtering: $y(n) = w(0)x(n) + w(1)x(n-1)$

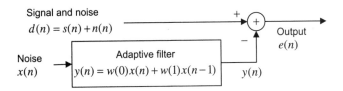

FIGURE 10.8

Noise cancellation in Example 10.3.

Computing the output: $e(n) = d(n) - y(n)$

Updating each weight to be used for the next coming sample:

$$w(i) = w(i) + 2\mu e(n)x(n-i), \quad \text{for} \quad i = 0, 1$$

or

$$w(0) = w(0) + 2\mu e(n)x(n)$$

$$w(1) = w(1) + 2\mu e(n)x(n-1)$$

b. We can see the adaptive filtering operations as follows:

For $n = 0$

Digital filtering:

$$y(0) = w(0)x(0) + w(1)x(-1) = 0 \times 1 + 0 \times 0 = 0$$

Computing the output:

$$e(0) = d(0) - y(0) = 2 - 0 = 2$$

Updating coefficients:

$$w(0) = w(0) + 2 \times 0.1 \times e(0)x(0) = 0 + 2 \times 0.1 \times 2 \times 1 = 0.4$$

$$w(1) = w(1) + 2 \times 0.1 \times e(0)x(-1) = 0 + 2 \times 0.1 \times 2 \times 0 = 0.0$$

For $n = 1$

Digital filtering:

$$y(1) = w(0)x(1) + w(1)x(0) = 0.4 \times 1 + 0 \times 1 = 0.4$$

Computing the output:

$$e(1) = d(1) - y(1) = 1 - 0.4 = 0.6$$

Updating coefficients:

$$w(0) = w(0) + 2 \times 0.1 \times e(1)x(1) = 0.4 + 2 \times 0.1 \times 0.6 \times 1 = 0.52$$

$$w(1) = w(1) + 2 \times 0.1 \times e(1)x(0) = 0 + 2 \times 0.1 \times 0.6 \times 1 = 0.12$$

For $n = 2$

Digital filtering:

$$y(2) = w(0)x(2) + w(1)x(1) = 0.52 \times (-1) + 0.12 \times 1 = -0.4$$

Computing the output:

$$e(2) = d(2) - y(2) = -2 - (-0.4) = -1.6$$

Updating coefficients:

$$w(0) = w(0) + 2 \times 0.1 \times e(2)x(2) = 0.52 + 2 \times 0.1 \times (-1.6) \times (-1) = 0.84$$

$$w(1) = w(1) + 2 \times 0.1 \times e(2)x(1) = 0.12 + 2 \times 0.1 \times (-1.6) \times 1 = -0.2$$

Hence, the adaptive filter outputs for the first three samples are listed as

$$e(0) = 2, \; e(1) = 0.6, \; e(2) = -1.6$$

Next we examine the MSE function assuming the following statistical data for the two-tap adaptive filter $y(n) = w(0)x(n) + w(1)x(n-1)$:

$$\sigma^2 = E\big[d^2(n)\big] = 4, \; E\big[x^2(n)\big] = E\big[x^2(n-1)\big] = 1, \; E[x(n)x(n-1)] = 0$$

$$E[d(n)x(n)] = 1, \text{ and } E[d(n)x(n-1)] = -1$$

We follow Equations (10.2) to (10.5) to achieve the minimum MSE function in two dimensions as

$$J = 4 + w^2(0) + w^2(1) - 2w(0) + 2w(1)$$

Figure 10.9 shows the MSE function versus the weights, where the optimal weights and the minimum MSE are $w^*(0) = 1$, $w^*(1) = -1$, and $J_{min} = 2$. If the adaptive filter continues to process the data, it will converge to the optimal weights, which locate the minimum MSE. The plot also indicates that the function is quadratic and that there exists only one minimum of the MSE surface.

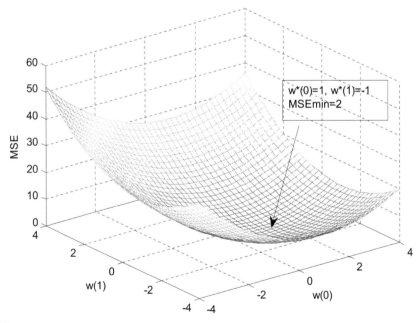

FIGURE 10.9

Plot of the MSE function versus two weights.

Next, a simulation example is given to illustrate this idea and its results. The noise cancellation system is assumed to have the following specifications:

- Sample rate = 8,000 Hz
- Original speech data: wen.dat
- Speech corrupted by Gaussian noise with a power of 1 delayed by 5 samples from the noise reference
- Noise reference containing Gaussian noise with a power of 1
- Adaptive FIR filter used to remove the noise
- Number of FIR filter taps = 21
- Convergence factor for the LMS algorithm is chosen to be 0.01 (<1/21).

The speech waveforms and spectral plots for the original, corrupted, and reference noise and for the cleaned speech are plotted in Figures 10.10A and Figure 10.10B. From the figures, it is observed that the enhanced speech waveform and spectrum are very close to the original ones. The LMS algorithm converges after approximately 400 iterations. The method is a very effective approach for noise canceling. The MATLAB implementation is detailed in Program 10.1.

Program 10.1. MATLAB program for adaptive noise cancellation.

```
close all; clear all
load wen.dat                  % Given by the instructor
fs=8000;                      % Sampling rate
t=0:1:length(wen)-1;          % Create index array
```

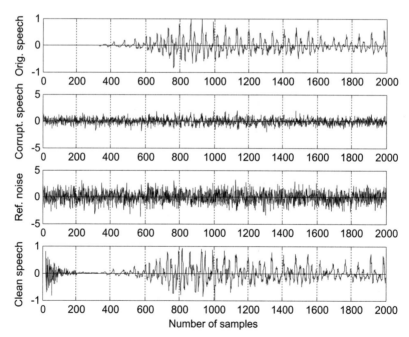

FIGURE 10.10A

Waveforms for original speech, corrupted speech, reference noise, and clean speech.

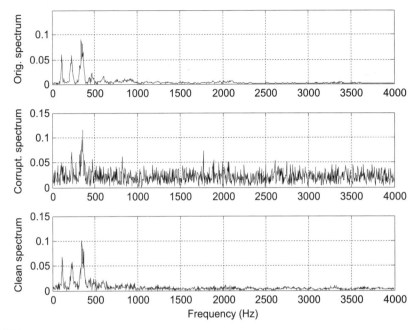

FIGURE 10.10B

Spectrum for original speech, corrupted speech, and clean speech.

```
t=t/fs;                          % Convert indices to time instant
x=randn(1,length(wen));          % Generate random noise
n=filter([ 0 0 0 0 0 0.5 ],1,x); % Generate the corruption noise
d=wen+n;                 % Generate signal plus noise
mu=0.01;                 % Initialize step size
w=zeros(1,21);           % Initialize adaptive filter coefficients
y=zeros(1,length(t));    % Initialize the adaptive filter output array
e=y;                     % Initialize the output array
% Adaptive filtering using LMS algorithm
for m=22:1:length(t)-1
  sum=0;
  for i=1:1:21
  sum=sum+w(i)*x(m-i);
  end
  y(m)=sum;
  e(m)=d(m)-y(m);
  for i=1:1:21
  w(i)=w(i)+2*mu*e(m)*x(m-i);
  end
end
% Calculate the single-sided amplitude spectrum for the original signal
WEN=2*abs(fft(wen))/length(wen);WEN(1)=WEN(1)/2;
```

```
% Calculate the single-sided amplitude spectrum for the corrupted signal
D=2*abs(fft(d))/length(d);D(1)=D(1)/2;
f=[0:1:length(wen)/2]*8000/length(wen);
% Calculate the single-sided amplitude spectrum for the noise-cancelled signal
E=2*abs(fft(e))/length(e);E(1)=E(1)/2;
% Plot signals and spectrums
subplot(4,1,1), plot(wen);grid; ylabel('Orig. speech');
subplot(4,1,2),plot(d);grid; ylabel('Corrupt. speech')
subplot(4,1,3),plot(x);grid;ylabel('Ref. noise');
subplot(4,1,4),plot(e);grid; ylabel('Clean speech');
xlabel('Number of samples');
figure
subplot(3,1,1),plot(f,WEN(1:length(f)));grid
ylabel('Orig. spectrum')
subplot(3,1,2),plot(f,D(1:length(f)));grid; ylabel('Corrupt. spectrum')
subplot(3,1,3),plot(f,E(1:length(f)));grid
ylabel('Clean spectrum'); xlabel('Frequency (Hz)');
```

Other interference cancellations include that of 60-Hz interference cancellation in electrocardiography (ECG) (Chapter 8) and echo cancellation in long-distance telephone circuits, which will be described in Section 10.4.

10.3.2 System Modeling

Another application of the adaptive filter is system modeling. The adaptive filter can keep tracking the behavior of an unknown system by using the unknown system input and output, as depicted in Figure 10.11.

As shown in the figure, after the adaptive filter converges, the adaptive filter output $y(n)$ will approach to the unknown system's output. Since both the unknown system and the adaptive filter respond to the same input, the transfer function of the adaptive filter approximates the transfer function of the unknown system.

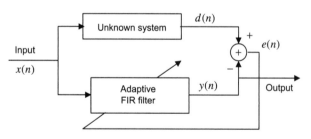

FIGURE 10.11

Adaptive filter for system modeling.

EXAMPLE 10.4

Given the system modeling described in this section and using a single-weight adaptive filter $y(n) = wx(n)$ to perform the system-modeling task,

a. set up the LMS algorithm to implement the adaptive filter assuming that initially $w = 0$ and $\mu = 0.5$;
b. perform adaptive filtering to obtain $y(0)$, $y(1)$, $y(2)$, and $y(3)$, given

$$d(0) = 1, \ d(1) = 2, \ d(2) = -2, \ d(3) = 2,$$

$$x(0) = 0.5, \ x(1) = 1, \ x(2) = -1, \ x(3) = 1$$

Solution:

a. Adaptive filtering equations are set up as

$$w = 0 \quad \text{and} \quad 2\mu = 2 \times 0.5 = 1$$

$$y(n) = wx(n)$$

$$e(n) = d(n) - y(n)$$

$$w = w + e(n)x(n)$$

b. Adaptive filtering:

$$n = 0, \qquad y(0) = wx(0) = 0 \times 0.5 = 0$$

$$e(0) = d(0) - y(0) = 1 - 0 = 1$$

$$w = w + e(0)x(0) = 0 + 1 \times 0.5 = 0.5$$

$$n = 1, \qquad y(1) = wx(1) = 0.5 \times 1 = 0.5$$

$$e(1) = d(1) - y(1) = 2 - 0.5 = 1.5$$

$$w = w + e(1)x(1) = 0.5 + 1.5 \times 1 = 2.0$$

$$n = 2, \qquad y(2) = wx(2) = 2 \times (-1) = -2$$

$$e(2) = d(2) - y(2) = -2 - (-2) = 0$$

$$w = w + e(2)x(2) = 2 + 0 \times (-1) = 2$$

$$n = 3, \qquad y(3) = wx(3) = 2 \times 1 = 2$$

$$e(3) = d(3) - y(3) = 2 - 2 = 0$$

$$w = w + e(3)x(3) = 2 + 0 \times 1 = 2$$

For this particular case, the system is actually a digital amplifier with a gain of 2.

Next, we assume the unknown system is a fourth-order bandpass IIR filter whose 3-dB lower and upper cutoff frequencies are 1,400 Hz and 1,600 Hz operating at 8,000 Hz. We use an input consisting of tones of 500, 1,500, and 2,500 Hz. The unknown system's frequency responses are shown in Figure 10.12.

The input waveform $x(n)$ with three tones is shown as the first plot in Figure 10.13. We can predict that the output of the unknown system will contain a 1,500 Hz tone only, since the other two tones are rejected by the unknown system. Now, let us look at adaptive filter results. We use an FIR adaptive filter with the number of taps being 21, and a convergence factor set to 0.01. In the time domain, the output waveforms of the unknown system $d(n)$ and adaptive filter output $y(n)$ are almost identical after 70 samples when the LMS algorithm converges. The error signal $e(n)$ is also plotted to show the adaptive filter keeps tracking the unknown system's output with no difference after the first 50 samples.

Figure 10.14 depicts the frequency domain comparisons. The first plot displays the frequency components of the input signal, which clearly shows 500 Hz, 1,500 Hz, and 2,500 Hz. The second plot shows the unknown system's output spectrum, which contains only a 1,500 Hz tone, while the third plot displays the spectrum of the adaptive filter output. As we can see, in the frequency domain, the adaptive filter tracks the characteristics of the unknown system. The MATLAB implementation is given in Program 10.2.

Program 10.2. MATLAB program for adaptive system identification.

```
close all; clear all
%Design unknown system
fs=8000; T=1/fs;          % Sampling rate and sampling period
```

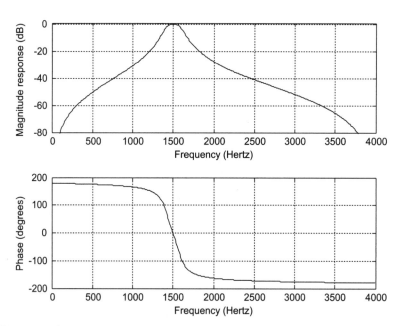

FIGURE 10.12

The unknown system's frequency responses.

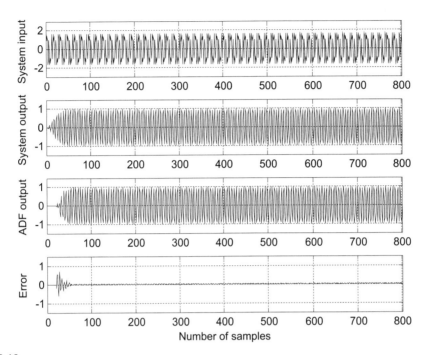

FIGURE 10.13

The waveforms for the unknown system's output, adaptive filter output, and error output.

```
% Bandpass filter design
% for the assumed unknown system using the bilinear transformation
%(BLT) method (see Chapter 8)
wd1=1400*2*pi; wd2=1600*2*pi;
wa1=(2/T)*tan(wd1*T/2); wa2=(2/T)*tan(wd2*T/2);
BW=wa2-wa1;
w0=sqrt(wa2*wa1);
[B,A]=lp2bp([1],[1 1.4141 1],w0,BW);
[b,a]=bilinear(B,A,fs);
freqz(b,a,512,fs); axis([0 fs/2 -80 1]); % Frequency response plots
figure
t=0:T:0.1;     % Generate the time vector
x=cos(2*pi*500*t)+sin(2*pi*1500*t)+cos(2*pi*2500*t+pi/4);
d=filter(b,a,x); % Produce unknown system output
mu=0.01;    % Convergence factor
w=zeros(1,21); y=zeros(1,length(t)); % Initialize the coefficients and output
e=y;              % Initialize the error vector
% Perform adaptive filtering using LMS algorithm
for m=22:1:length(t)-1
  sum=0;
  for i=1:1:21
```

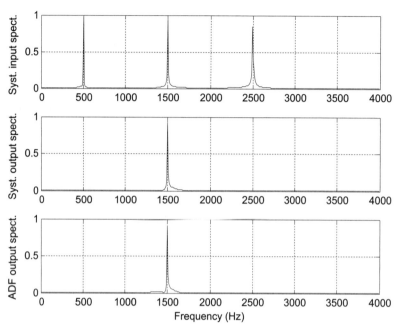

FIGURE 10.14

Spectrum for the input signal, unknown system output, and the adaptive filter output.

```
    sum=sum+w(i)*x(m-i);
    end
    y(m)=sum;
    e(m)=d(m)-y(m);
    for i=1:1:21
    w(i)=w(i)+2*mu*e(m)*x(m-i);
    end
end
% Calculate the single-sided amplitude spectrum for the input
X=2*abs(fft(x))/length(x);X(1)=X(1)/2;
% Calculate the single-sided amplitude spectrum for the unknown system output
D=2*abs(fft(d))/length(d);D(1)=D(1)/2;
% Calculate the single-sided amplitude spectrum for the adaptive filter output
Y=2*abs(fft(y))/length(y);Y(1)=Y(1)/2;
% Map the frequency index to its frequency in Hz
f=[0:1:length(x)/2]*fs/length(x);
% Plot signals and spectra
subplot(4,1,1), plot(x);grid; axis([0 length(x) -3 3]);
ylabel('System input');
subplot(4,1,2), plot(d);grid; axis([0 length(x) -1.5 1.5]);
ylabel('System output');
subplot(4,1,3),plot(y);grid; axis([0 length(y) -1.5 1.5]);
```

```
ylabel('ADF output')
subplot(4,1,4),plot(e);grid; axis([0 length(e) -1.5 1.5]);
ylabel('Error'); xlabel('Number of samples')
figure
subplot(3,1,1),plot(f,X(1:length(f)));grid; ylabel('Syst. input spect.')
subplot(3,1,2),plot(f,D(1:length(f)));grid; ylabel('Syst. output spect.')
subplot(3,1,3),plot(f,Y(1:length(f)));grid
ylabel('ADF output spect.'); xlabel('Frequency (Hz)');
```

10.3.3 Line Enhancement Using Linear Prediction

We study adaptive filtering via another application example: line enhancement. If the signal frequency content is very narrow compared with the bandwidth and changes with time, then the signal can efficiently be enhanced by the adaptive filter, which is line enhancement. Figure 10.15 shows line enhancement using the adaptive filter where the LMS algorithm is used. As illustrated in the figure, the signal $d(n)$ is the sine wave signal corrupted by the white Gaussian noise $n(n)$. The enhanced line consists of the delay element to delay the corrupted signal by Δ samples to produce an input to the adaptive filter. The adaptive filter is actually a linear predictor of the desired narrow band signal. A two-tap FIR adaptive filter can predict one sinusoid (proof is beyond the scope of this text). The value of Δ is usually determined by experiments or experience in practice to achieve the best enhanced signal.

Our simulation example has the following specifications:

- Sampling rate = 8,000 Hz
- Corrupted signal = 500 Hz tone with white Gaussian noise added to the unit amplitude
- Adaptive filter = FIR type, 21 taps
- Convergence factor = 0.001
- Delay value Δ = 7
- LMS algorithm is applied

Figure 10.16 shows the time domain results. The first plot is the noisy signal, while the second plot clearly demonstrates the enhanced signal. Figure 10.17 describes the frequency domain point of view. The spectrum of the noisy signal is shown in the top plot, where we can see the white noise is populated

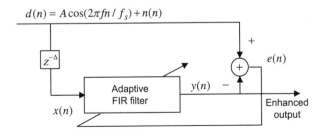

$d(n) = A\cos(2\pi fn / f_s) + n(n)$

$z^{-\Delta}$

Adaptive FIR filter $y(n)$

$x(n)$

$+$

$+$ $e(n)$

$-$

Enhanced output

FIGURE 10.15

Line enhancement using an adaptive filter.

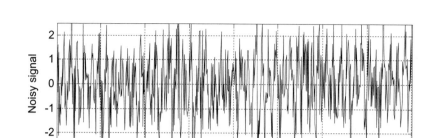

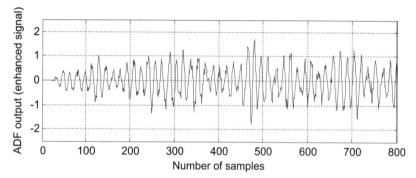

FIGURE 10.16

Noisy signal and enhanced signal.

over the entire bandwidth. The bottom plot is the enhanced signal spectrum. Since the method is adaptive, it is especially effective when the enhanced signal frequency is changing with time. Program 10.3 lists the MATLAB program for this simulation.

Program 10.3. MATLAB program for adaptive line enhancement.

```
close all; clear all
fs=8000; T=1/fs;                    % Sampling rate and sampling period
t=0:T:0.1;                          % 1 second time instant
n=randn(1,length(t));               % Generate Gaussian random noise
d=cos(2*pi*500*t)+n;                % Generate 500-Hz tone plus noise
x=filter([ 0 0 0 0 0 0 0 1 ],1,d);  % Delay filter
mu=0.001;                           % Initialize the step size for LMS algorithms
w=zeros(1,21);                      % Initialize the adaptive filter coefficients
y=zeros(1,length(t));               % Initialize the adaptive filter output
e=y;                                % Initialize the error vector
% Perform adaptive filtering using the LMS algorithm
for m=22:1:length(t)-1
  sum=0;
  for i=1:1:21
  sum=sum+w(i)*x(m-i);
```

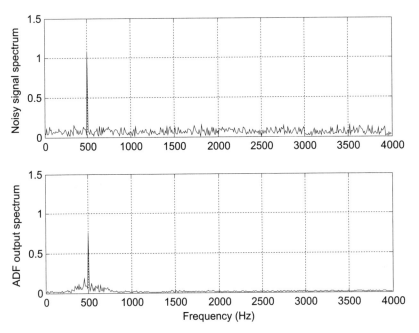

FIGURE 10.17

Spectrum plots for the noisy signal and enhanced signal.

```
   end
   y(m)=sum;
   e(m)=d(m)-y(m);
   for i=1:1:21
   w(i)=w(i)+2*mu*e(m)*x(m-i);
   end
end
% Calculate the single-sided amplitude spectrum for corrupted signal
D=2*abs(fft(d))/length(d);D(1)=D(1)/2;
% Calculate the single-sided amplitude spectrum for enhanced signal
Y=2*abs(fft(y))/length(y);Y(1)=Y(1)/2;
% Map the frequency index to its frequency in Hz
f=[0:1:length(x)/2]*8000/length(x);
% Plot the signals and spectra
subplot(2,1,1), plot(d);grid; axis([0 length(x) -2.5 2.5]); ylabel('Noisy signal');
subplot(2,1,2),plot(y);grid; axis([0 length(y) -2.5 2.5]);
ylabel('ADF output (enhanced signal)'); xlabel('Number of samples')
figure
subplot(2,1,1),plot(f,D(1:length(f)));grid; axis([0 fs/2 0 1.5]);
ylabel('Noisy signal spectrum')
subplot(2,1,2),plot(f,Y(1:length(f)));grid; axis([0 fs/2 0 1.5]);
ylabel('ADF output spectrum'); xlabel('Frequency (Hz)');
```

10.4 OTHER APPLICATION EXAMPLES

This section continues to explore other adaptive filter applications briefly, without showing computer simulations. The topics include periodic interference cancellation, ECG interference cancellation, and echo cancellation in long-distance telephone circuits. Detailed information can also be explored in Haykin (1991), Ifeachor and Jervis (2002), Stearns (2003), and Widrow and Stearns (1985).

10.4.1 Canceling Periodic Interferences Using Linear Prediction

An audio signal may be corrupted by periodic interference with no noise reference available. Such examples include the playback of speech or music with tape hum interference, turntable rumble, or vehicle engine or power line interference. We can use the modified line enhancement structure as shown in Figure 10.18.

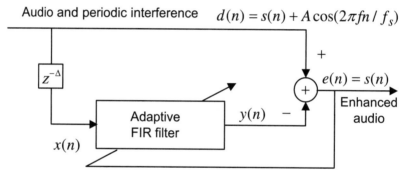

FIGURE 10.18

Canceling periodic interference using the adaptive filter.

The adaptive filter uses the delayed version of the corrupted signal $x(n)$ to predict the periodic interference. The number of delayed samples is selected through experiments that determine the performance of the adaptive filter. Note that a two-tap FIR adaptive filter can predict one sinusoid, as noted earlier. After convergence, the adaptive filter would predict the interference as

$$y(n) = \sum_{i=0}^{N-1} w(i)x(n-i) \approx A\cos(2\pi fn/f_s) \tag{10.16}$$

Therefore, the error signal contains only the desired audio signal

$$e(n) \approx s(n) \tag{10.17}$$

10.4.2 Electrocardiography Interference Cancellation

As we discussed in Chapters 1 and 8, in recording of electrocardiograms (ECG), there often exists unwanted 60-Hz interference, along with its harmonics, in the recorded data. This interference comes

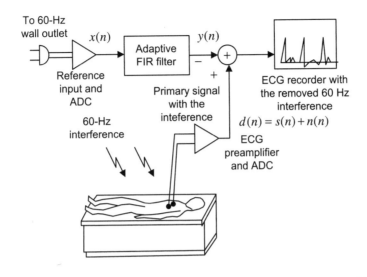

FIGURE 10.19

Illustration of canceling 60-Hz interference in ECG.

from the power line, including effects from magnetic induction, displacement currents in leads or in the body of the patient, and equipment interconnections and imperfections.

Figure 10.19 illustrates the application of adaptive noise canceling in ECG. The primary input is taken from the ECG preamplifier, while a 60-Hz reference input is taken from a wall outlet with proper attenuation. After proper signal conditioning, the digital interference $x(n)$ is acquired by the digital signal (DS) processor. The digital adaptive filter uses this reference input signal to produce an estimate, which approximates the 60-Hz interference $n(n)$ sensed from the ECG amplifier:

$$y(n) \approx n(n) \tag{10.18}$$

Here, an FIR adaptive filter with N taps and the LMS algorithm can be used for this application:

$$y(n) = w(0)x(n) + w(1)x(n-1) + \cdots + w(N-1)x(n-N+1) \tag{10.19}$$

Then after convergence of the adaptive filter, the estimated interference is subtracted from the primary signal of the ECG preamplifier to produce the output signal $e(n)$, in which the 60-Hz interference is cancelled:

$$e(n) = d(n) - y(n) = s(n) + n(n) - x(n) \approx s(n) \tag{10.20}$$

With enhanced ECG recording, doctors in clinics can give more accurate diagnoses for patients.

Canceling the maternal ECG in fetal monitoring is another important application. The block diagram is shown in Figure 10.20(a). Fetal ECG plays an important role in monitoring the condition of the baby before or during birth. However, the ECG acquired from the mother's abdomen is contaminated by noise such as muscle activity and fetal motion, as well as the mother's own ECG. In order to reduce the effect of the mother's ECG, four (or more) chest leads

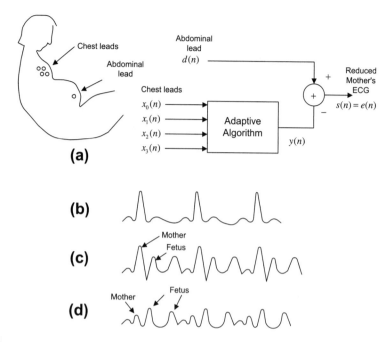

FIGURE 10.20

Canceling the maternal ECG in fetal monitoring.

(electrodes) are used to acquire the reference inputs: $x_0(n)$, $x_1(n)$, $x_2(n)$, and $x_3(n)$, with the assumption that these channels only contain the mother's ECG (see Figure 10.20(b)). One lead (electrode) placed on the mother's abdomen is used to capture the fetal information $d(n)$, which may be corrupted by the mother's ECG as shown in Figure 10.20(c). An adaptive filter uses its references to predict the mother ECG, which will be subtracted from the corrupted fetus signal. Then the fetal ECG with the reduced mother's ECG is obtained, as depicted in Figure 10.20(d). One possible LMS algorithm is listed below:

For $k = 0,1,2,3$

$$y_k(n) = w_k(0)x_k(n) + w_k(1)x_k(n-1) + \cdots + w_k(N-1)x_k(n-N+1)$$

$$y(n) = y_0(n) + y_1(n) + y_2(n) + y_3(n)$$

$$s(n) = e(n) = d(n) - y(n)$$

For $k = 0,1,2,3$

$$w_k(n-i) = w_k(n-i) + 2\mu e(n)x_k(n-i), \text{ for } i = 0,1,\cdots N-1$$

10.4.3 Echo Cancellation in Long-Distance Telephone Circuits

Long-distance telephone transmission often suffers from impedance mismatches. This occurs primarily at the hybrid circuit interface. Balancing electric networks within the hybrid can never perfectly match the hybrid to the subscriber loop due to temperature variations, degradation of transmission lines, and so on. As a result, a small portion of the received signal is leaked for transmission. For example, in Figure 10.21A, if speaker B talks, the speech indicated as $x_B(n)$ will pass the transmission line to reach user A, and a portion of $x_B(n)$ at site A is leaked and transmitted back to the user B, forcing caller B to hear his or her own voice. This is known as an echo for speaker B. A similar echo illustration can be conducted for speaker A. When the telephone call is made over a long distance (more than 1,000 miles, such as geostationary satellites), the echo can be delayed by as much as 540 milliseconds. The echo impairment can be annoying to the customer and increases with distance.

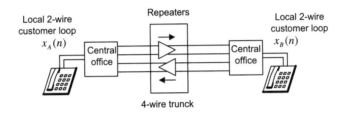

FIGURE 10.21A

Simplified long-distance circuit.

To circumvent the problem of echo in long-distance communications, an adaptive filter is applied at each end of the communication system, as shown in Figure 10.21B. Let us examine the adaptive filter installed at the speaker A site. The incoming signal is $x_B(n)$ from speaker B, while the outgoing signal contains the speech from the speaker A and a portion of leakage from the hybrid circuit

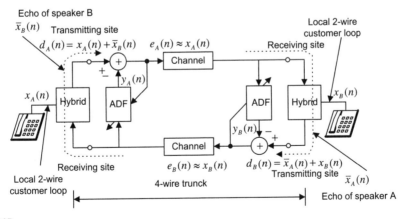

FIGURE 10.21B

Adaptive echo cancellers.

$d_A(n) = x_A(n) + \bar{x}_B(n)$. If the leakage $\bar{x}_B(n)$ returns back to speaker B, it becomes an annoying echo. To prevent the echo, the adaptive filter at the speaker A site uses the incoming signal from speaker B as an input and makes its output approximate to the leaked speaker B signal by adjusting its filter coefficients; that is,

$$y_A(n) = \sum_{i=0}^{N-1} w(i)x_B(n-i) \approx \bar{x}_B(n) \tag{10.21}$$

As shown in Figure 10.21B, the estimated echo $y_A(n) \approx \bar{x}_B(n)$ is subtracted from the outgoing signal, thus producing the signal that contains only speech A; that is, $e_A(n) \approx x_A(n)$. As a result, the echo of speaker B is removed. We can illustrate similar operations for the adaptive filter used at the speaker B site. In practice, an FIR adaptive filter with several hundred coefficients or more is commonly used to effectively cancel the echo. If nonlinearities are concerned in the echo path, a corresponding nonlinear adaptive canceller can be used to improve the performance of the echo cancellation.

Other forms of adaptive filters and other applications are beyond the scope of this book. The reader is referred to the references for further development.

10.5 LABORATORY EXAMPLES USING THE TMS320C6713 DSK

The implementation for system modeling in Section 10.4.3 is shown in Figure 10.22, where the input is fed from a function generator. The unknown system is a bandpass filter with a lower cutoff frequency of 1,400 Hz and upper cutoff frequency of 1,600 Hz. As shown in Figure 10.22, the left input channel (Left Line In [LCI]) is used for the input while the left output channel (Left Line Out [LCO]) and the right output channel (Right Line Out [RCO]) are designated as the system output and error output,

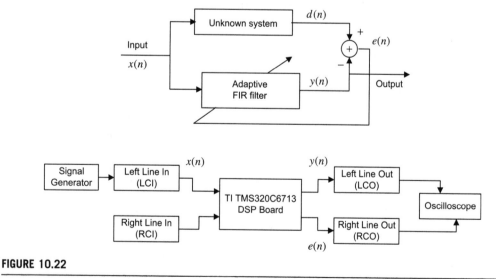

FIGURE 10.22

Setup for system modeling using the LMS adaptive filter.

respectively. Note that the right input channel (Right Line In [RCI]) is not used. When the input frequency is swept from 200 Hz to 3,000 Hz, the output shows a maximum peak when the input frequency is dialed to around 1,500 Hz. Hence, the adaptive filter acts like the unknown system. Program 10.4 gives the sample program segment.

Program 10.4. Program segment for system modeling.

```
/*Numerator coefficients */
/*for the bandpass filter (unknown system) with fL=1.4 kHz, fH=1.6 kHz*/
float b[5]={ 0.005542761540433, 0.000000000000002, -0.011085523080870,
      0.000000000000003 0.005542761540431};
/*Denominator coefficients */
/*for the bandpass filter (unknown system) with fL=1.4 kHz, fH=1.6 kHz*/
float a[5]={ 1.000000000000000, -1.450496619180500, 2.306093105231476,
      -1.297577189144526 0.800817049322883};
float x[40]={0,0,0,0,0,0,0,0,0,0,0,0,0,0,0,0,0,0,0,0,
      0,0,0,0,0,0,0,0,0,0,0,0,0,0,0,0,0,0,0,0}; /*Reference input buffer*/
float w[40]={0,0,0,0,0,0,0,0,0,0,0,0,0,0,0,0,0,0,0,0,
      0,0,0,0,0,0,0,0,0,0,0,0,0,0,0,0,0,0,0,0}; /*Adaptive filter coefficients*/
float d[1]={0.0}; /*Unknown system output */
float y[1]={0,0}; /*Adaptive filter output */
float e[1]={0.0}; /*Error signal */
float mu=0.000000000002; /*Adaptive filter convergence factor*/
interrupt void c_int11()
{
  float lc; /*left channel input */
  float rc; /*right channel input */
  float lcnew; /*left channel output */
  float rcnew; /*right channel output */
  int i;
//Left channel and right channel inputs
  AIC23_data.combo=input_sample();
  lc=(float) (AIC23_data.channel[LEFT]);
  rc= (float) (AIC23_data.channel[RIGHT]);
// Insert DSP algorithm below
  for(i=39;i>0;i--) /*Update the input buffer*/
  { x[i]=x[i-1]; }
  x[0]=lc;
  d[0]=b[0]*x[0]+ b[1]*x[1]+ b[2]*x[2]+ b[3]*x[3]+ b[4]*x[4]
    -a[1]*d[1]-a[2]*d[2]- a[3]*d[3]- a[4]*d[4]; /*Unknown system output*/
// Adaptive filter
  y[0]=0;
  for(i=0;i<40; i++)
  { y[0]=y[0]+w[i]*x[i];}
  e[0]=d[0]-y[0]; /* Error output */
  for(i=0;i<40; i++)
  { w[i]=w[i]+2*mu*e[0]*x[i];} /* LMS algorithm */
// End of the DSP algorithm
  lcnew=y[0]; /* Send the tracked output */
```

```
  rcnew=e[0]; /* Send the error signal*/
  AIC23_data.channel[LEFT]=(short) lcnew;
  AIC23_data.channel[RIGHT]=(short) rcnew;
  output_sample(AIC23_data.combo);
}
```

With the advantage of the stereo input and output channels, we can conduct system modeling for an unknown analog system illustrated in Figure 10.23, where RCI is used to feed the unknown analog system output to the DSK. The program segment is listed in Program 10.5.

Program 10.5. Program segment for modeling an analog system.

```
float x[40]={0,0,0,0,0,0,0,0,0,0,0,0,0,0,0,0,0,0,0,0,
      0,0,0,0,0,0,0,0,0,0,0,0,0,0,0,0,0,0,0,0}; /*Reference input buffer*/
float w[40]={0,0,0,0,0,0,0,0,0,0,0,0,0,0,0,0,0,0,0,0,
      0,0,0,0,0,0,0,0,0,0,0,0,0,0,0,0,0,0,0,0}; /*Adaptive filter coefficients*/
float d[1]={0.0}; /*Unknown system output */
float y[1]={0,0}; /*Adaptive filter output */
float e[1]={0.0}; /*Error signal */
float mu=0.000000000002; /*Adaptive filter convergence factor*/
interrupt void c_int11()
{
  float lc; /*left channel input */
  float rc; /*right channel input */
  float lcnew; /*left channel output */
  float rcnew; /*right channel output */
  int i;
//Left channel and right channel inputs
  AIC23_data.combo=input_sample();
  lc=(float) (AIC23_data.channel[LEFT]);
  rc= (float) (AIC23_data.channel[RIGHT]);
// Insert DSP algorithm below
  for(i=39;i>0;i--) /*Update the input buffer*/
  { x[i]=x[i-1]; }
```

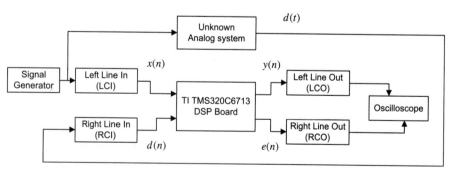

FIGURE 10.23

System modeling using an LMS adaptive filter.

```
  x[0]=lc;
  d[0]=rc; /*Unknown system output*/
// Adaptive filter
  y[0]=0;
  for(i=0;i<40; i++)
  { y[0]=y[0]+w[i]*x[i];}
  e[0]=d[0]-y[0]; /* Error output */
  for(i=0;i<40; i++)
  { w[i]=w[i]+2*mu*e[0]*x[i];} /* LMS algorithm */
// End of the DSP algorithm
  lcnew=y[0]; /* Send the tracked output */
  rcnew=e[0]; /* Send the error signal*/
  AIC23_data.channel[LEFT]=(short) lcnew;
  AIC23_data.channel[RIGHT]=(short) rcnew;
  output_sample(AIC23_data.combo);
}
```

Figure 10.24A shows an example of a tonal noise reduction system, and Figure 10.24B shows the details of the adaptive noise cancellation. The first DSP board is used to create the real-time corrupted signal, which is obtained by mixing the mono audio source (Left Line In [LCI1]) from any audio device and the tonal noise (Right Line In [RCI1]) generated from a function generator. The output (Left line Out [LCO1]) is the corrupted signal, which is fed to the second DSP board for noise cancellation application. The adaptive FIR filter in the second DSP board uses the reference input (Right Line In [RCI2]) to generate the output, which is used to cancel the tonal noise embedded in the corrupted signal (Left Line In [LCI2]). The output (Left Line Out [LCO2]) produces the clean mono audio signal (Jiang and Tan, 2012). Program 10.6 details the implementation.
Program 10.6. Program segments for noise cancellation.
   (a) Program segment for DSK 1 (generation of the corrupted signal).

```
float x[1]={0.0}; /* Tonal reference noise */
Float s[1]={0,0}; /* Audio signal */
float d[1]={0.0}; /* Corrupted signal*/
```

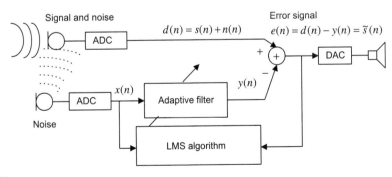

FIGURE 10.24A

Block diagram for tonal noise cancellation.

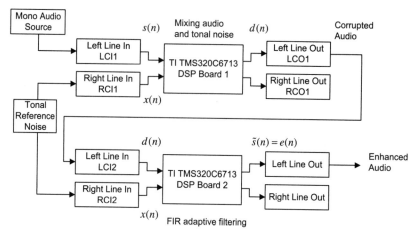

FIGURE 10.24B

Tonal noise cancellation with the adaptive filter.

```
interrupt void c_int11()
{
  float lc; /*left channel input */
  float rc; /*right channel input */
  float lcnew; /*left channel output */
  float rcnew; /*right channel output */
  int i;
//Left channel and right channel inputs
 AIC23_data.combo=input_sample();
 lc=(float) (AIC23_data.channel[LEFT]);
 rc= (float) (AIC23_data.channel[RIGHT]);
// Insert DSP algorithm below
  s[0]=lc;
  x[0]=rc;
  D[0]=s[0]+x[0];
// End of the DSP algorithm
  lcnew=d[0]; /* Send to DAC */
  rcnew=rc; /* keep the original data */
  AIC23_data.channel[LEFT]=(short) lcnew;
  AIC23_data.channel[RIGHT]=(short) rcnew;
  output_sample(AIC23_data.combo);
}
```

(b) Program segment for DSK 2 (LMS adaptive filter).

```
float x[20]={0,0,0,0,0,0,0,0,0,0,0,0,0,0,0,0,0,0,0,0}; /*Reference input buffer*/
float w[20]={0,0,0,0,0,0,0,0,0,0,0,0,0,0,0,0,0,0,0,0}; /*Adaptive filter
coefficients*/
float d[1]={0.0}; /* Corrupted signal*/
float y[1]={0,0}; /* Adaptive filter output */
```

```
float e[1]={0.0}; /* Enhanced signal */
float mu=0.000000000004; /*Adaptive filter convergence factor*/
interrupt void c_int11()
{
  float lc; /*left channel input */
  float rc; /*right channel input */
  float lcnew; /*left channel output */
  float rcnew; /*right channel output */
  int i;
//Left channel and right channel inputs
 AIC23_data.combo=input_sample();
 lc=(float) (AIC23_data.channel[LEFT]);
 rc= (float) (AIC23_data.channel[RIGHT]);
// Insert DSP algorithm below
 d[0]=lc; /*Corrupted signal*/
 for(i=19;i>0;i--) /*Update the reference noise buffer input buffer*/
 { x[i]=xn[i-1]; }
 x[0]=lc;
// Adaptive filter
 y[0]=0;
 for(i=0;i<20; i++)
 { y[0]=y[0]+w[i]*x[i];}
 e[0]=d[0]-y[0]; /* Enhanced output */
 for(i=0;i<20; i++)
 { w[i]=w[i]+2*mu*e[0]*x[i]; }/* LMS algorithm */
// End of the DSP algorithm
  lcnew=e[0]; /* Send to DAC */
  rcnew=rc; /* keep the original data */
  AIC23_data.channel[LEFT]=(short) lcnew;
  AIC23_data.channel[RIGHT]=(short) rcnew;
  output_sample(AIC23_data.combo);
}
```

Many other practical configurations can be implemented similarly.

10.6 SUMMARY

1. Adaptive filters can be applied to signal-changing environments, spectral overlap between noise and signal, and unknown, or time-varying, noise.
2. Wiener filter theory provides optimal weight solutions based on statistics. It involves collection of a large block of data, calculation of an autocorrelation matrix and a cross-correlation matrix, and inversion of a large autocorrelation matrix.
3. The steepest decent algorithm can find the optimal weight solution using an iterative method, so a large matrix inversion is not needed. But it still requires calculating an autocorrelation matrix and cross-correlation matrix.
4. The LMS is a sample-based algorithm, which does not need collection of data or computation of statistics and does not involve matrix inversion.

5. The convergence factor for the LMS algorithm is bounded by the reciprocal of the product of the number of filter coefficients and input signal power.
6. The LMS adaptive FIR filter can be effectively applied for noise cancellation, system modeling, and line enhancement.
7. Further exploration includes other applications such as cancellation of periodic interference, biomedical ECG signal enhancement, and adaptive telephone echo cancellation.

10.7 PROBLEMS

10.1. Given a quadratic MSE function for the Wiener filter

$$J = 50 - 40w + 10w^2$$

find the optimal solution for w^* to achieve the minimum MSE J_{min} and determine J_{min}.

10.2. Given a quadratic MSE function for the Wiener filter

$$J = 15 + 20w + 10w^2$$

find the optimal solution for w^* to achieve the minimum MSE J_{min} and determine J_{min}.

10.3. Given a quadratic MSE function for the Wiener filter

$$J = 100 + 20w + 2w^2$$

find the optimal solution for w^* to achieve the minimum MSE J_{min} and determine J_{min}.

10.4. Given a quadratic MSE function for the Wiener filter

$$J = 10 - 30w + 15w^2$$

find the optimal solution for w^* to achieve the minimum MSE J_{min} and determine J_{min}.

10.5. Given a quadratic MSE function for the Wiener filter

$$J = 50 - 40w + 10w^2$$

use the steepest descent method with an initial guess of $w_0 = 0$ and a convergence factor $\mu = 0.04$ to find the optimal solution for w^* and determine J_{min} by iterating three times.

10.6. Given a quadratic MSE function for the Wiener filter

$$J = 15 + 20w + 10w^2$$

use the steepest descent method with an initial guess of $w_0 = 0$ and a convergence factor $\mu = 0.04$ to find the optimal solution for w^* and determine J_{min} by iterating three times.

10.7. Given a quadratic MSE function for the Wiener filter

$$J = 100 + 20w + 2w^2$$

use the steepest descent method with an initial guess of $w_0 = -4$ and a convergence factor $\mu = 0.2$ to find the optimal solution for w^* and determine J_{min} by iterating three times.

10.8. Given a quadratic MSE function for the Wiener filter

$$J = 10 - 30w + 15w^2$$

use the steepest descent method with an initial guess of $w_0 = 2$ and a convergence factor $\mu = 0.02$ to find the optimal solution for w^* and determine J_{min} by iterating three times.

10.9. Consider the following DSP system used for noise cancellation applications (Figure 10.25), in which $d(0) = 3, d(1) = -2, d(2) = 1, x(0) = 3, x(1) = -1, x(2) = 2$, and there is an adaptive filter with two taps $y(n) = w(0)x(n) + w(1)x(n-1)$ with initial values $w(0) = 0, w(1) = 1$, and $\mu = 0.1$,

a. Determine the LMS algorithm equations

$$y(n) =$$
$$e(n) =$$
$$w(0) =$$
$$w(1) =$$

b. Perform adaptive filtering for each $n = 0, 1, 2$.

10.10. Given a DSP system with a sampling rate of 8,000 samples per second, implement an adaptive filter with five taps for system modeling.

As shown in Figure 10.26, assume the unknown system transfer function is

$$H(z) = \frac{0.25 + 0.25z^{-1}}{1 - 0.5z^{-1}}$$

Determine the DSP equations

$$y(n) =$$
$$e(n) =$$
$$w(i) =$$

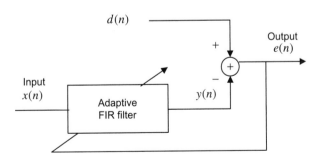

FIGURE 10.25

Noise cancellation in Problem 10.9.

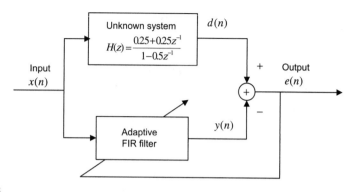

FIGURE 10.26

System modeling in Problem 10.10.

using the LMS algorithm for $i = 0, 1, 2, 3, 4$; that is, write the equations for all adaptive coefficients:

$$w(0) =$$
$$w(1) =$$
$$w(2) =$$
$$w(3) =$$
$$w(4) =$$

10.11. Consider the adaptive filter used for the noise cancellation application in Problem 10.9, in which $d(0) = 3$, $d(1) = -2$, $d(2) = 1$, $x(0) = 3$, $x(1) = -1$, $x(2) = 2$, and an adaptive filter with three taps $y(n) = w(0)x(n) + w(1)x(n-1) + w(2)x(n-2)$ with initial values $w(0) = 0$, $w(1) = 0$, $w(2) = 0$ and $\mu = 0.2$.

a. Determine the LMS algorithm equations

$$y(n) =$$
$$e(n) =$$
$$w(0) =$$
$$w(1) =$$
$$w(2) =$$

b. Perform adaptive filtering for each of $n = 0, 1, 2$.

10.12. Consider the DSP system with a sampling rate of 8,000 samples per second in Problem 10.10. Implement an adaptive filter with five taps for system modeling, assuming the unknown system transfer function is

$$H(z) = 0.2 + 0.3z^{-1} + 0.2z^{-2}$$

Determine the DSP equations

$$y(n) =$$
$$e(n) =$$
$$w(i) =$$

using the LMS algorithm for $i = 0, 1, 2, 3, 4$; that is, write the equations for all adaptive coefficients:

$$w(0) =$$
$$w(1) =$$
$$w(2) =$$
$$w(3) =$$
$$w(4) =$$

10.13. Consider the DSP system set up for noise cancellation applications with a sampling rate of 8,000 Hz shown in Figure 10.27. The desired 1,000 Hz tone is generated internally via a tone generator, and the generated tone is corrupted by the noise captured from a microphone. An FIR adaptive filter with 25 taps is applied to reduce the noise in the corrupted tone.

a. Determine the DSP equation for the channel noise $n(n)$.

b. Determine the DSP equation for signal tone $yy(n)$.

c. Determine the DSP equation for the corrupted tone $d(n)$.

d. Set up the LMS algorithm for the adaptive FIR filter.

10.14. Consider the DSP system for noise cancellation applications with two taps in Figure 10.28.

a. Set up the LMS algorithm for the adaptive filter.

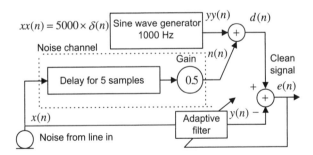

FIGURE 10.27

Noise cancellation in Problem 10.13.

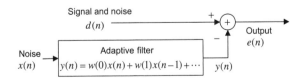

FIGURE 10.28

Noise cancellation in Problem 10.14.

b. Assume the inputs and outputs $x(0) = 1$, $x(1) = 1$, $x(2) = -1$, $x(3) = 2$, $d(0) = 0$, $d(1) = 2$, $d(2) = -1$, and $d(3) = 1$; initial weights $w(0) = w(1) = 0$; and convergence factor $\mu = 0.1$. Perform adaptive filtering to obtain outputs $e(n)$ for $n = 0, 1, 2$.

10.15. Again consider the DSP system for noise cancellation applications with three taps in Figure 10.28.

   a. Set up the LMS algorithm for the adaptive filter.

   b. Assume the inputs and outputs $x(0) = 1$, $x(1) = 1$, $x(2) = -1$, $x(3) = 2$, $d(0) = 0$, $d(1) = 2$, $d(2) = -1$, and $d(3) = 1$; initial weights $w(0) = w(1) = w(2) = 0$, and convergence factor $\mu = 0.1$. Perform adaptive filtering to obtain outputs $e(n)$ for $n = 0, 1, 2$.

10.16. For a line enhancement application using the FIR adaptive filter depicted in Figure 10.29.

   a. Set up the LMS algorithm for the adaptive filter using two filter coefficients and delay $\Delta = 2$.

   b. Assume the inputs and outputs $d(0) = -1$, $d(1) = 1$, $d(2) = -1$, $d(3) = 1$, $d(4) = -1$, $d(5) = 1$ and $d(6) = -1$; initial weights $w(0) = w(1) = 0$; and convergence factor $\mu = 0.1$. Perform adaptive filtering to obtain outputs $y(n)$ for $n = 0, 1, 2, 3, 4$.

10.17. Repeat Problem 10.16 using a three-tap FIR filter and $\Delta = 3$.

10.18. An audio playback application is described in Figure 10.30.

Due to the interference environment, the audio is corrupted by 15 different periodic interferences. The DSP engineer uses an FIR adaptive filter to remove such interferences as shown in Figure 10.30.

   a. What is the minimum number of filter coefficients?

   b. Set up the LMS algorithm for the adaptive filter using the number of taps obtained in (a).

10.19. Repeat Problem 10.18 for corrupted audio that contains five different periodic interferences.

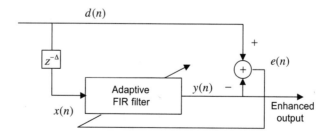

FIGURE 10.29

Line enhancement in Problem 10.16.

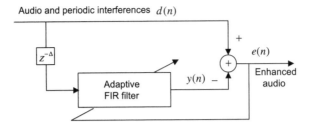

Audio and periodic interferences $d(n)$

FIGURE 10.30

Interference cancellation in Problem 10.18.

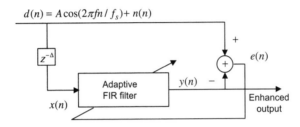

$d(n) = A\cos(2\pi fn / f_s) + n(n)$

FIGURE 10.31

A line enhancement system in Problem 10.26

10.20. In a noisy ECG acquisition environment, the DSP engineer uses an adaptive FIR filter with 20 coefficients to remove 60-Hz interference. The system is set up as shown in Figure 10.19, where the corrupted ECG and enhanced ECG are represented as $d(n)$ and $e(n)$, respectively; $x(n)$ is the captured reference signal from the 60-Hz interference; and $y(n)$ is the adaptive filter output. Determine all difference equations to implement the adaptive filter.

10.21. Given an application of the echo cancellation shown in Figure 10.21B, determine all difference equations to implement the adaptive filter with four adaptive coefficients at the speaker A site.

10.22. Given an application of the echo cancellation shown in Figure 10.21B,

  a. explain the concepts and benefits using the echo canceller;

  b. explain the operations of the adaptive filter at the speaker B site;

  c. determine all difference equations to implement the adaptive filter at the speaker A site.

10.7.1 Computer Problems with MATLAB
Use MATLAB to solve Problems 10.23 to 10.26.

10.23. Write a MATLAB program for minimizing the two-weight MSE (mean squared error) function

$$J = 100 + 100w_1^2 + 4w_2^2 - 100w_1 - 8w_2 + 10w_1w_2$$

by applying the steepest descent algorithm for 500 iterations. The derivatives are

$$\frac{dJ}{dw_1} = 200w_1 - 100 + 10w_2 \quad \text{and} \quad \frac{dJ}{dw_2} = 8w_2 - 8 + 10w_1$$

and the initial weights are assumed as $w_1(0) = 0$, $w_2(0) = 0$, $\mu = 0.001$. Plot $w_1(k)$, $w_2(k)$, and $J(k)$ versus the number of iterations, respectively. Summarize your results.

10.24. In Problem 10.10, the unknown system is assumed to be a fourth-order Butterworth bandpass filter with a lower cutoff frequency of 700 Hz and an upper cutoff frequency of 900 Hz. Design a bandpass filter by the bilinear transformation method for simulating the unknown system with a sampling rate of 8,000 Hz.

a. Generate the input signal for 0.1 second using a sum of three sinusoids with 100 Hz, 800 Hz, and 1,500 Hz and a sampling rate of 8,000 Hz.

b. Use the generated input as the unknown system input to produce the system output. The adaptive FIR filter is then applied to model the designed bandpass filter. The following parameters are assumed:

Adaptive FIR filter
Number of taps: 15 coefficients
Algorithm: LMS algorithm
Convergence factor: 0.01

c. Implement the adaptive FIR filter, and plot the system input, system output, adaptive filter output, and error signal, respectively.

d. Plot the input spectrum, system output spectrum, and adaptive filter output spectrum, respectively.

10.25. Use the following MATLAB code to generate reference noise and a signal of 300 Hz corrupted by the noise with a sampling rate of 8,000 Hz.

```
fs=8000; T=1/fs;                         % Sampling rate and sampling period

t=0:T:1;                                 % Create time instants

x=randn(1,length(t));                    % Generate reference noise

n=filter([ 0 0 0 0 0 0 0 0 0 0.8 ],1,x); % Generate the corruption noise

d=sin(2*pi*300*t)+n;                     % Generate the corrupted signal
```

a. Implement an adaptive FIR filter to remove the noise. The adaptive filter specifications are as follows:

Sample rate = 8,000 Hz
Signal corrupted by Gaussian noise delayed by nine samples from the reference noise
Reference noise: Gaussian noise with a power of 1
Number of FIR filter taps: 16
Convergence factor for the LMS algorithm: 0.01

b. Plot the corrupted signal, reference noise, and enhanced signal, respectively.

c. Compare the spectral plots between the corrupted signal and the enhanced signal.

10.26. A line enhancement system (Figure 10.31) has following specifications:

Sampling rate = 1,000 Hz
Corrupted signal: 100 Hz tone with the unit amplitude added with the unit power Gaussian noise
Adaptive filter: FIR type, 16 taps
Convergence factor: 0.001
Delay value Δ: to be decided according to the experiment
LMS algorithm

a. Write a MATLAB program to perform the line enhancement for a one-second corrupted signal. Run the developed program with a trail of delay value Δ and plot the noisy signal, the enhanced signals, and their spectra.

b. Run your simulation to find the delay value Δ that achieves the largest noise reduction.

10.7.2 MATLAB Projects

10.27. Active noise control:

The ANC (active noise control) system is based on the principle of superposition of the primary noise source and secondary source with its acoustic output being the same amplitude but the opposite phase of the primary noise source, as shown in Figure 10.32. The primary noise is captured using the reference microphone, which is located close to the noise source. The ANC system uses the sensed reference signal $x(n)$ to generate a canceling signal $y(n)$, which drives the secondary speaker to destructively attenuate the primary noise. An error microphone is used to detect the residue noise $e(n)$, which is fed back to the ANC system to monitor the system performance. The residue noise $e(n)$ together with the reference signal $x(n)$ are used by the linear adaptive controller whose coefficients are adjusted via an adaptive algorithm to minimize the measured error signal $e(n)$, or the residue acoustic noise. $P(z)$ designates the physical primary path between the reference sensor and the error sensor, and $S(z)$ designates the physical secondary path between the ANC adaptive filter output and the error sensor. To control the noise at the cancelling point, the instantaneous power $e^2(n)$ must be minimized. Note that

$$E(z) = D(z) - Y(z)S(z) = D(z) - [W(z)X(z)]S(z)$$

where $W(z)$ denotes the adaptive control filter. Exchange of the filter order (since the filters are linear filters) gives

$$E(z) = D(z) - W(z)[S(z)X(z)] = D(z) - Y(z)$$

Assuming that $\bar{S}(z)$ is the secondary path estimate and noticing that $U(z) = \bar{S}(z)X(z)$ and $Y(z) = W(z)X(z)$ are the filtered reference signal and adaptive filter output, applying the LMS algorithm gives the filtered-x LMS algorithm

$$w(i) = w(i) + 2\mu e(n)u(n-i)$$

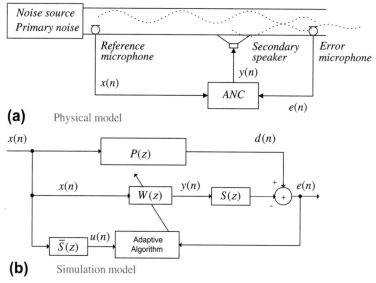

(a) Physical model

(b) Simulation model

FIGURE 10.32

An active noise control system.

Note that $e(n)$ is measured from the error microphone.

The completed filtered-x LMS algorithm is summarized below:

1. Initialize $w(0)$, $w(1)$, ... $w(N-1)$ to arbitrary values.

2. Read $x(n)$ and perform digital filtering:

$$y(n) = w(0)x(n) + w(1)x(n-1) + \cdots + w(N-1)x(n-N+1)$$

3. Compute the filtered inputs:

$$u(n) = \bar{s}(0)x(n) + \bar{s}(1)x(n-1) + \cdots + \bar{s}(M-1)x(n-M+1)$$

4. Read $e(n)$ and update each filter coefficient:

$$\text{for } i = 0, \cdots, N-1, \ w(i) = w(i) + 2\mu e(n)u(n-i),$$

Assuming the following:

Sampling rate = 8,000 Hz and simulation duration = 10 seconds
Primary noise: $x(n)$ = 500-Hz sine wave
Primary path: $P(z) = 0.2 + 0.25z^{-1} + 0.2z^{-2}$
Secondary path: $S(z) = 0.25 + 0.2z^{-1}$
Secondary path estimate: $\bar{S}(z) = S(z) = 0.2 + 0.2z^{-1}$ (can have slight error as compared to $S(z)$)

Residue error signal: $e(n) = d(n) -$ filtering $y(n)$ using coefficients $s(n) = d(n) - y(n)*$ $s(n)$, where the symbol "$*$" denotes the filter convolution.

Implement the ANC system and plot the residue sensor signal to verify the effectiveness. The primary noise at cancelling point $d(n)$, and filtered reference signal $u(n)$ can be generated in MATLAB as follows:

```
d = filter([0.2 0.25 0.2],1,x); % Simulate physical media
u=filter([0.2 0.2],1,x);
```

The residue error signal $e(n)$ should be generated sample by sample and embedded into the adaptive algorithm, that is,

```
e(n)=d(n)-(s(1)*y(n)+s(2)*y(n-1)); % Simulate the residue error
```

where $s(1) = 0.2$ and $s(2) = 0.2$. Details of active control systems can be found in the textbook by Kuo and Morgan (1996).

10.28. Frequency tracking:

An adaptive filter can be applied for real-time frequency tracking (estimation). In this application, a special second notch IIR filter structure, as shown in Figure 10.33, is preferred for simplicity.

The notch filter transfer function

$$H(z) = \frac{1 - 2\cos(\theta)z^{-1} + z^{-2}}{1 - 2r\cos(\theta)z^{-1} + r^2 z^{-2}}$$

has only one adaptive parameter θ. It has two zeros on the unit circle resulting in an infinite-depth notch. The parameter r controls the notch bandwidth. It requires $0 << r < 1$ for achieving a narrowband notch. When r is close to 1, the 3-dB notch filter bandwidth can be approximated as $BW \approx 2(1 - r)$ (see Chapter 8). The input sinusoid whose frequency f needs to be estimated and tracked is given below:

$$x(n) = A\cos(2\pi fn/f_s + \alpha)$$

where A and α are the amplitude and phase angle. The filter output is expressed as

$$y(n) = x(n) - 2\cos[\theta(n)]x(n-1) + x(n-2) + 2r\cos[\theta(n)]y(n-1) - r^2 y(n-2)$$

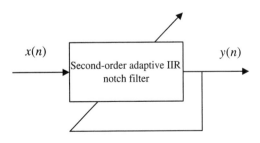

FIGURE 10.33

A frequency tracking system.

The objective is to minimize the filter instantaneous output power $y^2(n)$. Once the output power is minimized, the filter parameter $\theta = 2\pi f/f_s$ will converge to its corresponding frequency f Hz. The LMS algorithm to minimize the instantaneous output power $y^2(n)$ is given as

$$\theta(n+1) = \theta(n) - 2\mu y(n)\beta(n),$$

where the gradient function $\beta(n) = \partial y(n)/\partial \theta(n)$ can be derived as follows:

$$\beta(n) = 2\sin[\theta(n)]x(n-1) - 2r\sin[\theta(n)]y(n-1) + 2r\cos[\theta(n)]\beta(n-1) - r^2\beta(n-2)$$

μ is the convergence factor which controls speed of algorithm convergence.

In this project, plot and verify the notch frequency response by setting $f_s = 8,000$ Hz, $f = 1,000$ Hz, and $r = 0.95$. Then generate the sinusoid with a duration of 10 seconds, frequency of 1,000 Hz, and amplitude of 1. Implement the adaptive algorithm using an initial guess $\theta(0) = 2\pi \times 2000/f_s = 0.5\pi$ and plot the tracked frequency $f(n) = \theta(n)f_s/2\pi$ for tracking verification.

Notice that this particular notch filter only works for a single frequency tracking, since the mean squares error function $E[y^2(n)]$ has one global minimum (one best solution when the LMS algorithm converges). Details of the adaptive notch filter can be found in Tan and Jiang (2012). Notice that the general IIR adaptive filter suffers from local minima, that is, the LMS algorithm converges to local minimum and the nonoptimal solution results.

Waveform Quantization and Compression

CHAPTER OUTLINE

11.1 Linear Midtread Quantization 497
11.2 μ-law Companding 501
    11.2.1 Analog μ-Law Companding 501
    11.2.2 Digital μ-Law Companding...504
11.3 Examples of Differential Pulse Code Modulation (DPCM), Delta Modulation,
    and Adaptive DPCM G.721 509
    11.3.1 Examples of Differential Pulse Code Modulation and Delta Modulation.......................509
    11.3.2 Adaptive Differential Pulse Code Modulation G.721 512
            Simulation Example.. 517
11.4 Discrete Cosine Transform, Modified Discrete Cosine Transform,
    and Transform Coding in MPEG Audio 519
    11.4.1 Discrete Cosine Transform...519
    11.4.2 Modified Discrete Cosine Transform 522
    11.4.3 Transform Coding in MPEG Audio 525
11.5 Laboratory Examples of Signal Quantization Using the TMS320C6713 DSK...................... 528
11.6 Summary 533
11.7 MATLAB Programs 533

OBJECTIVES:

This chapter studies speech quantization and compression techniques such as signal companding, differential pulse code modulation, and adaptive differential pulse code modulation. The chapter continues to explore the discrete-cosine transform (DCT) and the modified DCT and shows how to apply the developed concepts to understand the MP3 audio format. The chapter also introduces industry standards that are widely used in the digital signal processing field.

11.1 LINEAR MIDTREAD QUANTIZATION

As we discussed in Chapter 2, in the digital signal processing (DSP) system, the first step is to sample and quantize the continuous signal. Quantization is the process of rounding off the sampled signal voltage to the predetermined levels that will be encoded by analog-to-digital conversion (ADC). We described the quantization process in Chapter 2, in which we studied unipolar and bipolar linear quantizers in detail. In

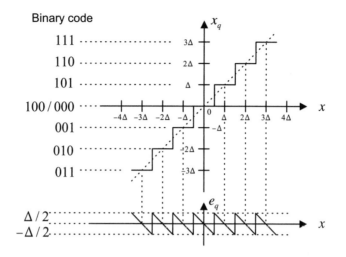

FIGURE 11.1

Characteristics of a 3-bit midtread quantizer.

this section, we focus on a linear midtread quantizer, which is used in digital communications (Roddy and Coolen, 1997; Tomasi, 2004), and its use to quantize speech waveforms. The linear midtread quantizer is similar to the bipolar linear quantizer discussed in Chapter 2 except that the midtread quantizer offers the same decoded magnitude range for both positive and negative voltages.

Let us look at a midtread quantizer. The characteristics and binary codes for a 3-bit midtread quantizer are depicted in Figure 11.1, where the code is in a sign magnitude format. Positive voltage is coded using a sign bit of logic 1, while negative voltage is coded by a sign bit of logic 0; the next two bits are the magnitude bits. The key feature of the linear midtread quantizer is noted as follows: when $0 \le x < \Delta/2$, the binary code of 100 is produced; when $-\Delta/2 \le x < 0$, the binary code of 000 is generated, where Δ is the quantization step size. However, the quantized values for both codes 100 and 000 are the same and equal to $x_q = 0$. We can also see details in Table 11.1. For the 3-bit midtread

Table 11.1 Quantization Table for the 3-bit Midtread Quantizer

Binary Code	Quantization Level x_q (V)	Input Signal Subrange (V)
0 1 1	-3Δ	$-3.5\Delta \le x < -2.5\Delta$
0 1 0	-2Δ	$-2.5\Delta \le x < -1.5\Delta$
0 0 1	$-\Delta$	$-1.5\Delta \le x < -0.5\Delta$
0 0 0	0	$-0.5 \le x < 0$
1 0 0	0	$0 \le x < 0.5\Delta$
1 0 1	Δ	$0.5\Delta \le x < 1.5\Delta$
1 1 0	2Δ	$1.5\Delta \le x < 2.5\Delta$
1 1 1	3Δ	$2.5\Delta \le x < 3.5\Delta$

Note: Step size $= \Delta = (x_{max} - x_{min})/(2^3 - 1)$; $x_{max} = $ maximum voltage; and $x_{min} = -x_{max}$. Coding format: a. sign bit: 1= plus, 0 = minus; b. 2 magnitude bits.

qunatizer, we expect seven quantized values instead of eight; that is; there are $2^n - 1$ quantization levels for the n-bit midtread quantizer. Notice that quantization signal range is $(2^n - 1)\Delta$ and the magnitudes of the quantized values are symmetric, as shown in Table 11.1. We apply the midtread quantizer particularly for speech waveform coding.

The following example serves to illustrate the coding principles of the 3-bit midtread quantizer.

EXAMPLE 11.1

For the 3-bit midtread quantizer described in Figure 11.1 and the analog signal with a range from -5 volts to 5 volts,

a. determine the quantization step size;
b. determine the binary codes, recovered voltages, and quantization errors when the input is -3.6 volts and 0.5 volt, respectively.

Solution:

a. The quantization step size is calculated as

$$\Delta = \frac{5 - (-5)}{2^3 - 1} = 1.43 \text{ volts}$$

b. For $x = -3.6$ volts, we have $x = \frac{-3.6}{1.43} = -2.52\Delta$. From quantization characteristics, it follows that the binary code $= 011$ and the recovered voltage is $x_q = -3\Delta = -4.29$ volts. Thus the quantization error is computed as

$$e_q = x_q - x = -4.29 - (-3.6) = -0.69 \text{ volt}$$

For $x = 0.5 = \frac{0.5}{1.43}\Delta = 0.35\Delta$, we get binary code $= 100$. Based on Figure 11.1, the recovered voltage and quantization error are found to be

$$x_q = 0 \quad \text{and} \quad e_q = 0 - 0.5 = -0.5 \text{ volt}$$

As discussed in Chapter 2, the linear midtread quantizer introduces quantization noise, as shown in Figure 11.1; the signal-to-noise power ratio (SNR) is given by

$$SNR = 10.79 + 20 \cdot \log_{10}\left(\frac{x_{rms}}{\Delta}\right) \text{ dB} \tag{11.1}$$

where x_{rms} designates the root mean squared value of the speech data to be quantized. The practical equation for estimating the SNR for the speech data sequence $x(n)$ of N data points is written as

$$SNR = \left(\frac{\sum_{n=0}^{N-1} x^2(n)}{\sum_{n=0}^{N-1}(x_q(n) - x(n))^2}\right) \tag{11.2}$$

$$SNR \ dB = 10 \cdot \log_{10}(SNR) \text{ dB} \tag{11.3}$$

Notice that $x(n)$ and $x_q(n)$ are the speech data to be quantized and the quantized speech data, respectively. Equation (11.2) gives the absolute SNR, and Equation (11.3) produces the SNR in terms of decibels (dB). Quantization error is the difference between the quantized speech data (or quantized voltage level) and speech data (or analog voltage), that is, $x_q(n) - x(n)$. Also note that from Equation (11.1), adding 1 bit to the linear quantizer would improve SNR by approximately 6 dB. Let us examine performance of the 5-bit linear midtread quantizer.

In the following simulation, we use a 5-bit midtread quantizer to quantize the speech data. After quantization, the original speech, quantized speech, and quantized error after quantization are plotted in Figure 11.2. Since the program calculates $x_{rms}/x_{max} = 0.203$, we yield x_{rms} as $x_{rms} = 0.203 \times x_{max} = 0.0203 \times 5 = 1.015$ and $\Delta = 10/(2^5 - 1) = 0.3226$. Applying Equation (11.1) gives $SNR = 21.02$ dB. The SNR using Equations (11.2) and (11.3) is approximately 21.6 dB.

The first plot in Figure 11.2 is the original speech, and the second plot shows the quantized speech. Quantization error is displayed in the third plot, where the error amplitude interval is uniformly distributed between -0.1613 and 0.1613, indicating the bounds of the quantized error ($\Delta/2$). The details of the MATLAB implementation are given in Program 11.1 in Section 11.7.

To improve the SNR, the number of bits must be increased. However, increasing the number of encoding bits will cost an expansive ADC device, larger storage media for storing the speech data, and more bandwidth for transmitting the digital data. To gain a more efficient quantization approach, we will study μ-law companding in the next section.

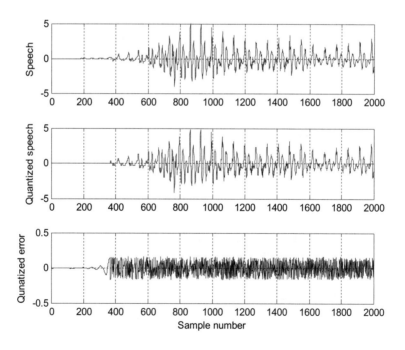

FIGURE 11.2

Plots of original speech, quantized speech, and quantization error.

11.2 μ-LAW COMPANDING

In this section, we will study analog μ-law companding, which takes an analog input signal; and digital μ-law companding, which deals with linear pulse code modulation (PCM) codes.

11.2.1 Analog μ-Law Companding

To reduce the number of bits required to encode each speech datum, μ-law companding, called log-PCM coding, is applied. μ-law companding (Roddy and Coolen, 1997; Tomasi, 2004) was first used in the United States and Japan in the telephone industry (G.711 standard). μ-law companding is a compression process. It explores the principle that the higher amplitudes of analog signals are compressed before ADC and expanded after digital-to-analog conversion (DAC). As studied in the linear quantizer, the quantization error is uniformly distributed. This means that the maximum quantization error stays the same no matter how big or small the speech samples are. μ-law companding can be employed to make the quantization error smaller when the sample amplitude is smaller and to make the quantization error bigger when the sample amplitude is bigger, using the same number of bits per sample. It is described in Figure 11.3.

As shown in Figure 11.3, x is the original speech sample, which is the input to the compressor, while y is the output from the μ-law compressor; then the output y is uniformly quantized. Assuming that the quantized sample y_q is encoded and sent to the μ-law expander, the expander will perform the reverse process to obtain the quantized speech sample x_q. The compression and decompression processes cause the maximum quantization error $|x_q - x|_{\max}$ to be small for the smaller sample amplitudes and large for the larger sample amplitudes.

The equation for the μ-law compressor is given by

$$ y = sign(x)\dfrac{\ln\left(1 + \mu\dfrac{|x|}{|x|_{\max}}\right)}{\ln(1 + \mu)} \tag{11.4} $$

where $|x|_{\max}$ is the maximum amplitude of the inputs, while μ is a positive parameter to control the degree of the compression. $\mu = 0$ corresponds to no compression, while $\mu = 255$ is adopted in the industry. The compression curve with $\mu = 255$ is plotted in Figure 11.4. Note that the sign function $sign(x)$ shown in Equation (11.4) is defined as

$$ sign(x) = \begin{cases} 1 & x \geq 0 \\ -1 & x < 0 \end{cases} \tag{11.5} $$

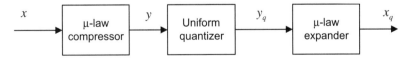

FIGURE 11.3

Block diagram for μ-law compressor and μ-law expander.

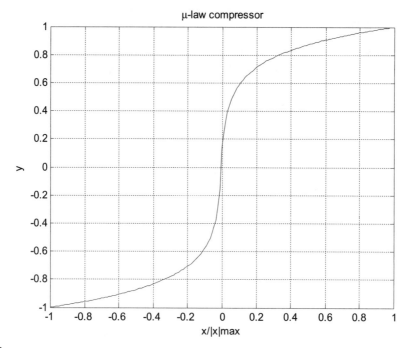

FIGURE 11.4

Characteristics for the μ-law compander.

Solving Equation (11.4) by substituting the quantized value, that is, $y = y_q$ we achieve the expander equation as

$$x_q = |x|_{\max} sign(y_q)\frac{(1+\mu)^{|y_q|}-1}{\mu} \tag{11.6}$$

For the case $\mu = 255$, the expander curve is plotted in Figure 11.5.

Let's look at Example 11.2 on μ-law compression.

EXAMPLE 11.2

For the μ-law compression and expansion process shown in Figure 11.3, with $\mu = 255$, the 3-bit midtread quantizer described in Figure 11.1, and an analog signal ranging from -5 to 5 volts, determine the binary codes, recovered voltages, and quantization errors when the input is

a. -3.6 volts
b. 0.5 volt

Solution:

a. For μ-law compression and $x = -3.6$ volts, we can determine the quantization input as

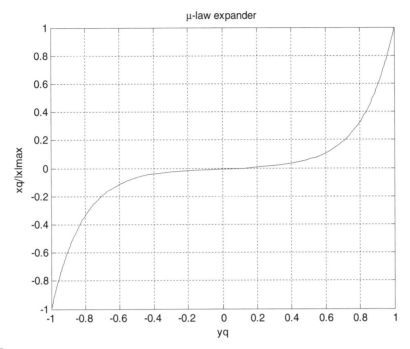

FIGURE 11.5

Characteristics for the μ-law expander.

$$y = sign(-3.6)\frac{\ln\left(1 + 255\frac{|-3.6|}{|5|_{max}}\right)}{\ln(1 + 255)} = -0.94$$

As shown in Figure 11.4, the range of y is 2, thus the quantization step size is calculated as

$$\Delta = \frac{2}{2^3 - 1} = 0.286 \quad \text{and} \quad y = \frac{-0.94}{0.286} = -3.28\Delta$$

From quantization characteristics, it follows that the binary code is 011 and the recovered signal is $y_q = -3\Delta = -0.858$.

Applying the μ-law expander leads to

$$x_q = |5|_{max}sign(-0.858)\frac{(1 + 255)^{|-0.858|} - 1}{255} = -2.264$$

Thus the quantization error is computed as

$$e_q = x_q - x = -2.264 - (-3.6) = 1.336 \text{ volts}$$

b. Similarly, for $x = 0.5$, we get

$$y = sign(0.5)\frac{\ln\left(1 + 255\frac{|0.5|}{|5|_{max}}\right)}{\ln(1 + 255)} = 0.591$$

In terms of the quantization step, we get

$$y = \frac{0.519}{0.286}\Delta = 2.1\Delta \quad \text{and binary code} = 110$$

Based on Figure 11.1, the recovered signal is

$$y_q = 2\Delta = 0.572$$

and the expander gives

$$x_q = |5|_{max} sign(0.572)\frac{(1+255)^{|0.572|}-1}{255} = 0.448 \text{ volt}$$

Finally, the quantization error is given by

$$e_q = 0.448 - 0.5 = -0.052 \text{ volt}$$

As we can see, with 3 bits per sample, the stronger signal is encoded with more quantization error, while the weak signal is encoded with less quantization error.

In the following simulation, we apply a 5-bit μ-law compander with $\mu = 255$ in order to quantize and encode the speech data used in the last section. Figure 11.6 is a block diagram of compression and decompression.

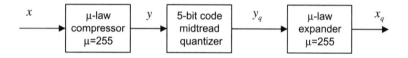

FIGURE 11.6

The 5-bit midtread uniform quantizer with $\mu = 255$ used for simulation.

Figure 11.7 shows the original speech data, the quantized speech data using μ-law compression, and the quantization error for comparisons. The quantized speech wave is very close to the original speech wave. From the plots in Figure 11.7, we can observe that the amplitude of the quantization error changes according to the amplitude of the speech being quantized. More quantization error is introduced when the amplitude of speech data is larger; on the other hand, a smaller quantization error is produced when the amplitude of speech data is smaller.

Compared with the quantized speech using the linear quantizer shown in Figure 11.2, the decompressed signal using the μ-law compander looks and sounds much better, since the quantized signal can better track the original large amplitude signal and original small amplitude signal as well. The MATLAB implementation is shown in Program 11.2 in Section 11.7.

11.2.2 Digital μ-Law Companding

In many multimedia applications, the analog signal is first sampled and then it is digitized into a linear PCM code with a larger number of bits per sample. Digital μ-law companding further compresses the linear PCM code using the compressed PCM code with a smaller number of bits per sample without losing sound quality. The block diagram of a digital μ-law compressor and expander is shown in Figure 11.8.

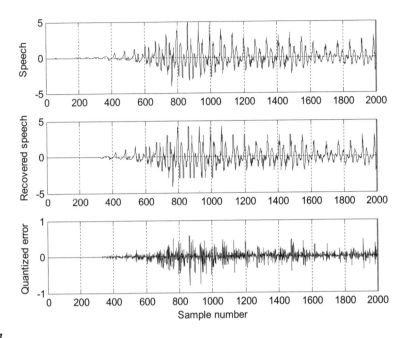

FIGURE 11.7

Plots of the original speech, quantized speech, and quantization error with the μ-law compressor and expander.

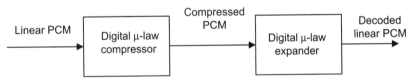

FIGURE 11.8

The block diagram for a μ-law compressor and expander.

The typical digital μ-law companding system compresses a 12-bit linear PCM code to an 8-bit compressed code. This companding characteristic is depicted in Figure 11.9, where it closely resembles an analog compression curve with $\mu = 255$ by approximating the curve using a set of eight straight-line segments. The slope of each successive segment is exactly one-half that of the previous segment. Figure 11.9 shows the 12-bit to 8-bit digital companding curve for the positive portion only. There are 16 segments, accounting for both positive and negative portions.

However, like the midtread quantizer discussed in the first section of this chapter, only 13 segments are used, since segments $+0$, -0, $+1$, and -1 form a straight line with a constant slope and are considered one segment. As shown in Figure 11.9, when the relative input is very small, such as in segment 0 or segment 1, there is no compression, while when the relative input is larger such as in segment 3 or segment 4, the compression occurs with compression ratios of 2:1 and 4:1, respectively.

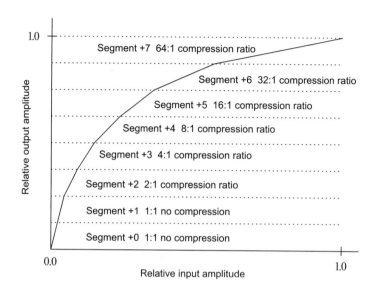

FIGURE 11.9

$\mu - 255$ compression characteristics (for positive portion only).

The format of the 12-bit linear PCM code is in the sign-magnitude form with the most significant bit (MSB) as the sign bit (1 = positive value and 0 = negative value) plus 11 magnitude bits. The compressed 8-bit code has the format shown in Table 11.2, where it consists of a sign bit, a 3-bit segment identifier, and a 4-bit quantization interval within the specified segment. Encoding and decoding procedures are very simple, as illustrated in Tables 11.3 and 11.4, respectively.

As shown in those two tables, the prefix "S" is used to indicate the sign bit, which could be either 1 or 0; A, B, C, and D, are transmitted bits; and the bit position with an "X" is the truncated bit during the compression and hence would be lost during decompression. For the 8-bit compressed PCM code in Table 11.3, the 3 bits between "S" and "ABCD" indicate the segment number that is obtained by subtracting the number of consecutive zeros (less than or equal to 7) after the "S" bit in the original 12-bit PCM code from 7. Similarly, to recover the 12-bit linear code in Table 11.4, the number of consecutive zeros after the "S" bit can be determined by subtracting the segment number in the 8-bit compressed code from 7. We will illustrate the encoding and decoding processes in Examples 11.3 and 11.4.

Table 11.2 The Format of 8-Bit Compressed PCM Code

Sign bit: 1 = + 0 = −	3-bit segment identifier: 000 to 111	4-bit quantization interval: A B C D 0000 to 1111

Table 11.3 $\mu - 255$ Encoding Table

Segment	12-Bit Linear Code	12-Bit Amplitude Range in Decimal	8-Bit Compressed Code
0	S0000000ABCD	0 to 15	S000ABCD
1	S0000001ABCD	16 to 31	S001ABCD
2	S000001ABCDX	32 to 63	S010ABCD
3	S00001ABCDXX	64 to 127	S011ABCD
4	S0001ABCDXXX	128 to 255	S100ABCD
5	S001ABCDXXXX	256 to 511	S101ABCD
6	S01ABCDXXXXX	512 to 1023	S110ABCD
7	S1ABCDXXXXXX	1023 to 2047	S111ABCD

Table 11.4 $\mu - 255$ Decoding Table

8-Bit Compressed Code	8-Bit Amplitude Range in Decimal	Segment	12-Bit Linear Code
S000ABCD	0 to 15	0	S0000000ABCD
S001ABCD	16 to 31	1	S0000001ABCD
S010ABCD	32 to 47	2	S000001ABCD1
S011ABCD	48 to 63	3	S00001ABCD10
S100ABCD	64 to 79	4	S0001ABCD100
S101ABCD	80 to 95	5	S001ABCD1000
S110ABCD	96 to 111	6	S01ABCD10000
S111ABCD	112 to 127	7	S1ABCD100000

EXAMPLE 11.3

In a digital companding system, encode each of the following 12-bit linear PCM codes into an 8-bit compressed PCM code.

a. 1 0 0 0 0 0 0 0 0 1 0 1
b. 0 0 0 0 1 1 1 0 1 0 1 0

Solution:

a. Based on Table 11.3, we identify the 12-bit PCM code as $S = 1$, $A = 0$, $B = 1$, $C = 0$, and $D = 1$, which is in segment 0. From the fourth column in Table 11.3, we get the 8-bit compressed code as

$$1 0 0 0 0 1 0 1$$

b. For the second 12 bit PCM code, we note that $S = 0$, $A = 1$, $B = 1$, $C = 0$, $D = 1$, and $XXX = 010$, and the code belongs to segment 4. Thus, from the fourth column in Table 11.3, we have

$$0 1 0 0 1 1 0 1$$

EXAMPLE 11.4

In a digital companding system, decode each of the following 8-bit compressed PCM codes into a 12-bit linear PCM code.

a. 1 0 0 0 0 1 0 1
b. 0 1 0 0 1 1 0 1

Solution:

a. Using Table 11.4, we notice that S = 1, A = 0, B = 1, C = 0, and D = 1, and the code is in segment 0. Decoding leads to 1 0 0 0 0 0 0 0 0 1 0 1, which is identical to the 12-bit PCM code in (a) in Example 11.3. We expect this result, since there is no compression for segment 0 and segment 1.

b. Applying Table 11.4, it follows that S = 0, A = 1, B = 1, C = 0, and D = 1, and the code resides in segment 4. Decoding achieves 0 0 0 0 1 1 1 0 1 1 0 0. As expected, this code is the approximation of the code in (b) in Example 11.3. Since segment 4 has compression, the last 3 bits in the original 12-bit linear code, that is, XXX = 010 = 2 in decimal, are discarded during transmission or storage. When we recover these three bits, the best guess should be the middle value: XXX = 100 = 4 in decimal for the 3-bit coding range from 0 to 7.

Now we apply the $\mu - 255$ compander to compress 12-bit speech data as shown in Figure 11.10(a). The 8-bit compressed code is plotted in Figure 11.10(b). Plots (c) and (d) in the figure show the 12-bit

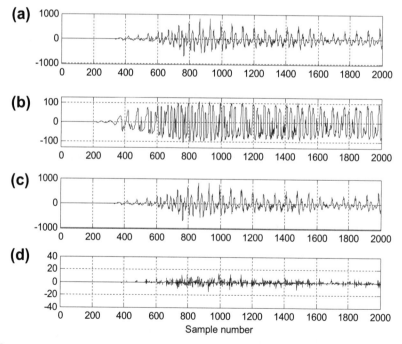

FIGURE 11.10

The $\mu - 255$ compressor and expander: (a) 12-bit speech data; (b) 8-bit compressed data; (c) 12-bit decoded speech; (d) quantization error.

speech after decoding and quantization error, respectively. We can see that the quantization error follows the amplitude of speech data relatively. The decoded speech sounds no different when compared with the original speech. Programs 11.8 to 11.10 in Section 11.7 show the details of the MATLAB implementation.

11.3 EXAMPLES OF DIFFERENTIAL PULSE CODE MODULATION (DPCM), DELTA MODULATION, AND ADAPTIVE DPCM G.721

Data compression can be further achieved using *differential pulse code modulation* (DPCM). The general idea is to use past recovered values as the basis to predict the current input data and then encode the difference between the current input and the predicted input. Since the difference has a significantly reduced signal dynamic range, it can be encoded with fewer bits per sample. Therefore, we obtain data compression. First, we study the principles of the DPCM concept that will help us understand adaptive DPCM in the next subsection.

11.3.1 Examples of Differential Pulse Code Modulation and Delta Modulation

Figure 11.11 shows a schematic diagram for the DPCM encoder and decoder. We denote the original signal $x(n)$; the predicted signal $\tilde{x}(n)$; the quantized or recovered signal $\hat{x}(n)$; the difference signal to be quantized $d(n)$; and the quantized difference signal $d_q(n)$. The quantizer can be chosen as a uniform quantizer, a midtread quantizer (e.g., see Table 11.5), or any others available. The encoding block produces a binary bit stream in the DPCM encoding. The predictor uses the past predicted signal and quantized difference signal to predict the current input value $x(n)$ as close as possible. The digital filter or adaptive filter can serve as the predictor. On the other hand, the decoder recovers the quantized difference signal, which can be added to the predictor output signal to produce the quantized and recovered signal, as shown in Figure 11.11(b).

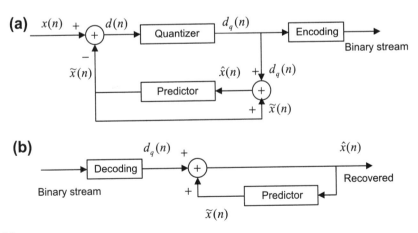

FIGURE 11.11

DPCM block diagram: (a) encoder; (b) decoder.

Table 11.5 Quantization Table for the 3-bit Quantizer in Example 11.5

Binary Code	Quantization Value $d_q(n)$	Subrange in $d(n)$
0 1 1	−11	$-15 \le d(n) < -7$
0 1 0	−5	$-7 \le d(n) < -3$
0 0 1	−2	$-3 \le d(n) < -1$
0 0 0	0	$-1 \le d(n) < 0$
1 0 0	0	$0 \le d(n) \le 1$
1 0 1	2	$1 < d(n) \le 3$
1 1 0	5	$3 < d(n) \le 7$
1 1 1	11	$7 < d(n) \le 15$

In Example 11.5, we examine a simple DPCM coding system via the process of encoding and decoding numerical actual data.

EXAMPLE 11.5

A DPCM system has the following specifications:

$$\text{Encoder scheme: } \tilde{x}(n) = \hat{x}(n-1); \text{ predictor}$$
$$d(n) = x(n) - \tilde{x}(n)$$
$$d_q(n) = Q[d(n)] = \text{quantizer in Table 11.5}$$
$$\hat{x}(n) = \tilde{x}(n) + d_q(n)$$

$$\text{Decoding scheme: } \tilde{x}(n) = \hat{x}(n-1); \text{ predictor}$$
$$d_q(n) = \text{quantizer in Table 11.5}$$
$$\hat{x}(n) = \tilde{x}(n) + d_q(n)$$

5-bit input data: $x(0) = 6$, $x(1) = 8$, $x(2) = 13$.

a. Perform DPCM encoding to produce the binary code for each input datum.
b. Perform DCPM decoding to recover the data using the binary code in (a).

Solution:

a. Let us perform encoding according to the encoding scheme.
   For $n = 0$, we have
   $\tilde{x}(0) = \hat{x}(-1) = 0$
   $d(0) = x(0) - \tilde{x}(0) = 6 - 0 = 6$
   $d_q(0) = Q[d(0)] = 5$
   $\hat{x}(0) = \tilde{x}(0) + d_q(0) = 0 + 5 = 5$
   Binary code $=110$

   For $n = 1$, it follows that
   $\tilde{x}(1) = \hat{x}(0) = 5$
   $d(1) = x(1) - \tilde{x}(1) = 8 - 5 = 3$
   $d_q(1) = Q[d(1)] = 2$
   $\hat{x}(1) = \tilde{x}(1) + d_q(1) = 5 + 2 = 7$
   Binary code $= 101$

   For $n = 2$, the results are
   $\tilde{x}(2) = \hat{x}(1) = 7$
   $d(2) = x(2) - \tilde{x}(2) = 13 - 7 = 6$

$d_q(2) = Q[d(2)] = 5$
$\hat{x}(2) = \tilde{x}(2) + d_q(2) = 7 + 5 = 12$
Binary code $= 110$

b. We conduct the decoding scheme as follows.
   For $n = 0$, we get
   Binary code $= 110$
   $d_q(0) = 5$; from Table 11.5
   $\tilde{x}(0) = \hat{x}(-1) = 0$
   $\hat{x}(0) = \tilde{x}(0) + d_q(0) = 0 + 5 = 5$ (recovered)

   For $n = 1$, decoding shows:
   Binary code $= 101$
   $d_q(1) = 2$; from Table 11.5
   $\tilde{x}(1) = \hat{x}(0) = 5$
   $\hat{x}(1) = \tilde{x}(1) + d_q(1) = 5 + 2 = 7$ (recovered)

   For $n = 2$, we have:
   Binary code $= 110$
   $d_q(2) = 5$; from Table 11.5
   $\tilde{x}(2) = \hat{x}(1) = 7$
   $\hat{x}(2) = \tilde{x}(2) + d_q(2) = 7 + 5 = 12$ (recovered)

From this example, we can verify that the 5-bit code is compressed to the 3-bit code. However, we can see that each piece of recovered data has a quantization error. Hence, DPCM is a lossy data compression scheme.

DPCM for which a single bit is used in the quantization table becomes *delta modulation* (DM). The quantization table contains two quantized values, A and $-A$, where A is the quantization step size. Delta modulation quantizes the difference of the current input sample and the previous input sample using a 1-bit code word. To conclude the idea, we list the equations for encoding and decoding as follows:

$$\text{Encoder scheme: } \tilde{x}(n) = \hat{x}(n-1); \text{ predictor}$$
$$d(n) = x(n) - \tilde{x}(n)$$
$$d_q(n) = \begin{cases} +A & d_q(n) \geq 0, \text{ output bit: } 1 \\ -A & d_q(n) \langle 0, \text{ output bit: } 0 \end{cases}$$
$$\hat{x}(n) = \tilde{x}(n) + d_q(n)$$

$$\text{Decoding scheme : } \tilde{x}(n) = \hat{x}(n-1); \text{ predictor}$$
$$d_q(n) = \begin{cases} +A & \text{input bit: } 1 \\ -A & \text{input bit: } 0 \end{cases}$$
$$\hat{x}(n) = \tilde{x}(n) + d_q(n)$$

Note that the predictor has a sample delay.

EXAMPLE 11.6

Consider a DM system with 5-bit input data
$$x(0) = 6, \; x(1) = 8, \; x(2) = 13$$

and a quantized constant of $A = 7$.

a. Perform DM encoding to produce the binary code for each input datum.
b. Perform DM decoding to recover the data using the binary code in (a).

Solution:

a. Applying encoding accordingly, we have

For $n = 0$,

$\tilde{x}(0) = \hat{x}(-1) = 0$, $d(0) = x(0) - \tilde{x}(0) = 6 - 0 = 6$

$d_q(0) = 7$, $\hat{x}(0) = \tilde{x}(0) + d_q(0) = 0 + 7 = 7$

Binary code $= 1$

For $n = 1$,

$\tilde{x}(1) = \hat{x}(0) = 7$, $d(1) = x(1) - \tilde{x}(1) = 8 - 7 = 1$

$d_q(1) = 7$, $\hat{x}(1) = \tilde{x}(1) + d_q(1) = 7 + 7 = 14$

Binary code $= 1$.

For $n = 2$,

$\tilde{x}(2) = \hat{x}(1) = 14$, $d(2) = x(2) - \tilde{x}(2) = 13 - 14 = -1$

$d_q(2) = -7$, $\hat{x}(2) = \tilde{x}(2) + d_q(2) = 14 - 7 = 7$

Binary code $= 0$

b. Applying the decoding scheme leads to

For $n = 0$,

Binary code $= 1$

$d_q(0) = 7$, $\tilde{x}(0) = \hat{x}(-1) = 0$

$\hat{x}(0) = \tilde{x}(0) + d_q(0) = 0 + 7 = 7$ (recovered)

For $n = 1$,

Binary code 1

$d_q(1) = 7$, $\tilde{x}(1) = \hat{x}(0) = 7$

$\hat{x}(1) = \tilde{x}(1) + d_q(1) = 7 + 7 = 14$ (recovered)

For $n = 2$,

Binary code 0

$d_q(2) = -7$, $\tilde{x}(2) = \hat{x}(1) = 14$

$\hat{x}(2) = \tilde{x}(2) + d_q(2) = 14 - 7 = 7$ (recovered)

We can see that coding causes a larger quantization error for each recovered sample. In practice, this can be solved by using a very high sampling rate (much larger than the Nyquist rate), and by making the quantization step size A adaptive. The quantization step size increases by a factor when the slope magnitude of the input sample curve becomes bigger, that is, the condition in which the encoder produces continuous logic 1s or continuous logic 0s in the coded bit stream. Similarly, the quantization step decreases by a factor when the encoder generates logic 1 and logic 0 alternatively. Hence, the resultant DM is called *adaptive* DM. In practice, the DM chip replaces the predictor, feedback path, and summer (see Figure 11.11) with an integrator for both the encoder and the decoder. Detailed information can be found in Li and Drew (2004), Roddy and Coolen (1997), and Tomasi (2004).

11.3.2 Adaptive Differential Pulse Code Modulation G.721

In this subsection, an efficient compression technique for speech waveform is described, that is, *adaptive DPCM* (ADPCM), per recommendation G.721 of the CCITT (the Comité Consultatif

International Téléphonique et Télégraphique). General discussion can be found in Li and Drew (2004), Roddy and Coolen (1997), and Tomasi (2004). The simplified block diagrams of the ADPCM encoder and decoder are shown in Figures 11.12A and 11.12B.

As shown in Figure 11.12A for the ADPCM encoder, first a difference signal $d(n)$ is obtained, by subtracting an estimate of the input signal $\tilde{x}(n)$ from the input signal $x(n)$. An adaptive 16-level quantizer is used to assign four binary digits $I(n)$ to the value of the difference signal for transmission to the decoder. At the same time, the inverse quantizer produces a quantized difference signal $d_q(n)$ from the same four binary digits $I(n)$. The adaptive quantizer and inverse quantizer operate based on the quantization table and the scale factor obtained from the quantizer adaptation to keep tracking the energy change of the difference signal to be quantized. The input signal estimate from the

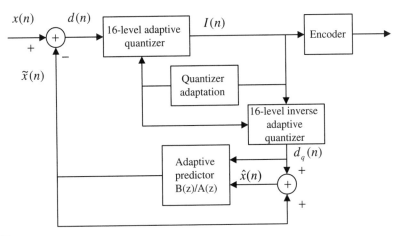

FIGURE 11.12A

ADPCM encoder.

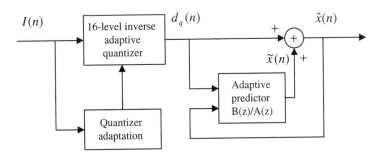

FIGURE 11.12B

ADPCM decoder.

adaptive predictor is then added to this quantized difference signal to produce the reconstructed version of the input $\hat{x}(n)$. Both the reconstructed signal and the quantized difference signal are operated on by an adaptive predictor, which generates the estimate of the input signal, thereby completing the feedback loop.

The decoder shown in Figure 11.12B includes a structure identical to the feedback part of the encoder as depicted in Figure 11.12A. It first converts the received 4-bit data $I(n)$ to the quantized difference signal $d_q(n)$ using the adaptive quantizer. Then, at the second stage, the adaptive predictor uses the recovered quantized difference signal $d_q(n)$ and recovered current output $\tilde{x}(n)$ to generate the next output. Notice that the adaptive predictors of both the encoder and the decoder change correspondingly based on the signal to be quantized. The details of the adaptive predictor will be discussed.

Now, let us examine the ADPCM encoder principles. As shown in Figure 11.12A, the difference signal is computed as

$$d(n) = x(n) - \tilde{x}(n) \tag{11.7}$$

A 16-level nonuniform adaptive quantizer is used to quantize the difference signal $d(n)$. Before quantization, $d(n)$ is converted to a base-2 logarithmic representation and scaled by $y(n)$, which is computed by the scale factor algorithm. Four binary codes $I(n)$ are used to specify the quantized signal level representing $d_q(n)$, and the quantized difference $d_q(n)$ is also fed to the inverse adaptive quantizer. Table 11.6 shows the quantizer normalized input and output characteristics.

The scaling factor for the quantizer and the inverse quantizer $y(n)$ is computed according to the 4-bit quantizer output $I(n)$ and the adaptation speed control parameter $a_l(n)$, the fast (unlocked) scale factor $y_u(n)$, the slow (locked) scale factor $y_l(n)$, and the discrete function $W(I)$, defined in Table 11.7:

$$y_u(n) = (1 - 2^{-5})y(n) + 2^{-5}W(I(n)) \tag{11.8}$$

where $1.06 \le y_u(n) \le 10.00$.

The slow scale factor $y_l(n)$ is derived from the fast scale factor $y_u(n)$ using a lowpass filter as follows:

$$y_l(n) = (1 - 2^{-6})y_l(n-1) + 2^{-6}y_u(n) \tag{11.9}$$

Table 11.6 Quantizer Normalized Input and Output Characteristics

Normalized Quantizer Input Range: $\log_2 \lvert d(n) \rvert - y(n)$	Magnitude: $\lvert I(n) \rvert$	Normalized Quantizer Output: $\log_2 \lvert d_q(n) \rvert - y(n)$
$[3.12, +\infty)$	7	3.32
$[2.72, 3.12)$	6	2.91
$[2.34, 2.72)$	5	2.52
$[1.91, 2.34)$	4	2.13
$[1.38, 1.91)$	3	1.66
$[0.62, 1.38)$	2	1.05
$[-0.98, 0.62)$	1	0.031
$(-\infty, -0.98)$	0	$-\infty$

TABLE 11.7 Discrete Function $W(I)$

$|I(n)|$	7	6	5	4	3	2	1	0
$W(I)$	70.13	22.19	12.38	7.00	4.0	2.56	1.13	−0.75

The fast and slow scale factors are then combined to compute the scale factor:

$$y(n) = a_l(n)y_u(n-1) + (1 - a_l(n))y_l(n-1) \tag{11.10}$$

Next the controlling parameter $0 \le a_l(n) \le 1$ tends toward unity for speech signals and toward zero for voice band data signals and tones. It is updated based on the following parameters: $d_{ms}(n)$, which is the relatively short-term average of $F(I(n))$; $d_{ml}(n)$, which is the relatively long-term average of $F(I(n))$; and the variable $a_p(n)$, where $F(I(n))$ is defined as in Table 11.8.

TABLE 11.8 Discrete Function $F(I(n))$

$|I(n)|$	7	6	5	4	3	2	1	0
$F(I(n))$	7	3	1	1	1	0	0	0

Hence, we have

$$d_{ms}(n) = \left(1 - 2^{-5}\right)d_{ms}(n-1) + 2^{-5}F(I(n)) \tag{11.11}$$

and

$$d_{ml}(n) = \left(1 - 2^{-7}\right)d_{ml}(n-1) + 2^{-7}F(I(n)) \tag{11.12}$$

while the variable $a_p(n)$ is given by

$$a_p(n) = \begin{cases} \left(1 - 2^{-4}\right)a_p(n-1) + 2^{-3} & \text{if } |d_{ms}(n) - d_{ml}(n)| \ge 2^{-3}d_{ml}(n) \\ \left(1 - 2^{-4}\right)a_p(n-1) + 2^{-3} & \text{if } y(n)\langle 3 \\ \left(1 - 2^{-4}\right)a_p(n) + 2^{-3} & \text{if } t_d(n) = 1 \\ 1 & \text{if } t_r(n) = 1 \\ \left(1 - 2^{-4}\right)a_p(n) & \text{otherwise} \end{cases} \tag{11.13}$$

$a_p(n)$ approaches 2 when the difference between $d_{ms}(n)$ and $d_{ml}(n)$ is large and approaches 0 when the difference is small. Also $a_p(n)$ approaches 2 for an idle channel (indicated by $y(n) < 3$) or partial band signals (indicated by $t_d(n) = 1$). Finally, $a_p(n)$ is set to 1 when a partial band signal transition is detected ($t_r(n) = 1$).

$a_l(n)$, which is used in Equation (11.10), is defined as

$$a_l(n) = \begin{cases} 1 & a_p(n-1) \rangle 1 \\ a_p(n-1) & a_p(n-1) \leq 1 \end{cases} \tag{11.14}$$

The partial band signal $t_d(n)$ and the partial band signal transition $t_r(n)$ that appear in Equation (11.13) will be discussed later.

The predictor computes the signal estimate $\tilde{x}(n)$ from the quantized difference signal $d_q(n)$. The predictor z-transfer function, which is suitable for a variety of input signals, is given by

$$\frac{B(z)}{A(z)} = \frac{b_0 + b_1 z^{-1} + b_2 z^{-2} + b_3 z^{-3} + b_4 z^{-4} + b_5 z^{-5}}{1 - a_1 z^{-1} - a_2 z^{-2}} \tag{11.15}$$

It consists of a fifth-order portion that models the zeros and a second-order portion that models the poles of the input signals. The input signal estimate is expressed in terms of the processed signal $\hat{x}(n)$ and the signal $x_z(n)$ processed by the finite impulse response (FIR) filter as follows:

$$\tilde{x}(n) = a_1(n)\hat{x}(n-1) + a_2(n)\hat{x}(n-2) + x_z(n) \tag{11.16}$$

where

$$\hat{x}(n-i) = \tilde{x}(n-i) + d_q(n-i) \tag{11.17}$$

$$x_z(n) = \sum_{i=0}^{5} b_i(n) d_q(n-i) \tag{11.18}$$

Both sets of predictor coefficients are updated using a simplified gradient algorithm:

$$a_1(n) = (1 - 2^{-8})a_1(n-1) + 3 \cdot 2^{-8} signn(p(n)) sign(p(n-1)) \tag{11.19}$$

$$a_2(n) = (1 - 2^{-7})a_2(n-1) + 2^{-7}\{signn(p(n))sign(p(n-2)) \\ - f(a_1(n-1))signn(p(n))sign(p(n-1))\} \tag{11.20}$$

where $p(n) = d_q(n) + x_z(n)$ and

$$f(a_1(n)) = \begin{cases} 4a_1(n) & |a_1(n)| \leq 2^{-1} \\ 2\, sign(a_1(n)) & |a_1(n)| > 2^{-1} \end{cases} \tag{11.21}$$

Note that the function $sign(x)$ is defined in Equation (11.5), while the function $signn(x)$ equals 1 when $x > 0$, equals 0 when $x = 0$; and equals $= -1$ when $x < 0$ with stability constrains

$$|a_2(n)| \leq 0.75 \text{ and } |a_1(n)| \leq 1 - 2^{-4} - a_2(n) \tag{11.22}$$

$$a_1(n) = a_2(n) = 0 \quad \text{if} \quad t_r(n) = 1 \tag{11.23}$$

Also, the equations for updating the coefficients for the zero-order portion are given by

$$b_i(n) = (1 - 2^{-8})b_i(n-1) + 2^{-7} signn(d_q(n)) sign(d_q(n-i)) \text{ for } i = 0, 1, 2, \cdots, 5 \tag{11.24}$$

with the following constrains:

$$b_0(n) = b_1(n) = b_2(n) = b_3(n) = b_4(n) = b_5(n) = 0 \quad \text{if} \quad t_r(n) = 1 \tag{11.25}$$

$$t_d(n) = \begin{cases} 1 & a_2(n) \langle -0.71875 \\ 0 & otherwise \end{cases} \tag{11.26}$$

$$t_r(n) = \begin{cases} 1 & a_2(n) \langle -0.71875 \quad and \quad |d_q(n)| > 24 \cdot 2^{y_l} \\ 0 & otherwise \end{cases} \tag{11.27}$$

$t_d(n)$ is the indicator that detects a partial band signal (tone). If a tone is detected ($t_d(n) = 1$), Equation (11.13) is invoked to drive the quantizer into the fast mode of adaptation. $t_r(n)$ is the indicator for a transition from a partial band signal. If it is detected ($t_r(n) = 1$), setting the predictor coefficients to be zero as shown in Equations (11.23) and (11.25) will force the quantizer into the fast mode of adaptation.

Simulation Example

To illustrate performance, we apply the ADPCM encoder to the speech data used in Section 11.1 and then operate the ADPCM decoder to recover the speech signal. As described, the ADPCM uses 4 bits to encode each speech sample. The MATLAB implementations for the encoder and decoder are listed in Programs 11.11 to 11.13 in Section 11.7. Figure 11.13 plots the original speech samples, decoded

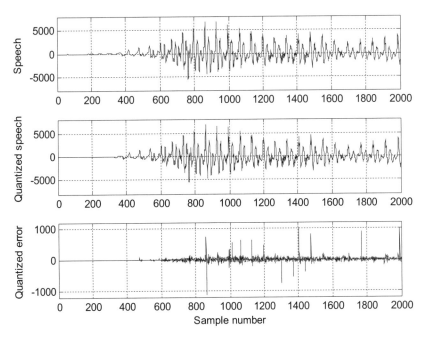

FIGURE 11.13

Original speech, quantized speech, and quantization error using ADPCM.

speech samples, and the quantization errors. From the figure, we see that the decoded speech data are very close to the original speech data; the quantization error is very small as compared with the speech sample, and its amplitude follows the change in amplitude of the speech data. In practice, we cannot tell the difference between the original speech and the decoded speech by listening to them. However, ADPCM encodes each speech sample using 4 bits per sample, while the original data are presented using 16 bits; thus the compression ratio (CR) of 4:1 is achieved.

In practical applications, data compression can reduce the storage media and bit rate for efficient digital transmission. To measure performance of data compression, we use

- the data CR, which is the ratio of original data file size to the compressed data file size, or ratio of the original code size in bits to the compressed code size in bits for fixed-length coding, and
- the bit rate, which is in terms of bits per second (bps) and can be calculated by

$$\text{bit rate} = m \times f_s \text{ (bps)} \tag{11.28}$$

where m — number of bits per sample (bits) and f_s = sampling rate (samples per second).

Now, let us look at an application example.

EXAMPLE 11.7

Speech is sampled at 8 kHz and each sample is encoded at 12 bits per sample. Using (1) no compression, (2) standard μ-law compression, and (3) standard ADPCM encoding,

a. determine the CR and the bit rate for each of the encoders and decoders;
b. determine the number of channels that the phone company can carry if a telephone system can transport the digital voice channel over a digital link with a capacity of 1.536 Mbps.

Solution:

a(1). For no compression:

$$\text{CR} = 1:1$$

$$\text{Bit rate} = 12\frac{bits}{sample} \times 8{,}000\frac{sample}{second} = 96 \text{ (kbps)}$$

b(1).

$$\text{Number of channels} = \frac{1.536}{96}\frac{MBPS}{KBPS} = 16$$

a(2). For standard μ-law compression, each sample is encoded using 8 bits per sample. Hence, we have

$$\text{CR} = \frac{12\ bits/sample}{8\ bits/sample} = 1.5:1$$

$$\text{Bit rate} = 8\frac{bits}{sample} \times 8{,}000\frac{sample}{second} = 64 \text{ (kbps)}$$

b(2).

$$\text{Number of channels} = \frac{1.536}{64}\frac{MBPS}{KBPS} = 24$$

a(3). For standard ADPCM with 4 bits per sample, it follows that

$$CR = \frac{12 \; bits/sample}{4 \; bits/sample} = 3:1$$

$$\text{Bit rate} = 4\frac{bits}{sample} \times 8{,}000\frac{sample}{second} = 32 \; (\text{kbps})$$

b(3).

$$\text{Number of channels} = \frac{1.536 \; MBPS}{32 \; KBPS} = 48.$$

11.4 DISCRETE COSINE TRANSFORM, MODIFIED DISCRETE COSINE TRANSFORM, AND TRANSFORM CODING IN MPEG AUDIO

This section introduces *discrete cosine transform* (DCT) and explains how to apply it in transform coding. This section will also show how to remove the block effects in transform coding using a modified DCT (MDCT). Finally, we will examine how MDCT coding is used in the MPEG (Motion Picture Experts Group) audio format, which is used as a part of MPEG audio, such as MP3 (MPEG-1 layer 3).

11.4.1 Discrete Cosine Transform

Given N data samples, we define the one-dimensional (1D) DCT pair as follows:

Forward transform:

$$X_{DCT}(k) = \sqrt{\frac{2}{N}} C(k) \sum_{n=0}^{N-1} x(n)\cos\left[\frac{(2n+1)k\pi}{2N}\right], \quad k = 0, 1, \cdots, N-1 \qquad (11.29)$$

Inverse transform:

$$x(n) = \sqrt{\frac{2}{N}} \sum_{k=0}^{N-1} C(k)X_{DCT}(k)\cos\left[\frac{(2n+1)k\pi}{2N}\right], \quad n = 0, 1, \cdots, N-1 \qquad (11.30)$$

$$C(k) = \begin{cases} \dfrac{\sqrt{2}}{2} & k = 0 \\ 1 & otherwise \end{cases} \qquad (11.31)$$

where $x(n)$ is the input data sample and $X_{DCT}(k)$ is the DCT coefficient. The DCT transforms the time domain signal to frequency domain coefficients. However, unlike the discrete Fourier transform (DFT), there are no complex number operations for both the forward and inverse transforms. Both the forward and inverse transforms use the same scale factor:

$$\sqrt{\frac{2}{N}}C(k)$$

In terms of transform coding, the DCT decomposes a block of data into a direct-current (DC) coefficient that corresponds to the average of the data samples and alternating-current (AC) coefficients that correspond to the frequency component (fluctuation). The terms "DC" and "AC" come from basic electrical engineering. In transform coding, we can quantize the DCT coefficients and encode them into binary information. The inverse DCT can transform the DCT coefficients back to the input data. Let us proceed to Examples 11.8 and 11.9.

EXAMPLE 11.8

Assume that the following input data can each be encoded by 5 bits, including a sign bit:

$$x(0) = 10,\ x(1) = 8,\ x(2) = 10 \text{ and } x(3) = 12$$

a. Determine the DCT coefficients.
b. Use the MATLAB function **dct()** to verify all the DCT coefficients.

Solution:

a. Using Equation (11.29) leads to

$$X_{DCT}(k) = \sqrt{\frac{1}{2}}C(k)\left[x(0)\cos\left(\frac{\pi k}{8}\right) + x(1)\cos\left(\frac{3\pi k}{8}\right) + x(2)\cos\left(\frac{5\pi k}{8}\right) + x(3)\cos\left(\frac{7\pi k}{8}\right)\right]$$

When $k = 0$, we see that the DC component is calculated as

$$X_{DCT}(0) = \sqrt{\frac{1}{2}}C(0)\left[x(0)\cos\left(\frac{\pi \times 0}{8}\right) + x(1)\cos\left(\frac{3\pi \times 0}{8}\right) + x(2)\cos\left(\frac{5\pi \times 0}{8}\right) + x(3)\cos\left(\frac{7\pi \times 0}{8}\right)\right]$$

$$= \sqrt{\frac{1}{2}} \times \frac{\sqrt{2}}{2}[x(0) + x(1) + x(2) + x(3)] = \frac{1}{2}(10 + 8 + 10 + 12) = 20$$

We clearly see that the first DCT coefficient is a scaled average value.
   For $k = 1$,

$$X_{DCT}(1) = \sqrt{\frac{1}{2}}C(1)\left[x(0)\cos\left(\frac{\pi \times 1}{8}\right) + x(1)\cos\left(\frac{3\pi \times 1}{8}\right) + x(2)\cos\left(\frac{5\pi \times 1}{8}\right) + x(3)\cos\left(\frac{7\pi \times 1}{8}\right)\right]$$

$$= \sqrt{\frac{1}{2}} \times 1\left[10 \times \cos\left(\frac{\pi}{8}\right) + 8 \times \cos\left(\frac{3\pi}{8}\right) + 10 \times \cos\left(\frac{5\pi}{8}\right) + 12 \times \cos\left(\frac{7\pi}{8}\right)\right] = -1.8478$$

Similarly, we have

$$X_{DCT}(2) = 2 \text{ and } X_{DCT}(3) = 0.7654$$

b. Using the MATLAB 1D-DCT function **dct()**, we can verify the DCT coeffcients:
    `>> dct([10 8 10 12])`
    ans = 20.0000 −1.8478 2.0000 0.7654

EXAMPLE 11.9

Assume the following DCT coefficients:

$$X_{DCT}(0) = 20,\ X_{DCT}(1) = -1.8478,\ X_{DCT}(0) = 2, \quad \text{and} \quad X_{DCT}(0) = 0.7654$$

a. Determine $x(0)$.
b. Use the MATLAB function **idct()** to verify all the recovered data samples.

Solution:

a. Applying Equations (11.30) and (11.31), we have

$$x(0) = \sqrt{\frac{1}{2}} \left[C(0) X_{DCT}(0) \cos\left(\frac{\pi}{8}\right) + C(1) X_{DCT}(1) \cos\left(\frac{3\pi}{8}\right) \right.$$
$$+ C(2) X_{DCT}(2) \cos\left(\frac{5\pi}{8}\right) + C(3) X_{DCT}(3) \cos\left(\frac{7\pi}{8}\right) \right]$$
$$= \sqrt{\frac{1}{2}} \left[\frac{\sqrt{2}}{2} \times 20 \times \cos\left(\frac{\pi}{8}\right) + 1 \times (-1.8478) \times \cos\left(\frac{3\pi}{8}\right) \right.$$
$$+ 1 \times 2 \times \cos\left(\frac{5\pi}{8}\right) + 1 \times 0.7654 \times \cos\left(\frac{7\pi}{8}\right) \right] = 10$$

b. With the MATLAB 1D inverse DCT function **idct()**, we obtain
    >> idct([20 −1.8478 2 0.7654])
    ans = 10.0000 8.0000 10.0000 12.0000

We verify that the input data samples are the same as those in Example 11.8.

In Example 11.9, we obtained an exact recovery of the input data from the DCT coefficients, since infinite precision of each DCT coefficient is preserved. However, in transform coding, each DCT coefficient is quantized using the number of bits per sample assigned by a bit allocation scheme. Usually the DC coefficient requires a larger number of bits to encode, since it carries more energy of the signal, while each AC coefficient requires a smaller number of bits to encode. Hence, the quantized DCT coefficients approximate the DCT coefficients in infinite precision, and the recovered input data with the quantized DCT coefficients will certainly have quantization errors.

EXAMPLE 11.10

Assuming the DCT coefficients

$$X_{DCT}(0) = 20, \; X_{DCT}(1) = -1.8478, \; X_{DCT}(2) = 2, \quad \text{and} \quad X_{DCT}(3) = 0.7654$$

in infinite precision, we recovered the exact data 10, 8, 10, and 12; this was verified in Example 11.9. If a bit allocation scheme quantizes the DCT coefficients using a scale factor of 4 in the form

$$X_{DCT}(0) = 4 \times 5 = 20, \; X_{DCT}(1) = 4 \times (-0) = 0, \; X_{DCT}(2) = 4 \times 1 = 4, \quad \text{and} \quad X_{DCT}(3) = 4 \times 0 = 0$$

we can code the scale factor of 4 with 3 bits (magnitude bits only), the scaled DC coefficient of 5 with 4 bits (including a sign bit), and the scaled AC coefficients of 0, 1, and 0 using 2 bits each. 13 bits in total are required. Use the MATLAB function **idct()** to recover the input data samples.

Solution:
Using the MATLAB function **idct()** and the quantized DCT coefficients, we obtain

>> idct([20 0 4 0])
ans = 12 8 8 12

As we see, the original sample requires 5 bits (4 magnitude bits and 1 sign bit) to encode each of 10, 8, 10, and 12 for a total of 20 bits. Hence, 7 bits are saved for coding this data block using the DCT. We expect many more bits to be saved in practice, in which a longer frame of the correlated data samples is used. However, quantization errors are introduced.

For comprehensive coverage of the topics on DCT, see Li and Drew (2004), Nelson (1992), Saywood (2000), and Stearns (2003).

11.4.2 Modified Discrete Cosine Transform

In the previous section, we observed how a 1D-DCT is adopted for coding a block of data. When we apply the 1D-DCT to audio coding, we first divide the audio samples into blocks and then transform each block of data with DCT. The DCT coefficients for each block are quantized according to the bit allocation scheme. However, when we decode DCT blocks, we encounter edge artifacts at boundaries of the recovered DCT blocks, since DCT coding is block based. This effect of edge artifacts produces periodic noise and is annoying in the decoded audio. To solve for such a problem, the windowed MDCT has been developed (described in Pan, 1995; Princen and Bradley, 1986). The principles are illustrated in Figure 11.14. As we shall see, the windowed MDCT is used in MPEG-1 MP3 audio coding.

We describe and discuss only the main steps for coding data blocks using the windowed MDCT (W-MDCT) based on Figure 11.14.

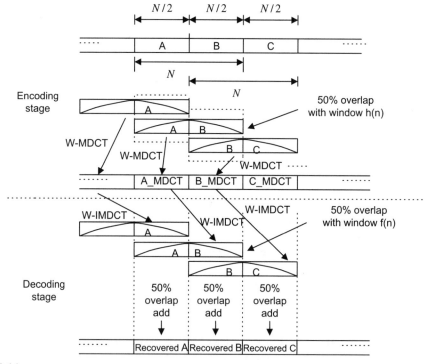

FIGURE 11.14

Modified discrete cosine transform (MDCT).

Encoding stage:

1. Divide the data samples into blocks that each have N (must be an even number) samples, and further divide each block into two subblocks, each with $N/2$ samples for data overlap purposes.
2. Apply the window function for the overlapped blocks. As shown in Figure 11.14, if one block contains the subblocks A and B, the next one would consist of subblocks B and C. The subblock B is the overlapped block. This procedure continues. A window function $h(n)$ is applied to each block with N samples to reduce possible edge effects. Next, the W-MDCT is applied. The W-MDCT is given by

$$X_{MDCT}(k) = 2 \sum_{n=0}^{N-1} x(n)h(n) \cos\left[\frac{2\pi}{N}(n + 0.5 + N/4)(k + 0.5)\right] \quad \text{for} \quad k = 0, 1, \cdots, N/2 - 1$$

(11.32)

Note that we need to compute and encode only half of the MDCT coefficients (since the other half can be reconstructed based on the first half of the MDCT coefficients).
3. Quantize and encode the MDCT coefficients.

Decoding stage:

1. Receive the $N/2$ MDCT coefficients, and use Equation (11.33) to recover the second half of the coefficients:

$$X_{MDCT}(k) = (-1)^{\frac{N}{2}+1} X_{MDCT}(N - 1 - k), \quad \text{for} \quad k = N/2, N/2 + 1, \cdots, N - 1 \qquad (11.33)$$

2. Apply the windowed inverse MDCT (W-IMDCT) to each N MDCT coefficient block using Equation (11.34) and then apply a decoding window function $f(n)$ to reduce the artifacts at the block edges:

$$x(n) = \frac{1}{N}f(n) \sum_{k=0}^{N-1} X_{MDCT}(k) \cos\left[\frac{2\pi}{N}(n + 0.5 + N/4)(k + 0.5)\right] \quad \text{for} \quad n = 0, 1, \cdots, N - 1$$

(11.34)

Note that the recovered sequence contains the overlap portion. As shown in Figure 11.14, if a decoded block has the decoded subblocks A and B, the next one would have subblocks B and C, where the subblock B is an overlapped block. The procedure continues.
3. Reconstruct the subblock B using the overlap and add operation, as shown in Figure 11.14, where two subblocks labeled B are overlapped and added to generate the recovered subblock B. Note that the first subblock B comes from the recovered block with N samples containing A and B, while the second subblock B belongs to the next recovered block with N samples consisting of B and C.

In order to obtain the perfect reconstruction, that is, the full cancellation of all aliasing introduced by the MDCT, the following two conditions must be met for selecting the window functions, in which one is used for encoding while the other is used for decoding (Princen and Bradley, 1986):

$$f\left(n + \frac{N}{2}\right)h\left(n + \frac{N}{2}\right) + f(n)h(n) = 1 \qquad (11.35)$$

$$f\left(n+\frac{N}{2}\right)h(N-n-1)-f(n)h\left(\frac{N}{2}-n-1\right) = 0 \qquad (11.36)$$

Here, we choose the following simple function for the W-MDCT:

$$f(n) = h(n) = \sin\left(\frac{\pi}{N}(n+0.5)\right) \qquad (11.37)$$

Equation (11.37) must satisfy the conditions described in Equations (11.35) and (11.36). This will be left for an exercise in the Problems section at the end of this chapter. The MATLAB functions **wmdcth()** and **wimdctf()** relate to this topic and are listed in Programs 11.14-11.5 in section (11.7). Now, let us examine the W-MDCT in Example 11.11.

EXAMPLE 11.11

Given the data 1, 2, −3, 4, 5, −6, 4, 5 ...,

a. determine the W-MDCT coefficients for the first three blocks using a block size of 4;
b. determine the first two overlapped subblocks, and compare the results with the original data sequence using the W-MDCT coefficients in (a).

Solution:

a. We divided the first two data blocks using the overlapping of 2 samples:
   First data block: 1 2 −3 4
   Second data block −3 4 5 −6
   Third data block: 5 −6 4 5
   We apply the W-MDCT to get
   >> wmdct([1 2 −3 4])
   ans = 1.1716 3.6569
   >> wmdct([−3 4 5 −6])
   ans = −8.0000 7.1716
   >> wmdct([5 −6 4 5])
   ans =−4.6569 −18.0711
b. The results from W-IWDCT are as follows:
   >> x1=wimdct([1.1716 3.6569])
   x1 =−0.5607 1.3536 −1.1465 −0.4749
   >> x2=wimdct([-8.0000 7.1716])
   x2 = −1.8536 4.4749 2.1464 0.8891
   >> x3=wimdct([-4.6569 -18.0711])
   x3 =2.8536 −6.8891 5.1820 2.1465
   Applying the overlap and add, we have
   >> [x1 0 0 0 0]+ [0 0 x2 0 0]+ [0 0 0 0 x3]
   ans = -0.5607 1.3536 -3.0000 4.0000 5.0000 −6.0000 5.1820 2.1465

The first two recovered subblocks contain the values −3, 4, 5 −6, which are consistent with the input data.

Figure 11.15 shows coding of speech data *we.dat* using the DCT transform and W-MDCT transform. To be able to see the block edge artifacts, the following parameters are used for both DCT and W-MDCT transform coding:

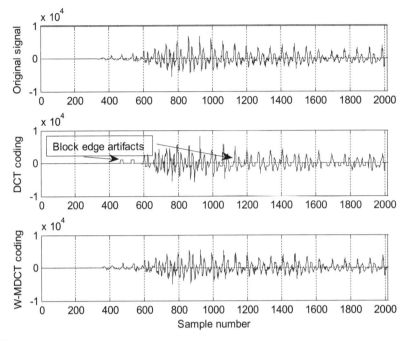

FIGURE 11.15

Waveform coding using DCT and W-MDCT.

Speech data: 16-bits per sample, 8,000 samples per second
Block size: 16 samples
Scale factor: 2-bit nonlinear quantizer
Coefficients: 3-bit linear quantizer

Note that we assume a lossless scheme will further compress the quantized scale factors and coeffi-cients. This stage does not affect the simulation results.

We use a 2-bit nonlinear quantizer with four levels to select the scale factor so that the block artifacts can be clearly displayed in Figure 11.15. We also apply a 3-bit linear quantizer to the scaled coefficients for both DCT and W-MDCT coding. As shown in Figure 11.15, the W-MDCT demon-strates significant improvement in smoothing out the block edge artifacts. The MATLAB simulation code is given in Programs 11.14 to 11.16 in Section 11.7, where Program 11.6 is the main program.

11.4.3 Transform Coding in MPEG Audio

With the DCT and MDCT concepts developed, we now explore the MPEG audio data format, where the DCT plays a key role. MPEG was established in 1988 to develop a standard for delivery of digital video and audio. Since MPEG audio compression contains so many topics, we focus here on examining its data format briefly, using the basic concepts developed in this book. Readers can further explore this subject by reading Pan's (1995) tutorial on MPEG audio compression, as well as Li and Drew (2004).

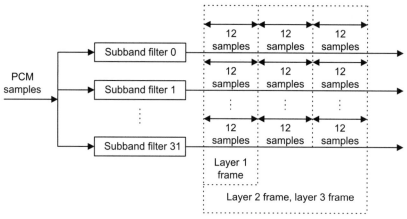

FIGURE 11.16

MPEG audio frame.

Figure 11.16 shows the MPEG audio frame. First, the input PCM samples—with possible sampling rates of 32 kHz, 44.1 kHz, and 48 kHz—are divided into 32 frequency subbands. All the subbands have equal bandwidths. The sum of their bandwidths covers up to the folding frequency, that is, $f_s/2$, which is the Nyquist limit in the DSP system. The subband filters are designed to minimize aliasing in the frequency domain. Each subband filter outputs one sample for every 32 input PCM samples continuously, and forms a data segment for every 12 output samples. The purpose of the filter banks is to separate the data into different frequency bands so that the psycho-acoustic model of the human auditory system (Yost, 1994) can be applied to activate the bit allocation scheme for a particular subband. The data frames are formed before quantization.

There are three types of data frames, as shown in Figure 11.16. Layer 1 contains 32 data segments, each coming from one subband with 12 samples, so the total frame has 384 data samples. As we see, layer 2 and layer 3 have the same size data frame, consisting of 96 data segments, where each filter outputs 3 data segments of 12 samples. Hence, layer 2 and layer 3 each have 1,152 data samples.

Next, let us examine briefly the content of each data frame, as show in Figure 11.17. Layer 1 contains 384 audio samples from 32 subbands, each having 12 samples. It begins with a header followed by a cyclic redundancy check (CRC) code. The numbers within parentheses indicate the possible number of bits to encode each field. The bit allocation informs the decoder of the number of bits used for each encoded sample in the specific band. Bit allocation can also be set to zero bits for a particular subband if analysis of the psycho-acoustic model finds that the data in the band can be discarded without affecting the audio quality. In this way, the encoder can achieve more data compression. Each scale factor is encoded with 6 bits. The decoder will multiply the scale factor by the decoded quantizer output to get the quantized subband value. Use of the scale factor allows for utilization of the full range of the quantizer. The field "ancillary data" is reserved for "extra" information.

The layer 2 encoder takes 1,152 samples per frame, with each subband channel having 3 data segments of 12 samples. These 3 data segments may have a bit allocation and up to to 3 scale factors.

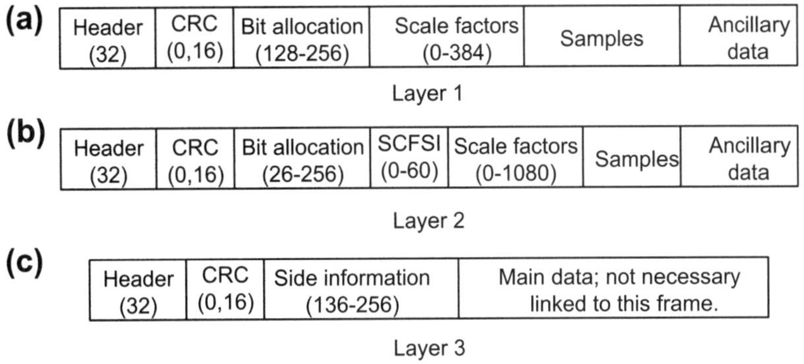

(a)

| Header (32) | CRC (0,16) | Bit allocation (128-256) | Scale factors (0-384) | Samples | Ancillary data |

Layer 1

(b)

| Header (32) | CRC (0,16) | Bit allocation (26-256) | SCFSI (0-60) | Scale factors (0-1080) | Samples | Ancillary data |

Layer 2

(c)

| Header (32) | CRC (0,16) | Side information (136-256) | Main data; not necessary linked to this frame. |

Layer 3

FIGURE 11.17

MPEG audio frame formats.

Using one scale factor for 3 data segments would be called for when values of the scale factors per subband are sufficiently close and the encoder applies temporal noise masking (a type of noise masking by the human auditory system) to hide any distortion. In Figure 11.17, the field "SCFSI" (scale-factor selection information) contains the information to inform the decoder. A different scale factor is used for each subband channel when avoidance of audible distortion is required. The bit allocation can also provide a possible single compact code word to represent three consecutive quantized values.

The layer 3 frame contains side information and main data that come from Huffman encoding (lossless coding with an exact recovery) of the W-MDCT coefficients to gain improvement over layer 1 and layer 2.

Figure 11.18 shows the MPEG-1 layer 1 and 2 encoder, and the layer 3 encoder. For MPEG-1 layer 1 and layer 2, the encoder examines the audio input samples using a 1,024-point fast Fourier transform (FFT). The psycho-acoustic model is analyzed based on the FFT coefficients. This includes possible frequency masking (hiding noise in frequency domain) and noise temporal masking (hiding noise in time domain). The result of the analysis of the psycho-acoustic model instructs the bit allocation scheme.

The major difference in layer 3, called MP3 (the most popular format in the multimedia industry), is that it adopts the MDCT. First, the encoder can gain further data compression by transforming the data segments from each subband channel using DCT and then quantizing the DCT coefficients, which, again, are losslessly compressed using Huffman encoding. As shown in Examples 11.8 to 11.11, since the DCT uses block-based processing, it produces block edge effects, where the beginning samples and ending samples show discontinuity and cause audible periodic noise. This periodic edge noise can be alleviated, as discussed in the previous section, by using the W-MDCT, in which there is 50% overlap between successive transform windows.

There are two sizes of windows. One has 36 samples and other 12 samples used in MPEG-1 layer 3 (MP3) audio. The larger block length offers better frequency resolution for low-frequency tonelike signals, hence it is used for the lowest two subbands. For the rest of the subbands, the shorter block is used, since it allows better time resolution for noiselike transient signals. Other improvements of MP3

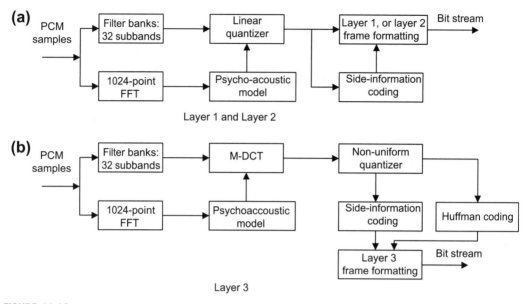

FIGURE 11.18

Encoder block diagrams for layers 1 and 2 and for layer 3.

over layers 1 and 2 include use of the scale-factor band, where the W-MDCT coefficients are regrouped from the original 32 uniformly divided subbands into 25 actual critical bands based on the human auditory system. Then the corresponding scale factors are assigned, and a nonlinear quantizer is used.

Finally, Huffman coding is applied to the quantizer outputs to obtain more compression. Particularly in CD-quality audio, MP3 (MPEG-1 layer 3) can achieve CRs varying from 12:1 to 8:1, corresponding to bit rates from 128 kbps to 192 kbps. Besides the use of DCT in MP3, MPEG-2 audio coding methods such as AC-2, AC-3, ATRAC, and PAC/MPAC also use W-MDCT coding. Readers can further explore these subjects in Brandenburg (1997) and Li and Drew (2004).

11.5 LABORATORY EXAMPLES OF SIGNAL QUANTIZATION USING THE TMS320C6713 DSK

Linear quantization can be implemented as shown in C Program 11.1. The program only demonstrates left channel coding; right channel coding can be easily extended from the program. The program consists of both an encoder and decoder. First, it converts the 16-bit 2's complement data to the sign-magnitude format with truncated magnitude bits as required. Then the decoder converts the compressed PCM code back to the 16-bit data. The encoding and decoding are explained in Example 11.12.

EXAMPLE 11.12

Given a 16-bit datum lc $= -2050$ (decimal), convert it to a 5-bit linear PCM code using the information in Program 11.1.

Solution:

a. Encoding:
   tmp $= 2050$ (decimal) $= $ 0x0802 (hex) $= $ 0000 1000 0000 0010 (binary)
   After shifting 11 bits, tmp $= $ 0x0001 (hex)
   PCMcode $= $ tmp&mask[5-1]$=$0x0001&0x000f$=$0x0001 (hex) // Get magnitude bits
   if lc >0 {
   PCMcode $= $ PCMcode | sign[5-1] // Add positive sign bit
   } // This line is not executed since lc < 0
   PCMcode $= $ <u>00001</u> (binary)

b. Decoding:
   dec $= $ PCMcode & mask[4] $= $ 0x0001&0x000f$=$0x0001 (hex)
   tmp $= $ PCMcode&sign[4] $= $ 0x0001&0x0010$=$0x0000 (hex) (the number is negative)
   dec $= $ dec$<<$11$=$0x0800
   dec $=$dec | rec[4]$=$0x0800| 0x0400$=$0x0C00 (hex)$=$3072 (decimal)
   lc $= -3072$ (recovered decimal)

As expected, the recovered PCM code is different from the original code, since a PCM is the lossy coding scheme. The same procedure can be followed for coding a positive decimal number.

C Program 11.1. Encoding and decoding using the TMS320C6713 DSK.

```
int PCMcode;
/* Sign-magnitude format: s magnitude bits, see Section 11.1, in Chapter 11 */
/* convert a 16-bit PCM code a 5-bit PCM code, and recover the 5-bit PCM code to the 16-bit PCM code */
// See the following example:
/* Convert a 16-bit sign-magnitude code 5-bit PCM code 16-bit recovered sign-magnitude code */
/* sADCDEFGHIJKLMNO     sABCDE      sABCD10000000000 */
/* Note that audio input/output data are in 2's complement form */
/* For encoder, convert the 2's complement input to the sign-magnitude form before
quantization          */
/* For decoder, convert the sign-magnitude form to the 2's complement output before DAC   */
int nofbits=5; // specify the number of quantization bits
int sign[16]={0x01,0x02,0x04,0x08,0x10,0x20,0x40,0x80,
     0x100,0x200,0x400,0x800,0x1000,0x2000,0x4000,0x8000}; // add sign bit (MSB)
int mask[16]={0x0,0x01,0x03,0x07,0x0f,0x1f,0x3f,0x7f,
     0xff,0x1ff,0x3ff,0x7ff,0xfff,0x1fff,0x3fff,0x7fff}; // mask for obtaining
magnitude bits
int rec[16]={0x4000,0x2000,0x1000,0x0800,
     0x0400,0x0200,0x0100,0x0080,
     0x0040,0x0020,0x0010,0x0008,
     0x0004,0x0002,0x0001,0x0000};// the midpoint of the quantization interval
interrupt void c_int11()
{
  int lc; /*left channel input */
  int rc; /*right channel input */
  int lcnew; /*left channel output */
```

```
    int rcnew; /*right channel output */
    int temp, dec;
//Left channel and right channel inputs
AIC23_data.combo=input_sample();
  lc=(int) (AIC23_data.channel[LEFT]);
  rc= (int) (AIC23_data.channel[RIGHT]);
// Insert DSP algorithm below
 /* Encoder :*/
 tmp =lc; // for the Left Line In channel
  if (tmp <0 )
  {
   tmp=-tmp; // Get magnitude bits to work with
  }
  tmp =(tmp>>(16-nofbits));
  PCMcode=tmp&mask[nofbits-1]; // Get magnitude bits
 if (lc>=0)
 {
   PCMcode= PCMcode | sign[nofbits-1]; // Add sign bit
 }
 /* PCM code (stored in the lower portion) */
 /* Decoder: */
  dec = PCMcode&mask[nofbits-1]; // Obtain magnitude bits
  tmp= PCMcode&sign[nofbits-1]; // Obtain the sign bit
  dec = (dec<<(16-nofbits)); // Scale to 15-bit magnitude
  dec = dec | rec[nofbits-1]; // Recover the midpoint of the quantization interval
 lc =dec;
 if (tmp == 0x00)
 {
  lc =-dec; // Back to 2's complement form (change the sign)
 }
// End of the DSP algorithm
  lcnew=lc; /* Send to DAC */
  rcnew=lc; /* Keep the original data */
  AIC23_data.channel[LEFT]=(short) lcnew;
  AIC23_data.channel[RIGHT]=(short) rcnew;
  output_sample(AIC23_data.combo);
}
```

C Program 11.2 demonstrates digital μ-law encoding and decoding. It converts a 12-bit linear PCM code to an 8-bit compressed PCM code using the principles discussed in Section 11.2. Note that the program only performs left-channel coding.

C Program 11.2. Digital μ-law encoding and decoding.

```
int ulawcode;
/* Digital mu-law definition*/
// Sign-magnitude format: s segment quantization
// s = 1 for the positive value, s = 0 for the negative value
// Segment defines compression
// quantization with 16 levels
```

```
// See Section 11.2, Chapter 11
/* Segment  12-bit PCM  4-bit quantization interval */
/* 0     s0000000ABCD     s000ABCD      */
/* 1     s0000001ABCD     s001ABCD      */
/* 2     s000001ABCDX     s010ABCD      */
/* 3     s00001ABCDXX     s011ABCD      */
/* 4     s0001ABCDXXX     s100ABCD      */
/* 5     s001ABCDXXXX     s101ABCD      */
/* 6     s01ABCDXXXXX     s110ABCD      */
/* 7     s1ABCDXXXXXX     s111ABCD      */
//
/* Segment Recovered 12-bit PCM         */
/* 0     s0000000ABCD                   */
/* 1     s0000001ABCD                   */
/* 2     s000001ABCD1                   */
/* 3     s00001ABCD10                   */
/* 4     s0001ABCD100                   */
/* 5     s001ABCD1000                   */
/* 6     s01ABCDq0000                   */
/* 7     s1ABCD100000                   */
/* Note that audio input/output data are in 2's complement form */
/* For encoder, convert the 2's complement form to the sign-magnitude form before
quantization        */
/* For decoder, convert the sign-magnitude form to the 2's complement output before DAC  */
interrupt void c_int11()
{
  int lc; /*left channel input */
  int rc; /*right channel input */
  int lcnew; /*left channel output */
  int rcnew; /*right channel output */
  int tmp,ulawcode, dec;
//Left channel and right channel inputs
  AIC23_data.combo=input_sample();
  lc=(int) (AIC23_data.channel[LEFT]);
  rc= (int) (AIC23_data.channel[RIGHT]);
// Insert DSP algorithm below
 /* Encoder :*/
 tmp =lc;
 if (tmp <0 )
 { tmp=-tmp; // Get magnitude bits to work with
 }
 tmp =(tmp>>4); // Linear scale down to 12 bits to use the u-255 law table
 if( (tmp&0x07f0)==0x0) // Segment 0
 { ulawcode= (tmp&0x000f); }
 if( (tmp&0x07f0)==0x0010) // Segment 1
 { ulawcode= (tmp&0x00f);
   ulawcode= ulawcode | 0x10; }
 if( (tmp&0x07E0)==0x0020) // Segment 2
```

```
{ ulawcode= (tmp&0x001f)>>1;
  ulawcode = ulawcode |0x20; }
if( (tmp&0x07c0)==0x0040) // Segment 3
{ ulawcode= (tmp&0x003f)>>2;
  ulawcode= ulawcode | 0x30; }
if( (tmp&0x0780)==0x0080) // Segment 4
{ ulawcode= (tmp&0x007f)>>3;
  ulawcode =ulawcode | 0x40;}
if( (tmp&0x0700)==0x0100) // Segment 5
{ ulawcode= (tmp&0x00ff)>>4;
  ulawcode = ulawcode | 0x50;}
if( (tmp&0x0600)==0x0200) // Segment 6
{ ulawcode= (tmp&0x01ff)>>5;
  ulawcode = ulawcode | 0x60; }
if( (tmp&0x0400)==0x0400) // Segment 7
{ ulawcode= (tmp&0x03ff)>>6;
  ulawcode=ulawcode | 0x70; }
if (lc>=0)
{
  ulawcode= ulawcode|0x80;
}
/* u-law code (8 bit compressed PCM code) for transmission and storage */
/* Decoder: */
  tmp = ulawcode&0x7f;
  tmp = (tmp>>4);
if ( tmp == 0x0) // Segment 0
{ dec = ulawcode&0xf; }
if ( tmp == 0x1) // Segment 1
{ dec = ulawcode&0xf | 0x10; }
if ( tmp == 0x2) // Segment 2
{ dec = ((ulawcode&0xf)<<1) | 0x20;
  dec= dec |0x01; }
if ( tmp == 0x3) // Segment 3
{ dec = ((ulawcode&0xf)<<2) | 0x40;
  dec = dec|0x02; }
if ( tmp == 0x4) // Segment 4
{ dec = ((ulawcode&0xf)<<3) | 0x80;
  dec = dec|0x04; }
if ( tmp == 0x5) // Segment 5
{ dec = ((ulawcode&0xf)<<4) | 0x0100;
  dec = dec|0x08; }
if ( tmp == 0x6) // Segment 6
{ dec = ((ulawcode&0xf)<<5) | 0x0200;
  dec =dec|0x10; }
if ( tmp == 0x7) // Segment 7
{ dec = ((ulawcode&0xf)<<6) | 0x0400;
  dec = dec |0x20; }
 tmp =ulawcode & 0x80;
```

```
lc =dec;
if (tmp == 0x00)
{ lc=-dec; // Back to 2's complement form
}
lc= (lc<<4); // Linear scale up to 16 bits
// End of the DSP algorithm
  lcnew=lc; /* Send to DAC */
  rcnew=lc; /* Keep the original data */
  AIC23_data.channel[LEFT]=(short) lcnew;
  AIC23_data.channel[RIGHT]=(short) rcnew;
  output_sample(AIC23_data.combo);
}
```

11.6 SUMMARY

1. The linear midtread quantizer used in PCM coding has an odd number of quantization levels, that is, $2^n - 1$. It accommodates the same decoded magnitude range for quantizing the positive and negative voltages.

2. Analog and digital μ-law compression improve coding efficiency. 8-bit μ-law compression of speech is equivalent to 12-bit linear PCM coding, with no difference in sound quality. These methods are widely used in the telecommunications industry and in multimedia system applications.

3. DPCM encodes the difference between the input sample and predicted sample using a predictor to achieve coding efficiency.

4. DM coding is essentially a 1-bit DPCM.

5. ADPCM is similar to DPCM except that the predictor transfer function has six zeros and two poles and is an adaptive filter. ADPCM is superior to 8-bit μ-law compression, since it provides the same sound quality with only 4 bits per code.

6. Data compression performance is measured in terms of the data compression ratio and the bit rate.

7. The DCT decomposes a block of data to the DC coefficient (average) and AC coefficients (fluctuation) so that different numbers of bits are assigned to encode DC coefficients and AC coefficients to achieve data compression.

8. W-MDCT alleviates the block effects introduced by the DCT.

9. MPEG-1 audio formats such as MP3 (MPEG-1, layer 3) include W-MDCT, filter banks, a psycho-acoustic model, bit allocation, a nonlinear quantizer, and Huffman lossless coding.

11.7 MATLAB PROGRAMS

Program 11.1 MATLAB program for the linear midtread quantizer.

```
clear all;close all
disp('load speech: We');
load we.dat;              % Provided by your instructor
sig = we;
```

```
lg=length(sig);              % Length of the speech data
t=[0:1:lg-1];                % Time index
sig=5*sig/max(abs(sig));     % Normalized signal to the range between -5 to 5
Emax = max(abs(sig));
Erms = sqrt( sum(sig .* sig) / length(sig))
k=Erms/Emax
disp('20*log10(k)=>');
k = 20*log10(k)
bits = input('input number of bits =>');
lg = length(sig);
% Encoding
for x=1:lg
 [indx(x) qy] = mtrdenc(bits, 5, sig(x));
end
disp('Finished and transmitted');
% Decoding
for x=1:lg
 qsig(x) = mtrddec(bits, 5, indx(x));
end
disp('decoding finished');
 qerr = sig-qsig;      % Calculate quantization errors
subplot(3,1,1);plot(t, sig);grid
ylabel('Speech');axis([0 length(we) -5 5]);
subplot(3,1,2);plot(t, qsig);grid
ylabel('Quantized speech');axis([0 length(we) -5 5]);
subplot(3,1,3);plot(qerr);grid
axis([0 length(we) -0.5 0.5]);
ylabel('Qunatized error');xlabel('Sample number');
disp('signal to noise ratio due to quantization noise')
snr(sig,qsig);        % Calculate signal to noise ratio due to quantization
```

Program 11.2. MATLAB program for μ-law encoding and decoding.

```
close all; clear all
disp('load speech file');
load we.dat;                 % Provided by your instructor
lg=length(we);               % Length of the speech data
we=5*we/max(abs(we));        % Normalize the speech data
we_nor=we/max(abs(we));      % Normalization
t=[0:1:lg-1];                % Time index
disp('mulaw companding')
mu=input('input mu =>');
for x=1:lg
 ymu(x) =mulaw(we_nor(x),1,mu);
end
disp('finished mu-law companding');
disp('start to quantization')
bits = input('input bits=>');
% Midtread quantization and encoding
```

```
for x=1:lg
 [indx(x) qy] = mtrdenc(bits, 1, ymu(x));
end
disp('finished and transmitted');
%
% Midtread decoding
for x=1:lg
 qymu(x) = mtrddec(bits, 1, indx(x));
end
disp('expander');
for x=1:lg
 dymu(x)= muexpand(qymu(x),1,mu)*5;
end
disp('finished')
qerr = dymu-we; % Quantization error
subplot(3,1,1);plot(we);grid
ylabel('Speech');axis([0 length(we) -5 5]);
subplot(3,1,2);plot(dymu);grid
ylabel('recovered speech');axis([0 length(we) -5 5]);
subplot(3,1,3);plot(qerr);grid
ylabel('Quantized error');xlabel('Sample number');
axis([0 length(we) -1 1]);
snr(we,dymu);       % Calculate signal to noise ratio due to quantization
```

Program 11.3. MATLAB function for μ-law companding.

```
Function qvalue = mulaw(vin, vmax, mu)
% This function performs mu-law companding
% Usage:
% function qvalue = mulaw(vin, vmax, mu)
% vin = input value
% vmax = input value
% mu = parameter for controlling the degree of compression which must be the same
% qvalue = output value from the mu-law compander
% as the mu-law expander
%
 vin = vin/vmax;     % Normalization
% mu-law companding formula
 qvalue = vmax*sign(vin)*log(1+mu*abs(vin))/log(1+mu);
```

Program 11.4. MATLAB program for μ-law expanding.

```
function rvalue = muexpand(y,vmax, mu)
% This function performs mu-law exoanding
% Usage:
% function rvalue = muexpand(y,vmax, mu)
% y = input signal
% vmax = maximum amplitude
% mu = parameter for controlling the degree of compression, which must be the same
% as the mu-law compander
```

```
% rvalue = output value from the mu-law expander
%
 y=y/vmax;      % Normalization
% mu-law expanding
 rvalue=sign(y)*(vmax/mu)*((1+mu)^abs(y) -1);
```

Program 11.5. MATLAB function for midread quantizer encoding.

```
function [indx, pq ] = mtrdenc(NoBits,Xmax,value)
% function pq = mtrdenc(NoBits, Xmax, value)
% This routine is created for simulation of midtread uniform quantizer.
%
% NoBits: number of bits used in quantization.
% Xmax: overload value.
% value: input to be quantized.
% pq: output of quantized value
% indx: integer index
%
% Note: the midtread method is used in this quantizer.
%
 if NoBits == 0
  pq = 0;
  indx=0;
 else
  delta = 2*abs(Xmax)/(2^NoBits-1);
  Xrmax=delta*(2^NoBits/2-1);
  if abs(value) >= Xrmax
   tmp = Xrmax;
 else
   tmp = abs(value);
 end
 indx=round(tmp/delta);
 pq =indx*delta;
 if value < 0
  pq = -pq;
  indx=-indx;
 end
end
```

Program 11.6. MATLAB function for midtread quantizer decoding.

```
function pq = mtrddec(NoBits,Xmax,indx)
% function pq = mtrddec(NoBits, Xmax, value)
% This routine is the dequantizer.
%
% NoBits: number of bits used in quantization.
% Xmax: overload value.
% pq: output of quantized value
% indx: integer index
%
```

```
% Note: the midtread method is used in this quantizer.
%
 delta = 2*abs(Xmax)/(2^NoBits-1);
 pq =indx*delta;
```

Program 11.7. MATLAB function for calculation of signal to quantization noise ratio (SNR).

```
function snr = calcsnr(speech, qspeech)
% function snr = calcsnr(speech, qspeech)
% This routine was created for calculation of SNR.
%
% speech: original speech waveform.
% qspeech: quantized speech.
% snr: output SNR in dB.
%
% Note: the midrise method is used in this quantizer.
%
 qerr = speech-qspeech;
 snr = 10*log10(sum(speech.*speech)/sum(qerr.*qerr))
```

Program 11.8. Main program for digital μ-law encoding and decoding.

```
load we12b.dat
for i=1:1:length(we12b)
  code8b(i)=dmuenc(12, we12b(i));  % Encoding
  qwe12b(i)=dmudec(code8b(i));   % Decoding
end
subplot(4,1,1),plot(we12b);grid
ylabel('a');axis([0 length(we12b) -1024 1024]);
subplot(4,1,2),plot(code8b);grid
ylabel('b');axis([0 length(we12b) -128 128]);
subplot(4,1,3),plot(qwe12b);grid
ylabel('c');axis([0 length(we12b) -1024 1024]);
subplot(4,1,4),plot(qwe12b-we12b);grid
ylabel('d');xlabel('Sample number');axis([0 length(we12b) -40 40]);
```

Program 11.9. The digital μ-law compressor.

```
function [cmp_code ] = dmuenc(NoBits, value)
% This routine is created for simulation of 12-bit mu law compression.
% function [cmp_code ] = dmuenc(NoBits, value)
% NoBits = number of bits for the data
% value = input value
% cmp_code = output code
%
 scale = NoBits-12;
 value=value*2^(-scale); % Scale to 12 bit
 if (abs(value) >=0) & (abs(value)<16)
   cmp_code=value;
 end
 if (abs(value) >=16) & (abs(value)<32)
```

```
  cmp_code=sgn(value)*(16+fix(abs(value)-16));
end
if (abs(value) >=32) & (abs(value)<64)
 cmp_code=sgn(value)*(32+fix((abs(value) -32)/2));
end
if (abs(value) >=64) & (abs(value)<128)
  cmp_code=sgn(value)*(48+fix((abs(value) -64)/4));
end
if (abs(value) >=128) & (abs(value)<256)
 cmp_code=sgn(value)*(64+fix((abs(value) -128)/8));
end
if (abs(value) >=256) & (abs(value)<512)
 cmp_code=sgn(value)*(80+fix((abs(value) -256)/16));
end
if (abs(value) >=512) & (abs(value)<1024)
 cmp_code=sgn(value)*(96+fix((abs(value) -512)/32));
end
if (abs(value) >=1024) & (abs(value)<2048)
   cmp_code=sgn(value)*(112+fix((abs(value) -1024)/64));
end
```

Program 11.10. The digital μ-law expander.

```
function [value ] = dmudec(cmp_code)
% This routine is created for simulation of 12-bit mu law decoding.
% Usage:
% unction [value ] = dmudec(cmp_code)
% cmp_code = input mu-law encoded code
% value = recovered output value
%
 if (abs(cmp_code) >=0) & (abs(cmp_code)<16)
   value=cmp_code;
end
if (abs(cmp_code) >=16) & (abs(cmp_code)<32)
 value=sgn(cmp_code)*(16+(abs(cmp_code)-16));
end
if (abs(cmp_code) >=32) & (abs(cmp_code)<48)
 value=sgn(cmp_code)*(32+(abs(cmp_code)-32)*2+1);
end
if (abs(cmp_code) >=48) & (abs(cmp_code)<64)
   value=sgn(cmp_code)*(64+(abs(cmp_code)-48)*4+2);
end
if (abs(cmp_code) >=64) & (abs(cmp_code)<80)
 value=sgn(cmp_code)*(128+(abs(cmp_code)-64)*8+4);
end
if (abs(cmp_code) >=80) & (abs(cmp_code)<96)
 value=sgn(cmp_code)*(256+(abs(cmp_code)-80)*16+8);
end
if (abs(cmp_code) >=96) & (abs(cmp_code)<112)
```

```
  value=sgn(cmp_code)*(512+(abs(cmp_code)-96)*32+16);
 end
 if (abs(cmp_code) >=112) & (abs(cmp_code)<128)
   value=sgn(cmp_code)*(1024+(abs(cmp_code)-112)*64+32);
 end
```

Program 11.11. Main program for ADPCM coding.

```
% This program is written for offline simulation.
% file: adpcm.m
clear all; close all
load we.dat                    % Provided by the instructor
speech=we;
desig= speech;
lg=length(desig);              % Length of speech data
enc = adpcmenc(desig);         % ADPCM encoding
%ADPCM finished
dec = adpcmdec(enc);           % ADPCM decoding
snrvalue = snr(desig,dec)      % Calculate signal to noise ratio due to quantization
subplot(3,1,1);plot(desig);grid;
ylabel('Speech');axis([0 length(we) -8000 8000]);
subplot(3,1,2);plot(dec);grid;
ylabel('Quantized speech');axis([0 length(we) -8000 8000]);
subplot(3,1,3);plot(desig-dec);grid
ylabel('Quantized error');xlabel('Sample number');
axis([0 length(we) -1200 1200]);
```

Program 11.12. MATLAB function for ADPCM encoding.

```
function iiout = adpcmenc(input)
% This function performs ADPCM encoding.
% function iiout = adpcmenc(input)
% Usage:
% input = input value
% iiout = output index
%
% Quantization tables
fitable = [0 0 0 1 1 1 1 3 7];
witable = [-0.75 1.13 2.56 4.00 7.00 12.38 22.19 70.13 ];
qtable = [ -0.98 0.62 1.38 1.91 2.34 2.72 3.12 ];
invqtable = [0.031 1.05 1.66 2.13 2.52 2.91 3.32 ];
lgth = length(input);
sr = zeros(1,2); pk = zeros(1,2);
a = zeros(1,2); b = zeros(1,6);
dq = zeros(1,6); ii= zeros(1,lgth);
y=0; ap = 0; al = 0; yu=0; yl = 0; dms = 0; dml = 0; tr = 0; td = 0;
for k = 1:lgth
 sl = input(k);
%
% predict zeros
```

```
%
 sez = b(1)*dq(1);
 for i=2:6
 sez = sez + b(i)*dq(i);
end
se = a(1)*sr(1)+a(2)*sr(2)+ sez;
 d = sl - se;
%
% Perform quantization
%
 dqq = log10(abs(d))/log10(2.0)-y;
  ik= 0;
  for i=1:7
   if dqq > qtable(i)
    ik = i;
   end
  end
 if d < 0
  ik = -ik;
 end
 ii(k) = ik;
 yu = (31.0/32.0)*y + witable(abs(ik)+1)/32.0;
 if yu > 10.0
  yu = 10.0;
 end
 if yu < 1.06
  yu = 1.06;
 end
 yl = (63.0/64.0)*yl+yu/64.0;
%
%Inverse quantization
%
 if ik == 0
 dqq = 2^(-y);
 else
 dqq = 2^(invqtable(abs(ik))+y);
 end
 if ik < 0
 dqq = -dqq;
 end
 srr = se + dqq;
 dqsez = srr+sez-se;
%
% Update state
%
pk1 = dqsez;
%
% Obtain adaptive predictor coefficients
```

```
%
  if tr == 1
    a = zeros(1,2); b = zeros(1,6);
    tr = 0;
    td = 0; % Set for the time being
  else
% Update predictor poles
% Update a2 first
  a2p = (127.0/128.0)*a(2);
  if abs(a(1)) <= 0.5
    fa1 = 4.0*a(1);
  else
    fa1 = 2.0*sgn(a(1));
  end
  a2p=a2p+(sign(pk1)*sgn(pk(1))-fa1*sign(pk1)*sgn(pk(2)))/128.0;
  if abs(a2p) > 0.75
    a2p = 0.75*sgn(a2p);
  end
   a(2) = a2p;
%
% Update a1
  a1p = (255.0/256.0)*a(1);
     a1p = a1p + 3.0*sign(pk1)*sgn(pk(2))/256.0;
   if abs(a1p) > 15.0/16.0-a2p
     a1p = 15.0/16.0 -a2p;
   end
  a(1) = a1p;
%
% Update b coefficients
%
 for i= 1:6
 b(i) = (255.0/256.0)*b(i)+sign(dqq)*sgn(dq(i))/128.0; % see Program 11.17 for sgn().
 end
 if a2p < -0.7185
 td = 1;
 else
 td = 0;
 end
 if a2p < -0.7185 & abs(dq(6)) > 24.0*2^(yl)
 tr = 1;
 else
 tr = 0;
 end
 for i=6:-1:2
 dq(i) = dq(i-1);
 end
 dq(1) = dqq; pk(2) = pk(1); pk(1) = pk1; sr(2) = sr(1); sr(1) = srr;
%
```

```
% Adaptive speed control
%
 dms = (31.0/32.0)*dms; dms = dms + fitable(abs(ik)+1)/32.0;
 dml = (127.0/128.0)*dml; dml = dml + fitable(abs(ik)+1)/128.0;
 if ap > 1.0
 al = 1.0;
 else
 al = ap;
 end
 ap = (15.0/16.0)*ap;
 if abs(dms-dml) >= dml/8.0
  ap = ap + 1/8.0;
 end
 if y < 3
  ap = ap +1/8.0;
 end
 if td == 1
  ap = ap + 1/8.0;
 end
 if tr == 1
  ap = 1.0;
 end
 y = al*yu + (1.0-al)*yl;
  end
end
iiout = ii;v
```

Program 11.13. MATLAB function for ADPCM decoding.

```
function iiout = adpcmdec(ii)
% This function performs ADPCM decoding.
% function iiout = adpcmdec(ii)
% Usage:
% ii = input ADPCM index
% iiout = decoded output value
%
% Quantization tables:
fitable = [0 0 0 1 1 1 1 3 7];
witable = [-0.75 1.13 2.56 4.00 7.00 12.38 22.19 70.13 ];
qtable = [ -0.98 0.62 1.38 1.91 2.34 2.72 3.12 ];
invqtable = [0.031 1.05 1.66 2.13 2.52 2.91 3.32 ];
lgth = length(ii);
sr = zeros(1,2); pk = zeros(1,2);
a = zeros(1,2); b = zeros(1,6);
dq = zeros(1,6); out= zeros(1,lgth);
y=0; ap = 0; al = 0; yu=0; yl = 0; dms = 0; dml = 0; tr = 0; td = 0;
for k = 1:lgth
%
 sez = b(1)*dq(1);
```

```
for i=2:6
 sez = sez + b(i)*dq(i);
end
se = a(1)*sr(1)+a(2)*sr(2)+ sez;
%
% Inverse quantization
%
 ik = ii(k);
 yu = (31.0/32.0)*y + witable(abs(ik)+1)/32.0;
 if yu > 10.0
 yu = 10.0;
 end
 if yu < 1.06
 yu = 1.06;
 end
 yl = (63.0/64.0)*yl+yu/64.0;
 if ik == 0
  dqq = 2^(-y);
 else
  dqq = 2^(invqtable(abs(ik))+y);
 end
 if ik < 0
  dqq = -dqq;
 end
 srr = se + dqq;
 dqsez = srr+sez-se;
 out(k) =srr;
%
% Update state
%
pk1 = dqsez;
%
% Obtain adaptive predictor coefficients
%
  if tr == 1
    a = zeros(1,2);
    b = zeros(1,6);
    tr = 0;
    td = 0; % Set for the time being
  else
% Update predictor poles
% Update a2 first
  a2p = (127.0/128.0)*a(2);
  if abs(a(1)) <= 0.5
    fa1 = 4.0*a(1);
  else
    fa1 = 2.0*sgn(a(1));
  end
```

```
   a2p=a2p+(sign(pk1)*sgn(pk(1))-fal*sign(pk1)*sgn(pk(2)))/128.0;
   if abs(a2p) > 0.75
    a2p = 0.75*sgn(a2p);
   end
    a(2) = a2p;
%
% Update a1
    a1p = (255.0/256.0)*a(1);
      a1p = a1p + 3.0*sign(pk1)*sgn(pk(2))/256.0;
    if abs(a1p) > 15.0/16.0-a2p
      a1p = 15.0/16.0-a2p;
    end
   a(1) = a1p;
%
% Update b coefficients
%
 for i= 1: 6
  b(i) = (255.0/256.0)*b(i)+sign(dqq)*sgn(dq(i))/128.0;
 end
 if a2p < -0.7185
  td = 1;
 else
  td = 0;
 end
 if a2p < -0.7185 & abs(dq(6)) > 24.0*2^(yl)
  tr = 1;
 else
  tr = 0;
 end
 for i=6:-1:2
  dq(i) = dq(i-1);
 end
 dq(1) = dqq; pk(2) = pk(1); pk(1) = pk1; sr(2) = sr(1); sr(1) = srr;
%
% Adaptive speed control
%
 dms = (31.0/32.0)*dms;
 dms = dms + fitable(abs(ik)+1)/32.0;
 dml = (127.0/128.0)*dml;
 dml = dml + fitable(abs(ik)+1)/128.0;
 if ap > 1.0
 al = 1.0;
 else
  al = ap;
 end
 ap = (15.0/16.0)*ap;
 if abs(dms-dml) >= dml/8.0
  ap = ap + 1/8.0;
```

```
end
if y < 3
 ap = ap +1/8.0;
end
if td == 1
 ap = ap + 1/8.0;
end
if tr == 1
 ap = 1.0;
end
y = al*yu + (1.0-al)*yl;
 end
end
iiout = out;
```

Program 11.14. W-MDCT function.

```
function [ tdac_coef ] = wmdct(ipsig)
%
% This function transforms the signal vector using the W-MDCT.
% Usage:
% ipsig: input signal block of N samples (N=even number)
% tdac_coe: W-MDCT coefficents (N/2 coefficients)
%
N = length(ipsig);
NN =N;
for i=1:NN
 h(i) = sin((pi/NN)*(i-1+0.5));
end
for k=1:N/2
 tdac_coef(k) = 0.0;
 for n=1:N
 tdac_coef(k) = tdac_coef(k) + ...
 h(n)*ipsig(n)*cos((2*pi/N)*(k-1+0.5)*(n-1+0.5+N/4));
 end
end
tdac_coef=2*tdac_coef;
```

Program 11.15. Inverse W-IMDCT function.

```
function [ opsig ] = wimdct(tdac_coef)
%
% This function transforms the W-MDCT coefficients back to the signal.
% Usage:
% tdac_coeff: N/2 W-MDCT coeffcients
% opsig: output signal black with N samples
%
N = length(tdac_coef);
tmp_coef = ((-1)^(N+1))*tdac_coef(N:-1:1);
tdac_coef = [ tdac_coef tmp_coef];
```

```
N = length(tdac_coef);
NN =N;
for i=1:NN
f(i) = sin((pi/NN)*(i-1+0.5));
end
for n=1:N
 opsig(n) = 0.0;
  for k=1:N
   opsig(n) = opsig(n) + ...
   tdac_coef(k)*cos((2*pi/N)*(k-1+0.5)*(n-1+0.5+N/4));
   end
 opsig(n) = opsig(n)*f(n)/N;
end
```

Program 11.16. Waveform coding using DCT and W-MDCT.

```
% Waveform coding using DCT and MDCT for a block size of 16 samples.
% Main program
close all; clear all
load we.dat          % Provided by the instructor
% Create a simple 3-bit scale fcator
scalef4bits=[1 2 4 8 16 32 64 128 256 512 1024 2048 4096 8192 16384 32768];
scalef3bits=[256 512 1024 2048 4096 8192 16384 32768];
scalef2bits=[4096 8192 16384 32768];
scalef1bit=[16384 32768];
scalef=scalef1bit;
nbits =3;
% Ensure the block size to be 16 samples.
x=[we zeros(1,16-mod(length(we),16))];
Nblock=length(x)/16;
DCT_code=[]; scale_code=[];
% DCT transform coding
% Encoder
 for i=1:Nblock
 xblock_DCT=dct(x((i-1)*16+1:i*16));
 diff=abs(scalef-(max(abs(xblock_DCT))));
 iscale(i)=min(find(diff<=min(diff))); % find a scale factor
 xblock_DCT=xblock_DCT/scalef(iscale(i)); % scale the input vector
  for j=1:16
  [DCT_coeff(j) pp]=biquant(nbits,-1,1,xblock_DCT(j));
  end
  DCT_code=[DCT_code DCT_coeff ];
end
% Decoder
Nblock=length(DCT_code)/16;
xx=[];
for i=1:Nblock
  DCT_coefR=DCT_code((i-1)*16+1:i*16);
 for j=1:16
```

```
   xrblock_DCT(j)=biqtdec(nbits,-1,1,DCT_coefR(j));
 end
   xrblock=idct(xrblock_DCT.*scalef(iscale(i)));
 xx=[xx xrblock];
end
% Transform coding using MDCT
xm=[zeros(1,8) we zeros(1,8-mod(length(we),8)), zeros(1,8)];
Nsubblock=length(x)/8;
MDCT_code=[];
% Encoder
 for i=1:Nsubblock
 xsubblock_DCT=wmdct(xm((i-1)*8+1:(i+1)*8));
 diff=abs(scalef-max(abs(xsubblock_DCT)));
 iscale(i)=min(find(diff<=min(diff))); % find a scale factor
 xsubblock_DCT=xsubblock_DCT/scalef(iscale(i)); % scale the input vector
 for j=1:8
  [MDCT_coeff(j) pp]=biquant(nbits,-1,1,xsubblock_DCT(j));
  end
  MDCT_code=[MDCT_code MDCT_coeff];
end
% Decoder
% Recover the first subblock
Nsubblock=length(MDCT_code)/8;
xxm=[];
MDCT_coeffR=MDCT_code(1:8);
for j=1:8
   xmrblock_DCT(j)=biqtdec(nbits,-1,1,MDCT_coeffR(j));
end
xmrblock=wimdct(xmrblock_DCT*scalef(iscale(1)));
xxr_pre=xmrblock(9:16) % recovered first block for overlap and add
for i=2:Nsubblock
 MDCT_coeffR=MDCT_code((i-1)*8+1:i*8);
 for j=1:8
   xmrblock_DCT(j)=biqtdec(nbits,-1,1,MDCT_coeffR(j));
  end
   xmrblock=wimdct(xmrblock_DCT*scalef(iscale(i)));
 xxr_cur=xxr_pre+xmrblock(1:8); % overlap and add
 xxm=[xxm xxr_cur];
 xxr_pre=xmrblock(9:16);       % set for the next overlap
end

subplot(3,1,1);plot(x,'k');grid; axis([0 length(x) -10000 10000])
ylabel('Original signal');
subplot(3,1,2);plot(xx,'k');grid;axis([0 length(xx) -10000 10000]);
ylabel('DCT coding')
subplot(3,1,3);plot(xxm,'k');grid;axis([0 length(xxm) -10000 10000]);
ylabel('W-MDCT coding');
xlabel('Sample number');
```

Program 11.17. Sign function.

```
function sgn = sgn(sgninp)
%
% Sign function
% if signp >=0 then sign=1
% else sign =-1
%
 if sgninp >= 0
  opt = 1;
 else
  opt = -1;
 end
 sgn = opt;
```

11.8 PROBLEMS

11.1. For the 3-bit midtread quantizer described in Figure 11.1, and an analog signal range from −2.5 to 2.5 volts, determine

  a. the quantization step size;

  b. the binary codes, recovered voltages, and quantization errors when each input is 1.6 volts and −0.2 volt.

11.2. For the 3-bit midtread quantizer described in Figure 11.1, and an analog signal range from −4 to 4 volts, determine

  a. the quantization step size;

  b. the binary codes, recovered voltages, and quantization errors when each input is −2.6 volts and 0.1 volt.

11.3. For the 3-bit midtread quantizer described in Figure 11.1, and an analog signal range from −5 to 5 volts, determine

  a. the quantization step size;

  b. the binary codes, recovered voltages, and quantization errors when each input is −2.6 volts and 3.5 volts.

11.4. For the 3-bit midtread quantizer described in Figure 11.1, and an analog signal range from −10 to 10 volts, determine

  a. the quantization step size;

  b. the binary codes, recovered voltages, and quantization errors when each input is −5 volts, 0 volts, and 7.2 volts.

11.5. For the μ-law compression and expanding process shown in Figure 11.3 with $\mu = 255$, a 3-bit midtread quantizer described in Figure 11.1, and an analog signal range from −2.5 to 2.5 volts, determine the binary codes, recovered voltages, and quantization errors when each input is 1.6 volts and −0.2 volt.

11.6. For the μ-law compression and expanding process shown in Figure 11.3 with $\mu = 255$, a 3-bit midtread quantizer described in Figure 11.1, and an analog signal range from -4 to 4 volts, determine the binary codes, recovered voltages, and quantization errors when each input is -2.6 volts and 0.1 volt.

11.7. For the μ-law compression and expanding process shown in Figure 11.3 with $\mu = 255$, a 3-bit midtread quantizer described in Figure 11.1, and an analog signal range from -5 to 5 volts, determine the binary codes, recovered voltages, and quantization errors when each input is -2.6 volts and 3.5 volts.

11.8. For the μ-law compression and expanding process shown in Figure 11.3 with $\mu = 255$, a 3-bit midtread quantizer described in Figure 11.1, and an analog signal range from -10 to 10 volts, determine the binary codes, recovered voltages, and quantization errors when each input is -5, 0, and 7.2 volts.

11.9. In a digital companding system, encode each of the following 12-bit linear PCM codes into 8-bit compressed PCM code:

a. 0 0 0 0 0 0 0 1 0 1 0 1

b. 1 0 1 0 1 1 1 0 1 0 1 0

11.10. In a digital companding system, decode each of the following 8-bit compressed PCM codes into 12-bit linear PCM code:

a. 0 0 0 0 0 1 1 1

b. 1 1 1 0 1 0 0 1

11.11. In a digital companding system, encode each of the following 12-linear PCM codes into the 8-bit compressed PCM code:

a. 0 0 1 0 1 0 1 0 1 0 1 0

b. 1 0 0 0 0 0 0 0 1 1 0 1

11.12. In a digital companding system, decode each of the following 8-bit compressed PCM codes into the 12-bit linear PCM code:

a. 0 0 1 0 1 1 0 1

b. 1 0 0 0 0 1 0 1

11.13. Consider a 3-bit DPCM encoding system with the following specifications (Figure 11.19):

$$\text{Encoder scheme}: \tilde{x}(n) = \hat{x}(n-1) \text{ (predictor)}$$
$$d(n) = x(n) - \tilde{x}(n)$$
$$d_q(n) = Q[d(n)] = \text{quantizer in Table 11.9}$$
$$\hat{x}(n) = \tilde{x}(n) + d_q(n)$$
$$\text{5-bit input data}: x(0) = -6, x(1) = -8, \text{ and } x(2) = -13$$

Perform DPCM encoding to produce the binary code for each input data.

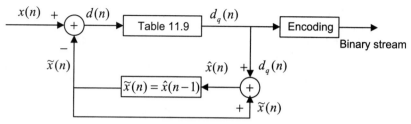

FIGURE 11.19

DPCM encoding in Problem 11.13.

Table 11.9 Quantization Table for the 3-bit Quantizer in Problem 11.13		
Binary Code	**Quantization Value $d_q(n)$**	**Subrange in $d(n)$**
0 1 1	−11	$-15 \leq d(n) < -7$
0 1 0	−5	$-7 \leq d(n) < -3$
0 0 1	−2	$-3 \leq d(n) < -1$
0 0 0	0	$-1 \leq d(n) < 0$
1 0 0	0	$0 \leq d(n) \leq 1$
1 0 1	2	$1 < d(n) \leq 3$
1 1 0	5	$3 < d(n) \leq 7$
1 1 1	11	$7 < d(n) \leq 15$

11.14. Consider a 3-bit DPCM decoding system with the following specifications (Figure 11.20):

$$\text{Decoding scheme}: \tilde{x}(n) = \hat{x}(n-1) \text{ (predictor)}$$
$$d_q(n) = \text{quantizer in Table 11.9}$$
$$\hat{x}(n) = \tilde{x}(n) + d_q(n)$$

Three received binary codes: 110, 100, 101

Perform DCPM decoding to recover each digital value using its binary code.

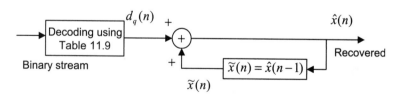

FIGURE 11.20

DPCM decoding in Problem 11.14.

11.15. For a 3-bit DPCM encoding system shown in Problem 11.13 and the given 5-bit input data $x(0) = 6, x(1) = 8$, and $x(2) = 13$, perform DPCM encoding to produce the binary code for each of input data.

11.16. For the 3-bit DPCM decoding system shown in Problem 11.14 and the received data 010, 000, 001, perform DCPM decoding to recover each digital value using its binary code.

11.17. Assuming that a speech waveform is sampled at 8 kHz and each sample is encoded by 16 bits, determine the compression ratio for each of the following encoding methods:

a. no compression

b. standard μ-law compression (8 bits per sample)

c. standard ADPCM encoding (4 bits per sample)

11.18. Assuming that a speech waveform is sampled at 8 kHz and each sample is encoded by 16 bits, determine the bit rate for each of the following encoding methods:

a. no compression

b. standard μ-law companding (8 bits per sample)

c. standard ADPCM encoding (4 bits per sample)

11.19. Assuming that an audio waveform is sampled at 44.1 kHz and each sample is encoded by 16 bits, determine the compression ratio for each of the following encoding methods:

a. no compression

b. standard μ-law compression (8 bits per sample)

c. standard ADPCM encoding (4 bits per sample)

11.20. Assuming that an audio waveform is sampled at 44.1 kHz and each sample is encoded by 12 bits, determine the bit rate for each of the encoding methods.

a. no compression

b. standard μ-law companding (8 bits per sample)

c. standard ADPCM encoding (4 bits per sample)

11.21. Speech is sampled at 8 kHz and each sample is encoded by 16 bits. A telephone system can transport the digital voice channel over a digital link with a capacity of 1.536 MBPS. Determine the number of channels that the phone company can carry for each of the following encoding methods:

a. no compression

b. standard 8-bit μ-law companding (8 bits per sample)

c. standard ADPCM encoding (4 bits per sample)

11.22. Given the input data

$$x(0) = 25, \ x(1) = 30, \ x(2) = 28, \quad \text{and} \quad x(3) = 25$$

determine the DCT coefficients.

11.23. Given the input data

$$x(0) = 25 \quad \text{and} \quad x(1) = 30$$

determine the DCT coefficients.

11.24. Given the input data

$$x(0) = 25, \ x(1) = 30, \ x(2) = 28, \ x(3) = 25,$$
$$x(4) = 10, \ x(5) = 0, \ x(6) = 0, \ \text{and} \ x(7) = 0$$

determine the DCT coefficients $X_{DCT}(0)$, $X_{DCT}(2)$, $X_{DCT}(4)$, and $X_{DCT}(6)$.

11.25. Given the input data

$$x(0) = 25, \ x(1) = 30, \ x(2) = 28, \ x(3) = 25,$$
$$x(4) = 10, \ x(5) = 0, \ x(6) = 0, \ \text{and} \ x(7) = 0$$

determine the DCT coefficients $X_{DCT}(1)$, $X_{DCT}(3)$, $X_{DCT}(5)$, and $X_{DCT}(7)$.

11.26. Assume the following DCT coefficients with infinite precision:

$$X_{DCT}(0) = 14, \ X_{DCT}(1) = 6, \ X_{DCT}(2) = -6, \ \text{and} \ X_{DCT}(3) = 8$$

a. Determine the input data using the MATLAB function **idct()**.

b. Recover the input data samples using the MATLAB function **idct()** if a bit allocation scheme quantizes the DCT coefficients as follows: 2 magnitude bits plus 1 sign bit (3 bits) for the DC coefficient, 1 magnitude bit plus 1 sign bit (2 bits) for each AC coefficient and a scale factor of 8, that is,

$$X_{DCT}(0) = 8 \times 2 = 16, \ X_{DCT}(1) = 8 \times 1 = 8, \ X_{DCT}(2) = 8 \times (-1)$$
$$= -8, \ \text{and} \ X_{DCT}(3) = 8 \times 1 = 8$$

c. Compute the quantized error in part (b) of this problem.

11.27. Assume the following DCT coefficients with infinite precision:

$$X_{DCT}(0) = 11, \ X_{DCT}(1) = 5, \ X_{DCT}(2) = 7, \ \text{and} \ X_{DCT}(3) = -3$$

a. Determine the input data using the MATLAB function **idct()**.

b. Recover the input data samples using the MATLAB function **idct()** if a bit allocation scheme quantizes the DCT coefficients as follows: 2 magnitude bits plus 1 sign bit (3 bits) for the DC coefficient, 1 magnitude bit plus 1 sign bit (2 bits) for each AC coefficient and a scale factor of 8, that is,

$$X_{DCT}(0) = 8 \times 1 = 8, \ X_{DCT}(1) = 8 \times 1 = 8, \ X_{DCT}(2) = 8 \times 1 = -8, \quad \text{and}$$

$$X_{DCT}(3) = 8 \times 0 = 0$$

c. Compute the quantized error in part (b) of this problem.

11.28. a. Verify that the window function

$$f(n) = h(n) = \sin\left(\frac{\pi}{N}(n + 0.5)\right)$$

used in MDCT is satisfied with Equations (11.35) and (11.36).

b. Verify the relation for W-MDCT coefficients

$$X_{MDCT}(k) = (-1)^{\frac{N}{2}+1}X_{MDCT}(N - 1 - k) \quad \text{for} \quad k = N/2, N/2 + 1, \cdots, N - 1$$

11.29. Given data 1, 2, 3, 4, 5, 4, 3, 2, ...,

a. determine the W-MDCT coefficients for the first three blocks using a block size of 4;

b. determine the first two overlapped subblocks and compare the results with the original data sequence using the W-MDCT coefficients in part (a).

11.30. Given data 1, 2, 3, 4, 5, 4, 3, 2, 1, 2, 3, 4, 5,...,

a. determine the W-MDCT coefficients for the first three blocks using a block size of 6;

b. determine the first two overlapped subblocks and compare the results with the original data sequence using the W-MDCT coefficients in part (a).

11.8.1 Computer Problems with MATLAB

Use the MATLAB programs in Section 11.7 for Problems 11.31 to 11.33.

11.31. Consider the data file "speech.dat" with 16 bits per sample and a sampling rate of 8 kHz.

a. Use PCM coding (midtread quantizer) to perform compression and decompression and apply the MATLAB function **sound()** to evaluate the sound quality in terms of "excellent", "good", "intelligent", and "unacceptable" for the following bit rates:

1. 4 bits/sample (32 kbits per second)

2. 6 bits/sample (48 kbits per second)

3. 8 bits/sample (64 kbits per second)

b. Use μ-law PCM coding to perform compression and decompression and apply the MATLAB function **sound()** to evaluate the sound quality for the following bit rates:

1. 4 bits/sample (32 kbits per second)

2. 6 bits/sample (48 kbits per second)

3. 8 bits/sample (64 kbits per second)

11.32. Given the data file "speech.dat" with 16 bits per sample, a sampling rate of 8 kHz, and ADCPM coding, perform compression and decompression and apply the MATLAB function **sound()** to evaluate the sound quality.

11.33. Given the data file "speech.dat" with 16 bits per sample, a sampling rate of 8 kHz, and DCT and M-DCT coding as described in the programs in Section 11.7, perform compression and decompression using the following specified parameters in Program 11.16 to compare the sound quality:

a. nbits=3, scalef=scalef2bits

b. nbits=3, scalef=scalef3bits

c. nbits=4, scalef=scalef2bits

d. nbits=4, scalef=scalef3bits

Multirate Digital Signal Processing, Oversampling of Analog-to-Digital Conversion, and Undersampling of Bandpass Signals

12

CHAPTER OUTLINE

12.1 **Multirate Digital Signal Processing Basics**.. 555
    12.1.1 Sampling Rate Reduction by an Integer Factor...556
    12.1.2 Sampling Rate Increase by an Integer Factor 562
    12.1.3 Changing the Sampling Rate by a Noninteger Factor *L/M*...567
    12.1.4 Application: CD Audio Player..571
    12.1.5 Multistage Decimation 574
12.2 **Polyphase Filter Structure and Implementation**.. 578
12.3 **Oversampling of Analog-to-Digital Conversion** 585
    12.3.1 Oversampling and Analog-to-Digital Conversion Resolution..586
    12.3.2 Sigma-Delta Modulation Analog-to-Digital Conversion 592
12.4 **Application Example: CD Player**.. 601
12.5 **Undersampling of Bandpass Signals** 603
    Simulation Example 608
12.6 **Sampling Rate Conversion Using the TMS320C6713 DSK** 608
12.7 **Summary** 613

OBJECTIVES:

This chapter investigates basics of multirate digital signal processing, illustrates how to change a sampling rate for speech and audio signals, and describes the polyphase implementation for the decimation filter and interpolation filter. Next, the chapter introduces the advanced analog-to-digital conversion system with the oversampling technique and sigma-delta modulation. Finally, the chapter explores the principles of undersampling of bandpass signals.

12.1 MULTIRATE DIGITAL SIGNAL PROCESSING BASICS

In many areas of digital signal processing (DSP) applications—such as communications, speech, and audio processing—rising or lowering of a sampling rate is required. The principles relating to changing the sampling rate belong essentially within the topic of *multirate signal processing* (Ifeachor and Jervis, 2002; Proakis and Manolakis,1996; Porat, 1997; Sorensen and Chen,1997). As an introduction, we will focus on the sampling rate conversion; that is, sampling rate reduction or increase.

12.1.1 Sampling Rate Reduction by an Integer Factor

The process of reducing the sampling rate by an integer factor is referred to as *downsampling* of a data sequence. We also refer downsampling as "decimation" (not taking one of ten). The term "decimation" has been accepted and used in many textbooks and fields. To downsample a data sequence $x(n)$ by an integer factor of M, we use the following notation:

$$y(m) = x(mM) \tag{12.1}$$

where $y(m)$ is the downsampled sequence, obtained by taking a sample from the data sequence $x(n)$ for every M samples (discarding M-1 samples for every M samples). As an example, if the original sequence with a sampling period $T = 0.1$ second (sampling rate $= 10$ samples per second) is given by

$$x(n): 8 \quad 7 \quad 4 \quad 8 \quad 9 \quad 6 \quad 4 \quad 2 \quad -2 \quad -5 \quad -7 \quad -7 \quad -6 \quad -4 \dots$$

and we downsample the data sequence by a factor of 3, we obtain the downsampled sequence as

$$y(m): 8 \quad 8 \quad 4 \quad -5 \quad -6 \dots$$

with the resultant sampling period $T = 3 \times 0.1 = 0.3$ second (the sampling rate now is 3.33 samples per second). Although the example is straightforward, there is a requirement to avoid aliasing noise. We will illustrate this next.

From the Nyquist sampling theorem, it is known that aliasing can occur in the downsampled signal due to the reduced sampling rate. After downsampling by a factor of M, the new sampling period becomes MT, and therefore the new sampling frequency is

$$f_{sM} = \frac{1}{MT} = \frac{f_s}{M} \tag{12.2}$$

where f_s is the original sampling rate.

Hence, the folding frequency after downsampling becomes

$$f_{sM}/2 = \frac{f_s}{2M} \tag{12.3}$$

This tells us that after downsampling by a factor of M, the new folding frequency will be decreased M times. If the signal to be downsampled has frequency components larger than the new folding frequency, $f > f_s/(2M)$, aliasing noise will be introduced into the downsampled data.

To overcome this problem, it is required that the original signal $x(n)$ be processed by a lowpass filter $H(z)$ before downsampling, which should have a stop frequency edge at $f_s/(2M)$ (Hz). The corresponding normalized stop frequency edge is then converted to

$$\Omega_{stop} = 2\pi \frac{f_s}{2M} T = \frac{\pi}{M} \text{ radians} \tag{12.4}$$

In this way, before downsampling, we can guarantee the maximum frequency of the filtered signal satisfies

$$f_{max} < \frac{f_s}{2M} \qquad (12.5)$$

such that no aliasing noise is introduced after downsampling. A general block diagram of decimation is given in Figure 12.1, where the filtered output in terms of the z-transform can be written as

$$W(z) = H(z)X(z) \qquad (12.6)$$

where $X(z)$ is the z-transform of the sequence to be decimated, $x(n)$, and $H(z)$ is the lowpass filter transfer function. After anti-aliasing filtering, the downsampled signal $y(m)$ takes its value from the filter output as

$$y(m) = w(mM) \qquad (12.7)$$

The process of reducing the sampling rate by a factor of 3 is shown in Figure 12.1. The corresponding spectral plots for $x(n)$, $w(n)$, and $y(m)$ in general are shown in Figure 12.2.

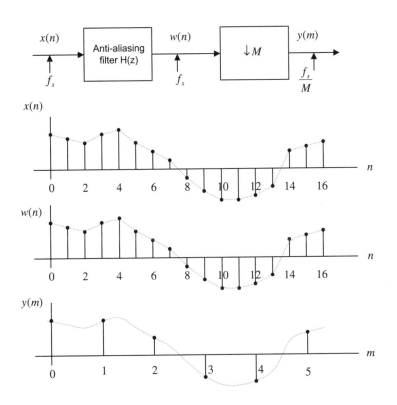

FIGURE 12.1

Block diagram of the downsampling process with $M = 3$.

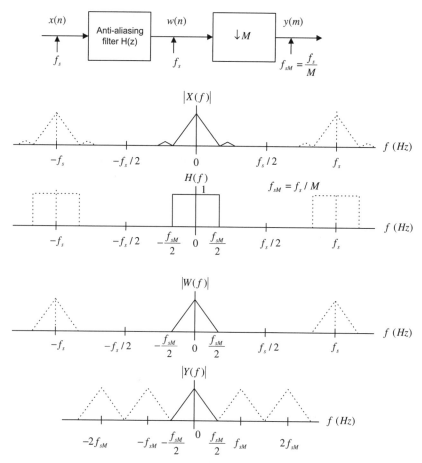

FIGURE 12.2

Spectrum after downsampling.

To verify this principle, let us consider a signal $x(n)$ generated by the following:

$$x(n) = 5 \sin\left(\frac{2\pi \times 1{,}000n}{8{,}000}\right) + \cos\left(\frac{2\pi \times 2{,}500}{8{,}000}\right) = 5 \sin\left(\frac{n\pi}{4}\right) + \cos\left(\frac{5n\pi}{8}\right) \qquad (12.8)$$

With a sampling rate of $f_s = 8{,}000$ Hz, the spectrum of $x(n)$ is plotted in the first graph in Figure 12.3A, where we observe that the signal has components at frequencies of 1,000 Hz and 2,500 Hz. Now we downsample $x(n)$ by a factor of 2, that is, $M = 2$. According to Equation (12.3), we know that the new folding frequency is $4{,}000/2 = 2{,}000$ Hz. Hence, without using the anti-aliasing lowpass filter, the spectrum would contain an aliasing frequency of 4 kHz – 2.5 kHz = 1.5 kHz introduced by 2.5 kHz, plotted in the second graph in Figure 12.3A.

Now we apply a finite impulse response (FIR) lowpass filter designed with a filter length of $N = 27$ and a cutoff frequency of 1.5 kHz to remove the 2.5 kHz signal before downsampling to avoid

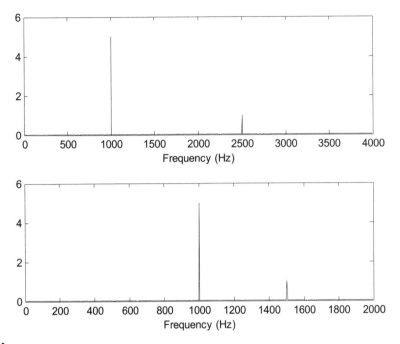

FIGURE 12.3A

Spectrum before downsampling and spectrum after downsampling without using the anti-aliasing filter.

aliasing. How to obtain such specifications will be discussed in the later example. The normalized cutoff frequency used for design is given by

$$\Omega_c = 2\pi \times 1,500 \times (1/8,000) = 0.375\pi$$

Thus, the aliasing noise is avoided. The spectral plots are given in Figure 12.3B, where the first plot shows the spectrum of $w(n)$ after anti-aliasing filtering, while the second plot describes the spectrum of $y(m)$ after downsampling. Clearly, we prevent aliasing noise in the downsampled data by sacrificing the original 2.5-kHz signal. Program 12.1 gives the details of the MATLAB implementation.

Program 12.1. MATLAB program for decimation.

```
close all; clear all;
% Downsampling filter (see Chapter 7 for FIR filter design)
B =[0.00074961181416 0.00247663033476 0.00146938649416 -0.00440446121505 ...
 -0.00910635730662 0.00000000000000 0.02035676831506 0.02233710562885...
 -0.01712963672810 -0.06376620649567 -0.03590670035210 0.10660980550088...
 0.29014909103794 0.37500000000000 0.29014909103794 0.10660980550088...
 -0.03590670035210 -0.06376620649567 -0.01712963672810 0.02233710562885...
 0.02035676831506 0.00000000000000 -0.00910635730662 -0.00440446121505...
 0.00146938649416 0.00247663033476 0.00074961181416];
```

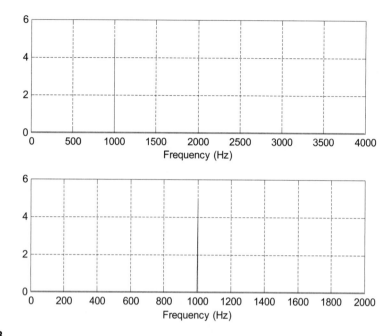

FIGURE 12.3B

Spectrum before downsampling and spectrum after downsampling using the anti-aliasing filter.

```
% Generate 2048 samples
fs=8000;                % Sampling rate
N=2048;                 % Number of samples
M=2;                    % Downsample factor
n=0:1:N-1;
x=5*sin(n*pi/4)+cos(5*n*pi/8);
% Compute single-sided amplitude spectrum
% AC component will be doubled, and DC component will be kept at the same value
X=2*abs(fft(x,N))/N;X(1)=X(1)/2;
% Map the frequency index up to the folding frequency in Hz
f=[0:1:N/2-1]*fs/N;
% Downsampling
y=x(1:M:N);
NM=length(y);           % Length of the downsampled data
% Compute the single-sided amplitude spectrum for the downsampled signal
Y=2*abs(fft(y,NM))/length(y);Y(1)=Y(1)/2;
% Map the frequency index to the frequency in Hz
fsM=[0:1:NM/2-1]*(fs/M)/NM;
subplot(2,1,1);plot(f,X(1:1:N/2));grid; xlabel('Frequency (Hz)');
subplot(2,1,2);plot(fsM,Y(1:1:NM/2));grid; xlabel('Frequency (Hz)');
figure
```

```
w=filter(B,1,x);        % Anti-aliasing filtering
% Compute the single-sided amplitude spectrum for the filtered signal
W=2*abs(fft(w,N))/N;W(1)=W(1)/2;
% Downsampling
y=w(1:M:N);
NM=length(y);
% Compute the single-sided amplitude spectrum for the downsampled signal
Y=2*abs(fft(y,NM))/NM;Y(1)=Y(1)/2;
% Plot spectra
subplot(2,1,1);plot(f,W(1:1:N/2));grid; xlabel('Frequency (Hz)');
subplot(2,1,2);plot(fsM,Y(1:1:NM/2));grid; xlabel('Frequency (Hz)');
```

Now we focus on how to design an anti-aliasing FIR filter, or decimation filter. We will discuss this topic via the following example.

EXAMPLE 12.1

Consider a DSP downsampling system with the following specifications:

Sampling rate = 6,000 Hz
Input audio frequency range = 0–800 Hz
Passband ripple = 0.02 dB
Stopband attenuation = 50 dB
Downsample factor $M = 3$
Determine the FIR filter length, cutoff frequency, and window type if the window method is used.

Solution:
The specifications are reorganized as

Anti-aliasing filter operating at the sampling rate = 6,000 Hz
Passband frequency range = 0–800 Hz
Stopband frequency range = 1–3 kHz
Passband ripple = 0.02 dB
Stopband attenuation = 50 dB
Filter type = FIR

The block diagram and specifications are depicted in Figure 12.4.

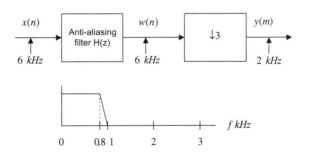

FIGURE 12.4

Filter specifications for Example 12.1.

The Hamming window is selected, since it provides 0.019 dB ripple and 53 dB stopband attenuation. The normalized transition band is given by

$$\Delta f = \frac{f_{stop} - f_{pass}}{f_s} = \frac{1,000 - 800}{6,000} = 0.033$$

The length of the filter and the cutoff frequency can be determined by

$$N = \frac{3.3}{\Delta f} = \frac{3.3}{0.033} = 100$$

We choose an odd number; that is, $N = 101$, and

$$f_c = \frac{f_{pass} + f_{stop}}{2} = \frac{800 + 1,000}{2} = 900 \text{ Hz}$$

12.1.2 Sampling Rate Increase by an Integer Factor

Increasing a sampling rate is a process of upsampling by an integer factor of L. This process is described as follows:

$$y(m) = \begin{cases} x\left(\dfrac{m}{L}\right) & m = nL \\ 0 & otherwise \end{cases} \tag{12.9}$$

where $n = 0, 1, 2, \cdots$, $x(n)$ is the sequence to be upsampled by a factor of L, and $y(m)$ is the upsampled sequence. As an example, suppose that the data sequence is given as follows:

$$x(n): \quad 8 \quad 8 \quad 4 \quad -5 \quad -6 \dots$$

After upsampling the data sequence $x(n)$ by a factor of 3 (adding $L - 1$ zeros for each sample), we have the upsampled data sequence $w(m)$ as

$$w(m): \quad 8\,0\,0 \quad 8\,0\,0 \quad 4\,0\,0 \quad -5\,0\,0 \quad -6\,0\,0 \dots$$

The next step is to smooth the upsampled data sequence via an interpolation filter. The process is illustrated in Figure 12.5A.

Similar to the downsampling case, assuming that the data sequence has the current sampling period of T, the Nyquist frequency is given by $f_{max} = f_s/2$. After usampling by a factor of L, the new sampling period becomes T/L, thus the new sampling frequency is changed to be

$$f_{sL} = Lf_s \tag{12.10}$$

This indicates that after upsampling, the spectral replicas originally centered at $\pm f_s$, $\pm 2f_s$, $\dots$ are included in the frequency range from 0 Hz to the new Nyquist limit $Lf_s/2$ Hz, as shown in Figure 12.5B. To remove those included spectral replicas, an interpolation filter with a stop frequency edge of $f_s/2$ in Hz must be attached, and the normalized stop frequency edge is given by

$$\Omega_{stop} = 2\pi\left(\frac{f_s}{2}\right) \times \left(\frac{T}{L}\right) = \frac{\pi}{L} \text{ radians} \tag{12.11}$$

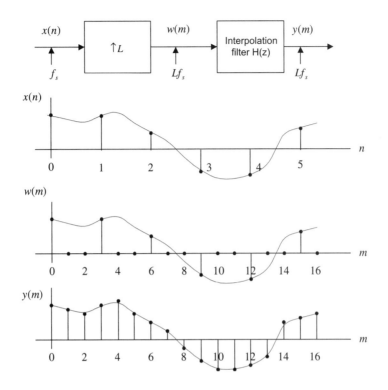

FIGURE 12.5A

Block diagram for the upsampling process with L = 3.

After filtering via the interpolation filter, we will achieve the desired spectrum for $y(n)$, as shown in Figure 12.5B. Note that since the interpolation is to remove the high-frequency images that are aliased by the upsampling operation, it is essentially an anti-aliasing lowpass filter.

To verify the upsampling principle, we generate the signal $x(n)$ with 1 kHz and 2.5 kHz as follows:

$$x(n) = 5 \sin \left(\frac{2\pi \times 1{,}000n}{8{,}000} \right) + \cos \left(\frac{2\pi \times 2{,}500n}{8{,}000} \right)$$

with a sampling rate of $f_s = 8{,}000$ Hz. The spectrum of $x(n)$ is plotted in Figure 12.6. Now we upsample $x(n)$ by a factor of 3, that is, $L = 3$. We know that the sampling rate is increased to be $3 \times 8{,}000 = 24{,}000$ Hz. Hence, without using the interpolation filter, the spectrum would contain the image frequencies originally centered at the multiple frequencies of 8 kHz. The top plot in Figure 12.6 shows the spectrum for the sequence after upsampling and before applying the interpolation filter.

Now we apply an FIR lowpass filter designed with a length of 53, a cutoff frequency of 3,250 Hz, and a new sampling rate of 24,000 Hz as the interpolation filter, whose normalized frequency should be

$$\Omega_c = 2\pi \times 3{,}250 \times \left(\frac{1}{24{,}000} \right) = 0.2708\pi$$

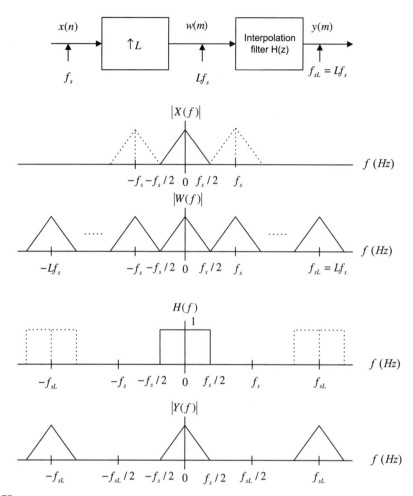

FIGURE 12.5B

Spectra before and after upsampling

The bottom plot in Figure 12.6 shows the spectrum for $y(m)$ after applying the interpolation filter, where only the original signals with frequencies of 1 kHz and 2.5 kHz are presented. Program 12.2 shows the implementation details in MATLAB.

Program 12.2. MATLAB program for interpolation.

```
close all; clear all
% Upsampling filter (see Chapter 7 for FIR filter design)
B =[-0.00012783931504 0.00069976044649 0.00123831516738 0.00100277549136...
 -0.00025059018468 -0.00203448515158 -0.00300830295487 -0.00174101657599...
 0.00188598835011 0.00578414933758 0.00649330625041 0.00177982369523...
```

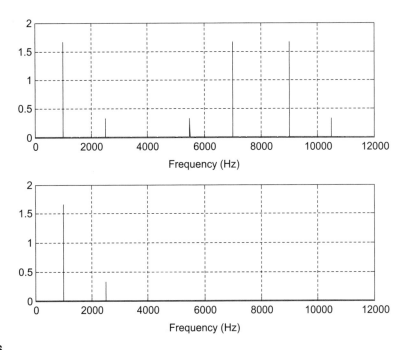

FIGURE 12.6

(Top) The spectrum after upsampling and before applying the interpolation filter; (bottom) spectrum after applying the interpolation filter.

```
 -0.00670672686935 -0.01319379342716 -0.01116855281442 0.00123034314117...
 0.01775600060894 0.02614700427364 0.01594155162392 -0.01235169936557...
 -0.04334322148505 -0.05244745563466 -0.01951094855292 0.05718573279009...
 0.15568416401644 0.23851539047347 0.27083333333333 0.23851539047347...
 0.15568416401644 0.05718573279009 -0.01951094855292 -0.05244745563466...
 -0.04334322148505 -0.01235169936557 0.01594155162392 0.02614700427364...
 0.01775600060894 0.00123034314117 -0.01116855281442 -0.01319379342716...
 -0.00670672686935 0.00177982369523 0.00649330625041 0.00578414933758...
 0.00188598835011 -0.00174101657599 -0.00300830295487 -0.00203448515158...
 -0.00025059018468 0.00100277549136 0.00123831516738 0.00069976044649...
 -0.00012783931504];
% Generate the 2048 samples with fs = 8000 Hz
fs=8000;            % Sampling rate
N=2048;             % Number of samples
L = 3;              % Upsampling factor
n=0:1:N-1;
x=5*sin(n*pi/4)+cos(5*n*pi/8);
% Upsampling by a factor of L
w=zeros(1,L*N);
```

```
for n=0:1:N-1
 w(L*n+1)=x(n+1);
end
NL = length(w);      % Length of the upsampled data
W=2*abs(fft(w,NL))/NL;W(1)=W(1)/2; % Compute one-sided amplitude spectrum
f=[0:1:NL/2-1]*fs*L/NL; % Map the frequency index to the frequency (Hz)
% Interpolation
y=filter(B,1,w);      % Apply interpolation filter
Y=2*abs(fft(y,NL))/NL;Y(1)=Y(1)/2; % Compute the one-sided amplitude spectrum
fsL=[0:1:NL/2-1]*fs*L/NL; % Map the frequency index to the frequency (Hz)
subplot(2,1,1);plot(f,W(1:1:NL/2));grid; xlabel('Frequency (Hz)');
subplot(2,1,2);plot(fsL,Y(1:1:NL/2));grid; xlabel('Frequency (Hz)');
```

Now let us study how to design an interpolation filter via Example 12.2.

EXAMPLE 12.2

Consider a DSP upsampling system with the following specifications:

Sampling rate = 6,000 Hz
Input audio frequency range = 0–800 Hz
Passband ripple = 0.02 dB
Stopband attenuation = 50 dB
Upsample factor $L = 3$
Determine the FIR filter length, cutoff frequency, and window type if the window design method is used.

Solution:
The specifications are reorganized as follows:

Interpolation filter operating at the sampling rate = 18,000 Hz
Passband frequency range = 0–800 Hz
Stopband frequency range = 3–9 kHz
Passband ripple = 0.02 dB
Stopband attenuation = 50 dB
Filter type: FIR filter

The block diagram and filter frequency specifications are given in Figure 12.7.

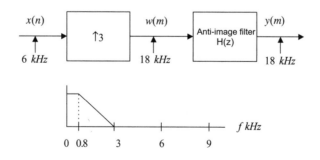

FIGURE 12.7

Filter frequency specifications for Example 12.2.

We choose the Hamming window for this application. The normalized transition band is

$$\Delta f = \frac{f_{stop} - f_{pass}}{f_{sL}} = \frac{3{,}000 - 800}{18{,}000} = 0.1222$$

The length of the filter and the cutoff frequency can be determined by

$$N = \frac{3.3}{\Delta f} = \frac{3.3}{0.1222} = 27$$

and the cutoff frequency is given by

$$f_c = \frac{f_{pass} + f_{stop}}{2} = \frac{3{,}000 + 800}{2} = 1{,}900 \text{ Hz}.$$

12.1.3 Changing the Sampling Rate by a Noninteger Factor *L/M*

With an understanding of the downsampling and upsampling processes, we now study sampling rate conversion by a noninteger L/M. This can be viewed as two sampling conversion processes. In step 1, we perform the upsampling process by a factor of integer L following application of an interpolation filter $H_1(z)$; in step 2, we continue filtering the output from the interpolation filter via an anti-aliasing filter $H_2(z)$, and finally execute downsampling. The entire process is illustrated in Figure 12.8.

Since the interpolation and anti-aliasing filters are in a cascaded form and operate at the same rate, we can select one of them. We choose the one with the lower stop frequency edge and choose the most demanding requirements for passband gain and stopband attenuation for the filter design. A lot of computational savings can be achieved by using one lowpass filter. We illustrate the procedure via the following simulation. Let us generate the signal $x(n)$ by

$$x(n) = 5\sin\left(\frac{2\pi \times 1{,}000n}{8{,}000}\right) + \cos\left(\frac{2\pi \times 2{,}500n}{8{,}000}\right)$$

with a sampling rate of $f_s = 8{,}000$ Hz and frequencies of 1 kHz and 2.5 kHz. Now we resample $x(n)$ to 3,000 Hz by a noninteger factor of 0.375, that is,

$$\left(\frac{L}{M}\right) = 0.375 = \frac{3}{8}$$

Upsampling is at a factor of $L = 3$ and the upsampled sequence is filtered by an FIR lowpass filter designed with a filter length $N = 53$ and a cutoff frequency of 3,250 Hz at a sampling rate of $3 \times 8{,}000 = 24{,}000$ Hz. The spectrum for the upsampled sequence and the spectrum after application of the interpolation filter are plotted in Figure 12.9A.

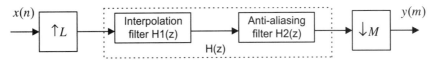

FIGURE 12.8

Block diagram for sampling rate conversion.

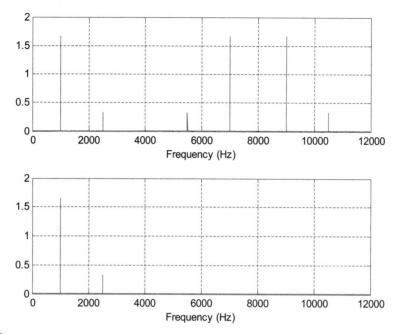

FIGURE 12.9A

(Top) Spectrum after upsampling, and (bottom) spectrum after interpolation filtering.

The sequence from step 1 can be filtered via another FIR lowpass filter with filter length $N = 159$ and a cutoff frequency of 1,250 Hz, followed by downsampling by a factor of $M = 8$. The spectrum after the anti-aliasing filter and the spectrum for the final output $y(m)$ are plotted in Figure 12.9B. Note that the anti-aliasing filter removes the frequency component of 2.5 kHz to avoid aliasing. This is because after downsampling, the Nyquist limit is 1.5 kHz. As we discussed previously, we can select one filter for implementation. We choose a FIR lowpass filter with $N = 159$ and a cutoff frequency of 1,250 Hz because its bandwidth is smaller than that of the interpolation filter. The MATLAB implementation is listed in Program 12.3.

Program 12.3 MATLAB program for changing sampling rate with a noninteger factor.

```
close all; clear all;clc;
% Downsampling filter
Bdown=firwd(159,1,2*pi*1250/24000,0,4);
% Generate 2048 samples with fs=8000 Hz
fs=8000;                          % Original sampling rate
N = 2048;                         % The number of samples
L=3;                              % Upsampling factor
M=8;                              % Downsampling factor
n=0:1:N-1;                        % Generate the time index
x=5*sin(n*pi/4)+cos(5*n*pi/8);    % Generate the test signal
% Upsampling by a factor of L
w1=zeros(1,L*N);
```

```
for n=0:1:N-1
 w1(L*n+1)=x(n+1);
end
NL= length(w1);               % Length of upsampled data
W1=2*abs(fft(w1,NL))/NL;W1(1)=W1(1)/2; % Compute the one-sided
                        % amplitude spectrum
f=[0:1:NL/2-1]*fs*L/NL;       % Map frequency index to its frequency in Hz
subplot(3,1,1);plot(f,W1(1:1:NL/2));grid
xlabel('Frequency (Hz)');
w2=filter(Bdown,1,w1);        % Perform the combined anti-aliasing filter
W2=2*abs(fft(w2,NL))/NL;W2(1)=W2(1)/2; % Compute the one-sided
                        % amplitude spectrum
y2=w2(1:M:NL);
NM=length(y2);                % Length of the downsampled data
Y2=2*abs(fft(y2,NM))/NM;Y2(1)=Y2(1)/2;% Compute the one-sided
                        %amplitude spectrum
% Map frequency index to its frequency in Hz before downsampling
fbar=[0:1:NL/2-1]*24000/NL;
% Map frequency index to its frequency in Hz
fsM=[0:1:NM/2-1]*(fs*L/M)/NM;
subplot(3,1,2);plot(f,W2(1:1:NL/2));grid; xlabel('Frequency (Hz)');
subplot(3,1,3);plot(fsM,Y2(1:1:NM/2));grid; xlabel('Frequency (Hz)');
```

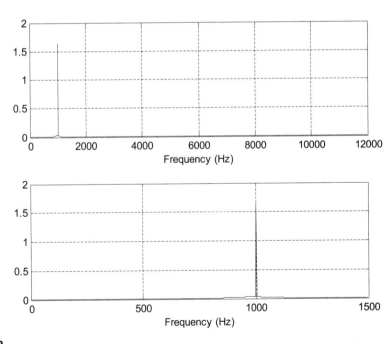

FIGURE 12.9B

(Top) Spectrum after anti-aliasing filtering, and (bottom) spectrum after downsampling.

Therefore, three steps are required to accomplish the process:

1. Upsampling by a factor of $L = 3$;
2. Filtering the upsampled sequence by an FIR lowpass filter designed with filter length $N = 159$ and a cutoff frequency of 1,250 Hz at a sampling rate of $3 \times 8,000 = 24,000$ Hz;
3. Downsampling by a factor of $M = 8$.

EXAMPLE 12.3

Consider a sampling conversion DSP system (Figure 12.10A) with the following specifications:

Audio input $x(n)$ is sampled at the rate of 6,000 Hz.
Audio output $y(m)$ is operated at the rate of 9,000 Hz.
Determine the filter length and cutoff frequency for the combined anti-aliasing filter $H(z)$, and the window types, respectively, if the window design method is used.

Solution:
The filter frequency specifications and corresponding block diagram are developed in Figure 12.10B.
Specifications for the interpolation filter $H_1(z)$:
Passband frequency range = 0–2,500 Hz
Passband ripples for $H_1(z)$ = 0.04 dB
Stopband frequency range = 3,000–9,000 Hz
Stopband attenuation = 42 dB
Specifications for the anti-aliasing filter $H_2(z)$:
Passband frequency range = 0–2,500 Hz

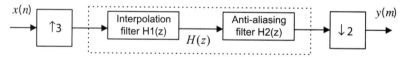

FIGURE 12.10A

Sampling conversion in Example 12.3.

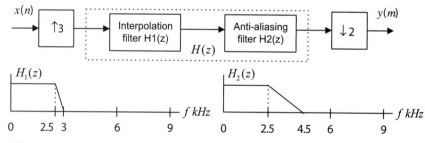

FIGURE 12.10B

Filter frequency specifications for Example 12.3.

Passband ripples for $H_2(z) = 0.02$ dB
Stopband frequency range $= 4{,}500-9{,}000$ Hz
Stopband attenuation $= 46$ dB
Combined specifications $H(z)$:
Passband frequency range $= 0-2{,}500$ Hz
Passband ripples for $H(z) = 0.02$ dB
Stopband frequency range $= 3{,}000-9{,}000$ Hz
Stopband attenuation $= 46$ dB

We use an FIR filter with a Hamming window. Since

$$\Delta f = \frac{f_{stop} - f_{pass}}{f_{sL}} = \frac{3{,}000 - 2{,}500}{18{,}000} = 0.0278$$

the length of the filter and the cutoff frequency can be determined by

$$N = \frac{3.3}{\Delta f} = \frac{3.3}{0.0278} = 118.8$$

We choose $N = 119$, and

$$f_c = \frac{f_{pass} + f_{stop}}{2} = \frac{3{,}000 + 2{,}500}{2} = 2{,}750 \text{ Hz}$$

12.1.4 Application: CD Audio Player

In this application example, we will discuss principles of the upsampling and interpolation-filter processes used in CD audio systems to help with reconstruction filter design.

Each raw digital sample recorded on a CD audio system contains 16 bits and is sampled at the rate of 44.1 kHz. Figure 12.11 describes a portion of one channel of the CD player in terms of a simplified block diagram.

Let us consider the situation without upsampling and application of a digital interpolation filter. We know that the audio signal has a bandwidth of 22.05 kHz, that is, the Nyquist frequency, and digital-to-analog conversion (DAC) produces the sample-and-hold signals that contain the desired audio band and images thereof. To achieve the audio band signal, we need to apply a *reconstruction filter* (also called a smooth filter or anti-image filter) to remove all image frequencies beyond the Nyquist frequency of 22.05 kHz. Due to the requirement of the sharp transition band, a higher-order analog filter design becomes a requirement.

The design of the higher-order analog filter is complex and expensive to implement. As shown in Figure 12.11, in order to relieve such design constraints, we can add the upsampling process

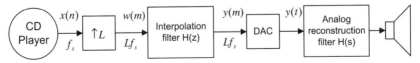

FIGURE 12.11

Sample rate conversion in the CD audio player system.

before DAC, followed by application of the digital interpolation filter (assume $L = 4$). Certainly, the interpolation filter design must satisfy the specifications studied in the previous section on increasing the sampling rate by an integer factor. Again, after digital interpolation, the audio band is kept the same, while the sampling frequency is increased by fourfold ($L = 4$), that is, $44.1 \times 4 = 176.4$ kHz.

Since the audio band of 22.05 kHz is now relatively low compared with the new folding frequency ($176.4/2 = 88.2$ kHz), the use of a simple first-order or second-order analog anti-image filter may be sufficient. Let us look the following simulation.

A test audio signal with a frequency of 16 kHz and a sampling rate of 44.1 kHz is generated using the formula

$$x(n) = \sin\left(\frac{2\pi \times 16{,}000n}{441{,}000}\right)$$

If we use an upsampling factor of 4, then the bandwidth would increase to 88.2 kHz. Based on the audio frequency of 16 kHz, the original Nyquist frequency of 22.05 kHz, and the new sampling rate of 176.4 kHz, we can determine the filter length as

$$\Delta f = \frac{22.05 - 16}{176.4} = 0.0343$$

Using the Hamming window for FIR filter design leads to

$$N = \frac{3.3}{\Delta f} = 96.2$$

We choose $N = 97$. The cutoff frequency therefore is

$$f_c = \frac{16 + 22.05}{2} = 19.025 \text{ kHz}$$

The spectrum of the interpolated audio test signal is shown in Figure 12.12, where the top plot illustrates that after the upsampling, the audio test signal has a frequency of 16 kHz, along with image frequencies coming from $44.1 - 16 = 28.1$ kHz, $44.1 + 16 = 60.1$ kHz, $88.2 - 16 = 72.2$ kHz, and so on. The bottom graph describes the spectrum after the interpolation filter. From lowpass FIR filtering, an interpolated audio signal with a frequency of 16 kHz is observed.

Let us examine the corresponding process in the time domain, as shown in Figure 12.13. The upper left plot shows the original samples. The upper right plot describes the upsampled signals. The lower left plot shows the signals after the upsampling process and digital interpolation filter. Finally, the lower right plot shows the sample-and-hold signals after DAC. Clearly, we can easily design a reconstruction filter to smooth the sample-and-hold signals and obtain the original audio test signal. The advantage of reducing hardware is illustrated. The MATLAB implementation can be seen in Program 12.4. Program 12.4. MATLAB program for CD player example.

```
close all; clear all; clc
% Generate the 2048 samples with fs = 44100 Hz
fs= 44100;               % Original sampling rate
T=1/fs;                  % Sampling period
```

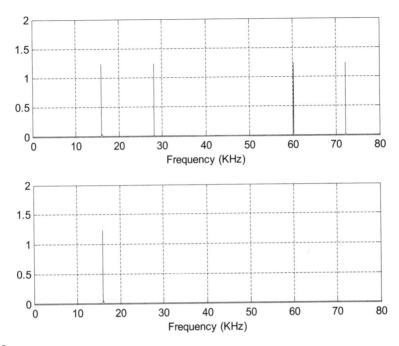

FIGURE 12.12

(Top) The spectrum after upsampling, and (bottom) the spectrum after applying the interpolation filter.

```
N=2048;                          % Number of samples
L=4;
fsL=fs*L;                        % Upsampling rate
%Upsampling filter (see Chapter 7 for FIR filter design)
Bup=firwd(97,1,2*19025*pi/fsL,0,4);
n=0:1:N-1;                       % Generate the time indices
x=5*sin(2*pi*16000*n*T);         % Generate the test signal
% Upsampling by a factor of L
w=zeros(1,L*N);
for n=0:1:N-1
 w(L*n+1)=x(n+1);
end
NL=length(w);                    % Number of the upsampled data
W=2*abs(fft(w,NL))/NL;W(1)=W(1)/2; % Compute the one-sided
                                 % amplitude spectrum
f=[0:1:NL/2-1]*fs*L/NL;          % Map the frequency index to its frequency in Hz
f=f/1000;                        % Convert to kHz
% Interpolation
y=filter(Bup,1,w);               % Perform the interpolation filter
Y=2*abs(fft(y,NL))/NL;Y(1)=Y(1)/2; % Compute the one-sided
                                 % amplitude spectrum
```

```
subplot(2,1,1);plot(f,W(1:1:NL/2));grid;
xlabel('Frequency (kHz)'); axis([0 f(length(f)) 0 2]);
subplot(2,1,2);plot(f,Y(1:1:NL/2));grid;
xlabel('Frequency (kHz)');axis([0 f(length(f)) 0 2]);
figure
subplot(2,2,1);stem(x(21:30));grid
xlabel('Number of Samples');ylabel('x(n)');
subplot(2,2,2);stem(w(81:120));grid
xlabel('Number of Samples'); ylabel('w(n)');
subplot(2,2,3);stem(y(81:120));grid
xlabel('Number of Samples'); ylabel('y(n)')
subplot(2,2,4);stairs([80:1:119]*1000*T,y(81:120));grid
xlabel('Time (ms)'); ylabel('y(t)')
```

12.1.5 **Multistage Decimation**

The multistage approach for downsampling rate conversion can be used to dramatically reduce the anti-aliasing filter length. Figure 12.14 describes a two-stage decimator.

As shown in Figure 12.14, a total decimation factor is $M = M_1 \times M_2$. Here, even though we develop a procedure for a two-stage case, a similar principle can be applied to general multistage cases.

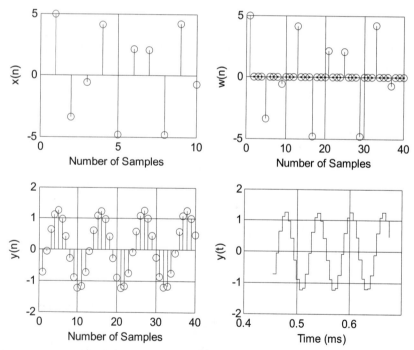

FIGURE 12.13

Plots of signals at each stage according to the block diagram in Figure 12.11.

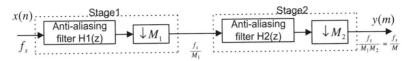

FIGURE 12.14

Multistage decimation.

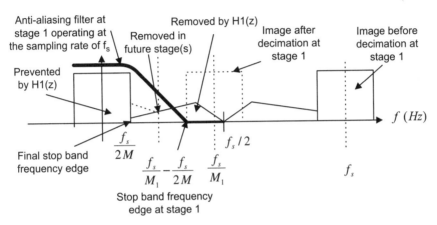

FIGURE 12.15

Stopband frequency edge for the anti-aliasing filter at stage 1 for two-stage decimation.

Using the two-stage decimation in Figure 12.15, the final Nyquist limit is $\dfrac{f_s}{2M}$ after final down-sampling. So our useful information bandwidth should stop at the frequency edge of $\dfrac{f_s}{2M}$. Next, we need to determine the stop frequency edge for the anti-aliasing lowpass filter at stage 1 before the first decimation process begins. This stop frequency edge is actually the lower frequency edge of the first image replica centered at the sampling frequency of $\dfrac{f_s}{M_1}$ after the stage 1 decimation. This lower frequency edge of the first image replica is then determined by

$$\frac{f_s}{M_1} - \frac{f_s}{2M}$$

After downsampling, we expect that some frequency components from $\dfrac{f_s}{2M_1}$ to $\dfrac{f_s}{M_1} - \dfrac{f_s}{2M}$ to be folded over to the frequency band between $\dfrac{f_s}{2M}$ and $\dfrac{f_s}{2M_1}$. However, these aliased frequency components do not affect the final useful band between 0 Hz to $\dfrac{f_s}{2M}$ and will be removed by the anti-aliasing filter(s) in the future stage(s). As illustrated in Figure 12.15, any frequency components beyond the edge $\dfrac{f_s}{M_1} - \dfrac{f_s}{2M}$

can fold over into the final useful information band to create aliasing distortion. Therefore, we can use this frequency as the lower stop frequency edge of the anti-aliasing filter to prevent the aliasing distortion at the final stage. The upper stopband edge (Nyquist limit) for the anti-image filter at stage 1 is clearly $\frac{f_s}{2}$, since the filter operates at f_s samples per second. So the stopband frequency range at stage 1 is therefore from $\frac{f_s}{M_1} - \frac{f_s}{2M}$ to $\frac{f_s}{2}$. The aliasing distortion, introduced into the frequency band from $\frac{f_s}{2M}$ to $\frac{f_s}{2M_1}$, will be filtered out after future decimation stage(s).

Similarly, for stage 2, the lower frequency edge of the first image developed after stage 2 down-sampling is

$$\frac{f_s}{M_1 M_2} - \frac{f_s}{2M} = \frac{f_s}{2M}$$

As is evident in our two-stage scheme, the stopband frequency range for the second anti-aliasing filter at stage 2 should be from $\frac{f_s}{2M}$ to $\frac{f_s}{2M_1}$.

We summarize the specifications for the two-stage decimation as follows:
Filter requirements for stage 1:

- Passband frequency range = 0 to f_p
- Stopband frequency range = $\frac{f_s}{M_1} - \frac{f_s}{2M}$ to $\frac{f_s}{2}$
- Passband ripple = $\delta_p/2$, where δ_p is the combined absolute ripple on the passband
- Stopband attenuation = δ_s

Filter requirements for stage 2:

- Passband frequency range = 0 to f_p
- Stopband frequency range = $\frac{f_s}{M_1 \times M_2} - \frac{f_s}{2M}$ to $\frac{f_s}{2M_1}$
- Passband ripple = $\delta_p/2$, where δ_p is the combined absolute ripple on the passband
- Stopband attenuation = δ_s

Example 12.4 illustrates the two-stage decimator design.

EXAMPLE 12.4

Determine the anti-aliasing FIR filter lengths and cutoff frequencies for the two-stage decimator with the following specifications and the block diagram in Figure 12.16A:

Original sampling rate: $f_s = 240$ kHz
Audio frequency range: 0–3,400 Hz
Passband ripple: $\delta_p = 0.05$ (absolute)
Stopband attenuation: $\delta_s = 0.005$ (absolute)
FIR filter design using the window method
New sampling rate: $f_{sM} = 8$ kHz

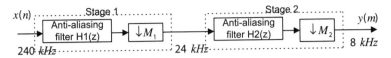

FIGURE 12.16A

Multistage decimation in Example 12.4.

Solution:

$$M = \frac{240 \ kHz}{8 \ kHz} = 30 = 10 \times 3$$

We choose $M_1 = 10$ and $M_2 = 3$; there could be other choices. Figure 12.16B shows the block diagram and filter frequency specifications.

Filter specifications for $H_1(z)$:

Passband frequency range: 0–3,400 Hz
Passband ripples: $0.05/2 = 0.025$ (δ_p dB $= 20 \log_{10}(1 + \delta_p) = 0.212$ dB)
Stopband frequency range: 20,000–120,000 Hz
Stopband attenuation: 0.005, δ_s dB $= -20 \times \log_{10}(\delta_s) = 46$ dB
Filter type: FIR, Hamming window

Note that the lower stopband edge can be determined as

$$f_{stop} = \frac{f_s}{M_1} - \frac{f_s}{2 \times M} = \frac{240,000}{10} - \frac{240,000}{2 \times 30} = 20,000 \ Hz$$

$$\Delta f = \frac{f_{stop} - f_{pass}}{f_s} = \frac{20,000 - 3,400}{240,000} = 0.06917$$

The length of the filter and the cutoff frequency can be determined by

$$N = \frac{3.3}{\Delta f} = 47.7$$

We choose $N = 49$, and

$$f_c = \frac{f_{pass} + f_{stop}}{2} = \frac{20,000 + 3,400}{2} = 11,700 \ Hz$$

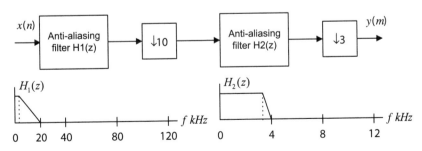

FIGURE 12.16B

Filter frequency specifications for Example 12.4.

Filter specifications for $H_2(z)$:

Passband frequency range: 0–3,400 Hz
Passband ripples: $0.05/2 = 0.025$ (0.212 dB)
Stopband frequency range: 4,000–12,000 Hz
Stopband attenuation: 0.005, $\delta_s dB = 46$ dB
Filter type: FIR, Hamming window

Note that

$$\Delta f = \frac{f_{stop} - f_{pass}}{f_{sM1}} = \frac{4,000 - 3,400}{24,000} = 0.025.$$

The length of the filter and the cutoff frequency can be determined by

$$N = \frac{3.3}{\Delta f} = 132$$

We choose $N = 133$, and

$$f_c = \frac{f_{pass} + f_{stop}}{2} = \frac{4,000 + 3,400}{2} = 3,700 \text{ Hz}$$

The reader can verify this case by using only one stage with a decimation factor of $M = 30$. Using the Hamming window for the FIR filter, the resulting number of taps is 1,321, and the cutoff frequency is 3,700 Hz. Thus, such a filter requires a huge number of computations and causes a large delay during implementation compared with the two-stage case.

The multistage scheme is very helpful for sampling rate conversion between audio systems. For example, to convert CD audio at a sampling rate of 44.1 kHz to MP3 or Digital Audio Tape (DAT), in which the sampling rate of 48 kHz is used, the conversion factor $L/M = 48/44.1 = 160/147$ is required. Using the single stage scheme may cause impractical FIR filter sizes for interpolation and downsampling. However, since $L/M = 160/147 = (4/3)(8/7)(5/7)$, we may design an efficient three-stage system, in which stages 1, 2, and 3 use the conversion factors $L/M = 8/7, L/M = 5/7$, and $L/M = 4/3$, respectively.

12.2 POLYPHASE FILTER STRUCTURE AND IMPLEMENTATION

Due to the nature of the decimation and interpolation processes, polyphase filter structures can be developed to efficiently implement the decimation and interpolation filters (using fewer multiplications and additions). As we will explain, these filters are all-pass filters with different phase shifts (Proakis and Manolakis, 1996), thus we call them *polyphase filters.*

Here, we skip their derivations and illustrate implementations of decimation and interpolation using simple examples. Consider the interpolation process shown in Figure 12.17, where $L = 2$.

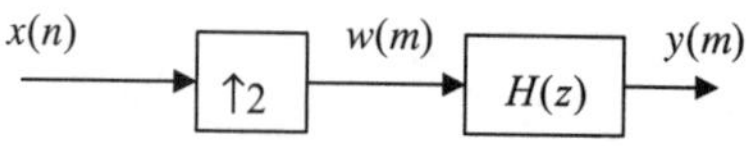

FIGURE 12.17

Upsampling by a factor of 2 and a four-tap interpolation filter.

Table 12.1 Results of the Direct Interpolation Process in Figure 12.17 (8 multiplications and 6 additions for processing each input sample $x(n)$)

n	$x(n)$	m	$w(m)$	$y(m)$
$n = 0$	$x(0)$	$m = 0$	$w(0) = x(0)$	$y(0) = h(0)x(0)$
		$m = 1$	$w(1) = 0$	$y(1) = h(1)x(0)$
$n = 1$	$x(1)$	$m = 2$	$w(2) = x(1)$	$y(2) = h(0)x(1) + h(2)x(0)$
		$m = 3$	$w(3) = 0$	$y(3) = h(1)x(1) + h(3)x(0)$
$n = 2$	$x(2)$	$m = 4$	$w(4) = x(2)$	$y(4) = h(0)x(2) + h(2)x(1)$
		$m = 5$	$w(5) = 0$	$y(5) = h(1)x(2) + h(3)x(1)$
...	...	...	...	...

We assume that the FIR interpolation filter has four taps, shown as

$$H(z) = h(0) + h(1)z^{-1} + h(2)z^{-2} + h(3)z^{-3}$$

and the filter output is

$$y(m) = h(0)w(m) + h(1)w(m-1) + h(2)w(m-2) + h(3)w(m-3)$$

For the purpose of comparison, the direct interpolation process shown in Figure 12.17 is summarized in Table 12.1, where $w(m)$ is the upsampled signal and $y(m)$ the interpolated output. Processing each input sample $x(n)$ requires applying the difference equation twice to obtain $y(0)$ and $y(1)$. Hence, for this example, we need eight multiplications and six additions.

The output results in Table 12.1 can be easily obtained by using the polyphase filters shown in Figure 12.18.

In general, there are L polyphase filters. With a designed interpolation filter $H(z)$ of N taps, we can determine each bank of filter coefficients as follows:

$$\rho_k(n) = h(k + nL) \quad \text{for} \quad k = 0, 1, \cdots, L-1 \quad \text{and} \quad n = 0, 1, \cdots, \frac{N}{L} - 1 \tag{12.12}$$

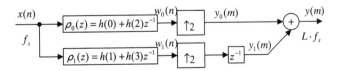

FIGURE 12.18

Polyphase filter implementation for the interpolation in Figure 12.17 (4 multiplications and 3 additions for processing each input sample $x(n)$).

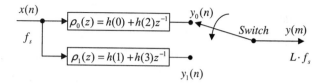

FIGURE 12.19

Commutative model for the polyphase interpolation filter.

For our example, $L = 2$ and $N = 4$, we have $L - 1 = 1$, and $N/L - 1 = 1$, respectively. Hence, there are two filter banks, $\rho_0(z)$ and $\rho_1(z)$, each having a length of 2, as illustrated in Figure 12.18. When $k = 0$ and $n = 1$, the upper limit of time index required for $h(k + nL)$ is $k + nL = 0 + 1 \times 2 = 2$. When $k = 1$ and $n = 1$, the upper limit of the time index for $h(k+nL)$ is 3. Hence, the first filter $\rho_0(z)$ has the coefficients $h(0)$ and $h(2)$. Similarly, the second filter $\rho_1(z)$ has coefficients $h(1)$ and $h(3)$. In fact, the filter coefficients of $\rho_0(z)$ are a decimated version of $h(n)$ starting at $k = 0$, while the filter coefficients of $\rho_1(z)$ are a decimated version of $h(n)$ starting at $k = 1$, and so on.

As shown in Figure 12.18, we can reduce the computational complexity from eight multiplications and six additions down to four multiplications and three additions for processing each input sample $x(n)$. Generally, the computation can be reduced by a factor of L as compared with the direct process.

The commutative model for the polyphase interpolation filter is given in Figure 12.19.

EXAMPLE 12.5

Verify $y(1)$ in Table 12.1 using the polyphase filter implementation in Figures 12.18 and 12.19, respectively.

Solution:

Applying the ployphase interpolation filter as shown in Figure 12.18 leads to

$$w_0(n) = h(0)x(n) + h(2)x(n-1)$$
$$w_1(n) = h(1)x(n) + h(3)x(n-1)$$

When $n = 0$,

$$w_0(0) = h(0)x(0)$$
$$w_1(0) = h(1)x(0)$$

After interpolation, we have

$$y_0(m): w_0(0) \quad 0 \quad \cdots$$

and

$$y_1(m): 0 \quad w_1(0) \quad 0 \quad \cdots$$

Note: there is a unit delay for the second filter bank. Hence

$$m = 0, \ y_0(0) = h(0)x(0), \ y_1(0) = 0$$
$$m = 1, \ y_0(1) = 0, \ y_1(1) = h(1)x(0)$$

Combining two channels, we finally get

$$m = 0, \ y(0) = y_0(0) + y_1(0) = h(0)x(0)$$
$$m = 1, \ y(1) = y_0(1) + y_1(1) = h(1)x(0)$$

Therefore, $y(1)$ matches that in the direct interpolation process given in Table 12.1.
Applying the polyphase interpolation filter using the commutative model in Figure 12.19, we have

$$y_0(n) = h(0)x(n) + h(2)x(n-1)$$
$$y_1(n) = h(1)x(n) + h(3)x(n-1)$$

When $n = 0$,

$$m = 0, \ y(0) = y_0(0) = h(0)x(0) + h(2)x(-1) = h(0)x(0)$$
$$m = 1, \ y(1) = y_1(0) = h(1)x(0) + h(3)x(-1) = h(1)x(0)$$

Clearly, $y(1) = h(1)x(0)$ matches the $y(1)$ result in Table 12.1.

Next, we will briefly explain the properties of polyphase filters (that is, they have all-pass gain and possible different phases). Each polyphase filter $p_k(n)$ operating at the original sampling rate f_s (assuming 8 kHz) is a downsampled version of the interpolation filter $h(n)$ operating at the upsampling rate Lf_s (32 kHz assuming an interpolation factor of $L = 4$). Considering that the designed interpolation FIR filter coefficients $h(n)$ are the impulse response sequence with a flat frequency spectrum up to a bandwidth of $f_s/2$ (assuming a bandwidth of 4 kHz with a perfect flat frequency magnitude response, theoretically) at a sampling rate of Lf_s (32 kHz), we then downsample $h(n)$ to obtain polyphase filters by a factor of $L = 4$ and operate them at a sampling rate of f_s (8 kHz).

The Nyquist frequency after downsampling should be $(Lf_s/2)/L = f_s/2$ (4 kHz); at the same time, each downsampled sequence $p_k(n)$ operating at f_s (8 kHz) has a flat spectrum up to $f_s/2$ (4 kHz) due to the $f_s/2$ (4 kHz) bandlimited sequence of $h(n)$ at the sampling rate of Lf_s (32 kHz). Hence, all of the ployphase filters are all-pass filters. Since each polyphase $p_k(n)$ filter has different coefficients, each may have a different phase. Therefore, these polyphase filters are the all-pass filters with possible different phases, theoretically.

Next, consider the decimation process in Figure 12.20. Assuming a three-tap decimation filter, we have

$$H(z) = h(0) + h(1)z^{-1} + h(2)z^{-2}$$
$$w(n) = h(0)x(n) + h(1)x(n-1) + h(2)x(n-2)$$

The direct decimation process is shown in Table 12.2 for the purpose of comparison. Obtaining each output $y(m)$ requires processing filter difference equations twice, resulting in six multiplications and four additions for this particular example.

FIGURE 12.20

Decimation by a factor of 2 and a three-tap anti-aliasing filter.

Table 12.2 Results of the Direct Decimation Process in Figure 12.20 (6 multiplications and 4 additions for obtaining each output $y(m)$)

n	$x(n)$	$w(n)$		m	$y(m)$
$n = 0$	$x(0)$	$w(0) = h(0)x(0)$		$m = 0$	$y(0) = h(0)x(0)$
$n = 1$	$x(1)$	$w(1) = h(0)x(1) + h(1)x(0)$ discard			
$n = 2$	$x(2)$	$w(2) = h(0)x(2) + h(1)x(1) + h(2)x(0)$		$m = 1$	$y(1) = h(0)x(2) +$ $h(1)x(1) + h(2)x(0)$
$n = 3$	$x(3)$	$w(3) = h(0)x(3) + h(1)x(2) + h(2)x(1)$ discard			
$n = 4$	$x(5)$	$w(4) = h(0)x(4) + h(1)x(3) + h(2)x(2)$		$m = 2$	$y(2) = h(0)x(4) +$ $h(1)x(3) + h(2)x(2)$
$n = 5$	$x(6)$	$w(5) = h(0)x(5) + h(1)x(4) + h(2)x(3)$ discard			
...	...	...		...	...

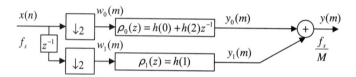

FIGURE 12.21

Polyphase filter implementation for the decimation in Figure 12.20. (3 multiplications and 1 addition for obtaining each output y(m)).

The efficient way to implement a polyphase filter is given in Figure 12.21.

Similarly, there are M polyphase filters. With the designed decimation filter $H(z)$ of N taps, we can obtain filter bank coefficients by

$$\rho_k(n) = h(k + nM) \quad \text{for} \quad k = 0, 1, \cdots, M - 1 \quad \text{and} \quad n = 0, 1, \cdots, \frac{N}{M} - 1 \qquad (12.13)$$

For our example, we see that $M - 1 = 1$ and $N/M - 1 = 1$(rounded up). Thus, we have two filter banks. Since $k = 0$ and $n = 1$, $k + nM = 0 + 1 \times 2 = 2$. The time index upper limit required for $h(k + nM)$ is 2 for the first filter bank $\rho_0(z)$. Hence $\rho_0(z)$ has filter coefficients $h(0)$ and $h(2)$.

However, when $k = 1$ and $n = 1$, $k + nM = 1 + 1 \times 2 = 3$, the time index upper limit required for $h(k + nM)$ is 3 for the second filter bank, and the corresponding filter coefficients are required to be $h(1)$ and $h(3)$. Since our direct interpolation filter $h(n)$ does not contain the coefficient $h(3)$, we set $h(3) = 0$ to get the second filter bank with one tap only, as shown in Figure 12.21. Also as shown in that figure, achieving each $y(m)$ requires three multiplications and one addition. In general, the number of multiplications can be reduced by a factor of M.

The commutative model for the polyphase decimator is shown in Figure 12.22.

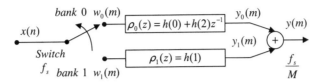

FIGURE 12.22

Commutative model for the polyphase decimation filter.

EXAMPLE 12.6

Verify $y(1)$ in Table 12.2 using the polyphase decimation filter implementation in Figure 12.21.

Solution:

Using Figure 12.21, we write the difference equations as

$$y_0(m) = h(0)w_0(m) + h(2)w_0(m-1)$$
$$y_1(m) = h(1)w_1(m)$$

Assuming $n = 0$, $n = 1$, $n = 2$, and $n = 3$, we have the inputs as $x(0)$, $x(1)$, $x(2)$ and $x(3)$, and

$$w_0(m): \quad x(0) \quad x(2) \quad \cdots$$

Delaying $x(n)$ by one sample and decimating it by a factor or 2 leads to

$$w_1(m): \quad 0 \quad x(1) \quad x(3) \quad \cdots$$

Hence, applying the filter banks yields the following: For $m = 0$, we have inputs for each filter as

$$w_0(0) = x(0) \text{ and } w_1(0) = 0$$

Then

$$y_0(0) = h(0)w_0(0) + h(2)w_0(-1) = h(0)x(0)$$

$$y_1(0) = h(1)w_1(0) = h(1) \times 0 = 0$$

Combining two channels, we obtain

$$y(1) = y_0(1) + y_1(1) = h(0)x(0) + 0 = h(0)x(0)$$

For $m = 1$, we get inputs for each filter as

$$w_0(1) = x(2) \quad \text{and} \quad w_1(1) = x(1)$$

Then

$$y_0(1) = h(0)w_0(1) + h(2)w_0(0) = h(0)x(2) + h(2)x(0)$$

$$y_1(1) = h(1)w_1(1) = h(1)x(1)$$

Combining two channels leads to

$$y(1) = y_0(1) + y_1(1) = h(0)x(2) + h(2)x(0) + h(1)x(1)$$

We note that $y(1)$ is the same as that shown in Table 12.2. Similar analysis can be done for the commutative model shown in Figure 12.22.

Program 12.5 demonstrates the polyphase implementation of decimation. The program is modified based on Program 12.1.

Program 12.5. Decimation using polyphase implementation.

```
close all; clear all;
% Downsampling filter (see Chapter 7 for FIR filter design)
B =[0.00074961181416 0.00247663033476  0.00146938649416  -0.00440446121505...
 -0.00910635730662  0.00000000000000  0.02035676831506  0.02233710562885...
 -0.01712963672810  -0.06376620649567 -0.03590670035210 0.10660980550088...
 0.29014909103794  0.37500000000000  0.29014909103794  0.10660980550088...
 -0.03590670035210  -0.06376620649567 -0.01712963672810  0.02233710562885...
 0.02035676831506  0.00000000000000  -0.00910635730662 -0.00440446121505...
 0.00146938649416  0.00247663033476  0.00074961181416];
% Generate 2048 samples
fs=8000;              % Sampling rate
N=2048;               % Number of samples
M=2;                  % Downsample factor
n=0:1:N-1;
x=5*sin(n*pi/4)+cos(5*n*pi/8);
% Compute the single-sided amplitude spectrum
% AC component will be doubled, and DC component will be kept the
% same value
X=2*abs(fft(x,N))/N;X(1)=X(1)/2;
% Map the frequency index up to the folding frequency in Hz
f=[0:1:N/2-1]*fs/N;
% Decimation
w0=x(1:M:N); p0=B(1:2:length(B)); % Downsampling
w1=filter([0 1],1,x); % Delay one sample
w1=w1(1:M:N); p1=B(2:M:length(B)) % Downsampling
y=filter(p0,1,w0)+filter(p1,1,w1);
NM=length(y);         % Length of the downsampled data
% Compute the single-sided amplitude spectrum for the downsampled
% signal
Y=2*abs(fft(y,NM))/NM;Y(1)=Y(1)/2;
% Map the frequency index to the frequency in Hz
fsM=[0:1:NM/2-1]*(fs/M)/NM;
% Plot spectra
subplot(2,1,1);plot(f,X(1:1:N/2));grid; xlabel('Frequency (Hz)');
subplot(2,1,2);plot(fsM,Y(1:1:NM/2));grid; xlabel('Frequency (Hz)');
```

Program 12.6 demonstrates polyphase implementation of interpolation using the information in Program 12.2.

Program 12.6. Interpolation using polyphase implementation.

```
close all; clear all
% Upsampling filter (see Chapter 7 for FIR filter design)
B =[ -0.00012783931504 0.00069976044649 0.00123831516738 0.00100277549136...
 -0.00025059018468 -0.00203448515158 -0.00300830295487 -0.00174101657599...
```

```
0.00188598835011 0.00578414933758 0.00649330625041 0.00177982369523...
-0.00670672686935 -0.01319379342716 -0.01116855281442 0.00123034314117...
0.01775600060894 0.02614700427364 0.01594155162392 -0.01235169936557...
-0.04334322148505 -0.05244745563466 -0.01951094855292 0.05718573279009...
0.15568416401644 0.23851539047347 0.27083333333333 0.23851539047347...
0.15568416401644 0.05718573279009 -0.01951094855292 -0.05244745563466...
-0.04334322148505 -0.01235169936557 0.01594155162392 0.02614700427364...
0.01775600060894 0.00123034314117 -0.01116855281442 -0.01319379342716...
-0.00670672686935 0.00177982369523 0.00649330625041 0.00578414933758...
0.00188598835011 -0.00174101657599 -0.00300830295487 -0.00203448515158...
-0.00025059018468 0.00100277549136 0.00123831516738 0.00069976044649...
-0.00012783931504];
% Generate 2048 samples with fs=8000 Hz
fs=8000;                          % Sampling rate
N=2048;                           % Number of samples
L = 3;                            % Upsampling factor
n=0:1:N-1;
x=5*sin(n*pi/4)+cos(5*n*pi/8);
p0=B(1:L:length(B)); p1=B(2:L:length(B)); p2=B(3:L:length(B));
% Interpolation
w0=filter(p0,1,x);
w1=filter(p1,1,x);
w2=filter(p2,1,x);
y0=zeros(1,L*N);y0(1:L:length(y0))=w0;
y1=zeros(1,L*N);y1(1:L:length(y1))=w1;
y1=filter([0 1],1,y1);
y2=zeros(1,L*N);y2(1:L:length(y2))=w2;
y2=filter([0 0 1],1,y2);
y=y0+y1+y2;                       % Interpolated signal
NL = length(y);                   % Length of the upsampled data
X=2*abs(fft(x,N))/N;X(1)=X(1)/2;  % Compute the one-sided amplitude
                                  % spectrum
f=[0:1:N/2-1]*fs/N; % Map the frequency index to the frequency (Hz)
Y=2*abs(fft(y,NL))/NL;Y(1)=Y(1)/2; % Compute the one-sided amplitude
                                  % spectrum
fsL=[0:1:NL/2-1]*fs*L/NL;         % Map the frequency index to the frequency (Hz)
subplot(2,1,1);plot(f,X(1:1:N/2));grid; xlabel('Frequency (Hz)');
subplot(2,1,2);plot(fsL,Y(1:1:NL/2));grid; xlabel('Frequency (Hz)');
```

Note that wavelet transform and subband coding are also topics in the area of multirate signal processing. We will pursue these subjects in Chapter 13.

12.3 OVERSAMPLING OF ANALOG-TO-DIGITAL CONVERSION

Oversampling of the analog signal has become more popular in the DSP industry to improve the resolution of analog-to-digital conversion (ADC). Oversampling uses a sampling rate that is much higher than the Nyquist rate. We can define an oversampling ratio as

$$\frac{f_s}{2f_{max}} \gg 1 \tag{12.14}$$

The benefits of an oversampling ADC include

1. helping to design a simple analog anti-aliasing filter before ADC, and
2. reducing the ADC noise floor with possible noise shaping so that a low-resolution ADC can be used.

12.3.1 Oversampling and Analog-to-Digital Conversion Resolution

To begin developing the relation between oversampling and ADC resolution, we first summarize the regular ADC and some useful definitions discussed in Chapter 2:

$$\text{Quantization noise power} = \sigma_q^2 = \frac{\Delta^2}{12} \tag{12.15}$$

$$\text{Quantization step} = \Delta = \frac{A}{2^n} \tag{12.16}$$

where A = full range of the analog signal to be digitized and n = number of bits per sample (ADC resolution).

Substituting Equation (12.16) into Equation (12.15), we have

$$\text{Quantization noise power} = \sigma_q^2 = \frac{A^2}{12} \times 2^{-2n} \tag{12.17}$$

The power spectral density of the quantization noise with an assumption of uniform probability distribution is shown in Figure 12.23. Note that this assumption is true for quantizing a uniformly

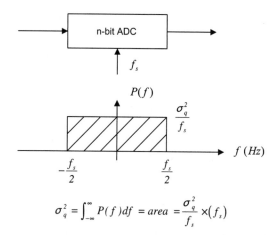

$$\sigma_q^2 = \int_{-\infty}^{\infty} P(f)df = area = \frac{\sigma_q^2}{f_s} \times (f_s)$$

FIGURE 12.23

Regular ADC system.

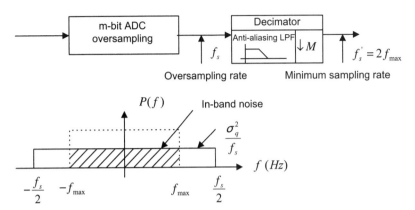

FIGURE 12.24

Oversampling ADC system.

distributed signal in a full range with a sufficiently long duration. It is not generally true in practice. See research papers authored by Lipshitz et al. (1992) and Maher (1992). However, using the assumption will guide us to some useful results for oversampling systems.

The quantization noise power is the area obtained from integrating the power spectral density function in the range of $-f_s/2$ to $f_s/2$. Now let us examine the oversampling ADC, where the sampling rate is much larger than that of the regular ADC; that is $f_s \gg 2f_{max}$. The scheme is shown in Figure 12.24.

As we can see, oversampling can reduce the level of noise power spectral density. After the decimation process with the decimation filter, only a portion of quantization noise power in the range from $-f_{max}$ and f_{max} is kept in the DSP system. We call this an *in-band frequency range*.

In Figure 12.24, the shaded area, which is the quantization noise power, is given by

$$Quantization\ noise\ power = \int_{-\infty}^{\infty} P(f)df = \frac{2f_{max}}{f_s} \times \sigma_q^2 = \frac{2f_{max}}{f_s} \frac{A^2}{12} \times 2^{-2m} \qquad (12.18)$$

Assuming that the regular ADC shown in Figure 12.23 and the oversampling ADC shown in Figure 12.24 are equivalent, we set their quantization noise powers to be the same to obtain

$$\frac{A^2}{12} \cdot 2^{-2n} = \frac{2f_{max}}{f_s} \cdot \frac{A^2}{12} \times 2^{-2m} \qquad (12.19)$$

Equation (12.19) leads to two useful equations for applications:

$$n = m + 0.5 \times \log_2\left(\frac{f_s}{2f_{max}}\right) \text{ and} \qquad (12.20)$$

$$f_s = 2f_{max} \times 2^{2(n-m)} \qquad (12.21)$$

where

f_s = sampling rate in the oversampling DSP system
f_{max} = maximum frequency of the analog signal
m = number of bits per sample in the oversampling DSP system
n = number of bits per sample in the regular DSP system using the minimum sampling rate

From Equation (12.20) and given the number of bits (m) used in the oversampling scheme, we can determine the number of bits per sample equivalent to the regular ADC. On the other hand, given the number of bits in the oversampling ADC, we can determine the required oversampling rate so that the oversampling ADC is equivalent to the regular ADC with the larger number of bits per sample (n). Let us look at the following examples.

EXAMPLE 12.7

Given an oversampling audio DSP system with maximum audio input frequency of 20 kHz and ADC resolution of 14 bits, determine the oversampling rate to improve the ADC resolution to 16-bit resolution.

Solution:
Based on the specifications, we have

$$f_{max} = 20\text{ kHz},\ m = 14\text{ bits}\quad \text{and}\quad n = 16\text{ bits}$$

Using Equation (12.21) leads to

$$f_s = 2f_{max} \times 2^{2(n-m)} = 2 \times 20 \times 2^{2(16-14)} = 640\text{ kHz}$$

Since $f_s/(2f_{max}) = 2^4$, we see that each doubling of the minimum sampling rate ($2f_{max} = 40$ kHz) will increase the resolution by a half bit.

EXAMPLE 12.8

Given an oversampling audio DSP system with a maximum audio input frequency of 4 kHz, and ADC resolution of 8 bits, an a sampling rate of 80 MHz, determine the equivalent ADC resolution.

Solution:
Since $f_{max} = 4$ kHz, $f_s = 80$ kHz, and $m = 8$ bits, appyling Equation (12.20) yields

$$n = m + 0.5 \times \log_2 \left(\frac{f_s}{2f_{max}}\right) = 8 + 0.5 \times \log_2 \left(\frac{80{,}000\text{ kHz}}{2 \times 4\text{ kHz}}\right) \approx 15\text{ bits}$$

The MATLAB program shown in Program 12.7 validates the oversampling technique. We consider the following signal,

$$x(t) = 1.5 \sin (2\pi \times 150t) + 0.9 \sin (2\pi \times 175t + \pi/6) + 0.6 \sin (2\pi \times 200t + \pi/4) \quad (12.22)$$

with a regular sampling rate of 1 kHz. The oversampling rate is 4 kHz and each sample is quantized using a 3-bit code. The anti-aliasing lowpass filter is designed with a cutoff frequency of $\Omega = 2\pi f_{max}T = 2\pi \times 500/4{,}000 = 0.25\pi$ radians. Figure 12.25 shows the frequency responses of

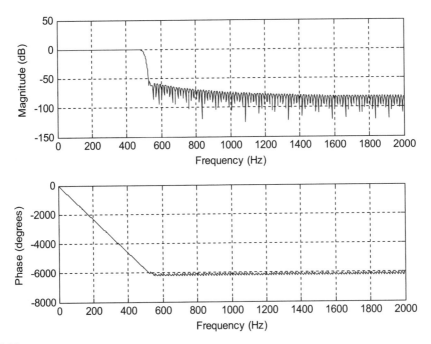

FIGURE 12.25

Frequency responses of the designed filter.

the designed filter while Figure 12.26 compares the signals in the time and frequency domains, respectively, where $x(t)$ denotes the continuous version, $x_q(n)$ is the quantized version using a regular sampling rate of 1 kHz, and $y_q(n)$ is the enhanced version using the oversampling system with $L = 4$. The detailed amplitude comparisons are given in Figure 12.27. The measured signal-to-noise ratios (SNRs) are 14.3 dB using the regular sampling system and 21.0 dB using the oversampling system. Since $L = 4$, the achieved signal is expected to have 4-bit quality ($0.5 \times \log_2 4 = 1$ bit improvement). From simulation, we achieve an SNR improvement of approximately 6dB. The improvement will stop when L increases due to the fact that when the sampling increases the quantization error may have correlation with the sinusoidal signal. The degradation performance can be cured using the dithering technique (Tan and Wang, 2011), which is beyond our scope.

Program 12.7. Oversampling implementation.

```
clear all; close all,clc
ntotal=512;
n=0:ntotal; % Number of samples
L=4; % Oversampling factor
nL=0:ntotal*L; % Number of samples for oversampling
numb=3; % Number of bits
A=2^(numb-1)-1; % Peak value
f1=150;C1=0.5*A;f2=175;C2=A*0.3;f3=200;C3=A*0.2; % Frequencies and amplitudes
```

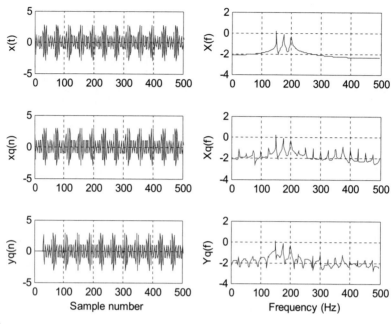

FIGURE 12.26

Signal comparisons in both the time and frequency domains.

```
fmax=500;fs=1000;T=1/fs; % Maximum frequency, sampling rate, sampling period
fsL=L*fs;TL=1/fsL;% Oversampling rate and oversampling period
% Sampling at fs=1000 Hz
x=C1*sin(2*pi*f1*n*T)+C2*sin(2*pi*f2*T*n+pi/6)+C3*sin(2*pi*f3*T*n+pi/4);
xq=round(x); % Quantized signal at the minimum sampling rate
NN=length(n);
f=[0:ntotal-1]*fs/NN;
M=32*L;nd=M/L; % Delay in terms of samples due to anti-aliasing filtering
B=firwd(2*M+1,1,2*pi*fmax/fsL,0,4); % Anti-aliasing filter design (5% transition
% bandwidth)
figure(1);
freqz(B,1,1000,fsL)
% Oversampling
xx=C1*sin(2*pi*f1*nL*TL)+C2*sin(2*pi*f2*nL*TL+pi/6)+C3*sin(2*pi*f3*nL*TL+pi/4);
xxq=round(xx); % Quantized signal
% Down sampling
y=filter(B,1,xxq);% Anti-aliasing filtering
yd=y(1:L:length(y));% Downsample
figure (2)
subplot(3,2,1);plot(n,x,'k');grid;axis([0 500 -5 5]);ylabel('x(t)')
Ak=2*abs(fft(x))/NN; Ak(1)=Ak(1)/2;
```

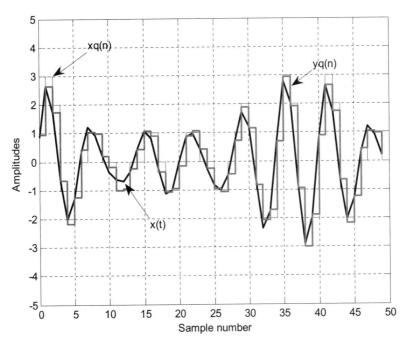

FIGURE 12.27

Comparisons of continuous, regular sampled, and oversampled signal amplitudes.

```
subplot(3,2,2);plot(f(1:NN/2),log10(Ak(1:NN/2)),'k');grid;ylabel('X(f)'); axis([0 500
-4 2])
subplot(3,2,3);plot(n,xq,'k');grid;axis([0 500 -5 5]);ylabel('xq(n)');
Ak=2*abs(fft(xq))/NN; Ak(1)=Ak(1)/2;
subplot(3,2,4);plot(f(1:NN/2),log10(Ak(1:NN/2)),'k');grid;ylabel('Xq(f)'); axis([0 500
-4 2])
subplot(3,2,5);plot(n,yd,'k');grid;axis([0 500 -5 5]);ylabel('yq(n)');
xlabel('Sample number');
Ak=2*abs(fft(yd))/NN; Ak(1)=Ak(1)/2;
subplot(3,2,6);plot(f(1:NN/2),log10(Ak(1:NN/2)),'k');grid;ylabel('Yq(f)'); axis([0 500
-4 2])
xlabel('Frequency (Hz)');
figure (3)
plot(n(1:50),x(1:50),'k','LineWidth',2); hold % Plot of first 50 samples
stairs(n(1:50),xq(1:50),'b');
stairs(n(1:50),yd(1+nd:50+nd),'r','LineWidth',2);grid
axis([0 50 -5 5]);xlabel('Sample number');ylabel('Amplitudes')
snr(x,xq);
snr(x(1:ntotal-nd),yd(1+nd:ntotal));
```

12.3.2 Sigma-Delta Modulation Analog-to-Digital Conversion

To further improve ADC resolution, *sigma-delta modulation* (SDM) ADC is used. The principles of the first-order SDM are described in Figure 12.28.

First, the analog signal is sampled to obtain the discrete-time signal $x(n)$. This discrete-time signal is subtracted by the analog output from the m-bit DAC, converting the m-bit oversampled digital signal $y(n)$. Then the difference is sent to the discrete-time analog integrator, which is implemented by the switched-capacitor technique, for example. The output from the discrete-time analog integrator is converted using an m-bit ADC to produce the oversampled digital signal. Finally, the decimation filter removes outband quantization noise. Further decimation processes can change the oversampling rate back to the desired sampling rate for the output digital signal $w(m)$.

To examine the SDM, we need to develop a DSP model for the discrete-time analog filter described in Figure 12.29.

As shown in Figure 12.29, the input signal $c(n)$ designates the amplitude at time instant n, while the output $d(n)$ is the area under the curve at time instant n, which can be expressed as a sum of the area under the curve at time instant $n - 1$ and an area increment:

$$d(n) = d(n - 1) + \text{area incremetal}$$

Using the extrapolation method, we have

$$d(n) = d(n - 1) + 1 \times c(n) \tag{12.23}$$

Applying the z-transform to Equation (12.23) leads to a transfer function of the discrete-time analog filter as

$$H(z) = \frac{D(z)}{C(z)} = \frac{1}{1 - z^{-1}} \tag{12.24}$$

Again, considering that the m-bit quantization requires one sample delay, we get the DSP model for the first-order SDM depicted in Figure 12.30, where $y(n)$ is the oversampling data encoded by m bits each, and $e(n)$ represents quantization error.

The SDM DSP model represents a feedback control system. Appling the z-transform leads to

$$Y(z) = \frac{1}{1 - z^{-1}} \left(X(z) - z^{-1}Y(z) \right) + E(z) \tag{12.25}$$

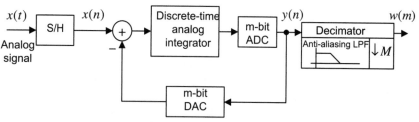

FIGURE 12.28

Block diagram of SDM ADC.

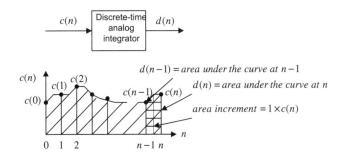

FIGURE 12.29

Illustration of discrete-time analog integrator.

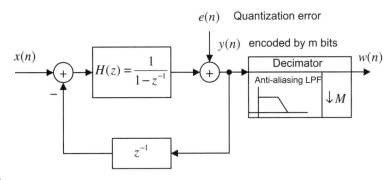

FIGURE 12.30

DSP model for first-order SDM ADC.

After simple algebra, we have

$$Y(z) = \underbrace{X(z)}_{\substack{\textit{Original} \\ \textit{digital signal} \\ \textit{transform}}} + \underbrace{(1 - z^{-1})}_{\substack{\textit{Highpass} \\ \textit{filter}}} \cdot \underbrace{E(z)}_{\substack{\textit{Quantization} \\ \textit{error} \\ \textit{transform}}} \qquad (12.26)$$

In Equation (12.26), the indicated highpass filter pushes quantization noise to the high-frequency range, where later the quantization noise can be removed by the decimation filter. Thus we call this highpass filter $(1 - z^{-1})$ the *noise shaping filter* (illustrated in Figure 12.31).

Shaped-in-band noise power after use of decimation filter can be estimated by the solid area under the curve. We have

$$\text{Shaped-in-band noise power} = \int\limits_{-\Omega_{\max}}^{\Omega_{\max}} \frac{\sigma_q^2}{2\pi} |1 - e^{-j\Omega}|^2 d\Omega \qquad (12.27)$$

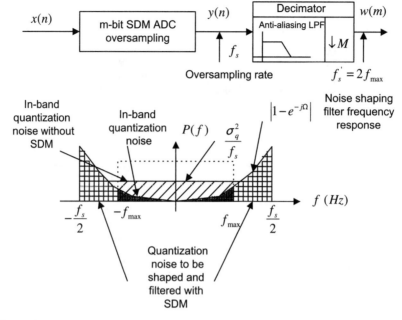

FIGURE 12.31

Noise shaping of quantization noise for SDM ADC.

Using the Maclaurin series expansion and neglecting the higher-order terms due to the small value of $\Omega_{\max}$, we yield

$$1 - e^{-j\Omega} = 1 - \left(1 + \frac{(-j\Omega)}{1!} + \frac{(-j\Omega)^2}{2!} + \cdots\right) \approx j\Omega$$

Applying this approximation to Equation (12.27) leads to

$$\text{Shaped-in-band noise power} \approx \int_{-\Omega_{\max}}^{\Omega_{\max}} \frac{\sigma_q^2}{2\pi} |j\Omega|^2 d\Omega = \frac{\sigma_q^2}{3\pi} \Omega_{\max}^3 \qquad (12.28)$$

After simple algebra, we have

$$\text{Shaped-in-band noise power} \approx \frac{\pi^2 \sigma_q^2}{3}\left(\frac{2f_{\max}}{f_s}\right)^3 = \frac{\pi^2}{3} \cdot \frac{A^2 2^{-2m}}{12}\left(\frac{2f_{\max}}{f_s}\right)^3 \qquad (12.29)$$

If we let the shaped-in-band noise power equal the quantization noise power from the regular ADC using a minimum sampling rate, we have

$$\frac{\pi^2}{3} \cdot \frac{A^2 2^{-2m}}{12}\left(\frac{2f_{\max}}{f_s}\right)^3 = \frac{A^2}{12} \cdot 2^{-2n} \qquad (12.30)$$

We modify Equation (12.30) into the following useful formats for applications:

$$n = m + 1.5 \times \log_2\left(\frac{f_s}{2f_{max}}\right) - 0.86 \qquad (12.31)$$

$$\left(\frac{f_s}{2f_{max}}\right)^3 = \frac{\pi^2}{3} \times 2^{2(n-m)} \qquad (12.32)$$

EXAMPLE 12.9

Given the following DSP system specifications, determine the equivalent ADC resolution.

Oversampling rate system
First-order SDM with 2-bit ADC
Sampling rate = 4 MHz
Maximum audio input frequency = 4 kHz

Solution:
Since $m = 2$ bits, and

$$\frac{f_s}{2f_{max}} = \frac{4{,}000 \ kHz}{2 \times 4 \ kHz} = 500$$

we calculate

$$n = m + 1.5 \times \log_2\left(\frac{f_s}{2f_{max}}\right) - 0.86 = 2 + 1.5 \times \log_2(500) - 0.86 \approx 15 \ \text{bits}$$

We can also extend the first-order SDM DSP model to the second-order SDM DSP model by cascading one section of the first-order discrete-time analog filter as depicted in Figure 12.32.

Similarly to the first-order SDM DSP model, applying the z-transform leads to the following relationship:

$$Y(z) = \underbrace{X(z)}_{\substack{Original \\ digital \ signal \\ transform}} + \underbrace{\left(1 - z^{-1}\right)^2}_{\substack{Highpass \\ noise \ shaping \\ filter}} \cdot \underbrace{E(z)}_{\substack{Quantization \\ error \\ transform}} \qquad (12.33)$$

Notice that the noise shape filter becomes a second-order highpass filter; hence, the more quantization noise is pushed to the high frequency range, the better ADC resolution is expected to be. In a similar analysis to the first-order SDM, we get the following useful formulas:

$$n = m + 2.5 \times \log_2\left(\frac{f_s}{2f_{max}}\right) - 2.14 \qquad (12.34)$$

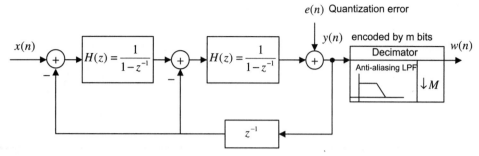

FIGURE 12.32

DSP model for the second-order SDM ADC.

$$\left(\frac{f_s}{2f_{\text{max}}}\right)^5 = \frac{\pi^4}{5} \times 2^{2(n-m)} \tag{12.35}$$

In general, the Kth-order SDM DSP model and ADC resolution formulas are given as

$$Y(z) = \underbrace{X(z)}_{\substack{Original \\ digital\ signal \\ transform}} + \underbrace{\left(1 - z^{-1}\right)^K}_{\substack{Highpass \\ noise\ shaping \\ filter}} \times \underbrace{E(z)}_{\substack{Quantization \\ error \\ transform}} \tag{12.36}$$

$$n = m + 0.5 \cdot (2K + 1) \times \log_2\left(\frac{f_s}{2f_{\text{max}}}\right) - 0.5 \times \log_2\left(\frac{\pi^{2K}}{2K + 1}\right) \tag{12.37}$$

$$\left(\frac{f_s}{2f_{\text{max}}}\right)^{2K+1} = \frac{\pi^{2K}}{2K + 1} \times 2^{2(n-m)} \tag{12.38}$$

EXAMPLE 12.10

Given an oversampling rate DSP system with the following specifications, determine the effective ADC resolution:

Second-order SDM = 1-bit ADC
Sampling rate = 1 MHz
Maximum audio input frequency = 4 kHz

Solution:

$$n = 1 + 2.5 \times \log_2\left(\frac{1{,}000\ kHz}{2 \times 4\ kHz}\right) - 2.14 \approx 16\ \text{bits}$$

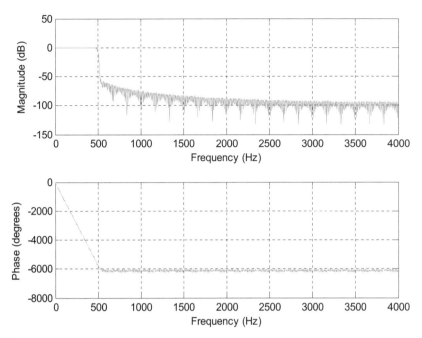

FIGURE 12.33

Frequency responses of the designed filter.

We implement the first-order SDM system using the same continuous signal in Equation (12.22). The continuous signal is originally sampled at 1 kHz and each sample is encoded using 3 bits. The SDM system uses an oversampling rate of 8 kHz ($L = 8$) and each sample is quantized using a 3-bit code. The anti-aliasing lowpass filter is designed with a cutoff frequency of $\Omega = 2\pi f_{\max}T = 2\pi \times 500/8,000 = \pi/8$ radians. Figure 12.33 shows the frequency responses of the designed filter while Figure 12.34 compares the time and frequency domain signals, where $x(t)$ designates the continuous version, $x_q(n)$ denotes the quantized version using a regular sampling rate ($L = 1$), and $y_q(n)$ is the enhanced version using $L = 8$. The detailed amplitude comparisons are given in Figure 12.35. The measured SNRs are 14.3 dB in the regular sampling system and 33.83 dB in the oversampling SDM system. We observe a significant SNR improvement of 19.5 dB. The detailed implementation using MATLAB is given in Program 12.8.

Program 12.8. First-order SDM oversampling implementation.

```
clear all; close all;clc
ntotal=512; % Number of samples
n=0:ntotal;
L=8; % Oversampling factor
nL=0:ntotal*L;numb=3;A=2^(numb-1)-1; % Peak value
f1=150;C1=0.5*A;f2=175;C2=A*0.3;f3=200;C3=A*0.2;% Frequencies and amplitudes
fmax=500;fs=1000; T=1/fs; % Sampling rate and sampling period
fsL=L*fs;TL=1/fsL; % Oversampling rate and oversampling period
```

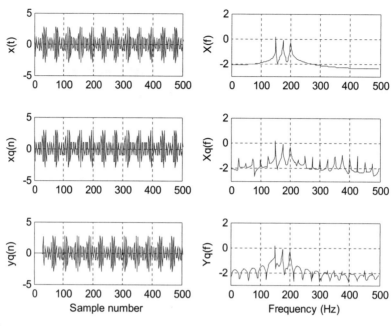

FIGURE 12.34

Signal comparisons in both the time and frequency domains.

```
% Sampling at fs = 1000 Hz
x=C1*sin(2*pi*f1*n*T)+C2*sin(2*pi*f2*T*n+pi/6)+C3*sin(2*pi*f3*T*n+pi/4);
xq=round(x); % Quantization
NN=length(n);
M=32*L;nd=M/L; % Delay in terms of samples for anti-aliasing filtering
B=firwd(2*M+1,1,2*pi*fmax/fsL,0,4); % Design of an anti-aliasing filter
figure(1)
freqz(B,1,1000,fsL);
% Oversampling
xx=C1*sin(2*pi*f1*nL*TL)+C2*sin(2*pi*f2*nL*TL+pi/6)+C3*sin(2*pi*f3*nL*TL+pi/4);
% The first-order SDM processing
 yq=zeros(1,ntotal*L+1+1); % Initializing the buffer
 y=yq;
for i=1:ntotal*L
  y(i+1)=(xx(i+1)-yq(i))+y(i);
  yq(i+1)=round(y(i+1));
end
xxq=yq(1:ntotal*L+1); % Signal quantization
% Downsampling
y=filter(B,1,xxq);
yd=y(1:L:length(y));
f=[0:ntotal-1]*fs/NN;
```

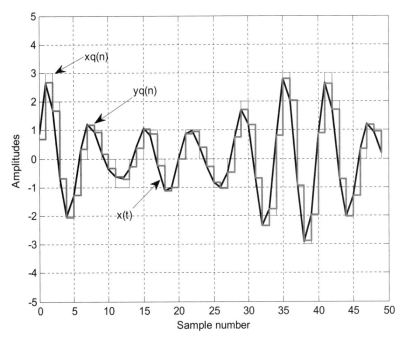

FIGURE 12.35

Comparison of continuous, regular sampled, and oversampled signal amplitudes.

```
figure (2)
subplot(3,2,1);plot(n,x,'k');grid;axis([0 500 -5 5]);ylabel('x(t)');
Ak=2*abs(fft(x))/NN; Ak(1)=Ak(1)/2;
subplot(3,2,2);plot(f(1:NN/2),log10(Ak(1:NN/2)),'k');grid;
axis([0 500 -3 2]);ylabel('X(f)');
subplot(3,2,3);plot(n,xq,'k');grid;axis([0 500 -5 5]);ylabel('xq(n)');
Ak=2*abs(fft(xq))/NN; Ak(1)=Ak(1)/2;
subplot(3,2,4);plot(f(1:NN/2),log10(Ak(1:NN/2)),'k');grid
axis([0 500 -3 2]);ylabel('Xq(f)');
subplot(3,2,5);plot(n,yd,'k');grid;axis([0 500 -5 5]);ylabel('yq(n)');
xlabel('Sample number');
Ak=2*abs(fft(yd))/NN; Ak(1)=Ak(1)/2;
subplot(3,2,6);plot(f(1:NN/2),log10(Ak(1:NN/2)),'k');grid
axis([0 500 -3 2]);ylabel('Yq(f)');xlabel('Frequency (Hz)');
figure (3)
plot(n(1:50),x(1:50),'k','LineWidth',2); hold
stairs(n(1:50),xq(1:50),'b');
stairs(n(1:50),yd(1+nd:50+nd),'r','LineWidth',2);
axis([0 50 -5 5]);grid;xlabel('Sample number');ylabel('Amplitudes');
snr(x,xq);
snr(x(1:ntotal-nd),yd(1+nd:ntotal));
```

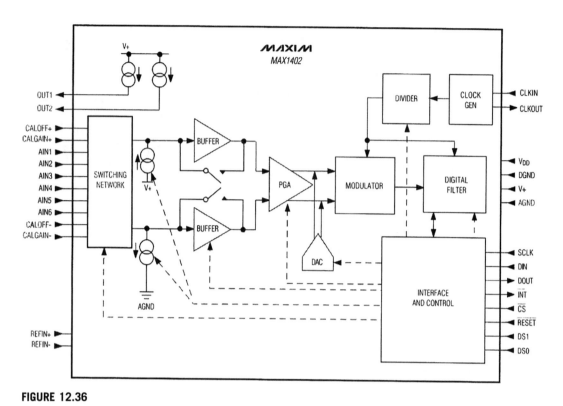

FIGURE 12.36

Functional diagram for the sigma-delta ADC.

Next, we review application of oversampling ADC in industry. Figure 12.36 illustrates a function diagram for the MAX1402 low-power, multichannel oversampling sigma-delta analog-to-digital converter used in industry. It applies a sigma-delta modulator with a digital decimation filter to achieve 16-bit accuracy. The device offers three fully differential input channels, which can be independently programmed. It can also be configured as five pseudo-differential input channels. It comprises two chopper buffer amplifiers and a programmable gain amplifier, a DAC unit with predicted input subtracted from the analog input to acquire the differential signal, and a second-order switched-capacitor sigma-delta modulator.

The chip produces a 1-bit data stream, which will be filtered by the integrated digital filter to complete ADC. The digital filter's user-selectable decimation factor offers flexibility as conversion resolution can be reduced in exchange for a higher data rate or *vice versa*. The integrated digital lowpass filter is a first-order or third-order Sinc infinite impulse response filter. Such a filter offers notches corresponding to its output data rate and its frequency harmonics, so it can effectively reduce the developed image noises in the frequency domain (the Sinc filter is beyond the scope of our discussion). The MAX1402 can provide 16-bit accuracy at 480 samples per second and 12-bit accuracy at 4,800 samples per second. The chip finds wide application in sensors and instrumentation. The MAX1402 data sheet contains detailed figure information (Maxim Integrated Products, 2007).

12.4 APPLICATION EXAMPLE: CD PLAYER

Figure 12.37 illustrates a CD playback system, also described earlier in this chapter. A laser optically scans the tracks on a CD to produce a digital signal. The digital signal is then demodulated, and parity bits are used to detect bit errors due to manufacturing defects, dust, and so on, and to correct them. The demodulated signal is again oversampled by a factor of 4 and hence the sampling rate is increased to 176.4 kHz for each channel. Each digital sample then passes through a 14-bit DAC, which produces the sample-and-hold voltage signals that pass the anti-image lowpass filter. The output from each analog filter is fed to its corresponding loudspeaker. Over-sampling relaxes the design requirements of the analog anti-image lowpass filter, which is used to smooth out the voltage steps.

The earliest system used a third-order Bessel filter with a 3-dB gain attenuation at 30 kHz. Notice that first-order sigma delta modulation (first-order SDM) is added to the 14-bit DAC unit to further improve the 14-bit DAC to 16-bit DAC.

Let us examine the single-channel DSP portion as shown in Figure 12.38.

The spectral plots for the oversampled and interpolated signal $\bar{x}(n)$, the 14-bit SDM output $y(n)$, and the final analog output audio signal are given in Figure 12.39. As we can see in plot (a) in the figure, the quantization noise is uniformly distributed, and only in-band quantization noise (0 to 22.05 kHz) is expected. Again, 14 bits for each sample are kept after oversampling. Without using the first-order SDM, we expect the effective ADC resolution due to oversampling to be

$$n = 14 + 0.5 \times \log_2 \left(\frac{176.4}{44.1} \right) = 15 \text{ bits}$$

which is fewer than 16 bits. To improve quality further, the first-order SDM is used. The in-band quantization noise is then shaped. The first-order SDM pushes quantization noise to the high-frequency range, as illustrated in plot (b) in Figure 12.39. The effective ADC resolution now becomes

$$n = 14 + 1.5 \times \log_2 \left(\frac{176.4}{44.1} \right) - 0.86 \approx 16 \text{ bits}$$

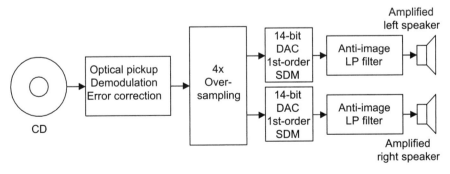

FIGURE 12.37

Simplified decoder of a CD recording system.

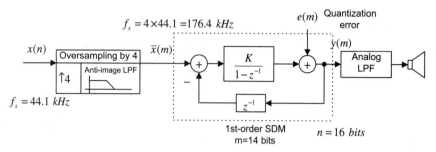

FIGURE 12.38

Illustration of oversampling and SDM ADC used in the decoder of a CD recording system.

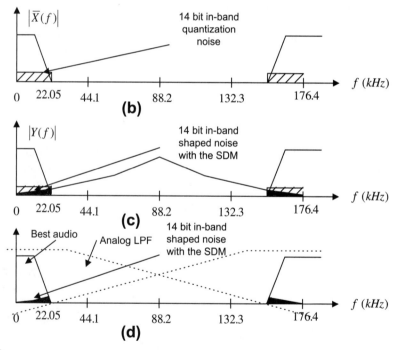

FIGURE 12.39

Spectral illustrations for oversampling and SDM ADC used in the decoder of a CD recording system.

Hence, 16-bit ADC audio quality is preserved. On the other hand, from plot (c) in Figure 12.39, the audio occupies a frequency range up to 22.05 kHz, while the DSP Nyquist limit is 88.2 kHz, so the low-order analog anti-image filter can satisfy the design requirement.

12.5 UNDERSAMPLING OF BANDPASS SIGNALS

As we discussed in Chapter 2, the sampling theorem requires that the sampling rate be twice as large as the highest frequency of the analog signal to be sampled. The sampling theorem ensures the complete reconstruction of the analog signal without aliasing distortion. In some applications, such as the modulated signals in communications systems, the signal exists in only a small portion of the band-width. Figure 12.40 shows an amplitude modulated (AM) signal in both the time domain and frequency domain. Assuming that the message signal has a bandwidth of 4 kHz and a carrier frequency of 96 kHz, the upper frequency edge of the AM signal is therefore 100 kHz ($fc+B$). Then the tradi-tional sampling process requires that the sampling rate be larger than 200 kHz $2(fc+B)$, resulting in at a high processing cost. Note that sampling the baseband signal of 4 kHz only requires a sampling rate of 8 kHz ($2B$).

If a certain condition is satisfied at the undersampling stage, we are able to make use of the aliasing signal to recover the message signal, since the aliasing signal contains the folded original message information (which we used to consider distortion). The reader is referred to the undersampling technique discussed in Ifeachor and Jervis (2002) and Porat (1997). Let the message to be recovered have a bandwidth of B, the theoretical minimum sampling rate be $f_s = 2B$, and the carrier frequency of the modulated signal be f_c. We discuss the following cases.

Case 1

If f_c = even integer $\times B$ and $f_c = 2B$, the sampled spectrum with all the replicas will be as shown in Figure 12.41(a).

As an illustrative example in the time domain for Case 1, suppose we have a bandpass signal with a carrier frequency of 20 Hz; that is,

$$x(t) = \cos(2\pi \times 20t)m(t) \tag{12.39}$$

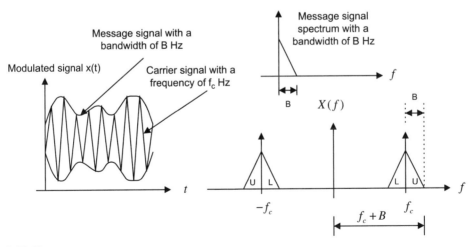

FIGURE 12.40

Message signal, modulated signal, and their spectra.

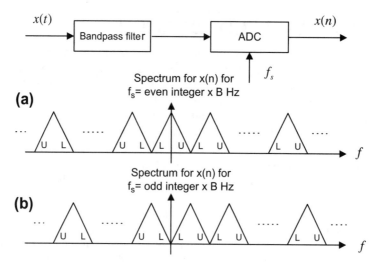

FIGURE 12.41

Spectrum of the undersampled signal.

where $m(t)$ is the message signal with a bandwidth of 2 Hz. Using a sampling rate of 4 Hz by substituting $t = nT$, where $T = 1/f_s$ into Equation (12.39), we get the sampled signal as

$$x(nT) = \cos(2\pi \times 20t)m(t)\big|_{t=nT} = \cos(2\pi \times 20n/4)m(nT) \tag{12.40}$$

Since $10n\pi = 5n(2\pi)$ is a multiple of 2π,

$$\cos(2\pi \times 20n/4) = \cos(10\pi n) = 1 \tag{12.41}$$

we obtain the undersampled signal as

$$x(nT) = \cos(2\pi \times 20n/4)m(nT) = m(nT) \tag{12.42}$$

which is a perfect digital message signal. Figure 12.42 shows the bandpass signal and its sampled signals when the message signal is 1 Hz, given as

$$m(t) = \cos(2\pi t) \tag{12.43}$$

Case 2

If $f_c = $ odd integer $\times B$ and $f_c = 2B$, the sampled spectrum with all the replicas will be as shown in Figure 12.41(b), where the spectral portions L and U are reversed. Hence, frequency reversal will occur. Then a further digital modulation in which the signal is multiplied by the digital oscillator with a frequency of B Hz can be used to adjust the spectrum to be the same as that in Case 1.

As another illustrative example for Case 2, let us sample the following the bandpass signal with a carrier frequency of 22 Hz, given by

$$x(t) = \cos(2\pi \times 22t)m(t) \tag{12.44}$$

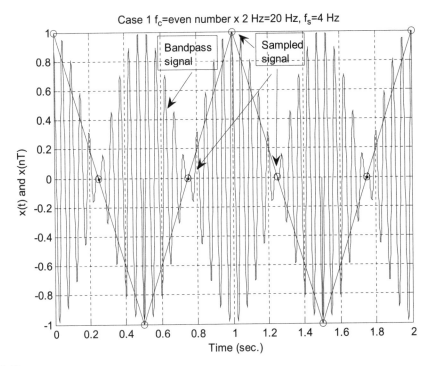

FIGURE 12.42

Plots of the bandpass signal and sampled signal for Case 1.

Applying undersampling using a sampling rate of 4 Hz, it follows that

$$x(nT) = \cos(2\pi \times 22n/4)m(nT) = \cos(11n\pi)m(nT) \qquad (12.45)$$

Since $11n\pi$ can be either an odd or an even integer multiple of π, we have

$$\cos(11\pi n) = \begin{cases} -1 & n = \text{odd} \\ 1 & n = \text{even} \end{cases} \qquad (12.46)$$

We see that Equation (12.46) causes the message samples to change sign alternatively with a carrier frequency of 22 Hz, which is the odd integer multiple of the message bandwidth of 2 Hz. This in fact will reverse the baseband message spectrum. To correct the spectrum reversal, we multiply an oscillator with a frequency of $B = 2$ Hz by the bandpass signal, that is

$$x(t)\cos(2\pi \times 2t) = \cos(2\pi \times 22t)m(t)\cos(2\pi \times 2t) \qquad (12.47)$$

Then the undersampled signal is given by

$$\begin{aligned} x(nT)\cos(2\pi \times 2n/4) &= \cos(2\pi \times 22n/4)m(nT)\cos(2\pi \times 2n/4) \\ &= \cos(11n\pi)m(nT)\cos(n\pi) \end{aligned} \qquad (12.48)$$

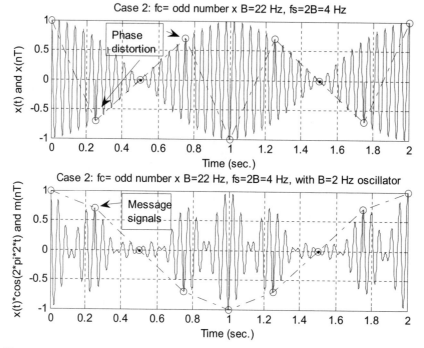

FIGURE 12.43

Plots of the bandpass signals and sampled signals for Case 2.

Since

$$\cos(11\pi n)\cos(\pi n) = 1 \qquad (12.49)$$

it follows that

$$x(nT)\cos(2\pi \times 2n/4) = \cos(\pi \times 11n)m(nT)\cos(\pi \times n) = m(nT) \qquad (12.50)$$

which is the recovered message signal. Figure 12.43 shows the sampled bandpass signals with the reversed message spectrum and the corrected message spectrum, respectively, for a message signal having a frequency of 0.5 Hz; that is,

$$m(t) = \cos(2\pi \times 0.5t) \qquad (12.51)$$

Case 3

If $f_c = $ noninterger $\times B$, we can extend the bandwidth B to $\overline{B}$ such that

$$f_c = \text{integer} \times \overline{B} \quad \text{and} \quad f_s = 2\overline{B} \qquad (12.52)$$

Then we can apply Case 1 or Case 2. An illustration of Case 3 is included in the following example.

EXAMPLE 12.11

Given a bandpass signal with the spectrum and carry frequency f_c shown in Figures 12.44A, 12.44B, and 12.44C, respectively, and assuming the baseband bandwidth $B = 4$ kHz, select the sampling rate and sketch the sampled spectrum ranging from 0 Hz to the carrier frequency for each of the following carrier frequencies:

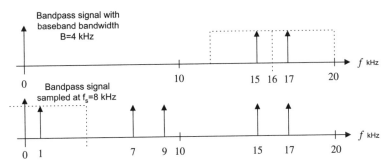

FIGURE 12.44A

Sampled signal spectrum for $f_c = 16$ kHz.

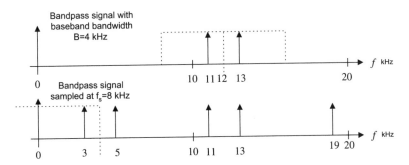

FIGURE 12.44B

Sampled signal spectrum for $f_c = 12$ kHz.

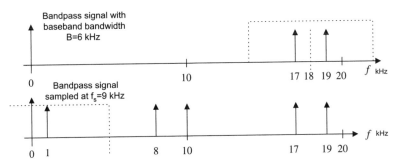

FIGURE 12.44C

Sampled signal spectrum for $f_c = 18$ kHz.

a. $f_c = 16$ kHz
b. $f_c = 12$ kHz
c. $f_c = 18$ kHz

Solution:

a. Since $f_c/B = 4$ is an even number, which is Case 1, we select $f_s = 8$ kHz and sketch the sampled spectrum shown in Figure 12.44A.
b. Since $f_c/B = 3$ is an odd number, we select $f_s = 8$ kHz and sketch the sampled spectrum shown in Figure 12.44B.
c. Now, $f_c/B = 4.5$, which is a noninteger. We extend the bandwidth $\overline{B} = 4.5$ kHz, so $f_c/\overline{B} = 4$ and $f_s = 2\overline{B} = 9$ kHz. The sketched spectrum is shown in Figure 12.44C.

Simulation Example

An AM with a 1-kHz message signal is given as

$$x(t) = [1 + 0.8 \times \sin(2\pi \times 1,000t)]\cos(2\pi \times f_c t) \qquad (12.53)$$

Assuming a message bandwidth of 4 kHz, determine the sampling rate, use MATLAB to sample the AM signal, and sketch the sampled spectrum up to the sampling frequency for each of the following carrier frequencies:

a. $f_c = 96$ kHz
b. $f_c = 100$ kHz
c. $f_c = 99$ kHz

a. For this case, $f_c/B = 24$ is an even number. We select $f_s = 8$ kHz. Figure 12.45A describes the simulation, where the upper left plot is the AM signal, the upper right plot is the spectrum of the AM signal, the lower left plot is the undersampled signal, and the lower right plot is the spectrum of the undersampled signal displayed from 0 to 8 kHz.
b. $f_c/B = 25$ is an odd number, so we choose $f_s = 8$ kHz, and a further process is needed. We can multiply the undersampled signal by a digital oscillator with a frequency of $B = 4$ kHz to achieve the 1-kHz baseband signal. The plots of the AM signal spectrum, undersampled signal spectrum, and the oscillator mixed signal and its spectrum are shown in Figure 12.45B.
c. For $f_c = 99$ kHz, $f_c/B = 24.75$. We extend the bandwidth to $\overline{B} = 4.125$ so that $f_c/\overline{B} = 24$. Hence, $f_s = 8.25$ kHz is used as the undersampling rate. Figure 12.45C shows the plots for the AM signal, the AM signal spectrum, the undersampled signal based on the extended baseband width, and the sampled signal spectrum ranging from 0 to 8.25 kHz, respectively.

This example verifies the principles of undersampling of bandpass signals.

12.6 SAMPLING RATE CONVERSION USING THE TMS320C6713 DSK

Downsampling by an integer factor of M using the TMS320C6713 is depicted in Figure 12.46. The idea is that we set up the DSK running at the original sampling rate and update the DAC channel once for M samples. The program (Tan and Jiang, 2008) is shown in Program 12.9.

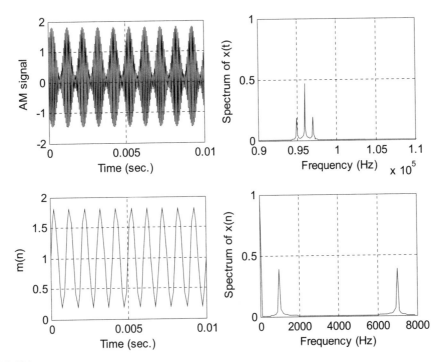

FIGURE 12.45A

Sampled AM signal and spectrum for $f_c = 96$ kHz.

Program 12.9. Downsampling implementation (anti-aliasing filter has 67 coefficients that are stored in an array b[67]).

```
int M=2;
float x[67];
float y[1]={0.0};
interrupt void c_int11()
{
  float lc; /*Left channel input */
  float rc; /*Right channel input */
  float lcnew; /*Left channel output */
  float rcnew; /*Right channel output */
  int i,j;
  float sum;
// Left channel and right channel inputs
  AIC23_data.combo=input_sample();
  lc=(float) (AIC23_data.channel[LEFT]);
  rc= (float) (AIC23_data.channel[RIGHT]);
// Insert DSP algorithm below
```

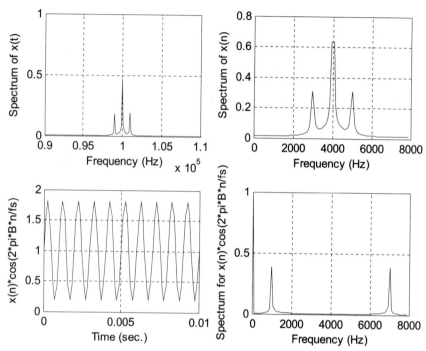

FIGURE 12.45B

Sampled AM signal and spectrum for $f_c = 100$ kHz.

```
for (j=0;j<M;j++)
{
 for(i=66; i>0; i--) // Update input buffer
 { x[i]=x[i-1]; }
 x[0]=lc; // Load new sample
 sum=0.0;
 for(i=0;i<67;i++)   // FIR filtering
 { sum=sum+x[i]*b[i]; }
 if (j== 0)
 { y[0]=sum; } // Update DAC with processed sample (decimation)
 }
// End of the DSP algorithm
 lcnew=y[0]; /* Send to DAC */
 rcnew=y[0];
 AIC23_data.channel[LEFT]=(short) lcnew;
 AIC23_data.channel[RIGHT]=(short) rcnew;
 output_sample(AIC23_data.combo);
}
```

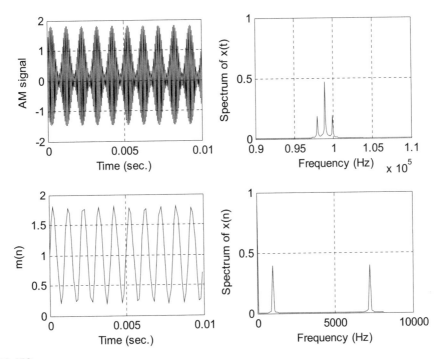

FIGURE 12.45C

Sampled AM signal and spectrum for $f_c = 99$ kHz.

Upsampling by an integer factor of L using the TMS320C6713 is depicted in Figure 12.47. Again, we set up the DSK to run at the upsampling rate f_{sL} and acquire a sample from the ADC channel once for L samples. The program (Tan and Jiang, 2008) is shown in Program 12.10.

Program 12.10. Upsampling implementation (anti-image filter has 67 coefficients that are stored in an array b[67]).

```
int L=2;
int Lcount=0;
float x[67];
float y[1]={0.0};
interrupt void c_int11()
{
  float lc; /*Left channel input */
  float rc; /*Right channel input */
  float lcnew; /*Left channel output */
  float rcnew; /*Right channel output */
  int i,j;
// Left channel and right channel inputs
  AIC23_data.combo=input_sample();
```

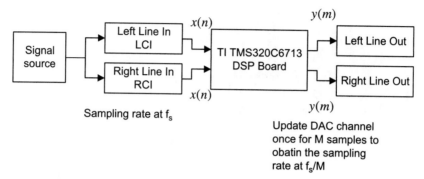

FIGURE 12.46

Downsampling using the TMS320C6713.

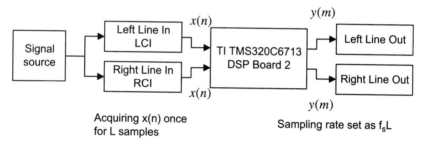

FIGURE 12.47

Upsampling using the TMS320C6713.

```
 lc=(float) (AIC23_data.channel[LEFT]);
 rc= (float) (AIC23_data.channel[RIGHT]);
// Insert DSP algorithm below
 Lcount++;
 for(i=66; i>0; i--) // Update input buffer with zeros
 { x[i]=x[i-1]; }
 x[0]=0;
 if (Lcount==L)
 {
  x[0]=lc; // Load new sample for every L samples
  Lcount =0;
 }
 y[0]=0.0;
 for(i=0;i<67;i++)   // FIR filtering
 { y[0]=y[0]+x[i]*b[i]; }
 y[0]=(float) L*y[0]
```

```
// End of the DSP algorithm
  lcnew=y[0]; /* Send to DAC */
  rcnew=y[0];
  AIC23_data.channel[LEFT]=(short) lcnew;
  AIC23_data.channel[RIGHT]=(short) rcnew;
  output_sample(AIC23_data.combo);
}
```

12.7 SUMMARY

1. Downsampling (decimation) by an integer factor of M means taking one sample from the data sequence $x(n)$ for every M samples and discarding the last $M - 1$ samples.
2. Upsampling (interpolation) by an integer factor of L means inserting $L - 1$ zeros for every sample in the data sequence $x(n)$.
3. Downsampling requires a decimation (anti-aliasing) filter to avoid frequency aliasing before downsampling.
4. Upsampling requires an interpolation (anti-image) filter to remove the images after interpolation.
5. Changing the sampling rate by a noninteger factor of L/M requires two stages: an interpolation stage and a downsampling stage.
6. Two-stage decimation can dramatically reduce the anti-aliasing filter length.
7. Polyphase implementations of the decimation filter and interpolation filter can reduce the complexity of the filter operations, that is, fewer multiplications and additions.
8. Using oversampling can improve the regular ADC resolution. Sigma-delta modulation ADC can achieve even higher ADC resolution, using the noise shaping effect for further reduction of quantization noise.
9. The audio CD player uses multirate signal processing and oversampling.
10. Undersampling can be used to sample the bandpass signal, leading to applications in communications.

12.8 PROBLEMS

12.1. Consider a single-stage decimator with the following specifications:
Original sampling rate = 1 kHz
Decimation factor $M = 2$
Frequency of interest = 0–100 Hz
Passband ripple = 0.015 dB
Stopband attenuation = 40 dB

   a. Draw the block diagram for the decimator.

   b. Determine the window type, filter length, and cutoff frequency if the window method is used for the anti-aliasing FIR filter design.

12.2. Consider a single-stage interpolator with the following specifications:
Original sampling rate = 1 kHz

Interpolation factor $L = 2$
Frequency of interest $= 0-150$ Hz
Passband ripple $= 0.02$ dB
Stopband attenuation $= 45$ dB

a. Draw the block diagram for the interpolator.

b. Determine the window type, filter length, and cutoff frequency if the window method is used for the anti-image FIR filter design.

12.3. Consider a single stage decimator with the following specifications:
Original sampling rate $= 8$ kHz
Decimation factor $M = 4$
Frequency of interest $= 0-800$ Hz
Passband ripple $= 0.02$ dB
Stopband attenuation $= 46$ dB,

a. Draw the block diagram for the decimator.

b. Determine the window type, filter length, and cutoff frequency if the window method is used for the anti-aliasing FIR filter design.

12.4. Consider a single-stage interpolator with the following specifications:
Original sampling rate $= 8$ kHz
Interpolation factor $L = 3$
Frequency of interest $= 0-3,400$ Hz
Passband ripple $= 0.02$ dB
Stopband attenuation $= 46$ dB

a. Draw the block diagram for the interpolator.

b. Determine the window type, filter length, and cutoff frequency if the window method is used for the anti-image FIR filter design.

12.5. Consider the sampling conversion from 4 kHz to 3 kHz with the following specifications:
Original sampling rate $= 4$ kHz
Interpolation factor $L = 3$
Decimation factor $M = 2$
Frequency of interest $= 0-400$ Hz
Passband ripple $= 0.02$ dB
Stopband attenuation $= 46$ dB

a. Draw the block diagram for the interpolator.

b. Determine the window type, filter length, and cutoff frequency if the window method is used for the combined FIR filter $H(z)$.

12.6. Consider the design of a two-stage decimator with the following specifications:
Original sampling rate $= 32$ kHz
Frequency of interest $= 0-250$ Hz
Passband ripple $= 0.05$ (absolute)

Stopband attenuation = 0.005 (absolute)

Final sampling rate = 1,000 Hz

a. Draw the decimation block diagram.

b. Specify the sampling rate for each stage.

c. Determine the window type, filter length, and cutoff frequency for the first stage if the window method is used for anti-aliasing FIR filter design ($H_1(z)$).

d. Determine the window type, filter length, and cutoff frequency for the second stage if the window method is used for the anti-aliasing FIR filter design ($H_2(z)$).

12.7. Consider the sampling conversion from 6 kHz to 8 kHz with the following specifications:

Original sampling rate = 6 kHz

Interpolation factor $L = 4$

Decimation factor $M = 3$

Frequency of interest = 0–2,400 Hz

Passband ripple = 0.02 dB

Stopband attenuation = 46 dB

a. Draw the block diagram for the processor.

b. Determine the window type, filter length, and cutoff frequency if the window method is used for the combined FIR filter $H(z)$.

12.8. Consider the design of a two-stage decimator with the following specifications:

Original sampling rate = 320 kHz

Frequency of interest = 0–3,400 Hz

Passband ripple = 0.05 (absolute)

Stopband attenuation = 0.005 (absolute)

Final sampling rate = 8,000 Hz

a. Draw the decimation block diagram.

b. Specify the sampling rate for each stage.

c. Determine the window type, filter length, and cutoff frequency for the first stage if the window method is used for anti-aliasing FIR filter design ($H_1(z)$).

d. Determine the window type, filter length, and cutoff frequency for the second stage if the window method is used for anti-aliasing FIR filter design ($H_2(z)$).

12.9. a. Given an interpolator filter

$$H(z) = 0.25 + 0.4z^{-1} + 0.5z^{-2}$$

draw the block diagram for interpolation polyphase filter implementation for the case of $L = 2$.

b. Given a decimation filter

$$H(z) = 0.25 + 0.4z^{-1} + 0.5z^{-2} + 0.6z^{-3}$$

draw the block diagram for decimation polyphase filter implementation for the case of $M = 2$.

12.10. Using the commutative models for the polyphase interpolation and decimation filters,

    a. draw the block diagram for interpolation polyphase filter implementation for the case of $L = 2$, and $H(z) = 0.25 + 0.4z^{-1} + 0.5z^{-2}$;

    b. draw the block diagram for decimation polyphase filter implementation for the case of $M = 2$, and $H(z) = 0.25 + 0.4z^{-1} + 0.5z^{-2} + 0.6z^{-3}$.

12.11. **a.** Given an interpolator filter

$$H(z) = 0.25 + 0.4z^{-1} + 0.5z^{-2} + 0.6z^{-3} + 0.7z^{-4} + 0.6z^{-5},$$

    draw the block diagram for interpolation polyphase filter implementation for the case of $L = 4$.

    b. Given a decimation filter

$$H(z) = 0.25 + 0.4z^{-1} + 0.5z^{-2} + 0.6z^{-3} + 0.5z^{-3} + 0.4z^{-4}$$

    draw the block diagram for decimation polyphase filter implementation for the case of $M = 4$.

12.12. Using the commutative models for the polyphase interpolation and decimation filters,

    a. draw the block diagram for interpolation polyphase filter implementation for the case of $L = 4$, and $H(z) = 0.25 + 0.4z^{-1} + 0.5z^{-2} + 0.6z^{-3} + 0.7z^{-4} + 0.6z^{-5}$;

    b. draw the block diagram for decimation polyphase filter implementation for the case of $M = 4$, and $H(z) = 0.25 + 0.4z^{-1} + 0.5z^{-2} + 0.6z^{-3} + 0.5z^{-3} + 0.4z^{-4}$.

12.13. Consider a speech system with the following specifications:
Speech input frequency range: 0–4 kHz
ADC resolution $= 16$ bits
Current sampling rate $= 8$ kHz

    a. Determine the oversampling rate if a 12-bit ADC chip is used to replace the speech system.

    b. Draw the block diagram.

12.14. Consider a speech system with the following specifications:
Speech input frequency range: 0–4 kHz
ADC resolution $= 6$ bits
Oversampling rate $= 4$ MHz

    a. Draw the block diagram.

    b. Determine the actual effective ADC resolution (number of bits per sample).

12.15. Consider an audio system with the following specifications:
Audio input frequency range: 0–15 kHz
ADC resolution $= 16$ bits
Current sampling rate $= 30$ kHz

a. Determine the oversampling rate if a 12-bit ADC chip is used to replace the audio system.

b. Draw the block diagram.

12.16. Consider an audio system with the following specifications:
Audio input frequency range: 0–15 kHz
ADC resolution = 6 bits
Oversampling rate = 45 MHz

a. Draw the block diagram.

b. Determine the actual effective ADC resolution (number of bits per sample).

12.17. Consider the following specifications of an oversampling DSP system:
Audio input frequency range: 0–4 kHz
First-order SDM with a sampling rate of 128 kHz
ADC resolution in SDM = 1 bit

a. Draw the block diagram using the DSP model.

b. Determine the equivalent (effective) ADC resolution.

12.18. Consider the following specifications of an oversampling DSP system:
Audio input frequency range: 0–20 kHz
Second-order SDM with a sampling rate of 160 kHz
ADC resolution in SDM = 10 bits

a. Draw the block diagram using the DSP model.

b. Determine the equivalent (effective) ADC resolution.

12.19. Consider the following specifications of an oversampling DSP system:
Signal input frequency range: 0–500 Hz
First-order SDM with a sampling rate of 128 kHz
ADC resolution in SDM = 1 bit

a. Draw the block diagram using the DSP model.

b. Determine the equivalent (effective) ADC resolution.

12.20. Consider the following specifications of an oversampling DSP system:
Signal input frequency range: 0–500 Hz
Second-order SDM with a sampling rate of 16 kHz
ADC resolution in SDM = 8 bits

a. Draw the block diagram using the DSP model.

b. Determine the equivalent (effective) ADC resolution.

12.21. Given a bandpass signal with a spectrum shown in Figure 12.48, and assuming the bandwidth $B = 5$ kHz, select the sampling rate and sketch the sampled spectrum ranging from 0 Hz to the carrier frequency for each of the following carrier frequencies:

a. $f_c = 30$ kHz

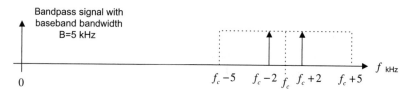

FIGURE 12.48

Spectrum of the bandpass signal in Problem 12.21.

b. $f_c = 25$ kHz

c. $f_c = 33$ kHz

12.22. Given a bandpass signal with a spectrum shown in Figure 12.48, and assuming $f_s = 10$ kHz, select the sampling rate and sketch the sampled spectrum ranging from 0 Hz to the carrier frequency for each of the following carrier frequencies:

a. $f_c = 15$ kHz

b. $f_c = 20$ kHz

12.23. Given a bandpass signal with a spectrum shown in Figure 12.48, and assuming $B = 5$ kHz, select the sampling rate and sketch the sampled spectrum ranging from 0 Hz to the carrier frequency for each of the following carrier frequencies:

a. $f_c = 35$ kHz

b. $f_c = 40$ kHz

c. $f_c = 22$ kHz

12.8.1 MATLAB Problems

Use MATLAB to solve Problems 12.24 to 12.30.

12.24. Generate a sinusoid with a frequency of 1,000 Hz for 0.05 second using a sampling rate of 8 kHz.

a. Design a decimator to change the sampling rate to 4 kHz with the specifications below:
Signal frequency range: 0–1,800 Hz
Hamming window required for FIR filter design

b. Write a MATLAB program to implement the downsampling scheme, and plot the original signal and the downsampled signal versus the sample number, respectively.

12.25. Generate a sinusoid with a frequency of 1,000 Hz for 0.05 second using a sampling rate of 8 kHz.

a. Design an interpolator to change the sampling rate to 16 kHz with the following specifications:
Signal frequency range: 0–3,600 Hz
Hamming window required for FIR filter design

b. Write a MATLAB program to implement the upsampling scheme, and plot the original signal and the upsampled signal versus the sample number, respectively.

12.26. Generate a sinusoid with a frequency of 500 Hz for 0.1 second using a sampling rate of 8 kHz.

a. Design an interpolation and decimation processing algorithm with the following specifications to change the sampling rate to 22 kHz:
Signal frequency range: 0–3,400 Hz
Hamming window required for FIR filter design

b. Write a MATLAB program to implement the scheme, and plot the original signal and the sampled signal at the rate of 22 kHz versus the sample number, respectively.

12.27. Repeat Problem 12.24 using the polyphase form for the decimator.

12.28. Repeat Problem 12.25 using the polyphase form for the interpolator.

12.29. a. Use MATLAB to create a 1-second sinusoidal signal using a sampling rate of 1 kHz,

$$x(t) = 1.8 \cos(2\pi \times 100t) + 1.0 \sin(2\pi \times 150t + \pi/4)$$

where each sample $x(t)$ can be rounded off using a 3-bit signed integer (directly round off the calculated $x(t)$). Evaluate the signal-to-quantization-noise ratio (SQNR).

b. Use MATLAB to design an oversampling system including an anti-aliasing filter with a selectable integer factor L using the same equation for the input $x(t)$.

c. Recover the signal using the quantized 3-bit signal and measure the SQNRs for the following integer factors: $L = 2, L = 4, L = 8, L = 16$, and $L = 32$. From the results, explain which one offers better quality for the recovered signals.

12.30. a. Use MATLAB to create a 1-second sinusoidal signal using a sampling rate of 1 kHz,

$$x(t) = 1.8 \cos(2\pi \times 100t) + 1.0 \sin(2\pi \times 150t + \pi/4)$$

where each sample $x(t)$ can be rounded off using a 3-bit signed integer (directly round off the calculated $x(t)$). Evaluate the signal-to-quantization-noise ratio (SQNR).

b. Use MATLAB to implement the first-order SDM system including an anti-aliasing filter with an oversampling factor of 16. Measure the SQNR (signal to noise ratio due to quantization).

c. Use MATLAB to implement the second-order SDM system including an anti-aliasing filter with an oversampling factor of 16. Measure the SQNR. Compare the SQNR with the one obtained in (b).

12.8.2 MATLAB Project

12.31. Audio rate conversion system:
Given a 16-bit stereo audio file ("No9seg.wav") with a sampling rate of 44.1 kHz, design a multistage conversion system and implement the designed system to convert the audio file from 44.1 to 48 kHz. Listen and compare the quality of the original audio with the converted audio.

Subband- and Wavelet-Based Coding

CHAPTER OUTLINE

13.1 Subband Coding Basics 621
13.2 Subband Decomposition and Two-Channel Perfect Reconstruction Quadrature Mirror Filter Bank..... 626
13.3 Subband Coding of Signals 635
13.4 Wavelet Basics and Families of Wavelets 638
13.5 Multiresolution Equations 650
13.6 Discrete Wavelet Transform 655
13.7 Wavelet Transform Coding of Signals.. 664
13.8 MATLAB Programs 668
13.9 Summary 672

OBJECTIVES

This chapter is a continuation of Chapter 12 and further studies basic principles of multirate digital signal processing, specifically for subband and wavelet transform coding. First, the chapter explains digital filter bank theory and develops subband coding techniques for compressing various signals, including speech and seismic data. Then the chapter focuses on wavelet basics with applications of waveform coding and signal denoising.

13.1 SUBBAND CODING BASICS

In many applications such as speech and audio analysis, synthesis, and compression, digital filter banks are often used. The filter bank system consists of two stages. The first stage, called the analysis stage, is in the form of filter bank decomposition, in which the signal is filtered into subbands along with a sampling rate decimation; the second stage interpolates the decimated subband signals to reconstruct the original signal. For the purpose of data compression, spectral information from each subband channel can be used to quantize the subband signal efficiently to achieve efficient coding.

Figure 13.1 illustrates the basic framework for a four-channel filter bank analyzer and synthesizer. At the analysis stage, the input signal $x(n)$ at the original sampling rate f_s is divided via the analysis filter bank into four channels, $x_0(m)$, $x_1(m)$, $x_2(m)$, and $x_3(m)$, each at the decimated sampling rate f_s/M, where $M = 4$. For the synthesizer, these four decimated signals are interpolated via a synthesis filter bank. The outputs from all four channels ($\bar{x}_0(n)$, $\bar{x}_1(n)$, $\bar{x}_2(n)$, and $\bar{x}_3(n)$) of the

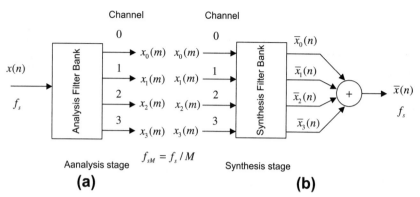

Channel   Channel
0   0

$x(n)$ Analysis Filter Bank
f_s

1
$x_0(m)$ $x_0(m)$ ⟶
$x_1(m)$ $x_1(m)$ ⟶
2
$x_2(m)$ $x_2(m)$ ⟶
3
$x_3(m)$ $x_3(m)$ ⟶

Synthesis Filter Bank

$\bar{x}_0(n)$
$\bar{x}_1(n)$
$\bar{x}_2(n)$
$\bar{x}_3(n)$

$+$ ⟶ $\bar{x}(n)$
f_s

$f_{sM} = f_s / M$

Aanalysis stage
(a)

Synthesis stage
(b)

FIGURE 13.1

Filter bank framework with an analyzer and synthesizer.

synthesis filter bank are then combined to reconstruct the original signal $\bar{x}(n)$ at the original sampling rate f_s. Each channel essentially generates a bandpass signal. The decimated signal spectrum for channel 0 can be achieved via a standard downsampling process, while the decimated spectra of other channels can be obtained using the principle of undersampling of bandpass signals with an integer band (discussed in Section 12.5), where the inherent frequency aliasing or image properties of decimation and interpolation are involved. The theoretical development will follow next. With a proper design of analysis and synthesis filter banks, we are able to achieve perfect reconstruction of the original signal.

Let us examine the spectral details of each band (subband). Figure 13.2 depicts the spectral information of the analysis and synthesis stages, as shown in Figure 13.2(a) and (b). $H_0(z)$ and $G_0(z)$ are the analysis and synthesis filters of channel 0, respectively. At the analyzer (Figure 13.2(c) to (e)), $x(n)$ is bandlimited by a lowpass filter $H_0(z)$ to get $w_0(n)$ and decimated by $M = 4$ to obtain $x_0(m)$. At the synthesizer (Figure 13.2(f) to (h)), $x_0(m)$ is upsampled by a factor of 4 to obtain $\bar{w}_0(n)$ and then goes through the anti-aliasing (synthesis) filter $G_0(z)$ to achieve the lowpass signal $\bar{x}_0(n)$.

Figure 13.3 depicts the analysis and synthesis stages for channel 1 (see Figure 13.3(a) and (b)). $H_1(z)$ and $G_1(z)$ are the bandpass analysis and synthesis filters, respectively. Similarly, at the analyzer (Figure 13.3(c) to (e)), $x(n)$ is filtered by a bandpass filter $H_1(z)$ to get $w_1(n)$ and decimated by $M = 4$ to obtain $x_1(m)$. Since the lower frequency edge of $W_1(z)$ is $f_c/B = 1 =$ odd number, where $f_c = f_s/(2M) = B$, f_c corresponds to the carrier frequency, and B is the baseband bandwidth as depicted in Section 12.5, the reversed spectrum in the baseband results in Figure 13.3(e). However, this is not a problem, since at the synthesizer as shown in Figures 13.3(f) and (g), the spectral reversal occurs again so that $\bar{W}_1(z)$ will have the same spectral components as $W_1(z)$ at the analyzer. After $\bar{w}_1(n)$ goes through the anti-aliasing (synthesis) filter $G_1(z)$, we achieve the reconstructed bandpass signal $\bar{x}_1(n)$.

Figure 13.4 describes the analysis and synthesis stages for channel 2. At the analyzer (Figure 13.4(c) to (e)), $x(n)$ is filtered by a bandpass filter $H_2(z)$ to get $w_2(n)$ and decimated by $M = 4$ to obtain $x_2(m)$. Similarly, considering the lower frequency edge of $W_2(z)$ as $f_c = 2(f_s/(2M)) = 2B$, $f_c/B = 2 =$ even. Therefore, we obtain the nonreversed spectrum in the baseband as shown in Figure 13.4(f). At the synthesizer shown in Figure 13.4(g), the spectrum $\bar{W}_2(z)$ has the same spectral

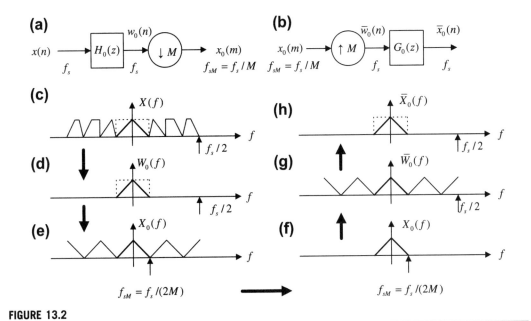

FIGURE 13.2

Analysis and synthesis stages for channel 0.

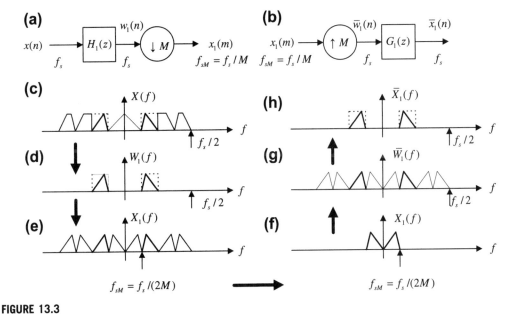

FIGURE 13.3

Analysis and synthesis stages for channel 1.

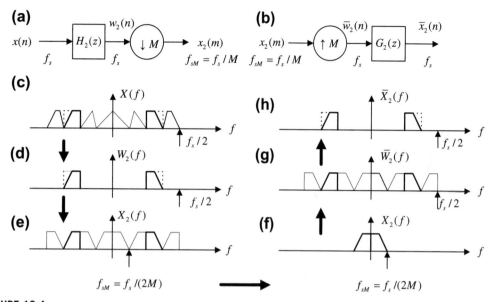

FIGURE 13.4

Analysis and synthesis stages for channel 2.

components as $W_2(z)$ at the analyzer. After $\overline{w}_2(n)$ is filtered by the synthesis bandpass filter, $G_2(z)$, we get the reconstructed bandpass signal $\overline{x}_2(n)$.

The process in channel 3 is similar to that in channel 1 with the spectral reversal effect and is illustrated in Figure 13.5.

Now let us examine the theory. Without quantization of subband channels, perfect reconstruction of the filter banks (see Figure 13.1) depends on the analysis and syntheses filter effects. To develop the perfect reconstruction required of the analysis and synthesis filters, consider a signal in a single channel flowing up to the synthesis filter in general as depicted in Figure 13.6.

As shown in Figure 13.6, $w(n)$ is the output signal from the analysis filter H(z) at the original sampling rate, that is,

$$W(z) = H(z)X(z) \tag{13.1}$$

$x_d(m)$ is the downsampled version of $w(n)$ while $\overline{w}(n)$ is the interpolated version of $w(n)$ prior to the synthesis filter and can be expressed as

$$\overline{w}(n) = \begin{cases} w(n) & n = 0, M, 2M, \dots \\ 0 & \text{otherwise} \end{cases} \tag{13.2}$$

Using a delta function $\delta(n)$, that is, $\delta(0) = 1$ for $n = 0$ and $\delta(n) = 0$ for $n \neq 0$, we can write $\overline{w}(n)$ as

$$\overline{w}(n) = \left[\sum_{k=0}^{\infty} \delta(n - kM) \right] w(n) = i(n)w(n) \tag{13.3}$$

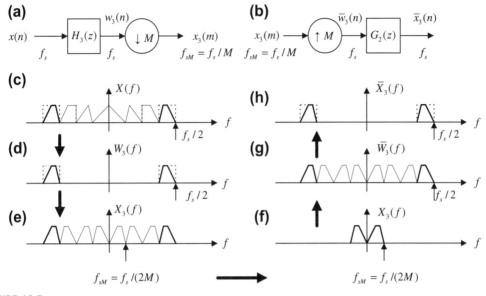

FIGURE 13.5

Analysis and synthesis stages for channel 3.

where $i(n)$ is defined as $i(n) = \sum_{k=0}^{\infty} \delta(n - kM) = \delta(n) + \delta(n - M) + \delta(n - 2M) + \cdots$

Clearly, $i(n)$ is a periodic function (impulse train with a period of M samples) as shown in Figure 13.7 where $M = 4$.

We can determine the discrete Fourier transform of the impulse train with a period of M samples as

$$I(k) = \sum_{n=0}^{M-1} i(n)e^{-j\frac{2\pi kn}{M}} = \sum_{n=0}^{M-1} \delta(n)e^{-j\frac{2\pi kn}{M}} = 1 \qquad (13.4)$$

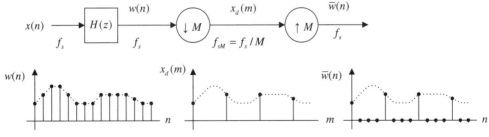

FIGURE 13.6

Signal flow in one channel.

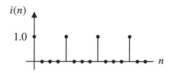

FIGURE 13.7

Impulse train with a period 4 samples.

Hence, using the inverse of discrete Fourier transform, $i(n)$ can be expressed as

$$i(n) = \frac{1}{M} \sum_{k=0}^{M-1} I(k) e^{j\frac{2\pi kn}{M}} = \frac{1}{M} \sum_{k=0}^{M-1} e^{j\frac{2\pi kn}{M}} \tag{13.5}$$

Substituting Equation (13.5) into Equation (13.3) leads to

$$\overline{w}(n) = \frac{1}{M} \sum_{k=0}^{M-1} w(n) e^{j\frac{2\pi kn}{M}} \tag{13.6}$$

Applying the z-transform in Equation (13.6), we achieve the fundamental relationship between $W(z)$ and $\overline{W}(z)$:

$$\overline{W}(z) = \frac{1}{M} \sum_{k=0}^{M-1} \sum_{n=0}^{\infty} w(n) e^{j\frac{2\pi kn}{M}} z^{-n} = \frac{1}{M} \sum_{k=0}^{M-1} \sum_{n=0}^{\infty} w(n) \left(e^{-j\frac{2\pi k}{M}} z \right)^{-n}$$

$$= \frac{1}{M} \sum_{k=0}^{M-1} W \left(e^{-j\frac{2\pi k}{M}} z \right) \tag{13.7}$$

$$= \frac{1}{M} \left[W \left(e^{-j\frac{2\pi \times 0}{M}} z \right) + W \left(e^{-j\frac{2\pi \times 1}{M}} z \right) + \cdots + W \left(e^{-j\frac{2\pi \times (M-1)}{M}} z \right) \right]$$

Equation (13.7) indicates that the signal spectrum $\overline{W}(z)$ before the synthesis filter is an average of the various modulated spectrum $W(z)$. Notice that both $\overline{W}(z)$ and $W(z)$ are at the original sampling rate f_s. We will use this result for further development in the next section.

13.2 SUBBAND DECOMPOSITION AND TWO-CHANNEL PERFECT RECONSTRUCTION QUADRATURE MIRROR FILTER BANK

To explore Equation (13.7), let us begin with a two-band case as illustrated in Figure 13.8.
Substituting $M = 2$ in Equation (13.7), it follows that

$$\overline{W}(z) = \frac{1}{2} \sum_{k=0}^{1} W(e^{-j\frac{2\pi k}{2}} z) = \frac{1}{2} [W(z) + W(-z)] \tag{13.8}$$

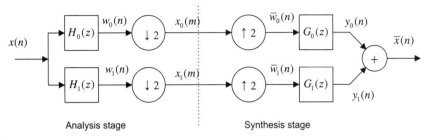

FIGURE 13.8

Two-band filter bank system.

Applying for each band in Figure 13.8 by substituting Equation (13.1) in Equation (13.8), we have

$$Y_0(z) = \frac{1}{2}G_0(z)(H_0(z)X(z) + H_0(-z)X(-z)) \tag{13.9}$$

$$Y_1(z) = \frac{1}{2}G_1(z)(H_1(z)X(z) + H_1(-z)X(-z)) \tag{13.10}$$

Since the synthesized signal $\overline{X}(z)$ is the sum of $Y_0(z)$ and $Y_1(z)$, it can be expressed as

$$\overline{X}(z) = \frac{1}{2}(G_0(z)H_0(z) + G_1(z)H_1(z))X(z)$$

$$+ \frac{1}{2}(G_0(z)H_0(-z) + G_1(z)H_1(-z))X(-z) \tag{13.11}$$

$$= A(z)X(z) + S(z)X(-z)$$

For perfect reconstruction, the recovered signal $\overline{x}(n)$ should be a scaled and delayed version of the original signal $x(n)$, that is, $\overline{x}(n) = cx(n - n_0)$. Hence, to achieve a perfect reconstruction, it is required that

$$S(z) = \frac{1}{2}(G_0(z)H_0(-z) + G_1(z)H_1(-z)) = 0 \tag{13.12}$$

$$A(z) = \frac{1}{2}(G_0(z)H_0(z) + G_1(z)H_1(z)) = cz^{-n_0} \tag{13.13}$$

where c is the constant while n_0 is the delay introduced by the analysis and synthesis filters.

Forcing $S(z) = 0$ leads to the following relationship:

$$\frac{G_0(z)}{G_1(z)} = -\frac{H_1(-z)}{H_0(-z)} \tag{13.14}$$

It follows that

$$G_0(z) = -H_1(-z) \tag{13.15}$$

$$G_1(z) = H_0(-z) \tag{13.16}$$

Substituting $G_0(z)$ and $G_1(z)$ in Equation (13.13) gives

$$A(z) = \frac{1}{2}(H_0(-z)H_1(z) - H_0(z)H_1(-z)) \tag{13.17}$$

Assume $H_0(z)$ and $H_1(z)$ are N-tap FIR filters, where N is even, and let

$$H_1(z) = z^{-(N-1)}H_0(-z^{-1}) \tag{13.18}$$

Notice that

$$H_1(-z) = -z^{-(N-1)}H_0(z^{-1}) \tag{13.19}$$

Substituting Equations (13.18) and (13.19), we can simplify Equation (13.17) as

$$A(z) = \frac{1}{2}z^{-(N-1)}(H_0(z)H_0(z^{-1}) + H_0(-z)H_0(-z^{-1})) \tag{13.20}$$

Finally, for perfect reconstruction, we require that

$$H_0(z)H_0(z^{-1}) + H_0(-z)H_0(-z^{-1}) = R(z) + R(-z) = \text{constant} \tag{13.21}$$

where

$$R(z) = H_0(z)H_0(z^{-1}) = a_{N-1}z^{N-1} + a_{N-2}z^{N-2} + \cdots + a_0z^0 + \cdots + a_{N-1}z^{-(N-1)} \tag{13.22}$$

$$R(-z) = H_0(-z)H_0(-z^{-1}) = -a_{N-1}z^{N-1} + a_{N-2}z^{N-2} + \cdots + a_0z^0 + \cdots - a_{N-1}z^{-(N-1)} \tag{13.23}$$

It is important to note that the sum of $R(z) + R(-z)$ only consists of even order of powers of z, since the terms with odd powers of z cancel each other. Using algebraic simplification, we conclude that the coefficients of $R(z) = H(z)H(z^{-1})$ are essentially samples of the autocorrelation function given by

$$\rho(n) = \sum_{k=0}^{N-1} h_0(k)h_0(k+n) = \rho(-n) = h_0(n) \odot h_0(n) \tag{13.24}$$

where $\odot$ denotes the correlation operation. Hence, we require $\rho(n) = 0$ for $n = $ even and $n \neq 0$, that is,

$$\rho(2n) = \sum_{k=0}^{N-1} h_0(k)h_0(k+2n) = 0 \tag{13.25}$$

For the normalization for $n = 0$, we require

$$\sum_{k=0}^{N-1}|h_0(k)|^2 = 0.5 \tag{13.26}$$

We then obtain the filter design constraint as

$$\rho(2n) = \sum_{k=0}^{N-1} h_0(k)h_0(k+2n) = \delta(n) \tag{13.27}$$

For a two-band filter bank, $h_0(k)$ and $h_1(k)$ are designed as lowpass and highpass filters, respectively, which are essentially the quadrature mirror filters. Their expected frequency responses must satisfy Equation (13.28) and are shown in Figure 13.9:

$$\left|H_0(e^{j\Omega})\right|^2 + \left|H_1(e^{j\Omega})\right|^2 = 1 \tag{13.28}$$

Equation (13.28) implies that

$$R(z) + R(-z) = 1 \tag{13.29}$$

To verify Equation (13.29), we use

$$\left|H(e^{j\Omega})\right|^2 = H(e^{j\Omega})H(e^{-j\Omega}) = \left.H(z)H(z^{-1})\right|_{z=e^{j\Omega}}$$

Equation (13.28) becomes

$$\left.H_0(z)H_0(z^{-1}) + H_1(z)H_1(z^{-1})\right|_{z=e^{j\Omega}} = 1$$

which is equivalent to

$$H_0(z)H_0(z^{-1}) + H_1(z)H_1(z^{-1}) = 1$$

From Equation (13.18), we can verify that

$$H_1(z)H_1(z^{-1}) = H_0(-z)H_0(-z^{-1})$$

Finally, we see that

$$\begin{aligned}
&H_0(z)H_0(z^{-1}) + H_1(z)H_1(z^{-1}) \\
&= H_0(z)H_0(z^{-1}) + H_0(-z)H_0(-z^{-1}) = R(z) + R(-z) = 1
\end{aligned}$$

Once the lowpass analysis filter $H_0(z)$ is designed, the highpass filter can be obtained using the developed relationship in Equation (13.18). The key equations are summarized as follows:

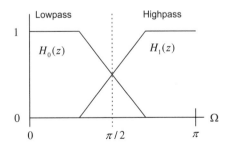

FIGURE 13.9

Frequency responses for quadrature mirror filters.

Filter design constraint equations for the lowpass filter $H_0(z)$:

$$R(z) = H_0(z)H_0(z^{-1})$$
$$R(z) + R(-z) = 1$$
$$\rho(2n) = 0.5\delta(n)$$

Equations for the other filters:

$$H_1(z) = z^{-(N-1)}H_0(-z^{-1})$$
$$G_0(z) = -H_1(-z)$$
$$G_1(z) = H_0(-z)$$

Design of the analysis and synthesis filters to satisfy the above conditions is very challenging. Smith and Barnwell (1984) were the first to show that perfect reconstruction in a two-band filter bank is possible when the linear phase of the FIR filter requirement is relaxed. The Smith–Barnwell filters are called the conjugate quadrature filters (PR-CQF). Eight- and 16-tap PR-CQF coefficients are listed in Table 13.1. As shown in Table 13.1, the filter coefficients are not symmetric; hence, the obtained analysis filter does not have a linear phase. The detailed design of Smith–Barnwell filters can be found in their research paper (Smith and Barnwell, 1984) and the design of other types of analysis and synthesis filters can be found in Akansu and Haddad (1992).

Now let us verify the filter constraint in the following example.

Table 13.1 Smith–Barnwell PR-CQF Filters

8 Taps	16 Taps
0.0348975582178515	0.02193598203004352
−0.01098301946252854	0.0015786164976663704
−0.06286453934951963	−0.06025449102875281
0.223907720892568	−0.0118906596205391
0.556856993531445	0.137537915636625
0.357976304997285	0.05745450056390939
−0.02390027056113145	−0.321670296165893
−0.07594096379188282	−0.528720271545339
	−0.295779674500919
	0.0002043110845170894
	0.0290669978946796
	−0.03533486088708146
	−0.006821045322743358
	0.02606678468264118
	0.001033363491944126
	−0.01435930957477529

EXAMPLE 13.1

Use the 8-tap PR-CQF coefficients (Table 13.1) and MATLAB to verify the following conditions:

$$p(2n) = \sum_{k=0}^{N-1} h_0(k)h_0(k+2n) = 0.5\delta(n)$$

$$R(z) + R(-z) = 1$$

Also, plot the magnitude frequency responses of the analysis and synthesis filters.

Solution:
Since $p(n) = \sum_{k=0}^{N-1} h_0(k)h_0(k+n)$, we obtain the following:
For $n = 0$,

$$p(0) = \sum_{k=0}^{8-1} h_0(k)h_0(k) = h_0^2(0) + h_0^2(1) + \cdots + h_0^2(7) = 0.5$$

For $n = 1$,

$$p(1) = \sum_{k=0}^{8-1} h_0(k)h_0(k+1) = h_0(0)h_0(1) + h_0(1)h_0(2) + \cdots + h_0(6)h_0(7) = 0.3035$$

For $n = -1$,

$$p(-1) = \sum_{k=0}^{8-1} h_0(k)h_0(k-1) = h_0(1)h_0(0) + h_0(2)h_0(1) + \cdots + h_0(7)h_0(6) = 0.3035$$

For $n = 2$,

$$p(2) = \sum_{k=0}^{8-1} h_0(k)h_0(k+2) = h_0(0)h_0(2) + h_0(1)h_0(3) + \cdots + h_0(5)h_0(7) = 0.0$$

For $n = -2$,

$$p(-2) = \sum_{k=0}^{8-1} h_0(k)h_0(k-2) = h_0(2)h_0(0) + h_0(3)h_0(1) + \cdots + h_0(7)h_0(5) = 0.0$$

We can easily verify that $p(n) = 0$ for $n \neq 0$ and $n =$ even number.
Next, we use the MATLAB built-in function **xcorr()** to compute the autocorrelation coefficients. The results are listed as

```
>>h0=[0.0348975582178515 −0.01098301946252854 −0.06286453934951963 ...
      0.223907720892568 0.556856993531445 0.357976304997285 ...
      −0.02390027056113145 −0.07594096379188282];
>>p=xcorr(h0,h0)
p = −0.0027 −0.0000 0.0175 0.0000 −0.0684 −0.0000 0.3035 0.50000
     0.3035 −0.0000 −0.0684 0.0000 0.0175 −0.0000 −0.0027
```

We observe that there are 15 coefficients. The middle one is $p(0) = 0.5$ and we also have $p(\pm2) = p(\pm4) = p(\pm6) = 0$ as well as $p(\pm1) = 0.3035$, $p(\pm3) = -0.0684$, $p(\pm5) = 0.0175$, and $p(\pm7) = -0.0027$.

Next, we write

$$R(z) = -0.0027z^7 - 0.0000z^6 + 0.0175z^5 + 0.0000z^4 - 0.0684z^3 - 0.0000z^2 + 0.3035z^1 + 0.5000z^0$$
$$+ 0.3035z^{-1} - 0.0000z^{-2} - 0.0684z^{-3} + 0.0000z^{-4} + 0.0175z^{-5} - 0.0000z^{-6} - 0.0027z^{-7}$$

Substituting $z = -z$ in $R(z)$ yields

$$R(-z) = 0.0027z^7 - 0.0000z^6 - 0.0175z^5 + 0.0000z^4 + 0.0684z^3 - 0.0000z^2 - 0.3035z^1 + 0.5000z^0$$
$$- 0.3035z^{-1} - 0.0000z^{-2} + 0.0684z^{-3} + 0.0000z^{-4} - 0.0175z^{-5} - 0.0000z^{-6} + 0.0027z^{-7}$$

Clearly, by adding the expressions $R(z)$ and $R(-z)$, we can verify that

$$R(z) + R(-z) = 1$$

Using MATLAB, the PR-CQF frequency responses are plotted and shown in Figure 13.10.

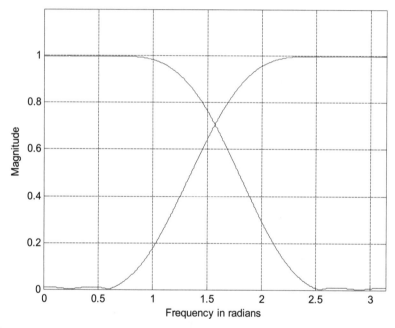

FIGURE 13.10

Magnitude frequency responses of the analysis and synthesis filters in Example 13.1.

Figure 13.11 shows the perfect reconstruction of the two-band system in Figure 13.8 using two-band CQF filters for speech data. The MATLAB program is listed in Program 13.1, in which the quantization is deactivated. Since the obtained signal-to-noise ratio (SNR) = 135.5803 dB, a perfect reconstruction is achieved. Notice that both $x_0(m)$ and $x_1(m)$ have half of the data samples, where $x_0(m)$ contains low-frequency components with more signal energy while $x_1(m)$ possesses high-frequency components with less signal energy.

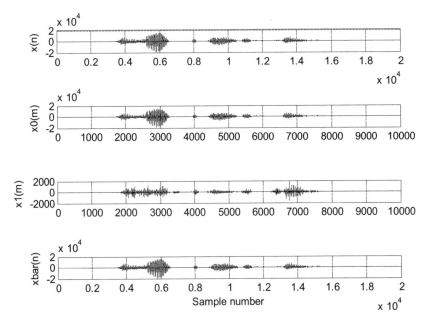

FIGURE 13.11

Two-band analysis and synthesis for speech data.

Program 13.1. Two-band subband system MATLAB implementation.

```
% This program is for implementing analysis and synthesis using two subbands.
close all; clear all;clc
% Smith-Barnwell PR-CQF 8-taps
h0=[0.0348975582178515 -0.01098301946252854 -0.06286453934951963 ...
 0.223907720892568 0.556856993531445 0.357976304997285 ...
 -0.02390027056113145 -0.075940963791188282];
% Read data file "orig.dat" with sampling rate of 8 kHz
load orig.dat; % Load speech data
M=2; % Downsample factor
N=length(h0); PNones=ones(1,N); PNones(2:2:N)=-1;
h1=h0.*PNones; h1=h1(N:-1:1);
g0=-h1.*PNones; g1=h0.*PNones;
disp('check R(z)+R(-z) =>');
xcorr(h0,h0)
sum(h0.*h0)
w=0:pi/1000:pi;
fh0=freqz(h0,1,w); fh1=freqz(h1,1,w);
plot(w,abs(fh0),'k',w,abs(fh1),'k');grid; axis([0 pi 0 1.2]);
xlabel('Frequency in radians');ylabel('Magnitude)')
figure
speech=orig;
```

```
% Analysis
sb_low=filter(h0,1,speech); sb_high=filter(h1,1,speech);
% Downsampling
sb_low=sb_low(1:M:length(sb_low)); sb_high=sb_high(1:M:length(sb_high));
% Quantization
%sb_low=round((sb_low/2^15)*2^9)*2^(15-9); %Quantization with 10 bits
%sb_high=round((sb_high/2^15)*2^5)*2^(15-5); % Quantization with 6 bits
% Synthesis
low_sp=zeros(1,M*length(sb_low)); % Upsampling
low_sp(1:M:length(low_sp))=sb_low;
high_sp=zeros(1,M*length(sb_high)); high_sp(1:M:length(high_sp))=sb_high;
low_sp=filter(g0,1,low_sp); high_sp=filter(g1,1,high_sp);
rec_sig=2*(low_sp+high_sp);
% Signal alignment for SNR calculations
speech=[zeros(1,N-1) speech]; % Align the signal
```

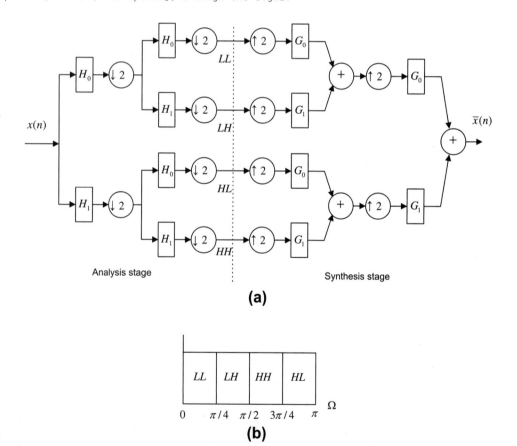

(a)

(b)

FIGURE 13.12

Four-band implementation based on a binary tree structure.

```
subplot(4,1,1);plot(speech);grid,ylabel('x(n)');axis([0 20000 -20000 20000]);
subplot(4,1,2);plot(sb_low);grid,ylabel('x0(m)'); axis([0 10000 -20000 20000]);
subplot(4,1,3);plot(sb_high);grid, ylabel('x1(m)'); axis([0 10000 -2000 2000]);
subplot(4,1,4);plot(rec_sig);grid, ylabel('xbar(n)'),xlabel('Sample number');
axis([0 20000 -20000 20000]);
NN=min(length(speech),length(rec_sig));
err=rec_sig(1:NN)-speech(1:NN);
SNR=sum(speech.*speech)/sum(err.*err);
disp('PR reconstruction SNR dB=>');
SNR=10*log10(SNR)
```

This two-band composition method can easily be extended to a multiband filter bank using a binary tree structure. Figure 13.12 describes a four-band implementation. As shown in Figure 13.12, the filter banks divide an input signal into two equal subbands, resulting the low (L) and high (H) bands using PR-QMF. This two-band PR-QMF again splits L and H into half bands to produce quarter bands: LL, LH, HL, and HH. The four-band spectrum is labeled in Figure 13.12(b). Note that the HH band is actually centered in $[\pi/2, 3\pi/4]$ instead of $[3\pi/4, \pi]$.

In signal coding applications, a dyadic subband tree structure is often used, as shown in Figure 13.13, where the PR-QMF bank splits only the lower half of the spectrum into two equal bands at any level. Through continuation of splitting, we can achieve a coarser-and-coarser version of the original signal.

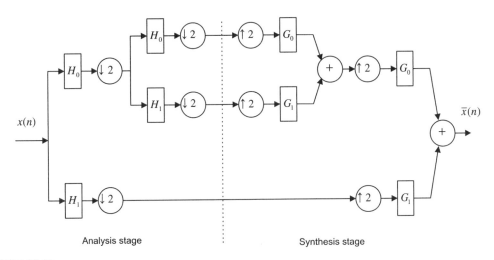

Analysis stage   Synthesis stage

FIGURE 13.13

Four-band implementation based on a dyadic tree structure.

13.3 SUBBAND CODING OF SIGNALS

Subband analysis and synthesis can be successfully applied to signal coding. Figure 13.14 presents an example of a two-band case. The analytical signals from each channel are filtered by the analysis filter, downsampled by a factor of 2, and quantized using quantizers Q_0 and Q_1 each with a assigned number

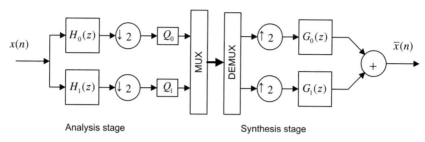

FIGURE 13.14

Two-band filter bank system used for signal compression.

of bits. The quantized codes are multiplexed for transmission or storage. At the synthesis stage, the received or recovered quantized signals are demultiplexed, upsampled by a factor of 2, and processed by the synthesis filters. Then the output signals from all the channels are added to reconstruct the original signal. Since the signal from each analytical channel is quantized, the resultant scheme is a lossy compression one. The coding quality can be measured using the SNR.

Figure 13.15 shows speech coding results using a subband coding system (two-band) with the following specifications:

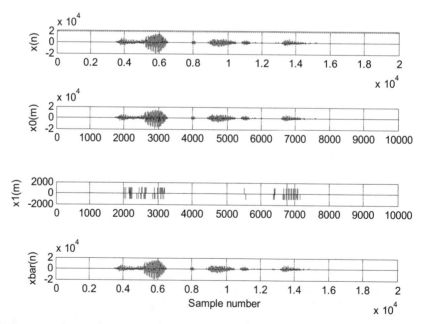

FIGURE 13.15

Two-subband compression for speech data.

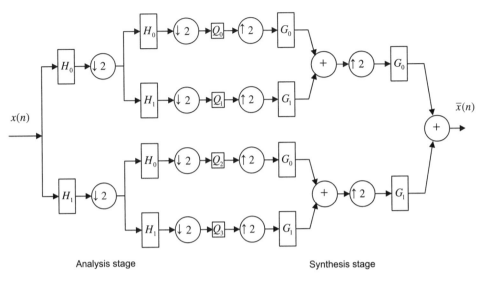

FIGURE 13.16

Four-subband compression for speech data.

Sampling rate = 8 ksps (kilosamples per second)
Sample size = 16 bits/sample
Original data rate = 8 kHz × 16 bits =128 kbps (kilobits per second)

We assign ten bits for Q_0 (since the low-band signal contains more energy) and six bits for Q_1. We obtain a new data rate of $(10 + 6)$ bits × 8 ksps/2 = 64 kbps. The MATLAB implementation is shown in Program 13.1 with the activated quantizers. Notice that $x_0(m)$ and $x_1(m)$ in Figure 13.15 are the quantized versions using Q_0 and Q_1. The measured SNR is 24.51 dB.

Figure 13.16 shows the results using a four-band system. We designate both Q_0 and Q_1 as 11 bits, Q_3 as 10 bits, and Q_2 as 0 bits (discarded). Note that the HL band contains the highest frequency components with the lowest signal energy level (see Figure 13.12(b)). Hence, we discard HL band information to increase the coding efficiency. Therefore, we obtain the data rate as $(11+11+10+0)$ bits × 8 ksps/4 = 64 kbps. The measured SNR is 27.06 dB. A four-band system offers a possibility of signal quality improvement over the two-band system. Plots for the original speech, reconstructed speech, and quantized signal version for each subband are displayed in Figure 13.17.

Figure 13.18 shows the results using a four-band system (Figure 13.16) for encoding seismic data. The seismic signal (provided by the US Geological Survey (USGS)) Albuquerque Seismological Laboratory) has a sampling rate of 15 Hz with 6,700 data samples, and each sample is encoded in 32 bits. For the four-band system, the bit allocations for all bands are as follows: Q_0 (LL) =21 bits, Q_1 (LH) = 21 bits, Q_3 (HL) = 0 (discarded), and Q_4 (HH) =19 bits. We achieve a compression ratio of 4:1 with SNR = 36.00 dB.

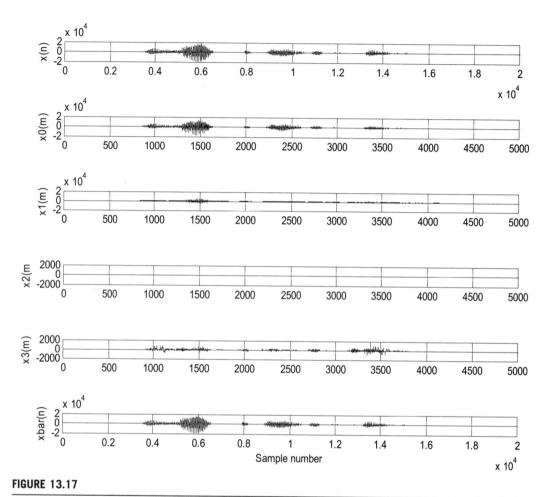

FIGURE 13.17

Four-subband compression for 16-bit speech data and SNR = 27.5 dB.

13.4 WAVELET BASICS AND FAMILIES OF WAVELETS

Wavelet transform has become a powerful tool for signal processing. It offers time–frequency analysis to decompose the signal in terms of a family of wavelets or a set of basic functions, which have a fixed shape but can be shifted and dilated in time. The wavelet transform can present a signal with a good time resolution or a good frequency resolution. There are two types of wavelet transforms: the continuous wavelet transform (CWT) and the discrete wavelet transform (DWT). Specifically, the DWT provides an efficient tool for signal coding. It operates on discrete samples of the signal and has a relation with the dyadic subband coding described in Section 13.2. The DWT resembles other discrete transforms, such as the discrete Fourier transform (DFT) or the discrete cosine transform (DCT). In this section, without getting too detailed with mathematics, we review the basics of the

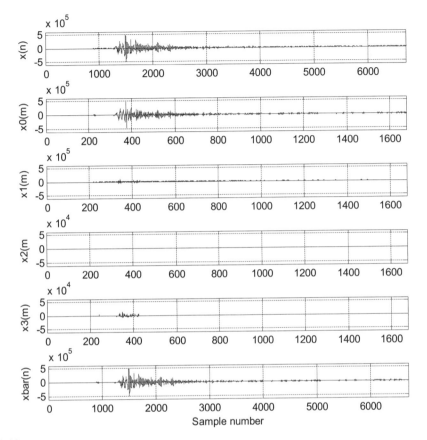

FIGURE 13.18

Four-subband compression for 32-bit seismic data and SNR = 36 dB.

CWT, which will lay out the foundation. Next, we emphasize the DWT for applications of signal coding.

Let us examine a signal sampled at 1 kHz with 1024×32 (32,678) samples given by

$$x(t) = 0.5\cos(2\pi \times 80t)[u(t) - u(t-8)] + \sin(2\pi \times 180t)[u(t-8) - u(t-16)]$$
$$+\sin(2\pi \times 250t)[u(t-16) - u(t-32)] + 0.1\sin(2\pi \times 0.8t)[u(t-8) - u(t-24)]$$

$$(13.30)$$

The signal contains four sinusoids: 80 Hz for $0 \le t < 8$ seconds, 180 Hz for $8 \le t < 16$ seconds, 350 Hz for $16 \le t \le 32$ seconds, and finally 0.8 Hz for $8 \le t \le 24$ seconds. All the signals are plotted separately in Figure 13.19 while Figure 13.20 shows the combined signal and its DFT spectrum.

Based on the traditional spectral analysis shown in Figure 13.20, we can identify the frequency components of 80, 180, and 350 Hz. However, the 0.8-Hz component and transient behaviors such as the start and stop time instants of the sinusoids (discontinuity) cannot be observed from the spectrum. Figure 13.21 depicts the wavelet transform of the same signal. The horizontal axis is time in seconds

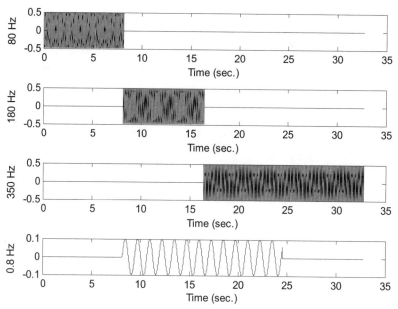

FIGURE 13.19

Individual signal components.

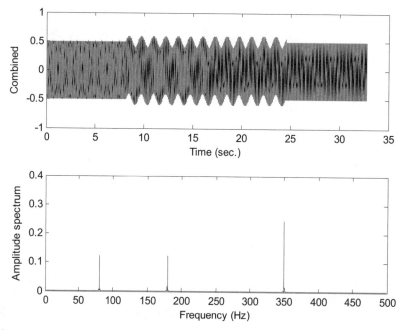

FIGURE 13.20

Combined signal and its spectrum.

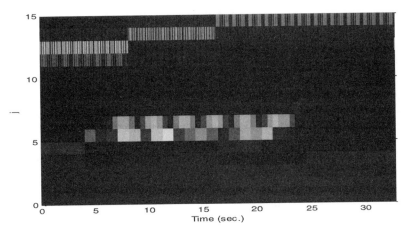

FIGURE 13.21

Wavelet transform amplitudes.

while the vertical axis is index j, which is inversely proportional to the scale factor ($a = 2^{-j}$). As will be discussed, the larger the scale factor (the smaller the index j), the smaller the frequency value. The amplitudes of the wavelet transform are displayed according to the intensity. The brighter the intensity, the larger the amplitude. The areas with brighter intensities indicate the strongest resonances between the signal and the wavelets of various frequency scales and time shifts. In Figure 13.21, the four different frequency components and the discontinuities of the sinusoids are displayed as well. We can further observe the fact that the finer the frequency resolution, the coarser the time resolution. For example, we can clearly identify the start and stop times for 80-, 180-, and 350-Hz frequency components, but frequency resolution is coarse, since index j has larger frequency spacing. However, for the 0.8-Hz sinusoid, we have fine frequency resolution (small frequency spacing so we can see the 0.8-Hz sinusoid) and coarse time resolution as evidenced by the way in which the start and stop times are blurred.

The CWT is defined as

$$W(a,b) = \int_{-\infty}^{\infty} f(t)\psi_{ab}(t)dt \qquad (13.31)$$

where $W(a, b)$ is the wavelet transform and $\psi_{ab}(t)$ is called the mother wavelet, which is defined as

$$\psi_{ab}(t) = \frac{1}{\sqrt{a}}\psi\left(\frac{t-b}{a}\right) \qquad (13.32)$$

The wavelet function consists of two important parameters: scaling a and translation b. A scaled version of the function $\psi(t)$ with a scale factor of a is defined as $\psi(t/a)$. Consider a base function $\psi(t) = \cos(\omega t)$ when $a = 1$. When $a > 1$, $\psi(t) = \cos(\omega t/a)$ is a scaled function with a frequency less than ω rad/s. When $a < 1$, $\psi(t) = \cos(\omega t/a)$ has a frequency larger than ω. Figure 13.22 shows the scaled wavelet functions.

A translated version of the function $\psi(t)$ with a shifted time constant b is defined as $\psi(t-b)$. Figure 13.23 shows several translated versions of the wavelet. A scaled and translated function $\psi(t)$ is given by $\psi((t-b)/a)$. This means that $\psi((t-b)/a)$ changes frequency and time shift. Several combined scaling and translated wavelets are displayed in Figure 13.24.

Besides these two properties, a wavelet function must satisfy admissibility and regularity conditions (vanishing moment up to a certain order). Admissibility requires that the wavelet (mother wavelet) have a bandpass-limited spectrum and a zero average in the time domain, which means that wavelets must be oscillatory. Regularity requires that wavelets have some smoothness and concentration in both time and frequency domains. This topic is beyond the scope of this book and the details can be found in Akansu and Haddad (1992). There exists a pair of wavelet functions: the father wavelet (also called the scaling function) and mother wavelet. Figure 13.25 shows a simplest pair of wavelets: the Haar father wavelet and mother wavelet.

To devise an efficient wavelet transform algorithm, we let the scale factor be a power of two, that is,

$$a = 2^{-j} \tag{13.33}$$

Note that the larger the index j, the smaller the scale factor $a = 2^{-j}$. The time shift becomes

$$b = k2^{-j} = ka \tag{13.34}$$

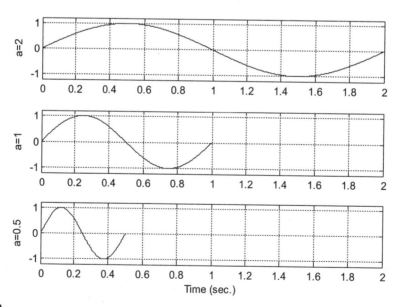

FIGURE 13.22

Scaled wavelet functions.

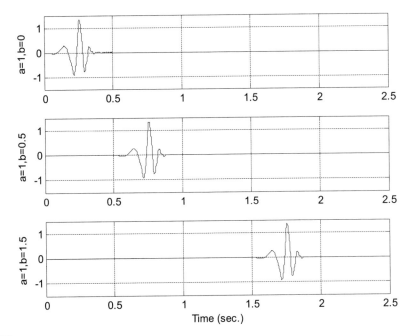

FIGURE 13.23

Translated wavelet functions.

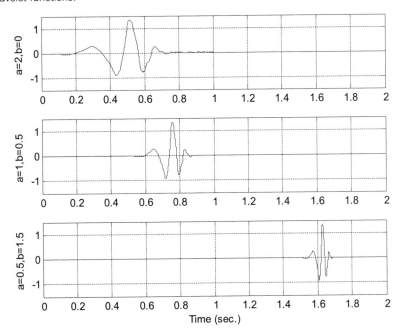

FIGURE 13.24

Scaled and translated wavelet functions.

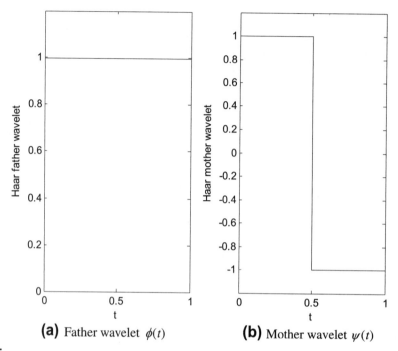

FIGURE 13.25

Haar (a) father and (b) mother wavelets.

Substituting Equations (13.33) and (13.34) into the base function gives

$$\psi\left(\frac{t-b}{a}\right) = \psi\left(\frac{t-kb}{a}\right) = \psi(a^{-1}t - k) = \psi(2^j t - k) \tag{13.35}$$

We can define a mother wavelet at scale j and translation k as

$$\psi_{jk}(t) = 2^{j/2}\psi(2^j t - k) \tag{13.36}$$

Similarly, a father wavelet (scaling function) at scale j and translation k is defined as

$$\phi_{jk}(t) = 2^{j/2}\phi(2^j t - k) \tag{13.37}$$

EXAMPLE 13.2

Sketch the Haar father wavelet families for four different scales, $j = 0, 1, 2, 3$, for a period of one second.

Solution:
Based on Equation (13.37), we can determine the wavelet at each required scale as follows:

For $j = 0$, $\phi_{0k}(t) = \phi(t - k)$, only $k = 0$ is required to cover a 1-second duration.
For $j = 1$, $\phi_{1k}(t) = \sqrt{2}\phi(2t - k)$, $k = 0$ and $k = 1$ are required.
For $j = 2$, $\phi_{2k}(t) = 2\phi(4t - k)$, we need $k = 0, 1, 2, 3$.
For $j = 3$, $\phi_{3k}(t) = 2\sqrt{2}\phi(8t - k)$, we need $k = 0, 1, 2, \cdots, 7$.

Using Figure 13.25(a), we obtain the plots shown in Figure 13.26.

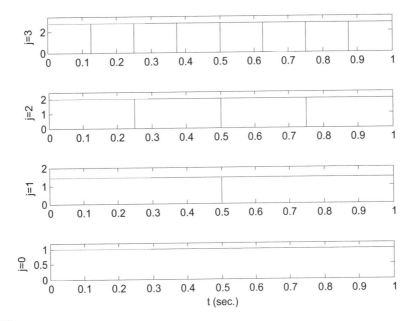

FIGURE 13.26

Haar father wavelets at different scales and translations.

EXAMPLE 13.3

Sketch the Haar mother wavelet families for four different scales, $j = 0, 1, 2, 3$, for a period of 1 second.

Solution:
Based on Equation (13.36), we have

For $j = 0$, $\psi_{0k} = \psi(t - k)$, $k = 0$ and $k = 1$ are required.
For $j = 1$, $\psi_{1k} = \sqrt{2}\psi(2t - k)$, $k = 0$ and $k = 1$ are required.
For $j = 2$, $\psi_{2k} = 2\psi(4t - k)$, we need $k = 0, 1, 2, 3$.
For $j = 3$, $\psi_{3k}(t) = 2\sqrt{2}\psi(8t - k)$, we need $k = 0, 1, 2, \cdots, 7$.

Using Figure 13.25(b), we obtain the plots shown in Figure 13.27.

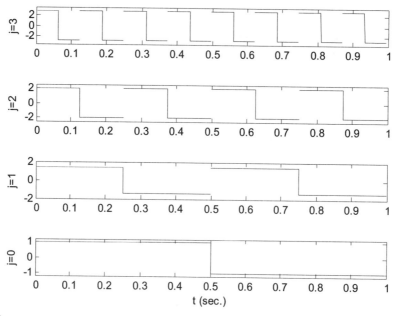

FIGURE 13.27

Haar mother wavelet at different scales and translations.

A signal can be expanded by the father wavelets (scaling function and its translations) at level j. More accuracy can be achieved by using the larger j. The expanded function is approximated by

$$f(t) \approx f_j(t) = \sum_{k=-\infty}^{\infty} c_j(k) 2^{j/2} \phi(2^j t - k) \qquad (13.38)$$

where the wavelet coefficients $c_j(k)$ can be determined by an inner product:

$$c_j(k) = \langle f(t) \phi_{jk}(t) \rangle = \int f(t) 2^{j/2} \phi(2^j t - k) dt. \qquad (13.39)$$

EXAMPLE 13.4

Approximate the following function using the Haar scaling function at level $j = 1$:

$$f(t) = \begin{cases} 2 & 0 \le t < 0.5 \\ 1 & 0.5 \le t \le 1 \end{cases}$$

Solution:

Substituting $j = 1$ in Equation (13.38) leads to

$$f(t) \approx f_1(t) = \sum_{k=-\infty}^{\infty} c_1(k)2^{1/2}\phi(2t - k)$$

We only need $k = 0$ and $k = 1$ to cover the range $0 \le t \le 1$, that is,

$$f(t) = c_1(0)2^{1/2}\phi(2t) + c_1(1)2^{1/2}\phi(2t - 1)$$

Notice that

$$\phi(2t) = \begin{cases} 1 & \text{for} \quad 0 \le t \le 0.5 \\ 0 & \text{elsewhere} \end{cases} \quad \text{and} \quad \phi(2t - 1) = \begin{cases} 1 & \text{for} \quad 0.5 \le t \le 1 \\ 0 & \text{elsewhere} \end{cases}$$

Applying Equation (13.39) yields

$$c_1(0) = \int_0^{1/2} f(t)2^{1/2}\phi(2t)dt = \int_0^{1/2} 2 \times 2^{1/2} \times 1 dt = 2^{1/2}$$

Similarly,

$$c_1(1) = \int_{1/2}^{1} f(t)2^{1/2}\phi(2t - 1)dt = \int_{1/2}^{1} 1 \times 2^{1/2} \times 1 dt = 0.5 \times 2^{1/2}$$

Then substituting the coefficients $c_1(0)$ and $c_1(1)$ leads to

$$f_1(t) = 2^{1/2} \times 2^{1/2}\phi(2t) + 0.5 \times 2^{1/2}2^{1/2}\phi(2t - 1) = 2\phi(2t) + \phi(2t - 1) = f(t)$$

Equation (13.39) can also be approximated numerically:

$$c_j(k) \approx \sum_{m=0}^{M-1} f(t_m)2^{j/2}\phi(2^j t_m - k)\Delta t$$

where $t_m = m\Delta t$ is the time instant, Δt denotes the time step, and M is the number of intervals. In this example, if we choose $\Delta t = 0.2$, then $M = 5$ and $t_m = m\Delta t$. The numerical calculations for Example 13.4 are done as follows:

$$c_1(0) \approx \sum_{m=0}^{4} f(t_m)2^{1/2}\phi(2t_m)\Delta t = 2^{1/2}[f(0) \times \phi(0) + f(0.2) \times \phi(0.4)$$
$$+ f(0.4) \times \phi(0.8) + f(0.6) \times \phi(1.2) + f(0.8) \times \phi(1.6)]\Delta t$$
$$= 2^{1/2}(2 \times 1 + 2 \times 1 + 2 \times 1 + 1 \times 0 + 1 \times 0) \times 0.2 = 1.2 \times 2^{1/2}$$

$$c_1(1) \approx \sum_{m=0}^{4} f(t_m)2^{1/2}\phi(2t_m - 1)\Delta t = 2^{1/2}[f(0) \times \phi(-1) + f(0.2) \times \phi(-0.6)$$
$$+ f(0.4) \times \phi(-0.2) + f(0.6) \times \phi(0.2) + f(0.8) \times \phi(0.6)]\Delta t$$
$$= 2^{1/2}(2 \times 0 + 2 \times 0 + 2 \times 0 + 1 \times 1 + 1 \times 1) \times 0.2 = 0.4 \times 2^{1/2}$$

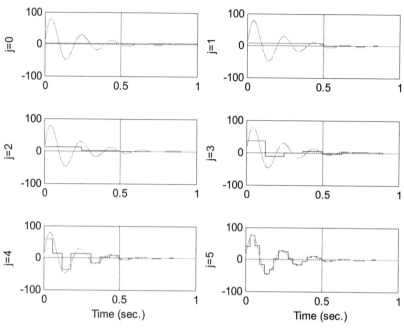

FIGURE 13.28

Signal expanded by Haar father wavelets.

Finally, we have

$$f_1(t) = 1.2 \times 2^{1/2} \times 2^{1/2}\phi(2t) + 0.4 \times 2^{1/2}2^{1/2}\phi(2t-1) = 2.4\phi(2t) + 0.8\phi(2t-1) \approx f(t)$$

It is clear that there is a numerical error. The error can be reduced when a smaller time interval Δt is adopted.

Figure 13.28 demonstrates the approximation of a sinusoidal delaying function using the scaling functions (Haar father wavelets) at different scales, i.e., $j = 0, 1, 2, 4, 5$.

Now, let us examine the function approximation at resolution $j = 1$:

$$f_1(t) \approx \sum_{k=-\infty}^{\infty} c_1(k)\sqrt{2}\phi(2t-k) = c_1(0)\sqrt{2}\phi(2t) + c_1(1)\sqrt{2}\phi(2t-1)$$

We also look at another possibility at a coarser scale with both the scaling functions (father wavelets) and mother wavelets, that is, $j = 0$:

$$f_1(t) \approx \sum_{k=-\infty}^{\infty} c_0(k)\phi_{0k}(t) + \sum_{k=-\infty}^{\infty} d_0(k)\psi_{0k}(t)$$
$$= c_0(0)\phi_{00}(t) + d_0(0)\psi_{00}(t) = c_0(0)\phi(t) + d_0(0)\psi(t)$$

Furthermore, we see that

$$f_1(t) \approx c_0(0)\phi(t) + d_0(0)\psi(t)$$

$$= c_0(0)\left(\frac{1}{\sqrt{2}}\phi(2t) + \frac{1}{\sqrt{2}}\phi(2t-1)\right) + d_0(0)\left(\frac{1}{\sqrt{2}}\phi(2t) - \frac{1}{\sqrt{2}}\phi(2t-1)\right)$$

$$= \frac{1}{\sqrt{2}}(c_0(0) + d_0(0))\phi(2t) + \frac{1}{\sqrt{2}}(c_0(0) - d_0(0))\phi(2t-1)$$

We observe that

$$c_1(0) = \frac{1}{2}(c_0(0) + d_0(0))$$

$$c_1(1) = \frac{1}{2}(c_0(0) - d_0(0))$$

This means that

$$S_1 = S_0 \cup W_0$$

where S_1 contains functions in terms of basis scaling functions at $\phi_{1k}(t)$, and the function can also be expanded using the scaling functions $\phi_{0k}(t)$ and wavelet functions $\psi_{0k}(t)$ at a coarser level $j - 1$. In general, the following statement is true:

$$\begin{aligned} S_j &= S_{j-1} \cup W_{j-1} = [S_{j-2} \cup W_{j-2}] \cup W_{j-1} \\ &= \{[S_{j-3} \cup W_{j-3}] \cup W_{j-2}\} \cup W_{j-1} \\ &\cdots = S_0 \cup W_0 \cup W_1 \cup \cdots \cup W_{j-1}. \end{aligned} \qquad (13.40)$$

Hence, the approximation of $f_j(t)$ can be expressed as

$$f(t) \approx f_j(t) = \sum_{k=-\infty}^{\infty} c_j(k)\phi_{jk}(t)$$

$$= \sum_{k=-\infty}^{\infty} c_{j-1}(k)\phi_{(j-1)k}(t) + \sum_{k=-\infty}^{\infty} d_{j-1}(k)\psi_{(j-1)k}(t)$$

$$= \sum_{k=-\infty}^{\infty} c_{(j-1)}(k)2^{(j-1)/2}\phi(2^{(j-1)}t - k) + \sum_{k=-\infty}^{\infty} d_{(j-1)}(k)2^{(j-1)/2}\psi(2^{(j-1)}t - k)$$

Repeating the expansion of the first sum leads to

$$f(t) \approx f_J(t) = \sum_{k=-\infty}^{\infty} c_0(k)\phi_{0k}(t) + \sum_{j=0}^{J-1} \sum_{k=-\infty}^{\infty} d_j(k)\psi_{jk}(t)$$

$$= \sum_{k=-\infty}^{\infty} c_0(k)\phi(t-k) + \sum_{j=0}^{J-1} \sum_{k=-\infty}^{\infty} d_j(k)2^{j/2}\psi(2^j t - k)$$

$$(13.41)$$

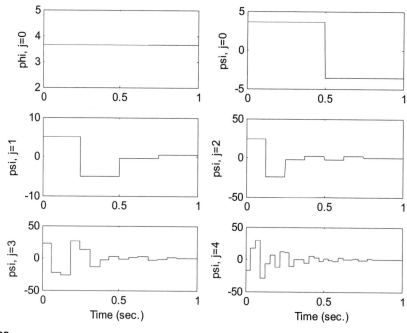

FIGURE 13.29

Approximations using Haar scaling functions and mother wavelets.

where the mother wavelet coefficients $d_j(k)$ can also be determined by the inner product:

$$d_j(k) = \langle f(t)\psi_{jk}(t)\rangle = \int f(t)2^{j/2}\psi(2^j t - k)dt \qquad (13.42)$$

Figure 13.29 demonstrates the function approximation (Figure 13.28) with the base scaling function at resolution $j = 0$, and mother wavelets at scales $j = 0, 1, 2, 3, 4$. The combined approximation ($J = 5$) using Equation (13.41) is shown in Figure 13.30.

13.5 MULTIRESOLUTION EQUATIONS

There are two very important equations for multiresolution analysis. Each scaling function can be constructed by a linear combination of translations with the doubled frequency of a base scaling function $\phi(2t)$, that is,

$$\phi(t) = \sum_{k=-\infty}^{\infty} \sqrt{2}h_0(k)\phi(2t - k) \qquad (13.43)$$

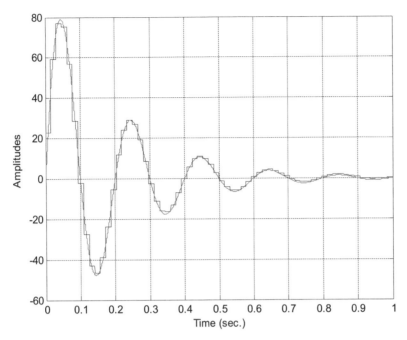

FIGURE 13.30

Signal coded using the wavelets at resolution $J = 5$.

where $h_0(k)$ is a set of scaling function coefficients (wavelet filter coefficients). The mother wavelet function can also be built by a sum of translations with the double frequency of the base scaling function $\phi(2t)$, that is,

$$\psi(t) = \sum_{k=-\infty}^{\infty} \sqrt{2}h_1(k)\phi(2t - k) \tag{13.44}$$

where $h_1(k)$ is another set of wavelet filter coefficients. Let us verify these two relationships via Example 13.5 below.

EXAMPLE 13.5

Determine $h_0(k)$ for the Haar father wavelet.

Solution:

From Equation (13.43), we can express

$$\phi(t) = \sqrt{2}h_0(0)\phi(2t) + \sqrt{2}h_0(1)\phi(2t - 1)$$

Then we deduce that

$$h_0(0) = h_0(1) = 1/\sqrt{2}$$

Figure 13.31 shows that the Haar father wavelet is the sum of two scaling functions at scale $j = 1$.

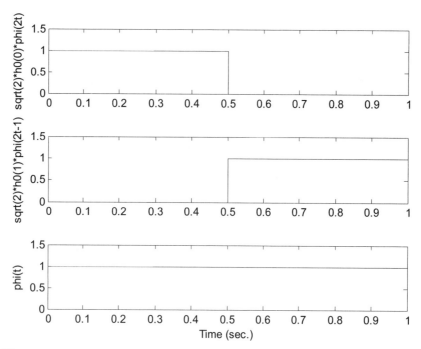

FIGURE 13.31

Haar wavelets in Example 13.5.

EXAMPLE 13.6

Determine $h_1(k)$ for the Haar mother wavelet.

Solution:

From Equation (13.44), we can write

$$\psi(t) = \sqrt{2}h_1(0)\phi(2t) + \sqrt{2}h_1(1)\phi(2t-1)$$

Hence, we deduce that

$$h_1(0) = 1/\sqrt{2} \quad \text{and} \quad h_1(1) = -1/\sqrt{2}$$

Figure 13.32 shows that the Haar mother wavelet is the difference of two scaling functions at scale $j = 1$.

Notice that the relation between $H_0(z)$ and $H_1(z)$ exists and is given by

$$h_1(k) = (-1)^k h_0(N-1-k) \qquad (13.45)$$

We can verify Equation (13.45) for the Haar wavelet:

$$h_1(k) = (-1)^k h_0(1-k)$$

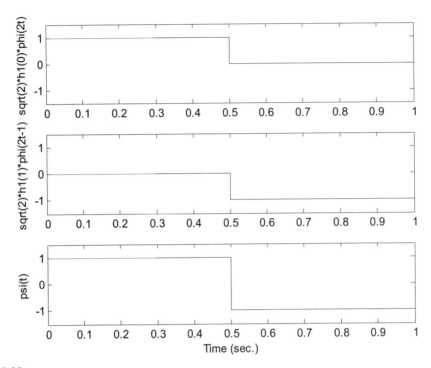

FIGURE 13.32

Haar wavelets in Example 13.6.

$$h_1(0) = (-1)^0 h_0(1-0) = h_0(1) = 1/\sqrt{2}$$

$$h_1(1) = (-1)^1 h_0(1-1) = -h_0(0) = -1/\sqrt{2}$$

This means that once we obtain the coefficients of $h_0(k)$, the coefficients $h_1(k)$ can be determined via Equation (13.45). We do not aim to obtain wavelet filter coefficients here. The topic is beyond the scope of this book and the details are given in Akansu and Haddad (1992). Instead, some typical filter coefficients for Haar and Daubechies are given in Table 13.2.

We can apply the Daubechies-4 filter coefficients to examine multiresolution Equations (13.43) and (13.44). From Table 13.2, we have

$$h_0(0) = 0.4830,\ h_0(1) = 0.8365,\ h_0(2) = 0.2241,\ h_0(3) = -0.1294$$

We then expand Equation (13.43) as

$$\phi(t) = \sqrt{2}h_0(0)\phi(2t) + \sqrt{2}h_0(1)\phi(2t-1) + \sqrt{2}h_0(2)\phi(2t-2) + \sqrt{2}h_0(3)\phi(2t-3)$$

Table 13.2 Typical Wavelet Filter Coefficients $h_0(k)$

Haar	Daubechies 4	Daubechies 6	Daubechies 8
0.707106781186548	0.482962913144534	0.332670552950083	0.230377813308896
0.707106781186548	0.836516303737808	0.806891509311093	0.714846570552915
	0.224143868042013	0.459877502118492	0.630880767929859
	−0.129409522551260	−0.135011020010255	−0.027983769416859
		−0.085441273882027	−0.187034811719093
		0.035226291885710	0.030841381835561
			0.032883011666885
			−0.010597401785069

Figure 13.33 shows each component at resolution $j = 1$ and the constructed scaling function $\phi(t)$. The original scaling function $\phi(t)$ is also included as shown in the last plot for comparison.

With the given coefficients $h_0(k)$ and applying Equation (13.45), we can obtain the wavelet coefficients $h_1(k)$ as

$$h_1(0) = -0.1294, \quad h_1(1) = -0.2241, \quad h_1(2) = 0.8365, \quad \text{and} \quad h_1(2) = -0.4830$$

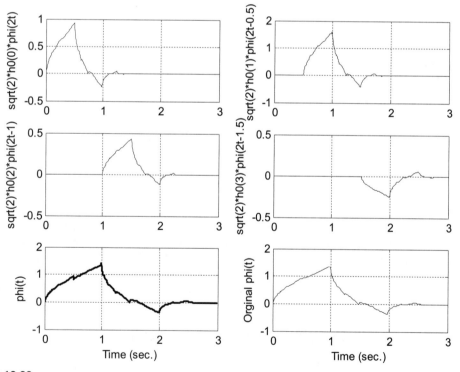

FIGURE 13.33

Constructed 4-tap Daubechies father wavelet.

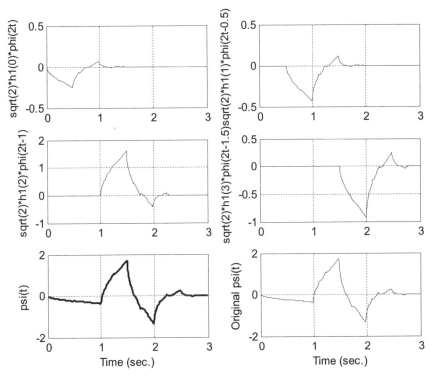

FIGURE 13.34

Constructed 4-tap Daubechies mother wavelet.

Expanding Equation (13.44) leads to

$$\psi(t) = \sqrt{2}h_1(0)\phi(2t) + \sqrt{2}h_1(1)\phi(2t-1) + \sqrt{2}h_1(2)\phi(2t-2) + \sqrt{2}h_1(3)\phi(2t-3)$$

Similarly, Figure 13.34 displays each component at resolution $j = 1$ and the constructed mother wavelet function $\psi(t)$. The last plot displays the original mother wavelet function $\psi(t)$ for comparison.

13.6 DISCRETE WAVELET TRANSFORM

Now let us examine the discrete wavelet transform (DWT). We begin with coding a signal using a wavelet expansion as follows:

$$f(t) \approx f_{j+1}(t) = \sum_{k=-\infty}^{\infty} c_j(k)2^{j/2}\phi(2^j t - k) + \sum_{k=-\infty}^{\infty} d_j(k)2^{j/2}\psi(2^j t - k) \qquad (13.46)$$

By applying and continuing to apply Equation (13.46), $f(t)$ can be coded at any level we wish. Furthermore, by recursively applying Equation (13.46) until $j = 0$, we can obtain signal expansion using all the mother wavelets plus one scaling function at scale $j = 0$, that is,

$$f(t) \approx f_J(t) = \sum_{k=-\infty}^{\infty} c_0(k)\phi(t-k) + \sum_{j=0}^{J-1}\sum_{k=-\infty}^{\infty} d_j(k)2^{j/2}\psi(2^j t - k) \qquad (13.47)$$

All $c_j(k)$ and all $d_j(k)$ are called the wavelet coefficients. They are essentially weights for the scaling function(s) and wavelet functions (mother wavelets). The DWT computes these wavelet coefficients. On the other hand, given the wavelet coefficients, we are able to reconstruct the original signal by applying the inverse discrete wavelet transform (IDWT).

Based on the wavelet theory without proof (see Appendix F), we can perform the DWT using the analysis equations as follows:

$$c_j(k) = \sum_{m=-\infty}^{\infty} c_{j+1}(m)h_0(m-2k) \qquad (13.48)$$

$$d_j(k) = \sum_{m=-\infty}^{\infty} c_{j+1}(m)h_1(m-2k) \qquad (13.49)$$

where $h_0(k)$ are the lowpass wavelet filter coefficients listed in Table 13.2, while $h_1(k)$, the highpass filter coefficients, can be determined by

$$h_1(k) = (-1)^k h_0(N-1-k) \qquad (13.50)$$

These lowpass and highpass filters are called the quadrature mirror filters (QMF). As an example, the frequency responses of the 4-tap Daubechies wavelet filters are plotted in Figure 13.35.

Next, we need to determine the filter inputs $c_{j+1}(k)$ in Equations (13.48) and (13.49). In practice, since j is a large number, the function $\phi(2^j t - k)$ appears to be close to an impulse-like function, that is, $\phi(2^j t - k) \approx 2^{-j}\delta(t - k2^{-j})$. For example, the Haar scaling function can be expressed as $\phi(t) = u(t) - u(t-1)$, where $u(t)$ is the step function. We can easily get $\phi(2^5 t - k) = u(2^5 t - k) - u(2^5 t - 1 - k) = u(t - k2^{-5}) - u(t - (k+1)2^{-5})$ for $j = 5$, which is a narrow pulse with a unit height and a width 2^{-5} located at $t = k2^{-5}$. The area of the pulse is therefore 2^{-5}. When j approaches a larger positive integer, $\phi(2^j t - k) \approx 2^{-j}\delta(t - k2^{-j})$. Therefore, $f(t)$ approximated by the scaling function at level j is rewritten as

$$
\begin{aligned}
f(t) \approx f_j(t) &= \sum_{k=-\infty}^{\infty} c_j(k)2^{j/2}\phi(2^j t - k) \\
&= \cdots + c_j(0)2^{j/2}\phi(2^j t) + c_j(1)2^{j/2}\phi(2^j t - 1) + c_j(2)2^{j/2}\phi(2^j t - 2) + \cdots \\
&\approx \cdots + c_j(0)2^{-j/2}\delta(t) + c_j(1)2^{-j/2}\delta(t - 1 \times 2^{-j}) + c_j(2)2^{-j/2}\delta(t - 2 \times 2^{-j}) + \cdots
\end{aligned}
\qquad (13.51)
$$

On the other hand, if we sample $f(t)$ using the same sample interval $T_s = 2^{-j}$ (time resolution), the discrete-time function can be expressed as

$$f(n) = f(nT_s) = \cdots + f(0T_s)T_s\delta(t - T_s) + f(T_s)T_s\delta(t - T_s) + f(2T_s)T_s\delta(n - 2T_s) + \cdots \qquad (13.52)$$

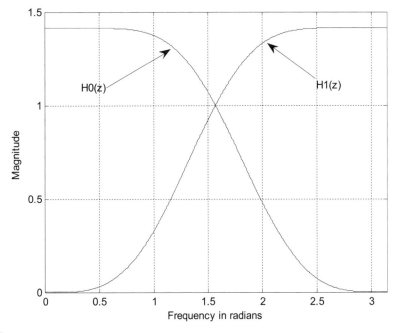

FIGURE 13.35

Frequency responses for 4-tap Daubechies filters.

Hence, comparing Equation (13.51) with the discrete-time version in Equation (13.52), it follows that

$$c_j(k)2^{-j/2} = f(k)T_s \tag{13.53}$$

Substituting $T_s = 2^{-j}$ in Equation (13.53) leads to

$$c_j(k) = 2^{-j/2}f(k) \tag{13.54}$$

With the obtained sequence $c_j(k)$ using sample values $f(k)$, we can perform the DWT using Equations (13.48) and (13.49). Furthermore, Equations (13.48) and (13.49) can be implemented using a dyadic tree structure similar to the subband coding case. Figure 13.36 depicts the case for $j = 2$.

Note that the reversed sequences $h_0(-k)$ and $h_1(-k)$ are used in the analysis stage. Similarly, the IDWT (synthesis equation) can be developed (see Appendix F) and expressed as

$$c_{j+1}(k) = \sum_{m=-\infty}^{\infty} c_j(m)h_0(k - 2m) + \sum_{m=-\infty}^{\infty} d_j(m)h_1(k - 2m) \tag{13.55}$$

Finally, the signal amplitude can be rescaled by

$$f(k) = 2^{j/2}c_j(k) \tag{13.56}$$

An implementation for $j = 2$ using the dyadic subband coding structure is illustrated in Figure 13.37.

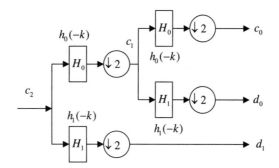

FIGURE 13.36

Analysis using the dyadic subband coding structure.

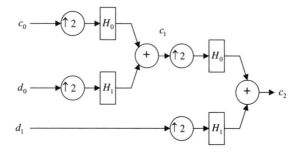

FIGURE 13.37

Synthesis using the dyadic subband coding structure.

Now, let us study the DWT and IDWT in the following examples.

EXAMPLE 13.7

Given the sample values [4 2 −1 0], use the Haar wavelets to determine the wavelet coefficients.

Solution:

Form the filter inputs:

$$c_2(k) = 2^{-2/2} \times [4 \quad 2 \quad -1 \quad 0] = \left[2 \quad 1 \quad -\frac{1}{2} \quad 0\right]$$

The acquired Haar wavelet filter coefficients are listed as

$$h_0(k) = \left[\frac{1}{\sqrt{2}} \quad \frac{1}{\sqrt{2}}\right] \quad \text{and} \quad h_1(k) = \left[\frac{1}{\sqrt{2}} \quad -\frac{1}{\sqrt{2}}\right]$$

The function is expanded by the scaling functions as

$$f(t) \approx f_2(t) = \sum_{k=-\infty}^{\infty} c_j(k) 2^{j/2} \phi(2^j t - k)$$

$$= 4 \times \phi(4t) + 2 \times \phi(4t - 1) - 1 \times \phi(4t - 2) + 0 \times \phi(4t - 3).$$

We will verify this expression later. Applying the wavelet analysis equations, we have

$$c_1(k) = \sum_{m=-\infty}^{\infty} c_2(m)h_0(m-2k)$$

$$d_1(k) = \sum_{m=-\infty}^{\infty} c_2(m)h_1(m-2k)$$

Specifically,

$$c_1(0) = \sum_{m=-\infty}^{\infty} c_2(m)h_0(m) = c_2(0)h_0(0) + c_2(1)h_0(1) = 2 \times \frac{1}{\sqrt{2}} + 1 \times \frac{1}{\sqrt{2}} = \frac{3\sqrt{2}}{2}$$

$$c_1(1) = \sum_{m=-\infty}^{\infty} c_2(m)h_0(m-2) = c_2(2)h_0(0) + c_2(3)h_0(1) = \left(-\frac{1}{2}\right) \times \frac{1}{\sqrt{2}} + 0 \times \frac{1}{\sqrt{2}} = -\frac{1}{2\sqrt{2}}$$

$$d_1(0) = \sum_{m=-\infty}^{\infty} c_2(m)h_1(m) = c_2(0)h_1(0) + c_2(1)h_1(1) = 2 \times \frac{1}{\sqrt{2}} + 1 \times \left(-\frac{1}{\sqrt{2}}\right) = \frac{1}{\sqrt{2}}$$

$$d_1(1) = \sum_{m=-\infty}^{\infty} c_2(m)h_1(m-2) = c_2(2)h_1(0) + c_2(3)h_1(1) = \left(-\frac{1}{2}\right) \times \frac{1}{\sqrt{2}} + 0 \times \left(-\frac{1}{\sqrt{2}}\right) = -\frac{1}{2\sqrt{2}}$$

Using the subband coding method in Figure 13.36 yields

```
>> x0=rconv([1 1]/sqrt(2),[2 1 -0.5 0])
x0 = 2.1213 0.3536 -0.3536 1.4142
>> c1=x0(1:2:4)
c1 = 2.1213 -0.3536
>> x1=rconv([1 -1]/sqrt(2),[2 1 -0.5 0])
x1 = 0.7071 1.0607 -0.3536 -1.4142
>> d1=x1(1:2:4)
d1 = 0.7071 -0.3536
```

where the MATLAB function **rconv()** for filter operations with the reversed filter coefficients is listed in Section 13.8. Repeating for the next level, we have

$$c_0(k) = \sum_{m=-\infty}^{\infty} c_1(m)h_0(m-2k)$$

$$d_0(k) = \sum_{m=-\infty}^{\infty} c_1(m)h_1(m-2k)$$

Thus

$$c_0(0) = \sum_{m=-\infty}^{\infty} c_1(m)h_0(m) = c_1(0)h_0(0) + c_1(1)h_0(1) = \frac{3\sqrt{2}}{2} \times \frac{1}{\sqrt{2}} + \left(-\frac{1}{2\sqrt{2}}\right) \times \frac{1}{\sqrt{2}} = \frac{5}{4}$$

$$d_0(0) = \sum_{m=-\infty}^{\infty} c_1(m)h_1(m) = c_1(0)h_1(0) + c_1(1)h_1(1) = \frac{3\sqrt{2}}{2} \times \frac{1}{\sqrt{2}} + \left(-\frac{1}{2\sqrt{2}}\right) \times \left(-\frac{1}{\sqrt{2}}\right) = \frac{7}{4}$$

The MATLAB verifications follow:

```
>> xx0=rconv([1 1]/sqrt(2),c1)
xx0 = 1.2500 1.2500
>> c0=xx0(1:2:2)
c0 = 1.2500
>> xx1=rconv([1 −1]/sqrt(2),c1)
xx1 = 1.7500 −1.7500
>> d0=xx1(1:2:2)
d0 = 1.7500
```

Finally, we pack the wavelet coefficients $w_2(k)$ at $j = 2$ together as

$$w_2(k) = [c_0(0)\, d_0(0)\, d_1(0)\, d_1(1)] = \left[\frac{5}{4}\ \frac{7}{4}\ \frac{1}{\sqrt{2}}\ -\frac{1}{2\sqrt{2}}\right]$$

Then the function can be expanded using one scaling function and three mother wavelet functions:

$$f(t) \approx f_2(t) = \sum_{k=-\infty}^{\infty} c_0(k)\phi(t-k) + \sum_{j=0}^{1} \sum_{k=-\infty}^{\infty} d_j(k)2^{j/2}\psi(2^j t - k)$$

$$= \frac{5}{4}\phi(t) + \frac{7}{4}\psi(t) + \psi(2t) - \frac{1}{2}\psi(2t-1)$$

Figure 13.38 shows the plots for each function and the combined function to verify that $f(t)$ does have amplitudes of 4, 2, −1, and 0.

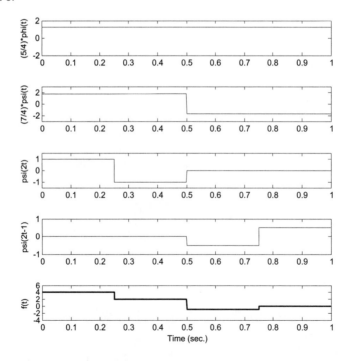

FIGURE 13.38

The signal reconstructed using the Haar wavelets in Example 13.7.

We can use the MATLAB function **dwt()** provided in Section 13.8 to compute the DWT coefficients.
dwt.m
```
function w = dwt(h0,c,kLevel)
% h0 = wavelet filter coefficients (lowpass filter)
% c = input vector
% kLevel = level
% w= wavelet coefficients
```
The results are verified as follows:

>> w=dwt([1/sqrt(2) 1/sqrt(2)],[4 2 −1 0]/2,2)'
w = 1.2500 1.7500 0.7071 −0.3536

From Example 13.7, we can create a time–frequency plot of the DWT amplitudes in two dimensions as shown in Figure 13.39. Assuming the sampling frequency is f_s, we have the smallest frequency resolution as $f_s/N = f_s/4$, where $N = 4$. When j ($j = 0$) is small, we achieve a small frequency resolution $\Delta f = f_s/4$ and each wavelet presents four samples. In this case, we have a good frequency resolution but a poor time resolution. Similarly, when j ($j = 1$) is a large value, the frequency resolution becomes $\Delta f = 2f_s/4$ and each wavelet presents two samples (more details in the time domain). Hence, we achieve a good time resolution but a poor frequency resolution. Note that

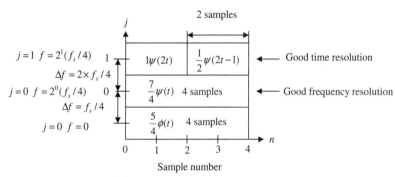

(a) Time–frequency plane

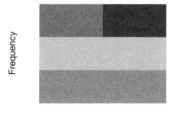

(b) Time–frequency plot

FIGURE 13.39

Time–frequency plot of the DWT amplitudes.

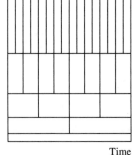

High frequency
(Large j, small scale=2^{-j})

Low frequency
(Small j, large scale=2^{-j})

Time

FIGURE 13.40

Time–frequency plane.

the DWT cannot achieve good resolutions in both frequency and time at the same time. The time–frequency plot of the DWT amplitudes in terms of their intensity is shown in Figure 13.39(b), and the time and frequency plane for the DWT is shown in Figure 13.40.

EXAMPLE 13.8

Given the wavelet coefficients obtained using the Haar wavelet filters

$$[c_0(0)\, d_0(0)\, d_1(0)\, d_1(1)] = \left[\frac{5}{4}\, \frac{7}{4}\, \frac{1}{\sqrt{2}}\, -\frac{1}{2\sqrt{2}}\right]$$

perform the IDWT.

Solution:
From Equation (13.55) we get

$$c_1(k) = \sum_{m=-\infty}^{\infty} c_0(m)h_0(k-2m) + \sum_{m=-\infty}^{\infty} d_0(m)h_1(k-2m)$$

Then we recover coefficients $c_1(k)$ as

$$c_1(0) = \sum_{m=-\infty}^{\infty} c_0(m)h_0(-2m) + \sum_{m=-\infty}^{\infty} d_0(m)h_1(-2m)$$

$$= c_0(0)h_0(0) + d_0(0)h_1(0) = \frac{5}{4} \times \frac{1}{\sqrt{2}} + \frac{7}{4} \times \frac{1}{\sqrt{2}} = \frac{3\sqrt{2}}{2}$$

$$c_1(1) = \sum_{m=-\infty}^{\infty} c_0(m)h_0(1-2m) + \sum_{m=-\infty}^{\infty} d_0(m)h_1(1-2m)$$

$$= c_0(0)h_0(1) + d_0(0)h_1(1) = \frac{5}{4} \times \frac{1}{\sqrt{2}} + \frac{7}{4} \times \left(-\frac{1}{\sqrt{2}}\right) = -\frac{1}{2\sqrt{2}}$$

MATLAB verification using Figure 13.37 is given as

```
>> c1=fconv([1 1]/sqrt(2),[5/4 0])+fconv([1 -1]/sqrt(2),[7/4 0])
c1 = 2.1213 -0.3536
```

where the MATLAB function **fconv()** for filter operations with the forward filter coefficients is listed in Section 13.8. Again, from Equation (13.55), we obtain

$$c_2(k) = \sum_{m=-\infty}^{\infty} c_1(m) h_0(k - 2m) + \sum_{m=-\infty}^{\infty} d_1(m) h_1(k - 2m)$$

Substituting the achieved wavelet coefficients $c_2(k)$, we yield

$$c_2(0) = \sum_{m=-\infty}^{\infty} c_1(m) h_0(-2m) + \sum_{m=-\infty}^{\infty} d_1(m) h_1(-2m)$$

$$= c_1(0) h_0(0) + d_1(0) h_1(0) = \frac{3\sqrt{2}}{2} \times \left(\frac{1}{\sqrt{2}}\right) + \frac{1}{\sqrt{2}} \times \frac{1}{\sqrt{2}} = 2$$

$$c_2(1) = \sum_{m=-\infty}^{\infty} c_1(m) h_0(1 - 2m) + \sum_{m=-\infty}^{\infty} d_1(m) h_1(1 - 2m)$$

$$= c_1(0) h_0(1) + d_1(0) h_1(1) = \frac{3\sqrt{2}}{2} \times \frac{1}{\sqrt{2}} + \frac{1}{\sqrt{2}} \left(-\frac{1}{\sqrt{2}}\right) = 1$$

$$c_2(2) = \sum_{m=-\infty}^{\infty} c_1(m) h_0(2 - 2m) + \sum_{m=-\infty}^{\infty} d_1(m) h_1(2 - 2m)$$

$$= c_1(1) h_0(0) + d_1(1) h_1(0) = \left(-\frac{1}{2\sqrt{2}}\right) \times \frac{1}{\sqrt{2}} + \left(-\frac{1}{2\sqrt{2}}\right) \times \frac{1}{\sqrt{2}} = -\frac{1}{2}$$

$$c_2(3) = \sum_{m=-\infty}^{\infty} c_1(m) h_0(3 - 2m) + \sum_{m=-\infty}^{\infty} d_1(m) h_1(3 - 2m)$$

$$= c_1(1) h_0(1) + d_1(1) h_1(1) = \left(-\frac{1}{2\sqrt{2}}\right) \times \frac{1}{\sqrt{2}} + \left(-\frac{1}{2\sqrt{2}}\right) \times \left(-\frac{1}{\sqrt{2}}\right) = 0$$

We can verify the results using the MATLAB program as follows:

```
>> c2=fconv([1 1]/sqrt(2),[3*sqrt(2)/2 0 −1/(2*sqrt(2)) 0])+fconv([1 −1]/sqrt(2),[1/sqrt(2) 0 −1/(2*sqrt(2)) 0])
c2 = 2.0000 1.0000 −0.5000 0
```

Scaling the wavelet coefficients, we finally recover the original sample values as

$$f(k) = 2^{2/2}[2\ 1 - 0.5\ 0] = [4\ 2 - 1\ 0]$$

Similarly, we can use the MATLAB function **idwt()** provided in Section 13.8 to perform the IDWT.

```
idwt.m
function c = idwt(h0,w,kLevel)
% h0 = wavelet filter coefficients (lowpass filter)
% w= wavelet coefficients
% kLevel = level
% c = input vector
```

Appling the MATLAB function **idwt()** leads to

```
>> f=2*idwt([1/sqrt(2) 1/sqrt(2)],[5/4 7/4 1/sqrt(2) −1/(2*sqrt(2))],2)'
f = 4.0000 2.0000 −1.0000 0.0000
```

Since $2^{j/2}$ scales signal amplitudes down in the analysis stage and scales them back up in the synthesis stage, we can omit $2^{j/2}$ by using $c(k) = f(k)$ directly in practice.

EXAMPLE 13.9

Given the sample values [4 2 −1 0], use the provided MATLAB DWT (dwt.m) and IDWT (idwt.m) and specified wavelet filter to perform the DWT and IWDT without using the scale factor $2^{j/2}$.

a. Haar wavelet filter
b. 4-tap Daubechies wavelet filter

Solution:

a. From Table 13.2, the Haar wavelet filter coefficients are

$$h_0 = \left(\frac{1}{\sqrt{2}} \frac{1}{\sqrt{2}} \right)$$

Applying the MATLAB functions **dwt()** and **idwt()**, we have

```
>> w=dwt([1/sqrt(2) 1/sqrt(2)],[4 2 −1 0],2)'
w = 2.5000 3.5000 1.4142 −0.7071

>> f=idwt([1/sqrt(2) 1/sqrt(2)],w,2)'
f = 4.0000 2.0000 −1.0000 0
```

b. From Table 13.2, the 4-Tap Duabechies wavelet filter coefficients are

h0=[0.482962913144534 0.836516303737808 0.224143868042013 −0.129409522551260]

MATLAB program verification is demonstrated as follows:

```
>> w=dwt([0.482962913144534  0.836516303737808  0.224143868042013  −0.129409522551260],
[4 2 −1 0],2)'
w = 2.5000 2.2811 −1.8024 2.5095

>> f=idwt([0.482962913144534 0.836516303737808 0.224143868042013 −0.129409522551260],
w,2)'
f = 4.0000 2.0000 −1.0000 0
```

13.7 WAVELET TRANSFORM CODING OF SIGNALS

We can apply the DWT and IWDT for data compression and decompression. The compression and decompression involves two stages, that is, the analysis stage and the synthesis stage. At the analysis stage, the wavelet coefficients are quantized based on their significance. Usually, we assign more bits to the coefficient in a coarser scale, since the corresponding subband has larger signal energy and low frequency components. We assign a small number of bits to a coefficient that resides in a finer scale, since the corresponding subband has lower signal energy and high frequency components. The quantized coefficients can be efficiently transmitted. The DWT coefficients are laid out in a format described in Figure 13.41. The coarse coefficients are placed towards the left side. For example, in Example 13.7, we organized the DWT coefficient vector as

$$w_2(k) = [c_0(0)d_0(0)d_1(0)d_1(1)] = \left[\frac{5}{4} \frac{7}{4} \frac{1}{\sqrt{2}} - \frac{1}{2\sqrt{2}} \right]$$

Let us look at the following simulation examples.

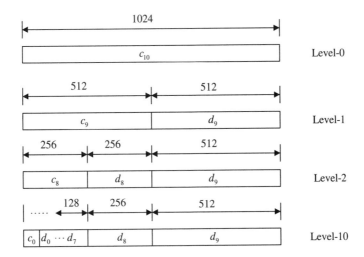

FIGURE 13.41

DWT coefficient layout.

EXAMPLE 13.10

Consider a 40-Hz sinusoidal signal plus random noise sampled at 8,000 Hz with 1,024 samples:

$$x(n) = 100 \cos (2\pi \times 40nT) + 10 \times randn$$

where $T = 1/8,000$ seconds and $randn$ is a random noise generator with a unit power and Gaussian distribution. Use a 16-bit code for each wavelet coefficient and write a MATLAB program to perform data compression for each of the following ratios: 2:1, 4:1, 8:1, and 16:1. Plot the reconstructed waveforms.

Solution:
We use the 8-tap Daubechies filter as listed in Table 13.2. We achieve the data compression by dropping the high subband coefficients for each level consecutively and coding each wavelet coefficient in the lower subband using 16 bits. For example, we achieve the 2:1 compression ratio by omitting 512 high frequency coefficients at the first level, 4:1 by omitting 512 high frequency coefficients at the first level, and 256 high frequency coefficients at the second level, and so on. The recovered signals are plotted in Figure 13.42. SNR = 21 dB is achieved for the 2:1 compression ratio. As we can see, when more and more higher frequency coefficients are dropped, the reconstructed signal contains less and less details. The recovered signal with 16:1 compression presents the least details but shows the smoothest signal. On the other hand, omitting the high frequency wavelet coefficients can be very useful for a signal denoising application, in which the high frequency noise contaminating the clean signal is removed. A complete MATLAB program is given in Program 13.2.

Program 13.2. Wavelet data compression.

```
close all; clear all;clc
t=0:1:1023;t=t/8000;
x=100*cos(40*2*pi*t)+10*randn(1,1024);
h0=[0.230377813308896 0.714846570552915 0.630880767929859 ...
  -0.027983769416859 -0.187034811719092 0.030841381835561 ....
  0.032883011666885 -0.010597401785069];
N=1024; nofseg=1
rec_sig=[]; rec_sig2t1=[]; rec_sig4t1=[]; rec_sig8t1=[]; rec_sig16t1=[];
for i=1:nofseg
```

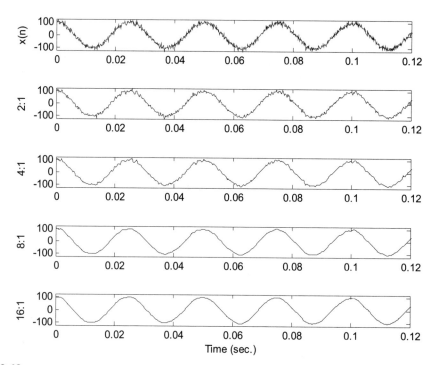

FIGURE 13.42

Reconstructed signal at various compression ratios.

```
  sp=x((i-1)*1024+1:i*1024);
  w=dwt(h0,sp,10);
% Quantization
  wmax=round(max(abs(w)));
  wcode=round(2^15*w/wmax); % 16-bit code for storage
  w=wcode*wmax/2^15; % Recovered wavelet coefficients
  w(513:1024)=zeros(1,512); % 2:1 compression ratio
  sig_rec2t1=idwt(h0,w,10);
  rec_sig2t1=[rec_sig2t1 sig_rec2t1'];
  w(257:1024)=0; % 4:1 compression ratio
  sig_rec4t1=idwt(h0,w,10);
  rec_sig4t1=[rec_sig4t1 sig_rec4t1'];
  w(129:1024)=0; % 8:1 compression ratio
  sig_rec8t1=idwt(h0,w,10);
  rec_sig8t1=[rec_sig8t1 sig_rec8t1'];
  w(65:1024)=0; % 16:1 compression ratio
  sig_rec16t1=idwt(h0,w,10);
  rec_sig16t1=[rec_sig16t1 sig_rec16t1'];
end
subplot(5,1,1),plot(t,x,'k'); axis([0 0.12 -120 120]);ylabel('x(n)');
subplot(5,1,2),plot(t,rec_sig2t1,'k'); axis([0 0.12 -120 120]);ylabel('2:1');
subplot(5,1,3),plot(t,rec_sig4t1,'k'); axis([0 0.12 -120 120]);ylabel(4:1);
subplot(5,1,4),plot(t,rec_sig8t1,'k'); axis([0 0.12 -120 120]);ylabel('8:1');
```

```
subplot(5,1,5),plot(t,rec_sig16t1,'k'); axis([0 0.12 -120 120]);ylabel('16:1');
xlabel('Time (sec.)')
NN=min(length(x),length(rec_sig2t1)); axis([0 0.12 -120 120]);
err=rec_sig2t1(1:NN)-x(1:NN);
SNR=sum(x.*x)/sum(err.*err);
disp('PR reconstruction SNR dB=>');
SNR=10*log10(SNR)
```

Figure 13.43 shows the wavelet compression for 16-bit speech data sampled at 8 kHz. The original speech data is divided into speech segments, each with 1,024 samples. After applying the DWT to each segment, the coefficients, which correspond to high frequency components indexed from 513 to 1,024, are discarded in order to achieve coding efficiency. The reconstructed speech data has a compression ratio 2:1 with SNR = 22 dB. The MATLAB program is given in Program 13.3.

Program 13.3. Wavelet data compression for speech segments.

```
close all; clear all;clc
load orig.dat ; % Load speech data
h0=[0.230377813308896 0.714846570552915 0.630880767929859 ...
  -0.027983769416859 -0.187034811719092 0.030841381835561 ....
  0.032883011666885 -0.010597401785069];
N=length(orig);
nofseg=ceil(N/1024);
speech=zeros(1,nofseg*1024);
```

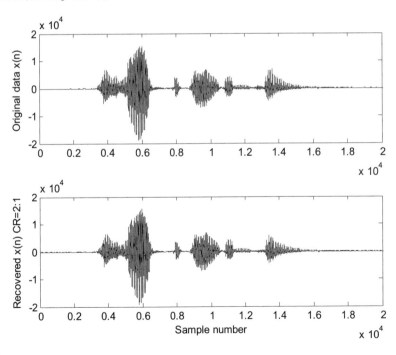

FIGURE 13.43

Reconstructed speech signal with compression ratio of 2 and SNR = 22 dB.

```
speech(1:N)=orig(1:N); % Making the speech length a multiple of 1024 samples
rec_sig=[];
for i=1:nofseg
  sp=speech((i-1)*1024+1:i*1024);
  w=dwt(h0,sp,10);
% Quantization
  w=(round(2^15*w/2^15))*2^(15-15);
  w(513:1024)=zeros(1,512); % Omitting the high frequency coefficients
  sp_rec=idwt(h0,w,10);
  rec_sig=[rec_sig sp_rec'];
end
subplot(2,1,1),plot([0:length(speech)-1],speech,'k');axis([0 20000 -20000 20000]);
ylabel('Original data x(n)');
subplot(2,1,2),plot([0:length(rec_sig)-1],rec_sig,'k');axis([0 20000 -20000 20000]);
xlabel('Sample number');ylabel('Recovered x(n) CR=2:1');
NN=min(length(speech),length(rec_sig));
err=rec_sig(1:NN)-speech(1:NN);
SNR=sum(speech.*speech)/sum(err.*err);
disp('PR reconstruction SNR dB=>');
SNR=10*log10(SNR)
```

Figure 13.44 displays the wavelet compression for 16-bit ECG data using Program 13.3. The reconstructed ECG data has a compression ratio of 2:1 with SNR = 33.8 dB.

Figure 13.45 illustrates an application of signal denoising using the DWT with a coefficient threshold. During the analysis stage, an obtained DWT coefficient (quantization is not necessary) is set to zero if its value is less than the predefined threshold depicted in Figure 13.45. This simple technique is called the hard threshold. Usually, the small wavelet coefficients are related to the high frequency components in signals. Therefore, setting high frequency components to zero is the same as lowpass filtering.

An example is shown in Figure 13.46. The first plot depicts a 40-Hz noisy sinusoidal signal (sine wave plus noise with SNR = 18 dB) and the clean signal with a sampling rate of 8,000 Hz. The second plot shows that after zero threshold operations, 67% of coefficients are set to zero and the recovered signal has SNR = 19 dB. Similarly, the third and fourth plots illustrate that 93% and 97% of coefficients are set to zero after threshold operations and the recovered signals have SNR = 23 and 28 dB, respectively. As an evidence that the signal is smoothed, that is, high frequency noise is attenuated, the wavelet denoising technique is equivalent to lowpass filtering.

13.8 MATLAB PROGRAMS

In this section, four key MATLAB programs are listed. **rconv()** and **fconv()** perform circular convolutions with the reversed filter coefficients and the forward filter coefficients, respectively. **dwt()** and **idwt()** are the programs to compute the DWT coefficients and IDWT coefficients. The resolution level can be specified.

Program 13.4. Circular convolution with the reversed filter coefficients (rconv.m).

```
function [y] = rconv(h,c)
% Circular convolution using the reversed filter coefficients h(-k)
```

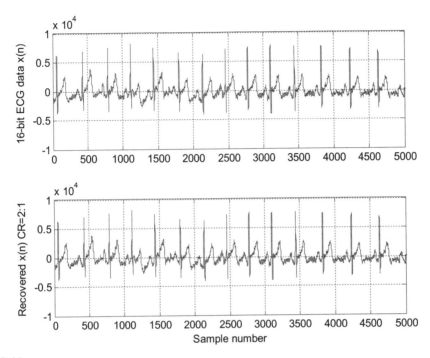

FIGURE 13.44

Reconstructed ECG signal with compression ratio of 2 and SNR = 33.8 dB.

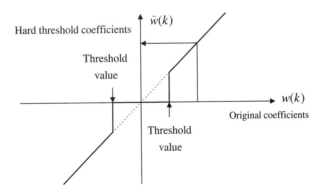

FIGURE 13.45

Hard threshold for the DWT coefficients.

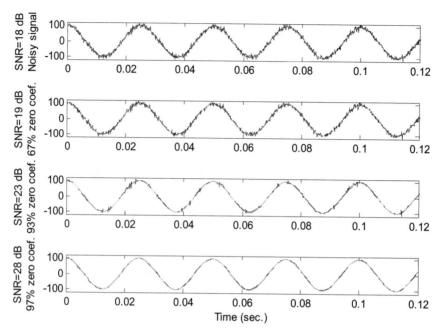

FIGURE 13.46

Signal denoising using wavelet transform coding.

```
% h = filter coefficients
% c = input vector
% y = output vector
N=length(c); M=length(h);
xx=zeros(1,M+N-1);
xx(1:N)=c;
xx(N+1:N+M-1)=c(1:M-1); % Use periodized input
for n=1:N;
y(n)=0;
 for m=1:M
  y(n)=y(n)+h(m)*xx(n+m-1);
  end
end
```

Program 13.5. Circular convolution with the forward filter coefficients (fconv.m).

```
function [y] = fconv(h,c)
% Circular convolution using the forward filter coefficients h(k)
% h = filter coefficients
% c = input vector
% y = output vector
N=length(c); M=length(h);
```

```
x(1:N+M-1) = zeros(1,N+M-1);
  for j = 1:N
  x(j:M+(j-1)) = x(j:M+(j-1)) + c(j)*h;
  end
  for i = N+M-1:-1:N+1
  x(i-N) = x(i-N) + x(i); % Circular convolution
  end
  y=x(1:N);
```

Program 13.6. DWT coefficients (dwt.m).

```
function w = dwt(h0,c,kLevel)
% w = dwt(h,c,k)
% Computes wavelet transform coefficients for a vector c using the
% orthonormal wavelets defined by the coefficients h.
% h = wavelet coefficients
% c =input vector
% kLelvel= level
% w = wavelet coefficients
n=length(c); m = length(h0);
h1 = h0(m:-1:1); h1(2:2:m)=-h1(2:2:m);
h0 = h0(:)'; h1 = h1(:)';
c = c(:); w = c;
x = zeros(n+m-2,1);
% Perform decomposition through k levels
% at each step, x = periodized version of x coefficients
for j = 1:kLevel
  x(1:n) = w(1:n);
  for i = 1:m-2
  x(n+i) = x(i);
  end
  for i = 1:n/2
  w(i) = h0 * x(1 + 2*(i-1):m + 2*(i-1));
  w(n/2 + i) = h1* x(1 + 2*(i-1):m + 2*(i-1));
  end
  n = n/2;
end
```

Program 13.7. IDWT coefficients (idwt.m).

```
function c = idwt(h0,w,kLevel)
% c = idwt(h0,w,kLevel)
% Computes the inverse fast wavelet transform from data W using the
% orthonormal wavelets defined by the coefficients.
% h0 = wavelet filter coefficients
% w = wavelet coefficients
% kLevel = level
% c = IDWT coefficients
n=length(w); m = length(h0);
h1 = h0(m:-1:1); h1(2:2:m)=-h1(2:2:m);
```

```
h0 = h0(:); h1 = h1(:);
w = w(:); c = w;
x = zeros(n+m-2,1);
% Perform the reconstruction through k levels
% x = periodized version of x coefficients
n = n/2^kLevel;
for i = 1:kLevel
  x(1:2*n+m-2) = zeros(2*n+m-2,1);
  for j = 1:n
  x(1+2*(j-1):m+2*(j-1)) = x(1+2*(j-1):m+2*(j-1)) + c(j)*h0 + w(n+j)*h1;
  end
  for i = 2*n+m-2:-1:2*n+1
  x(i-2*n) = x(i-2*n) + x(i);
  end
  c(1:2*n) = x(1:2*n);
  n = 2 * n;
end
```

13.9 SUMMARY

1. A signal can be decomposed using a filter bank system. The filter bank contains two stages: the analysis stage and the synthesis stage. The analysis stage applies analysis filters to decompose the signal into multiple channels. The signal from each channel is downsampled and coded. At the synthesis stage, the recovered signal from each channel is upsampled and processed using its synthesis filter. Then the outputs from all the synthesis filters are combined to produce the recovered signal.

2. Perfect reconstruction conditions for the two-band case are derived to design the analysis and synthesis filters. The conditions consist of a half-band filter requirement and normalization. Once the lowpass analysis filter coefficients are obtained, the coefficients for other filters can be achieved from the derived relationships.

3. In a binary tree structure, a filter bank divides an input signal into two equal subbands, resulting in the low and high bands. Each band again splits into low and high bands to produce quarter bands. The process continues in this form.

4. The dyadic structure implementation of the filter bank first splits the input signal to low and high bands and then continues to split the low band only each time.

5. By quantizing each subband channel using the assigned number bits based on the signal significance (more bits assigned to code the samples for channels with large signal energy while less bits assigned to code the samples for channels with small signal energy), the subband coding method demonstrates efficiency for data compression.

6. The wavelet transform can identify the frequencies in a signal and the times when the signal elements occur and end.

7. The wavelet transform provides either good frequency resolution or good time resolution, but not both.

8. The wavelet has two important properties: scaling and translation. The scaling process is related to changes of wavelet frequency (oscillation) while translation is related to the time localization.

9. A family of wavelets contains a father wavelet and a mother wavelet, and their scaling and translation versions. The father wavelet and its scaling and translation are called the scaling function while the mother wavelet and its scaling and translation are called the wavelet function. Each scaling function and wavelet function can be presented using the scaling functions in the next finer scale.

10. A signal can be approximated from a sum of weighted scaling functions and wavelet functions. The weights are essentially the DWT coefficients. A signal can also be coded at any desired level using smaller-scale wavelets.

11. Implementation of DWT and IDWT consists of the analysis and synthesis stages, which are similar to the subband coding scheme. The implementation uses the dyadic structure but with analysis filter coefficients in a reversed format.

12. The DWT and IDWT are very effective for data compression or signal denoising by eliminating smaller DWT coefficients, which correspond to higher frequency components.

13.10 PROBLEMS

13.1. Given the downsampling systems in Figure 13.47(a) and (b) and input spectrum $W(f)$, sketch the downsampled spectrum $X(f)$.

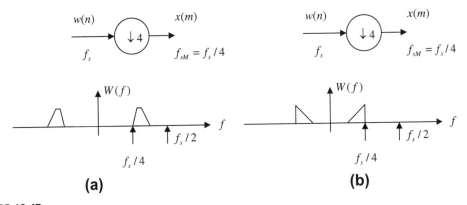

FIGURE 13.47

Downsampling systems in Problem 13.1.

13.2. Given the upsampling systems in Figure 13.48(a) and (b) and input spectrum $X(f)$, sketch the upsampled spectrum $\overline{W}(f)$. Note that the sampling rate for input $x(m)$ $f_{sM} = f_s/4$ and the output sampling rate $\overline{w}(n)$ is f_s.

13.3. Given the down- and upsampling systems in Figure 13.49(a) and (b) and the input spectrum $W(f)$, sketch the output spectrum $\overline{Y}(f)$ and express $\overline{Y}(f)$ in terms of $W(f)$.

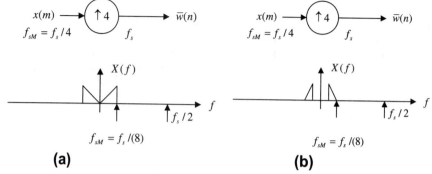

(a)   **(b)**

FIGURE 13.48

Upsampling systems in Problem 13.2.

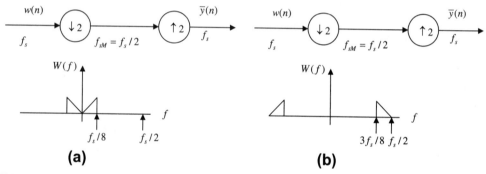

(a)   **(b)**

FIGURE 13.49

Down- and upsampling systems in Problem 13.3.

13.4. Given the down- and upsampling systems in Figure 13.50(a) and (b) and the input spectrum $W(f)$, sketch the output spectrum $\overline{Y}(f)$ and express $\overline{Y}(f)$ in terms of $W(f)$.

13.5. Given $H_0(z) = \dfrac{1}{\sqrt{2}} + \dfrac{1}{\sqrt{2}}z^{-1}$, determine $H_1(z)$, $G_0(z)$, and $G_1(z)$.

13.6. Given $H_0(z) = \dfrac{1}{\sqrt{2}} + \dfrac{1}{\sqrt{2}}z^{-1}$, verify the following conditions:

$$p(2n) = \sum_{k=0}^{N-1} h_0(k)h_0(k+2n) = \delta(n)$$
$$R(z) + R(-z) = 2$$

Also, plot the magnitude frequency responses of the analysis and synthesis filters.

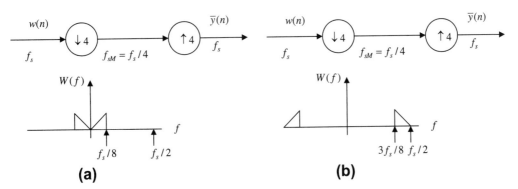

FIGURE 13.50

Down- and upsampling systems in Problem 13.4.

13.7. Given

$$H_0(z) = 0.483 + 0.837z^{-1} + 0.224z^{-2} - 0.129z^{-3}$$

determine $H_1(z)$, $G_0(z)$, and $G_1(z)$.

13.8. Given

$$H_0(z) = 0.483 + 0.837z^{-1} + 0.224z^{-2} - 0.129z^{-3}$$

verify the following conditions:

$$\rho(2n) = \sum_{k=0}^{N-1} h_0(k)h_0(k+2n) = \delta(n)$$
$$R(z) + R(-z) = 2$$

13.9. Draw a four-band dyadic tree structure of a subband system including the analyzer and synthesizer.

13.10. Draw an eight-band dyadic tree structure of a subband system including the analyzer and synthesizer.

13.11. Consider the function in Figure 13.51.

Sketch

a. $f(4t)$

b. $f(t-2)$

c. $f(2t-3)$

d. $f(t/2)$

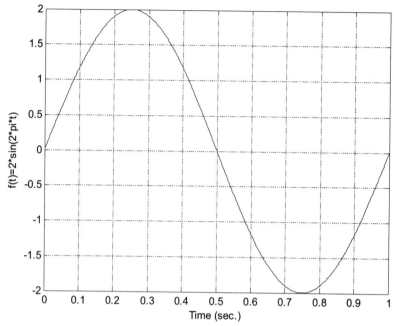

FIGURE 13.51

A sine function in Problem 13.11.

      e. $f(t/4 - 0.5)$

13.12. Given a father wavelet (base scaling function) in base scale plotted in Figure 13.52(a), determine a and b for each of the wavelets plotted in Figure 13.52(b) and (c).

13.13. Consider the signal in Figure 13.53.

    Sketch

    a. $f(4t)$

    b. $f(t - 2)$

    c. $f(2t - 3)$

    d. $f(t/2)$

    e. $f(t/4 - 0.5)$

13.14. Consider the signal in Figure 13.54.

    Sketch

    a. $f(4t)$

    b. $f(t - 2)$

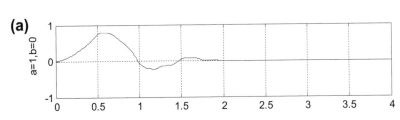

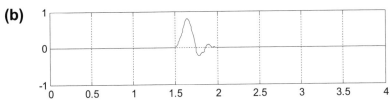

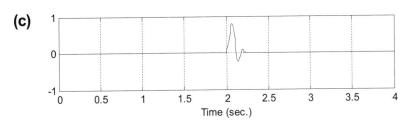

FIGURE 13.52

Wavelets in Problem 13.12.

c. $f(2t - 3)$

d. $f(t/2)$

e. $f(t/4 - 1)$

13.15. Sketch the Haar father wavelet families for three different scales, $j = 0, 1, 2$ for a period of 2 seconds.

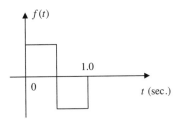

FIGURE 13.53

The function in Problem 13.13.

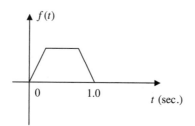

FIGURE 13.54

A trapezoidal function in Problem 13.14.

13.16. Sketch the Haar mother wavelet families for three different scales, $j = 0, 1, 2$ for a period of 2 seconds.

13.17. Use the Haar wavelet family to expand the signal depicted in Figure 13.55.

   a. Use only scaling functions $\phi(2t - k)$.

   b. Use scaling functions and wavelets $\phi(t)$ and $\psi(t - k)$

13.18. Use the Haar wavelet family to expand the signal depicted in Figure 13.56.

   a. Use only scaling functions $\phi(4t - k)$.

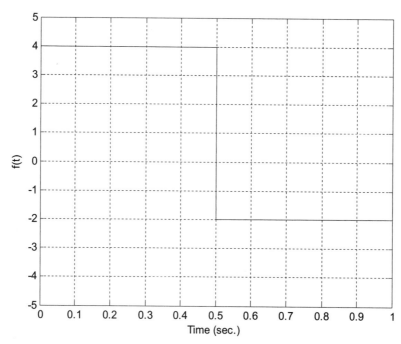

FIGURE 13.55

A gate function in Problem 13.17.

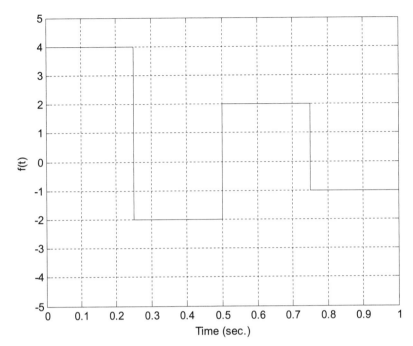

FIGURE 13.56

A piecewise function in Problem 13.18.

    b. Use scaling functions and wavelets $\phi(2t - k)$ and $\psi(2t - k)$.

    c. Use scaling functions and wavelets $\phi(t)$, $\psi(2t - k)$, and $\psi(t - k)$.

13.19. Use the Haar wavelet family to expand the signal

$$x(t) = \sin(2\pi t) \quad \text{for} \quad 0 \le t \le 1$$

    a. Use only scaling functions $\phi(2t - k)$.

    b. Use scaling functions and wavelets $\phi(t)$ and $\psi(t - k)$.

13.20. Use the Haar wavelet family to expand the signal

$$x(t) = e^{-5t} \quad \text{for} \quad 0 \le t \le 1$$

    a. Use only scaling functions $\phi(2t - k)$.

    b. Use scaling functions and wavelets $\phi(t)$ and $\psi(t - k)$.

13.21. Verify the following equations using the Haar wavelet families:

    a. $\phi(2t) = \sum_{k=-\infty}^{\infty} \sqrt{2} h_0(k) \phi(4t - k)$

    b. $\psi(2t) = \sum_{k=-\infty}^{\infty} \sqrt{2} h_1(k) \phi(4t - k)$

13.22. Given the 4-tap Daubechies wavelet coefficients

$$h_0(k) = [0.483\ 0.837\ 0.224\ -0.129]$$

determine $h_1(k)$ and plot magnitude frequency responses for both $h_0(k)$ and $h_1(k)$.

13.23. Given the sample values $[8\ -2\ 4\ 1]$, use the Haar wavelet to determine the level-2 wavelet coefficients.

13.24. Given the sample values $[8\ -2\ 4\ 3\ 0\ -1\ -2\ 0]$, use the Haar wavelet to determine the level-3 wavelet coefficients.

13.25. Given the level-2 wavelet coefficients $[4\ 2\ -1\ 2]$, use the Haar wavelet to determine the sampled signal vector $f(k)$.

13.26. Given the level-3 wavelet coefficients $[4\ 2\ -1\ 2\ 0\ 0\ 0\ 0]$, use the Haar wavelet to determine the sampled signal vector $f(k)$.

13.27. Given the level-1 wavelet coefficients $[4\ 2\ -1\ 2]$, use the Haar wavelet to determine the sampled signal vector $f(k)$.

13.28. The four-level DWT coefficients are given as follows:

$$W = [100\ 20\ 16\ -5\ -3\ 4\ 2\ -6\ 4\ 6\ 1\ 2\ -3\ 0\ 2\ -1]$$

List the wavelet coefficients to achieve each of the following compression ratios:

a. 2:1

b. 4:1

c. 8:1

d. 16:1

13.10.1 MATLAB Problems

Use MATLAB to solve Problems 13.29 to 13.31.

13.29. Use the 16-tap PR-CQF coefficients and MATLAB to verify the following conditions:

$$p(2n) = \sum_{k=0}^{N-1} h_0(k)h_0(k+2n) = \delta(n)$$

$$R(z) + R(-z) = 2$$

Plot the frequency responses for $h_0(k)$ and $h_1(k)$.

13.30. Use the MATLAB functions provided in Section 13.8 [**dwt()**, **idwt()**] to verify Problems 13.23–13.27.

13.31. Consider a 20-Hz sinusoidal signal plus random noise sampled at 8,000 Hz with 1,024 samples:

$$x(n) = 100\cos(2\pi \times 20nT) + 50 \times randn$$

where $T = 1/8,000$ seconds and *randn* is a random noise generator with a unit power and Gaussian distribution.

a. Use a 16-bit code for each wavelet coefficient and write a MATLAB program to perform data compression with the following ratios: 2:1, 4:1, 8:1, 16:1, and 32:1.
b. Measure the SNR in dB for each case.
c. Plot the reconstructed waveform for each case.

13.10.2 MATLAB Projects

13.32. Data compression using subband coding:

Given 16-bit speech data ("speech.dat") and using the four-band subband coding method, write a MATLAB program to compress a speech signal with the following specifications:

a. 16 bits for each of the subband coefficients, code the LL, LH, HL, HH subbands, and measure the SNR in dB.
b. 16 bits for each of the subband coefficients, code the LL band, discard the LH, HL, and HH subbands.
c. 16 bits for each of the subband coefficients, code the LL and LH bands, discard the HL and HH subbands.
d. 16 bits for each of subband coefficients, code the LL, LH, and HL bands, discard the HH subband.
e. Measure SNR in dB for (a), (b), (c), and (d).
f. Determine the achieved compression ratios for (a), (b), (c), and (d).
g. Repeat (a) to (f) for seismic data ("seismic.dat") in which each sample is encoded using 32 bits instead of 16 bits.

13.33. Wavelet-based data compression:

Given 16-bit speech data ("speech.dat") and using the wavelet coding method with 16-tap Daubechies wavelet filters, write a MATLAB program to compress a speech signal with the following specifications:

a. 16 bits for each of the wavelet coefficients, compression 2:1.
b. 16 bits for each of the wavelet coefficients, compression 4:1.
c. 16 bits for each of the wavelet coefficients, compression 8:1.
d. 16 bits for each of the wavelet coefficients, compression 16:1.
e. 16 bits for each of the wavelet coefficients, compression 32:1.
f. Measure SNR in dB for (a), (b), (c), (d) and (e).
g. Repeat (a) to (f) for seismic data ("seismic.dat") in which each sample is encoded using 32 bits instead of 16 bits.

Image Processing Basics

CHAPTER OUTLINE

14.1 Image Processing Notation and Data Formats..684
    14.1.1 8-Bit Gray Level Images 684
    14.1.2 24-bit Color Images 686
    14.1.3 8-Bit Color Images...687
    14.1.4 Intensity Images 688
    14.1.5 Red, Green, and Blue Components and Grayscale Conversion 688
    14.1.6 MATLAB Functions for Format Conversion 690
14.2 Image Histogram and Equalization...692
    14.2.1 Grayscale Histogram and Equalization 692
    14.2.2 24-Bit Color Image Equalization 695
    14.2.3 8-Bit Indexed Color Image Equalization 700
    14.2.4 MATLAB Functions for Equalization 702
14.3 Image Level Adjustment and Contrast 704
    14.3.1 Linear Level Adjustment..704
    14.3.2 Adjusting the Level for Display...707
    14.3.3 MATLAB Functions for Image Level Adjustment 707
14.4 Image Filtering Enhancement 707
    14.4.1 Lowpass Noise Filtering 709
    14.4.2 Median Filtering 712
    14.4.3 Edge Detection..715
    14.4.4 MATLAB Functions for Image Filtering 718
14.5 Image Pseudo-Color Generation and Detection 722
14.6 Image Spectra..725
14.7 Image Compression by Discrete Cosine Transform 728
    14.7.1 Two-Dimensional Discrete Cosine Transform 729
    14.7.2 Two-Dimensional JPEG Grayscale Image Compression Example................................731
    14.7.3 JPEG Color Image Compression 735
        RGB to YIQ Transformation 735
        DCT on Image Blocks..735
        Quantization...735
        Differential Pulse Code Modulation on Direct-Current Coefficients.................................736
        Run-Length Coding on Alternating-Current Coefficients 736
        Lossless Entropy Coding..737

Digital Signal Processing.
Copyright © 2013 Elsevier Inc. All rights reserved.

683

    Coding DC Coefficients...737
    Coding AC Coefficients...737
    14.7.4 Image Compression Using Wavelet Transform Coding 738
14.8 Creating a Video Sequence by Mixing Two Images 745
14.9 Video Signal Basics 746
    14.9.1 Analog Video 747
        PAL Video..752
        SECAM Video..752
    14.9.2 Digital Video 753
14.10 Motion Estimation in Video 755
14.11 Summary 757

OBJECTIVES:

In today's modern computers, media information such as audio, images, and video have become necessary for daily business operations and entertainment. In this chapter, we will study the digital image and its processing techniques. This chapter introduces the basics of image processing, including image enhancement using histogram equalization and filtering methods, and proceeds to study pseudo-color generation for object detection and recognition. Finally, the chapter investigates image compression techniques and the basics of video signals.

14.1 IMAGE PROCESSING NOTATION AND DATA FORMATS

The digital image is picture information in a digital form. The image can be filtered to remove noise to enhance it. It can also be transformed to extract features for pattern recognition. The image can be compressed for storage and retrieval, as well as transmitted via a computer network or a communication system.

The digital image consists of pixels. The position of each pixel is specified in terms of an index for the number of columns and another for the number of rows. Figure 14.1 shows that a pixel $p(2, 8)$ has a level of 86 and is located in the second row, eighth column. We express it in notation as

$$p(2, 8) = 86 \qquad (14.1)$$

The number of pixels in the presentation of a digital image is its *spatial resolution*, which relates to the image quality. The higher the spatial resolution, the better quality the image has. The resolution can be fairly high, for instance, as high as $1,600 \times 1,200$ (1,920,000 pixels = 1.92 megapixels), or as low as 320×200 (64,000 pixels = 64 kilopixels). In notation, the number to the left of the multiplication symbol represents the width, and that to the right of the symbol represents the height. Image quality also depends on the numbers of bits used in encoding each pixel level, which will be discussed in next section.

14.1.1 8-Bit Gray Level Images

If a pixel is encoded on a gray scale from 0 to 255, where $0 =$ black and $255 =$ white, the numbers in between represent levels of gray forming a *grayscale image*. For a 640×480 8-bit image, 307.2

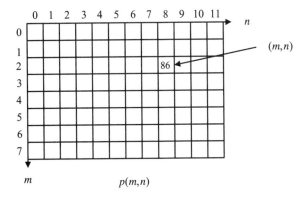

FIGURE 14.1

Image pixel notation.

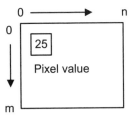

FIGURE 14.2

Grayscale image format.

kilobytes are required for storage. Figure 14.2 shows a grayscale image format. As shown in the figure, the pixel indicated in the box has an 8-bit value of 25.

The image of a cruise ship with a spatial resolution of 320×240 using an 8-bit grayscale format is shown in Figure 14.3.

FIGURE 14.3

Grayscale image (8-bit 320×240).

14.1.2 **24-bit Color Images**

In a 24-bit color image representation, each pixel is recoded with red, green, and blue (RGB) components. With each component value encoded in 8 bits, resulting in 24 bits in total, we achieve a full color RGB image. With such an image, we can have $2^{24} = 16.777216 \times 10^6$ different colors. A 640×480 24-bit color image requires 921.6 kilobytes for storage. Figure 14.4 shows the format for the 24-bit color image where the indicated pixel has 8-bit RGB components.

Figure 14.5 shows a 24-bit color image of the Grand Canyon, along with grayscale displays for the 8-bit RGB component images. The full color picture at the upper left is included in the color insert.

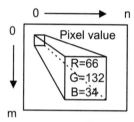

FIGURE 14.4

The 24-bit color image format.

(a) RGB   **(b)** Red

(c) Green   **(d)** Blue

FIGURE 14.5

The 24-bit color image and its respective RGB components. See color image on book's companion site.

14.1.3 **8-Bit Color Images**

The 8-bit color image is also a popular image format. Its pixel value is a color index that points to a color lookup table containing RGB components. We call this a *color indexed image*, and its format is shown in Figure 14.6. As an example in the figure, the color indexed image has a pixel index value of 5, which is the index for the entry of the color table, called the *color map*. At location 5 in the color table, there are three color components with RGB values of 66, 132, and 34, respectively. Each color component is encoded in 8 bits. There are only 256 different colors in the image. A 640×480 8-bit color image requires 307.2 kilobytes for data storage and $3 \times 256 = 768$ bytes for color map storage. The 8-bit color image for the cruise ship shown in Figure 14.3 is displayed in Figure 14.7.

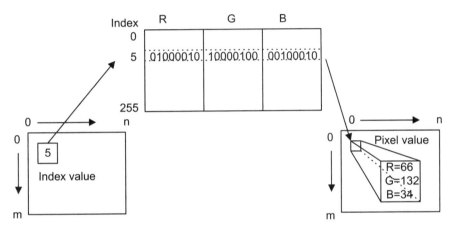

FIGURE 14.6

The 8-bit color indexed image format.

FIGURE 14.7

The 8-bit color indexed image. See color image on book's companion site.

14.1.4 Intensity Images

As we noted in the first section, the grayscale image uses a pixel value ranging from 0 to 255 to present luminance, or the light intensity. A pixel value of 0 designates black, and a value of 255 represents white.

In some processing environments such as MATLAB (*matrix laboratory*), floating-point operations are used. The grayscale image has an intensity value that is normalized to the range from 0 to 1.0, where 0 represents black and 1 represents white. We often change the pixel value to the normalized range to get the grayscale intensity image before processing it, then scale it back to the standard 8-bit range after processing for display. With the intensity image in floating point format, the digital filter implementation can be easily applied. Figure 14.8 shows the format of the grayscale intensity image, where the indicated pixel shows the intensity value of 0.5988.

14.1.5 Red, Green, and Blue Components and Grayscale Conversion

In some applications, we need to convert a color image to a grayscale image so that storage space can be saved. As an example, fingerprint images are stored in grayscale format in a database system. As another example, in color image compression, the transformation converts the RGB color space to the YIQ color space (Li and Drew, 2004; Rabbani and Jones, 1991), where Y is the luminance (Y) channel representing light intensity while the I (in-space) and Q (quadrature) chrominance channels represent color details.

The luminance $Y(m, n)$ carries grayscale information with most of the signal energy (as much as 93%), and the chrominance channels $I(m, n)$ and $Q(m, n)$ carry color information with much less energy (as little as 7%). The transformation in terms of the standard matrix notion is given by

$$\begin{bmatrix} Y(m,n) \\ I(m,n) \\ Q(m,n) \end{bmatrix} = \begin{bmatrix} 0.299 & 0.587 & 0.114 \\ 0.596 & -0.274 & -0.322 \\ 0.212 & -0.523 & 0.311 \end{bmatrix} \begin{bmatrix} R(m,n) \\ G(m,n) \\ B(m,n) \end{bmatrix} \tag{14.2}$$

As an example of data compression, after transformation, we can encode $Y(m, n)$ with a higher resolution using a larger number of bits, since it contains most of the signal energy, while we encode chrominance channels $I(m, n)$ and $Q(m, n)$ with less resolution using a smaller number of bits. Inverse transformation can be solved as

$$\begin{bmatrix} R(m,n) \\ G(m,n) \\ B(m,n) \end{bmatrix} = \begin{bmatrix} 1.000 & 0.956 & 0.621 \\ 1.000 & -0.272 & -0.647 \\ 1.000 & -1.106 & 1.703 \end{bmatrix} \begin{bmatrix} Y(m,n) \\ I(m,n) \\ Q(m,n) \end{bmatrix} \tag{14.3}$$

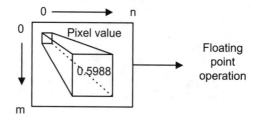

FIGURE 14.8

The grayscale intensity image format.

To obtain the grayscale image, we simply convert each RGB pixel to the YIQ pixel, and then keep its luminance channel and discard IQ channel chrominance. The conversion formula is hence given by

$$Y(m, n) = 0.299 \cdot R(m, n) + 0.587 \cdot G(m, n) + 0.114 \cdot B(m, n) \qquad (14.4)$$

Note that $Y(m, n)$, $I(m, n)$, and $Q(m, n)$ can be matrices that represent the luminance image and two color component images, respectively. Similarly, $R(m, n)$, $G(m, n)$, and $B(m, n)$ can be matrices for the RGB component images.

EXAMPLE 14.1

Given a pixel in an RGB image

$$R = 200, G = 10, B = 100$$

convert the pixel values to the YIQ values.

Solution:
Applying Equation (14.2), it follows that

$$\begin{bmatrix} Y \\ I \\ Q \end{bmatrix} = \begin{bmatrix} 0.299 & 0.587 & 0.114 \\ 0.596 & -0.274 & -0.322 \\ 0.212 & -0.523 & 0.311 \end{bmatrix} \begin{bmatrix} 200 \\ 10 \\ 100 \end{bmatrix}$$

Carrying out the matrix operations leads to

$$\begin{bmatrix} Y \\ I \\ Q \end{bmatrix} = \begin{bmatrix} 0.299 \times 200 & 0.587 \times 10 & 0.114 \times 100 \\ 0.596 \times 200 & -0.274 \times 10 & -0.322 \times 100 \\ 0.212 \times 200 & -0.523 \times 10 & 0.311 \times 100 \end{bmatrix} = \begin{bmatrix} 77.07 \\ 84.26 \\ 68.27 \end{bmatrix}$$

Rounding the values to integers, we have

$$\begin{bmatrix} Y \\ I \\ Q \end{bmatrix} = round \begin{bmatrix} 77.07 \\ 84.26 \\ 68.27 \end{bmatrix} = \begin{bmatrix} 77 \\ 84 \\ 68 \end{bmatrix}$$

Now let us study the following example to convert the YIQ values back to the RGB values.

EXAMPLE 14.2

Given a pixel of an image in the YIQ color format

$$Y = 77, I = 84, Q = 68$$

convert the pixel values back to the RGB values.

Solution:
Applying Equation (14.3) yields

$$\begin{bmatrix} R \\ G \\ B \end{bmatrix} = \begin{bmatrix} 1.000 & 0.956 & 0.621 \\ 1.000 & -0.272 & -0.647 \\ 1.000 & -1.106 & 1.703 \end{bmatrix} \begin{bmatrix} 77 \\ 84 \\ 68 \end{bmatrix} = \begin{bmatrix} 199.53 \\ 10.16 \\ 99.90 \end{bmatrix}$$

After rounding, it follows that

$$
\begin{bmatrix} R \\ G \\ B \end{bmatrix} = round \begin{bmatrix} 199.53 \\ 10.16 \\ 99.9 \end{bmatrix} = \begin{bmatrix} 200 \\ 10 \\ 100 \end{bmatrix}
$$

EXAMPLE 14.3

Given the 2×2 RGB image

$$
R = \begin{bmatrix} 100 & 50 \\ 200 & 150 \end{bmatrix} \quad G = \begin{bmatrix} 10 & 25 \\ 20 & 50 \end{bmatrix} \quad B = \begin{bmatrix} 10 & 5 \\ 20 & 15 \end{bmatrix}
$$

convert the RGB color image into a grayscale image.

Solution:
Since only Y components are kept in the grayscale image, we apply Equation (14.4) to each pixel in the 2 x 2 image and round the results to integers as follows:

$$
Y = 0.299 \times \begin{bmatrix} 100 & 50 \\ 200 & 150 \end{bmatrix} + 0.587 \times \begin{bmatrix} 10 & 25 \\ 20 & 50 \end{bmatrix} + 0.114 \times \begin{bmatrix} 10 & 5 \\ 20 & 15 \end{bmatrix} = \begin{bmatrix} 37 & 30 \\ 74 & 76 \end{bmatrix}
$$

Figure 14.9 shows the grayscale image converted from the 24-bit color image in Figure 14.5 using the RGB-to-YIQ transformation, where only the luminance information is retained.

14.1.6 MATLAB Functions for Format Conversion

The following list summarizes MATLAB functions for image format conversion:

imread = read image data file with the specified format
X = 8-bit grayscale image, 8-bit indexed image, or 24-bit RGB color image

FIGURE 14.9

Grayscale image converted from the 24-bit color image in Figure 14.5 using RGB-to-YIQ transformation.

map = color map table for the indexed image (256 entries)
imshow(X,map) = 8-bit image display
imshow(X) = 24-bit RGB color image display if image X is in a 24-bit RGB color format; grayscale image display if image X is in an 8-bit grayscale format
ind2gray = 8-bit indexed color image to 8-bit grayscale image conversion
ind2rgb = 8-bit indexed color image to 24-bit RGB color image conversion
rgb2ind = 24-bit RGB color image to 8-bit indexed color image conversion
rgb2gray = 24-bit RGB color image to 8-bit grayscale image conversion
im2double = 8-bit image to intensity image conversion
mat2gray = image data to intensity image conversion
im2uint8 = intensity image to 8-bit grayscale image conversion

Image format and conversion

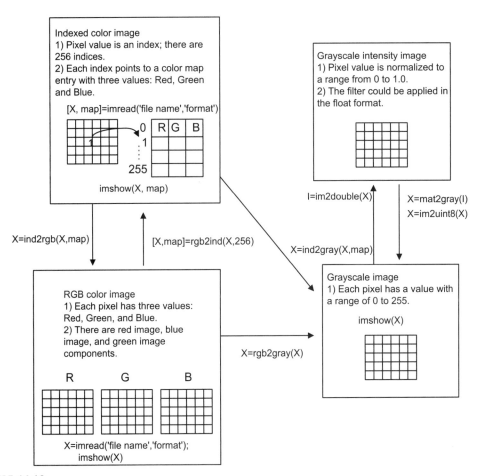

FIGURE 14.10

Outlines the applications of image format conversions.

14.2 IMAGE HISTOGRAM AND EQUALIZATION

An image histogram is a graph to show how many pixels are at each scale level, or at each index for the indexed color image. The histogram contains information needed for image equalization, where the image pixels are stretched to give a reasonable contrast.

14.2.1 Grayscale Histogram and Equalization

We can obtain a histogram by plotting pixel value distribution over the full grayscale range.

EXAMPLE 14.4

Produce a histogram given the following image (a matrix filled with integers) with the grayscale value ranging from 0 to 7, that is, with each pixel encoded into 3 bits:

$$\begin{bmatrix} 0 & 1 & 2 & 2 & 6 \\ 2 & 1 & 1 & 2 & 1 \\ 1 & 3 & 4 & 3 & 3 \\ 0 & 2 & 5 & 1 & 1 \end{bmatrix}$$

Solution:

Since the image is encoded using 3 bits for each pixel, the pixel value ranges from 0 to 7. The count for each grayscale is listed in Table 14.1.

Table 14.1 Pixel Count Distribution	
Pixel $p(m,n)$ Level	**Number of Pixels**
0	2
1	7
2	5
3	3
4	1
5	1
6	1
7	0

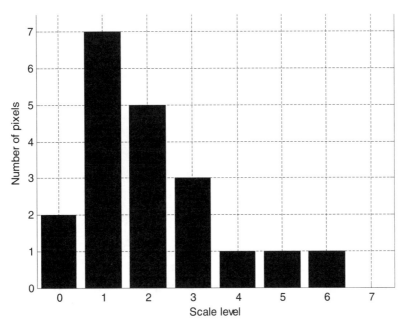

FIGURE 14.11

Histogram in Example 14.4.

Based on the grayscale distribution counts, the histogram is created as shown in Figure 14.11.
As we can see, the image has pixels whose levels are more concentrated in the dark scale in this example.

With the histogram, the equalization technique can be developed. Equalization stretches the scale range of the pixel levels to the full range to improve the contrast of the given image. To utilize this technique, the equalized new pixel value is redefined as

$$p_{eq}(m,n) = \frac{\text{Number of pixels with scale level} \leq p(m,n)}{\text{Total number of pixels}} \times (\text{maximum scale level}) \qquad (14.5)$$

The new pixel value is reassigned using the value obtained by multiplying the maximum scale level by the scaled ratio of the accumulative counts up to the current image pixel value over the total number of pixels. Clearly, since the accumulative counts can range from 0 up to the total number of pixels, the equalized pixel value can vary from 0 to the maximum scale level. It is due to this accumulation procedure that the pixel values are spread over the whole range from 0 to the maximum scale level (255). Let us look at a simplified equalization example.

EXAMPLE 14.5

Consider the following image (matrix filled with integers) with a grayscale value ranging from 0 to 7, that is, with each pixel encoded using 3 bits:

$$\begin{bmatrix} 0 & 1 & 2 & 2 & 6 \\ 2 & 1 & 1 & 2 & 1 \\ 1 & 3 & 4 & 3 & 3 \\ 0 & 2 & 5 & 1 & 1 \end{bmatrix}$$

Table 14.2 Image Equalization in Example 14.5.

Pixel $p(m,n)$ Level	Number of Pixels	Number of Pixels $\leq p(m,n)$	Equalized Pixel Level
0	2	2	1
1	7	9	3
2	5	14	5
3	3	17	6
4	1	18	6
5	1	19	7
6	1	20	7
7	0	20	7

Perform equalization using the histogram in Example 14.4, and plot the histogram for the equalized image.

Solution:

Using the histogram result in Table 14.1, we can compute an accumulative count for each grayscale level as shown in Table 14.2. The equalized pixel level using Equation (14.5) is given in the last column.

To see how the old pixel level $p(m,n) = 4$ is equalized to the new pixel level $p_{eq}(m,n) = 6$, we apply Equation (14.5):

$$p_{eq}(m,n) = round\left(\frac{18}{20} \times 7\right) = 6$$

The equalized image using Table 14.2 is finally obtained by replacing each old pixel value in the old image with its corresponding equalized new pixel value:

$$\begin{bmatrix} 1 & 3 & 5 & 5 & 7 \\ 5 & 3 & 3 & 5 & 3 \\ 3 & 6 & 6 & 6 & 6 \\ 1 & 5 & 7 & 3 & 3 \end{bmatrix}$$

To see how the histogram is changed, we compute the pixel level counts according to the equalized image. The result is given in Table 14.3, and Figure 14.12 shows the new histogram for the equalized image.

Table 14.3 Pixel Level Distribution Counts of the Equalized Image in Example 14.5

Pixel $p(m,n)$ Level	Number of Pixels
0	0
1	2
2	0
3	6
4	0
5	5
6	4
7	2

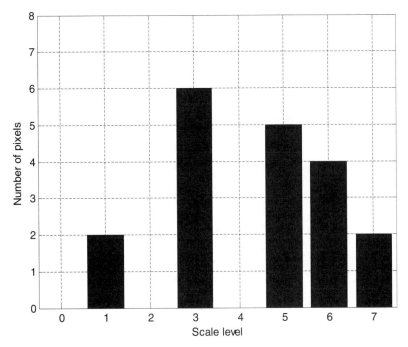

FIGURE 14.12

Histogram for the equalized image in Example 14.5.

As we can see, the pixel levels in the equalized image are stretched to the larger scale levels. This technique works for underexposed images.

Next, we apply image histogram equalization to enhance a biomedical image of a human neck in Figure 14.13A, while Figure 14.13B shows the original image histogram. (The purpose of the arrow in Figure 14.13A will be explained later.) We see that there are many pixel counts residing at the lower scales in the histogram. Hence, the image looks rather dark, and may be underexposed.

Figure 14.14A and Figure 14.14B show the equalized grayscale image using the histogram method and its histogram, respectively. As shown in the histogram, the equalized pixels reside on a larger scale, and hence the equalized image has improved contrast.

14.2.2 24-Bit Color Image Equalization

For equalizing the RGB image, we first transform RGB values to YIQ values since the Y channel contains most of the signal energy, about 93%. Then Y channel is equalized just like the grayscale equalization to enhance the luminance. We leave the I and Q channels as they are, since these contain color information only and we do not equalize them. Next, we can repack the equalized Y channel back to the YIQ format. Finally, the YIQ values are transformed back to the RGB values for display. Figure 14.15 shows the procedure.

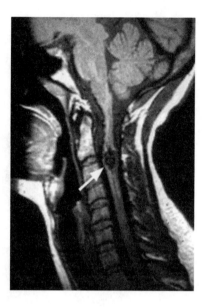

FIGURE 14.13A

Original grayscale image.

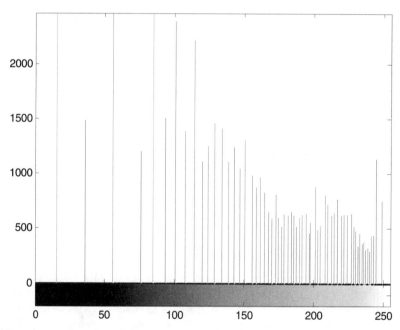

FIGURE 14.13B

Histogram for the original grayscale image.

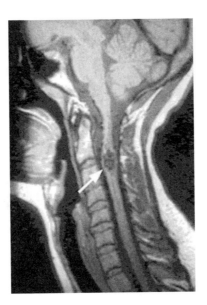

FIGURE 14.14A

Grayscale equalized image.

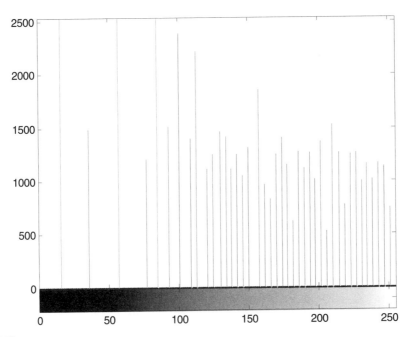

FIGURE 14.14B

Histogram for the grayscale equalized image.

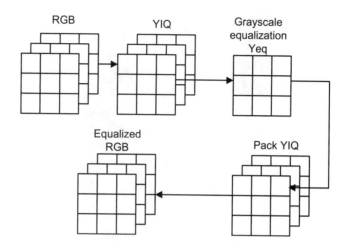

FIGURE 14.15

Color image equalization.

Figure 14.16A shows an original RGB color outdoors scene that is underexposed. Figure 14.16B shows the equalized RGB image using the method of equalizing the Y channel only. We can verify significant improvement with the equalized image showing much detailed information. The color print of the image is included in the color insert.

We can also use the histogram equalization method to equalize each of the R, G , and B channels, or their possible combinations. Figure 14.17 illustrates such a procedure.

Some color effects can be observed. Equalization of the R channel only would make the image look redder since the red pixel values are stretched out to the full range. Similar

FIGURE 14.16A

Original RGB color image. See color image on book's companion site.

FIGURE 14.16B

Equalized RGB color image. See color image on book's companion site.

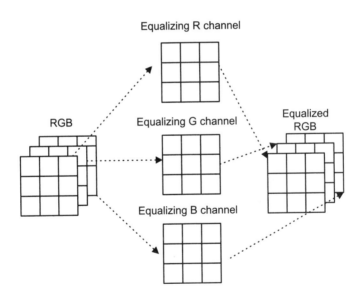

FIGURE 14.17

Equalizing RGB channels.

observations can be made for equalizing the G channel, or the B channel only. The equalized images for the R, G, and B channels, respectively, are shown in Figure 14.18. The image from equalizing the R, G, and B channels simultaneously is shown in the upper left corner, which offers improved image contrast.

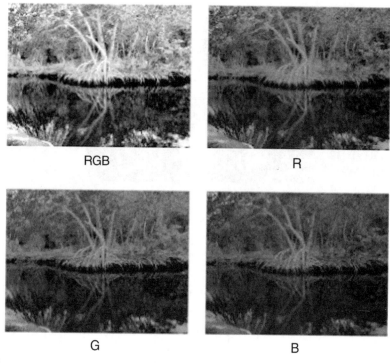

RGB

R

G

B

FIGURE 14.18

Equalization effects for RGB channels. See color image on book's companion site.

14.2.3 8-Bit Indexed Color Image Equalization

Equalization of the 8-bit color indexed image is more complicated. This is due to the fact that the pixel value is the index for color map entries, and there are three RGB color components for each entry. We expect that after equalization, the index for each pixel will not change from its location on the color map table. Instead, the RGB components in the color map are equalized and changed. The procedure is described in the following is shown in Figure 14.19.

Step 1. The RGB color map is converted to the YIQ color map. Note that there are only 256 color table entries. Since the image contains the index values, which point to locations on the color table containing RGB components, it is natural to convert the RGB color table to the YIQ color table.

Step 2. The grayscale image is generated using the Y channel value, so that grayscale equalization can be performed.

Step 3. Grayscale equalization is executed.

Step 4. The equalized 256 Y values are divided by their corresponding old Y values to obtain the relative luminance scale factors.

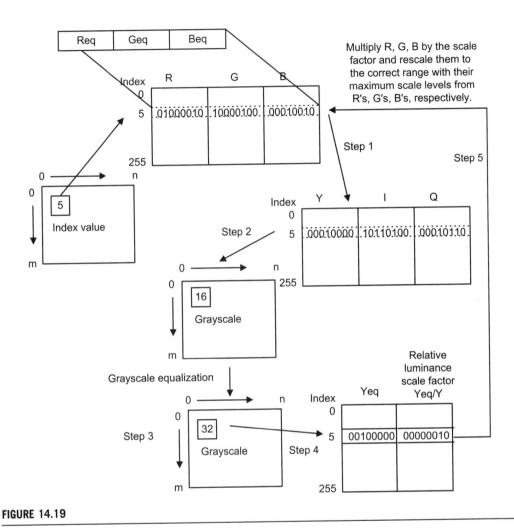

FIGURE 14.19

Equalization of 8-bit indexed color image.

Step 5. Finally, the R, G, B values are each scaled in the old RGB color table with the corresponding relative luminance scale factor and are normalized as new RGB channels in the color table in the correct range. Then the new RGB color map is the output.

Note that original index values are not changed; only the color map content is.

Using the previous outdoors picture for the condition of underexposure, Figure 14.20 shows the equalized indexed color image. We see that the equalized image displays much more detail. Its color version is reprinted in the color insert.

FIGURE 14.20

Equalized indexed 8-bit color image. See color image on book's companion site.

14.2.4 **MATLAB Functions for Equalization**

Figure 14.21 lists MATLAB functions for performing equalization for the different image formats. The MATLAB functions are explained as follows:

> **histeq** = grayscale histogram equalization, or 8-bit indexed color histogram equalization
> **imhist** = histogram display
> **rgb2ntsc** = 24-bit RGB color image to 24-bit YIQ color image conversion
> **ntsc2rgb** = 24-bit YIQ color image to 24-bit RGB color image conversion

Examples using the MATLAB functions for image format conversion and equalization are given in Program 14.1.

Program 14.1. Examples of image format conversion and equalization.

```
disp('Read the RGB image');
 XX=imread('trees','JPEG');                % Provided by the instructor
 figure, imshow(XX); title('24-bit color');
disp('The grayscale image and histogram');
 Y=rgb2gray(XX);                           % RGB to grayscale conversion
 figure, subplot(1,2,1);imshow(Y);
 title('original');subplot(1,2,2);imhist(Y, 256);

disp('Equalization in grayscale domain');
Y=histeq(Y);
figure, subplot(1,2,1); imshow(Y);
title('EQ in grayscale domain'); subplot(1,2,2); imhist(Y, 256);

disp('Equalization of Y channel for RGB color image');
figure
subplot(1,2,1); imshow(XX);
title('EQ in RGB color');
subplot(1,2,2); imhist(rgb2gray(XX),256);

Z1=rgb2ntsc(XX);                           % Conversion from RGB to YIQ
```

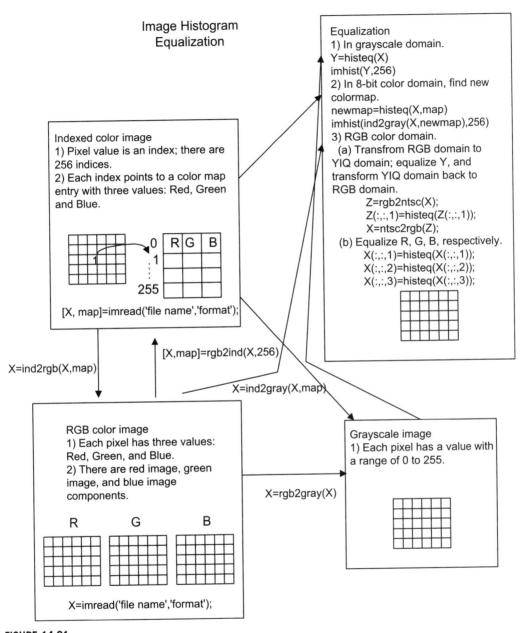

Image Histogram
Equalization

Equalization
1) In grayscale domain.
Y=histeq(X)
imhist(Y,256)
2) In 8-bit color domain, find new colormap.
newmap=histeq(X,map)
imhist(ind2gray(X,newmap),256)
3) RGB color domain.
 (a) Transfrom RGB domain to YIQ domain; equalize Y, and transform YIQ domain back to RGB domain.
   Z=rgb2ntsc(X);
   Z(:,:,1)=histeq(Z(:,:,1));
   X=ntsc2rgb(Z);
 (b) Equalize R, G, B, respectively.
   X(:,:,1)=histeq(X(:,:,1));
   X(:,:,2)=histeq(X(:,:,2));
   X(:,:,3)=histeq(X(:,:,3));

Indexed color image
1) Pixel value is an index; there are 256 indices.
2) Each index points to a color map entry with three values: Red, Green and Blue.

0 R G B
1
...
255

[X, map]=imread('file name','format');

X=ind2rgb(X,map)

[X,map]=rgb2ind(X,256)

X=ind2gray(X,map)

RGB color image
1) Each pixel has three values: Red, Green, and Blue.
2) There are red image, green image, and blue image components.

R   G   B

X=imread('file name','format');

X=rgb2gray(X)

Grayscale image
1) Each pixel has a value with a range of 0 to 255.

FIGURE 14.21

MATLAB functions for image equalization.

```
Z1(:,:,1)=histeq(Z1(:,:,1));              % Equalizing Y channel
ZZ=ntsc2rgb(Z1);                          % Conversion from YIQ to RGB

figure
subplot(1,2,1); imshow(ZZ);
title('EQ for Y channel for RGB color image');
subplot(1,2,2); imhist(im2uint8(rgb2gray(ZZ)),256);

ZZZ=XX;
ZZZ(:,:,1)=histeq(ZZZ(:,:,1));            %Equalizing R channel
ZZZ(:,:,2)=histeq(ZZZ(:,:,2));            %Equalizing G channel
ZZZ(:,:,3)=histeq(ZZZ(:,:,3));            %Equalizing B channel
figure
subplot(1,2,1); imshow(ZZZ);
title('EQ for RGB channels');
subplot(1,2,2); imhist(im2uint8(rgb2gray(ZZZ)),256);

disp('Equalization in 8-bit indexed color');
[Xind, map]=rgb2ind(XX, 256);             % RGB to 8-bit index image conversion
newmap=histeq(Xind,map);
figure
subplot(1,2,1); imshow(Xind,newmap);
title('EQ in 8-bit indexed color');
subplot(1,2,2); imhist(ind2gray(Xind,newmap),256);
```

14.3 IMAGE LEVEL ADJUSTMENT AND CONTRAST

Image level adjustment can be used to linearly stretch the pixel level in an image to increase contrast and shift the pixel level to change viewing effects. Image level adjustment is also a requirement for modifying results from image filtering or other operations to an appropriate range for display. We will study this technique in the following subsections.

14.3.1 Linear Level Adjustment

Sometimes, if the pixel range in an image is small, we can adjust the image pixel level to make use of a full pixel range. Hence, contrast of the image is enhanced. Figure 14.22 illustrates linear level adjustment.

The linear level adjustment is given by the following formula:

$$p_{adjust}(m,n) = Bottom + \frac{p(m,n) - L}{H - L} \times (\text{Top-Bottom}) \tag{14.6}$$

where $p(m,n) =$ original image pixel
$p_{adjust}(m,n) =$ desired image pixel
$H =$ maximum pixel level in the original image
$L =$ minimum pixel level in the original image
$Top =$ maximum pixel level in the desired image
$Bottom =$ minimum pixel level in the desired image

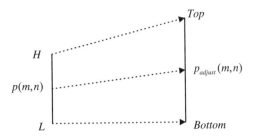

FIGURE 14.22

Linear level adjustment.

Besides adjusting the image level to a full range, we can also apply the method to shift the image pixel levels up or down.

EXAMPLE 14.6

Consider the following image (matrix filled with integers) with a grayscale value ranging from 0 to 7, that is, with each pixel encoded in 3 bits:

$$\begin{bmatrix} 3 & 4 & 4 & 5 \\ 5 & 3 & 3 & 3 \\ 4 & 4 & 4 & 5 \\ 3 & 5 & 3 & 4 \end{bmatrix}$$

a. Perform level adjustment to the full range.
b. Shift the level to the range from 3 to 7.
c. Shift the level to the range from 0 to 3.

Solution:

a. From the given image, we set the following for level adjustment to the full range:

$$H = 5, L = 3, \text{Top} = 2^3 - 1 = 7, \text{Bottom} = 0$$

Applying Equation (14.6) yields the second column in Table 14.4.

Table 14.4 Image Adjustment Results in Example 14.6.

Pixel $p(m,n)$ Level	Full Range	Range [3–7]	Range [0–3]
3	0	3	0
4	4	5	2
5	7	7	3

b. For the shift-up operation, it follows that

$$H = 5, L = 3, \text{Top} = 7, \text{Bottom} = 3$$

c. For the shift-down operation, we set

$$H = 5, L = 3, \text{Top} = 3, \text{Bottom} = 0$$

The results for (b) and (c) are listed in the third and fourth column, respectively, of Table 14.4.

According to Table 14.4, we have three images:

$$\begin{bmatrix} 0 & 4 & 4 & 7 \\ 7 & 0 & 0 & 0 \\ 4 & 4 & 4 & 7 \\ 0 & 7 & 0 & 4 \end{bmatrix} \begin{bmatrix} 3 & 5 & 5 & 7 \\ 7 & 3 & 3 & 3 \\ 5 & 5 & 5 & 7 \\ 3 & 7 & 3 & 5 \end{bmatrix} \begin{bmatrix} 0 & 2 & 2 & 3 \\ 3 & 0 & 0 & 0 \\ 2 & 2 & 2 & 3 \\ 0 & 3 & 0 & 2 \end{bmatrix}$$

Next, applying the level adjustment for the neck image of Figure 14.13A, we get the results shown in Figure 14.23: the original image, the full range stretched image, the level shift-up image, and the level shift-down image. As we can see, the stretching operation increases image contrast while the shift-up operation lightens the image and the shift-down operation darkens the image.

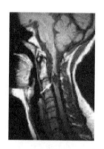 

Original   Stretching [0 0.5] to [0 1]

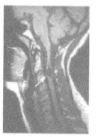 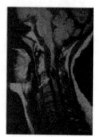

Shift-up   Shift-down

FIGURE 14.23

Image level adjustment.

14.3.2 Adjusting the Level for Display

When two 8-bit images are added together or undergo other mathematical operations, the sum of two pixel values could be as low as 0 and as high as 510. We can apply the linear adjustment to scale the range back to 0 to 255 for display. The following addition of two 8-bit images yields a sum that is out of the 8-bit range:

$$
\begin{bmatrix} 30 & 25 & 5 & 170 \\ 70 & 210 & 250 & 30 \\ 225 & 125 & 50 & 70 \\ 28 & 100 & 30 & 50 \end{bmatrix} + \begin{bmatrix} 30 & 255 & 50 & 70 \\ 70 & 3 & 30 & 30 \\ 50 & 200 & 50 & 70 \\ 30 & 70 & 30 & 50 \end{bmatrix} = \begin{bmatrix} 60 & 280 & 55 & 240 \\ 140 & 213 & 280 & 60 \\ 275 & 325 & 100 & 140 \\ 58 & 179 & 60 & 100 \end{bmatrix}
$$

To scale the combined image, modify Equation (14.6) as follows:

$$
p_{scaled}(m, n) = \frac{p(m, n) - \text{Minimum}}{\text{Maximum} - \text{Minimum}} \times (\text{Maximum scale level}) \tag{14.7}
$$

Note that in the image to be scaled,

  Maximum $= 325$
  Minimum $= 55$
  Maximum scale level $= 255$
  After scaling we have

$$
\begin{bmatrix} 5 & 213 & 0 & 175 \\ 80 & 149 & 213 & 5 \\ 208 & 255 & 43 & 80 \\ 3 & 109 & 5 & 43 \end{bmatrix}
$$

14.3.3 MATLAB Functions for Image Level Adjustment

Figure 14.24 lists applications of the MATLAB level adjustment function, which is defined as follows:

J = imajust(I, [bottom level, top level],[adjusted bottom, adjusted top], gamma)
I = input intensity image
J = output intensity image
gamma = 1 (linear interpolation function as we discussed in Section 14.3.1)
0 < gamma < 1 lightens image; gamma > 1 darkens image

14.4 IMAGE FILTERING ENHANCEMENT

As with one-dimensional digital signal processing, we can design a digital image filter such as low-pass, highpass, bandpass, and notch to process the image to obtain the desired effect. In this section, we

Image Level Adjustment

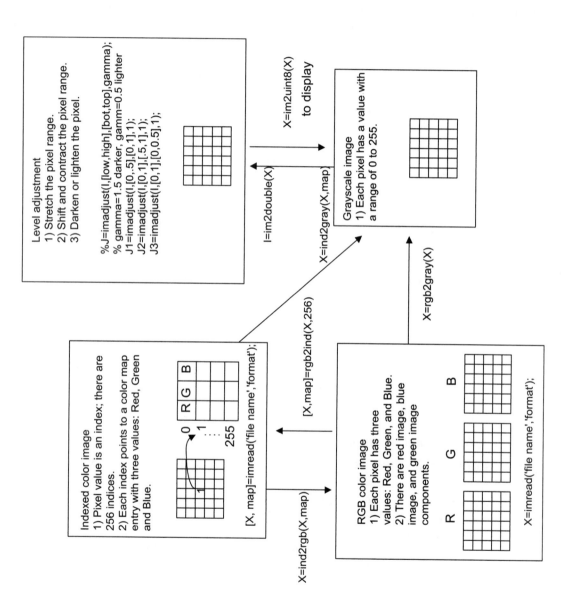

FIGURE 14.24

MATLAB functions for image level adjustment.

discuss the most common ones: lowpass filters to remove noise, median filters to remove impulse noise, and edge detection filters to discover the boundaries of objects in images. More advanced treatment of this subject can be explored in the well-known text by Gonzalez and Wintz (1987).

14.4.1 Lowpass Noise Filtering

One of the simplest lowpass filters is the average filter. The noisy image is filtered using the average convolution kernel with a size 3×3 block, 4×4 block, 8×8 block, and so on, in which the elements in the block have the same filter coefficients. The 3×3, 4×4, and 8×8 average kernels are as follows:

3×3 average kernel:

$$\frac{1}{9} \begin{bmatrix} 1 & 1 & 1 \\ 1 & 1 & 1 \\ 1 & 1 & 1 \end{bmatrix} \tag{14.8}$$

4×4 average kernel:

$$\frac{1}{16} \begin{bmatrix} 1 & 1 & 1 & 1 \\ 1 & 1 & 1 & 1 \\ 1 & 1 & 1 & 1 \\ 1 & 1 & 1 & 1 \end{bmatrix} \tag{14.9}$$

8×8 average kernel:

$$\frac{1}{64} \begin{bmatrix} 1 & 1 & 1 & 1 & 1 & 1 & 1 & 1 \\ 1 & 1 & 1 & 1 & 1 & 1 & 1 & 1 \\ 1 & 1 & 1 & 1 & 1 & 1 & 1 & 1 \\ 1 & 1 & 1 & 1 & 1 & 1 & 1 & 1 \\ 1 & 1 & 1 & 1 & 1 & 1 & 1 & 1 \\ 1 & 1 & 1 & 1 & 1 & 1 & 1 & 1 \\ 1 & 1 & 1 & 1 & 1 & 1 & 1 & 1 \\ 1 & 1 & 1 & 1 & 1 & 1 & 1 & 1 \end{bmatrix} \tag{14.10}$$

Each of the elements in the average kernel is 1 and the scale factor is the reciprocal of the total number of elements in the kernel. The convolution operates to modify each pixel in the image as follows. By passing the center of a convolution kernel through each pixel in the noisy image, we can sum each product of the kernel element and the corresponding image pixel value and multiply the sum by the scale factor to get the processed pixel. To understand the filter operation with the convolution kernel, let us study the following example.

EXAMPLE 14.7

Perform digital filtering on the noisy image using a 2×2 convolutional average kernel, and compare the enhanced image with the original one given the following 8-bit grayscale original and corrupted (noisy) images:

$$4 \times 4 \text{ original image:} \begin{bmatrix} 100 & 100 & 100 & 100 \\ 100 & 100 & 100 & 100 \\ 100 & 100 & 100 & 100 \\ 100 & 100 & 100 & 100 \end{bmatrix}$$

$$4 \times 4 \text{ corrupted image:} \begin{bmatrix} 99 & 107 & 113 & 96 \\ 92 & 116 & 84 & 107 \\ 103 & 93 & 86 & 108 \\ 87 & 109 & 106 & 107 \end{bmatrix}$$

$$2 \times 2 \text{ average kernel:} \frac{1}{4}\begin{bmatrix} 1 & 1 \\ 1 & 1 \end{bmatrix}$$

Solution:

In the following diagram, we pad edges with zeros in the last row and column before processing at the point where the first kernel and the last kernel are shown in the dotted-line boxes, respectively:

99	107	113	96	0
92	116	84	107	0
103	93	86	108	0
87	109	106	107	0
0	0	0	0	0

To process the first element, we know that the first kernel covers the image elements as $\begin{bmatrix} 99 & 107 \\ 92 & 116 \end{bmatrix}$. Summing each product of the kernel element and the corresponding image pixel value, multiplying by a scale factor of ¼, and rounding the result, it follows that

$$\frac{1}{4}(99 \times 1 + 107 \times 1 + 92 \times 1 + 116 \times 1) = 103.5$$

$$round(103.5) = 104$$

In the processing of the second element, the kernel covers $\begin{bmatrix} 107 & 113 \\ 116 & 84 \end{bmatrix}$. Similarly, we have

$$\frac{1}{4}(107 \times 1 + 113 \times 1 + 116 \times 1 + 84 \times 1) = 105$$

$$round(105) = 105$$

The process continues for the rest of image pixels. To process the last element of the first row, 96, since the kernel covers only $\begin{bmatrix} 96 & 0 \\ 107 & 0 \end{bmatrix}$, we assume that the last two elements are zeros. Then

$$\frac{1}{4}(96 \times 1 + 107 \times 1 + 0 \times 1 + 0 \times 1) = 50.75$$

$$round(50.75) = 51$$

Finally, we yield the following filtered image:

$$\begin{bmatrix} 104 & 105 & 100 & 51 \\ 101 & 95 & 96 & 54 \\ 98 & 98 & 102 & 54 \\ 49 & 54 & 53 & 27 \end{bmatrix}$$

As we know, due to zero padding for boundaries, the last-row and last-column values are in error. However, for a large image, these errors at the boundaries can be neglected without affecting image quality. The first 3×3 elements in the processed image have values that are close to those of the original image. Hence, the image is enhanced.

Figure 14.25 shows the noisy image and enhanced images using the 3×3, 4×4, 8×8 average lowpass filter kernels, respectively. The average kernel removes noise. However, it also blurs the image. When using a large-sized kernel, the quality of the processed image becomes unacceptable.

The sophisticated large-size kernels are used for noise filtering. Although it is beyond the scope of the text, the Gaussian filter kernel with a standard deviation $\sigma = 0.9$, for instance, is given by the following:

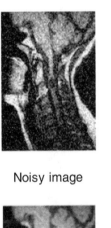
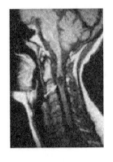

Noisy image   3x3 kernel

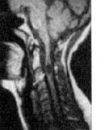
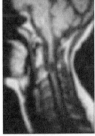

4x4 kernel   8x8 kernel

FIGURE 14.25

Noise filtering using the lowpass average kernels.

$$\frac{1}{25}\begin{bmatrix} 0 & 2 & 4 & 2 & 0 \\ 2 & 15 & 27 & 15 & 2 \\ 4 & 27 & 50 & 27 & 4 \\ 2 & 15 & 27 & 15 & 2 \\ 0 & 2 & 4 & 2 & 0 \end{bmatrix} \tag{14.11}$$

In this kernel the center pixel is weighted the most and the weights become lower and lower as we move away from the center. In this way, the blurring effect can be reduced when filtering the noise. The plot of kernel values in the special domain looks like the bell shape. The steepness of shape is controlled by the standard deviation of the Gaussian distribution function. The larger the standard deviation, the flatter the kernel; with a flatter kernel, the blurring effect will be more pronounced.

Figure 14.26A shows the noisy image, while Figure 14.26B shows the enhanced image using a 5×5 Gaussian filter kernel. Clearly, the majority of the noise has been filtered, while the blurring effect is significantly reduced.

14.4.2 Median Filtering

The median filter is one type of nonlinear filter. It is very effective at removing impulse noise, the "pepper and salt" noise, in an image. The principle of the median filter is to replace the gray level of each pixel by the median of the gray levels in a neighborhood of the pixels, instead of using the average operation. For median filtering, we specify the kernel size, list the pixel values covered by the kernel, and determine the median level. If the kernel covers an even number of pixels, the average of two median values is used. Before beginning median filtering, zeros must be padded around the row edge

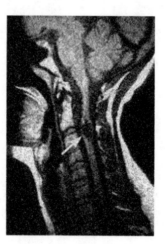

FIGURE 14.26A

Noisy image for a human neck.

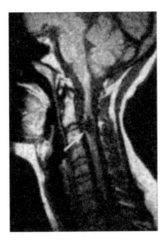

FIGURE 14.26B

Enhanced image using a Gaussian lowpass filter.

and the column edge. Hence, edge distortion is introduced at image boundary. Let us look at Example 14.8.

EXAMPLE 14.8

Consider a 3×3 median filter kernel and the following 8-bit grayscale original and corrupted (noisy) images:

$$4 \times 4 \text{ original image}: \begin{bmatrix} 100 & 100 & 100 & 100 \\ 100 & 100 & 100 & 100 \\ 100 & 100 & 100 & 100 \\ 100 & 100 & 100 & 100 \end{bmatrix}$$

$$4 \times 4 \text{ corrupted image by impulse noise}: \begin{bmatrix} 100 & 255 & 100 & 100 \\ 100 & 255 & 100 & 100 \\ 255 & 100 & 100 & 0 \\ 100 & 100 & 100 & 100 \end{bmatrix}$$

$$3 \times 3 \text{ median filter kernel}: \begin{bmatrix} \\ \\ \end{bmatrix}_{3\times 3}$$

Perform digital filtering, and compare the filtered image with the original one.

Solution:

Step 1: The 3×3 kernel requires zero padding $3/2 = 1$ column of zeros at the left and right edges and $3/2 = 1$ row of zeros at the upper and bottom edges:

```
0    0    0    0    0    0
0   100  255  100  100   0
0   100  255  100  100   0
0   255  100  100   0    0
0   100  100  100  100   0
0    0    0    0    0    0
```

Step 2: To process the first element, we cover the 3×3 kernel with the center pointing to the first element to be processed. The sorted data within the kernel are listed in terms of thier value as

$$0, \ 0, \ 0, \ 0, \ 0, \ 100, \ 100, \ 255, \ 255$$

The median value $=$ median(0, 0, 0, 0, 0, 100, 100, 255, 255) $= 0$. Zero will replace 100.

Step 3: Continue for each element until the last is replaced. Let us review the element at location (1,1):

```
0    0    0    0    0    0
0   100  255  100  100   0
0   100  255  100  100   0
0   255  100  100   0    0
0   100  100  100  100   0
0    0    0    0    0    0
```

The values covered by the kernel are

$$100, \ 100, \ 100, \ 100, \ 100, \ 100, \ 255, \ 255, \ 255$$

The median value $=$ median(100, 100, 100, 100, 100, 100, 255, 255, 255) $= 100$. The final processed image is

$$\begin{bmatrix} 0 & 100 & 100 & 0 \\ 100 & 100 & 100 & 100 \\ 0 & 100 & 100 & 0 \\ 100 & 100 & 100 & 100 \end{bmatrix}$$

Some boundary pixels are distorted due to the zero padding effect. However, for a large image, the portion of the boundary pixels (outmost image edges) is significant small so that their distortion can be omitted versus the overall quality of the image. The 2×2 middle portion matches the original image exactly. The effectiveness of the median filter is verified via this example.

The image in Figure 14.27A is corrupted by "pepper and salt" noise. The median filter with a 3×3 kernel is used to filter the impulse noise. The enhanced image shown in Figure 14.27B has a significant quality improvement. Note that a larger size kernel is not appropriate for median filtering, because for a larger set of pixels the median value deviates from the pixel value.

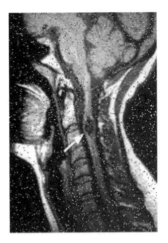

FIGURE 14.27A

Noisy image (corrupted by "pepper and salt" noise).

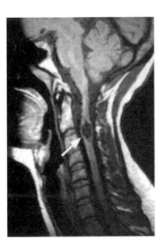

FIGURE 14.27B

The enhanced image using a 3 $\times$ 3 median filter.

14.4.3 Edge Detection

In many applications, such as pattern recognition and fingerprint and iris biometric identification, image edge information is required. To obtain the edge information, a differential convolution kernel is used. Of these kernels, Sobel convolution kernels are used for horizontal and vertical edge detection. They are listed in the following:

Horizontal Sobel edge detector:

$$\begin{bmatrix} -1 & -2 & -1 \\ 0 & 0 & 0 \\ 1 & 2 & 1 \end{bmatrix} \tag{14.12}$$

The kernel subtracts the first row in the kernel from the third row to detect the horizontal difference.

Vertical Sobel edge detector:

$$\begin{bmatrix} -1 & 0 & 1 \\ -2 & 0 & 2 \\ -1 & 0 & 1 \end{bmatrix} \tag{14.13}$$

The kernel subtracts the first column in the kernel from the third column to detect the vertical difference.

A Laplacian edge detector is devised to tackle both vertical and horizontal edges. It is described in the following:

Laplacian edge detector:

$$\begin{bmatrix} 0 & 1 & 0 \\ 1 & -4 & 1 \\ 0 & 1 & 0 \end{bmatrix} \tag{14.14}$$

EXAMPLE 14.9

Given the following 8-bit grayscale image, use the Sobel horizontal edge detector to detect horizontal edges:

$$5 \times 4 \text{ original image:} \begin{bmatrix} 100 & 100 & 100 & 100 \\ 110 & 110 & 110 & 110 \\ 100 & 100 & 100 & 100 \\ 100 & 100 & 100 & 100 \\ 100 & 100 & 100 & 100 \end{bmatrix}$$

Solution:

We pad the image with zeros before processing as follows:

```
  0    0    0    0    0    0
  0  100  100  100  100    0
  0  110  110  110  110    0
  0  100  100  100  100    0
  0  100  100  100  100    0
  0  100  100  100  100    0
  0    0    0    0    0    0
```

After processing using the Sobel horizontal edge detector, we have

$$
\begin{bmatrix}
330 & 440 & 440 & 330 \\
0 & 0 & 0 & 0 \\
-30 & -40 & -40 & -30 \\
0 & 0 & 0 & 0 \\
-300 & -400 & -400 & -300
\end{bmatrix}
$$

Adjusting the scale level leads to

$$
\begin{bmatrix}
222 & 255 & 255 & 222 \\
121 & 121 & 121 & 121 \\
112 & 109 & 109 & 112 \\
121 & 121 & 121 & 121 \\
30 & 0 & 0 & 30
\end{bmatrix}
$$

Disregarding the first row and column and the last row and column, since they are at image boundaries, we identify a horizontal line of 109 in the third row.

Figure 14.28 shows the results from edge detection.

Figure 14.29 shows the edge detection for the grayscale image of the cruise ship in Figure 14.3. Sobel edge detection can tackle only the horizontal edge or the vertical edge, as shown in Figure 14.29, where the edges of the image have both horizontal and vertical features. We can simply combine the two horizontal and vertical edge-detected images and then rescale the resultant image in the full range. Figure 14.29(c) shows that the edge detection result is equivalent to that of the Laplacian edge detector.

Next, we apply a more sophisticated Laplacian of Gaussian filter for edge detection, which is a combined Gaussian lowpass filter and Laplacian derivative operator (highpass filter). The filter *smoothes* the image to suppress noise using the lowpass Gaussian filter, then uses the Laplacian derivative operation for edge detection, since the noisy image is very sensitive to the Laplacian derivative operation. As we discussed for the Gaussian lowpass filter, the standard deviation in the Gaussian distribution function controls the degree of noise filtering before the Laplacian derivative operation. A larger value of the standard deviation may blur the image; hence, some edge boundaries could be lost. Its selection should be based on the particular noisy image. The filter kernel with a standard deviation of $\sigma = 0.8$ is given by

$$
\begin{bmatrix}
4 & 13 & 16 & 13 & 4 \\
13 & 9 & -25 & 9 & 13 \\
16 & -25 & -124 & -25 & 16 \\
13 & 9 & -25 & 9 & 13 \\
4 & 13 & 16 & 13 & 4
\end{bmatrix}
\tag{14.15}
$$

The processed edge detection using the Laplacian of Gaussian filter in Equation (14.15) is shown in Figure 14.30. We can further use a threshold value to convert the processed image to a black and white image, where the contours of objects can be clearly displayed.

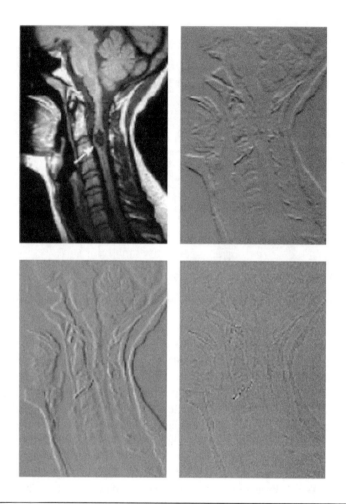

FIGURE 14.28

Image edge detection. (Upper left) original image; (upper right) result from Sobel horizontal edge detector; (lower left) result from Sobel vertical edge detector; (lower right) result from Laplacian edge detector.

14.4.4 MATLAB Functions for Image Filtering

MATLAB image filter design and implementation functions are summarized in Figure 14.31. The MATLAB functions are explained in the following:

X = image to be processed

fspecial('filter type', kernel size, parameter) = convolution kernel generation

H = FSPECIAL('gaussian',HSIZE,SIGMA) = returns a rotationally symmetric Gaussian lowpass filter of size HSIZE with standard deviation SIGMA (positive)

H = FSPECIAL('log',HSIZE,SIGMA) = returns a rotationally symmetric Laplacian of Gaussian filter of size HSIZE with standard deviation SIGMA (positive)

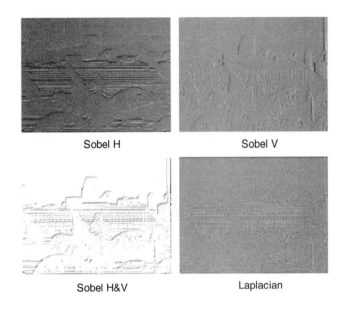

FIGURE 14.29

Edge detection (H, horizontal; V, vertical; H&V, horizontal and vertical).

Laplacian filter 5x5 kernel

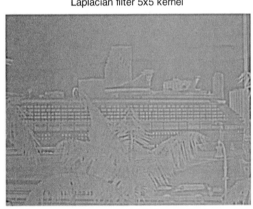

FIGURE 14.30

Image edge detection using a Laplacian of Gaussian filter.

$\mathbf{Y = filter2([convolution\ kernel], X)}$ = two-dimensional filter using the convolution kernel
$\mathbf{Y = medfilt2(X, [row\ size, column\ size])}$ = two-dimensional median filter

Program 14.2 lists the sample MATLAB codes for filtering applications. Figure 14.31 outlines the applications of the MATLAB functions.

Image filtering

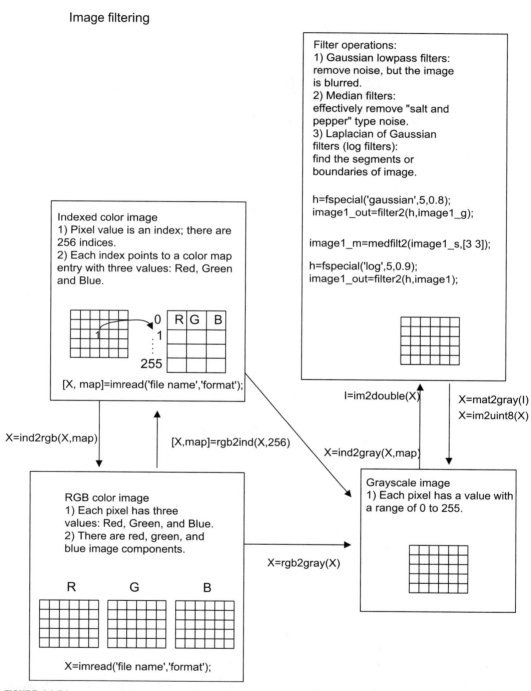

FIGURE 14.31

MATLAB functions for filter design and implementation.

Program 14.2. Examples of Gaussian filtering, media filtering, and Laplacian of Gaussian filtering.

```
close all;clear all; clc;
X=imread('cruise','jpeg');           % Provided by the instructor
Y=rgb2gray(X);                       % Convert the RGB image to the grayscale image
I=im2double(Y);                      % Get the intensity image
image1_g=imnoise(I,'gaussian');      % Add random noise to the intensity image
ng=mat2gray(image1_g);               % Adjust the range
ng=im2uint8(ng);                     % 8-bit corrupted image

% Linear filtering
K_size= 5;                           % Kernel size = 5×5
sigma =0.8;                          % sigma (the bigger, the smoother the image)
h=fspecial('gaussian',K_size,sigma); % Determine Gaussian filter coefficients
% This command will construct a Gaussian filter
% of size 5×5 with a main lobe width of 0.8.
image1_out=filter2(h,image1_g);      % Perform filtering
image1_out=mat2gray(image1_out);     % Adjust the range
image1_out=im2uint8(image1_out);     % Get the 8-bit image
subplot(1,2,1); imshow(ng),title('Noisy image');
subplot(1,2,2); imshow(image1_out);

title('5×5 Gaussian kernel');
% Median filtering
image1_s=imnoise(I,'salt & pepper'); % Add "salt and pepper" noise to the image
mn=mat2gray(image1_s);               % Adjust the range
mn=im2uint8(mn);                     % Get the 8-bit image

K_size=3;                            % Kernel size
image1_m=medfilt2(image1_s,[K_size, K_size]); % Perform median filtering
image1_m=mat2gray(image1_m);         % Adjust the range
image1_m=im2uint8(image1_m);         % Get the 8-bit image
figure, subplot(1,2,1);imshow(mn)
title('Median noisy');
subplot(1,2,2);imshow(image1_m);
title('3×3 median kernel');

% Laplacian of Gaussian filtering
K_size =5;                           % Kernel size
sigma =0.9;                          % Sigma parameter
h=fspecial('log',K_size,alpha);      % Determine the Laplacian of Gaussian %filter
kernel
image1_out=filter2(h,I);             % Perform filtering
image1_out=mat2gray(image1_out);     % Adjust the range
image1_out=im2uint8(image1_out);     % Get the 8-bit image
figure,subplot(1,2,1); imshow(Y)
title('Original');
subplot(1,2,2); imshow(image1_out);

title('Laplacian filter 5×5 kernel');
```

14.5 IMAGE PSEUDO-COLOR GENERATION AND DETECTION

We can apply certain transformations to the grayscale image so that it becomes a color image, and a wider range of pseudo-color enhancement can be obtained. In object detection, pseudo-color generation can produce the specific color for the object that is to be detected, say, red. This would significantly increase the accuracy of the identification. To do so, we choose three transformations of the grayscale level to the RGB components, as shown in Figure 14.32.

As a simple choice, we choose three sine functions for RGB transformations, as shown in Figure 14.33A. The phase and period of one sine function can be easily changed so that the

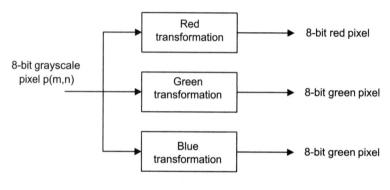

FIGURE 14.32

Block diagram for transforming a grayscale pixel to a pseudo-color pixel.

grayscale pixel level of the object to be detected is aligned to the desired color with its component value as large as possible, while the other two functions transform the same grayscale level such that their color component values are as small as possible. Hence, the single specified color object can be displayed in the image for identification. By carefully choosing the phase and period of each sine function, certain object(s) can be transformed to the red, green, or blue with a favorable choice.

EXAMPLE 14.10

In the grayscale image in Figure 14.13A, the area pointed to by the arrow has a grayscale value approximately equal to 60. The background has a pixel value approximately equal to 10. Make the background to as close to blue as possible, and make the area pointed to by the arrow as close to red as possible.

Solution:

The transformation functions are chosen as shown in Figure 14.33A, where the red value is largest at 60 and the blue and green values approach zero. At the grayscale of 10, the blue value is dominant. Figure 14.33B shows the processed pseudo-color image; it is included in the color insert.

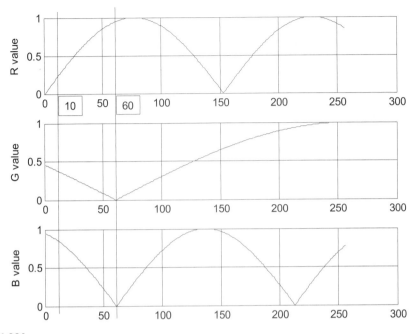

FIGURE 14.33A

Three sine functions for grayscale transformation.

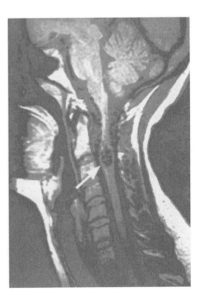

FIGURE 14.33B

The pseudo-color image. See color image on book's companion site.

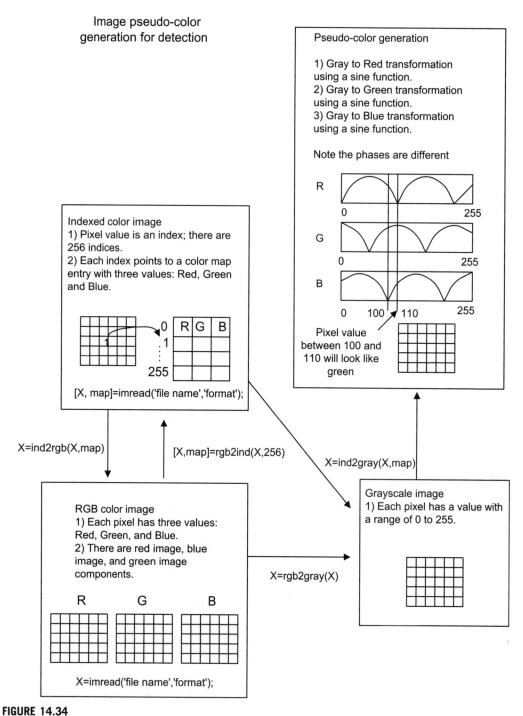

FIGURE 14.34

Illustrative procedure for pseudo-color generation.

Program 14.3 lists the sample MATLAB codes for pseudo-color generation for a grayscale image. Program 14.3. Program examples for pseudo-color generation.

```
close all; clear all;clc
disp('Convert the grayscale image to the pseudo-color image');
[X, map]=imread('clipim2','gif'); % Read 8-bit index image, provided by the
                                  % instructor
  Y=ind2gray(X,map);             % 8-bit color image to the grayscale conversion
% Apply pseudo-color functions using sinusoids
  C_r =304;                      % Cycle change for the red channel
  P_r=0;                         % Phase change for the red channel
  C_b=804;                       % Cycle change for the blue channel
  P_b=60;                        % Phase change for the blue channel
  C_g=304;                       % Cycle change for the green channel
  P_g=60;                        % Phase change for the green channel
  r=abs(sin(2*pi*[-P_r:255-P_r]/C_r));
  g=abs(sin(2*pi*[-P_b:255-P_b]/C_b));
  b=abs(sin(2*pi*[-P_g:255-P_g]/C_g));
  figure, subplot(3,1,1);plot(r,'r');grid;ylabel('R value')
  subplot(3,1,2);plot(g,'g');grid;ylabel('G value');
  subplot(3,1,3);plot(b,'b');grid;ylabel('B value');
  figure, imshow(Y);
  map=[r;g;b;]';                 % Construct the color map
  figure, imshow(Y,map);         % Display the pseudo-color image
```

14.6 IMAGE SPECTRA

In one-dimensional signal processing such as for speech and audio, we need to examine the frequency contents, check filtering effects, and perform feature extraction. Image processing is similar. However, we need apply a two-dimensional discrete Fourier transform (2D-DFT) instead of a one-dimensional (1D) DFT. The spectrum including the magnitude and phase is also in two dimensions. The equations of the 2D-DFT are given by

$$X(u, v) = \sum_{m=0}^{M-1} \sum_{n=0}^{N-1} p(m, n) W_M^{um} W_N^{vn} \tag{14.16}$$

where $W_M = e^{-j\frac{2\pi}{M}}$ and $W_N = e^{-j\frac{2\pi}{N}}$

m and n = pixel locations

u and v = frequency indices

Taking the absolute value of the 2D-DFT coefficients $X(u, v)$ and dividing the absolute value by $(M \times N)$, we get the magnitude spectrum as

$$A(u, v) = \frac{1}{(N \times M)} |X(u, v)| \tag{14.17}$$

Instead of going through the details of the 2D-DFT, we focus on application results via examples.

EXAMPLE 14.11

Determine the 2D-DFT coefficients and magnitude spectrum for the following 2×2 image:

$$\begin{bmatrix} 100 & 50 \\ 100 & -10 \end{bmatrix}$$

Solution:

Since $M = N = 2$, applying Equation (14.16) leads to

$$X(u,v) = p(0,0)e^{-j\frac{2\pi u \times 0}{2}} \times e^{-j\frac{2\pi v \times 0}{2}} + p(0,1)e^{-j\frac{2\pi u \times 0}{2}} \times e^{-j\frac{2\pi v \times 1}{2}}$$
$$+ p(1,0)e^{-j\frac{2\pi u \times 1}{2}} \times e^{-j\frac{2\pi v \times 0}{2}} + p(1,1)e^{-j\frac{2\pi u \times 1}{2}} \times e^{-j\frac{2\pi v \times 1}{2}}$$

For $u = 0$ and $v = 0$, we have

$$X(0,0) = 100e^{-j0} \times e^{-j0} + 50e^{-j0} \times e^{-j0} + 100e^{-j0} \times e^{-j0} - 10e^{-j0} \times e^{-j0}$$
$$= 100 + 50 + 100 - 10 = 240$$

For $u = 0$ and $v = 1$, we have

$$X(0,1) = 100e^{-j0} \times e^{-j0} + 50e^{-j0} \times e^{-j\pi} + 100e^{-j0} \times e^{-j0} - 10e^{-j0} \times e^{-j\pi}$$
$$= 100 + 50 \times (-1) + 100 - 10 \times (-1) = 160$$

Following similar operations,

$$X(1,0) = 60, \text{ and } X(1,1) = -60$$

Thus, we have the following DFT coefficients:

$$X(u,v) = \begin{bmatrix} 240 & 160 \\ 60 & -60 \end{bmatrix}$$

Using Equation (14.17), we can calculate the magnitude spectrum as

$$A(u,v) = \begin{bmatrix} 60 & 40 \\ 15 & 15 \end{bmatrix}$$

We can use the MABLAB function **fft2()** to verify the calculated DFT coefficients:

```
>> X=fft2([100 50;100 -10])
X =
   240 160
   60  -60
```

EXAMPLE 14.12

Given the 200×200 grayscale image shown in Figure 14.35A with a white rectangle (11×3 pixels) at its center and a black background, we can compute its magnitude spectrum (which ranges from 0 to 255). We can display the spectrum in terms of the grayscale. Figure 14.35B shows the spectrum image.

The displayed spectrum has four quarters. The left upper quarter corresponds to the frequency components, and the other three quarters are the image counterparts. In the spectrum image, the upper left corner area in the left upper quarter is white and hence has a highest scale value. Therefore, the image signal has low-frequency

FIGURE 14.35A

A square image.

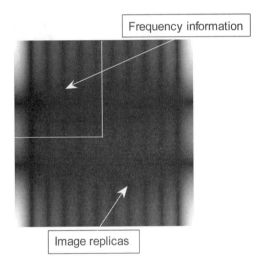

FIGURE 14.35B

Magnitude spectrum for the square image.

dominant components. The spectrum exhibits horizontal and vertical null lines (dark lines). The first vertical null line location can be estimated as 200/11=18 pixels from the left side, while the first horizontal null line happens at 200/3 = 67 pixels from the top. Next, let us apply the 2D spectrum to understand image filtering effects in image enhancement.

EXAMPLE 14.13

Figure 14.36(a) is a biomedical image corrupted by random noise. Before we apply lowpass filtering, its 2D-DFT coefficients are calculated. We then compute its magnitude spectrum and scale it to the range from 0 to 255. To

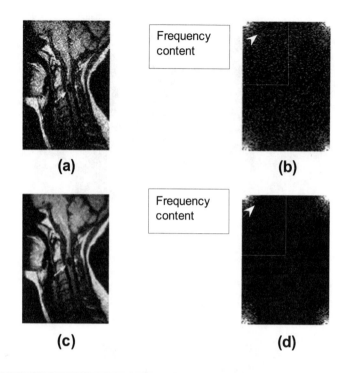

FIGURE 14.36

Magnitude spectrum plots for the noisy image and the noise-filtered image: (a) the noisy image; (b) magnitude spectrum of the noisy image; (c) noise-filtered image; (d) magnitude spectrum of the noise-filtered image.

see noise spectral components, the spectral magnitude is further multiplied by a factor of 100. Once the spectral value is larger than 255, it is clipped to 255. The resultant spectrum is displayed in Figure 14.36(b), where we can see that noise occupies the entirety of the image.

To enhance the image, we apply a Gaussian lowpass filter. The enhanced image is shown in Figure 14.36(c), in which the enhancement is easily observed. Figure 14.36(d) displays the spectra for the enhanced image with the same scaling process described just above. As we can see, the noise is significantly reduced compared with Figure 14.36(b).

14.7 IMAGE COMPRESSION BY DISCRETE COSINE TRANSFORM

Image compression is a must in our modern media systems, such as digital still and video cameras and computer systems. The purpose of compression is to reduce information storage or transmission bandwidth without losing image quality or at least without losing it significantly. Image compression can be classified as lossless compression or lossy compression. Here we focus on lossy compression using the discrete cosine transform (DCT).

The DCT is a core compression technology used in the industry standards JPEG (Joint Photographic Experts Group) for still-image compression and MPEG (Motion Picture Experts Group) for video compression, achieving a compression ratio of 20:1 without noticeable quality degradation. JPEG standard image compression is used everyday in real life.

The principle of the DCT is to transform the original image pixels to an identical number of DCT coefficients, where the DCT coefficients have a nonuniform distribution of direct-current (DC) terms representing the average values, and alternate-current (AC) terms representing fluctuations. The compression is achieved by applying the advantages of encoding DC terms (with a large dynamic range) with a large number of bits and low-frequency AC terms (a few, with a reduced dynamic range) with a reduced number of bits, and neglecting some high-frequency AC terms that have small dynamic ranges (most of them do not affect the visual quality of the picture).

14.7.1 Two-Dimensional Discrete Cosine Transform

Image compression uses 2D-DCT, whose transform pairs are defined as follows:
Forward DCT:

$$F(u,v) = \frac{2C(u)C(v)}{\sqrt{MN}} \sum_{i=0}^{M-1} \sum_{j=0}^{N-1} p(i,j) \cos\left(\frac{(2i+1)u\pi}{2M}\right) \cos\left(\frac{(2j+1)v\pi}{2N}\right) \qquad (14.18)$$

Inverse DCT:

$$p(i,j) = \sum_{u=0}^{M-1} \sum_{v=0}^{N-1} \frac{2C(u)C(v)}{\sqrt{MN}} F(u,v) \cos\left(\frac{(2i+1)u\pi}{2M}\right) \cos\left(\frac{(2j+1)v\pi}{2N}\right) \qquad (14.19)$$

where

$$C(m) = \begin{cases} \dfrac{\sqrt{2}}{2} & \text{if } m = 0 \\ 1 & \text{otherwise} \end{cases} \qquad (14.20)$$

$p(i,j) = $ pixel level at the location (i,j)
$F(u,v) = $ DCT coefficient at the frequency indices (u,v)

JPEG divides an image into 8×8 image subblocks and applies DCT for each subblock individually. Hence, we simplify the general 2D-DCT in terms of 8×8 size. The equation for 2D 8×8 DCT is modified as

$$F(u,v) = \frac{C(u)C(v)}{4} \sum_{i=0}^{7} \sum_{j=0}^{7} p(i,j) \cos\left(\frac{(2i+1)u\pi}{16}\right) \cos\left(\frac{(2j+1)v\pi}{16}\right) \qquad (14.21)$$

The inverse of 2D 8×8 DCT is expressed as

$$p(i,j) = \sum_{u=0}^{7} \sum_{v=0}^{7} \frac{C(u)C(v)}{4} F(u,v) \cos\left(\frac{(2i+1)u\pi}{16}\right) \cos\left(\frac{(2j+1)v\pi}{16}\right) \qquad (14.22)$$

To become familiar with the 2D-DCT formulas, we study Example 14.14.

EXAMPLE 14.14

Determine the 2D-DCT coefficients for the following image:

$$\begin{bmatrix} 100 & 50 \\ 100 & -10 \end{bmatrix}$$

Solution:

Applying $N = 2$ and $M = 2$ to Equation (14.18) yields

$$F(u, v) = \frac{2C(u)C(v)}{\sqrt{2} \times 2} \sum_{i=0}^{1} \sum_{j=0}^{1} p(i, j) \cos\left(\frac{(2i+1)u\pi}{4}\right) \cos\left(\frac{(2j+1)v\pi}{4}\right)$$

For $u = 0$ and $v = 0$, we achieve

$$F(0,0) = c(0)c(0) \sum_{i=0}^{1} \sum_{j=0}^{1} p(i, j) \cos(0) \cos(0)$$

$$= \left(\frac{\sqrt{2}}{2}\right)^2 [p(0,0) + p(0,1) + p(1,0) + p(1,1)]$$

$$= \frac{1}{2}(100 + 50 + 100 - 10) = 120$$

For $u = 0$ and $v = 1$, we achieve

$$F(0,1) = c(0)c(1) \sum_{i=0}^{1} \sum_{j=0}^{1} p(i, j) \cos(0) \cos\left(\frac{(2j+1)\pi}{4}\right)$$

$$= \left(\frac{\sqrt{2}}{2}\right) \times 1 \times \left(p(0,0) \cos\frac{\pi}{4} + p(0,1) \cos\frac{3\pi}{4} + p(1,0) \cos\frac{\pi}{4} + p(1,1) \cos\frac{3\pi}{4}\right)$$

$$= \frac{\sqrt{2}}{2}\left(100 \times \frac{\sqrt{2}}{2} + 50\left(-\frac{\sqrt{2}}{2}\right) + 100 \times \frac{\sqrt{2}}{2} - 10\left(-\frac{\sqrt{2}}{2}\right)\right) = 80$$

Similarly,

$$F(1,0) = 30, \text{ and } F(1,1) = -30$$

Finally, we get

$$F(u, v) = \begin{bmatrix} 120 & 80 \\ 30 & -30 \end{bmatrix}$$

Applying the MATLAB function **dct2()** verifies the DCT coefficients as follows:

```
>> F = dct2([100 50;100 -10])
F =
   120.0000   -80.0000
    30.0000   -30.0000
```

EXAMPLE 14.15

Given the following DCT coefficients from a 2× 2 image, determine the pixel $p(0,0)$:

$$F\left(u, v\right) = \begin{bmatrix} 120 & 80 \\ 30 & -30 \end{bmatrix}$$

Solution:

Applying Equation (14.19) of the inverse 2D-DCT with $N = M = 2$, $i = 0$, and $j = 0$, it follows that

$$p(0,0) = \sum_{u=0}^{1} \sum_{v=0}^{1} c(u)c(v)F(u, v) \cos\left(\frac{u\pi}{4}\right) \cos\left(\frac{v\pi}{4}\right)$$

$$= \left(\frac{\sqrt{2}}{2}\right) \times \left(\frac{\sqrt{2}}{2}\right) \times F(0,0) + \left(\frac{\sqrt{2}}{2}\right) \times F(0,1) \times \left(\frac{\sqrt{2}}{2}\right)$$

$$+ \left(\frac{\sqrt{2}}{2}\right) \times F(1,0) \times \left(\frac{\sqrt{2}}{2}\right) + F(0,1)\left(\frac{\sqrt{2}}{2}\right) \times \left(\frac{\sqrt{2}}{2}\right)$$

$$= \frac{1}{2} \times 120 + \frac{1}{2} \times 80 + \frac{1}{2} \times 30 + \frac{1}{2}(-30) = 100$$

We apply the MATLAB function **idct2()** to verify the inverse DCT and get the following pixel values:
```
>> p = idct2([120 80; 30 −30])
p =
   100.0000    50.0000
   100.0000   −10.0000
```

14.7.2 Two-Dimensional JPEG Grayscale Image Compression Example

To understand JPEG image compression, we examine an 8 × 8 grayscale subblock. Table 14.5 shows a subblock of the grayscale image in Figure 14.37 that is to be compressed. Applying 2D-DCT leads to Table 14.6.

These DCT coefficients have a big DC component of 1198 but small AC component values. These coefficients are further normalized (quantized) with a quality factor Q, defined in Table 14.7.

Table 14.5 8 × 8 Subblock

150	148	140	132	150	155	155	151
155	152	143	136	152	155	155	153
154	149	141	135	150	150	150	150
156	150	143	139	154	152	152	155
156	151	145	140	154	152	152	155
154	152	146	139	151	149	150	151
156	156	151	142	154	154	154	154
151	154	149	139	151	153	154	153

FIGURE 14.37

Original image.

After normalization, as shown in Table 14.8, the DC coefficient is reduced to 75, and a few small AC coefficients exist, while most are zero. We can encode and transmit only nonzero DCT coefficients and omit transmitting zeros, since they do not carry any information. They can be easily recovered by resetting coefficients to zero during decoding. By this principle we achieve data compression.

As shown in Table 14.8, most nonzero coefficients reside in the upper left corner. Hence, the order of encoding for each value is based on the zigzag path in which the order is numbered, as in Table 14.9.

According to the order, we record the nonzero DCT coefficients as a JPEG vector, shown as

$$\text{JPEG vector}: [75 \ -1 \ -1 \ 0 \ -1 \ 3 \ 2 \ \mathbf{EOB}]$$

where "EOB" = end of block coding. The JPEG vector can further be compressed by encoding the difference of DC values between subblocks, in *differential pulse code modulation* (DPCM), as discussed in Chapter 11, as well as by run-length coding of AC values and Huffman coding, which both belong to lossless compression techniques. We will pursue this in the next section.

Table 14.6 DCT Coefficients for the Subblock Image in Table 14.5

1198	−10	26	24	−5	−16	0	12
−8	−6	3	8	0	0	0	0
0	−3	0	0	−8	0	0	0
0	0	0	0	0	0	0	0
0	−4	0	0	0	0	0	0
0	0	−1	0	0	0	0	0
−10	1	0	0	0	0	0	0
0	0	0	0	0	0	0	0

Table 14.7 The Quality Factor

16	11	10	16	24	40	51	61
12	12	14	19	26	58	60	55
14	13	16	24	40	57	69	56
14	17	22	29	51	87	80	62
18	22	37	56	68	109	103	77
24	35	55	64	81	104	113	92
49	64	78	87	103	121	120	101
72	92	95	98	112	100	103	99

Table 14.8 Normalized DCT Coefficients

75	−1	3	2	0	0	0	0
−1	−1	0	0	0	0	0	0
0	0	0	0	0	0	0	0
0	0	0	0	0	0	0	0
0	0	0	0	0	0	0	0
0	0	0	0	0	0	0	0
0	0	0	0	0	0	0	0
0	0	0	0	0	0	0	0

During the decoding stage, the JPEG vector is recovered first. Then the quantized DCT coefficients are recovered according to the zigzag path. Next, the recovered DCT coefficients are multiplied by a quality factor to obtain the estimate of the original DCT coefficients. Finally, we apply the inverse DCT to achieve the recovered image subblock, which is shown in Table 14.10.

For comparison, the errors between the recovered image and original image are calculated and listed in Table 14.11.

The original and compressed images are displayed in Figures 14.37 and 14.38, respectively. We do not see any noticeable difference between these two grayscale images.

Table 14.9 DCT Coefficient Scan Order

0	1	5	6	14	15	27	28
2	4	7	13	16	26	29	42
3	8	12	17	25	30	41	43
9	11	18	24	31	40	44	53
10	19	23	32	39	45	52	54
20	22	33	38	46	51	55	60
21	34	37	47	50	56	59	61
35	36	48	49	57	58	62	63

Table 14.10 The Recovered Image Subblock

153	145	138	139	147	154	155	153
153	145	138	139	147	154	155	153
155	147	139	140	148	154	155	153
157	148	141	141	148	155	155	152
159	150	142	142	149	155	155	152
161	152	143	143	149	155	155	152
162	153	144	144	150	155	154	151
163	154	145	144	150	155	154	151

Table 14.11 The Coding Error of the Image Subblock

3	−3	−2	7	−3	−1	0	2
−2	−7	−5	3	−5	−1	0	0
1	−2	−2	5	−2	4	5	3
1	−2	−2	2	−6	3	3	−3
3	−1	−3	2	−5	3	3	−3
7	0	−3	4	−2	6	5	1
6	−3	−7	2	−4	1	0	−3
12	0	−4	5	−1	2	0	−2

FIGURE 14.38

JPEG compressed image.

14.7.3 JPEG Color Image Compression

This section is devoted to reviewing JPEG standard compression and examines the steps briefly. We focus on the encoder, since the decoder is just the reverse process of the encoding. The block diagram for the JPEG encoder is in Figure 14.39.

The JPEG encoder has the following main steps:

1. Transform RGB to YIQ or YUV (U and V = chrominance components).
2. Perform DCT on blocks.
3. Perform quantization.
4. Perform zigzag ordering, DPCM, and run-length encoding.
5. Perform entropy encoding (Huffman coding).

RGB to YIQ Transformation

The first transformation is of the RGB image to a YIQ or YUV image. Transformation from RGB to YIQ has previously been discussed. The principle is that in YIQ format, the luminance channel carries more signal energy, up to 93%, while the chrominance channels carry up to only 7% of signal energy. After transformation, more effort can be spent on coding the luminance channel.

DCT on Image Blocks

Each image is divided into 8×8 blocks. 2D-DCT is applied to each block to obtain the 8×8 DCT coefficient block. Note that there are three blocks, Y, I, and Q.

Quantization

The quantization is operated using the 8×8 quantization matrix. Each DCT coefficient is quantized, that is, divided by the corresponding value given in the quantization matrix. In this way, a smaller

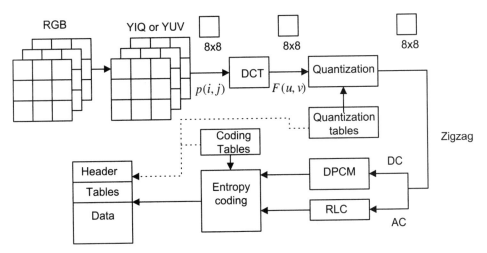

FIGURE 14.39

Block diagram for JPEG encoder.

Table 14.12 The Quality Factor for Luminance

16	11	10	16	24	40	51	61
12	12	14	19	26	58	60	55
14	13	16	24	40	57	69	56
14	17	22	29	51	87	80	62
18	22	37	56	68	109	103	77
24	35	55	64	81	104	113	92
49	64	78	87	103	121	120	101
72	92	95	98	112	100	103	99

Table 14.13 The Quality Factor for Chrominance

17	18	24	47	99	99	99	99
18	21	26	66	99	99	99	99
24	26	56	99	99	99	99	99
47	66	99	99	99	99	99	99
99	99	99	99	99	99	99	99
99	99	99	99	99	99	99	99
99	99	99	99	99	99	99	99
99	99	99	99	99	99	99	99

number of bits can be used for encoding the DCT coefficients. There are two different quantization tables, one for luminance (which is the same as the one in the last section and listed here again for comparison) and the other for chrominance (Tables 14.12 and 14.13).

We can see that the chrominance table has numbers with larger values, so that small values of DCT coefficients will result and hence fewer bits are required for encoding each DCT coefficient. Zigzag ordering to produce the JPEG vector is similar to the grayscale case, except that there are three JPEG vectors.

Differential Pulse Code Modulation on Direct-Current Coefficients

Since each 8×8 image block has only one DC coefficient, which can be a very large number and varies slowly, we make use of DPCM for coding DC coefficients. As an example for the first five image blocks, the DC coefficients are 200, 190, 180, 160, and 170. DPCM with a coding rule of $d(n) = DC(n) - DC(n-1)$ with initial condition $d(0) = DC(0)$ produces a DPCM sequence of

$$200, -10, -10, -20, 10$$

Hence, the reduced signal range of these values is feasible for entropy coding.

Run-Length Coding on Alternating-Current Coefficients

The run-length method encodes the pair of

- the number of zeros to skip and
- the next nonzero value.

The zigzag scan of the 8×8 matrix makes up a vector of 64 values with a long runs of zeros. For example, the quantized DCT coefficients are scanned as

$$[75, \ -1, \ 0, \ -1, \ 0, \ 0, \ -1, \ 3, \ 0, \ 0, \ 0, \ 2, \ 0, \ 0, \ ..., \ 0]$$

with one run, two runs, and three runs of zeros in the middle and 52 extra zeros towards the end. The run-length method encodes AC coefficients by producing a pair (run length, value), where the run length is the number of zeros in the run, while the value is the next nonzero coefficient. A special pair $(0, 0)$ indicates EOB. Here is the result from a run-length encoding of AC coefficients:

$$(0, -1), \ (1, -1), \ (2, -1), \ (0, 3), \ (3, 2), \ (0, 0)$$

Lossless Entropy Coding

The DC and AC coefficients are further compressed using entropy coding. JPEG allows Huffman coding and arithmetic coding. We focus on the Huffman coding here.

Coding DC Coefficients

Each DPCM-coded DC coefficient is encoded by a pair of symbols (size, amplitude) with the size (4-bit code) designating the number of bits for the coefficient as shown in Table 14.14, while the amplitude is encoded by the actual bits. For the negative number of the amplitude, 1's complement is used.

For example, we can code the DPCM-coded DC coefficients $200, -10, -10, -20, 10$ as

$$(8, \ 11001000), \ (4, \ 0101), \ (4, 0101), \ (5, 01011), \ (4, 1010)$$

Since there needs to be 4 bits for encoding each size, we can use 45 bits in total for encoding the DC coefficients for these five subblocks.

Coding AC Coefficients

The run-length AC coefficients have the format (run length, value). The value can be further compressed using the Huffman coding method, similar to coding the DPCM-coded DC coefficients. The run length and the size are each encoded by 4 bits and packed into a byte.

Table 14.14 Huffman Coding Table

Size	Amplitude
1	$-1, 1$
2	$-3, -2, 2, 3$
3	$-7, .., -4, 4, ..., 7$
4	$-15, ..., -8, 8, ..., 15$
5	$-31, ..., -16, 16, ..., 31$
.	.
.	.
.	.
10	$-1023, ..., -512, 512, ..., 1023$

FIGURE 14.40

JPEG compressed color image. See color image on book's companion site.

Symbol 1: (run length, size)
Symbol 2: (amplitude)

The 4-bit run length can only tackle runs for zeros from 1 to 15. If the run length of zeros is larger than 15, then a special code (15, 0) is used for Symbol 1. Symbol 2 is the amplitude in Huffman coding as shown in Table 14.14, while the encoded Symbol 1 is kept in its format:

(run length, size, amplitude)

Let us code the following run-length code of AC coefficients:

$$(0, -1), (1, -1), (2, -1), (0, 3), (3, 2), (0, 0)$$

We can produce a bit stream for AC coefficients:

$$(0000, 0001, 0), (0001, 0001, 0), (0010, 0001, 0),$$

$$(0000, 0010, 11), (0011, 0010, 10), (0000, 0000)$$

There are 55 bits in total. Figure 14.40 shows a JPEG compressed color image (included in the color insert). The decompressed image is indistinguishable from the original image after comparison.

14.7.4 Image Compression Using Wavelet Transform Coding

We can extend the one-dimensional discrete wavelet transform (DWT) discussed in Chapter 13 to the two-dimensional DWT. The procedure is described as follows. Given an N × N image, the 1D-DWT using level 1 is applied to each row of the image; and after all the rows are transformed, the level-1 1D-DWT is applied again to each column. After the rows and columns are transformed, we achieve four first-level subbands labeled LL, HL, LH, and HH as shown in Figure 14.41(a). The same procedure repeats for the LL band only and results in the second-level subbands: LL2, HL2, LH2, and HH2

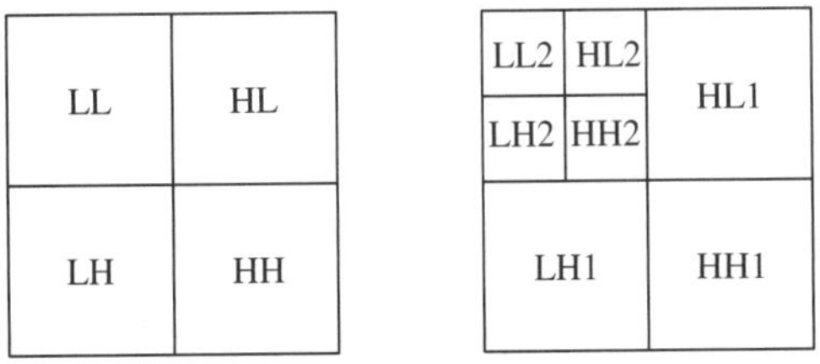

(a) Level-one transformation   **(b)** Level-two transformation

FIGURE 14.41

The two-dimensional DWT for level 1 and level 2.

(Figure 14.41(b)). The process proceeds to higher levels as desired. With the obtained wavelet transform, we can quantize coefficients to achieve the compression requirement. For example, for the second-level coefficients, we can omit HL1, LH1, HH1 to simply achieve a 4:1 compression ratio. Decompression reverses the process, that is, we inversely transform columns and then rows of the wavelet coefficients. We can apply the IDWT to the recovered LL band with the column and row inverse transform processes, and continue until the inverse transform at level 1 is completed. Let us look at an illustrative example.

EXAMPLE 14.16

Consider the following 4×4 image:

114	135	122	109
102	116	119	124
105	148	138	122
141	102	140	132

a. Perform 2D-DWT using the 2-tap Haar wavelet.
b. Using the result in (a), perform 2D-IDWT using the 2-tap Haar wavelet.

Solution:

a. The MATLAB function **dwt()** is applied to each row. The result for the first row is displayed below:
```
>> dwt([1 1]/sqrt(2), [114  135  122  109],1)'   % Row vector coefficients
ans =176.0696   163.3417   −14.8492   9.1924
```

The completed row transform is listed below:

176.0696	163.3417	−14.8492 9	9.1924
154.1493	171.8269	−9.8995	3.5355
178.8980	183.8478	−30.4056	11.3137
171.8269	192.3330	27.5772	5.6569

Next, the result for the first column is presented:
>> dwt([1 1]/sqrt(2),[176.0696 154.1493 178.8980 171.8269],1)'
ans = 233.5000 248.0000 15.5000 5.0000

Applying the transform to each column completes the level-1 transformation:

233.5000	237.0000	−17.5000	4.0000
248.0000	266.0000	−2.0000	12.0000
15.5000	−6.0000	−3.5000	9.0000
5.0000	−6.0000	−41.0000	4.0000

Now, we perform the level-2 transformation. Applying the transform to the first row for the LL1 band yields
>> dwt([1 1]/sqrt(2),[233.5000 237.0000],1)'
ans = 332.6937 −2.4749

After completing the row transformation, the result is given by

332.6937	−2.4749	−17.5000	4.0000
363.4529	−12.7279	−2.0000	12.0000
15.5000	−6.0000	−3.5000	9.0000
5.0000	−6.0000	−41.0000	4.0000

Similarly, the first-column MATLAB result is listed below:
>> dwt([1 1]/sqrt(2),[332.6937 363.4529],1)'
ans = 492.2500 −21.7500

Finally, we achieve the completed level-2 DWT as

492.2500	−10.7500	−17.5000	4.0000
−21.7500	7.2500	−2.0000	12.0000
15.5000	−6.0000	−3.5000	9.0000
5.0000	−6.0000	−41.0000	4.0000

b. Recovering the LL2 band first, the first column reconstruction is given as
>> idwt([1 1]/sqrt(2),[492.2500 −21.7500],1)'
ans = 332.6937 363.4529

Completing the inverse of the second column in the LL2 band gives

332.6937	−2.4749	−17.5000	4.0000
363.4529	−12.7279	−2.0000	12.0000
15.5000	−6.0000	−3.5000	9.0000
5.0000	−6.0000	−41.0000	4.0000

Now, we show the first row result for the LL2 band in MATLAB as follows:

>> idwt([1 1]/sqrt(2),[332.6937 −2.4749],1)'
ans = 233.5000 237.0000

The recovered LL1 band is shown in the following:

233.5000	237.0000	−17.5000	4.0000
248.0000	266.0000	−2.0000	12.0000
15.5000	−6.0000	−3.5000	9.0000
5.0000	−6.0000	−41.0000	4.0000

Now we are at the level-1 inverse process. For simplicity, the first column result in MATLAB and the completed results are listed below, respectively.

>> idwt([1 1]/sqrt(2),[233.5000 248.0000 15.5000 5.0000],1)'
ans = 176.0696 154.1493 178.8980 171.8269

176.0696	163.3417	−14.8492	9.1924
154.1493	171.8269	−9.8995	−3.5355
178.8980	183.8478	−30.4056	11.3137
171.8269	192.3330	27.5772	5.6569

Finally, we perform the inverse of the row transform at level 1. The first row result in MATLAB is listed below:

>> idwt([1 1]/sqrt(2),[176.0696 163.3417 −14.8492 9.1924],1)'
ans = 114.0000 135.0000 122.0000 109.0000

The final inverse DWT is obtained as

114.0000	135.0000	122.0000	109.0000
102.0000	116.0000	119.0000	124.0000
105.0000	148.0000	138.0000	122.0000
141.0000	102.0000	140.0000	132.0000

Since there is no quantization for each coefficient, we obtain a perfect reconstruction.

Figure 14.42 shows 8-bit grayscale image compression by applying the one-level wavelet transform, in which a 16-tap Daubechies wavelet is used. The wavelet coefficients (each is coded using 8 bits) are shown in Figure 14.42(b). By discarding the HL, LH, and HH band coefficients, we can achieve 4:1 compression. The decoded image is displayed in Figure 14.42(c). The MATLAB program is listed in Program 14.4.

Figure 14.43 illustrates two-level wavelet transform and compression results. By discarding the HL2, LH2, HH2, HL1, LH1, and HH1 subbands, we achieve 16:1 compression. However, as shown in

(a) Wavelet coefficients

(b) Original image

(c) 4:1 Compression

FIGURE 14.42

(a) Wavelet coefficients; (b) original image; (c) 4:1 compression.

Figure 14.43(c), we can observe a noticeable degradation of image quality. Since the high-frequency details are discarded, the compressed image shows a significant smoothing effect. In addition, there are many advanced methods to quantize and compress the wavelet coefficients. Of these compression techniques, the embedded zerotree wavelet (EZW) method is the most efficient one; it can be found in Li and Drew (2004).

Program 14.4. One-level wavelet transform and compression.

```
close all; clear all; clc
X=imread('cruise','JPEG');
Y=rgb2gray(X); % Convert the image into grayscale
h0 =[0.054415842243144 0.312871590914520 0.675630736297712 ...
  0.585354683654425 -0.015829105256675 -0.284015542962009 ...
  0.000472484573805 0.128747426620538 -0.017369301001845 ...
  -0.044088253930837 0.013981027917411 0.008746094047413 ...
  -0.004870352993456 -0.000391740373377 0.000675449406451 ...
  -0.000117476784125];
M= length(h0);
h1(1:2:M-1) = h0(M:-2:2);h1(2:2:M) = -h0(M-1:-2:1);% Obtain QMF highpass filter
[m n]=size(Y);
% Level-1 transform
[m n]=size(Y);
```

(a) Wavelet coefficients

(b) Original image

(c) 16:1 Compression

FIGURE 14.43

(a) Wavelet coefficients; (b) original image; (c) 16:1 compression.

```
for i=1:m
  W1(i,:)=dwt(h0,double(Y(i,:)),1)';
end
for i=1:n
  W1(:,i)=dwt(h0,W1(:,i),1); % Wavelet coefficients at level 1
end
% Quantization using 8 bits
wmax=double(max(max(abs(W1)))); % Scale factor
W1=round(double(W1)*2^7/wmax); % Get 8-bit data
W1=double(W1)*wmax/2^7;% Recover the wavelet
figure (1); imshow(uint8(W1));xlabel('Wavelet coefficients');
% 8-bit quantization
[m, n]=size(W1);
WW=zeros(m,n);
WW(1:m/2,1:n/2)=W1(1:m/2,1:n/2);
W1=WW;
% Decoding from level 1 using W1
[m, n]=size(W1);
for i=1:n
  Yd1(:,i)=idwt(h0,double(W1(:,i)),1);
end
```

```
for i=1:m
  Yd1(i,:)=idwt(h0,double(Yd1(i,:)),1)';
end
YY1=uint8(Yd1);
figure (2),imshow(Y);xlabel('Original image');
figure (3),imshow(YY1);xlabel('4:1 Compression');
```

Program 14.5. Two-level wavelet compression.

```
close all; clear all; clc
X=imread('cruise','JPEG');
Y=rgb2gray(X);
h0 =[0.054415842243144 0.312871590914520 0.675630736297712 ...
  0.585354683654425 -0.015829105256675 -0.284015542962009 ...
  0.000472484573805 0.128747426620538 -0.017369301001845 ...
  -0.044088253930837 0.013981027917411 0.008746094047413 ...
  -0.004870352993456 -0.000391740373377 0.000675449406451 ...
  -0.000117476784125];
M= length(h0);
h1(1:2:M-1) = h0(M:-2:2);h1(2:2:M) = -h0(M-1:-2:1);% Obtain QMF highpass filter
[m n]=size(Y);
% Level-1 transform
[m n]=size(Y);
for i=1:m
  W1(i,:)=dwt(h0,double(Y(i,:)),1)';
end
for i=1:n
  W1(:,i)=dwt(h0,W1(:,i),1); % Wavelet coefficients at level-1 transform
end
% Level-2 transform
Y1=W1(1:m/2,1:n/2); % Obtain LL subband
[m n]=size(Y1);
for i=1:m
  W2(i,:)=dwt(h0,Y1(i,:),1)';
end
for i=1:n
  W2(:,i)=dwt(h0,W2(:,i),1);
end
W22=W1; W22(1:m,1:n)=W2; % Wavelet coefficients at level-2 transform
wmax=max(max(abs(W22)));
% 8-bit quantization
W22=round(W22*2^7/wmax);
W22=double(W22)*wmax/2^7;
figure(1), imshow(uint8(W22));xlabel('Wavelet coefficients');
[m, n]=size(W22); WW=zeros(m,n);
WW(1:m/4,1:n/4)=W22(1:m/4,1:n/4);
W22=WW; % Discard HL2,LH2, HH2, HL1, LH1, HH1 subbands
% Decoding from level-2 transform
[m,n]=size(W22); Wd2=W22(1:m/2,1:n/2);
```

```
% Level 2
[m n]=size(Wd2);
for i=1:n
  Wd1(:,i)=idwt(h0,double(Wd2(:,i)),1);
end
for i=1:m
  Wd1(i,:)=idwt(h0,double(Wd1(i,:))',1);
end
% Level 1
[m, n]=size(W22);Yd11=W22;
Yd11(1:m/2,1:n/2)=Wd1;
for i=1:n
  Yd(:,i)=idwt(h0,Yd11(:,i),1);
end
for i=1:m
  Yd(i,:)=idwt(h0,double(Yd(i,:)),1)';
end
figure (2),imshow(Y),xlabel('Original image');
Y11=uint8(Yd); figure (3),imshow(Y11);xlabel('16:1 compression');
```

14.8 CREATING A VIDEO SEQUENCE BY MIXING TWO IMAGES

In this section, we introduce a method to mix two images to generate an image (video) sequence. Applications of mixing the two images may include fading in and fading out images, blending two images, or overlaying text on an image.

In mixing two images in a video sequence, a smooth transition is required from fading out one image of interest to fading in another image of interest. We want to fade out the first image and gradually fade in the second. This cross-fade scheme is implemented using the following operation:

$$\text{Mixed image} = (1 - \alpha) \times \text{image}_1 + \alpha \times \text{image}_2 \tag{14.23}$$

where α = fading in proportionally to the weight of the second image (value between 0 and 1), and $(1 - \alpha)$ = fade out proportionally to the weight of the second image.

The video sequence in Figure 14.44A consisting of six frames is generated using $\alpha = 0$, $\alpha = 0.2$, $\alpha = 0.4$, $\alpha = 0.6$, $\alpha = 0.8$, and $\alpha = 1.0$, respectively, for two images. The equations for generating these frames are listed as follows:

$$\text{Mixed image}_1 = 1.0 \times \text{image}_1 + 0.0 \times \text{image}_2$$
$$\text{Mixed image}_2 = 0.8 \times \text{image}_1 + 0.2 \times \text{image}_2$$
$$\text{Mixed image}_3 = 0.6 \times \text{image}_1 + 0.4 \times \text{image}_2$$
$$\text{Mixed image}_4 = 0.4 \times \text{image}_1 + 0.6 \times \text{image}_2$$
$$\text{Mixed image}_5 = 0.2 \times \text{image}_1 + 0.8 \times \text{image}_2$$
$$\text{Mixed image}_6 = 0.0 \times \text{image}_1 + 1.0 \times \text{image}_2$$

The sequence begins with the Grand Canyon image and fades in with the cruise ship image. At frame 4, 60% of the cruise ship is faded in, and the image begins to be discernible as such. The sequence ends

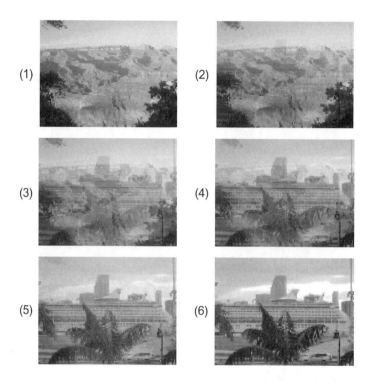

FIGURE 14.44A

Grayscale video sequence.

with the cruise ship in 100% fade-in. Figure 14.44A displays the generated grayscale sequence. Figure 14.44B shows the RGB color video sequence (also given in the color insert).

14.9 VIDEO SIGNAL BASICS

Video signals generally can be classified as component video, composite video, and S-video. In *component video*, three video signals—such as the red, green, and blue channels or the Y, I, and Q channels—are used. Three wires are required for connection to a camera or TV. Most computer systems use component video signals. *Composite video* has intensity (luminance) and two-color (chrominance) components that modulate the carrier wave. This signal is used in broadcast color TV. The standard by the US-based National Television System Committee (NTSC) combines channel signals into a chroma signal, which is modulated to a higher frequency for transmission. Connecting TVs or VCRs requires only one wire, since both video and audio are mixed into the same signal. *S-video* sends luminance and chrominance separately, since the luminance presenting black-and-white intensity contains most of the signal information for visual perception.

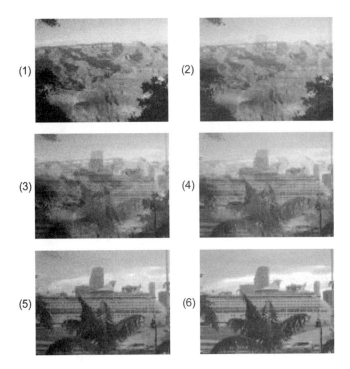

FIGURE 14.44B

The RGB color video sequence.

14.9.1 Analog Video

In computer systems, progressive scanning traces a whole picture, called *frame via row-wise*. A higher-resolution computer uses 72 frames per second (fps). The video is usually played at a frame rate varying from 15 frames to 30 frames.

In TV reception and some monitors, *interlaced scanning* is used in a cathode-ray tube display, or raster. The odd-numbered lines are traced first, and the even-numbered lines are traced next. We then get the odd-field and even-field scans per frame. The interlaced scheme is illustrated in Figure 14.45, where the odd lines are traced, such as A to B, then C to D, and so on, ending in the middle at E. The even field begins at F in the middle of the first line of the even field and ends at G. The purpose of using interlaced scanning is to transmit a full frame quickly to reduce flicker. Trace jumping from B to C is called horizontal retrace, while trace jumping from E to F or G to A is called vertical retrace.

The video signal is amplitude modulated. The modulation levels for NTSC video are shown in Figure 14.46. In the United States, negative modulation is used, meaning that the bright and dark intensities are inverted before modulation. In the negative modulated video signal, less amplitude comes from a brighter scene while more amplitude comes from a darker one. Since most pictures contain more white than black levels, the video signal level is reduced. Hence, with negative modulation, possible power efficiency can be achieved for transmission. The reverse process will apply for display at the receiver.

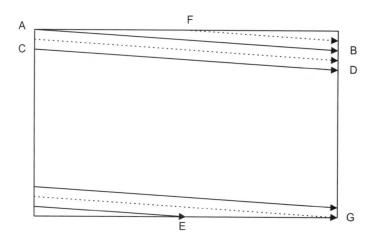

FIGURE 14.45

Interlaced raster scanning.

The horizontal synchronizing pulse controls the timing for the horizontal retrace. The blanking levels are used for synchronizing as well. The "back porch" (Figure 14.46) of the blanking also contains the color subcarrier burst for color demodulation.

The demodulated electrical signal can be seen in Figure 14.47, where a typical electronic signal for one scan line is depicted. The white intensity has a peak value of 0.714 volt, and the black has a voltage level of 0.055 volt, which is close to zero. The blank corresponds to zero voltage, and the synchronizing pulse is at the level of -0.286 volt. Synchronizing takes 10.9 microseconds, while video

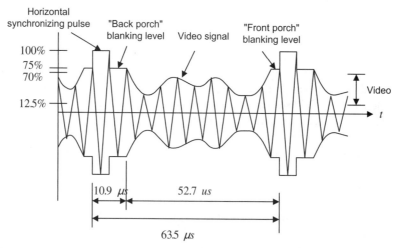

FIGURE 14.46

Video-modulated waveform.

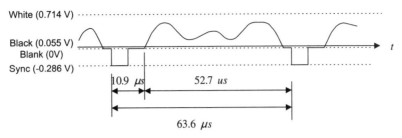

FIGURE 14.47

The demodulated signal level for one NTSC scan line.

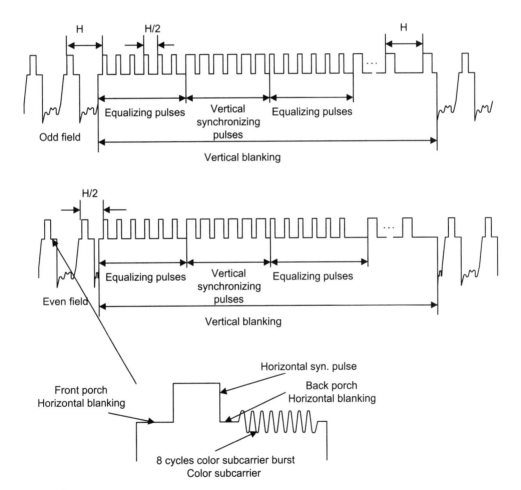

FIGURE 14.48

Vertical synchronization for each field and the color subcarrier burst.

occupies 52.7 microseconds, and one entire scan line occupies 63.6 microseconds. Hence, the line scan rate can be determined as 15.75 kHz.

Figure 14.48 describes vertical synchronization. A pulse train is generated at the end of each field. The pulse train contains 6 equalizing pulses, 6 vertical synchronizing pulses, and another 6 equalizing pulses at the rate of twice the size of the line scan rate (31.5 kHz), so that the timing for sweeping half the width of the field is feasible. In NTSC, the vertical retrace takes the time interval of 20 horizontal lines designated for control information at the beginning of each field. The 18 pulses of the vertical blanking occupy a time interval that is equivalent to 9 lines. This leaves lines 10 to 20 for other uses.

A color subcarrier resides on the back porch, as shown in Figure 14.48. The eight cycles of the color subcarrier are recovered via a delayed gating circuit trigged by the horizontal sync pulse. Synchronization includes the color burst frequency and phase information. The color subcarrier is then applied to demodulate the color (chrominance).

Let us summarize NTSC video signals. The NTSC TV standard uses an aspect ratio of 4:3 (ratio of picture width to height), and 525 scan lines per frame at 30 fps. Each frame has an odd field and an even field. So there are $525/2=262.5$ lines per field. NTSC actually uses 29.97 fps. The horizontal sweep frequency is $525\times 29.97=15{,}734$ lines per second, and each line takes $1/15{,}734=63.6\ \mu$ sec. Horizontal retrace takes $10.9\ \mu$ sec, while the line signal takes $52.7\ \mu$ sec. for one line of image display. Vertical retrace and sync are also needed so that the first 20 lines for each field are reserved. The active video lines per frame are 485. The layout of the video data, retrace, and sync data is shown in Figure 14.49.

Blanking regions can be used for V-chip information, stereo audio channel data, and subtitles in various languages. The active line is then sampled for display. A pixel clock divides each horizontal line of video into samples. For example, vertical helical scan (VHS) uses 240 samples per line, while Super VHS uses 400–425 samples per line.

Figure 14.50 shows the NTSC video signal spectra. The NTSC standard assigns a bandwidth of 4.2 MHz for luminance Y, 1.5 MHz for I, and 0.5 MHz for Q, due to the human perception of color

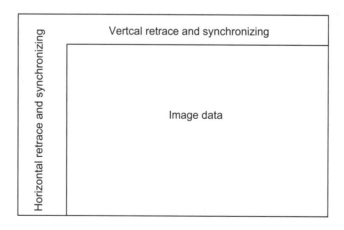

FIGURE 14.49

Video data, retrace, and sync layout.

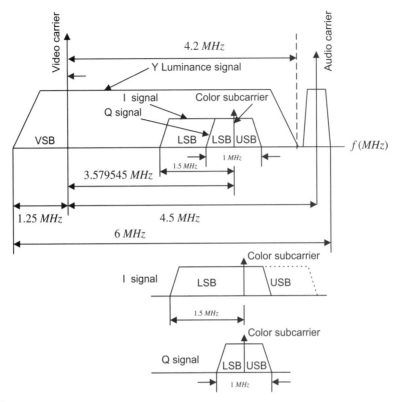

FIGURE 14.50

NTSC Y, I, and Q spectra.

information. There is a wider bandwidth for I because the human eye has higher resolution to the I color component than the Q color component.

As shown in Figure 14.50, *vestigial sideband modulation* (VSB) is employed for the luminance, with a picture carrier of 1.25 MHz relative to the VSB left edge. The space between the picture carrier and the audio carrier is 4.5 MHz.

The audio signal containing the frequency range from 50 Hz to 15 kHz is stereo frequency modulated (FM), using a peak frequency deviation of 25 kHz. Therefore, stereo FM audio requires a transmission bandwidth of 80 kHz, with an audio carrier located at 4.5 MHz relative to the picture carrier.

The color burst carrier is centered at 3.58 MHz above the picture carrier. The two color components I and Q undergo *quadrature amplitude modulation* (QAM) with modulated component I output, which is VSB filtered to remove two-thirds of the upper sideband, so that all chroma signals fall within a 4.2 MHz video bandwidth. The color burst carrier of 3.58 MHz is chosen such that the chroma signal and luminance are interleaved in the frequency domain to reduce interference between them.

Generating a chroma signal with QAM gives

$$C = I \cos\left(2\pi f_{sc}t\right) + Q \sin\left(2\pi f_{sc}t\right) \tag{14.24}$$

where C = chroma component and f_{sc} = color subcarrier = 3.58 MHz. The NTSC signal is further combined into a composite signal:

$$Composite = Y + C = Y + I \cos\left(2\pi f_{sc}t\right) + Q \sin\left(2\pi f_{sc}t\right) \tag{14.25}$$

At decoding, the chroma signal is obtained by separating Y and C first. Generally, the lowpass filters located at the lower end of the channel can be used to extract Y. Comb filters may be employed to cancel interferences between the modulated luminance signal and the chroma signal (Li and Drew, 2004). Then we perform demodulation for I and Q as follows:

$$\begin{aligned} C \times 2 \cos\left(2\pi f_{sc}t\right) &= I2 \cos^2\left(2\pi f_{sc}t\right) + Q \times 2 \sin\left(2\pi f_{sc}t\right) \cos\left(2\pi f_{sc}t\right) \\ &= I + I \times \cos\left(2 \times 2\pi f_{sc}t\right) + Q \sin\left(2 \times 2\pi f_{sc}t\right) \end{aligned} \tag{14.26}$$

Applying a lowpass filter yields the I component. Similar operation applying a carrier signal of $2\sin(2\pi f_{sc}t)$ for demodulation recovers the Q component.

PAL Video

The phase alternative line (PAL) system uses 625 scan lines per frame at 25 fps, with an aspect ratio of 4:3. It is widely used in Western Europe, China, and India. PAL uses the YUV color model, with an 8-MHz channel in which Y has 5.5 MHz and U and V each have 1.8 MHz with the color subcarrier frequency of 4.43 MHz relative to the picture carrier. U and V are the color difference signals (chroma signals) of the B-Y signal and R-Y signal, respectively. The chroma signals have alternate signs (e.g., +V and –V) in successive scan lines. Hence, in consecutive lines, the signal and its sign-reversed counterpart are averaged to cancel out phase errors that could be displayed as color errors.

SECAM Video

The SECAM (Séquentiel Couleur à Mémoire) system uses 625 scan lines per frame at 25 fps, with an aspect ratio of 4:3 and interlaced fields. The YUV color model is employed, and U and V signals are modulated using separate color subcarriers of 4.25 and 4.41 MHz, respectively. The U and V signals are sent on each line alternatively. In this way, quadrature multiplexing and the possible cross-coupling of color signals can be avoided by halving the color resolution in the vertical dimension. Table 14.15 includes a summary of analog broadband TV systems.

Table 14.15 Analog Broadband TV Systems

TV System	Frame Rate (fps)	Number of Scan Lines	Total Bandwidth (MHz)	Y Bandwidth (MHz)	U or I Bandwidth (MHz)	V or Q Bandwidth (MHz)
NTSC	29.97	525	6.0	4.2	1.6	0.6
PAL	25	625	8.0	5.5	1.8	1.8
SECAM	25	625	8.0	6.0	2.0	2.0

Source: Li and Drew, 2004.

14.9.2 **Digital Video**

Digital video has become dominant over the long-standing analog method in modern systems and devices because it offers: high image quality; flexibility of storage, retrieval, and editing capabilities; digital enhancement of video images; encryption; channel noise tolerance; and multimedia system applications.

Digital video formats are developed by the Consultative Committee for International Radio (CCIR). One of the most important standards is CCIR-601, which became ITU-R-601, an international standard for professional video applications.

In CCIR-601, chroma subsampling is carried for digital video. Each pixel is in the YCbCr color space, where Y is the luminance, and Cb and Cr are the chrominance. Subsampling schemes include 4:4:4 (no chroma subsampling), 4:2:2, 4:1:1, and 4:2:0 as illustrated in Figure 14.51.

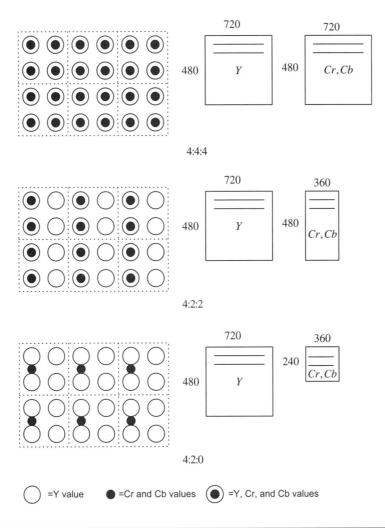

FIGURE 14.51

Chroma subsampling.

Table 14.16 Digital Video Specifications

	CCIR 601 525/60 NTSC	CCR 601 625/50 PAL/SECAM	CIF	QCIF
Luminance resolution	720 × 480	720 × 576	352 × 288	176 × 144
Chrominance resolution	360 × 480	360 × 576	176 × 144	88 × 72
Color subsampling	4:2:2	4:2:2	4:2:0	4:2:0
Aspect ratio	4:3	4:3	4:3	4:3
Fields/sec	60	50	30	30
Interlaced	Yes	Yes	No	No

CCR, comparison category rating; CIF, common intermediate format; QCIF, quarter-CIF.

Source: Li and Drew, 2004.

In each frame in a 4:4:4 video format, the number of values for each chrominance component, Cb or Cr, is the same as that for luminance, Y, both horizontally and vertically. This format finds applications in the computer graphics, in which the chrominance resolution is required for both horizontal and vertical dimension. The format is not widely used in video applications due to a huge storage requirement.

As shown in Figure 14.51, for each frame in the 4:2:2 video format, the number of chrominance components for Cr or Cb is half the number of luminance components for Y. The resolution is full vertically, and the horizontal resolution is downsampled by a factor of 2. Considering the first line of six pixels, transmission occurs in the following form: (Y0, Cb0), (Y1, Cr0), (Y2,Cb2), (Y3,Cr2), (Y4,Cb4), (Y5, Cr4), and so on. Six Y values are sent for every two Cb and Cr values that are sent.

In the 4:2:0 video format, the number of values for each chrominance Cb and Cr is half the number of luminance Y values for both the horizontal and vertical directions. That is, the chroma is down-sampled horizontally and vertically by a factor of 2. The location for both Cb and Cr is shown in Figure 14.51. Digital video specifications are given in Table 14.16.

CIF was specified by the Comité Consultatif International Téléphonique et Télégraphique (CCITT), which is now the International Telecommunications Union (ITU). CIF produces low bit rate video and supports progressive scan. QCIF produces video with an even lower bit rate. Neither format supports interlaced scan mode.

Table 14.17 outlines the high-definition TV (HDTV) formats supported by the Advanced Television System Committee (ATSC), where "I" means interlaced scan and "P" indicates progressive scan. MPEG compressions of video and audio are employed.

Table 14.17 High-Definition TV (HDTV) Formats

Number of Active Pixels per Line	Number of Active Lines	Aspect Ratio	Picture Rate
1920	1080	16:9	60I 30P 24P
1280	720	16:9	60P 30P 24 P
704	480	16:9 and 4:3	60I 60P 30P 24P
640	480	4:3	60I 60P 30P 24P

Source: Li and Drew, 2004.

14.10 MOTION ESTIMATION IN VIDEO

In this section, we study motion estimation since this technique is widely used in MPEG video compression. A video contains a time-ordered sequence of frames. Each frame consists of image data. When the objects in an image are still, the pixel values do not change under constant lighting conditions. Hence, there is no motion between the frames. However, if the objects are moving, then the pixels are moved. If we can find the motions, which are the pixel displacements, with *motion vectors*, the frame data can be recovered from the reference frame by copying and pasting at locations specified by the motion vector. To explore such an idea, let us look at Figure 14.52.

As shown in Figure 14.52, the reference frame is displayed first, and the next frame is the target frame containing a moving object. The image in the target frame is divided into N $\times$ N macroblocks (20 macroblocks). A macroblock match is searched within the search window in the reference frame to find the closest match between a macroblock under consideration in the target frame and the macroblock in the reference frame. The differences between two locations (motion vectors) for the matched macroblocks are encoded.

The criteria for finding the best match can be chosen using the mean absolute difference (MAD) between the reference frame and the target frame:

$$MAD(i, j) = \frac{1}{N^2} \sum_{k=0}^{N-1} \sum_{l=0}^{N-1} |T(m+k, n+l) - R(m+k+i, n+l+j)| \qquad (14.27)$$

$$u = i, v = j \; for \; MAD(i, j) = \text{minimum, and} -p \leq i, j \leq p \qquad (14.28)$$

There are many search methods for finding the motion vectors, including optimal, sequential, or brute force searches, and suboptimal searches such as 2D-logarithmic and hierarchical searches. Here we examine sequential search to understand the basic idea.

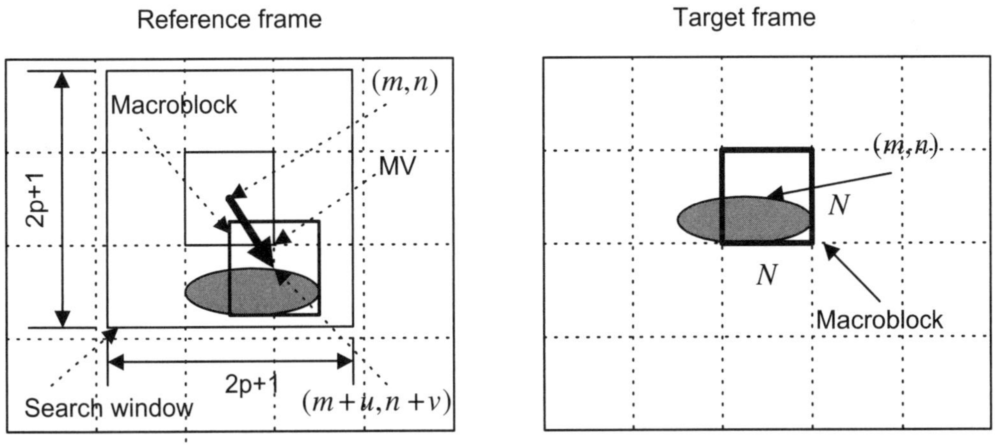

FIGURE 14.52

Macroblocks and motion vectors in the reference frame and target frame.

The sequential search for finding the motion vectors employs methodical "brute force" to search the entire $(2p+1) \times (2p+1)$ search window in the reference frame. The macroblock in the target frame compares each macroblock centered at each pixel in the search window in the reference frame. Comparison using Equation (14.27) proceeds pixel by pixel to find the best match in which the vector (i, j) produces the smallest MAD. Then the motion vector $(MV(u, v))$ is found to be $u = i$, and $v = j$. The algorithm is described as follows:

```
min_MAD=large value
for i=-p, ..., p
  for j=-p,...,p
    cur_MAD=MDA(i,j);
    if cur_MAD < min_MAD
       min_MAD= cur_MAD;
       u=i;
       v=j;
  end
 end
end
```

Sequential search provides the best match with the least MAD. However, it requires a huge amount of computations. Other suboptimal search methods can be employed to reduce the computational requirement, but with sacrifices of image quality. These topics are beyond our scope.

EXAMPLE 14.17

An 80 × 80 reference frame, target frame, and their difference are displayed in Figure 14.53. A macroblock with a size of 16 × 16 is used, and the search window has a size of 32 × 32. The target frame is obtained by moving the reference frame to the right by 6 pixels and to the bottom by 4 pixels. The sequential search method is applied to find all the motion vectors. The reconstructed target frame using the motion vectors and reference image is given in Figure 14.53.

Since 80 × 80/(16 × 16) = 25, there are 25 macroblocks in the target frame and 25 motion vectors in total. The motion vectors are

Horizontal direction =

$$
\begin{matrix}
-6 & -6 & -6 & -6 & -6 & -6 & -6 & -6 & -6 & -6 & -6 & -6 & -6 & -6 \\
-6 & -6 & -6 & -6 & -6 & -6 & -6 & -6 & -6 & -6 & -6 &&&
\end{matrix}
\tag{14.29}
$$

Vertical direction =

$$
\begin{matrix}
-4 & -4 & -4 & -4 & -4 & -4 & -4 & -4 & -4 & -4 & -4 & -4 & -4 & -4 & -4 \\
-4 & -4 & -4 & -4 & -4 & -4 & -4 & -4 & -4 & -4 &&&&
\end{matrix}
\tag{14.30}
$$

The motion vector comprises the pixel displacements from the target frame to the reference frame. Hence, given the reference frame, directions specified in the motion vector should be switched to indicate the motion towards the target frame. As indicated by the obtained motion vectors, the target image is a version of the reference image moving to the right by 6 pixels and down by 4 pixels.

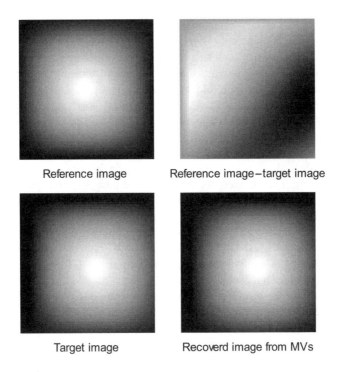

Reference image   Reference image–target image

Target image   Recoverd image from MVs

FIGURE 14.53

Reference frame, target frame, their difference, and the reconstructed frame by the motion vectors.

14.11 SUMMARY

1. A digital image consists of pixels. For a grayscale image, each pixel is assigned a grayscale level that presents the luminance of the pixel. For an RGB color image, each pixel is assigned a red component, a green component, and a blue component. For an indexed color image, each pixel is assigned an address that is the location of the color table (map) made up of the red, green, and blue components.

2. Common image data formats are 8-bit grayscale image, 24-bit color, and 8-bit indexed color.

3. The larger the number of pixels in an image, or the larger the numbers of the RGB components, the finer is the spatial resolution in the image. Similarly, the more scale levels used for each pixel, the better the scale-level image resolution. The more pixels and more bits used for the scale levels in the image, the more storage is required.

4. RGB color pixels can be converted to YIQ color pixels. The Y component is the luminance occupying 93% of the signal energy, while the I and Q components represent the color information of the image, occupying the remainder of the energy.

5. The histogram for a grayscale image shows the number of pixels at each grayscale level. The histogram can be modified to enhance the image. Image equalization using the histogram can

improve the image contrast and effectively enhances contrast for image underexposure. Color image equalization can be done only in the luminance channel or RGB channels.

6. Image enhancement techniques such as average lowpass filtering can filter out random noise in the image; however, it also blurs the image. The degree of blurring depends on the kernel size. The bigger the kernel size, the more blurring occurs.

7. Median filtering effectively removes the "pepper and salt" noise in an image.

8. The edge detection filter with Sobel convolution, Laplacian, and Laplacian of Gaussian kernels can detect image boundaries.

9. The grayscale image can be made into a facsimile of the color image by pseudo-color image generation, using the red, green, and blue transformation functions.

10. RGB-to-YIQ transformation is used to obtain the color image in YIQ space, or *vice versa*. It can also be used for color-to-grayscale conversion, that is, keeping only the luminance channel after the transformation.

11. 2D spectra can be calculated and are used to examine filtering effects.

12. JPEG compression uses the 2D-DCT transform for both grayscale and color images in the YIQ color space. JPEG uses different quality factors to normalize DCT coefficients for the luminance (Y) channel and the chrominance (IQ) channels.

13. The mixing of two images, in which two pixels are linearly interpolated using the weights $1 - \alpha$ and α, can produce video sequences that have effects such as fading in and fading out of images, blending of two images, and overlaying of text on an image.

14. Analog video uses interlaced scanning. A video frame contains odd and even fields. Analog video standards include NTSC, PAL, SECAM.

15. Digital video carries the modulated information for each pixel in the YCbCr color space, where Y is the luminance and Cb and Cr are the chrominance. Chroma subsampling creates various digital video formats. The industry standards include CCIR601, CCR601, CIF, and QCIF.

16. Motion compensation of a video sequence produces motion vectors for all the image blocks in the target video frame, which contain displacements of these image blocks relative to the reference video frame. Recovering the target frame involves simply copying each image block of the reference frame to the target frame at the location specified in the motion vector. Motion compensation is a key element in MPEG video.

14.12 PROBLEMS

14.1. Determine the memory storage requirement for each of the following images:

    a. 320 × 240 8-bit grayscale

    b. 640 × 480 24-bit color image

    c. 1600 × 1200 8-bit indexed image

14.2. Determine the number of colors for each of the following images:

    a. 320 × 240 16-bit indexed image

    b. 200 × 100 24-bit color image

14.3. Given a pixel in an RGB image

$$R = 200, G = 120, B = 100$$

convert the RGB values to YIQ values.

14.4. Given a pixel of an image in YIQ color format

$$Y = 141, I = 46, Q = 5$$

convert the YIQ values back to RGB values.

14.5. Given the 2×2 RGB image,

$$R = \begin{bmatrix} 100 & 50 \\ 100 & 50 \end{bmatrix} \quad G = \begin{bmatrix} 20 & 40 \\ 10 & 30 \end{bmatrix} \quad B = \begin{bmatrix} 100 & 50 \\ 200 & 150 \end{bmatrix}$$

convert the image into grayscale.

14.6. Produce a histogram of the following image, which has a grayscale value ranging from 0 to 7, that is, each pixel is encoded in 3 bits.

$$\begin{bmatrix} 0 & 1 & 2 & 2 & 0 \\ 2 & 1 & 1 & 2 & 1 \\ 1 & 1 & 4 & 2 & 3 \\ 0 & 2 & 5 & 6 & 1 \end{bmatrix}$$

14.7. Consider the following image with a grayscale value ranging from 0 to 7, that is, each pixel is encoded in 3 bits:

$$\begin{bmatrix} 0 & 1 & 2 & 2 & 0 \\ 2 & 1 & 1 & 2 & 1 \\ 1 & 1 & 4 & 2 & 3 \\ 0 & 2 & 5 & 6 & 1 \end{bmatrix}$$

Perform equalization using the histogram in Problem 14.6, and plot the histogram for the equalized image.

14.8. Consider the following image with a grayscale value ranging from 0 to 7, that is, each pixel is encoded in 3 bits:

$$\begin{bmatrix} 2 & 4 & 4 & 2 \\ 2 & 3 & 3 & 3 \\ 4 & 4 & 4 & 2 \\ 3 & 2 & 3 & 4 \end{bmatrix}$$

Perform level adjustment to the full range, shift the level to the range from 3 to 7, and shift the level to the range from 0 to 3.

14.9. Consider the following 8-bit grayscale original and noisy images and 2×2 convolution average kernel:

$$4 \times 4 \text{ original image}: \begin{bmatrix} 100 & 100 & 100 & 100 \\ 100 & 100 & 100 & 100 \\ 100 & 100 & 100 & 100 \\ 100 & 100 & 100 & 100 \end{bmatrix}$$

$$4 \times 4 \text{ corrupted image}: \begin{bmatrix} 93 & 116 & 109 & 96 \\ 92 & 107 & 103 & 108 \\ 84 & 107 & 86 & 107 \\ 87 & 113 & 106 & 99 \end{bmatrix}$$

$$2 \times 2 \text{ average kernel}: \frac{1}{4}\begin{bmatrix} 1 & 1 \\ 1 & 1 \end{bmatrix}$$

Perform digital filtering on the noisy image, and compare the enhanced image with the original image.

14.10. Consider the following 8-bit grayscale original and noisy image, and 3×3 median filter kernel:

$$4 \times 4 \text{ original image}: \begin{bmatrix} 100 & 100 & 100 & 100 \\ 100 & 100 & 100 & 100 \\ 100 & 100 & 100 & 100 \\ 100 & 100 & 100 & 100 \end{bmatrix}$$

$$4 \times 4 \text{ corrupted image by impulse noise}: \begin{bmatrix} 100 & 255 & 100 & 100 \\ 0 & 255 & 255 & 100 \\ 100 & 0 & 100 & 0 \\ 100 & 255 & 100 & 100 \end{bmatrix}$$

$$3 \times 3 \text{ average kernel}: \begin{bmatrix} & \\ & \end{bmatrix}$$

Perform digital filtering, and compare the filtered image with the original image.

14.11. Given the 8-bit 5 × 4 original grayscale image

$$\begin{bmatrix} 110 & 110 & 110 & 110 \\ 110 & 100 & 100 & 110 \\ 110 & 100 & 100 & 110 \\ 110 & 110 & 110 & 110 \\ 110 & 110 & 110 & 110 \end{bmatrix}$$

apply the following edge detectors to the image:

a. Sobel vertical edge detector

b. Laplacian edge detector

Scale the resultant image pixel value to the range of 0 to 255.

14.12. In Example 14.10, if we switch the transformation functions between the red function and the green function, what is the expected color for the area pointed to by the arrow, and what is the expected background color?

14.13. In Example 14.10, if we switch the transformation functions between the red function and the blue function, what is the expected color for the area pointed to by the arrow, and what is the expected background color?

14.14. Consider the following grayscale image $p(i, j)$:

$$\begin{bmatrix} 100 & -50 & 10 \\ 100 & 80 & 100 \\ 50 & 50 & 40 \end{bmatrix}$$

Determine the 2D-DFT coefficient $X(1, 2)$ and the magnitude spectrum $A(1, 2)$.

14.15. Consider the following grayscale image $p(i, j)$:

$$\begin{bmatrix} 10 & 100 \\ 200 & 150 \end{bmatrix}$$

Determine the 2D-DFT coefficients $X(u, v)$ and magnitude $A(u, v)$.

14.16. Consider the following grayscale image $p(i, j)$:

$$\begin{bmatrix} 10 & 100 \\ 200 & 150 \end{bmatrix}$$

Apply the 2D-DCT to determine the DCT coefficients.

14.17. Consider the following DCT coefficients $F(u, v)$:

$$\begin{bmatrix} 200 & 10 \\ 10 & 0 \end{bmatrix}$$

Apply the inverse 2D-DCT to determine the 2D data.

14.18. In JPEG compression, DCT DC coefficients from several blocks are 400, 390, 350, 360, and 370. Use DPCM to produce the DPCM sequence, and use the Huffman table to encode the DPCM sequence.

14.19. In JPEG compression, DCT coefficients from the an image subblock are

$$[175, \ -2, \ 0, \ 0, \ 0, \ 4, \ 0, \ 0, \ -3\,7, \ 0, \ 0, \ 0, \ 0, \ -2, \ 0, \ 0, \ ..., \ 0]$$

a. Generate the run-length codes for AC coefficients.

b. Perform entropy coding for the run-length codes using the Huffman table.

14.20. Consider the following grayscale image $p(i, j)$:

$$\begin{bmatrix} 10 & 100 \\ 200 & 150 \end{bmatrix}$$

Apply the 2D-DWT using the Haar wavelet to determine the level-1 DWT coefficients.

14.21. Consider the following level-1 IDWT coefficients $W(u, v)$ obtained using the Haar wavelet:

$$\begin{bmatrix} 200 & 10 \\ 10 & 0 \end{bmatrix}$$

Apply the IDWT to determine the 2D data.

14.22. Consider the following grayscale image $p(i, j)$:

$$\begin{bmatrix} 100 & 150 & 60 & 80 \\ 80 & 90 & 50 & 70 \\ 110 & 120 & 100 & 80 \\ 90 & 50 & 40 & 90 \end{bmatrix}$$

Apply the 2D-DWT using the Haar wavelet to determine the level-1 DWT coefficients.

14.23. Consider the following level-1 IDWT coefficients $W(u, v)$ obtained using the Haar wavelet:

$$\begin{bmatrix} 250 & 50 & -30 & -20 \\ 30 & 20 & 10 & -20 \\ 10 & 20 & 0 & 0 \\ 20 & 15 & 0 & 0 \end{bmatrix}$$

Apply the IDWT to determine the 2D data.

14.24. Explain the difference between horizontal retrace and vertical retrace. Which one would take more time?

14.25. What is the purpose of using interlaced scanning in a traditional NTSC TV system?

14.26. What is the bandwidth in the traditional NTSC TV broadcast system? What is the bandwidth to transmit luminance Y, and what are the required bandwidths to transmit Q and I channels, respectively?

14.27. What type of modulation is used for transmitting audio signals in the NTSC TV system?

14.28. Given the composite NTSC signal

$$Composite = Y + C = Y + I\cos(2\pi f_{sc}t) + Q\sin(2\pi f_{sc}t)$$

show demodulation for the Q channel.

14.29. Where does the color subcarrier burst reside? What is the frequency of the color subcarrier, and how many cycles does the color burst have?

14.30. Compare differences of the NTSC, PAL and SECAM video systems in terms of the number of scan lines, frame rates, and total bandwidths required for transmission.

14.31. In the NTSC TV system, what is the horizontal line scan rate? What is the vertical synchronizing pulse rate?

14.32. Explain which of the following digital video formats achieves the most data transmission efficiency:

  a. 4:4:4

  b. 4:2:2

  c. 4:2:0

14.33. What is the difference between interlaced scan and progressive scan? Which of the following video systems use progressive scan?

  a. CCIR-601

  b. CIF

14.34. In motion compensation, which of the following would require more computation? Explain.

  a. Finding the motion vector using sequential search

  b. Recovering the target frame with the motion vectors and reference frame

14.35. Given a reference frame and target frame of size 80 × 80, a macroblock size of 16 × 16, and a search window size of 32 × 32, estimate the number of subtractions, absolute value calculations, and additions for searching all the motion vectors using the sequential search method.

14.12.1 MATLAB Problems
Use MATLAB to solve Problems 14.36 to 14.42.

14.36. Given the image data "trees.jpg", use MATLAB functions to perform each of the following processes:

  a. Use MATLAB to read and display the image.

  b. Convert the image to grayscale.

c. Perform histogram equalization for the grayscale image in (b) and display the histogram plots for both the original grayscale image and equalized grayscale image.

d. Perform histogram equalization for the color image in (a) and display the histogram plots of the Y channel for both the original color image and equalized color image.

14.37. Given the image data "cruise.jpg", perform the following linear filtering:

a. Convert the image to grayscale and then create an 8-bit noisy image by adding Gaussian noise using the following code:

noise_image = imnoise(I,'gaussian');

where I is the intensity image obtained from normalizing the grayscale image.

b. Process the noisy image using a Gaussian filter with the following parameters: convolution kernel size = 4, SIGMA = 0.8. Compare the filtered image with the noisy image.

14.38. Given the image data "cruise.jpg", perform the following filtering process:

a. Convert the image to grayscale and then create an 8-bit noisy image by adding "pepper and salt" noise using the following code:

noise_image = imnoise(I,'salt & pepper');

where I is the intensity image obtained from normalizing the grayscale image.

b. Process the noisy image using median filtering with a convolution kernel size of 4 × 4.

14.39. Given the image data "cruise.jpg", convert the image to the grayscale and detect the image boundaries using Laplacian of Gaussian filtering with the following parameters:

a. Kernel size = 4 and SIGMA = 0.9

b. Kernel size = 10 and SIGMA = 10

Compare the results.

14.40. Given the image data "clipim2.gif", perform the following process:

a. Convert the indexed image to grayscale.

b. Adjust the color transformation functions (sine functions) to make the object indicated by the arrow in the image red and the background color green.

14.41. Given the image data "cruiseorg.tiff", perform JPEG compression by completing the following steps:

a. Convert the image to grayscale.

b. Write a MATLAB program for encoding with the following features: (1) divide the image into 8 × 8 blocks; (2) transform each block using the discrete-cosine transform; (3) scale and round DCT coefficients with the standard quality factor. Note that using lossless compression with the quantized DCT coefficients is omitted here for a simple simulation.

c. Continue and write a MATLAB program for decoding with the following features: (1) invert the scaling process for quantized DCT coefficients; (2) perform the inverse DCT for each 8 × 8 image block; (3) recover the image.

d. Run the developed MATLAB program to examine the image quality using

  I. The quality factor

  II. The quality factor × 5

  III. The quality factor × 10

14.42. Given the image data "cruiseorg.tiff", perform wavelet-based compression by completing the following steps:

a. Convert the image to grayscale.

b. Write a MATLAB program for encoding with the following features: (1) use an 8-tap Daubechies filter; (2) apply the two-level DWT; (3) perform 8-bit quantization for subbands LL2, LH2, HL2, HH2, LH1, HL1, and HH1 for simulation.

c. Write the MATLAB program for the decoding process.

d. Run the developed MATLAB program to examine the image quality using the following methods:

  I. Reconstruct the image using all subband coefficients.

  II. Reconstruct the image using LL2 subband coefficients.

  III. Reconstruct the image using LL2, HL2, LH2, and HH2 subband coefficients.

  IV. Reconstruct the image using LL2, HL2, and LH2 subband coefficients.

  V. Reconstruct the image using LL2, HL2, LH2, HH2, LH1, and HL1 subband coefficients.

Appendix A: Introduction to the MATLAB Environment

Matrix Laboratory (MATLAB) is used extensively in this book for simulations. The goal here is to help students acquire familiarity with MATLAB and build basic skills in the MATLAB environment. Hence, Appendix A serves the following objectives:

1. Learn to use the help system to study basic MATLAB commands and syntax.
2. Learn array indexing.
3. Learn basic plotting utilities.
4. Learn to write script m-files.
5. Learn to write functions in MATLAB.

A.1 BASIC COMMANDS AND SYNTAX

MATLAB has an accessible help system through the help command. By issuing the MATLAB help command following the topic or function (routine), MATLAB will return information on the topic and show how to use the function. For example, by entering **help sum** at the MATLAB prompt, we see information (listed partially here) on how to use the function **sum()**.

```
» help sum
SUM   Sum of the elements.
    For vectors, SUM(X) is the sum of the elements of X.
    For matrices, SUM(X) is a row vector with the sum over
    each column.
»
```

The following examples are given to demonstrate the usage of **sum()**:

```
» x=[ 1 2 3 1.5 -1.5 -2];      % Initialize an array
» sum(x)                        % Call MATLAB function sum
ans =
   4                            % Display the result
»
» x=[1 2 3; -1.5 1.5 2.5; 4 5 6] % Initialize 3 × 3 matrix
x =                              % Display the contents of 3 × 3 matrix
   1.0000 2.0000 3.0000
  -1.5000 1.5000 2.5000
   4.0000 5.0000 6.0000
```

```
» sum(x)                      % Call MATLAB function sum
ans =
  3.5000 8.5000 11.5000       % Display the results

»
```

MATLAB can be used like a calculator to work with numbers, variables, and expressions in the command window. The following are basic syntax examples:

```
» sin(pi/4)
ans =
  0.7071
» pi*4
ans =
  12.5664
```

In MATLAB, variable names can store values, vectors and matrices. See the following examples.

```
» x=cos(pi/8)
x =
  0.9239

» y=sqrt(x)-2^2
y =
  -3.0388

» z=[1 -2 1 2]
z =
  1 -2 1 2

» zz=[1 2 3; 4 5 6]
zz =
  1 2 3
  4 5 6
```

Complex numbers are natural in MATLAB. See the following examples.

```
» z=3+4i                      % Complex number
z =
  3.0000 + 4.0000i

» conj(z)                     % Complex conjugate of z
ans =
  3.0000 - 4.0000i

» abs(z)                      % Magnitude of z
ans =
  5
» angle(z)                    % Angle of z (radians)
ans =
```

```
  0.9273
» real(z)                    % Real part of a complex number z
ans =
  3
» imag(z)                    % Imaginary part of a complex number z
ans =
  4
» exp(j*pi/4)                % Polar form of a complex number
ans =
  0.7071 + 0.7071i
```

The following shows examples of array operations. Krauss, Shure, and Little (1994) and Stearns (2003) give detailed explanation for each operation.

```
» x=[1 2; 3 4]               % Initialize 2 × 2 matrixes
x =
  1 2
  3 4

» y=[-4 3; -2 1]
y =
  -4 3
  -2 1

» x+y                        % Add two matrixes
ans =
  -3 5
  1 5
» x*y                        % Matrix product
ans =
  -8 5
  -20 13

» x.*y                       % Array element product
ans =
  -4 6
  -6 4

» x'                         % Matrix transpose
ans =
  1 3
  2 4

» 2.^x                       % Exponentiation: matrix x contains each exponent
ans =
  2 4
  8 16
```

```
» x.^3                   % Exponentiation: power of 3 for each element in matrix x
ans =
  1 8
  27 64

» y.^x                   % Exponentiation: each element in y has a power
                         % specified by a corresponding element in matrix x
ans =
  -4 9
  -8 1

» x=[0 1 2 3 4 5 6]      % Initialize a row vector
x =
  0 1 2 3 4 5 6

» y=x.*x-2*x             % Use array element product to compute a quadratic function
y =
  0 -1 0 3 8 15 24

» z=[1 3]'               % Initialize a column vector
z =
  1
  3

» w=x\z                  % Invert matrix x, then multiply it by the column vector z
w =
  1
  0
```

A.2 MATLAB ARRAYS AND INDEXING

Let us look at the syntax to create an array as follows:

Basic syntax: **x=begin: step: end**

An array x is created with the first element value of **begin**. The value increases by a value of **step** for the next element in the array and stops when the next stepped value is beyond the specified end value of **end**. In simulation, we may use this operation to create the sample indexes or array of time instants for digital signals. The **begin**, **step,** and **end** can be assigned to integers or floating-point numbers.

The following examples are given for illustrations:

```
» n=1:3:8                % Create a vector n=[1 4 7]
n =
  1 4 7

» m=9:-1:2               % Create a vector m=[9 8 7 6 5 4 3 2 ]
m =
  9 8 7 6 5 4 3 2
```

```
» x=2:(1/4):4          % Create x=[2 2.25 2.5 2.75 3 3.25 3.5 3.75 4]
x =
 Columns 1 through 7
 2.0000 2.2500 2.5000 2.7500 3.0000 3.2500 3.5000
 Columns 8 through 9
 3.7500 4.0000

» y=2:-0.5:-1  % Create y=[2 1.5 1 0.5 0 -0.5 -1]
y =
 2.0000 1.5000 1.0000 0.5000 0 -0.5000 -1.0000
```

Next, we examine creating a vector and extracting numbers in a vetcor:

```
» xx= [ 1 2 3 4 5 [5:-1:1] ]  % Create xx=[ 1 2 3 4 5 5 4 3 2 1]
xx =
 1 2 3 4 5 5 4 3 2 1
» xx(3:7)                      % Show elements to 7, that is, [3 4 5 5 4]
ans =
 3 4 5 5 4
» length(xx)                   % Return of the number of elements in vetcor xx
ans =
 10
» yy=xx(2:2:length(xx))        % Display even indexed numbers in array xx
yy =
 2 4 5 3 1
```

A.3 PLOT UTILITIES: SUBPLOT, PLOT, STEM, AND STAIR

The following are common MATLAB plot functions for digital signal processing (DSP) simulation:

subplot opens subplot windows for plotting.
plot produces an x–y plot. It can also create multiple plots.
stem produces discrete-time signals.
stair produces staircase (sample-and-hold) signals.

The following program contains different MATLAB plot functions:

```
t=0:0.01:0.4;  % Create time vector for time instants from 0 to 0.4 second
xx=4.5*sin(2*pi*10*t+pi/4); % Calculate a sine function with a frequency of 10 Hz
yy=4.5*cos(2*pi*5*t-pi/3); % Calculate cos function with a frequency of 5 Hz
subplot(4,1,1), plot(t,xx);grid        % Plot a sine function in window 1
subplot(4,1,2), plot(t,xx, t,yy, ' -. ' );grid; % Plot sine and cos functions in window 2
subplot(4,1,3), stem(t,xx);grid        % Plot a sine function in the discrete-time form
subplot(4,1,4), stairs(t,yy);grid      % Plot a cos function in the sample-and-hold form
xlabel( ' Time (sec.) ' );
```

Each plot is shown in Figure A.1. Notice that dropping the semicolon at the end of the MATLAB syntax will display values on the MATLAB prompt.

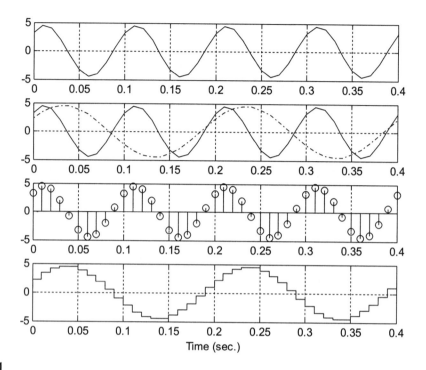

FIGURE A.1

Illustration for the MATLAB plot functions.

A.4 MATLAB SCRIPT FILES

We can create a MATLAB script file using the built-in MATLAB editor (or Windows Notepad) to write MATLAB source code. The script file is named "filename.m" and can be run by typing the file name at the MATLAB prompt and hitting the return key. The script file **test.m** is described here for illustration. Figure A.2 illustrates the plot produced by **test.m**.

At MATLAB prompt, run the program

>>which test   % show the folder where test.m resides

Go to the folder that contains test.m, and run your script from MATLAB.

>>test   % run the test.m

>>type test   % display the contents of test.m

test.m

```
t=0:0.01:1;
x=sin(2*pi*2*t);
y=0.5*cos(2*pi*5*t-pi/4);
plot(t,x), grid on
title( ' Test plots of sinusoids ' )
ylabel( ' Signal amplitudes ' );
```

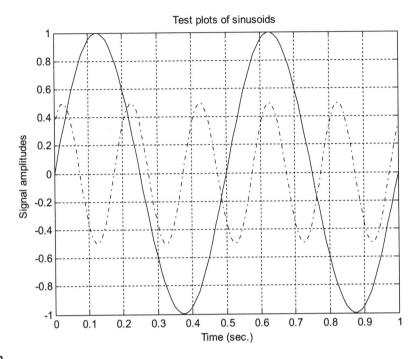

FIGURE A.2

Illustration of MATLAB script file test.m.

```
xlabel( ' Time (sec.) ' ); hold on
plot(t,y, ' -. ' );
```

A.5 MATLAB FUNCTIONS

A MATLAB function is often used to replace the repetitive portions of MATLAB code. It is created using a MATLAB script file. However, the code begins with the keyword **function**, followed by the function declaration, comments for the help system, and program code. A function **sumsub.m** that computes the addition and subtraction of two numbers is listed here for illustration.

sumsub.m

```
function [sum, sub]=sumsub(x1,x2)
%sumsub: Function to add and subtract two numbers
% Usage:
% [sum, sub] = sumsub(x1,x2)
% x1 = the first number
% x2= the second number
% sum = x1+x2;
```

```
% sub = x1-x2
sum = x1+x2;   % Add two numbers
sub= x1-x2;    % Subtract x2 from x1
```

To use the MATLAB function, go to the folder that contains **sumsub.m**. At the MATLAB prompt, try the following:

```
» help sumsub  % Display usage information on MATLAB prompt
sumsub: Function to add and subtract two numbers
usage:
[sum, sub] = sumsub(x1,x2)
x1 = the first number
x2= the second number
sum = x1+x2;
sub = x1-x2
```

Run the function as follows:

```
» [x1, x2]=sumsub(3, 4-3i);  % Call function sumsub
» x1                          % Display the result of sum
x1 =
  7.0000 - 3.0000i
» x2                          % Display the result of subtraction
x2 =
 -1.0000 + 3.0000i
```

MATLAB functions can also be used inside an m-file. More MATLAB exercises for introduction to DSP can be explored in McClellan, Schafer, and Yoder (1998) and Stearns (2003).

Appendix B: Review of Analog Signal Processing Basics

B.1 FOURIER SERIES AND FOURIER TRANSFORM

Electronics applications require familiarity with some periodic signals such the square wave, rectangular wave, triangular wave, sinusoid, sawtooth wave, and so on. These periodic signals can be analyzed in the frequency domain with the help of the Fourier series expansion. According to Fourier theory, a periodic signal can be represented by a Fourier series that contains the sum of a series of sine and/or cosine functions (harmonics) plus a direct current (DC) term. There are three forms of Fourier series: (1) sine-cosine, (2) amplitude-phase, and (3) complex exponential. We will review each of them individually in the following text. Comprehensive treatments can be found in Ambardar (1999), Soliman and Srinath (1998), and Stanley (2003).

B.1.1 Sine-Cosine Form

The Fourier series expansion of a periodic signal $x(t)$ with a period of T via the sine-cosine form is given by

$$x(t) = a_0 + \sum_{n=1}^{\infty} a_n \cos(n\omega_0 t) + \sum_{n=1}^{\infty} b_n \sin(n\omega_0 t) \tag{B.1}$$

where $\omega_0 = 2\pi/T_0$ is the fundamental angular frequency in radians per second, while the fundamental frequency in terms of Hz is $f_0 = 1/T_0$. The Fourier coefficients of a_0, a_n, and b_n may be found according to the following integral equations:

$$a_0 = \frac{1}{T_0} \int_{T_0} x(t)dt \tag{B.2}$$

$$a_n = \frac{2}{T_0} \int_{T_0} x(t) \cos(n\omega_0 t)dt \tag{B.3}$$

$$b_n = \frac{2}{T_0} \int_{T_0} x(t) \sin(n\omega_0 t)dt \tag{B.4}$$

Notice that the integral is performed over one period of the signal to be expanded. From Equation (B.1), the signal $x(t)$ consists of a DC term and sums of sine and cosine functions with their corresponding harmonic frequencies. Again, note that $n\omega_0$ is the nth harmonic frequency.

B.1.2 Amplitude-Phase Form

From the sine-cosine form, we notice that there is a sum of two terms with the same frequency. The term in the first sum is $a_n \cos(n\omega_0 t)$ while the other is $b_n \sin(n\omega_0 t)$. We can combine these two terms and modify the sine-cosine form into the amplitude-phase form:

$$x(t) = A_0 + \sum_{n=1}^{\infty} A_n \cos(n\omega_0 t + \phi_n) \tag{B.5}$$

The DC term is same as before, that is,

$$A_0 = a_0 \tag{B.6}$$

and the amplitude and phase are given by

$$A_n = \sqrt{a_n^2 + b_n^2} \tag{B.7}$$

$$\phi_n = \tan^{-1}\left(\frac{-b_n}{a_n}\right) \tag{B.8}$$

respectively. The amplitude-phase form provides very useful information for spectral analysis. With the calculated amplitude and phase for each harmonic frequency, we can create the spectral plots. One depicts a plot of the amplitude versus its corresponding harmonic frequency (the amplitude spectrum), while the other plot shows each phase versus its harmonic frequency (the phase spectrum). Note that the spectral plots are one-sided, since amplitudes and phases are plotted versus the positive harmonic frequencies. We will illustrate these in Example B.1.

B.1.3 Complex Exponential Form

The complex exponential form is developed based on expanding sine and cosine functions in the sine-cosine form into their exponential expressions using Euler's formula and regrouping these exponential terms. Euler's formula is given by

$$e^{\pm jx} = \cos(x) \pm j \sin(x)$$

which can be written as two separate forms:

$$\cos(x) = \frac{e^{jx} + e^{-jx}}{2}$$

$$\sin(x) = \frac{e^{jx} - e^{-jx}}{2j}$$

We will focus on interpretation and application rather than the derivation of this form. Thus the complex exponential form is expressed as

$$x(t) = \sum_{n=-\infty}^{\infty} c_n e^{jn\omega_0 t} \tag{B.9}$$

where c_n represents the complex Fourier coefficients, which may be found from

$$c_n = \frac{1}{T_0} \int_{T_0} x(t) e^{-jn\omega_0 t} dt \qquad (B.10)$$

The relationship between the complex Fourier coefficients c_n and the coefficients of the sine-cosine form are

$$c_0 = a_0 \qquad (B.11)$$

$$c_n = \frac{a_n - jb_n}{2}, \quad \text{for} \quad n > 0 \qquad (B.12)$$

Considering a real signal $x(t)$ ($x(t)$ is not a complex function) in Equation (B.10), c_{-n} is equal to the complex conjugate of c_n, that is, $\bar{c}_n$. It follows that

$$c_{-n} = \bar{c}_n = \frac{a_n + jb_n}{2}, \quad \text{for} \quad n > 0 \qquad (B.13)$$

Since c_n is a complex value that can be written in the magnitude-phase form, we obtain

$$c_n = |c_n| \angle \phi_n \qquad (B.14)$$

where $|c_n|$ is the magnitude and ϕ_n is the phase of the complex Fourier coefficient. Similar to the magnitude-phase form, we can create the spectral plots for $|c_n|$ and ϕ_n. Since the frequency index n goes from $-\infty$ to ∞, the plots of the resultant spectra are two-sided.

EXAMPLE B.1

Consider the square waveform $x(t)$ shown in Figure B.1, where T_0 represents a period. Find the Fourier series expansions in terms of (a) the sine-cosine form, (b) the amplitude-phase form, and (c) the complex exponential form.

Solution:
From Figure B.1, we notice that $T_0 = 1$ second and $A = 10$. The fundamental frequency is

$$f_0 = 1/T_0 = 1 \text{ Hz} \quad \text{or} \quad \omega_0 = 2\pi \times f_0 = 2\pi \text{ rad/sec}$$

a. Using Equations (B.1) to (B.3) yields

$$a_0 = \frac{1}{T_0} \int_{-T_0/2}^{T_0/2} x(t) dt = \frac{1}{1} \int_{-0.25}^{0.25} 10 dt = 5$$

$$a_n = \frac{2}{T_0} \int_{-T_0/2}^{T_0/2} x(t) \cos(n\omega_0 t) dt$$

$$= \frac{2}{1} \int_{-0.25}^{0.25} 10 \cos(n 2\pi t) dt$$

$$= \frac{2}{1} \frac{10 \times \sin(n 2\pi t)}{n 2\pi} \Big|_{-0.25}^{0.25} = 10 \frac{\sin(0.5\pi n)}{0.5\pi n}$$

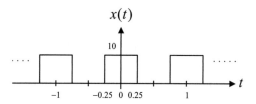

FIGURE B.1

Square waveform in Example B.1.

$$b_n = \frac{2}{T_0} \int_{-T_0/2}^{T_0/2} x(t)\sin(n\omega_0 t)\,dt$$

$$= \frac{2}{1} \int_{-0.25}^{0.25} 10 \times \sin(n2\pi t)\,dt$$

$$= \frac{2}{1} \frac{-10\cos(n2\pi t)}{n2\pi}\bigg|_{-0.25}^{0.25} = 0$$

Thus, the Fourier series expansion in terms of the sine-cosine form is written as

$$x(t) = 5 + \sum_{n=1}^{\infty} 10\frac{\sin(0.5\pi n)}{0.5\pi n}\cos(n2\pi t)$$

$$= 5 + \frac{20}{\pi}\cos(2\pi t) - \frac{20}{3\pi}\cos(6\pi t) + \frac{4}{\pi}\cos(10\pi t) - \frac{20}{7\pi}\cos(14\pi t) + \cdots$$

b. Making use of the relations between the sine-cosine form and the amplitude-phase form, we obtain

$$A_0 = a_0 = 5$$

$$A_n = \sqrt{a_n^2 + b_n^2} = |a_n| = 10 \times \left|\frac{\sin(0.5\pi n)}{0.5\pi n}\right|$$

Again, noting that $-\cos(x) = \cos(x + 180°)$, the Fourier series expansion in terms of the amplitude-phase form is

$$x(t) = 5 + \frac{20}{\pi}\cos(2\pi t) + \frac{20}{3\pi}\cos(6\pi t + 180°) + \frac{4}{\pi}\cos(10\pi t) + \frac{20}{7\pi}\cos(14\pi t + 180°) + \cdots$$

c. First let us find the complex Fourier coefficients using the formula, that is,

$$c_n = \frac{1}{T_0} \int_{-T_0/2}^{T_0/2} x(t)e^{-jn\omega_0 t}\,dt$$

$$= \frac{1}{1} \int_{-0.25}^{0.25} Ae^{-jn2\pi t}\,dt$$

$$= 10 \times \frac{e^{-jn2\pi t}}{-jn2\pi}\bigg|_{-0.25}^{0.25} = 10 \times \frac{\left(e^{-j0.5\pi n} - e^{j0.5\pi n}\right)}{-jn2\pi}$$

Applying Euler's formula yields

$$c_n = 10 \times \frac{\cos 0.5\pi n - j\sin(0.5\pi n) - [\cos(0.5\pi n) + j\sin(0.5\pi n)]}{-jn2\pi} = 5\frac{\sin(0.5\pi n)}{0.5\pi n}$$

Second, using the relationship between the sine-cosine form and the complex exponential form, it follows that

$$c_n = \frac{a_n - jb_n}{2} = \frac{a_n}{2} = 5\frac{\sin(0.5n\pi)}{(0.5n\pi)}$$

Certainly, the result is identical to the one obtained directly from the formula. Note that c_0 cannot be evaluated directly by substituting $n = 0$, since we have the indeterminate term $\frac{0}{0}$. Using L'Hospital's rule, described in Appendix G, leads to

$$c_0 = \lim_{n\to 0} 5\frac{\sin(0.5n\pi)}{(0.5n\pi)} = \lim_{n\to 0} 5\frac{\dfrac{d(\sin(0.5n\pi))}{dn}}{\dfrac{d(0.5n\pi)}{dn}}$$

$$= \lim_{n\to 0} 5\frac{0.5\pi\cos(0.5n\pi)}{0.5\pi} = 5$$

Finally, the Fourier expansion in terms of the complex exponential form is shown as follows:

$$x(t) = \cdots + \frac{10}{\pi}e^{-j2\pi t} + 5 + \frac{10}{\pi}e^{j2\pi t} - \frac{10}{3\pi}e^{j6\pi t} + \frac{2}{\pi}e^{j10\pi t} - \frac{10}{7\pi}e^{j14\pi t} + \cdots$$

B.1.4 Spectral Plots

As previously discussed, the magnitude-phase form can provide information to create a one-sided spectral plot. The amplitude spectrum is obtained by plotting A_n versus the harmonic frequency $n\omega_0$, and the phase spectrum is obtained by plotting ϕ_n versus $n\omega_0$, both for $n \geq 0$. Similarly, if the complex exponential form is used, the two-sided amplitude and phase spectral plots of $|c_n|$ and ϕ_n versus $n\omega_0$ for $-\infty < n < \infty$ can be achieved, respectively. We illustrate this by the following example.

EXAMPLE B.2

Based on the solution to Example B.1, plot the one-sided amplitude spectrum and two-sided amplitude spectrum, respectively.

Solution:
Based on the solution for A_n, the one-sided amplitude spectrum is shown in Figure B.2.

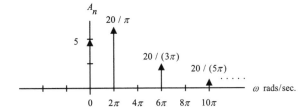

FIGURE B.2

One-sided spectrum of the square waveform in Example B.2.

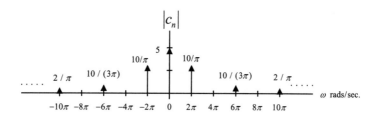

FIGURE B.3

Two-sided spectrum of the square waveform in Example B.2.

According to the solution of the complex exponential form, the two-sided amplitude spectrum is demonstrated in Figure B.3.

A general pulse train $x(t)$ with a period T_0 seconds and a pulse width τ seconds is shown in Figure B.4.

The Fourier series expansions for sine-cosine and complex exponential forms can be derived similarly and are given as follows:

Sine-cosine form:

$$x(t) = \frac{\tau A}{T_0} + \frac{2\tau A}{T_0}\left(\frac{\sin(\omega_0\tau/2)}{(\omega_0\tau/2)}\cos(\omega_0 t) + \frac{\sin(2\omega_0\tau/2)}{(2\omega_0\tau/2)}\cos(2\omega_0 t)\right.$$
$$\left. + \frac{\sin(3\omega_0\tau/2)}{(3\omega_0\tau/2)}\cos(3\omega_0 t) + \cdots\right) \tag{B.15}$$

Complex exponential form:

$$x(t) = \cdots + \frac{\tau A}{T_0}\frac{\sin(\omega_0\tau/2)}{(\omega_0\tau/2)}e^{-j\omega_0 t} + \frac{\tau A}{T_0} + \frac{\tau A}{T_0}\frac{\sin(\omega_0\tau/2)}{(\omega_0\tau/2)}e^{j\omega_0 t} + \frac{\tau A}{T_0}\frac{\sin(2\omega_0\tau/2)}{(2\omega_0\tau/2)}e^{j2\omega_0 t} + \cdots \tag{B.16}$$

where $\omega_0 = 2\pi f_0 = 2\pi/T_0$ is the fundamental angle frequency of the periodic waveform. The reader can derive the one-sided amplitude spectrum A_n and the two-sided amplitude spectrum $|c_n|$. The expressions for the one-sided amplitude and two-sided amplitude spectra are given by the following:

$$A_0 = \frac{\tau}{T_0}A \tag{B.17}$$

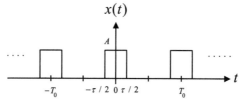

FIGURE B.4

Rectangular waveform (pulse train).

$$A_n = \frac{2\tau}{T_0} A \left| \frac{\sin(n\omega_0\tau/2)}{(n\omega_0\tau/2)} \right|, \quad for \quad n = 1, 2, 3... \tag{B.18}$$

$$|c_n| = \frac{\tau}{T_0} A \left| \frac{\sin(n\omega_0\tau/2)}{(n\omega_0\tau/2)} \right|, \quad -\infty < n < \infty \tag{B.19}$$

EXAMPLE B.3

In Figure B.4, if $T_0 = 1$ ms, $\tau = 0.2$ ms, and $A = 10$, use Equations (B.17) to (B.19) to derive the amplitude one-sided spectrum and two-sided spectrum for each of the first four harmonic frequency components.

Solution:

The fundamental frequency is

$$\omega_0 = 2\pi f_0 = 2\pi \times (1/0.001) = 2,000\pi \text{ rad/sec}$$

Using Equations (B.17) and (B.18) yields the one-sided spectrum as

$$A_0 = \frac{\tau}{T_0} A = \frac{0.0002}{0.001} \times 10 = 2, \quad for \quad n = 0, n\omega_0 = 0$$

For $n = 1$, $n\omega_0 = 2,000\pi$ rad/sec:

$$A_1 = \frac{2 \times 0.0002}{0.001} \times 10 \times \left| \frac{\sin(1 \times 2,000\pi \times 0.0002/2)}{(1 \times 2,000\pi \times 0.0002/2)} \right| = 4 \frac{\sin(0.2\pi)}{(0.2\pi)} = 3.7420$$

For $n = 2$, $n\omega_0 = 4,000\pi$ rad/sec:

$$A_2 = \frac{2 \times 0.0002}{0.001} \times 10 \times \left| \frac{\sin(2 \times 2,000\pi \times 0.0002/2)}{(2 \times 2,000\pi \times 0.0002/2)} \right| = 4 \frac{\sin(0.4\pi)}{(0.4\pi)} = 3.0273$$

For $n = 3$, $n\omega_0 = 6,000\pi$ rad/sec:

$$A_3 = \frac{2 \times 0.0002}{0.001} \times 10 \times \left| \frac{\sin(3 \times 2,000\pi \times 0.0002/2)}{(3 \times 2,000\pi \times 0.0002/2)} \right| = 4 \frac{\sin(0.6\pi)}{(0.6\pi)} = 2.0182$$

For $n = 4$, $n\omega_0 = 8,000\pi$ rad/sec:

$$A_4 = \frac{2 \times 0.0002}{0.001} \times 10 \times \left| \frac{\sin(4 \times 2,000\pi \times 0.0002/2)}{(4 \times 2,000\pi \times 0.0002/2)} \right| = 4 \frac{\sin(0.8\pi)}{(0.8\pi)} = 0.9355$$

The one-sided amplitude spectrum is plotted in Figure B.5.

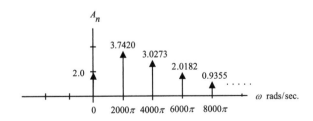

FIGURE B.5

One-sided spectrum in Example B.3.

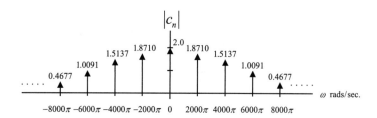

FIGURE B.6

Two-sided spectrum in Example B.3.

Similarly, applying Equation (B.19) leads to

$$|c_0| = \frac{0.0002}{0.001} \times 10 \times \left| \lim_{n \to 0} \frac{\sin(n \times 2,000\pi \times 0.0002/2)}{(n \times 2,000\pi \times 0.0002/2)} \right| = 2 \times |1| = 2$$

Note: We use the fact that $\lim_{x \to 0} \frac{\sin(x)}{x} = 1.0$ (see L'Hospital's rule in Appendix G).

$$|c_1| = |c_{-1}| = \frac{0.0002}{0.001} \times 10 \times \left| \frac{\sin(1 \times 2,000\pi \times 0.0002/2)}{(1 \times 2,000\pi \times 0.0002/2)} \right| = 2 \times \left| \frac{\sin(0.2\pi)}{0.2\pi} \right| = 1.8710$$

$$|c_2| = |c_{-2}| = \frac{0.0002}{0.001} \times 10 \times \left| \frac{\sin(2 \times 2,000\pi \times 0.0002/2)}{(2 \times 2,000\pi \times 0.0002/2)} \right| = 2 \times \left| \frac{\sin(0.4\pi)}{0.4\pi} \right| = 1.5137$$

$$|c_3| = |c_{-3}| = \frac{0.0002}{0.001} \times 10 \times \left| \frac{\sin(3 \times 2,000\pi \times 0.0002/2)}{(3 \times 2,000\pi \times 0.0002/2)} \right| = 2 \times \left| \frac{\sin(0.6\pi)}{0.6\pi} \right| = 1.0091$$

$$|c_4| = |c_{-4}| = \frac{0.0002}{0.001} \times 10 \times \left| \frac{\sin(4 \times 2,000\pi \times 0.0002/2)}{(4 \times 2,000\pi \times 0.0002/2)} \right| = 2 \times \left| \frac{\sin(0.8\pi)}{0.8\pi} \right| = 0.4677$$

Figure B.6 shows the two-sided amplitude spectral plot.

The following example illustrates the use of table information to determine the Fourier series expansion of the periodic waveform. Table B.1 consists of the Fourier series expansions for common periodic signals in the sine-cosine form while Table B.2 shows the expansions in the complex exponential form.

Table B.1 Fourier Series Expansions for Some Common Waveform Signals in the Sine-Cosine Form

Time Domain Signal $x(t)$	Fourier Series Expansion
Positive square wave 	$$x(t) = \frac{A}{2} + \frac{2A}{\pi}\left(\sin \omega_0 t + \frac{1}{3}\sin 3\omega_0 t + \frac{1}{5}\sin 5\omega_0 t + \frac{1}{7}\sin 7\omega_0 t + \cdots\right)$$
Square wave 	$$x(t) = \frac{4A}{\pi}\left(\cos \omega_0 t - \frac{1}{3}\cos 3\omega_0 t + \frac{1}{5}\cos 5\omega_0 t - \frac{1}{7}\cos 7\omega_0 t + \cdots\right)$$
Triangular wave 	$$x(t) = \frac{8A}{\pi^2}\left(\cos \omega_0 t + \frac{1}{9}\cos 3\omega_0 t + \frac{1}{25}\cos 5\omega_0 t + \frac{1}{49}\cos 7\omega_0 t + \cdots\right)$$
Sawtooth wave 	$$x(t) = \frac{2A}{\pi}\left(\sin \omega_0 t - \frac{1}{2}\sin 2\omega_0 t + \frac{1}{3}\sin 3\omega_0 t - \frac{1}{4}\sin 4\omega_0 t + \cdots\right)$$
Rectangular wave (Pulse train) Duty cycle $= d = \dfrac{\tau}{T_0}$	$$x(t) = Ad + 2Ad\left(\frac{\sin \pi d}{\pi d}\right)\cos \omega_0 t$$ $$+ 2Ad\left(\frac{\sin 2\pi d}{2\pi d}\right)\cos 2\omega_0 t + 2Ad\left(\frac{\sin 3\pi d}{3\pi d}\right)\cos 3\omega_0 t + \cdots$$
Ideal impulse train 	$$x(t) = \frac{1}{T_0} + \frac{2}{T_0}\left(\cos \omega_0 t + \cos 2\omega_0 t + \cos 3\omega_0 t + \cos 4\omega_0 t + \cdots\right)$$

TABLE B.2 Fourier Series Expansions for Some Common Waveform Signals in the Complex Exponential Form

Time Domain Signal $x(t)$	Fourier Series Expansion
Positive square wave 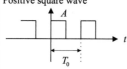	$$x(t) = \left(\cdots - \frac{A}{j3\pi} e^{-j3\omega_0 t} - \frac{A}{j\pi} e^{-j\omega_0 t} + \frac{A}{2} + \frac{A}{j\pi} e^{j\omega_0 t} + \frac{A}{j3\pi} e^{j3\omega_0 t} \right.$$ $$\left. + \frac{A}{j5\pi} e^{j5\omega_0 t} + \cdots \right)$$
Square wave 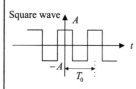	$$x(t) = \frac{2A}{\pi} \left(\cdots + \frac{1}{5} e^{-j5\omega_0 t} - \frac{1}{3} e^{-j3\omega_0 t} + e^{-j\omega_0 t} + e^{j\omega_0 t} - \frac{1}{3} e^{j3\omega_0 t} \right.$$ $$\left. + \frac{1}{5} e^{j5\omega_0 t} - \cdots \right)$$
Triangular wave 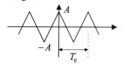	$$x(t) = \frac{4A}{\pi^2} \left(\cdots + \frac{1}{25} e^{-j5\omega_0 t} + \frac{1}{9} e^{-j3\omega_0 t} + e^{-j\omega_0 t} + e^{j\omega_0 t} \right.$$ $$\left. + \frac{1}{9} e^{j3\omega_0 t} + \frac{1}{3} e^{j5\omega_0 t} + \cdots \right)$$
Sawtooth wave 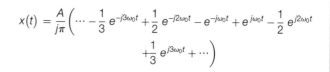	$$x(t) = \frac{A}{j\pi} \left(\cdots - \frac{1}{3} e^{-j3\omega_0 t} + \frac{1}{2} e^{-j2\omega_0 t} - e^{-j\omega_0 t} + e^{j\omega_0 t} - \frac{1}{2} e^{j2\omega_0 t} \right.$$ $$\left. + \frac{1}{3} e^{j3\omega_0 t} + \cdots \right)$$
Rectangular wave (Pulse train) 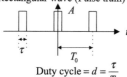 Duty cycle $= d = \dfrac{\tau}{T_0}$	$$x(t) = \cdots + Ad \left(\frac{\sin \pi d}{\pi d} \right) e^{-j\omega_0 t} + Ad \left(\frac{\sin \pi d}{\pi d} \right) e^{j\omega_0 t}$$ $$+ Ad \left(\frac{\sin 2\pi d}{2\pi d} \right) e^{j2\omega_0 t} + Ad \left(\frac{\sin 3\pi d}{3\pi d} \right) e^{j3\omega_0 t} + \cdots$$
Ideal impulse train 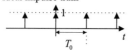	$$x(t) = \frac{1}{T_0} \left(\cdots + e^{-j3\omega_0 t} + e^{-j2\omega_0 t} + e^{-j\omega_0 t} + 1 + e^{j\omega_0 t} + e^{j2\omega_0 t} \right.$$ $$\left. + e^{j3\omega_0 t} + \cdots \right)$$

EXAMPLE B.4

In the sawtooth waveform shown in Table B.1 and reprinted in Figure B.7, if $T_0 = 1$ ms and $A = 10$, use the formula in the table to determine the Fourier series expansion in a magnitude-phase form, and determine the frequency f_3 and amplitude value of A_3 for the third harmonic. Write the Fourier series expansion in a complex exponential form also, and determine $|c_3|$ and $|c_{-3}|$ for the third harmonic.

Solution:

a. Based on the information in Table B.1, we have

$$x(t) = \frac{2A}{\pi}\left(\sin \omega_0 t - \frac{1}{2}\sin 2\omega_0 t + \frac{1}{3}\sin 3\omega_0 t - \frac{1}{4}\sin 4\omega_0 t + \cdots\right)$$

Since $T_0 = 1$ ms, the fundamental frequency is

$$f_0 = 1/T_0 = 1,000 \text{ Hz}, \quad \text{and} \quad \omega_0 = 2\pi f_0 = 2,000\pi \text{ rad/sec}$$

Then, the expansion is determined as

$$x(t) = \frac{2 \times 10}{\pi}\left(\sin 2,000\pi t - \frac{1}{2}\sin 4,000\pi t + \frac{1}{3}\sin 6,000\pi t - \frac{1}{4}\sin 8,000\pi t + \cdots\right)$$

Using the trigonometric identities

$$\sin x = \cos(x - 90°) \text{ and } -\sin x = \cos(x + 90°)$$

and simple algebra, we finally obtain

$$x(t) = \frac{20}{\pi}\cos(2,000\pi t - 90°) + \frac{10}{\pi}\cos(4,000\pi t + 90°)$$

$$+\frac{20}{3\pi}\cos(6,000\pi t - 90°) + \frac{5}{\pi}\cos(8,000\pi t + 90°) + \cdots$$

From the magnitude-phase form, we then determine f_3 and A_3 as follows:

$$f_3 = 3 \times f_0 = 3,000 \text{ Hz}, \quad \text{and} \quad A_3 = \frac{20}{3\pi} = 2.1221$$

b. From Table B.2, the complex exponential form is

$$x(t) = \frac{10}{j\pi}\left(\cdots - \frac{1}{3}e^{-j6,000\pi t} + \frac{1}{2}e^{-j4,000\pi t} - e^{-j2,000\pi t} + e^{j2,000\pi t} - \frac{1}{2}e^{j4,000\pi t} + \frac{1}{3}e^{j6,000\pi t} + \cdots\right)$$

From the expression, we have

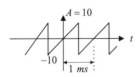

FIGURE B.7

Sawtooth waveform for Example B.4.

$$|c_3| = \left| \frac{10}{j\pi} \times \frac{1}{3} \right| = \left| \frac{1.061}{j} \right| = 1.061 \text{ and}$$

$$|c_{-3}| = \left| -\frac{10}{j\pi} \times \frac{1}{3} \right| = \left| -\frac{1.061}{j} \right| = 1.061$$

B.1.5 FOURIER TRANSFORM

The Fourier transform is a mathematical function that provides frequency spectral analysis for a nonperiodic signal. The Fourier transform pair is defined as

Fourier transform:

$$X(\omega) = \int_{-\infty}^{\infty} x(t)e^{-j\omega t}dt \tag{B.20}$$

Inverse Fourier transform:

$$x(t) = \frac{1}{2\pi} \int_{-\infty}^{\infty} X(\omega)e^{j\omega t}d\omega \tag{B.21}$$

where $x(t)$ is a nonperiodic signal and $X(\omega)$ is a two-sided continuous spectrum versus the continuous frequency variable ω, where $-\infty < \omega < \infty$. Again, the spectrum is a complex function that can be further written as

$$X(\omega) = |X(\omega)| \angle \phi(\omega) \tag{B.22}$$

where $|X(\omega)|$ is the continuous amplitude spectrum, while $\angle \phi(\omega)$ designates the continuous phase spectrum.

EXAMPLE B.5

Let $x(t)$ be a single rectangular pulse, shown in Figure B.8, where the pulse width is $\tau = 0.5$ second. Find its Fourier transform and sketch the amplitude spectrum.

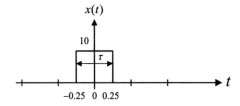

FIGURE B.8

Rectangular pulse in Example B.5.

Solution:

Applying Equation (B.21) and using Euler's formula, we have

$$X(\omega) = \int_{-\infty}^{\infty} x(t)e^{-j\omega t}\,dt = \int_{-0.25}^{0.25} 10e^{-j\omega}\,dt$$

$$= 10\left.\frac{e^{-j\omega t}}{-j\omega}\right|_{-0.25}^{0.25} = 10 \times \frac{\left(e^{-j0.25\omega} - e^{j0.25\omega}\right)}{-j\omega}$$

$$= 10 \times \frac{\cos(0.25\omega) - j\sin(0.25\omega) - [\cos(0.25\omega) + j\sin(0.25\omega)]}{-j\omega}$$

$$= 5\frac{\sin(0.25\omega)}{0.25\omega}$$

where the amplitude spectrum is expressed as

$$|X(\omega)| = 5 \times \left|\frac{\sin(0.25\omega)}{0.25\omega}\right|$$

Using $\omega = 2\pi f$, we can express the spectrum in terms of Hz as

$$|X(f)| = 5 \times \left|\frac{\sin(0.5\pi f)}{0.5\pi f}\right|$$

The amplitude spectrum is shown in Figure B.9. Note that the first null point is at $\omega = 2\pi/0.5 = 4\pi$ rad/sec, and the spectrum is symmetric.

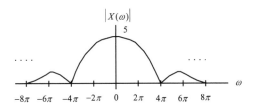

FIGURE B.9

Amplitude spectrum for Example B.5.

EXAMPLE B.6

Let $x(t)$ be an exponential function given by

$$x(t) = 10e^{-2t}u(t) = \begin{cases} 10e^{-2t} & t \geq 0 \\ 0 & t < 0 \end{cases}$$

Find its Fourier transform.

Solution:

According to the definition of the Fourier transform,

$$X(\omega) = \int_0^\infty 10e^{-2t}u(t)e^{-j\omega t}dt = \int_0^\infty 10e^{-(2+j\omega)t}dt$$

$$= \frac{10e^{-(2+j\omega)t}}{-(2+j\omega)}\bigg|_0^\infty = \frac{10}{2+j\omega}$$

$$X(\omega) = \frac{10}{\sqrt{2^2 + \omega^2}} \angle -\tan^{-1}\left(\frac{\omega}{2}\right)$$

Using $\omega = 2\pi f$, we get

$$X(f) = \frac{10}{2+j2\pi f} = \frac{10}{\sqrt{2^2 + (2\pi f)^2}} \angle -\tan^{-1}(\pi f)$$

The Fourier transforms for some common signals are listed in Table B.3. Some useful properties of the Fourier transform are summarized in Table B.4.

EXAMPLE B.7

Find the Fourier transforms of the following functions:

a. $x(t) = \delta(t)$, where $\delta(t)$ is an impulse function defined by

$$\delta(t) = \begin{cases} \neq 0 & t = 0 \\ 0 & \text{elsewhere} \end{cases}$$

with a property given as

$$\int_{-\infty}^\infty f(t)\delta(t-\tau)dt = f(\tau)$$

b. $x(t) = \delta(t-\tau)$

Solution:

a. We first use the Fourier transform definition and then apply the delta function property,

$$X(\omega) = \int_{-\infty}^\infty \delta(t)e^{-j\omega t}dt = e^{-j\omega t}\bigg|_{t=0} = 1$$

b. Similar to (a), we obtain

Table B.3 Fourier Transforms for Some Common Signals

Time Domain Signal $x(t)$	Fourier Spectrum $X(f)$
Rectangular pulse 	$X(f) = A\tau \dfrac{\sin \pi f \tau}{\pi f \tau}$
Triangular pulse 	$X(f) = A\tau \left(\dfrac{\sin \pi f \tau}{\pi f \tau}\right)^2$
Cosine pulse 	$X(f) = \dfrac{2A\tau}{\pi}\left(\dfrac{\cos \pi f \tau}{1 - 4f^2\tau^2}\right)$
Sawtooth pulse 	$X(f) = \dfrac{jA}{2\pi f}\left(\dfrac{\sin \pi f \tau}{\pi f \tau}e^{-j\pi f \tau} - 1\right)$
Exponential function $\alpha = \dfrac{1}{\tau}$ 	$X(f) = \dfrac{A}{\alpha + j2\pi f}$
Impulse function 	$X(f) = A$

$$X(\omega) = \int_{-\infty}^{\infty} \delta(t - \tau)e^{-j\omega t}\,dt = e^{-j\omega t}\Big|_{t=\tau} = e^{-j\omega \tau}$$

Example B.8 shows how to use the table information to determine the Fourier transform of a nonperiodic signal.

Table B.4 Properties of the Fourier Transform

Line	Time Function	Fourier Transform
1	$\alpha x_1(t) + \beta x_2(t)$	$\alpha X_1(f) + \beta X_2(f)$
2	$\dfrac{dx(t)}{dt}$	$j2\pi f X(f)$
3	$\displaystyle\int_{-\infty}^{t} x(t)dt$	$\dfrac{X(f)}{j2\pi f}$
4	$x(t-\tau)$	$e^{-j2\pi f\tau}X(f)$
5	$e^{j2\pi f_0 t}x(t)$	$X(f-f_0)$
6	$x(at)$	$\dfrac{1}{a}X\left(\dfrac{f}{a}\right)$

EXAMPLE B.8

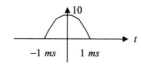

FIGURE B.10

Cosine pulse in Example B.8.

Use Table B.3 to determine the Fourier transform for the cosine pulse in Figure B.10.

Solution:
According to the graph, we can identify

$$\frac{\tau}{2} = 1 \; ms, \text{ and } A = 1$$

τ is given by

$$\tau = 2 \times 1 \; ms = 0.002 \text{ second}$$

Applying the formula from Table B.3 gives

$$X(f) = \frac{2 \times 10 \times 0.002}{\pi}\left(\frac{\cos \pi f 0.002}{1 - 4f^2 0.002^2}\right) = \frac{0.04}{\pi}\left(\frac{\cos 0.002\pi f}{1 - 4 \times 0.002^2 f^2}\right)$$

B.2 LAPLACE TRANSFORM

In this section, we will review Laplace transform and its applications.

B.2.1 Laplace Transform and Its Table

The Laplace transform plays an important role in the analysis of continuous signals and systems. We define the Laplace transform pairs as

$$X(s) = L\{x(t)\} = \int_0^\infty x(t)e^{-st}dt \tag{B.23}$$

$$x(t) = L^{-1}\{X(s)\} = \frac{1}{2\pi j} \int_{\gamma-j\infty}^{\gamma+j\infty} X(s)e^{st}ds \tag{B.24}$$

Notice that the symbol $L\{\}$ denotes the forward Laplace operation, while the symbol $L^{-1}\{\}$ indicates the inverse Laplace operation. Some common Laplace transform pairs are listed in Table B.5.

TABLE B.5 Laplace Transform Table

Line	Time Function $x(t)$	Laplace Transform $X(s) = L(x(t))$
1	$\delta(t)$	1
2	1 or $u(t)$	$\dfrac{1}{s}$
3	$tu(t)$	$\dfrac{1}{s^2}$
4	$e^{-at}u(t)$	$\dfrac{1}{s+a}$
5	$\sin(\omega t)u(t)$	$\dfrac{\omega}{s^2+\omega^2}$
6	$\cos(\omega t)u(t)$	$\dfrac{s}{s^2+\omega^2}$
7	$\sin(\omega t+\theta)u(t)$	$\dfrac{s\sin(\theta)+\omega\cos(\theta)}{s^2+\omega^2}$
8	$e^{-at}\sin(\omega t)u(t)$	$\dfrac{\omega}{(s+a)^2+\omega^2}$
9	$e^{-at}\cos(\omega t)u(t)$	$\dfrac{s+a}{(s+a)^2+\omega^2}$

(continued)

TABLE B.5 Laplace Transform Table *(continued)*

Line	Time Function $x(t)$	Laplace Transform $X(s) = L(x(t))$
10	$\left(A\cos(\omega t) + \dfrac{B - aA}{\omega}\sin(\omega t)\right)e^{-at}u(t)$	$\dfrac{As + B}{(s + a)^2 + \omega^2}$
11a	$t^n u(t)$	$\dfrac{n!}{s^{n+1}}$
11b	$\dfrac{1}{(n-1)!}t^{n-1}u(t)$	$\dfrac{1}{s^n}$
12a	$e^{-at}t^n u(t)$	$\dfrac{n!}{(s+a)^{n+1}}$
12b	$\dfrac{1}{(n-1)!}e^{-at}t^{n-1}u(t)$	$\dfrac{1}{(s+a)^n}$
13	$(2\text{Real}(A)\cos(\omega t) - 2\text{Imag}(A)\sin(\omega t))e^{-\alpha t}u(t)$	$\dfrac{A}{s + \alpha - j\omega} + \dfrac{A^*}{s + \alpha + j\omega}$
14	$\dfrac{dx(t)}{dt}$	$sX(s) - x(0^-)$
15	$\displaystyle\int_0^t x(t)dt$	$\dfrac{X(s)}{s}$
16	$x(t - a)u(t - a)$	$e^{-as}X(s)$
17	$e^{-at}x(t)u(t)$	$X(s + a)$

In Example B.9, we examine the Laplace transform in light of its definition.

EXAMPLE B.9

Derive the Laplace transform of the unit step function.

Solution:
By the definition in Equation (B.23),

$$X(s) = \int_0^\infty u(t)e^{-st}dt$$

$$= \int_0^\infty e^{-st}dt = \left.\frac{e^{-st}}{-s}\right|_0^\infty = \frac{e^{-\infty}}{-s} - \frac{e^0}{-s} = \frac{1}{s}$$

The answer is consistent with the result listed in Table B.5. Now we use the results in Table B.5 to find the Laplace transform of a function.

EXAMPLE B.10

Perform the Laplace transform for each of the following functions.

a. $x(t) = 5\sin(2t)u(t)$

b. $x(t) = 5e^{-3t}\cos(2t)u(t)$

Solution:

a. Using line 5 in Table B.5 and noting that $\omega = 2$, the Laplace transform immediately follows:

$$X(s) = 5L\{2\sin(2t)u(t)\}$$
$$= \frac{5\times 2}{s^2 + 2^2} = \frac{10}{s^2 + 4}$$

b. Applying line 9 in Table B.5 with $\omega = 2$ and $a = 3$ yields

$$X(s) = 5L\{e^{-3t}\cos(2t)u(t)\}$$
$$= \frac{5(s+3)}{(s+3)^2 + 2^2} = \frac{5(s+3)}{(s+3)^2 + 4}$$

B.2.2 Solving Differential Equations Using the Laplace Transform

One of the important applications of the Laplace transform is to solve differential equations. Using the differential property in Table B.5, we can transform a differential equation from the time domain to the Laplace domain. This will change the differential equation into an algebraic equation, and we then solve the algebraic equation. Finally, the inverse Laplace operation is processed to yield the time domain solution.

EXAMPLE B.11

Solve the following differential equation using the Laplace transform:

$$\frac{dy(t)}{dt} + 10y(t) = x(t) \text{ with an initial condition } y(0) = 0,$$

where the input $x(t) = 5u(t)$.

Solution:

Applying the Laplace transform on both sides of the differential equation and using the differential property (line 14 in Table B.5), we get

$$sY(s) - y(0) + 10Y(s) = X(s)$$

Note that

$$X(s) = L\{5u(t)\} = \frac{5}{s}$$

Substituting the initial condition yields

$$Y(s) = \frac{5}{s(s+10)}$$

Then we use a partial fraction expansion by writing

$$Y(s) = \frac{A}{s} + \frac{B}{s+10}$$

where

$$A = sY(s)|_{s=0} = \frac{5}{s+10}\Big|_{s=0} = 0.5$$

and

$$B = (s+10)Y(s)|_{s=-10} = \frac{5}{s}\Big|_{s=-10} = -0.5$$

Hence,

$$Y(s) = \frac{0.5}{s} - \frac{0.5}{s+10}$$

$$y(t) = L^{-1}\left\{\frac{0.5}{s}\right\} - L^{-1}\left\{\frac{0.5}{s+10}\right\}$$

Finally, applying the inverse of the Laplace transform leads to using the results listed in Table B.5, and we obtain the time domain solution as

$$y(t) = 0.5u(t) - 0.5e^{-10t}u(t)$$

B.2.3 Transfer Function

A linear analog system can be described using the Laplace transfer function. The transfer function relating the input and output of the linear system is depicted as

$$Y(s) = H(s)X(s) \tag{B.25}$$

where $X(s)$ and $Y(s)$ are the system input and response (output), respectively, in the Laplace domain, and the transfer function is defined as a ratio of the Laplace response of the system to the Laplace input given by

$$H(s) = \frac{Y(s)}{X(s)} \tag{B.26}$$

The transfer function will allow us to study the system behavior. Considering an impulse function as the input to a linear system, that is, $x(t) = \delta(t)$, whose Laplace transform is $X(s) = 1$, we then find the system output due to the impulse function to be

$$Y(s) = H(s)X(s) = H(s) \tag{B.27}$$

Therefore, the response in the time domain $y(t)$ is called the impulse response of the system and can be expressed as

$$h(t) = L^{-1}\{H(s)\} \tag{B.28}$$

The analog impulse response can be sampled and transformed to obtain a digital filter transfer function. This topic is covered in Chapter 8.

EXAMPLE B.12

Consider a linear system described by the differential equation shown in Example B.11. $x(t)$ and $y(t)$ designate the system input and system output, respectively. Derive the transfer function and the impulse response of the system.

Solution:

Taking the Laplace transform on both sides of the differential equation yields

$$L\left\{\frac{dy(t)}{dt}\right\} + L\{10y(t)\} = L\{x(t)\}$$

Applying the differential property and substituting the initial condition, we have

$$Y(s)(s + 10) = X(s)$$

Thus, the transfer function is given by

$$H(s) = \frac{Y(s)}{X(s)} = \frac{1}{s + 10}$$

The impulse response can be found by taking the inverse Laplace transform as

$$h(t) = L^{-1}\left\{\frac{1}{s + 10}\right\} = e^{-10t}u(t)$$

B.3 POLES, ZEROS, STABILITY, CONVOLUTION, AND SINUSOIDAL STEADY-STATE RESPONSE

This section is a review of analog system analysis.

B.3.1 Poles, Zeros, and Stability

To study system behavior, the transfer function is written in a general form given by

$$H(s) = \frac{N(s)}{D(s)} = \frac{b_m s^m + b_{m-1} s^{m-1} + \cdots + b_0}{a_n s^n + a_{n-1} s^{n-1} + \cdots + a_0} \tag{B.29}$$

It is a ratio of the numerator polynomial of degree m to the denominator polynomial of degree n. The numerator polynomial is expressed as

$$N(s) = b_m s^m + b_{m-1} s^{m-1} + \cdots + b_0 \tag{B.30}$$

while the denominator polynomial is given by

$$D(s) = a_n s^n + a_{n-1} s^{n-1} + \cdots + a_0 \tag{B.31}$$

Again, the roots of $N(s)$ are called zeros, while the roots of $D(s)$ are called poles of the transfer function $H(s)$. Notice that zeros and poles could be real numbers or complex numbers.

Given a system transfer function, the poles and zeros can be found. Further, a pole-zero plot could be created on the s-plane. With the pole-zero plot, the stability of the system is determined by the following rules:

1. The linear system is stable if the rightmost pole(s) is/are on the left-hand half plane (LHHP) on the s-plane.
2. The linear system is marginally stable if the rightmost pole(s) is/are simple-order (first-order) on the $j\omega$ axis, including the origin on the s-plane.
3. The linear system is unstable if the rightmost pole(s) is/are on the right-hand half plane (RHHP) of the s-plane or if the rightmost pole(s) is/are multiple-order on the $j\omega$ axis on the s-plane.
4. Zeros do not affect system stability.

EXAMPLE B.13

Determine whether each of the following transfer functions is stable, marginally stable, and unstable:

a. $H(s) = \dfrac{s+1}{(s+1.5)(s^2 + 2s + 5)}$

b. $H(s) = \dfrac{(s+1)}{(s+2)(s^2 + 4)}$

c. $H(s) = \dfrac{s+1}{(s-1)(s^2 + 2s + 5)}$

Solution:

a. A zero is found at $s = -1$. The poles are calculated as $s = -1.5$, $s = -1 + j2$, $s = -1 - j2$. The pole-zero plot is shown in Figure B.11A. Since all the poles are located on the LHHP, the system is stable.

b. A zero is found at $s = -1$. The poles are calculated as $s = -2$, $s = j2$, $s = -j2$. The pole-zero plot is shown in Figure B.11B. Since the first-order poles $s = \pm j2$ are located on the $j\omega$ axis, the system is marginally stable.

c. A zero is found at $s = -1$. The poles are calculated as $s = 1$, $s = -1 + j2$, $s = -1 - j2$. The pole-zero plot is shown in Figure B.11C. Since there is a pole $s = 1$ located on the RHHP, the system is unstable.

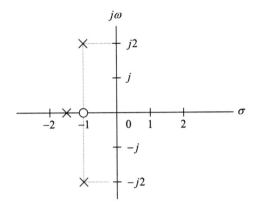

FIGURE B.11A

Pole-zero plot for (a).

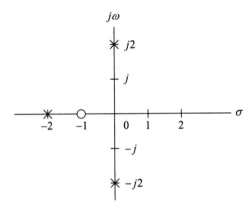

FIGURE B.11B

Pole-zero plot for (b).

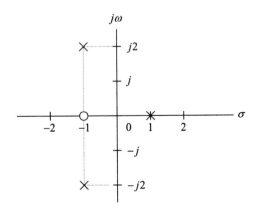

FIGURE B.11C

Pole-zero plot for (c).

B.3.2 Convolution

As we discussed before, the input and output relationship of a linear system in the Laplace domain is

$$Y(s) = H(s)X(s) \tag{B.32}$$

It is apparent that in the Laplace domain, the system output is the product of the Laplace input and transfer function. But in the time domain, the system output is given as

$$y(t) = h(t) * x(t) \tag{B.33}$$

where $*$ denotes linear convolution of the system impulse response $h(t)$ and the system input $x(t)$. The linear convolution is further expressed as

$$y(t) = \int_0^\infty h(\tau)x(t - \tau)d\tau \tag{B.34}$$

EXAMPLE B.14

As you have seen in Examples B.11 and B.12, for a linear system, the impulse response and the input are given, respectively, by

$$h(t) = e^{-10t}u(t) \quad \text{and} \quad x(t) = 5u(t)$$

Determine the system response $y(t)$ using the convolution method.

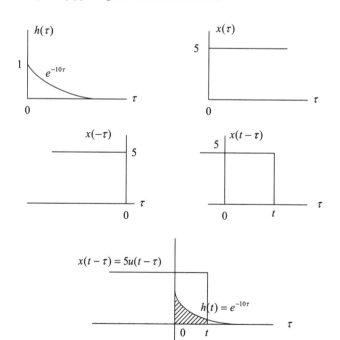

FIGURE B.12

Convolution illustration for Example B.14.

Solution:
Two signals $h(\tau)$ and $x(\tau)$ that are involved in the convolution integration are displayed in Figure B.12. To evaluate the convolution, the time-reversed signal $x(-\tau)$ and the shifted signal $x(t - \tau)$ are also plotted for reference. Figure B.12 shows an overlap of $h(\tau)$ and $x(t - \tau)$. According to the overlapped (shaded) area, the lower limit and the upper limit of the convolution integral are determined to be 0 and t, respectively. Hence,

$$y(t) = \int_0^t e^{-10\tau} \cdot 5 d\tau = \left. \frac{5}{-10} e^{-10\tau} \right|_0^t$$

$$= -0.5 e^{-10t} - (-0.5 e^{-10 \times 0})$$

Finally, the system response is found to be

$$y(t) = 0.5 u(t) - 0.5 e^{-10t} u(t)$$

The solution is the same as that obtained using the Laplace transform method described in Example B.11.

B.3.3 Sinusoidal Steady-State Response

For linear analog systems, if the input to a system is a sinusoid of radian frequency ω, the steady-state response of the system will also be a sinusoid of the same frequency. Therefore, the transfer function, which provides the relationship between a sinusoidal input and a sinusoidal output, is called the steady-state transfer function. The steady-state transfer function is obtained from the Laplace transfer function by substituting $s = j\omega$, as shown in the following:

$$H(j\omega) = H(s)|_{s=j\omega} \tag{B.35}$$

Thus we have a system relationship in a sinusoidal steady state as

$$Y(j\omega) = H(j\omega) X(j\omega) \tag{B.36}$$

Since $H(j\omega)$ is a complex function, we may write it in the phasor form:

$$H(j\omega) = A(\omega) \angle \beta(\omega) \tag{B.37}$$

where the quantity $A(\omega)$ is the amplitude response of the system defined as

$$A(\omega) = |H(j\omega)| \tag{B.38}$$

and the phase angle $\beta(\omega)$ is the phase response of the system. The following example is presented to illustrate the application.

EXAMPLE B.15

Consider a linear system described by the differential equation shown in Example B.12, where $x(t)$ and $y(t)$ designate the system input and system output, respectively. The transfer function has been derived as

$$H(s) = \frac{10}{s + 10}$$

a. Derive the steady-state transfer function.
b. Derive the amplitude response and phase response.
c. If the input is given as a sinusoid, that is, $x(t) = 5\sin(10t + 30°)u(t)$, find the steady-state response $y_{ss}(t)$.

Solution:

a. By substituting $s = j\omega$ into the transfer function in terms of a suitable form, we get the steady-state transfer function as

$$H(j\omega) = \frac{1}{\dfrac{s}{10} + 1} = \frac{1}{\dfrac{j\omega}{10} + 1}$$

b. The amplitude response and phase response are found to be

$$A(\omega) = \frac{1}{\sqrt{\left(\dfrac{\omega}{10}\right)^2 + 1}}$$

$$\beta(\omega) = \angle -\tan^{-1}\left(\frac{\omega}{10}\right)$$

c. When $\omega = 10$ rad/sec, the input sinusoid can be written in terms of the phasor form as

$$X(j10) = 5\angle 30°$$

For the amplitude and phase of the steady-state transfer function at $\omega = 10$, we have

$$A(10) = \frac{1}{\sqrt{\left(\dfrac{10}{10}\right)^2 + 1}} = 0.7071$$

$$\beta(10) = -\tan^{-1}\left(\frac{10}{10}\right) = -45°$$

Hence, we yield

$$H(j10) = 0.7071\angle - 45°$$

Using Equation (B.36), the system output in phasor form is obtained as

$$Y(j10) = H(j10)X(j10) = (1.4141\angle - 45°)(5\angle 30°)$$

$$Y(j10) = 3.5355\angle - 15°$$

Converting the output in phasor form back to the time domain results in the steady-state system output:

$$y_{ss}(t) = 3.5355\sin(10t - 15°)u(t)$$

B.4 PROBLEMS

B.1. Develop equations for the amplitude spectra, that is, A_n (one-sided) and $|c_n|$ (two-sided), of the pulse train $x(t)$ displayed in Figure B.13, where $\tau = 10~\mu sec$.

   a. Plot and label the one-sided amplitude spectrum up to 4 harmonic frequencies including DC.

   b. Plot and label the two-sided amplitude spectrum up to 4 harmonic frequencies including DC.

B.2. In the waveform shown in Figure B.14, $T_0 = 1$ ms and $A = 10$. Use the formula in Table B.1 to write a Fourier series expansion in magnitude-phase form. Determine the frequency f_3 and amplitude value of A_3 for the third harmonic.

B.3. In the waveform shown in Figure B.15, $T_0 = 1$ ms, $\tau = 0.2$ ms, and $A = 10$.

   a. Use the formula in Table B.1 to write a Fourier series expansion in magnitude-phase form.

   b. Determine the frequency f_2 and amplitude value of A_2 for the second harmonic.

B.4. Find the Fourier transform $X(\omega)$ and sketch the amplitude spectrum for the rectangular pulse $x(t)$ displayed in Figure B.16.

B.5. Use Table B.3 to determine the Fourier transform for the pulse in Figure B.17.

B.6. Use Table B.3 to determine the Fourier transform for the pulse in Figure B.18.

B.7. Determine the Laplace transform $X(s)$ for each of the following time domain functions using the Laplace transform in Table B.5.

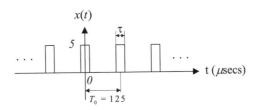

FIGURE B.13

Pulse train in Problem B.1.

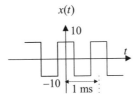

FIGURE B.14

Square wave in Problem B.2.

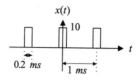

FIGURE B.15

Rectangular wave in Problem B.3.

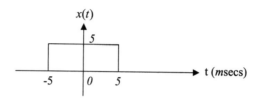

FIGURE B.16

Rectangular pulse in Problem B.4.

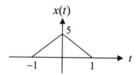

FIGURE B.17

Triangular pulse in Problem B.5.

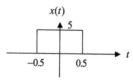

FIGURE B.18

Rectangular pulse in Problem B.6.

    a. $x(t) = 10\delta(t)$
    b. $x(t) = -100tu(t)$
    c. $x(t) = 10e^{-2t}u(t)$
    d. $x(t) = 2u(t-5)$
    e. $x(t) = 10\cos(3t)u(t)$
    f. $x(t) = 10\sin(2t + 45^\circ)u(t)$

g. $x(t) = 3e^{-2t}\cos(3t)u(t)$

h. $x(t) = 10t^5 u(t)$

B.8. Determine the inverse transform of the analog signal $x(t)$ for each of the following functions using Table B.5 and partial fraction expansion.

a. $X(s) = \dfrac{10}{s+2}$

b. $X(s) = \dfrac{100}{(s+2)(s+3)}$

c. $X(s) = \dfrac{100s}{s^2 + 7s + 10}$

d. $X(s) = \dfrac{25}{s^2 + 4s + 29}$

B.9. Solve the following differential equation using the Laplace transform method:

$$2\frac{dx(t)}{dt} + 3x(t) = 15u(t) \text{ with } x(0) = 0$$

a. Determine $X(s)$.

b. Determine the continuous signal $x(t)$ by taking the inverse Laplace transform of $X(s)$.

B.10. Solve the following differential equation using the Laplace transform method:

$$\frac{d^2x(t)}{dt^2} + 3\frac{dx(t)}{dt} + 2x(t) = 10u(t) \text{ with } x'(0) = 0 \text{ and } x(0) = 0$$

a. Determine $X(s)$.

b. Determine $x(t)$ by taking the inverse Laplace transform of $X(s)$.

B.11. Determine the locations of all finite zeros and poles in the following functions. In each case, make an s-plane plot of the poles and zeros, and determine whether the given transfer function is stable, unstable, or marginally stable.

a. $H(s) = \dfrac{(s-3)}{(s^2 + 4s + 4)}$

b. $H(s) = \dfrac{s(s^2 + 5)}{(s^2 + 9)(s^2 + 2s + 4)}$

c. $H(s) = \dfrac{(s^2 + 1)(s+1)}{s(s^2 + 7s - 8)(s+3)(s+4)}$

B.12. Given the transfer function of a system

$$H(s) = \frac{5}{s+5}$$

and the input $x(t) = u(t)$,

a. determine the system impulse response $h(t)$;

b. determine the system Laplace output based on $Y(s) = H(s)X(s)$;

c. determine the system response $y(t)$ in the time domain by taking the inverse Laplace transform of $Y(s)$.

B.13. Given the transfer function of a system

$$H(s) = \frac{5}{s+5}$$

a. determine the steady-state transfer function;

b. determine the amplitude response and phase response in terms of the frequency ω;

c. determine the steady-state response of the system output $y_{ss}(t)$ in time domain using the results that you obtained in (b), given an input to the system of $x(t) = 5\sin(2t)u(t)$.

B.14. Given the transfer function of a system

$$H(s) = \frac{5}{s+5}$$

and the input $x(t) = u(t)$, determine the system output $y(t)$ using the convolution method; that is, $y(t) = h(t) * x(t)$.

Appendix C: Normalized Butterworth and Chebyshev Functions

C.1 NORMALIZED BUTTERWORTH FUNCTION

The normalized Butterworth squared magnitude function is given by

$$|P_n(\omega)|^2 = \frac{1}{1 + \varepsilon^2 (\omega)^{2n}} \tag{C.1}$$

where n is the order and ε is the specified ripple on the filter passband. The specified ripple in dB is expressed as $\varepsilon_{dB} = 10 \cdot \log_{10}(1 + \varepsilon^2)$ dB.

To develop the transfer function $P_n(s)$, we first let $s = j\omega$ and then substitute $\omega^2 = -s^2$ into Equation (C.1) to obtain

$$P_n(s)P_n(-s) = \frac{1}{1 + \varepsilon^2 (-s^2)^n} \tag{C.2}$$

Equation (C.2) has $2n$ poles, and $P_n(s)$ has n poles on the left-hand half plane (LHHP) on the s-plane, while $P_n(-s)$ has n poles on the right-hand half plane (RHHP) on the s-plane. Solving for poles leads to

$$(-1)^n s^{2n} = -1/\varepsilon^2 \tag{C.3}$$

If n is an odd number, Equation (C.3) becomes

$$s^{2n} = 1/\varepsilon^2$$

and the corresponding poles are solved as

$$p_k = \varepsilon^{-1/n} e^{j\frac{2\pi k}{2n}} = \varepsilon^{-1/n}[\cos(2\pi k/2n) + j\sin(2\pi k/2n)] \tag{C.4}$$

where $k = 0, 1, \cdots, 2n - 1$. Thus in phasor form, we have

$$r = \varepsilon^{-1/n}, \quad \text{and} \quad \theta_k = 2\pi k/(2n) \quad \text{for } k = 0, 1, \cdots, 2n - 1 \tag{C.5}$$

When n is an even number, it follows that

$$s^{2n} = -1/\varepsilon^2$$

$$p_k = \varepsilon^{-1/n} e^{j\frac{2\pi k + \pi}{2n}} = \varepsilon^{-1/n}[\cos((2\pi k + \pi)/2n) + j\sin((2\pi k + \pi)/2n)] \tag{C.6}$$

where $k = 0, 1, \cdots, 2n - 1$. Similarly, the phasor form is given by

$$r = \varepsilon^{-1/n}, \quad \text{and} \quad \theta_k = (2\pi k + \pi)/(2n) \quad \text{for} \quad k = 0, 1, \cdots, 2n - 1 \tag{C.7}$$

When n is an odd number, we can identify the poles on the LHHP as

$$p_k = -r, \ k = 0 \text{ and}$$

$$p_k = -r\cos(\theta_k) + jr\sin(\theta_k), \ k = 1, \cdots, (n-1)/2 \tag{C.8}$$

Using complex conjugate pairs, we have

$$p_k^* = -r\cos(\theta_k) - jr\sin(\theta_k)$$

Notice that

$$(s - p_k)(s - p_k^*) = s^2 + (2r\cos(\theta_k))s + r^2$$

and from a factor from the real pole $(s + r)$, it follows that

$$P_n(s) = \frac{K}{(s + r)\prod_{k=1}^{(n-1)/2}(s^2 + (2r\cos(\theta_k))s + r^2)} \tag{C.9}$$

and

$$\theta_k = 2\pi k/(2n) \quad \text{for} \quad k = 1, \cdots, (n-1)/2$$

Setting $P_n(0) = 1$ for the unit passband gain leads to

$$K = r^n = 1/\varepsilon$$

When n is an even number, we can identify the poles on the LHHP as

$$p_k = -r\cos(\theta_k) + jr\sin(\theta_k), \ k = 0, 1, \cdots, n/2 - 1 \tag{C.10}$$

Using complex conjugate pairs, we have

$$p_k^* = -r\cos(\theta_k) - jr\sin(\theta_k)$$

The transfer function is given by

$$P_n(s) = \frac{K}{\prod_{k=1}^{n/2}(s^2 + (2r\cos(\theta_k))s + r^2)} \tag{C.11}$$

$$\theta_k = (2\pi k + \pi)/(2n) \quad \text{for} \quad k = 0, 1, \cdots, n/2 - 1$$

Setting $P_n(0) = 1$ for the unit passband gain, we have

$$K = r^n = 1/\varepsilon$$

Let us examine the following examples.

EXAMPLE C.1

Compute the normalized Butterworth transfer function for the following specifications:

Ripple $= 3$ dB
$n = 2$

Solution:

$n/2 = 1$
$\theta_k = (2\pi \times 0 + \pi)/(2 \times 2) = 0.25\pi$
$\varepsilon^2 = 10^{0.1\times3} - 1$
$r = 1$ and $K = 1$

Applying Equation (C.11) leads to

$$P_2(s) = \frac{1}{s^2 + 2 \times 1 \times \cos(0.25\pi)s + 1^2} = \frac{1}{s^2 + 1.4141s + 1}$$

EXAMPLE C.2

Compute the normalized Butterworth transfer function for the following specifications:

Ripple $= 3$ dB
$n = 3$

Solution:

$(n-1)/2 = 1$
$\varepsilon^2 = 10^{0.1\times3} - 1$
$r = 1$ and $K = 1$
$\theta_k = (2\pi \times 1)/(2 \times 3) = \pi/3$

From Equation (C.9), we have

$$P_3(s) = \frac{1}{(s+1)(s^2 + 2 \times 1 \times \cos(\pi/3)s + 1^2)}$$

$$= \frac{1}{(s+1)(s^2 + s + 1)}$$

For the unfactored form, we get

$$P_3(s) = \frac{1}{s^3 + 2s^2 + 2s + 1}$$

EXAMPLE C.3

Compute the normalized Butterworth transfer function for the following specifications:

Ripple = 1.5 dB
$n = 3$

Solution:

$(n - 1)/2 = 1$
$\varepsilon^2 = 10^{0.1 \times 1.5} - 1,$
$r = 1.1590$ and $K = 1.5569$
$\theta_k = (2\pi \times 1)/(2 \times 3) = \pi/3$

Applying Equation (C.9), we achieve the normalized Butterworth transfer function:

$$P_3(s) = \frac{1}{(s + 1.1590)(s^2 + 2 \times 1.1590 \times \cos(\pi/3)s + 1.1590^2)}$$

$$= \frac{1}{(s + 1)(s^2 + 1.1590s + 1.3433)}$$

For the unfactored form, we obtain

$$P_3(s) = \frac{1.5569}{s^3 + 2.3180s^2 + 2.6866s + 1.5569}$$

C.2 NORMALIZED CHEBYSHEV FUNCTION

Similar to analog Butterworth filter design, the transfer function is derived from the normalized Chebyshev function, and the result is usually listed in a table for design reference. The Chebyshev magnitude response function with an order of n and the normalized cutoff frequency $\omega = 1$ radian per second is given by

$$|B_n(\omega)| = \frac{1}{\sqrt{1 + \varepsilon^2 C_n^2(\omega)}}, \quad n \geq 1 \tag{C.12}$$

where the function $C_n(\omega)$ is defined as

$$C_n(\omega) = \begin{cases} \cos\left(n \cos^{-1}(\omega)\right) & \omega \leq 1 \\ \cos h(n \cos h^{-1}(\omega)) & \omega > 1 \end{cases} \tag{C.13}$$

where ε is the ripple specification on the filter passband. Notice that

$$\cos h^{-1}(x) = \ln(x + \sqrt{x^2 - 1}) \tag{C.14}$$

To develop the transfer function $B_n(s)$, we let $s = j\omega$ and substitute $\omega^2 = -s^2$ into Equation (C.12) to obtain

$$B_n(s)B_n(-s) = \frac{1}{1 + \epsilon^2 C_n^2(s/j)} \tag{C.15}$$

The poles can be found from

$$1 + \epsilon^2 C_n^2(s/j) = 0$$

or

$$C_n(s/j) = \cos\left(n\cos^{-1}(s/j)\right) = \pm j1/\epsilon \tag{C.16}$$

If we introduce a complex variable $v = \alpha + j\beta$ such that

$$v = \alpha + j\beta = \cos^{-1}(s/j) \tag{C.17}$$

we can then write

$$s = j\cos(v) \tag{C.18}$$

Substituting Equation (C.17) into Equation (C.16) and using trigonometric identities, it follows that

$$
\begin{aligned}
C_n(s/j) &= \cos\left(n\cos^{-1}(s/j)\right) \\
&= \cos(nv) = \cos(n\alpha + jn\beta) \\
&= \cos(n\alpha)\cos h(n\beta) - j\sin(n\alpha)\sin h(n\beta) = \pm j1/\epsilon
\end{aligned}
\tag{C.19}
$$

To solve Equation (C.19), the following conditions must be satisfied:

$$\cos(n\alpha)\cos h(n\beta) = 0 \tag{C.20}$$

$$-\sin(n\alpha)\sin h(n\beta) = \pm 1/\epsilon \tag{C.21}$$

Since $\cos h(n\beta) \geq 1$ in Equation (C.20), we must let

$$\cos(n\alpha) = 0 \tag{C.22}$$

which therefore leads to

$$\alpha_k = (2k+1)\pi/(2n), \quad k = 0, 1, 2, \cdots, 2n-1 \tag{C.23}$$

With Equation (C.23), we have $\sin(n\alpha_k) = \pm 1$. Then Equation (C.21) becomes

$$\sin h(n\beta) = 1/\epsilon \tag{C.24}$$

Solving Equation (C.24) gives

$$\beta = \sin h^{-1}(1/\epsilon)/n \tag{C.25}$$

Again from Equation (C.18),

$$s = j\cos(v) = j[\cos(\alpha_k)\cos h(\beta) - j\sin(\alpha_k)\sin h(\beta)] \quad \text{for} \quad k = 0, 1, \cdots, 2n-1 \tag{C.26}$$

The poles can be found from Equation (C.26):

$$p_k = \sin(\alpha_k)\sin h(\beta) + j\cos(\alpha_k)\cos h(\beta) \quad \text{for} \quad k = 0, 1, \cdots, 2n - 1 \tag{C.27}$$

Using Equation (C.27), if n is an odd number, the poles on the left side are

$$p_k = -\sin(\alpha_k)\sin h(\beta) + j\cos(\alpha_k)\cos h(\beta), \quad k = 0, 1, \cdots, (n-1)/2 - 1 \tag{C.28}$$

Using complex conjugate pairs, we have

$$p_k^* = -\sin(\alpha_k)\sin h(\beta) - j\cos(\alpha_k)\cos h(\beta) \tag{C.29}$$

and a real pole

$$p_k = -\sin h(\beta), \quad k = (n-1)/2 \tag{C.30}$$

Notice that

$$(s - p_k)(s - p_k^*) = s^2 + b_k s + c_k \tag{C.31}$$

and from a factor from the real pole $[s + \sin h(\beta)]$, it follows that

$$B_n(s) = \frac{K}{[s + \sin h(\beta)]\prod_{k=0}^{(n-1)/2-1}(s^2 + b_k s + c_k)} \tag{C.32}$$

$$b_k = 2\sin(\alpha_k)\sin h(\beta) \tag{C.33}$$

$$c_k = [\sin(\alpha_k)\sin h(\beta)]^2 + [\cos(\alpha_k)\cos h(\beta)]^2 \tag{C.34}$$

where

$$\alpha_k = (2k+1)\pi/(2n) \quad \text{for} \quad k = 0, 1, \cdots, (n-1)/2 - 1 \tag{C.35}$$

For the unit passband gain and the filter order as an odd number, we set $B_n(0) = 1$. Then

$$K = \sin h(\beta)\prod_{k=0}^{(n-1)/2-1} c_k \tag{C.36}$$

$$\beta = \sin h^{-1}(1/\varepsilon)/n \tag{C.37}$$

$$\sin h^{-1}(x) = \ln(x + \sqrt{x^2 + 1}) \tag{C.38}$$

Following a similar procedure for when n is even, we have

$$B_n(s) = \frac{K}{\prod_{k=0}^{n/2-1}(s^2 + b_k s + c_k)} \tag{C.39}$$

$$b_k = 2\sin(\alpha_k)\sin h(\beta) \tag{C.40}$$

$$c_k = [\sin(\alpha_k)\sin h(\beta)]^2 + [\cos(\alpha_k)\cos h(\beta)]^2 \tag{C.41}$$

where

$$\alpha_k = (2k+1)\pi/(2n) \quad \text{for} \quad k = 0, 1, \cdots, n/2 - 1 \tag{C.42}$$

For the unit passband gain and the filter order as an even number, we require that $B_n(0) = 1/\sqrt{1+\varepsilon^2}$, so that the maximum magnitude of the ripple on the passband equals 1. Then we have

$$K = \prod_{k=0}^{n/2-1} c_k / \sqrt{1+\varepsilon^2} \tag{C.43}$$

$$\beta = \sin h^{-1}(1/\varepsilon)/n \tag{C.44}$$

$$\sin h^{-1}(x) = \ln(x + \sqrt{x^2 + 1}) \tag{C.45}$$

Equations (C.32) to (C.45) are applied to compute the normalized Chebyshev transfer function. Now let us look at the following illustrative examples.

EXAMPLE C.4

Compute the normalized Chebyshev transfer function for the following specifications:

Ripple = 0.5 dB
$n = 2$

Solution:

$n/2 = 1$

Applying Equations (C.39) to (C.45), we obtain

$\alpha_0 = (2 \times 0 + 1)\pi/(2 \times 2) = 0.25\pi$
$\varepsilon^2 = 10^{0.1 \times 0.5} - 1 = 0.1220, \; 1/\varepsilon = 2.8630$
$\beta = \sin h^{-1}(2.8630)/n = \ln(2.8630 + \sqrt{2.8630^2 + 1})/2 = 0.8871$
$b_0 = 2\sin(0.25\pi)\sin h(0.8871) = 1.4256$
$c_0 = [\sin(0.25\pi)\sin h(0.8871)]^2 + [\cos(0.25\pi)\cos h(0.8871)]^2 = 1.5162$
$K = 1.5162/\sqrt{1 + 0.1220} = 1.4314$

Finally, the transfer function is derived as

$$B_2(s) = \frac{1.4314}{s^2 + 1.4256s + 1.5162}$$

EXAMPLE C.5

Compute the normalized Chebyshev transfer function for the following specifications:

Ripple $= 1$ dB
$n = 3$

Solution:

$(n-1)/2 = 1$

Applying Equations (C.32) to (C.38) leads to

$\alpha_0 = (2 \times 0 + 1)\pi/(2 \times 3) = \pi/6$
$\varepsilon^2 = 10^{0.1 \times 1} - 1 = 0.2589, 1/\varepsilon = 1.9653$
$\beta = \sinh^{-1}(1.9653)/n = \ln(1.9653 + \sqrt{1.9653^2 + 1})/3 = 0.4760$
$b_0 = 2\sin(\pi/6)\sinh(0.4760) = 0.4942$
$c_0 = [\sin(\pi/6)\sinh(0.4760)]^2 + [\cos(\pi/6)\cosh(0.4760)]^2 = 0.9942$
$\sinh(\beta) = \sinh(0.4760) = 0.4942$
$K = 0.4942 \times 0.9942 = 0.4913$

We can derive the transfer function as

$$B_3(s) = \frac{0.4913}{(s + 0.4942)(s^2 + 0.4942s + 0.9942)}$$

Finally, the unfactored form is found to be

$$B_3(s) = \frac{0.4913}{s^3 + 0.9883s^2 + 1.2384s + 0.4913}$$

Appendix D: Sinusoidal Steady-State Response of Digital Filters

D.1 SINUSOIDAL STEADY-STATE RESPONSE

Analysis of the sinusoidal steady-state response of digital filters will lead to the development of the magnitude and phase responses of digital filters. Let us look at the following digital filter with a digital transfer function $H(z)$ and a complex sinusoidal input

$$x(n) = Ve^{j(\Omega n + \varphi_x)} \tag{D.1}$$

where $\Omega = \omega T$ is the normalized digital frequency, while T is the sampling period and $y(n)$ denotes the digital output, as shown in Figure D.1.

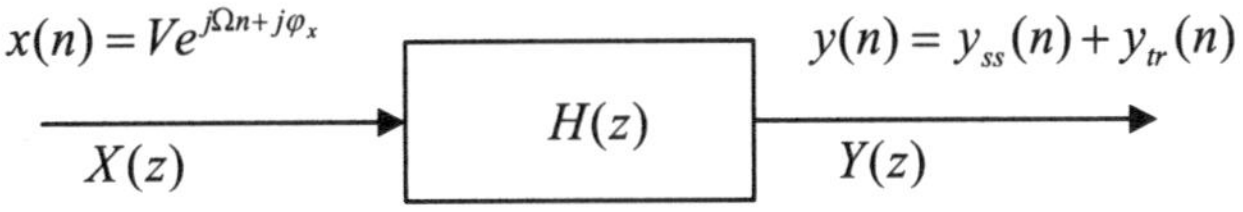

FIGURE D.1

Steady-state response of the digital filter.

The z-transform output from the digital filter is then given by

$$Y(z) = H(z)X(z) \tag{D.2}$$

Since $X(z) = \dfrac{Ve^{j\varphi_x}z}{z - e^{j\Omega}}$, we have

$$Y(z) = \frac{Ve^{j\varphi_x}z}{z - e^{j\Omega}}H(z) \tag{D.3}$$

Based on the partial fraction expansion, $Y(z)/z$ can be expanded to the following form:

$$\frac{Y(z)}{z} = \frac{Ve^{j\varphi_x}}{z - e^{j\Omega}}H(z) = \frac{R}{z - e^{j\Omega}} + \text{sum of the rest of partial fractions} \tag{D.4}$$

Multiplying the factor $(z - e^{j\Omega})$ on both sides of Equation (D.4) yields

$$Ve^{j\phi_x}H(z) = R + (z - e^{j\Omega})(\text{sum of the rest of partial fractions}) \tag{D.5}$$

813

Substituting $z = e^{j\Omega}$, we get the residue as

$$R = V e^{j\phi_x} H(e^{j\Omega})$$

Then substituting $R = V e^{j\phi_x} H(e^{j\Omega})$ back into Equation (D.4) results in

$$\frac{Y(z)}{z} = \frac{V e^{j\phi_x} H(e^{j\Omega})}{z - e^{j\Omega}} + \text{sum of the rest of partial fractions} \qquad \text{(D.6)}$$

and multiplying z on both sides of Equation (D.6) leads to

$$Y(z) = \frac{V e^{j\phi_x} H(e^{j\Omega}) z}{z - e^{j\Omega}} + z \times \text{sum of the rest of partial fractions} \qquad \text{(D.7)}$$

Taking the inverse z-transform leads to two parts of the solution:

$$y(n) = V e^{j\phi_x} H(e^{j\Omega}) e^{j\Omega n} + Z^{-1}(z \times \text{sum of the rest of partial fractions}) \qquad \text{(D.8)}$$

From Equation (D.8), we have the steady-state response

$$y_{ss}(n) = V e^{j\phi_x} H(e^{j\Omega}) e^{j\Omega n} \qquad \text{(D.9)}$$

and the transient response

$$y_{tr}(n) = Z^{-1}(z \times \text{sum of the rest of partial fractions}) \qquad \text{(D.10)}$$

Note that since the digital filter is a stable system, and the locations of its poles must be inside the unit circle on the z-plane, the transient response will settle to zero eventually. To develop the filter magnitude and phase responses, we write the digital steady-state response as

$$y_{ss}(n) = V|H(e^{j\Omega})| e^{j\Omega + j\phi_x + \angle H(e^{j\Omega})} \qquad \text{(D.11)}$$

Comparing Equation (D.11) and Equation (D.1), it follows that

$$\text{Magnitude response} = \frac{\text{Amplitude of the steady-state output}}{\text{Amplitude of the sinusoidal input}}$$

$$\qquad \text{(D.12)}$$

$$= \frac{V|H(e^{j\Omega})|}{V} = |H(e^{j\Omega})|$$

$$\text{Phase response} = \frac{e^{j\phi_x + j\angle H(e^{j\Omega})}}{e^{j\phi_x}} = e^{j\angle H(e^{j\Omega})} = \angle H(e^{j\Omega}) \qquad \text{(D.13)}$$

Thus we conclude that

$$\text{Frequency response: } H(e^{j\Omega}) = H(z)|_{z=e^{j\Omega}} \qquad \text{(D.14)}$$

Since $H(e^{j\Omega}) = |H(e^{j\Omega})| \angle H(e^{j\Omega})$

$$\text{Magnitude response: } |H(e^{j\Omega})| \qquad \text{(D.15)}$$

$$\text{Phase response: } \angle H(e^{j\Omega}) \qquad \text{(D.16)}$$

D.2 Properties of the Sinusoidal Steady-State Response

From Euler's identity and trigonometric identity, we know that

$$
\begin{aligned}
e^{j(\Omega+k2\pi)} &= \cos{(\Omega + k2\pi)} + j\sin{(\Omega + k2\pi)} \\
&= \cos\Omega + j\sin\Omega = e^{j\Omega}
\end{aligned}
\tag{D.17}
$$

where k is an integer taking values of $k = 0, \pm 1, \pm 2, \cdots$. Then

$$
\text{Frequency response: } H(e^{j\Omega}) = H(e^{j(\Omega+k2\pi)})
\tag{D.18}
$$

$$
\text{Magnitude frequency response: } |H(e^{j\Omega})| = \left|H(e^{j(\Omega+k2\pi)})\right|
\tag{D.19}
$$

$$
\text{Phase response: } \angle H(e^{j\Omega}) = \angle H(e^{j\Omega+2k\pi})
\tag{D.20}
$$

Clearly, the frequency response is periodic, with a period of 2π. Next, let us develop the symmetric properties. Since the transfer function is written as

$$
H(z) = \frac{Y(z)}{X(z)} = \frac{b_0 + b_1 z^{-1} + \cdots + b_M z^{-M}}{1 + a_1 z^{-1} + \cdots + a_N z^{-N}}
\tag{D.21}
$$

substituting $z = e^{j\Omega}$ into Equation (D.21) yields

$$
H(e^{j\Omega}) = \frac{b_0 + b_1 e^{-j\Omega} + \cdots + b_M e^{-jM\Omega}}{1 + a_1 e^{-j\Omega} + \cdots + a_N e^{-jN\Omega}}
\tag{D.22}
$$

Using Euler's identity, $e^{-j\Omega} = \cos\Omega - j\sin\Omega$, we have

$$
H(e^{j\Omega}) = \frac{(b_0 + b_1\cos\Omega + \cdots + b_M\cos M\Omega) - j(b_1\sin\Omega + \cdots + b_M\sin M\Omega)}{(1 + a_1\cos\Omega + \cdots + a_N\cos N\Omega) - j(a_1\sin\Omega + \cdots + a_N\sin N\Omega)}
\tag{D.23}
$$

Similarly,

$$
H(e^{-j\Omega}) = \frac{(b_0 + b_1\cos\Omega + \cdots + b_M\cos M\Omega) + j(b_1\sin\Omega + \cdots + b_M\sin M\Omega)}{(1 + a_1\cos\Omega + \cdots + a_N\cos N\Omega) + j(a_1\sin\Omega + \cdots + a_N\sin N\Omega)}
\tag{D.24}
$$

Then the magnitude response and phase response can be expressed as

$$
|H(e^{j\Omega})| = \frac{\sqrt{(b_0 + b_1\cos\Omega + \cdots + b_M\cos M\Omega)^2 + (b_1\sin\Omega + \cdots + b_M\sin M\Omega)^2}}{\sqrt{(1 + a_1\cos\Omega + \cdots + a_N\cos N\Omega)^2 + (a_1\sin\Omega + \cdots + a_N\sin N\Omega)^2}}
\tag{D.25}
$$

$$
\angle H(e^{j\Omega}) = \tan^{-1}\left(\frac{-(b_1\sin\Omega + \cdots + b_M\sin M\Omega)}{b_0 + b_1\cos\Omega + \cdots + b_M\cos M\Omega}\right)
$$

$$
- \tan^{-1}\left(\frac{-(a_1\sin\Omega + \cdots + a_N\sin N\Omega)}{1 + a_1\cos\Omega + \cdots + a_N\cos N\Omega}\right)
\tag{D.26}
$$

Based on Equation (D.24), we also obtain the magnitude and phase response for $H(e^{-j\Omega})$ as

$$|H(e^{-j\Omega})| = \frac{\sqrt{(b_0 + b_1 \cos \Omega + \cdots + b_M \cos M\Omega)^2 + (b_1 \sin \Omega + \cdots + b_M \sin M\Omega)^2}}{\sqrt{(1 + a_1 \cos \Omega + \cdots + a_N \cos N\Omega)^2 + (a_1 \sin \Omega + \cdots + a_N \sin N\Omega)^2}} \qquad (D.27)$$

$$\angle H(e^{-j\Omega}) = \tan^{-1}\left(\frac{b_1 \sin \Omega + \cdots + b_M \sin M\Omega}{b_0 + b_1 \cos \Omega + \cdots + b_M \cos M\Omega}\right)$$

$$-\tan^{-1}\left(\frac{a_1 \sin \Omega + \cdots + a_N \sin N\Omega}{1 + a_1 \cos \Omega + \cdots + a_N \cos N\Omega}\right) \qquad (D.28)$$

Comparing Equation (D.25) with (D.27), and Equation (D.26) with (D.28), respectively, we obtain the symmetric properties as

$$|H(e^{-j\Omega})| = |H(e^{j\Omega})| \qquad (D.29)$$

$$\angle H(e^{-j\Omega}) = -\angle H(e^{j\Omega}) \qquad (D.30)$$

Appendix E: Finite Impulse Response Filter Design Equations by the Frequency Sampling Design Method

Recall in Section 7.5 in Chapter 7 on the "Frequency Sampling Design Method" that we obtained

$$h(n) = \frac{1}{N} \sum_{k=0}^{N-1} H(k) W_N^{-kn} \tag{E.1}$$

where $h(n)$, $0 \leq n \leq N-1$, is the causal impulse response that approximates the finite impulse response (FIR) filter, $H(k)$, $0 \leq k \leq N-1$, represents the corresponding coefficients of the discrete Fourier transform (DFT), and $W_N = e^{-j\frac{2\pi}{N}}$. We further write the DFT coefficients, $H(k)$, $0 \leq k \leq N-1$, in polar form:

$$H(k) = H_k e^{j\varphi_k}, \ 0 \leq k \leq N-1 \tag{E.2}$$

where H_k and φ_k are the kth magnitude and the phase angle, respectively. The frequency response of the FIR filter is expressed as

$$H(e^{j\Omega}) = \sum_{n=0}^{N-1} h(n) e^{-jn\Omega} \tag{E.3}$$

Substituting (E.1) into (E.3) yields

$$H(e^{j\Omega}) = \sum_{n=0}^{N-1} \frac{1}{N} \sum_{k=0}^{N-1} H(k) W_N^{-kn} e^{-j\Omega n} \tag{E.4}$$

Interchanging the order of the summation in Equation (E.4) leads to

$$H(e^{j\Omega}) = \frac{1}{N} \sum_{k=0}^{N-1} H(k) \sum_{n=0}^{N-1} (W_N^{-k} e^{-j\Omega})^n \tag{E.5}$$

Since $W_N^{-k} e^{-j\Omega} = (e^{-j2\pi/N})^{-k} e^{-j\Omega} = e^{-(j\Omega - 2\pi k/N)}$ and using the identity $\sum_{n=0}^{N-1} r^n = 1 + r + r^2 + \cdots + r^{N-1} = \dfrac{1 - r^N}{1 - r}$, we can write the second summation in Equation (E.5) as

$$\sum_{n=0}^{N-1}(W_N^{-k}e^{-j\Omega})^n = \frac{1-e^{-j(\Omega-2\pi k/N)N}}{1-e^{-j(\Omega-2\pi k/N)}} \tag{E.6}$$

Using the Euler formula, Equation (E.6) becomes

$$\sum_{n=0}^{N-1}(W_N^{-k}e^{-j\Omega})^n = \frac{e^{-jN(\Omega-2\pi k/N)/2}\left(e^{jN(\Omega-2\pi k/N)/2} - e^{-jN(\Omega-2\pi k/N)/2}\right)/2j}{e^{-j(\Omega-2\pi k/N)/2}\left(e^{j(\Omega-2\pi k/N)/2} - e^{-j(\Omega-2\pi k/N)/2}\right)/2j}$$

$$= \frac{e^{-jN(\Omega-2\pi k/N)/2}\sin\left[N(\Omega-2\pi k/N)/2\right]}{e^{-j(\Omega-2\pi k/N)/2}\sin\left[(\Omega-2\pi k/N)/2\right]} \tag{E.7}$$

Substituting Equation (E.7) into Equation (E.5) leads to

$$H(e^{j\Omega}) = \frac{1}{N}e^{-j(N-1)\Omega/2}\sum_{k=0}^{N-1}H(k)e^{j(N-1)k\pi/N}\frac{\sin\left[N(\Omega-2\pi k/N)/2\right]}{\sin\left[(\Omega-2\pi k/N)/2)\right]} \tag{E.8}$$

Let $\Omega = \Omega_m = \dfrac{2\pi m}{N}$, and substitute it into Equation (E.8) to get

$$H(e^{j\Omega_m}) = \frac{1}{N}e^{-j(N-1)2\pi m/(2N)}\sum_{k=0}^{N-1}H(k)e^{j(N-1)k\pi/N}\frac{\sin\left[N(2\pi m/N-2\pi k/N)/2\right]}{\sin\left[(2\pi m/N-2\pi k/N)/2)\right]} \tag{E.9}$$

Clearly, when $m \neq k$, the last term of the summation in Equation (E.9) becomes

$$\frac{\sin\left[N(2\pi m/N-2\pi k/N)/2\right]}{\sin\left[(2\pi m/N-2\pi k/N)/2)\right]} = \frac{\sin\left(\pi(m-k)\right)}{\sin\left(\pi(m-k)/N\right)} = \frac{0}{\sin\left(\pi(m-k)/N\right)} = 0$$

When $m = k$, using L'Hospital's rule we have

$$\frac{\sin\left[N(2\pi m/N-2\pi k/N)/2\right]}{\sin\left[(2\pi m/N-2\pi k/N)/2)\right]} = \frac{\sin\left(N\pi(m-k)/N\right)}{\sin\left(\pi(m-k)/N\right)} = \lim_{x\to 0}\frac{\sin(Nx)}{\sin(x)} = N$$

Then Equation (E.9) is simplified to

$$H(e^{j\Omega_k}) = \frac{1}{N}e^{-j(N-1)\pi k/N}H(k)e^{j(N-1)k\pi/N}N = H(k)$$

that is,

$$H(e^{j\Omega_k}) = H(k), \ 0 \le k \le N-1 \tag{E.10}$$

where $\Omega_k = \dfrac{2\pi k}{N}$, corresponding to the kth DFT frequency component. The fact is that if we specify the desired frequency response, $H(\Omega_k)$, $0 \le k \le N-1$, at the equally spaced sampling frequency determined by $\Omega_k = \dfrac{2\pi k}{N}$, they are actually the DFT coefficients; that is, $H(k)$, $0 \le k \le N-1$, via Equation (E.10). Furthermore, the inverse of the DFT calculated using (E.10) will give the desired impulse response, $h(n)$, $0 \le n \le N-1$.

To devise the design procedure, we substitute Equation (E.2) in Equation (E.8) to obtain

$$H(e^{j\Omega}) = \frac{1}{N}e^{-j(N-1)\Omega/2}\sum_{k=0}^{N-1}H_k e^{j\varphi_k + j(N-1)k\pi/N}\frac{\sin[N(\Omega - 2\pi k/N)/2]}{\sin[(\Omega - 2\pi k/N)/2)]} \tag{E.11}$$

It is required that the frequency response of the designed FIR filter expressed in Equation (E.11) be linear phase. This can easily be accomplished by setting

$$\varphi_k + (N-1)k\pi/N = 0, \ 0 \le k \le N-1 \tag{E.12}$$

in Equation (E.11) so that the summation part becomes a real value, thus resulting in the linear phase of $H(e^{j\Omega})$, since only one complex term, $e^{-j(N-1)\Omega/2}$, is left, which presents the constant time delay of the transfer function. Second, the sequence $h(n)$ must be real. To proceed, let $N = 2M + 1$, and due to the properties of DFT coefficients, we have

$$\overline{H}(k) = H(N-k), \ 1 \le k \le M \tag{E.13}$$

where the bar indicates complex conjugate. Note the fact that

$$\overline{W}_N^{-k} = W_N^{-(N-k)}, \ 1 \le k \le M \tag{E.14}$$

From Equation (E.1), we write

$$h(n) = \frac{1}{N}\left(H(0) + \sum_{k=1}^{M}H(k)W_N^{-kn} + \sum_{k=M+1}^{2M}H(k)W_N^{-kn}\right) \tag{E.15}$$

Equation (E.15) is equivalent to

$$h(n) = \frac{1}{N}\left(H(0) + \sum_{k=1}^{M}H(k)W_N^{-kn} + \sum_{k=1}^{M}H(N-k)W_N^{-(N-k)n}\right)$$

Using Equations (E.13) and (E.14) in the last summation term leads to

$$h(n) = \frac{1}{N}\left(H(0) + \sum_{k=1}^{M}H(k)W_N^{-kn} + \sum_{k=1}^{M}\overline{H}(k)\overline{W}_N^{-kn}\right)$$

$$= \frac{1}{2M+1}\left(H(0) + \sum_{k=1}^{M}(H(k)W_N^{-kn} + \overline{H}(k)\overline{W}_N^{-kn})\right)$$

Combining the last two summation terms, we achieve

$$h(n) = \frac{1}{2M+1}\left\{H(0) + 2\text{Re}\left(\sum_{k=1}^{M}H(k)W_N^{-kn}\right)\right\}, \ 0 \le n \le N-1 \tag{E.16}$$

Solving Equation (E.12) gives

$$\varphi_k = -(N-1)k\pi/N, \ 0 \le k \le N-1 \tag{E.17}$$

Again, note that Equation (E.13) is equivalent to

$$H_k e^{-j\varphi_k} = H_{N-k} e^{j\varphi_{N-k}}, \quad 1 \leq k \leq M \tag{E.18}$$

Substituting (E.17) in (E.18) yields

$$H_k e^{j(N-1)k\pi/N} = H_{N-k} e^{-j(N-1)(N-k)\pi/N}, \quad 1 \leq k \leq M \tag{E.19}$$

Simplification of Equation (E.19) leads to the following result:

$$H_k = H_{N-k} e^{-j(N-1)\pi} = (-1)^{N-1} H_{N-k}, \quad 1 \leq k \leq M \tag{E.20}$$

Since we constrain the filter length to be $N = 2M + 1$, Equation (E.20) can be further reduced to

$$H_k = (-1)^{2M} H_{2M+1-k} = H_{2M+1-k}, \quad 1 \leq k \leq M \tag{E.21}$$

Finally, by substituting (E.21) and (E.17) into (E.16), we obtain a very simple design equation:

$$h(n) = \frac{1}{2M+1} \left\{ H_0 + 2 \sum_{k=1}^{M} H_k \cos\left(\frac{2\pi k(n-M)}{2M+1}\right) \right\}, \quad 0 \leq n \leq 2M \tag{E.22}$$

Thus the design procedure is simply summarized as follows: Given the filter length, $2M + 1$, and the specified frequency response, H_k at $\Omega_k = \dfrac{2\pi k}{(2M+1)}$ for $k = 0, 1, \cdots, M$, FIR filter coefficients can be calculated via Equation (E.22).

Appendix F: Wavelet Analysis and Synthesis Equations

F.1 BASIC PROPERTIES

The inner product of two functions is defined as

$$< x, y >= \int x(t)y(t)dt \tag{F.1}$$

Two functions are orthogonal if

$$< x(t), x(t-k) >= \begin{cases} A & \text{for} \quad k = 0 \\ 0 & \text{for} \quad k \neq 0 \end{cases} \tag{F.2}$$

Two functions are orthonormal if

$$< x(t), x(t-k) >= \begin{cases} 1 & \text{for } k = 0 \\ 0 & \text{for } k \neq 0 \end{cases} \tag{F.3}$$

The signal energy is defined as

$$E = \int x^2(t)dt \tag{F.4}$$

Many wavelet families are designed to be orthonormal:

$$E = \int \psi^2(t)dt = 1 \tag{F.5}$$

$$E = \int \psi_{jk}^2(t)dt = \int [2^{j/2}\psi(2^j t - k)]^2 dt = \int 2^j \psi^2(2^j t - k)dt \tag{F.6}$$

Let $u = 2^j t - k$. Then $du = 2^j dt$. Equation (F.6) becomes

$$E = \int 2^j \psi^2(u)2^{-j}du = 1 \tag{F.7}$$

Both father and mother wavelets are orthonormal at scale j:

$$\int \phi_{jk}(t)\phi_{jn}(t)dt = \begin{cases} 1 & k = n \\ 0 & \text{otherwise} \end{cases} \tag{F.8}$$

$$\int \psi_{jk}(t)\psi_{jn}(t)dt = \begin{cases} 1 & k = n \\ 0 & \text{otherwise} \end{cases} \tag{F.9}$$

F.2 ANALYSIS EQUATIONS

When a function $f(t)$ is approximated using the scaling functions only at scale $j+1$, it can be expressed as

$$f(t) = \sum_{k=-\infty}^{\infty} c_j(k)2^{j/2}\phi(2^j t - k)$$

Using the inner product,

$$c_j(k) = <f(t), \phi_{jk}(t)> = \int f(t)2^{j/2}\phi(2^j t - k)dt \tag{F.10}$$

Note that

$$\phi(t) = \sum_{n=-\infty}^{\infty} \sqrt{2}h_0(n)\phi(2t - n) \tag{F.11}$$

Substituting Equation (F.11) into Equation (F.10) leads to

$$c_j(k) = <f(t), \phi_{jk}(t)> = \int f(t)2^{j/2} \sum_{n=-\infty}^{\infty} \sqrt{2}h_0(n)\phi[2(2^j t - k) - n]dt$$

$$c_j(k) = \sum_{n=-\infty}^{\infty} \int f(t)2^{(j+1)/2}h_0(n)\phi(2^{(j+1)}t - 2k - n)dt$$

Let $m = n + 2k$. Interchange of the summation and integral leads to

$$c_j(k) = \sum_{m=-\infty}^{\infty} \int f(t)2^{(j+1)/2}h_0(m - 2k)\phi(2^{(j+1)}t - m)dt$$

$$c_j(k) = \sum_{m=-\infty}^{\infty} \left(\int f(t)\phi_{(j+1)m}(t)dt \right) h_0(m - 2k) \tag{F.12}$$

Using the inner product definition for the DWT coefficient again in (F.12), we achieve

$$c_j(k) = \sum_{m=-\infty}^{\infty} <f(t), \phi_{(j+1)m}(t)> h_0(m-2k) = \sum_{m=-\infty}^{\infty} c_{j+1}(m) h_0(m-2k) \qquad \text{(F.13)}$$

Similarly, notice that

$$\psi(t) = \sum_{k=-\infty}^{\infty} \sqrt{2} h_1(k) \phi(2t-k)$$

Using the inner product gives

$$d_j(k) = <f(t), \psi_{jk}(t)> = \int f(t) 2^{j/2} \sum_{n=-\infty}^{\infty} \sqrt{2} h_1(n) \phi[2(2^j t - k) - n] dt$$

$$d_j(k) = \sum_{n=-\infty}^{\infty} \int f(t) 2^{(j+1)/2} h_1(n) \phi(2^{(j+1)} t - 2k - n) dt \qquad \text{(F.14)}$$

Let $m = n + 2k$. Interchange of the summation and integral leads to

$$d_j(k) = \sum_{m=-\infty}^{\infty} \int f(t) 2^{(j+1)/2} h_1(m-2k) \phi(2^{(j+1)} t - m) dt$$

$$d_j(k) = \sum_{m=-\infty}^{\infty} \left(\int f(t) \phi_{(j+1)m}(t) dt \right) h_1(m-2k) \qquad \text{(F.15)}$$

Finally, applying the inner product definition for the wavelet discrete transform (WDT) coefficient, we obtain

$$d_j(k) = \sum_{m=-\infty}^{\infty} <f(t), \phi_{(j+1)m}(t)> h_1(m-2k) = \sum_{m=-\infty}^{\infty} c_{j+1}(m) h_1(m-2k) \qquad \text{(F.16)}$$

F.2 WAVELET SYNTHESIS EQUATIONS

We begin with

$$f(t) = \sum_{k=-\infty}^{\infty} c_j(k) 2^{j/2} \phi(2^j t - k) + \sum_{k=-\infty}^{\infty} d_j(k) 2^{j/2} \psi(2^j t - k)$$

Taking an inner product using the scaling function at scale level $j + 1$ gives

$$c_{j+1}(k) = <f(t), \phi_{(j+1)k}(t)> = \sum_{m=-\infty}^{\infty} c_j(m) 2^{j/2} \int \phi(2^j t - m) \phi_{(j+1)k}(t) dt$$

$$+ \sum_{m=-\infty}^{\infty} d_j(m) 2^{j/2} \int \psi(2^j t - m) \phi_{(j+1)k}(t) dt$$

$$c_{j+1}(k) = \sum_{m=-\infty}^{\infty} c_j(m) 2^{j/2} \int \sum_{n=-\infty}^{\infty} \sqrt{2} h_0(n) \phi(2^{j+1}t - 2m - n) \phi_{(j+1)k}(t) dt$$

$$+ \sum_{m=-\infty}^{\infty} d_j(m) 2^{j/2} \int \sum_{n=-\infty}^{\infty} \sqrt{2} h_1(n) \phi(2^{j+1}t - 2m - n) \phi_{(j+1)k}(t) dt$$

(F.17)

Interchange of the summation and integral yields

$$c_{j+1}(k) = \sum_{m=-\infty}^{\infty} \sum_{n=-\infty}^{\infty} c_j(m) h_0(n) \int 2^{(j+1)/2} \phi(2^{j+1}t - 2m - n) \phi_{(j+1)k}(t) dt$$

$$+ \sum_{m=-\infty}^{\infty} \sum_{n=-\infty}^{\infty} d_j(m) h_1(n) \int 2^{(j+1)/2} \phi(2^{j+1}t - 2m - n) \phi_{(j+1)k}(t) dt$$

Using the inner product, we get

$$c_{j+1}(k) = \sum_{m=-\infty}^{\infty} \sum_{n=-\infty}^{\infty} c_j(m) h_0(n) < \phi_{(j+1)(2m+n)}(t), \phi_{(j+1)k}(t) >$$

$$+ \sum_{m=-\infty}^{\infty} \sum_{n=-\infty}^{\infty} d_j(m) h_1(n) < \phi_{(j+1)(2m+n)}(t), \phi_{(j+1)k}(t) >$$

(F.18)

From the wavelet orthonormal property, we have

$$< \phi_{(j+1)(2m+n)}(t), \phi_{(j+1)k}(t) > = \begin{cases} 1 & n = k - 2m \\ 0 & \text{otherwise} \end{cases}$$

(F.19)

Substituting Equation (F.19) into Equation (F.18), we finally obtain

$$c_{j+1}(k) = \sum_{m=-\infty}^{\infty} c_j(m) h_0(k - 2m) + \sum_{m=-\infty}^{\infty} d_j(m) h_1(k - 2m)$$

(F.20)

Appendix G: Some Useful Mathematical Formulas

Form of a complex number:
Rectangular form:

$$a + jb, \text{ where } j = \sqrt{-1} \tag{G.1}$$

Polar form:

$$Ae^{j\theta} \tag{G.2}$$

Euler formula:

$$e^{\pm jx} = \cos x \pm j \sin x \tag{G.3}$$

Conversion from the polar form to the rectangular form:

$$Ae^{j\theta} = A \cos\theta + jA \sin\theta = a + jb \tag{G.4}$$

where $a = A \cos\theta$, and $b = A \sin\theta$.
Conversion from the rectangular form to the polar form:

$$a + jb = Ae^{j\theta} \tag{G.5}$$

where $A = \sqrt{a^2 + b^2}$. We usually specify the principal value of the angle such that $-180° < \theta \leq 180°$. The angle value can be determined as

$$\theta = \tan^{-1}\left(\frac{b}{a}\right) \quad \text{if} \quad a \geq 0$$

(that is, the complex number is in the first or fourth quadrant in the rectangular coordinate system);

$$\theta = 180° + \tan^{-1}\left(\frac{b}{a}\right) \quad \text{if} \quad a < 0 \quad \text{and} \quad b \geq 0$$

(that is, the complex number is in the second quadrant in the rectangular coordinate system); and

$$\theta = -180° + \tan^{-1}\left(\frac{b}{a}\right) \quad \text{if} \quad a < 0 \quad \text{and} \quad b \leq 0$$

(that is, the complex number is in the third quadrant in the rectangular coordinate system). Note that

$$\theta \ radian = \frac{\theta \ degree}{180°} \times \pi$$

$$\theta \ degree = \frac{\theta \ radian}{\pi} \times 180°$$

Complex numbers:

$$e^{\pm j\pi/2} = \pm j \tag{G.6}$$

$$e^{\pm j2n\pi} = 1 \tag{G.7}$$

$$e^{\pm j(2n+1)\pi} = -1 \tag{G.8}$$

Complex conjugate of $a + jb$:

$$(a + jb)^* = con \ j(a + jb) = a - jb \tag{G.9}$$

Complex conjugate of $Ae^{j\theta}$:

$$\left(Ae^{j\theta}\right)^* = con \ j\left(Ae^{j\theta}\right) = Ae^{-j\theta} \tag{G.10}$$

Complex number addition and subtraction:

$$(a_1 + jb_1) \pm (a_2 + jb_2) = (a_1 \pm a_2) + j(b_1 \pm b_2) \tag{G.11}$$

Complex number multiplication:
Rectangular form:

$$(a_1 + jb_1) \times (a_2 + jb_2) = a_1a_2 - b_1b_2 + j(a_1b_2 + a_2b_1) \tag{G.12}$$

$$(a + jb) \cdot con \ j(a + jb) = (a + jb)(a - jb) = a^2 + b^2 \tag{G.13}$$

Polar form:

$$A_1e^{j\theta_1}A_2e^{j\theta_2} = A_1A_2e^{j(\theta_1+\theta_2)} \tag{G.14}$$

Complex number division:
Rectangular form:

$$\frac{a_1 + jb_1}{a_2 + jb_2} = \frac{(a_1 + jb_1)(a_2 - jb_2)}{(a_2 + jb_2)(a_2 - jb_2)}$$

$$= \frac{(a_1a_2 + b_1b_2) + j(a_2b_1 - a_1b_2)}{(a_2)^2 + (b_2)^2} \tag{G.15}$$

Polar form:

$$\frac{A_1 e^{j\theta_1}}{A_2 e^{j\theta_2}} = \left(\frac{A_1}{A_2}\right) e^{j(\theta_1 - \theta_2)} \tag{G.16}$$

Trigonometric identities:

$$\sin x = \frac{e^{jx} - e^{-jx}}{2j} \tag{G.17}$$

$$\cos x = \frac{e^{jx} + e^{-jx}}{2} \tag{G.18}$$

$$\sin(x \pm 90°) = \pm \cos x \tag{G.19}$$

$$\cos(x \pm 90°) = \mp \sin x \tag{G.20}$$

$$\sin x \cos x = \frac{1}{2} \sin 2x \tag{G.21}$$

$$\sin^2 x + \cos^2 x = 1 \tag{G.22}$$

$$\sin^2 x = \frac{1}{2}\left(1 - \cos 2x\right) \tag{G.23}$$

$$\cos^2 x = \frac{1}{2}\left(1 + \cos 2x\right) \tag{G.24}$$

$$\sin(x \pm y) = \sin x \cos y \pm \cos x \sin y \tag{G.25}$$

$$\cos(x \pm y) = \cos x \cos y \mp \sin x \sin y \tag{G.26}$$

$$\sin x \cos y = \frac{1}{2}(\sin(x+y) + \sin(x-y)) \tag{G.27}$$

$$\sin x \sin y = \frac{1}{2}(\cos(x-y) - \cos(x+y)) \tag{G.28}$$

$$\cos x \cos y = \frac{1}{2}(\cos(x-y) + \cos(x+y)) \tag{G.29}$$

Series of exponentials:

$$\sum_{k=0}^{N-1} a^k = \frac{1 - a^N}{1 - a}, \; a \neq 1 \tag{G.30}$$

$$\sum_{k=0}^{\infty} a^k = \frac{1}{1 - a}, \; |a| < 1 \tag{G.31}$$

$$\sum_{k=0}^{\infty} k a^k = \frac{1}{(1 - a)^2}, \; |a| < 1 \tag{G.32}$$

$$\sum_{k=0}^{N-1} e^{\left(j \frac{2\pi nk}{N}\right)} = \begin{cases} 0 & 1 \leq n \leq N - 1 \\ N & n = 0, N \end{cases} \tag{G.33}$$

L'Hospital's rule:

If $\lim\limits_{x \to a} \dfrac{f(x)}{g(x)}$ results in the undetermined form $\dfrac{0}{0}$ or $\dfrac{\infty}{\infty}$, then

$$\lim_{x \to a} \frac{f(x)}{g(x)} = \lim_{x \to a} \frac{f'(x)}{g'(x)} \tag{G.34}$$

where $f'(x) = \dfrac{df(x)}{dx}$ and $g'(x) = \dfrac{dg(x)}{dx}$.

Solution of the quadratic equation:

For a quadratic equation expressed as

$$ax^2 + bx + c = 0 \tag{G.35}$$

the solution is given by

$$x = \frac{-b \pm \sqrt{b^2 - 4ac}}{2a} \tag{G.36}$$

Solution of simultaneous equations:

Simultaneous linear equations are listed below:

$$\begin{aligned} a_{11}x_1 + a_{12}x_2 + \cdots + a_{1n}x_n &= b_1 \\ a_{21}x_1 + a_{22}x_2 + \cdots + a_{2n}x_n &= b_2 \\ &\cdots \\ a_{n1}x_1 + a_{n2}x_2 + \cdots + a_{nn}x_n &= b_n \end{aligned} \tag{G.37}$$

The solution is given by Cramer's rule, that is

$$x_1 = \frac{D_1}{D}, \; x_2 = \frac{D_2}{D}, \cdots, x_n = \frac{D_n}{D} \tag{G.38}$$

where $D, D_1, D_2, ..., D_n$ are the $n \times n$ determinants. Each is defined below:

$$D = \begin{vmatrix} a_{11} & a_{12} & \cdots & a_{1n} \\ a_{21} & a_{22} & \cdots & a_{2n} \\ \vdots & \vdots & \ddots & \vdots \\ a_{n1} & a_{n2} & \cdots & a_{nn} \end{vmatrix} \tag{G.39}$$

$$D_1 = \begin{vmatrix} b_1 & a_{12} & \cdots & a_{1n} \\ b_2 & a_{22} & \cdots & a_{2n} \\ \vdots & \vdots & \ddots & \vdots \\ b_n & a_{n2} & \cdots & a_{nn} \end{vmatrix} \tag{G.40}$$

$$D_2 = \begin{vmatrix} a_{11} & b_1 & \cdots & a_{1n} \\ a_{21} & b_2 & \cdots & a_{2n} \\ \vdots & \vdots & \ddots & \vdots \\ a_{n1} & b_n & \cdots & a_{nn} \end{vmatrix} \tag{G.41}$$

$$\cdots$$

$$D_n = \begin{vmatrix} a_{11} & a_{12} & \cdots & b_1 \\ a_{21} & a_{22} & \cdots & b_2 \\ \vdots & \vdots & \ddots & \vdots \\ a_{n1} & a_{n2} & \cdots & b_n \end{vmatrix} \tag{G.42}$$

$$D = (-1)^{1+1}a_{11}M_{11} + (-1)^{1+2}a_{12}M_{12} + \cdots (-1)^{1+n}a_{1n}M_{1n} \tag{G.43}$$

where M_{ij} is an $(n-1) \times (n-1)$ determinant obtained from D by crossing out the ith row and jth column. D can also be expanded by any row or column. As an example, using the second column,

$$D = (-1)^{1+2}a_{12}M_{12} + (-1)^{2+2}a_{22}M_{12} + \cdots (-1)^{n+2}a_{n2}M_{n2} \tag{G.44}$$

2×2 determinant:

$$D = \begin{vmatrix} a_{11} & a_{12} \\ a_{21} & a_{22} \end{vmatrix} = a_{11}a_{22} - a_{12}a_{21} \tag{G.45}$$

3×3 determinant:

$$
\begin{aligned}
D &= \begin{vmatrix} a_{11} & a_{12} & a_{13} \\ a_{21} & a_{22} & a_{23} \\ a_{31} & a_{32} & a_{33} \end{vmatrix} \\
&= (-1)^{1+1} a_{11} \begin{vmatrix} a_{22} & a_{23} \\ a_{32} & a_{33} \end{vmatrix} + (-1)^{1+2} a_{12} \begin{vmatrix} a_{21} & a_{23} \\ a_{31} & a_{33} \end{vmatrix} + (-1)^{1+3} a_{13} \begin{vmatrix} a_{21} & a_{22} \\ a_{31} & a_{32} \end{vmatrix} \\
&= a_{11}(a_{22}a_{33} - a_{23}a_{32}) - a_{12}(a_{21}a_{33} - a_{23}a_{31}) + a_{13}(a_{21}a_{32} - a_{22}a_{31})
\end{aligned}
$$

(G.46)

Solution for two simultaneous linear equations:

$$
\begin{aligned}
ax + by &= e \\
cx + dy &= f
\end{aligned}
$$

(G.47)

The solution is given by

$$
x = \frac{D_1}{D} = \frac{\begin{vmatrix} e & b \\ f & d \end{vmatrix}}{\begin{vmatrix} a & b \\ c & d \end{vmatrix}} = \frac{ed - bf}{ad - bc}
$$

(G.48)

$$
y = \frac{D_2}{D} = \frac{\begin{vmatrix} a & e \\ c & f \end{vmatrix}}{\begin{vmatrix} a & b \\ c & d \end{vmatrix}} = \frac{af - ec}{ad - bc}
$$

(G.49)

Answers to Selected Problems

CHAPTER 2

2.1. Hint:

   b.

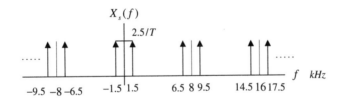

2.2. Hint:

   a.

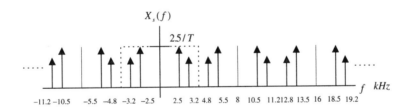

2.5. Hint:

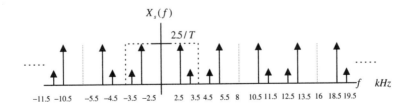

   c. The aliasing frequency = 3.5 kHz

2.9.

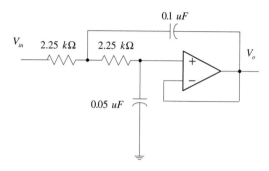

2.10. % aliasing level $= 8.39\%$

2.13. **a.** % aliasing level $= 57.44\%$

   b. % aliasing level $= 20.55\%$

2.17. **a.** % distortion $= 24.32\%$

   b. % distortion $= 5.68\%$

2.18. $f_c = 4,686$ Hz

2.21. $b1b0 = 01$

2.22. $V_0 = 1.25$ Volts

2.25. **a.** $L = 2^4 = 16$ levels

   b. $\Delta = \dfrac{x_{\max} - x_{\min}}{L} = \dfrac{5}{16} = 0.3125$

   c. $x_q = 3.125$

   d. binary code $= 1010$

   e. $e_q = -0.075$

2.27. **a.** $L = 2^3 = 8$ levels

   b. $\Delta = \dfrac{x_{\max} - x_{\min}}{L} = \dfrac{5}{8} = 0.625$

   c. $x_q = -2.5 + 2 \times 0.625 = -1.25$

   d. binary code $= 010$

   e. $e_q = -0.05$

2.29. **a.** $L = 2^6 = 64$ levels

   b. $\Delta = \dfrac{x_{\max} - x_{\min}}{L} = \dfrac{20}{64} = 0.3125$

   c. $SNR_{dB} = 1.76 + 6.02 \times 6 = 37.88$ dB

CHAPTER 3

3.1.

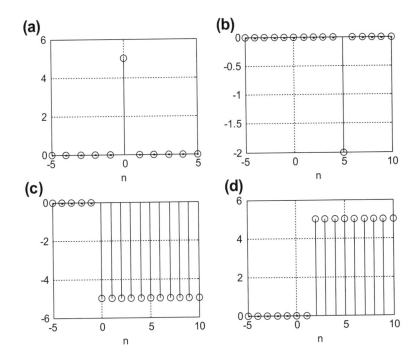

3.2. Hint:

a.

n	0	1	2	3	4	5	6	7
$x(n)$	1.000	0.5000	0.2500	0.1250	0.0625	0.0313	0.0156	0.0078

d.

n	0	1	2	3	4	5	6	7
$x(n)$	0.0000	1.1588	1.6531	1.7065	1.5064	1.1865	0.8463	0.5400

3.5.

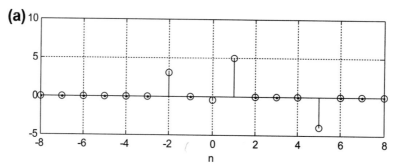

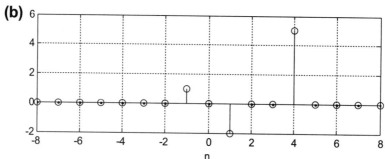

3.6. a. $x(n) = 3\delta(n) + \delta(n-1) + 2\delta(n-2) + \delta(n-3) + \delta(n-5)$
   b. $x(n) = \delta(n-1) - \delta(n-2) + \delta(n-4) - \delta(n-5)$
3.9. a. $x(n) = e^{-0.5n}u(n) = (0.6065)^{n}u(n)$
   b. $x(n) = 5\sin(0.2\pi n)u(n)$
   c. $x(n) = 10\cos(0.4\pi n + \pi/6)u(n)$
   d. $x(n) = 10e^{-n}\sin(0.15\pi n)u(n) = 10(0.3679)^{n}\sin(0.15\pi n)u(n)$
3.10. a. nonlinear system
   b. linear system
   c. nonlinear system
3.13. a. time-invariant
3.15. a. causal system
   b. noncausal system
   c. causal system
3.16. a. $h(n) = 0.5\delta(n) - 0.5\delta(n-2)$
   b. $h(n) = (0.75)^{n},\ n \geq 0$
   c. $h(n) = 1.25\delta(n) - 1.25(-0.8)^{n},\ n \geq 0$
3.19. a. $h(n) = 5\delta(n-10)$
   b. $h(n) = \delta(n) + 0.5\delta(n-1)$
3.20. Since $h(n) = 0.5\delta(n) + 100\delta(n-2) - 20\delta(n-10)$ and $S = 0.5 + 100 + 20 = 120.5 =$ finite number, the system is stable.
3.23. a. $h(n) = (0.75)^{n}u(n), S = \sum_{k=0}^{\infty}(0.75)^{k} = 1/(1-0.75) = 4 =$ finite, the system is stable.
   b. $h(n) = (2)^{n}u(n),\ S = \sum_{k=0}^{\infty}(2)^{k} = 1 + 2 + 2^{2} + \cdots = \infty =$ infinite, the system is unstable.

3.25.

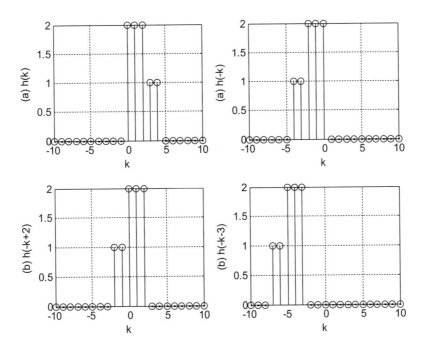

3.27. $y(0) = 4$, $y(1) = 6$, $y(2) = 8$, $y(3) = 6$, $y(4) = 5$, $y(5) = 2$, $y(6) = 1$,
$y(n) = 0$ for $n \geq 7$

3.29. $y(0) = 0$, $y(1) = 1$, $y(2) = 2$, $y(3) = 1$, $y(4) = 0$
$y(n) = 0$ for $n \geq 4$

CHAPTER 4

4.1. $X(0) = 1$, $X(1) = 2 - j$, $X(2) = -1$, $X(3) = 2 + j$

4.5. $x(0) = 4$, $x(1) = 3$, $x(2) = 2$, $x(3) = 1$

4.6. $X(0) = 10$, $X(1) = 3.5 - 4.3301j$, $X(2) = 2.5 - 0.8660j$, $X(3) = 2$,
$X(4) = 2.5 + 0.8660j$, $X(5) = 3.5 + 4.3301j$

4.9. $\bar{x}(0) = 4$, $\bar{x}(4) = 0$

4.10. $\Delta f = 2.5$ Hz and $f_{\max} = 10$ kHz

4.13. $N = 4096$, $\Delta f = 0.488$ Hz

4.15. a. w = [0.0800 0.2532 0.6424 0.9544 0.9544 0.6424 0.2532 0.0800]
  b. w = [0 0.1883 0.6113 0.9505 0.9505 0.6113 0.1883 0]

4.16. a. xw = [0 0.4000 0 −0.8000 0 0]
  b. xw = [0 0.3979 0 −0.9121 0 0.0800]
  c. xw = [0 0.3455 0 −0.9045 0 0]

4.19. a. $A_0 = 0.1667, A_1 = 0.3727, A_2 = 0.5, A_3 - 0.3727$
$\varphi_0 = 0^0, \varphi_1 = 154.43^0, \varphi_2 = 0^0, \varphi_3 = -154.43^0$
$P_0 = 0.0278, P_1 = 0.1389, P_2 = 0.25, P_3 = 0.1389$
b. $A_0 = 0.2925, A_1 = 0.3717, A_2 = 0.6375, A_3 = 0.3717$
$\varphi_0 = 0^0, \varphi_1 = 145.13^0, \varphi_2 = 0^0, \varphi_3 = -145.13^0$
$P_0 = 0.0586, P_1 = 0.1382, P_2 = 0.4064, P_3 = 0.1382$
c. $A_0 = 0.6580, A_1 = 0.3302, A_2 = 0.9375, A_3 = 0.3302$
$\varphi_0 = 0^0, \varphi_1 = 108.86^0, \varphi_2 = 0^0, \varphi_3 = -108.86^0$
$P_0 = 0.4330, P_1 = 0.1091, P_2 = 0.8789, P_3 = 0.1091$
4.21. $X(0) = 10, X(1) = 2 - 2j, X(2) = 2, X(3) = 2 + 2j$, 4 complex multiplications
4.22. $x(0) = 4, x(1) = 3, x(2) = 2, x(3) = 1$, 4 complex multiplications
4.25. $X(0) = 10, X(1) = 2 - 2j, X(2) = 2, X(3) = 2 + 2j$, 4 complex multiplications
4.26. $x(0) = 4, x(1) = 3, x(2) = 2, x(3) = 1$, 4 complex multiplications

CHAPTER 5

5.1. a. $X(z) = \dfrac{4z}{z-1}$,

b. $X(z) = \dfrac{z}{z+0.7}$

c. $X(z) = \dfrac{4z}{z - e^{-2}} = \dfrac{4z}{z - 0.1353}$,

d. $X(z) = \dfrac{4z[z - 0.8 \times \cos(0.1\pi)]}{z^2 - [2 \times 0.8z\cos(0.1\pi)] + 0.8^2} = \dfrac{4z(z - 0.7608)}{z^2 - 1.5217z + 0.64}$

e. $X(z) = \dfrac{4e^{-3}\sin(0.1\pi)z}{z^2 - 2e^{-3}z\cos(0.1\pi) + e^{-6}} = \dfrac{0.06154z}{z^2 - 0.0947z + 0.00248}$

5.2. a. $X(z) = \dfrac{z}{z-1} + \dfrac{z}{z-0.5}$

b. $X(z) = \dfrac{z^{-4}z[z - e^{-3}\cos(0.1\pi)]}{z^2 - [2e^{-3}\cos(0.1\pi)]z + e^{-6}} = \dfrac{z^{-3}(z - 0.0474)}{z^2 - 0.0948z + 0.0025}$

5.3. c. $X(z) = \dfrac{5z^{-2}}{z - e^{-2}}$

e. $X(z) = \dfrac{4e^{-3}\sin(0.2\pi)}{z^2 - 2e^{-3}\cos(0.2\pi)z + e^{-6}} = \dfrac{0.1171}{z^2 - 0.0806z + 0.0025}$

5.5. a. $X(z) = 15z^{-3} - 6z^{-5}$
b. $x(n) = 15\delta(n-3) - 6\delta(n-5)$

5.9. a. $X(z) = -25 + \dfrac{5z}{z - 0.4} + \dfrac{20z}{z + 0.1}$, $x(n) = -25\delta(n) + 5(0.4)^n u(n) + 20(-0.1)^n u(n)$

b. $X(z) = \dfrac{1.6667z}{z - 0.2} - \dfrac{1.6667z}{z + 0.4}$, $x(n) = 1.6667(0.2)^n u(n) - 1.6667(-0.4)^n u(n)$

c. $X(z) = \dfrac{1.3514z}{z+0.2} + \dfrac{Az}{z-P} + \dfrac{A^*z}{z-P^*}$

where $P = 0.5+0.5j = 0.707\angle 45^0, A = 1.1625\angle -125.54^0, x(n) = 1.3514(-0.2)^n u(n) + 2.325(0.707)^n \cos(45^0 \times n - 125.54^0)$

d. $X(z) = \dfrac{4.4z}{z-0.6} + \dfrac{-0.4z}{z-0.1} + \dfrac{-1.2z}{(z-0.1)^2}$,

$x(n) = 4.4(0.6)^n u(n) - 0.4(0.1)^n u(n) - 12n(0.1)^n u(n)$

5.10. $Y(z) = \dfrac{-4.3333z}{z-0.5} + \dfrac{5.333z}{z-0.8}$, $y(n) = -4.3333(0.5)^n u(n) + 5.3333(0.8)^n u(n)$

5.13. $Y(z) = \dfrac{9.84z}{z-0.2} + \dfrac{-29.46z}{z-0.3} + \dfrac{20z}{z-0.4}$

$y(n) = 9.84(0.2)^n u(n) - 29.46(0.3)^n u(n) + 20(0.4)^n u(n)$

5.14. a. $Y(z) = \dfrac{-4z}{z-0.2} + \dfrac{5z}{z-0.5}$, $y(n) = -4(0.2)^n u(n) + 5(0.5)^n u(n)$

b. $Y(z) = \dfrac{5z}{z-1} + \dfrac{-5z}{z-0.5} + \dfrac{z}{z-0.2}$,

$y(n) = 5u(n) - 5(0.5)^n u(n) + (0.2)^n u(n)$

5.17. a. $Y(z) = \dfrac{Az}{z-P} + \dfrac{A^*z}{z-P^*}$, $P = 0.2+0.5j = 0.5385\angle 68.20^0, A = 0.8602\angle -54.46^0$

$y(n) = 1.7204(0.5382)^n \cos(n \times 68.20^0 - 54.46^0)$

b. $Y(z) = \dfrac{1.6854z}{z-1} + \dfrac{Az}{z-P} + \dfrac{A^*z}{z-P^*}$, where $P = 0.2+0.5j = 0.5385\angle 68.20^0$, $A = 0.4910\angle -136.25^0$

$y(n) = 1.6845u(n) + 0.982(0.5382)^n \cos(n \times 68.20^0 - 136.25^0)$

CHAPTER 6

6.1. a. $y(0) = 0.5, y(1) = 0.25, y(2) = 0.125, y(3) = 0.0625, y(4) = 0.03125$
  b. $y(0) = 1, y(1) = 0, y(2) = 0.25, y(3) = 0, y(4) = 0.0625$
6.3. a. $y(0) = -2, y(1) = 2.3750, y(2) = -1.0312, y(3) = 0.7266, y(4) = -0.2910$
  b. $y(0) = 0, y(1) = 1, y(2) = -0.2500, y(3) = 0.3152, y(4) = -0.0781$
6.4. a. $H(z) = 0.5+0.5z^{-1}$
  b. $y(n) = 2\delta(n) + 2\delta(n-1), y(n) = -5\delta(n) + 10u(n)$

6.5. a. $H(z) = \dfrac{1}{1+0.5z^{-1}}$
  b. $y(n) = (-0.5)^n u(n), y(n) = 0.6667u(n) + 0.3333(-0.5)^n u(n)$

6.9. $H(z) = 1 - 0.3z^{-1} + 0.28z^{-2}$, $A(z) = 1$, $N(z) = 1 - 0.3z^{-1} + 0.28z^{-2}$

6.12. a. $y(n) = x(n) - 0.25x(n-2) - 1.1y(n-1) - 0.18y(n-2)$

    b. $y(n) = x(n-1) - 0.1x(n-2) + 0.3x(n-3)$

6.13. b. $H(z) = \dfrac{(z+0.4)(z-0.4)}{(z+0.2)(z+0.5)}$

6.15. a. zero: $z = 0.5$, poles: $z = -0.25$ ($|z| = 0.25$), $z = -0.5 \pm 0.7416j$ ($|z| = 0.8944$), stable

    b. zeros: $z = \pm 0.5j$, poles: $z = 0.5$ ($|z| = 0.5$), $z = -2 \pm 1.7321j$ ($|z| = 2.6458$), unstable

    c. zero: $z = -0.95$, poles: $z = 0.2$ ($|z| = 0.2$), $z = -0.7071 \pm 0.7071j$ ($|z| = 1$), marginally stable

    d. zeros: $z = -0.5$, $z = -0.5$, poles: $z = 1$ ($|z| = 1$), $z = -1$, $z = -1$ ($|z| = 1$), $z = 0.36$ ($|z| = 0.36$), unstable

6.17. $H(z) = 0.5z^{-1} + 0.5z^{-2}$, $H(e^{j\Omega}) = 0.5e^{-j\Omega} + 0.5e^{-j2\Omega}$

$$\left|H(e^{j\Omega})\right| = 0.5\sqrt{(1 + \cos\Omega)^2 + (\sin\Omega)^2}, \quad \angle H\left(e^{j\Omega}\right) = \tan^{-1}\left(\frac{-\sin\Omega - \sin 2\Omega}{\cos\Omega + \cos 2\Omega}\right)$$

6.19. $H(z) = \dfrac{1}{1 + 0.5z^{-2}}$, $H\left(e^{j\Omega}\right) = \dfrac{1}{1 + 0.5e^{-j2\Omega}}$

$$\left|H(e^{j\Omega})\right| = \frac{1}{\sqrt{(1 + 0.5\cos 2\Omega)^2 + (0.5\sin 2\Omega)^2}}, \quad \angle H\left(e^{j\Omega}\right) = -\tan^{-1}\left(\frac{-0.5\sin 2\Omega}{1 + 0.5\cos 2\Omega}\right)$$

6.21. a. $H(z) = 0.5 + 0.5z^{-1}$, $H(e^{j\Omega}) = 0.5 + 0.5e^{-j\Omega}$

$$\left|H(e^{j\Omega})\right| = \sqrt{(0.5 + 0.5\cos\Omega)^2 + (0.5\sin\Omega)^2}, \quad \angle H(e^{j\Omega}) = \tan^{-1}\left(\frac{-0.5\sin\Omega}{0.5 + 0.5\cos\Omega}\right)$$

    b. $H(z) = 0.5 - 0.5z^{-1}$, $H(e^{j\Omega}) = 0.5 - 0.5e^{-j\Omega}$

$$\left|H(e^{j\Omega})\right| = \sqrt{(0.5 - 0.5\cos\Omega)^2 + (0.5\sin\Omega)^2}, \quad \angle H(e^{j\Omega}) = \tan^{-1}\left(\frac{0.5\sin\Omega}{0.5 - 0.5\cos\Omega}\right)$$

    c. $H(z) = 0.5 + 0.5z^{-2}$, $H(e^{j\Omega}) = 0.5 + 0.5e^{-j2\Omega}$

$$\left|H(e^{j\Omega})\right| = \sqrt{(0.5 + 0.5\cos 2\Omega)^2 + (0.5\sin 2\Omega)^2}, \quad \angle H(e^{j\Omega}) = \tan^{-1}\left(\frac{-0.5\sin 2\Omega}{0.5 + 0.5\cos 2\Omega}\right)$$

    d. $H(z) = 0.5 - 0.5z^{-2}$, $H(e^{j\Omega}) = 0.5 - 0.5e^{-j2\Omega}$

$$\left|H(e^{j\Omega})\right| = \sqrt{(0.5 - 0.5\cos 2\Omega)^2 + (0.5\sin 2\Omega)^2}, \quad \angle H(e^{j\Omega}) = \tan^{-1}\left(\frac{0.5\sin 2\Omega}{0.5 - 0.5\cos 2\Omega}\right)$$

6.23. $H(z) = \dfrac{0.5}{1 + 0.7z^{-1} + 0.1z^{-2}}$,

$$y(n) = 0.5556u(n) - 0.111(-0.2)^n u(n) + 0.5556(-0.5)^n u(n)$$

6.25. a. $y(n) = x(n) - 0.9x(n-1) - 0.1x(n-2) - 0.3y(n-1) + 0.04y(n-2)$

    b. $w(n) = x(n) - 0.3w(n-1) + 0.04w(n-2)$

        $y(n) = w(n) - 0.9w(n-1) - 0.1w(n-2)$

c. Hint: $H(z) = \dfrac{(z-1)(z+0.1)}{(z+0.4)(z-0.1)}$

$w_1(n) = x(n) - 0.4w_1(n-1)$
$y_1(n) = w_1(n) - w_1(n-1)$
$w_2(n) = y_1(n) + 0.1w_2(n-1)$
$y(n) = w_2(n) + 0.1w_2(n-1)$

d. Hint: $H(z) = 2.5 + \dfrac{2.1z}{z+0.4} - \dfrac{3.6z}{z-0.1}$

$y_1(n) = 2.5x(n)$
$w_2(n) = x(n) - 0.4w_2(n-1)$
$y_2(n) = 2.1w_2(n)$
$w_3(n) = x(n) + 0.1w_3(n-1)$
$y_3(n) = -3.6w_3(n)$
$y(n) = y_1(n) + y_2(n) + y_3(n)$

6.26. a. $y(n) = x(n) - 0.5x(n-1)$
b. $y(n) = x(n) - 0.7x(n-1)$
c. $y(n) = x(n) - 0.9x(n-1)$
The filter in (c) emphasizes high frequencies most.

CHAPTER 7

7.1. a. $H(z) = 0.2941 + 0.3750z^{-1} + 0.2941z^{-2}$
b. $H(z) = 0.0235 + 0.3750z^{-1} + 0.0235z^{-2}$
7.3. a. $H(z) = 0.1514 + 0.1871z^{-1} + 0.2000z^{-2} + 0.1871z^{-3} + 0.1514z^{-4}$
b. $H(z) = 0.0121 + 0.1010z^{-1} + 0.2000z^{-2} + 0.1010z^{-3} + 0.0121z^{-4}$
7.5. a. $H(z) = -0.0444 + 0.0117z^{-1} + 0.0500z^{-2} + 0.0117z^{-3} - 0.0444z^{-4}$
b. $H(z) = -0.0035 + 0.0063z^{-1} + 0.0500z^{-2} + 0.0063z^{-3} - 0.0035z^{-4}$
7.7. a. Hanning window
b. filter length $=63$
c. cutoff frequency $= 1,000$ Hz
7.9. a. Hamming window
b. filter length $=45$
c. lower cutoff frequency $= 1500$ Hz, upper cutoff frequency $= 2,300$ Hz
7.11. Hint:
a. $y(n) = 0.25x(n) - 0.5x(n-1) + 0.25x(n-2)$
b. $y(n) = 0.25[x(n) + x(n-2)] - 0.5x(n-1)$
7.13. $N = 3$, $\Omega_c = 3\pi/10$, $\Omega_0 = 0$, $H_0 = 1$, $\Omega_1 = 2\pi/3$, $H_1 = 0$
$H(z) = 0.3333 + 0.3333z^{-1} + 0.3333z^{-2}$
7.15. $H(z) = -0.1236 + 0.3236z^{-1} + 0.6z^{-2} + 0.3236z^{-3} - 0.1236z^{-4}$

7.17. $H(z) = 0.1718 - 0.2574z^{-1} - 0.0636z^{-2} + 0.2857z^{-3} - 0.0636z^{-4} - 0.2574z^{-5}$
$+0.1781z^{-6}$

7.19. $W_p = 1$, $W_s = 12$
7.21. $W_p = 1$, $W_s = 122$
7.23. Hamming window, filter length $= 33$, lower cutoff frequency $= 3{,}500$ Hz
7.24. Hamming window, filter length $= 53$,
   lower cutoff frequency $= 1{,}250$ Hz, upper cutoff frequency $= 2{,}250$ Hz
7.25. Lowpass filter: Hamming window, filter length $= 91$, cutoff frequency $= 2{,}000$ Hz
   Highpass filter: Hamming window, filter length $= 91$, cutoff frequency $= 2{,}000$ Hz

---

CHAPTER 8

8.1. $H(z) = \dfrac{0.3333 + 0.3333z^{-1}}{1 - 0.3333z^{-1}}$

$y(n) = 0.3333x(n) + 0.3333x(n-1) + 0.3333y(n-1)$

8.3. a. $H(z) = \dfrac{0.6625 - 0.6625z^{-1}}{1 - 0.3249z^{-1}}$

$y(n) = 0.6225x(n) - 0.6225x(n-1) + 0.3249y(n-1)$

8.5. a. $H(z) = \dfrac{0.2113 - 0.2113z^{-2}}{1 - 0.8165z^{-1} + 0.5774z^{-2}}$

$y(n) = 0.2113x(n) - 0.2113x(n-2) + 08165y(n-1) - 0.5774y(n-2)$

8.7. a. $H(z) = \dfrac{0.1867 + 0.3734z^{-1} + 0.1867z^{-2}}{1 - 0.4629z^{-1} + 0.2097z^{-2}}$

$y(n) = 0.1867x(n) + 0.3734x(n-1) + 0.1867x(n-2)$
$\qquad + 0.4629y(n-1) - 0.2097y(n-2)$

8.9. a. $H(z) = \dfrac{0.0730 - 0.0730z^{-2}}{1 + 0.8541z^{-2}}$

$y(n) = 0.0730x(n) - 0.0730x(n-2) - 0.8541y(n-2)$

8.11. a. $H(z) = \dfrac{0.5677 + 0.5677z^{-1}}{1 + 0.1354z^{-1}}$

$y(n) = 0.5677x(n) + 0.5677x(n-1) - 0.1354y(n-1)$

8.13. a. $H(z) = \dfrac{0.1321 - 0.3964z^{-1} + 0.3964z^{-2} - 0.1321z^{-3}}{1 + 0.3432z^{-1} + 0.6044z^{-2} + 0.2041z^{-3}}$

$y(n) = 0.1321x(n) - 0.3964x(n-1) + 0.3964x(n-2) - 0.1321x(n-3)$
$\qquad - 0.3432y(n-1) - 0.6044y(n-2) - 0.2041y(n-3)$

8.15. a. $H(z) = \dfrac{0.9609 + 0.7354z^{-1} + 0.9609z^{-2}}{1 + 0.7354z^{-1} + 0.9217z^{-2}}$

$y(n) = 0.9609x(n) + 0.7354x(n-1) + 0.9609x(n-2)$
$\qquad -0.7354y(n-1) - 0.9217y(n-2)$

8.17. a. $H(z) = \dfrac{0.0242 + 0.0968z^{-1} + 0.1452z^{-2} + 0.0968z^{-3} + 0.0242z^{-4}}{1 - 1.5895z^{-1} + 1.6690z^{-2} - 0.9190z^{-3} + 0.2497z^{-4}}$

$y(n) = 0.0242x(n) + 0.0968x(n-1) + 0.1452x(n-2)$
$\qquad + 0.0968x(n-3) + 0.0242x(n-4) + 1.5895y(n-1) - 1.6690y(n-2)$
$\qquad + 0.9190y(n-3) - 0.2497y(n-4)$

8.19. a. $H(z) = \dfrac{1}{1 - 0.3679z^{-1}}$

$y(n) = x(n) + 0.3679y(n-1)$

8.21. a. $H(z) = \dfrac{0.1 - 0.09781z^{-1}}{1 - 1.6293z^{-1} + 0.6703z^{-2}}$

$y(n) = 0.1x(n) - 0.0978x(n-1) + 1.6293y(n-1) - 0.6703y(n-2)$

8.23. $H(z) = \dfrac{0.9320 - 1.3180z^{-1} + 0.9320z^{-2}}{1 - 1.3032z^{-1} + 0.8492z^{-2}}$

$y(n) = 0.9320x(n) - 1.3180x(n-1) + 0.9329x(n-2) + 1.3032y(n-1) - 0.8492y(n-2)$

8.25. $H(z) = \dfrac{0.9215 + 0.9215z^{-1}}{1 + 0.8429z^{-1}}$

$y(n) = 0.9215x(n) + 0.9215x(n-1) - 0.8429y(n-1)$

8.27. $H(z) = \dfrac{0.9607 - 0.9607z^{-1}}{1 - 0.9215z^{-1}}$

$y(n) = 0.9607x(n) - 0.9607x(n-1) + 0.9215y(n-1)$

8.29. a.

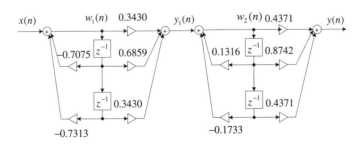

b. For section 1: $w_1(n) = x(n) - 0.7075w_1(n-1) - 0.7313w_1(n-2)$

$$y_1(n) = 0.3430w_1(n) + 0.6859w_1(n-1) + 0.3430w_1(n-2)$$

For section 2: $w_2(n) = y_1(n) + 0.1316w_2(n-1) - 0.1733w_2(n-2)$
$$y_2(n) = 0.4371w_2(n) + 0.8742w_2(n-1) + 0.4371w_2(n-2)$$

8.30. $H(z) = \dfrac{0.9511z^{-1}}{1.0000 - 0.6180z^{-1} + z^{-2}}$, $y(n) = 0.9511x(n-1) + 0.618y(n-1) - y(n-2)$

8.32. **a.** $H_{852}(z) = \dfrac{0.6203z^{-1}}{1 - 1.5687z^{-1} + z^{-2}}$, $H_{1477}(z) = \dfrac{0.9168z^{-1}}{1 - 0.7986z^{-1} + z^{-2}}$

b. $y_{852}(n) = 0.6203x(n-1) + 1.5678y_{852}(n-1) - y_{852}(n-2)$

$y_{1477}(n) = 0.9168x(n-1) + 0.7986y_{1477}(n-1) - y_{1477}(n-2)$

$y_9(n) = y_{1477}(n) + y_{852}(n)$

8.34. $X(0) = 2$, $|X(0)|^2 = 4$, $A_0 = 0.5$ (single side)

$X(1) = 1 - j3$, $|X(1)|^2 = 10$, $A_1 = 1.5811$ (single side)

8.36. $A_0 = 2.5$, $A_2 = 0.5$

8.39. Chebyshev notch filter 1: order $= 2$

$$H(z) = \frac{0.9915 - 1.9042z^{-1} + 0.9915z^{-2}}{1.0000 - 1.9042z^{-1} + 0.9830z^{-2}}$$

Chebyshev notch filter 2: order $= 2$

$$H(z) = \frac{0.9917 - 1.3117z^{-1} + 0.9917z^{-2}}{1.0000 - 1.3117z^{-1} + 0.9835z^{-2}}$$

8.41. Filter order $= 4$

$$H(z) = \frac{0.1103 + 0.4412z^{-1} + 0.6618z^{-2} + 0.4412z^{-3} + 0.1103z^{-4}}{1.0000 + 0.1509z^{-1} + 0.8041z^{-2} - 0.1619z^{-3} + 0.1872z^{-4}}$$

8.43. Filter order $= 4$

$$H(z) = \frac{0.0300 - 0.0599z^{-2} + 0.0300z^{-4}}{1.0000 - 0.6871z^{-1} + 1.5741z^{-2} - 0.5177z^{-3} + 0.5741z^{-4}}$$

8.45. $H(z) = \dfrac{0.5878z^{-1}}{1 - 1.6180z^{-1} + z^2}$

$y(n) = 0.5878x(n-1) + 1.6180y(n-1) - y(n-2)$

8.47. $X(0) = 1$, $|X(0)|^2 = 1$, $A_0 = 0.25$ (single side)

$X(1) = 1 - j2$; $|X(1)|^2 = 5$; $A_1 = 1.12$ (single side)

CHAPTER 9

9.1. 0.2560123 (decimal) $= 0.0100000110000101\,(Q - 15)$

9.2. $-0.2160123 \times 2^{15} = -7078_{10} = 1110010001011010$

-0.2160123 (decimal) $= 1.1100100010111010\,(Q - 15)$

9.5. $0.1010100010111110 = 0.6591186$

$1.0101011101000010 (Q - 15) = -0.6591186$

9.6. $0.0010001111101110 (Q - 15) = 0.1400756$

9.9. $1.1010101111000001 + 0.0100011111011010$

$= 1.1111001100110011$

9.13. **a.** $1101\ 011100001100$

b. $0100\ 101011001001$

9.15. $1101\ 011100011011$ (floating) $= 0.8881835 \times 2^{-3}$ (decimal)

$0100\ 101111100101$ (floating) $= -0.5131835 \times 2^{4}$ (decimal)

0.8881835×2^{-3} (decimal) $= 0.0069389 \times 2^{4}$ (decimal)

$= 0100\ 000000001110$ (floating)

$0100\ 101111100101$ (floating) $+ 0100\ 000000001110$ (floating)

$= 0100\ 101111110011$ (floating) $= -8.1016$ (decimal)

9.18. $(-1)^{1} \times 1.025 \times 2^{160-127} = -8.8047 \times 10^{9}$

9.20. $(-1)^{0} \times 1.625 \times 2^{1,536-1,023} = 4.3575 \times 10^{154}$

9.25. $B = 2,\ S = 2$

$$x_s(n) = \frac{x(n)}{2}$$

$$y_s(n) = -0.18x_s(n) + 0.8x_s(n-1) + 0.18x_s(n-2)$$

$$y(n) = 4y_s(n)$$

9.26. $S = 1,\ C = 2$

$$x_s(n) = x(n),\ y_s(n) = 0.75x_s(n) + 0.15y_f(n-1),\ y_f(n) = 2y_s(n),\ y(n) = y_f(n)$$

9.29. $S = 8,\ A = 2,\ B = 4$

$$x_s(n) = x(n)/8,\ w_s(n) = 0.5x_s(n) - 0.675w(n-1) - 0.25w(n-2),$$

$$w(n) = 2w_s(n)$$

$$y_s(n) = 0.18w(n) + 0.355w(n-1) + 0.18w(n-2),\ y(n) = 32y_s(n)$$

CHAPTER 10

10.1. $w^* = 2$ and $J_{min} = 10$

10.3. $w^* = -5$ and $J_{min} = 50$

10.5. $w^* \approx w_3 = 1.984$ and $J_{min} = 10.0026$

10.7. $w^* \approx w_3 = -4.992$ and $J_{min} = 5.0001$

10.9. **a.** $y(n) = w(0)x(n) + w(1)x(n-1)$

$e(n) = d(n) - y(n)$

$w(0) = w(0) + 0.2 \times e(n)x(n)$

$w(1) = w(1) + 0.2 \times e(n)x(n-1)$

b. For $n = 0$:

$$y(0) = 0$$

$$e(0) = 3$$

$$w(0) = 1.8$$

$$w(1) = 1$$

For $n = 1$:

$$y(1) = 1.2$$

$$e(1) = -3.2$$

$$w(0) = 2.44$$

$$w(1) = -0.92$$

For $n = 2$:

$$y(2) = 5.8$$

$$e(2) = -4.8$$

$$w(0) = 0.52$$

$$w(1) = 0.04$$

10.13. **a.** $n(n) = 0.5 \cdot x(n-5)$

b. $xx(n) = 5{,}000 \cdot \delta(n)$, $yy(n) = 0.7071xx(n-1) + 1.4141yy(n-1) - yy(n-2)$

c. $d(n) = yy(n) - n(n)$

d. For $i = 0, \cdots, 24$, $w(i) = 0$

$$y(n) = \sum_{i=0}^{24} w(i)x(n-i)$$

$$e(n) = d(n) - y(n)$$

For $i = 0, \cdots, 24$

$$w(i) = w(i) + 2\mu e(n)x(n - i)$$

10.15. a. $w(0) = w(1) = w(2) = 0$, $\mu = 0.1$
$y(n) = w(0)x(n) + w(1)x(n - 1) + w(2)x(n - 2)$
$e(n) = d(n) - y(n)$
$w(0) = w(0) + 0.2e(n)x(n)$
$w(1) = w(1) + 0.2e(n)x(n - 1)$
$w(2) = w(2) + 0.2e(n)x(n - 2)$
b. For $n = 0 : y(0) = 0$, $e(0) = 0$, $w(0) = 0$, $w(1) = 0$, $w(2) = 0$
For $n = 1 : y(1) = 0$, $e(1) = 2$, $w(0) = 0.4$, $w(1) = 0.4$, $w(2) = 0$
For $n = 2 : y(2) = 0$, $e(2) = -1$, $w(0) = 0.6$, $w(1) = 0.2$, $w(2) = -0.2$

10.17. a. $w(0) = w(1) = 0$, $\mu = 0.1$
$x(n) = d(n - 3)$
$y(n) = w(0)x(n) + w(1)x(n - 1)$
$e(n) = d(n) - y(n)$
$w(0) = w(0) + 0.2e(n)x(n)$
$w(1) = w(1) + 0.2e(n)x(n - 1)$
b. For $n = 0 : x(0) = 0$, $y(0) = 0$, $e(0) = -1$, $w(0) = 0$, $w(1) = 0$
For $n = 1 : x(1) = 0$, $y(1) = 0$, $e(1) = 1$, $w(0) = 0$, $w(1) = 0$
For $n = 2 : x(2) = 0$, $y(2) = 0$, $e(2) = -1$, $w(0) = 0$, $w(1) = 0$
For $n = 3 : x(3) = -1$, $y(3) = 0$, $e(3) = 1$, $w(0) = -0.2$, $w(1) = 0$
For $n = 4 : x(4) = 1$, $y(4) = -0.2$, $e(4) = -0.8$, $w(0) = -0.36$, $w(1) = 0.16$

10.18. a. 30 coefficients
10.20. For $i = 0, \cdots, 19$, $w(i) = 0$

$$y(n) = \sum_{i=0}^{19} w(i)x(n - i)$$

$$e(n) = d(n) - y(n)$$

For $i = 0, \cdots, 19$

$$w(i) = w(i) + 2\mu e(n)x(n - i)$$

CHAPTER 11

11.1. a. $\Delta = 0.714$
b. For $x = 1.6$ volts, binary code $= 110$, $x_q = 1.428$ volts, and $e_q = -0.172$ volts
For $x = -0.2$ volts, binary code $= 000$, $x_q = 0$ volts, and $e_q = 0.2$ volts

11.5. For $x = 1.6$ volts, binary code $= 111$, $x_q = 1.132$ volts, and $e_q = -0.468$ volts

For $x = -0.2$ volts, binary code $= 010$, $x_q = -0.224$ volts, and $e_q = -0.024$ volts

11.9. a. 0 0 0 1 0 1 0 1

b. 1 1 1 0 0 1 1 1

11.10. a. 0 0 0 0 0 0 0 0 0 1 1 1

b. 1 0 1 1 0 0 1 1 0 0 0 0

11.13. 010

001

010

11.14. For binary code $= 110$, $\hat{x}(0) = \tilde{x}(0) + d_q(0) = 0 + 5 = 5$

For binary code $= 100$, $\hat{x}(1) = \tilde{x}(1) + d_q(1) = 5 + 0 = 5$

For binary code $= 110$, $\hat{x}(2) = \tilde{x}(2) + d_q(2) = 5 + 2 = 7$

11.17. a. 1:1

b. 2:1

c. 4:1

11.18. a. 128 KBPS

b. 64 KBPS

c. 32 KBPS

11.21. a. 12 channels

b. 24 channels

c. 48 channels

11.22. $X_{DCT}(0) = 54$, $X_{DCT}(1) = 0.5412$, $X_{DCT}(2) = -4$, $X_{DCT}(3) = -1.3066$

11.25. $X_{DCT}(1) = 33.9730$, $X_{DCT}(3) = -10.4308$, $X_{DCT}(5) = 1.2001$, $X_{DCT}(7) = -1.6102$

11.26. a. Inverse DCT: 10.0845 6.3973 13.6027 −2.0845

b. Recovered inverse DCT: 11.3910 8.9385 15.0615 −3.3910

c. Quantization error: 1.3066 2.5412 1.4588 −1.3066

11.29. a. −9.0711 −0.5858

−13.3137 − 0.0000

−7.8995 0.5858

b. 3, 4, 5, 4

CHAPTER 12

12.3 a.

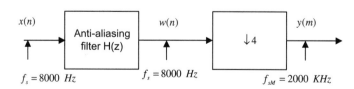

b. Hamming window, $N = 133$, $f_c = 900$ Hz

12.4. a.

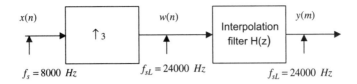

b. Hamming window, $N = 133$, $f_c = 3{,}700$ Hz

12.7. a.

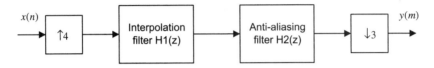

b. Combined filter $H(z)$: Hamming window, $N = 133$, $f_c = 2{,}700$ KHz

12.8. a.

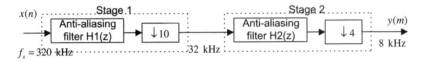

b. $M_1 = 10$ and $M_2 = 4$
c. Filter specification for $H_1(z)$: Hamming window, $N = 43$, $f_c = 15{,}700$ Hz
d. Filter specification for $H_2(z)$: Hamming window, $N = 177$, $f_c = 3{,}700$ Hz

12.9.

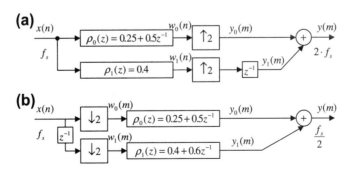

12.10.

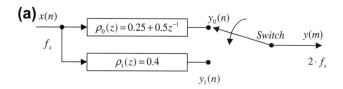

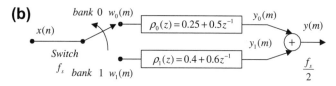

12.15. a. $f_s = 2f_{max}2^{2(n-m)} = 2 \times 15 \times 2^{2 \times (16-12)} = 7680$ KHz

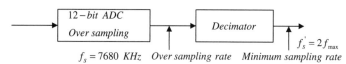

12.17. a.

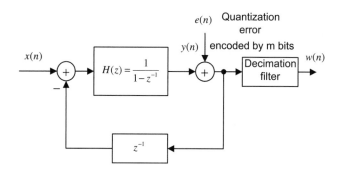

b. $n = 1 + 1.5 \times \log_2 \left(\dfrac{128}{2 \times 4} \right) - 0.86 \approx 6$ bits

12.18. a.

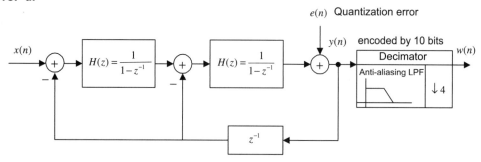

b. $n = m + 2.5 \times \log_2\left(\dfrac{f_s}{2f_{\max}}\right) - 2.14 = 10 + 2.5 \times \log_2\left(\dfrac{160}{2 \times 20}\right) - 2.14$

$= 12.86 \approx 13$ bits

12.21. a. $f_c/B = 6$ is an even number, which is case 1, so we select $f_s = 10$ kHz.

(a)

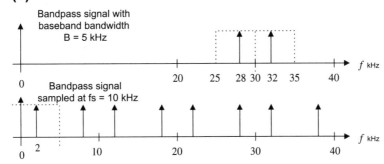

b. Since $f_c/B = 5$ is an odd number, we select $f_s = 10$ kHz.

(b)

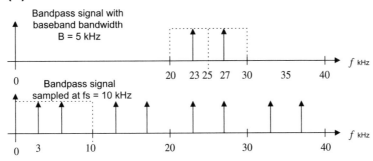

c. Now, $f_c/B = 6.6$, which is a noninteger. We extend the bandwidth to $\overline{B} = 5.5$ kHz, so $f_c/\overline{B} = 6$ and $f_s = 2\overline{B} = 11$ kHz.

(c)

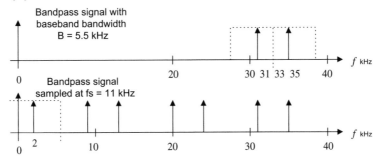

CHAPTER 13

13.1. a. $B = f_{sM}/2 = f_s/(2M)$, $f_c = 2(f_s/(2M)) = 2B$, $f_c/B = 2 = $ even

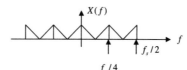

b. $B = f_{sM}/2 = f_s/(2M)$, $f_c = f_s/(2M) = B$, $f_c/B = 1 = $ odd

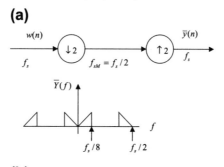

13.3. From Equation (13.7), $\bar{Y}(z) = \frac{1}{2}(W(z) + W(e^{-j\pi}z))$, $\bar{Y}(e^{j\Omega}) = \frac{1}{2}(W(e^{j\Omega}) + W(e^{j(\Omega-\pi)}))$, $W(e^{j(\Omega-\pi)})$ is the shifted version of $W(e^{j\Omega})$ by $f_s/2$.

(a)

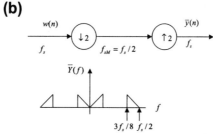

(b)

13.5. $H_1(z) = -\dfrac{1}{\sqrt{2}} + \dfrac{1}{\sqrt{2}}z^{-1}$, $G_0(z) = \dfrac{1}{\sqrt{2}} + \dfrac{1}{\sqrt{2}}z^{-1}$, $G_1(z) = \dfrac{1}{\sqrt{2}} - \dfrac{1}{\sqrt{2}}z^{-1}$

13.7. $H_1(z) = 0.129 + 0.224z^{-1} - 0.837z^{-2} + 0.483z^{-3}$

$G_0(z) = -0.129 + 0.224z^{-1} + 0.837z^{-2} + 0.483z^{-3}$

$G_1(z) = 0.483 - 0.837z^{-1} + 0.224z^{-2} + 0.129z^{-3}$

13.9.

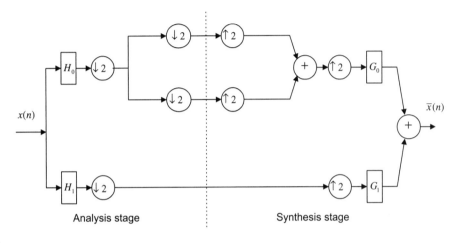

13.11.

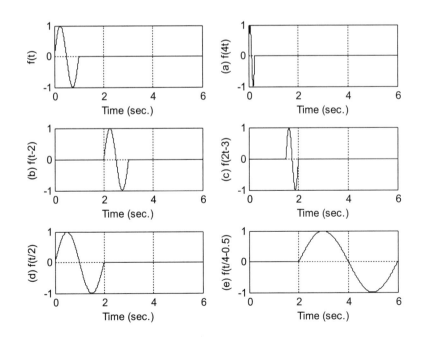

13.13.

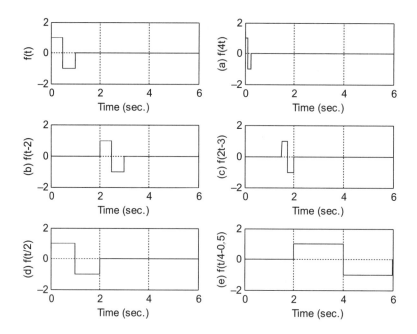

13.15.

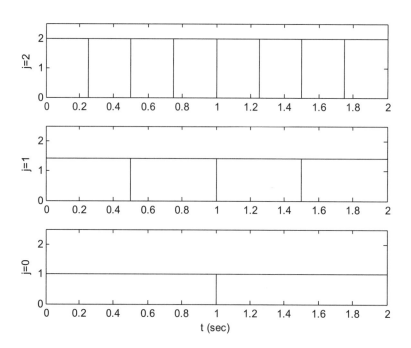

13.17. **a.** $f(t) = 4\phi(2t) - 2\phi(2t - 1)$
     b. $f(t) = \phi(t) + 3\psi(t)$

13.19. **a.** $f(t) = (2/\pi)\phi(2t) - (2/\pi)\phi(2t - 1)$
     b. $f(t) = (2/\pi)\psi(t)$

13.21. **a.** $\displaystyle\sum_{k=-\infty}^{\infty} \sqrt{2}h_0(k)\phi(4t - k) = \sqrt{2}h_0(0)\phi(4t) + \sqrt{2}h_0(1)\phi(4t - 1)$

     b. $= \sqrt{2} \times 0.707\phi(4t) + \sqrt{2} \times 0.707\phi(4t - 1) = \phi(4t) + \phi(4t - 1) = \phi(2t)$

        $\displaystyle\sum_{k=-\infty}^{\infty} \sqrt{2}h_1(k)\phi(4t - k) = \sqrt{2}h_1(0)\phi(4t) + \sqrt{2}h_1(1)\phi(4t - 1)$

        $= \sqrt{2} \times 0.707\phi(4t) + \sqrt{2}(-0.707)\phi(4t - 1) = \phi(4t) - \phi(4t - 1) = \psi(2t)$

13.23. $w(k) = [5.5000 \quad 0.5000 \quad 7.0711 \quad 2.1213]$

13.25. $c(k) = [2.2929 \quad 3.7071 \quad 2.4142 \quad -0.4142]$

13.27. $c(k) = [2.1213 \quad 3.5355 \quad 2.8284 \quad 0]$

CHAPTER 14

14.1. **a.** 76.8 K bytes
     b. 921.6 K bytes
     c. 1920.768 K bytes

14.3. $Y = 142, I = 54, Q = 11$

14.5.

$$\begin{bmatrix} 53 & 44 \\ 59 & 50 \end{bmatrix}$$

14.7.

$$\begin{bmatrix} 1 & 4 & 6 & 6 & 1 \\ 6 & 4 & 4 & 6 & 4 \\ 4 & 4 & 6 & 6 & 6 \\ 1 & 6 & 7 & 7 & 4 \end{bmatrix}$$

14.9.

$$\begin{bmatrix} 102 & 109 & 104 & 51 \\ 98 & 101 & 101 & 54 \\ 98 & 103 & 100 & 51 \\ 50 & 55 & 51 & 25 \end{bmatrix}$$

14.10.

$$
\begin{bmatrix}
0 & 100 & 100 & 0 \\
0 & 100 & 100 & 100 \\
0 & 100 & 100 & 100 \\
0 & 100 & 0 & 0
\end{bmatrix}
$$

14.11. a.

Vertical Sobel detector:
$$
\begin{bmatrix}
-1 & 0 & 1 \\
-2 & 0 & 2 \\
-1 & 0 & 1
\end{bmatrix}
$$
Processed image:
$$
\begin{bmatrix}
225 & 125 & 130 & 33 \\
249 & 119 & 136 & 6 \\
249 & 119 & 136 & 6 \\
255 & 125 & 130 & 0 \\
255 & 128 & 128 & 30
\end{bmatrix}
$$

b.

Laplacian edge detector:
$$
\begin{bmatrix}
0 & 1 & 0 \\
1 & -4 & 1 \\
0 & 1 & 0
\end{bmatrix}
$$
Processed image:
$$
\begin{bmatrix}
0 & 106 & 106 & 0 \\
106 & 255 & 255 & 106 \\
106 & 255 & 255 & 106 \\
117 & 223 & 223 & 117 \\
0 & 117 & 117 & 0
\end{bmatrix}
$$

14.13. Blue is dominant in the area pointed to by the arrow; red is dominant in the background.

14.15.

$$
X(u, v) = \begin{bmatrix} 460 & -40 \\ -240 & -140 \end{bmatrix} \text{ and } A(u, v) = \begin{bmatrix} 115 & 10 \\ 60 & 35 \end{bmatrix}
$$

14.16.

Forward DCT: $F(u, v) = \begin{bmatrix} 230 & -20 \\ -120 & -70 \end{bmatrix}$

14.17.

Inverse DCT: $p(i, j) = \begin{bmatrix} 110 & 100 \\ 100 & 90 \end{bmatrix}$

14.19. a. $(0, -2)$ $(3, 4)$, $(2, -3)$, $(0, 7)$, $(4, -2)$, $(0, 0)$
   b. $(0000, 0010, 01)$, $(0011, 0011, 100)$, $(0010, 0010, 00)$,
   $(0000, 0011, 111)$, $(0100, 0010, 01)$, $(0000, 0000)$

14.19. a. $(0, -2)$ $(3, 4)$, $(2, -3)$, $(0, 7)$, $(4, -2)$, $(0, 0)$
   b. $(0000, 0010, 01)$, $(0011, 0011, 100)$, $(0010, 0010, 00)$,
   $(0000, 0011, 111)$, $(0100, 0010, 01)$, $(0000, 0000)$

14.20. $w = 230.0000 - 20.0000$
$-120.0000 - 70.0000$

14.21. $f = 110.0000 \ 100.0000$
$100.0000 \ 90.0000$

14.23. $f = 115.0000 \quad 145.0000 \quad 25.0000 \quad 45.0000$

$\quad 105.0000 \quad 135.0000 \quad 5.0000 \quad 25.0000$

$\quad 30.0000 \quad 20.0000 \quad 7.5000 \quad 27.5000$

$\quad 10.0000 \quad -0.0000 \quad -7.5000 \quad 12.5000$

14.28. Hint :

$$Composite \times 2\sin(2\pi f_{sc}t) = Y \times 2\sin(2\pi f_{sc}t) + I\cos(2\pi f_{sc}t) \times 2\sin(2\pi f_{sc}t)$$
$$+ Q \times 2\sin^2(2\pi f_{sc}t) = Y \times 2\sin(2\pi f_{sc}t) + I\sin(2 \times 2\pi f_{sc}t)$$
$$+ Q - Q\cos(2 \times 2\pi f_{sc}t)$$

Then apply lowpass filtering.

14.35. $\dfrac{80 \times 80}{16 \times 16}(16^2 \times 32^2 \times 3) = 19.661 \times 10^6$ operations

---

APPENDIX B

B.1. $A_0 = 0.4, \ A_1 = 0.7916, \ A_2 = 0.7667, \ A_3 = 0.7263, \ A_4 = 0.6719$
$|c_0| = 0.4, \ |c_1| = |c_{-1}| = 0.3958,$

$|c_2| = |c_{-2}| = 0.3834, \ |c_3| = |c_{-3}| = 0.3632, \ |c_4| = |c_{-4}| = 0.3359$

B.3. $x(t) = 2 + 3.7420 \times \cos(2000\pi t) + 3.0273 \times \cos(4000\pi t)$
$+ 2.0182 \times \cos(6000\pi t) + 0.9355 \times \cos(8000\pi t) + \cdots$

$f_2 = 2000 \ \text{Hz}, \ A_2 = 3.0273$

B.5. $X(f) = 5\left(\dfrac{\sin \pi f}{\pi f}\right)^2$

B.7. **a.** $X(s) = 10$
b. $X(s) = -100/s^2$

c. $X(s) = \dfrac{10}{s+2}$

d. $X(s) = \dfrac{2e^{-5s}}{s}$

e. $X(s) = \dfrac{10s}{s^2+9}$

f. $X(s) = \dfrac{14.14 + 7.07s}{s^2 + 9}$

g. $X(s) = \dfrac{3(s + 2)}{(s + 2)^2 + 9}$

h. $X(s) = \dfrac{12,000}{s^6}$

B.9. a. $X(s) = \dfrac{7.5}{s(s + 1.5)}$

     b. $x(t) = 5u(t) - 5e^{-1.5t}u(t)$

B.11. a. zero: $s = 3$, poles: $s = -2$, $s = -2$, stable

       b. zeros: $s = 0$, $s = \pm 2.236j$, poles: $s = \pm 3j$, $s = -1 \pm 1.732j$, marginally stable

       c. zeros: $s = \pm j$, $s = -1$, poles: $s = 0$, $s = -3$, $s = -4$, $s = -8$, $s = 1$, unstable

B.13. a. $H(j\omega) = \dfrac{1}{\dfrac{j\omega}{5} + 1}$

       b. $A(\omega) = \dfrac{1}{\sqrt{1 + \left(\dfrac{\omega}{5}\right)^2}}$, $\beta(\omega) = \angle - \tan\left(\dfrac{\omega}{5}\right)$

       c. $Y(j2) = 4.6424 \angle - 21.80^0$ that is, $y_{ss}(t) = 4.6424 \sin(2t - 21.80^0)u(t)$

References

Ahmed, N., & Natarajan, T. (1983). *Discrete-Time Signals and Systems*. Englewood Cliffs, New Jersey: Prentice Hall.

Akansu, Ali N., & Haddad, Richard A. (1992). *Multiresolution Signal Decomposition: Transforms, Subbands, and Wavelets*. Boston: Academic Press, Inc.

Ambardar, A. (1999). *Analog and Digital Signal Processing* (second ed.). Pacific Grove, CA: Brooks/Cole Publishing Company, ITP.

Alkin, O. (1993). *Digital Signal Processing: A Laboratory Approach Using PC-DSP*. Englewood Cliffs, New Jersey: Prentice Hall.

Brandenburg, K. (1997). Overview of MPEG Audio: Current and Future Standards for Low-Bit-Rate Audio Coding. *Journal of Audio Engineering Society, 45*(1/2), 4–21.

Carr, J. J., & Brown, J. M. (2001). *Introduction to Biomedical Equipment Technology* (fourth ed.). Upper Saddle River, New Jersey: Prentice Hall.

Chassaing, R., & Reay, D. (2008). *Digital Signal Processing and Applications with the TMS320C6713 and TMS320C6416 DSK* (second ed.). Hoboken, New Jersey: John Wiley & Sons.

Chen, W. (1986). *Passive and Active Filters-Theory and Implementations*. New York: John Wiley & Sons.

Dahnoun, N. (2000). *Digital Signal Processing Implementation Using the TMS320C6000TM DSP Platform*. Englewood Cliffs, New Jersey: Prentice Hall.

Deller, J. R., Proakis, J. G., & Hansen, J. H. L. (1993). *Discrete-Time Processing of Speech Signals*. New York: Macmillian Publishing Company.

El-Sharkawy, M. (1996). *Digital Signal Processing Applications with Motorola's DSP56002 Processor*. Upper Saddle River, New Jersey: Prentice Hall.

Embree, P. M. (1995). *C Algorithms for Real-Time DSP*. Upper Saddle River, New Jersey: Prentice Hall.

Gonzalez, R. C., & Wintz, P. (1987). *Digital Image Processing* (second ed.). Reading, Massachusetts: Addison-Wesley Publishing Company.

Grover, D., & Deller, J. R. (1998). *Digital Signal Processing and the Microcontroller*. Upper Saddle River, New Jersey: Prentice Hall.

Haykin, S. (1991). *Adaptive Filter Theory* (second ed.). Englewood Cliffs, New Jersey: Prentice Hall.

Ifeachor, E. C., & Jervis, B. W. (2002). *Digital Signal Processing: A Practical Approach* (second ed.). Upper Saddle River, New Jersey: Prentice Hall.

Jiang, J., & Tan, L. (June 2012). *Teaching Adaptive Filters and Applications in Electrical and Computer Engineering Technology Program*. San Antonio, Texas: 2012 Proceedings of the American Society for Engineering Education.

Kehtaranavaz, N., & Simsek, B. (2000). *C6x-Based Digital Signal Processing*. Upper Saddle River, New Jersey: Prentice Hall.

Krauss, T. P., Shure, L., & Little, J. N. (1994). *Signal Processing TOOLBOX for Use with MATLAB*. Natick, Massachusetts: The Math Works Inc.

Kuo, S. M., & Morgan, D. R. (1996). *Active Noise Control Systems: Algorithms and DSP Implementations*. New York: John Wiley & Sons.

Li, Z. N., & Drew, M. S. (2004). *Fundamentals of Multimedia*. Upper Saddle River, New Jersey: Prentice Hall.

Lipshiz, S. P., Wannamaker, R. A., & Vanderkooy, J. (May 1992). Quantization and Dither: A Theoretical Survey. *Journal of Audio Engineering Society, 40*(5), 355–375.

Lynn, P. A., & Fuerst, W. (1999). *Introductory Digital Signal Processing with Computer Applications* (second ed.). Chichester and New York: John Wiley & Sons.

Maher, Robert C. (January/February 1992). On the Nature of Granulation Noise in Uniform Quantization Systems. *Journal of Audio Engineering Society, 40*(1/2), 12–20.

McClellan, J. H., Oppenheim, A. V., Shafer, R. W., Burrus, C. S., Parks, T. W., & Schuessler, H. (1998). *Computer Based Exercises for Signal Processing Using MATLAB 5*. Upper Saddle River, New Jersey: Prentice Hall.

McClellan, J. H., Shafer, R. W., & Yoder, M. A. (1998). *DSP First – A Multimedia Approach*. Upper Saddle River, New Jersey: Prentice Hall.

Nelson, M. (1992). *The Data Compression Book*. Redwood City, California: M&T Publishing, Inc.

Oppenheim, A. V., & Shafer, R. W. (1975). *Discrete-Time Signal Processing*. Englewood Cliffs, New Jersey: Prentice Hall.

Oppenheim, A. V., Shafer, R. W., & Buck, J. R. (1998). *Discrete-Time Signal Processing* (second ed.). Upper Saddle River, New Jersey: Prentice Hall.

Pan, David (1995). A Tutorial on MPEG/Audio Compression. *IEEE Multimedia, 2*(2), 60–74.

Phillips, C. L., & Harbor, R. D. (1991). *Feedback Control Systems* (second ed.). Englewood Cliffs, New Jersey: Prentice Hall.

Phillips, C. L., & Nagle, H. T. (1995). *Digital Control System Analysis and Design* (third ed.). Englewood Cliffs, New Jersey: Prentice Hall.

Porat, B. (1997). *A Course in Digital Signal Processing*. New York: John Wiley & Sons.

Princen, J., & Bradley, A. B. (October 1986). Analysis/Synthesis Filter Bank Design Based on Time Domain Aliasing Cancellation. *IEEE Transactions on Acoustics, Speech, and Signal Processing, Vol. ASSP 34, No. 5*, 1153–1161.

Proakis, J. G., & Manolakis, D. G. (1996). *Digital Signal Processing: Principles, Algorithms, and Applications* (third ed.). Upper Saddle River, New Jersey: Prentice Hall.

Rabbani, M., & Jones, P. W. (1991). *Digital Image Compression Techniques*. Bellingham, Washington: The Society of Photo-Optical Instrumentation Engineers (SPIE).

Rabiner, L. R., & Schafer, R. W. (1978). *Digital Processing of Speech Signals*. Englewood Cliffs, New Jersey: Prentice Hall.

Randall, R. B. (2011). *Vibration-Based Conditioning Monitoring: Industrial, Aerospace, and Automotive Applications*. Chichester: John Wiley & Sons.

Roddy, D., & Coolen, J. (1997). *Electronic Communications* (fourth ed.). Englewood Cliffs, New Jersey: Prentice Hall.

Sayood, K. (2000). *Introduction to Data Compression* (second ed.). San Francisco: Academic Press.

Smith, M. J. T., & Barnwell, T. P. (1984). A Procedure for Designing Exact Reconstruction Filter Banks for Tree-Structured Sub-band Coders. *San Diego, California: Proceedings of the IEEE International Conference on Acoustics, Speech, and Signal Processing*, 27.1.1–27.1.4.

Soliman, S. S., & Srinath, M. D. (1998). *Continuous and Discrete Signals and Systems* (second ed.). Upper Saddle River, New Jersey: Prentice Hall.

Sorensen, H. V., & Chen, J. P. (1997). *A Digital Signal Processing Laboratory Using TMS320C30*. Upper Saddle River, New Jersey: Prentice Hall.

Stanley, W. D. (2003). *Network Analysis with Applications* (fourth ed.). Upper Saddle River, New Jersey: Prentice Hall.

Stearns, S. D., & David, R. A. (1996). *Signal Processing Algorithms in MATLAB*. Upper Saddle River, New Jersey: Prentice Hall.

Stearns, S. D., & Hush, D. R. (1990). *Digital Signal Analysis* (second ed.). Englewood Cliff, New Jersey: Prentice Hall.

Stearns, S. D. (2003). *Digital Signal Processing with Examples in MATLAB*. Boca Raton, Florida: CRC Press LLC.

Tan, L., & Jiang, J. (June 2010). *Improving Digital Signal Processing Course with Real-Time Processing Experiences for Electrical and Computer Engineering Technology Students*. Louisville, Kentucky: 2010 Proceedings of the American Society for Engineering Education.

Tan, L., & Wang, L. (November 2011). Oversampling Technique for Obtaining Higher-Order Derivative of Low-Frequency Signals. *IEEE Transactions on Instrumentation and Measurement, 60*(11), 3677–3684.

Tan, L., & Jiang, J. (Fall 2008). A Simple DSP Project for Teaching Real-time Signal Rate Conversions. *The Technology Interface Journal, 9*(1).

Tan, L., & Jiang, J. (2012). Novel Adaptive IIR Notch Filters for Frequency Estimation and Tracking, Chapter 20. *Streamlining Digital Signal Processing: A Tricks of the Trade Guidebook*. Edited by Richard G. Lyons, Hoboken, New Jersey: IEEE Press/Wiley & Sons, 197–205.

Tomasi, W. (2004). *Electronic Communications Systems: Fundamentals Through Advanced* (fifth ed.). Upper Saddle River, New Jersey: Prentice Hall.

Texas Instruments. (1991). *TMS320C3x User's Guide*. Dallas, Texas: Texas Instruments.

Texas Instruments. (1998). *TMS320C6x CPU and Instruction Set Reference Guide, Literature ID# SPRU 189C*. Dallas, Texas: Texas Instruments.

Texas Instruments. (2001). *Code Composer Studio: Getting Started Guide*. Dallas, Texas: Texas Instruments.

Vegte, J. (2002). *Fundamentals of Digital Signal Processing*. Upper Saddle River, New Jersey: Prentice Hall.

Vetterli, M., & Kovacevic, J. (1995). *Wavelets and Subband Coding*. Englewood Cliffs, New Jersey: Prentice Hall PTR.

Webster, J. G. (1998). *Medical Instrumentation: Application and Design* (third ed.). New York: John Wiley & Sons, Inc.

Yost, William A. (1997). *Fundamentals of Hearing: An Introduction* (third ed.). San Diego, California: Academic Press.

Index

Note: Page Numbers with "f" denote figures; "t" tables

A

Adaptive filters
  adaptive echo cancellers, 479–480, 479f
  electrocardiography interference cancellation,
    476–478, 477f, 478f
  least mean square adaptive FIR filters
    corrupted signal and noise reference, 455–457, 456f,
      456t
    desired signal spectrum, 453, 454f
    noise canceller, 454, 454f
    one-tap FIR filter, 455, 457
  linear prediction
    line enhancement, 473–475, 473f, 474f, 475f
    periodic interference cancellation, 476, 476f
  long-distance telephone circuit, 479, 479f
  noise cancellation, *see* Noise cancellation
  system modeling, 468, 468f
    MATLAB program, 470
    spectrum for, 470, 472f
    unknown system's frequency responses, 469, 470f
    waveforms for, 470, 471f
  TMS320C6713 DSK, *see* TMS320C6713 DSK
  Wiener filter theory
    autocorrelation and cross-correlation, 459
    LMS algorithm, 461–462
    mean square error quadratic function, 457–458,
      458f
    noise cancellation, 457, 457f
    statistical expectation, 457–458, 461–462
    steepest descent algorithm, 459, 460f, 461–462
Adaptive differential pulse code modulation (ADPCM)
  decoder, 512–514, 513f
  discrete function, 514–515, 515t
  encoder, 512–514, 513f
  FIR filter, 516–517
  input and output characteristics, 514, 514t
  16-level nonuniform adaptive quantizer, 514
  performance measurement, 518
  predictor z-transfer function, 516–517
  scale factor, 514–515
  speech samples, 517–518, 517f
ADC, *see* Analog-to-digital conversion (ADC)
Address generators
  circular buffering, 409, 410f
    equivalent FIFO, 410f, 411
  FIR filtering, 411
Aliasing level, 28

Amplitude modulation (AM), 603, 603f
Amplitude spectrum
  DFT, 87, 88f, 97–101
  Fourier series, 89, 89f
Analog-to-digital conversion (ADC)
  binary codes, 40, 41f
  2-bit flash ADC unit, 36, 36f
  implementation, 35–36
  oversampling, 587, 587f
    benefits of, 586
    continuous *vs.* regular sampled *vs.* oversampled signal
      amplitudes, 588–589, 591f
    frequency response, 588–589, 589f
    in-band frequency range, 587
    MATLAB program, 589
    oversampling ratio, definition, 585–586
    quantization noise power, 586–588
    regular ADC system, 586–587, 586f
    time *vs.* frequency domains, 588–589,
      590f
  quantization
    bipolar quantizer, 38–40, 39f, 40t
    definition, 35, 36f
    error, 37
    notations and rules, 38
    process, 37
    SNR, 47
    unipolar quantizer, 38, 38f, 39t
  SDM ADC, *see* Sigma-delta modulation analog-
    to-digital conversion (SDM ADC)
Analog filters
  lowpass prototype transformation, 305,
    306t
    bandpass filter, 305, 306f
    bandstop filter, 305, 306f
    cutoff frequency, 304–305
    highpass filter, 305, 305f
    lowpass filter, 304, 304f
    magnitude response, 304–305
    MATLAB function, 307
  steady-state frequency response, 178
Analog μ-law companding
  characteristics, 502, 502f
  compressor, 501, 501f
  expander, 501–502, 501f
  original speech data, 504, 505f
  quantization error, 501

Analog signal processing
  convolution, 798–799
  Fourier series
    amplitude and phase spectrum, 779
    amplitude-phase form, 776
    complex exponential form, 776–782, 784t
    Fourier transform, 786–791, 789t, 790t
    rectangular waveform, 780, 780f
    sine-cosine form, 775, 780, 783t
  Laplace transform
    differential equations, 793–794
    and table, 791–793, 791t
    transfer function, 794–795
  poles, zeros and stability, 795–796
  sinusoidal steady-state response, 799–804
Analog system program, 482
Analog video
  "back porch", 748
  electrical signal demodulation, 748–750, 749f
  frame via row-wise scanning, 747
  frequency modulation, 751
  interlaced raster scanning, 747, 748f
  NTSC TV standard, 750, 751f
  PAL system, 752
  QAM, 751–752
  SECAM system, 752, 752t
  vertical synchronization, 749f, 750
  video data, retrace and sync layout, 750, 750f
  video-modulated waveform, 747, 748f
Analysis filter
  channel 0, 622, 623f
  channel 1, 622, 623f
  channel 2, 622–624, 624f
  channel 3, 624, 625f
  4-channel filter bank, 621–622, 622f
Anti-aliasing filter
  aliasing level percentage, 28
  Butterworth magnitude frequency response,
    25–26
  Sallen-Key lowpass filter, 26–27, 26f
  sampled analog signal, 25, 26f
Anti-image filter
  DAC unit, 30–31, 30f
  sample-and-hold effect
    digital equalizer, 32, 33f
    and distortion percentage, 31f, 32
    lowpass filtering effect, 30–31, 31f
    shaping effect, 32, 33f
  transfer function, 30–31
Application-specific integrated circuit (ASIC), 412
Auxiliary register arithmetic units (ARAUs),
  428–429

B

Bandpass filters
  amplitude spectra, 203, 205f
  analog lowpass filters, 321, 322f
  design specifications, 389
  digital Chebyshev lowpass prototype functions, 321, 322f,
    331–337
  digital fourth-order bandpass Butterworth filter, 203
  frequency responses, 203, 203f
  lowpass prototype transformation, 305, 306f
  MATLAB program, 204
  normalized filter, 187, 187f
  original and filtered speech plots, 203, 204f
  second-order bandpass filter, 352–354, 353f
Bandpass signals, undersampling, 603–608, 603f, 604f,
  605f, 606f
Bandstop filters
  digital Butterworth lowpass prototype functions,
    331–337
  digital Chebyshev lowpass prototype functions,
    331–337
  lowpass prototype transformation, 305, 306f
  normalized filter, 188, 188f
Bartlett window, 230, 231f
Bilinear transformation (BLT) design method, 391
  analog filters, lowpass prototype transformation,
    305, 306t
    bandpass filter, 305, 306f
    bandstop filter, 305, 306f
    cutoff frequency, 304–305
    highpass filter, 305, 305f
    lowpass filter, 304, 304f
    magnitude response, 304–305
    MATLAB function, 307
  design procedure, 303–304, 303f, 314–318, 316t
  frequency warping, 312, 313f
    digital frequency, 312
    digital integration method, 308, 309f
    graphical representation, 313, 314f
    Laplace transfer function, 309–310
    mapping properties, 310, 310f
    s-plane vs. z-plane, frequency mapping, 312, 312f
    z-transform, 309–310
Bipolar quantizer, 38–40, 39f, 40t
Blackman window, 230, 231f
BLT design method, see Bilinear transformation (BLT)
    design method
Bounded-in and bounded-out (BIBO) stability,
  71–72, 71f
Butterworth filters, 338–340, 338t, 339t
  see also Digital Butterworth lowpass prototype functions
Butterworth magnitude frequency response, 25–26

C

Cascade (series) realization method, 192, 195, 196f
Causal system, 66–67
CD recording system, *see* Compact-disc (CD) recording
      system
Chebyshev filters, 388–389
      see also Digital Chebyshev lowpass prototype functions
Chebyshev polynomial approximation, 269
Chrominance channels, 688–689
Circular convolution
      forward filter coefficients, 670
      reversed filter coefficients, 668
Comité Consultatif International Téléphonique et
      Télégraphique (CCITT), 754
Compact-disc (CD) recording system
      decoder, 8f, 9
      encoder, 7–9, 8f
Companding
      analog μ-law companding, *see* Analog μ-law companding
      digital μ-law companding, *see* Digital μ-law companding
Component video, 746
Composite video, 746
Compression, *see* Discrete cosine transform (DCT)
      see also image compression
Conjugate quadrature filter (CQF), 630
Continuous wavelet transform (CWT), 638, 641
Convolution, 72–80, 798–799
      impulse response, 69
      linear, 142–143
Cyclic redundancy check (CRC) code, 526

D

DAC, *see* Digital-to-analog conversion (DAC)
DCT, *see* Discrete cosine transform (DCT)
Decimation, 556
Decimation filter, 581, 581f, 582t
      commutative model, 582–583, 583f
      filter bank coefficients, 582
      implementation, 582, 582f, 584
      three-tap decimation filter, 581
Decimation-in-frequency method
      bit reversal process, 126, 127f
      eight-point FFT
            12 complex multiplications, 124–126, 125f
            first iteration, 123–124, 125f
            inverse of, 127–128, 127f
            second iteration, 124, 125f
      graphical operations, 123–124, 125f
      index mapping for, 126, 126t
      inverse FFT, definition, 126
      twiddle factor, 123–124
Decimation-in-time method

eight-point FFT algorithm, 128–129, 130f
eight-point IFFT, 129–131, 131f
first iteration, 128–129, 130f
frequency bins, 128–129
second iteration, 128–129, 130f
Decomposition, *see* Two-channel perfect reconstruction
      quadrature mirror filter bank
Delta modulation (DM), 511
Denoise, 668, 670f
DFT, *see* Discrete Fourier transform (DFT)
Difference equation, 67–68, 79
      DSP system, input and output, 162, 162f
      filter() function, 165
      filtic() function, 165
      nonzero initial conditions, 165
      transfer function
            impulse response, 169
            step response, 169
            system response, 169–172
            z-transfer function, 166–167, 166f
      zero initial conditions, 165
Differential pulse code modulation (DPCM)
      3-bit quantizer, 509, 510t
      direct-current coefficients, 736
      encoder and decoder, 509, 509f
      quantization step size, 512
Digital-to-analog conversion (DAC), 47
      anti-image filter and equalizer, 30–31, 30f
      process, 40, 41f
      quantization error, 40–42
      quantization noise, 42
      quantized *vs.* original signal, 44f
      R-2R ladder DAC, 36–37, 37f
      SNR, 42
Digital audio equalizer, 341f
      audio spectrum, 343–344, 343f
      audio test signal, 343–344
      filter banks design, 342, 342t
      magnitude frequency responses, 342, 342f
      MATLAB program, 344
      specifications for, 341, 341t
Digital Butterworth lowpass prototype functions, 318, 319t
      magnitude response function, 318, 320f
      prototype filter order, 318–320
Digital Chebyshev lowpass prototype functions, 318, 319t,
      320t
      analog filter specification conversion, 321, 322t
      analog lowpass and bandpass filters, 321, 322f
      lowpass prototype order, 321
      magnitude response function, 320–321, 321f
Digital convolution, 72–80
Digital crossover design

Digital crossover design (*Continued*)
  lowpass and highpass filters
    impulse responses, 260, 261f
    magnitude frequency responses, 260, 260f, 261f
  speaker drivers, 258–259, 259f
  specifications, 259–260
Digital filtering system
  analog filter steady-state frequency response, 178
  difference equation, *see* Difference equation
  digital filters, *see* Digital filters
  Euler's formula, 179
  FIR and IIR systems, 186
  frequency response properties, 180
  inverse z-transform, 179
  magnitude frequency response, 178, 181
  normalized digital frequency, 178
  signal enhancement
    biomedical signals, 199
    ECG signal, notch filtering, 205–206, 206f, 207f, 208f
    speech signals, *see* Speech signals
  sinusoidal inputs, system response, 180, 181f
  steady-state frequency responses, 178, 178f, 180
  system transient response, 178, 178f
  types
    freqz() function, 188
    normalized bandpass filter, 187, 187f
    normalized bandstop filter, 188, 188f
    normalized highpass filter, 187, 187f
    normalized lowpass filter, 186f, 187
    passband, stopband and transition band, 186
  z-plane pole-zero plot, 172f
    analog-to-digital conversion, 174
    bounded-in/bounded-out stability, 175
    features, 172
    Laplace shift property, 174
    Laplace *vs.* z-transform, 173, 174f
    s-plane *vs.* z-plane mapping, 175, 175f
    stability rules, 175, 176f
Digital filters
  cascade (series) realization method, 192, 195, 196f
  direct-form II realization method, 192–195, 195f
  direct-form I realization method, 192–193, 194f
  parallel realization method, 192, 196–199, 196f
  sinusoidal steady-state response, 813f
    inverse z-transform, 814
    magnitude and phase response, 814
    properties of, 815–816
    z-transform output, 813
Digital μ-law companding
  8-bit compressed PCM code format, 505–506, 506t, 508–509, 508f
  characteristics, 505, 506f

  compressor and expander, 504, 505f
  decoding table, 506–508, 507t
  encoding table, 505–506, 507t
Digital signal processing (DSP), 1, 2f
  aliasing distortion, 2
  analog input signal, 2
  audio signals and spectrums, 3, 5f
  digital filtering, 3, 3f, 4f
  DS processor, 2
  real-world applications, 12, 12t
    CD recording system, 7–9, 8f
    data compressor, 7, 8f
    data expander, 7, 8f
    digital image enhancement, 9–12, 12f
    interference cancellation, electrocardiography, 5–7, 7f
    software audio players, 9
    two-band digital crossover, 5, 6f
    vibration signature analysis, 9, 10f, 11f
  signal frequency (spectrum) analysis, 3, 4f
  speech samples and spectrum, 4, 6f
Digital signal (DS) processor
  adder output, 429
  ASIC, 412
  features, 406, 411
  FIR filter, direct-form I implementation, 430, 430f
  fixed-point format, 411–412
    3-bit 2's complement number, 412–413, 412t, 413t
    computational units, 427
    C program, 445–446, 446t, 447f, 448f
    fractional binary 2's complement system, 414
    program control unit, 427
    Q-30 format, 418, 418f
    Q-format number, 415, 415f, 418
    TMSC320C54x family architecture, 426–427, 426f
  floating-point format, 411–412, 419, 419f
    advantages, 427
    ARAUs, 428–429
    C programs, 445, 445f
    IEEE format, 423–426, 423f, 425f
    overflow, 422
    rules for, 420
    speech quality applications, 429
    TMS320C3x processor, 427–428, 428f
    underflow, 423
  hardware units
    address generators, *see* Address generators
    MAC, 408–409, 409f
    shifters, 409
  Harvard architecture, 407, 407f
    execution cycle, 407, 408f
    pipelining operation, 408

IIR filter
    direct-form II implementation, 432, 432f
    transfer function, 433
linear buffering, *see* Linear buffering
manufactures, 411
real-time processing
    input and output sample clock, 438, 439f
    program segment, 438, 440f
    TMS320C6713 DSK setup, 438, 440f
scale factor, 429
second-order section filters, 434
TMS320C67x DSK, 436f, 437
    C6713 DSK board, 434–436, 435f
    memory and internal buses, 438
    peripherals, 438
    registers of, 437, 437f
    software tool, 438
    Texas Instruments Veloci™ architecture, 437
    TMS320C6713 DSK, 438, 439f
Von Neumann architecture, 406, 406f
    applications, 408
    execution cycles, 407, 408f
    opcode and operand, 406
Digital signals
    BIBO stability, 71–72, 71f, 80
    causal system, 66–67
    difference equation format, 67–68, 79
    digital convolution, 72–80
    digital samples, 58, 58f
    digital sequences, 61, 61f
        analog signal function, 62, 79
        exponential function, 60, 60f, 61t
        sampling rate, 61
        shifted unit-impulse and unit-step sequences, 59, 59f
        sinusoidal function, 60, 60f, 60t
        unit-impulse sequence, 58–59, 59f
        unit-step sequence, 59, 59f, 62
    DS processor, 58
    impulse response
        digital convolution sum, 69
        FIR system, 69
        IIR system, 71
        unit-impulse response, 68, 68f
    linear system, 63–65, 64f
    notation of, 57–58, 58f
    time-invariant system, 65–66, 65f
Direct-form II realization method, 192–195, 195f
Direct-form I realization method, 192–193, 194f
Discrete cosine transform (DCT), 519–522, 524–525, 525f
    coefficients, 731, 732t

scan order, 732, 733t
    image compression
        2D-DCT, 729–731
        JPEG image compression, *see* JPEG image compression
        lossless/lossy compression, 728
        principle of, 729
        wavelet transform, *see* Wavelet transform
Discrete Fourier transform (DFT), 625
    amplitude spectrum, 87, 88f, 97–101
    data window time, 97–101
    definition, 88
    FFT
        applications of, 97, 97f
        data sequence, 101–102
        decimation-in-frequency method, *see* Decimation-in-frequency method
        decimation-in-time method, *see* Decimation-in-time method
        digital sequence sample, 123
        interpolated spectrum, 102–103
        zero-padding effect, 102–103, 102f
    fft() and ifft() MATLAB functions, 93, 93t
    formula development, 91, 92f
    Fourier series, 132
    see also Fourier series
        amplitude spectral components, 90
        coefficients, 88–89
        periodic digital signal, 88, 89f
        two-side line amplitude spectrum, 89, 89f
    frequency bin, 95
    frequency resolution, 96–101
    inverse of, 93
    phase spectrum, 97–101
    power spectrum, 97–101
    signal amplitude *vs.* sampling time instant, 87
    spectral estimation, window functions
        Hamming window, 109–110, 111f
            see also Hamming window function
        Hanning window, 109–110, 111f
        periodic, continuous and band limited data, 107, 107f
        rectangular window, 109–110, 111f
        signal samples and spectra, 107–108, 108f
        spectral leakage, 108
        triangular window, 109–110, 111f
        window operation, 108–109, 109f, 110f
    twiddle factor, 92–93
Discrete wavelet transform (DWT)
    discrete time function, 656–657
    dwt() function, 671
    dyadic subband coding structure, 657, 658f
    IDWT, 656

Discrete wavelet transform (DWT) (*Continued*)
   idwt() function, 671
  lowpass and highpass filter coefficients, 656
  signal amplitude, 657
  4-tap Daubechies filters, frequency response, 656, 657f
  time-frequency plane, 661–664, 662f
  time-frequency plot, 661–664, 661f
  wavelet coefficients, 656
  wavelet expansion, 655
Downsampling, 557f
  data sequence, 556
  definition, 556
  MATLAB program, 559, 609
  normalized stop frequency edge, 556–557
  Nyquist sampling theorem, 556
  spectral plots, 556–557, 558f
  TMS320C6713 DSK, 608, 612f
  using anti-aliasing filter, spectral plots, 558–559, 560f
  without using anti-aliasing filter, spectral plots, 558, 559f
  z-transform, 556–557
DPCM, *see* Differential pulse code modulation (DPCM)
DSP, *see* Digital signal processing (DSP)
Dual-tone multifrequency (DTMF) tone generator, 442
  Goertzel algorithm, 392
    advantages, 380
    DFT algorithm, 377
    DFT coefficient, 378–379
    Euler's identity, 378–379
    MATLAB function, 382
    modified second-order Goertzel IIR filter,
      380–381, 381f
    second-order Goertzel IIR filter, 310–311, 377
    transfer function, 377
  MATLAB program, 377
  modified Goertzel algorithm, 384f
  see also Modified Goertzel algorithm
    ASCII code, 385–386
    design principles, 383
    frequency bins, 383, 384t
    MATLAB simulation, 385–386, 386f
  telephone touch keypads, 373–375, 373f, 376f
DWT, *see* Discrete wavelet transform (DWT)

E

Echo cancellation, 479–480, 479f
Edge detection, 717, 718f
  differential convolution kernel, 715–716
  grayscale image, 717, 719f
  horizontal Sobel edge detector, 716
  Laplacian edge detector, 716–717
  Laplacian of Gaussian filter, 717, 719f
  MATLAB functions, 718–721, 720f
  vertical Sobel edge detector, 716
Electrocardiography (ECG)
  60-Hz hum eliminator and heart rate detection, 392
    cascaded frequency responses, 365, 366f
    characteristics of, 362, 363f
    design specifications, 364–365
    harmonics, 364
    heart rate, definition, 367–368
    MATLAB program, 368
    QRS complex, 362–364
    signal enhancement system, 364, 364f
    signal processing results, 366, 367f
    signal spectrum, 362, 363f
    transfer function and difference equation, 365
    zero-crossing algorithm, 366–367, 368f
  interference cancellation, 476–478, 477f, 478f
Equalizer, *see* Anti-image filter
Euler's identity, 378–379
Exponent, floating-point format, 419

F

Fast Fourier transform (FFT), 3–4
  applications of, 97, 97f
  data sequence, 101–102
  decimation-in-frequency method, *see* Decimation-in-
    frequency method
  decimation-in-time method, *see* Decimation-in-time
    method
  digital sequence sample, 123
  interpolated spectrum, 102–103
  zero-padding effect, 102–103, 102f
Father wavelet, 642, 644f
fconv() function, 670
fft() and ifft() MATLAB functions, 93, 93t
Finite impulse response (FIR) filter design, 69, 286t, 287
  coefficient accuracy effects, 282–285
  Fourier transform design, 221t, 222, 290
    coefficient symmetry, 220
    desired impulse response, 220, 221f
    Fourier coefficients, 219
    Gibbs oscillatory behavior, 224, 229
    ideal lowpass filter, 219, 219f
    ideal lowpass frequency response, 219, 219f
    linear phase response, 223, 224f, 226–227, 226f, 227f
    magnitude and phase frequency responses, 224, 225f
    nonlinear phase response, 226, 227f
    periodic frequency response, 219
    sinusoidal sequence, 225
    symmetric coefficients, 224–225
    17-tap FIR lowpass filter coefficients, 224, 225t
    z-transfer function, 220
  frequency sampling, 286, 817–820

design procedure, 263
desired filter frequency response, 262, 262f
DFT, 262
features, 262
IDFT, 262
magnitude frequency response, 269
input-output relationship, 217
linear phase form, 281, 282f
noise reduction
clean signal and spectrum, 254, 255f
data acquisition process, 253
MATLAB program, 255
noise signal and spectrum, 254, 254f
passband frequency range, 254
speech noise reduction, 256–257, 256f, 257f
stopband frequency range, 254
vibration signals, 257–258, 258f, 259f
optimal design method, 286–287
see also Parks-McClellan algorithm
transfer function, 218
transversal form, 280–281, 280f
two-band digital crossover
impulse responses, lowpass and highpass filters, 260,
261f
magnitude frequency responses, lowpass and highpass
filters, 260, 260f, 261f
speaker drivers, 258–259, 259f
specifications, 259–260
window method, 285–286
see also Window method
Finite precision, 35, 40–42
First-order IIR filter transfer function, 369–371
Fixed-point DS processor, 411–412
3-bit 2's complement number, 412, 412t
fractional representation, 413, 413t
computational units, 427
direct-form II implementation, C code, 445–446, 446t,
447f, 448f
fractional binary 2's complement system, 414
program control unit, 427
Q-30 format, 418, 418f
Q-format number, 415, 415f, 418
TMSC320C54x family architecture, 426–427, 426f
Floating-point DS processor, 411–412, 419, 419f
advantages, 427
ARAUs, 428–429
direct-form I implementation, C code, 445, 445f
IEEE format
double precision format, 425, 425f
single precision format, 423–424, 423f
overflow, 422
rules for, 420

speech quality applications, 429
TMS320C3x processor, 427–428, 428f
underflow, 423
Folding frequency, 20, 47
Fourier series, 132
amplitude-phase form, 776
amplitude and phase spectrum, 779
amplitude spectral components, 90
coefficients, 88–89
complex exponential form, 776–782
waveform signals, 784t
Fourier transform, 786–791
properties, 790t
waveform signals, 789t
periodic digital signal, 88, 89f
rectangular waveform, 780, 780f
sine-cosine form, 775, 780
waveform signals, 783t
two-side line amplitude spectrum, 89, 89f
Frequency modulation (FM), 751
Frequency resolution, 96–101
Frequency sampling method, 286
design procedure, 263
desired filter frequency response, 262, 262f
DFT, 262, 817
Euler formula, 817
features, 262
frequency response, 819–820
IDFT, 262
L'Hospital's rule, 817
magnitude frequency response, 269
Frequency warping effect, 312, 313f
digital frequency, 312
digital integration method, 308, 309f
graphical representation, 313, 314f
Laplace transfer function, 309–310
mapping properties, 310, 310f
s-plane *vs.* z-plane, frequency mapping, 312, 312f
z-transform, 309–310

G

Gaussian filter kernel, 711–712
Gibbs effect, 224
Goertzel algorithm, 392
advantages, 380
DFT algorithm, 377
DFT coefficient, 378–379
Euler's identity, 378–379
MATLAB function, 382
modified second-order Goertzel IIR filter, 380–381, 381f
second-order Goertzel IIR filter, 310–311, 377
transfer function, 377

Grayscale histogram and equalization
    equalized grayscale image, human neck, 695, 697f
    new pixel value, 693–695
    original grayscale image, human neck, 695, 696f
    pixel value distribution, 692–693

H

Hamming window function, 230, 231f
    ECG data, 118f, 119
    seismic data, 116, 118f
    speech data, 116, 117f
    vibration signal, 119, 119f, 121f, 122f
    vibration signature analysis, gearbox, 119–120,
        120f
Hanning window, 230, 231f
Harvard architecture, 407, 407f
    execution cycle, 407, 408f
    pipelining operation, 408
High-definition TV (HDTV) formats, 754, 754t
Highpass filters
    coefficients, 656
    digital Butterworth lowpass prototype functions,
        322–331
    digital Chebyshev lowpass prototype functions,
        322–331, 322f
    impulse responses, 260, 261f
    lowpass prototype transformation, 305, 305f
    magnitude frequency responses, 260, 260f, 261f
Histogram equalization, 9–12
Horizontal Sobel edge detector, 716
Huffman coding, 528, 737, 737t

I

IDWT, *see* Inverse discrete wavelet transform (IDWT)
IEEE floating-point format
    double precision format, 425, 425f
    single precision format, 423–424, 423f
IIR filter design, *see* Infinite impulse response (IIR) filter
    design
Image processing
    24-bit color image equalization, 695, 698f
        equalized RGB color image, 698, 699f
        histogram equalization method, 698, 699f
        original RGB color image, 698, 698f
        RGB channels, equalization effects, 698–699, 700f
    8-bit indexed color image equalization, 700–701, 701f,
        702f
    compression, DCT
        2D-DCT, 729–731
        JPEG image compression, *see* JPEG image compression
        lossless/lossy compression, 728
        principle of, 729

    wavelet transform coding, *see* Wavelet transform coding
    2D-DFT, 725
    definition, histogram, 692
    edge detection, 717, 718f
        differential convolution kernel, 715–716
        grayscale image, 717, 719f
        horizontal Sobel edge detector, 716
        Laplacian edge detector, 716–717
        Laplacian of Gaussian filter, 717, 719f
        MATLAB functions, 718–721, 720f
        vertical Sobel edge detector, 716
    grayscale histogram and equalization
        equalized grayscale image, human neck, 695, 697f
        new pixel value, 693–695
        original grayscale image, human neck, 695, 696f
        pixel value distribution, 692–693
    image level adjustment
        display level adjustment, 707
        linear level adjustment, 704–706, 705f, 706f
        MATLAB functions, 707, 708f
    lowpass noise filtering
        average convolution kernel, 709
        Gaussian filter kernel, 711–712
        noisy and enhanced image, 711–712, 711f, 712f,
         713f
    MATLAB functions, equalization, 702–704, 703f
    median filtering
        enhanced image, 714, 715f
        "pepper and salt" noise, 714, 715f
        principle of, 712–714
    notation and data formats
        8-bit color image, 687, 687f
        24-bit color image, 686, 686f
        8-bit grayscale image, 684–685, 685f
        chrominance channels, 688–689
        format conversion, 690–691, 691f
        grayscale image conversion, RGB-to-YIQ
         transformation, 690, 690f
        image pixel notation, 684, 685f
        intensity image, 688, 688f
        luminance channel, 688–689
        spatial resolution, 684
        transformation and inverse transformation,
         688–689
    pseudo-color generation and detection
        grayscale to pseudo-color pixel, 722, 722f
        MATLAB code, 725
        procedure for, 724f
        sine functions, RGB transformations, 722, 723f
    video sequence creation, 745–746, 746f, 747f
    video signals, *see* Video signals
Impulse function, 374

Impulse-invariant design method, 345f, 389, 392
    filter DC gain, 348
    inverse Laplace transform, analog impulse function, 345–
        346
    rectangular approximation, 346–348
    sampling interval effect, 348, 349f
    scaled magnitude frequency response, 347f, 348
    second-order filter design, 348–351
Impulse response system
    digital convolution sum, 69
    FIR system, 69
    IIR system, 71
    unit-impulse response, 68, 68f
Infinite impulse response (IIR) filter design, 71, 389, 390t,
        391f
    bandpass filter design specifications, 389
    BLT design method, 388
    see also Bilinear transformation (BLT) design method
    C code
        direct-form II implementation, 445–446, 446t, 447f,
            448f
        direct-form I structure, 445, 445f
    difference equation, 302–303
    digital audio equalizer, 341f
        audio spectrum, 343–344, 343f
        audio test signal, 343–344
        filter banks design, 342, 342t
        magnitude frequency responses, 342, 342f
        MATLAB program, 344
        specifications for, 341, 341t
    digital Butterworth lowpass prototype functions, 318, 319t
        bandpass and bandstop filter, 331–337
        lowpass and highpass filters, 322–331
        magnitude response function, 318, 320f
        prototype filter order, 318–320
    digital Chebyshev lowpass prototype functions, 318, 319t,
        320t
        analog filter specification conversion, 321, 322t
        bandpass filters, 321, 322f, 331–337
        bandstop filters, 331–337
        highpass filters, 322–331, 322f
        lowpass filters, 321–331, 322f
        lowpass prototype order, 321
        magnitude response function, 320–321, 321f
    direct-form I and direct-form II, realization structure, 358–
        360
    DTMF tone generator
        Goertzel algorithm, *see* Goertzel algorithm
        MATLAB program, 377
        modified Goertzel algorithm, *see* Modified Goertzel
            algorithm
        telephone touch keypads, 373–375, 373f, 376f

    first-order IIR filter transfer function, 369–371
    fixed-point system, 432, 432f
    format of, 302–303
    higher order IIR filter design, cascade method, 338–340,
        338t, 339t
        realization structure, 361–362
    60-Hz hum eliminator and heart rate detection,
        electrocardiography, 392
        cascaded frequency responses, 365, 366f
        characteristics of, 362, 363f
        design specifications, 364–365
        harmonics, 364
        heart rate, definition, 367–368
        MATLAB program, 368
        QRS complex, 362–364
        signal enhancement system, 364, 364f
        signal processing results, 366, 367f
        signal spectrum, 362, 363f
        transfer function and difference equation, 365
        zero-crossing algorithm, 366–367, 368f
    impulse-invariant design method, 345f, 389, 392
        filter DC gain, 348
        inverse Laplace transform, analog impulse function,
            345–346
        rectangular approximation, 346–348
        sampling interval effect, 348, 349f
        scaled magnitude frequency response, 347f, 348
        second-order filter design, 348–351
    pole-zero placement method, 389, 392
        first-order highpass filter, 357–358, 357f
        first-order lowpass filter, 355–357, 355f, 356f
        magnitude response, 351, 352f
        Nyquist limit, 351–352
        second-order bandpass filter, 352–354, 353f
        second-order bandstop (notch) filter, 354–355, 354f
    second-order IIR filter transfer function, 369–371
    single-tone generator, 374–375, 374f, 375f
    transfer function, 433
Infinite precision, 35, 282
Interlaced raster scan, 747, 748f
Interpolation filter, 579, 579t
    commutative model, 580, 580f
    filter bank coefficients, 579–580
    four-tap interpolation filter, 578, 578f
    implementation, 579, 579f, 584
Inverse discrete cosine transform (IDCT), 729
Inverse discrete Fourier transform (IDFT), 262
Inverse discrete wavelet transform (IDWT),
        656, 671
Inverse fast Fourier transform (IFFT)
    definition, 126
    eight-point IFFT, 129–131, 131f

Inverse z-transform
  definition, 144
  partial fraction expansion method, 144–145
    constant(s) formulas, 145, 146t
    MATLAB function residue(), 150–152

J

JPEG image compression
  alternating-current coefficients, 738f
    bit stream, 738
    run-length coding, 736–738
  direct-current coefficients
    DPCM, 736
    Huffman coding, 737, 737t
  encoder, 735, 735f
  image blocks, 735
  lossless entropy coding, 737
  quantization, 735–736, 736t
  RGB to YIQ transformation, 735
  two-dimensional grayscale image, 733, 734f
    coding error, 733, 734t
    DCT coefficients, 731, 732t
    DCT coefficient scan order, 732, 733t
    JPEG vector, 732–733
    normalized DCT coefficients, 732, 733t
    original image, 731, 732f
    quality factor, 731, 733t
    recovered image subblock, 733, 734t
    8×8 subblock, 731, 731t

K

Kaiser window, 230
Kernel
  average convolution, 709
  differential convolution, 715–716
  Gaussian filter, 711–712

L

Laplace shift property, 174
Laplace transform
  differential equations, 793–794
  and table, 791–793, 791t
  transfer function, 794–795
Laplacian edge detector, 716–717
Laplacian of Gaussian filter, 717, 719f
Least mean square (LMS) algorithm, 461–462
  adaptive FIR filters
    corrupted signal and noise reference, 455–457, 456f, 456t
    desired signal spectrum, 453, 454f
    noise canceller, 454, 454f
    one-tap FIR filter, 455, 457

Linear buffering
  FIR filter, 441–442, 441f
  IIR filter, 442, 443f
    coefficient buffer, 444, 444f
    digital oscillation, 442
Linear convolution, 142–143
Linear midtread quantizer, 533
Linear phase response, 223, 224f, 226–227, 226f, 227f
Linear systems, 63, 64f
  digital amplifier, 64
  system output, 65
Linear time invariant system
  difference equation, 67
  FIR system, 69
  stability criterion, 71
  unit-impulse response, 68–69, 68f, 72
Lowpass filters, 304, 304f
  analog filters, 321, 322f
  coefficients, 656
  digital Butterworth lowpass prototype functions, 322–331
  digital Chebyshev lowpass prototype functions, 321–331, 322f
  impulse responses, 260, 261f
  magnitude frequency responses, 260, 260f, 261f
  Sallen-Key lowpass filter, 26–27, 26f
  17-tap FIR lowpass filter coefficients, 224, 225t
Lowpass noise filtering
  average convolution kernel, 709
  Gaussian filter kernel, 711–712
  noisy and enhanced image, 711–712, 711f, 712f, 713f
Luminance channel, 688–689

M

MAC, *see* Multiplier and accumulator (MAC)
Maclaurin series expansion, 593–595
Macroblocks, 755, 755f
Mathematical formulas
  complex conjugate, 826
  complex number form, 825–826
    addition and subtraction, 826
    division, 826–828
    multiplication, 826
  L'Hospital's rule, 828
  quadratic equation solution, 828
  simultaneous equation solution, 828
  simultaneous linear equation solution, 830
Matrix Laboratory (MATLAB) programs
  ADPCM coding, 539
    decoding, 542
    encoding, 539
  analog filters, lowpass prototype transformation, 307

arrays and indexing, 770–771
CD audio player, 572
commands and syntax
    array operations, 769
    complex numbers, 768
    numbers, variables and expressions, 768
    sum() function, 767
    variable names, 768
DCT waveform coding, 546
digital audio equalizer, 344
digital μ-law compressor, 537
digital μ-law encoding and decoding, 537
digital μ-law expander, 538
downsampling, 559, 609
DTMF tone generator, 377
edge detection, 718–721, 720f
equalization, 702–704, 703f
fft() and ifft() function, 93, 93t
FIR filter design
    noise reduction, 255
    window method, 237–240, 237t, 288
first-order SDM, 597
Goertzel algorithm, 382
60-Hz hum eliminator and heart rate detection,
        electrocardiography, 368
image level adjustment, 707, 708f
μ-law companding, 535
μ-law encoding and decoding, 534
μ-law expanding, 535
midtread quantizer
    decoding, 536
    encoding, 536
    linear, 533
modified Goertzel algorithm, 385–386, 386f
noise cancellation, 466
noninteger factor L/M, 568
one-level wavelet transform and compression, 742
oversampling, 589
plot functions, 771–772, 772f
pseudo-color generation and detection, 725
residue() function, 150–152
script files, 164f, 772–773
signal to quantization noise ratio, 48, 537
sign function, 548
speech signals
    bandpass filtering, 204
    pre-emphasis of, 201
sumsub.m function, 773–774
system modeling, adaptive filters, 470
two-channel perfect reconstruction quadrature mirror filter
        bank, 633
two-level wavelet transform and compression, 744

uniform quantization decoding, 48
uniform quantization encoding, 48
upsampling, 564, 611
wavelet data compression, 667
W-MDCT function, 545
    inverse function, 545
    waveform coding, 546
MDCT, *see* Modified discrete cosine transform (MDCT)
Mean square error quadratic function, 457–458, 458f
Median filtering
    enhanced image, 714, 715f
    "pepper and salt" noise, 714, 715f
    principle of, 712–714
Modified discrete cosine transform (MDCT)
    1D-DCT, 522
    decoding stage, 523
    encoding stage, 523
    W-MDCT, 522, 522f
        waveform coding, 524–525, 525f
    wmdeth() and wimdetf() functions, 523–524
Modified Goertzel algorithm, 384f
    ASCII code, 385–386
    design principles, 383
    frequency bins, 383, 384t
    MATLAB simulation, 385–386, 386f
Mother wavelet, 642, 644, 644f
Motion estimation, 755–756, 755f
Motion vector, 755
MPEG audio
    audio frame formats, 526, 527f
    data frame types, 526, 526f
    DCT, 519–522, 524–525, 525f
    encoder, 527, 528f
    Huffman coding, 528
    MDCT, *see* Modified discrete cosine transform
        (MDCT)
Multiplier and accumulator (MAC), 407–409, 409f
Multirate digital signal processing, 555–556
    CD audio player
        interpolation filter design, 571–572, 573f
        MATLAB program, 572
        sample rate conversion, 571, 571f
        signal plots, 572, 574f
    multistage decimation, *see* Multistage decimation approach
    sampling rate, integer factor, *see* Sampling rate
Multiresolution analysis, 650–651
Multistage decimation approach
    sampling rate conversion, 578
    two-stage decimator, 574, 575f
        filter requirements, 576
        stopband frequency edge, anti-aliasing filter, 575–576,
        575f

N

Noise cancellation
  MATLAB program, 466
  MSE function *vs.* weights, 465, 465f
  one-tap adaptive filter, 462, 463f
  specifications, 466
  speech waveforms and spectral plots, 466,
    466f, 467f
  two-tap adaptive filter, 463–465
Noise reduction systems
  clean signal and spectrum, 254, 255f
  data acquisition process, 253
  MATLAB program, 255
  noise signal and spectrum, 254, 254f
  passband frequency range, 254
  speech noise reduction, 256–257, 256f, 257f
  stopband frequency range, 254
  vibration signals, 257–258, 258f, 259f
Noncausal FIR filter coefficients, 233
Noncausal sequence, 141, 233
Normalized bandpass filter, 187, 187f
Normalized bandstop filter, 188, 188f
Normalized Butterworth function, 805–808
Normalized Chebyshev function, 808–812
Normalized highpass filter, 187, 187f
Normalized lowpass filter, 186f, 187
Notch filter, 354–355, 354f
NTSC TV standard, 750, 751f
Nyquist frequency, 20, 47
Nyquist limit, 351–352, 562–563, 564f

O

Optimal design method, 286–287
  see also Parks-McClellan algorithm
Overflow, 422
Output digital signal, 2, 592

P

Parallel realization method, 192, 196–199, 196f
Parks-McClellan algorithm
  alternation theorem, 277–278
  approximation error, 269
  Chebyshev polynomial approximation, 269
  Chebyshev real magnitude function, 277
  design procedure, 270–279
  disadvantages, 279
  magnitude frequency response, 269–270, 270f
  minimax filters, 269
  Remez exchange algorithm, 269
Partial fraction expansion method, 144–145
  constant(s) formulas, 145, 146t
  MATLAB function residue(), 150–152

Perfect reconstruction, *see* Two-channel perfect
    reconstruction quadrature mirror filter bank
Phase alternative line (PAL) system, 752
Plot functions, 771
Pole-zero placement method, 389, 392
  first-order highpass filter, 357–358, 357f
  first-order lowpass filter, 355–357, 355f, 356f
  magnitude response, 351, 352f
  Nyquist limit, 351–352
  second-order bandpass filter, 352–354, 353f
  second-order bandstop (notch) filter, 354–355, 354f
Pole-zero plot, *see* Z-plane pole-zero plot
Polyphase filters
  direct decimation process, 581, 581f, 582t
    commutative model, 582–583, 583f
    filter bank coefficients, 582
    implementation, 582, 582f, 584
    three-tap decimation filter, 581
  direct interpolation filter, 579, 579t
    commutative model, 580, 580f
    filter bank coefficients, 579–580
    four-tap interpolation filter, 578, 578f
    implementation, 579, 579f, 584
  properties, 581
Power spectrum, 97–101
Progressive scan, 754

Q

Quadrature amplitude modulation (QAM), 751–752
Quantization, 735–736, 736t
  see also Waveform quantization and compression
  bipolar quantizer, 38–40, 39f, 40t
  definition, 35, 36f
  error, 37
  notations and rules, 38
  process, 37
  SNR, 47
  unipolar quantizer, 38, 38f, 39t
Quantization error
  analog μ-law companding, 501
  DAC, 40–42

R

Radix-2 FFT algorithm, 123
  decimation-in-frequency method, *see* Decimation-
    in-frequency method
  decimation-in-time method
    eight-point FFT algorithm, 128–129, 130f
    eight-point IFFT, 129–131, 131f
    first iteration, 128–129, 130f
    frequency bins, 128–129
    second iteration, 128–129, 130f

rconv() function, 668
Realization structure
    direct-form I and direct-form II, 358–360
    higher order IIR filter design, cascade method, 361–362
Real-time processing
    input and output sample clock, 438, 439f
    program segment, 438, 440f
    TMS320C6713 DSK setup, 438, 440f
Rectangular window, 230
Reference frame, 755, 755f
Remez exchange algorithm, 269
RGB components, 686, 686f
RGB-to-YIQ transformation, 690, 690f
Root mean square (RMS), 42, 499
Rounded off error, 282
Run-length coding, 736–738

S

Sallen-Key lowpass filter, 26–27, 26f
Sampling rate
    downsampling, 557f
        data sequence, 556
        definition, 556
        MATLAB program, 559, 609
        normalized stop frequency edge, 556–557
        Nyquist sampling theorem, 556
        spectral plots, 556–557, 558f
        TMS320C6713 DSK, 608, 612f
        using anti-aliasing filter, spectral plots, 558–559, 560f
        without using anti-aliasing filter, spectral plots, 558, 559f
        z-transform, 556–557
    noninteger factor L/M, 570–571
        anti-aliasing filter, 568, 569f
        interpolation filter, 567, 568f
        MATLAB program, 568
        sampling rate conversion, 567, 567f
    upsampling
        definition, 562
        interpolation filter, 563, 565f
        MATLAB program, 564, 611
        normalized stop frequency edge, 562–563
        Nyquist limit, 562–563, 564f
        process of, 562, 563f
        sampling frequency, 562–563
        TMS320C6713 DSK, 611, 612f
Scaling functions, 649–650, 650f
    multiresolution analysis, 650–651
SECAM system, *see* Séquentiel Couleur á; Mémoire (SECAM) system
Second-order bandpass filter, 352–354, 353f
Second-order bandstop (notch) filter, 354–355, 354f
Second-order Butterworth lowpass filter, 25–26

Second-order IIR filter transfer function, 369–371
Séquentiel Couleur á Mémoire (SECAM) system, 752, 752t
Sequential search method, 755–756
Shannon sampling theorem, 20
Shaped-in-band noise power, 593–595
Shifters, 409
Sigma-delta modulation analog-to-digital conversion (SDM ADC)
    ADC resolution, 595–596
    CD player, 601–602, 601f, 602f
    continuous *vs.* regular sampled *vs.* oversampled signal amplitudes, 597, 599f
    discrete-time analog filter, 592, 593f
    DSP model, second-order SDM, 595, 596f
    extrapolation method, 592
    feedback control system, 592–593
    first-order SDM
        DSP model, 592, 593f
        MATLAB program, 597
        principles, 592, 592f
    frequency responses, 597, 597f
    MAX1402, functional diagram, 600, 600f
    noise shaping filter, 592–593, 594f
    shaped-in-band noise power, 593–595
    time *vs.* frequency domains, 597, 598f
Signal denoising, 668, 670f
Signal-to-noise power ratio, 499
Signal reconstruction
    aliasing frequency component, 23–25
    anti-aliasing filtering, 35
        aliasing level percentage, 28
        Butterworth magnitude frequency response, 25–26
        Sallen-Key lowpass filter, 26–27, 26f
        sampled analog signal, 25, 26f
    anti-image filter and equalizer, *see* Anti-image filter and equalizer
    signal notations, 21–22, 22f
    signal spectrum recovery, 22–23, 22f, 23f
Signal sampling
    ADC
        see also Analog-to-digital conversion (ADC)
        sample-and-hold analog voltage, 15, 16f
    analog (continuous) signal and digital samples *vs.* time instants, 15, 16f
    anti-image filter, 18
    DAC, *see* Digital-to-analog conversion (DAC)
    DSP, 15, 16f
    lowpass reconstruction filter, 20
    MATLAB function
        signal to quantization noise ratio calculation, 48
        uniform quantization decoding, 48

Signal sampling (*Continued*)
  uniform quantization encoding, 48
  Nyquist frequency/folding frequency, 20, 47
  sampling process, 18, 18f, 47
  sampling rate, 16–17
  sampling theorem condition, 17–18, 17f, 20, 47
  Shannon sampling theorem, 20
  signal reconstruction, *see* Signal reconstruction
  spectral analysis, 18–19, 19f
Single-tone generator, 374–375, 374f, 375f
Smith-Barnwell PR-CQF filters, 630, 630t
Spectral leakage, 108
Speech coding
  four-band compression, 637, 637f, 638f
    seismic data, 637, 639f
  two-band compression, 636–637, 636f
Speech noise reduction, 256–257, 256f, 257f
Speech signals
  bandpass filtering
    amplitude spectra, 203, 205f
    digital fourth-order bandpass Butterworth filter, 203
    frequency responses, 203, 203f
    MATLAB program, 204
    original and filtered speech plots, 203, 204f
  pre-emphasis of
    amplitude spectral plots, 201, 202f
    magnitude and phase responses, 200, 200f
    MATLAB program, 201
    speech waveforms, 200, 201f
    transfer function, 200
Stair functions, 771
Steepest descent algorithm, 459, 460f, 461–462
Stem functions, 771
Step response, 169
Subband coding
  analysis and synthesis stages
    channel 0, 622, 623f
    channel 1, 622, 623f
    channel 2, 622–624, 624f
    channel 3, 624, 625f
  4-channel filter bank analyzer and synthesizer, 621–622, 622f
  decomposition, *see* Two-channel perfect reconstruction quadrature mirror filter bank
  delta function, 624
  discrete Fourier transform, 625
  filter bank system, 621
  impulse train, 625, 626f
  signal flow, 624, 625f
  speech coding, *see* Speech coding

two-band filter bank system, signal compression, 635–636, 636f
  z-transform, 626
Subplot functions, 771
S-video, 746
Synthesis filter
  channel 0, 622, 623f
  channel 1, 622, 623f
  channel 2, 622–624, 624f
  channel 3, 624, 625f
  4-channel filter bank, 621–622, 622f

T

17-Tap FIR lowpass filter coefficients, 224, 225t
Target frame, 755, 755f
Time-invariant system, 65–66, 65f
TMS320C6713 DSK
  analog system program, 482
  downsampling, 608, 612f
  system modeling
    LMS adaptive filter, 480–482, 480f, 482f
    program segment, 481
  tonal noise cancellation, 483, 483f, 484f
    DSK1 program, 483
    DSK2 program, 484
  upsampling, 611, 612f
Transition band, 186
Translated function, 642, 643f
Transversal FIR filter, 280–281, 280f
Twiddle factor, 92–93, 123–124
Two-band digital crossover design
  lowpass and highpass filters
    impulse responses, 260, 261f
    magnitude frequency responses, 260, 260f, 261f
  speaker drivers, 258–259, 259f
  specifications, 259–260
Two-channel perfect reconstruction quadrature mirror filter bank, 626, 627f
  analysis and synthesis filters, 627
  autocorrelation function, 628
  four-band implementation
    binary tree structure, 634f, 635
    dyadic tree structure, 635, 635f
  frequency response, 629, 629f
  lowpass filter equations, 630
  MATLAB program, 633
  N-tap FIR filters, 628
  Smith-Barnwell PR-CQF filters, 630, 630t
  two-band analysis and synthesis, 632, 633f
Two-dimensional discrete cosine transform (2D-DCT), 729–731
Two-dimensional discrete Fourier transform (2D-DFT), 725

U

Unipolar quantizer, 38, 38f, 39t
Unit circle, 175
Unit-impulse sequence, 58–59, 59f
Unit-step sequence, 59, 59f, 62
Underflow, 423
Unstable system, 175
Upsampling
    definition, 562
    interpolation filter, 563, 565f
    MATLAB program, 564, 611
    normalized stop frequency edge, 562–563
    Nyquist limit, 562–563, 564f
    process of, 562, 563f
    sampling frequency, 562–563
    TMS320C6713 DSK, 611, 612f

V

Vertical retrace, 747, 750
Vertical Sobel edge detector, 716
Vibration signature analysis, 9, 10f, 11f
Video signals
    analog video
        "back porch", 748
        electrical signal demodulation, 748–750, 749f
        frame via row-wise scanning, 747
        frequency modulation, 751
        interlaced raster scanning, 747, 748f
        NTSC TV standard, 750, 751f
        PAL system, 752
        QAM, 751–752
        SECAM system, 752, 752t
        vertical synchronization, 749f, 750
        video data, retrace and sync layout, 750, 750f
        video-modulated waveform, 747, 748f
    component video, 746
    composite video, 746
    digital video
        CCIR-601, chroma subsampling, 753, 753f
        HDTV formats, 754, 754t
        specifications, 754, 754t
    motion estimation, 755–756, 755f
    S-video, 746
Von Neumann architecture, 406, 406f
    applications, 408
    execution cycles, 407, 408f
    opcode and operand, 406

W

Waveform coding, 7
Waveform quantization and compression
    analog μ-law companding

        characteristics, 502, 502f
        compressor, 501, 501f
        expander, 501–502, 501f
        original speech data, 504, 505f
        quantization error, 501
    digital μ-law companding
        8-bit compressed PCM code format, 505–506, 506t, 508–509, 508f
        characteristics, 505, 506f
        compressor and expander, 504, 505f
        decoding table, 506–508, 507t
        encoding table, 505–506, 507t
    DM, 511
    DPCM
        3-bit quantizer, 509, 510t
        encoder and decoder, 509, 509f
        quantization step size, 512
    G.721 modulation, *see* Adaptive differential pulse code modulation (ADPCM)
    linear midtread quantization
        characteristics of, 498–499, 498f
        quantization, definition, 497–498
        quantization error, 500
        quantized values, 498–499, 498t
        signal-to-noise power ratio, 499
        speech data plot, 500, 500f
    MATLAB programs, *see* MATLAB programs
    MPEG audio
        audio frame formats, 526, 527f
        data frame types, 526, 526f
        DCT, 519–522, 524–525, 525f
        encoder, 527, 528f
        Huffman coding, 528
        MDCT, *see* Modified discrete cosine transform (MDCT)
    TMS320C6713 DSK
        digital μ-law encoding and decoding, 530
        encoding and decoding, linear quantization, 528–529
Wavelet analysis
    analysis equations, 822–823
    properties, 821–822
    scaled function, 641, 642f, 643f
    synthesis equations, 823–824
    translated function, 642, 643f
Wavelet transform
    amplitudes, 639–641, 641f
    analysis and synthesis stage, 664
    combined signal and spectrum, 639–641, 640f
    CWT, 638, 641
    Daubechies-4 filter coefficients, 653, 654t
    DWT, 638
    see also Discrete wavelet transform (DWT)

Wavelet transform (*Continued*)
    coefficient layout, 664, 665f
    hard threshold, 668, 669f
    signal denoising, 668, 670f
    Haar father and mother wavelets, 642, 644, 644f,
       652–653
    individual signal components, 639, 640f
    mother wavelet, definition, 641
    one-level wavelet transform and compression, 741,
       742f
       MATLAB program, 742
    scaled wavelet function, 641, 642f, 643f
    scaling functions, 649–650, 650f
       multiresolution analysis, 650–651
    signal coding, 650, 651f
    sinusoidal delaying function, 648, 648f
    4-tap Daubechies father wavelet, 654, 654f
    4-tap Daubechies mother wavelet, 655, 655f
    translated wavelet function, 642, 643f
    two-dimensional DWT, 738–741, 739f
    two-level wavelet transform and compression,
       741–742, 743f
       MATLAB program, 744
    types, 638
    wavelet coefficients, 646–647
    wavelet data compression
       16-bit ECG data, 668, 669f
       16-bit speech data, 667, 667f
       MATLAB program, 667
Wiener filter theory
    autocorrelation and cross-correlation, 459
    LMS algorithm, 461–462
    mean square error quadratic function, 457–458, 458f
    noise cancellation, 457, 457f
    statistical expectation, 457–458, 461–462
    steepest descent algorithm, 459, 460f, 461–462
Windowed modified discrete cosine transform (W-MDCT),
       522–523, 522f, 545
    inverse function, 545
    waveform coding, 524–525, 525f, 546
Window method
    Blackman window, 230, 231f
    cutoff frequency, 242
    design procedure, 233
    Gibbs oscillations, 230
    Hamming window, 230, 231f
    Hanning window, 230, 231f
    Kaiser window, 230
    length estimation, 241, 241t
    magnitude frequency response, 240
    MATLAB function, 237–240, 237t, 288
    passband ripple, 241–242, 241f
    rectangular window, 230
    stopband attenuation, 241–242, 241f
    triangular (Bartlett) window, 230, 231f

Y

YCbCr color space, 753
YIQ, 690, 690f
YUV color model, 752

Z

Zero-crossing algorithm, 366–367, 368f
Zigzag scan, 737
Z-plane pole-zero plot, 172f
    analog-to-digital conversion, 174
    bounded-in/bounded-out stability, 175
    features, 172
    Laplace shift property, 174
    Laplace *vs.* z-transform, 173, 174f
    s-plane *vs.* z-plane mapping, 175, 175f
    stability rules, 175, 176f
Z-transform
    definition, 137–138
    difference equations, 152–156
    exponential sequence, 138
    inverse z-transform
       definition, 144
       partial fraction expansion method, *see* Partial fraction
          expansion method
    lookup table, 156
    one-sided/unilateral transform, 137–138
    properties of, 144t
       causal sequence, 141–143
       linear convolution, 142–143
       linearity, 140–141
       time-shifted sequence, 141
    region of convergence, 138
    sequences for, 138–140, 139t